Applications of Essential Oils in Food Industry

Applications of Essential Oils in the Food Industry

Edited by

Charles Oluwaseun Adetunji
Applied Microbiology, Biotechnology and Nanotechnology Laboratory,
Department of Microbiology, Edo State University Uzairue, Iyamho,
Edo State, Nigeria

Javad Sharifi-Rad
Facultad de Medicina, Universidad del Azuay, Cuenca, Ecuador

ELSEVIER

ACADEMIC PRESS
An imprint of Elsevier

Academic Press is an imprint of Elsevier
125 London Wall, London EC2Y 5AS, United Kingdom
525 B Street, Suite 1650, San Diego, CA 92101, United States
50 Hampshire Street, 5th Floor, Cambridge, MA 02139, United States

Notices

Knowledge and best practice in this field are constantly changing. As new research and experience broaden our understanding, changes in research methods, professional practices, or medical treatment may become necessary.

Practitioners and researchers must always rely on their own experience and knowledge in evaluating and using any information, methods, compounds, or experiments described herein. In using such information or methods they should be mindful of their own safety and the safety of others, including parties for whom they have a professional responsibility.

To the fullest extent of the law, neither the Publisher nor the authors, contributors, or editors, assume any liability for any injury and/or damage to persons or property as a matter of products liability, negligence or otherwise, or from any use or operation of any methods, products, instructions, or ideas contained in the material herein.

ISBN: 978-0-323-98340-2

For Information on all Academic Press publications
visit our website at https://www.elsevier.com/books-and-journals

Publisher: Nikki P. Levy
Acquisitions Editor: Nina Bandeira
Editorial Project Manager: Kyle Gravel
Production Project Manager: Rashmi Manoharan
Cover Designer: Greg Harris

Typeset by MPS Limited, Chennai, India

Working together
to grow libraries in
developing countries

www.elsevier.com • www.bookaid.org

Contents

16. Combinatory effect of essential oil and lactic acid in formulations of different food products   199

Babatunde Oluwafemi Adetuyi, Charles Oluwaseun Adetunji, Juliana Bunmi Adetunji, Abel Inobeme, Osarenkhoe Omorefosa Osemwegie, Mohammed Bello Yerima, M.L. Attanda and Oluwabukola Atinuke Popoola

17. Application of essential oil in aromatherapy: current trends   207

Babatunde Oluwafemi Adetuyi, Peace Abiodun Olajide, Charles Oluwaseun Adetunji, Juliana Bunmi Adetunji, Oluwabukola Atinuke Popoola, Oloruntoyin Ajenifujah-Solebo, Yovwin D. Godwin, Olatunji Matthew Kolawole, Olalekan Akinbo and Abel Inobeme

18. Application of starter culture bacteria in dairy product   223

Babatunde Oluwafemi Adetuyi, Charles Oluwaseun Adetunji, Juliana Bunmi Adetunji, Abel Inobeme, Oluwabukola Atinuke Popoola, Oloruntoyin Ajenifujah-Solebo, Yovwin D. Godwin, Olatunji Matthew Kolawole, Olalekan Akinbo and Mohammed Bello Yerima

19. Toxicity and safety of essential oil   235

Olulope Olufemi Ajayi

30. Antihyperlipidemic and antioxidant properties of medicinal attributes of essential oil 327

Muhammad Akram, Rabia Zahid, Babatunde Oluwafemi Adetuyi, Olalekan Akinbo, Juliana Bunmi Adetunji, Charles Oluwaseun Adetunji, Mojisola Christiana Owoseni, Majolagbe Olusola Nathaniel, Ismail Ayoade Odetokun, Oluwabukola Atinuke Popoola, Joan Imah Harry, Olatunji Matthew Kolawole and Mohammed Bello Yerima

List of contributors

Benjamin Olusola Abere Department of Economics, Edo State University Uzairue, Iyamho, Edo State, Nigeria

Charles Oluwaseun Adetunji Applied Microbiology, Biotechnology and Nanotechnology Laboratory, Department of Microbiology, Edo State University Uzairue, Iyamho, Edo State, Nigeria

Juliana Bunmi Adetunji Department of Biochemistry, Osun State University, Osogbo, Osun State, Nigeria

Babatunde Oluwafemi Adetuyi Department of Natural Sciences, Faculty of Pure and Applied Sciences, Precious Cornerstone University, Ibadan, Oyo State, Nigeria

Adeyemi Ayotunde Adeyanju Department of Food Science and Microbiology, Landmark University, Omu-Aran, Kwara State, Nigeria

Olulope Olufemi Ajayi Department of Biochemistry, Edo State University Uzairue, Auchi, Edo State, Nigeria

Oloruntoyin Ajenifujah-Solebo Genetics, Genomics and Bioinformatics Department, National Biotechnology Development Agency, Abuja, FCT, Nigeria

Olalekan Akinbo Centre of Excellence in Science, Technology, and Innovation, AUDA-NEPAD, Johannesburg, Gauteng, South Africa

Muhammad Akram Department of Eastern Medicine, Government College University, Faisalabad, Pakistan

Christiana Eleojo Aruwa Department of Microbiology, School of Life Sciences, Federal University of Technology, Akure (FUTA), Akure, Ondo State, Nigeria

M.L. Attanda Department of Agricultural Engineering, Bayero University Kano, Kano, Kano State, Nigeria

Fisayo Yemisi Daramola Department of Agriculture, Cape Peninsula University of Agriculture, Cape Town, Western Cape, South Africa

Christianah Oluwakemi Erinle Department of Agricultural and Biosystems Engineering, Landmark University, Omu-Aran, Kwara State, Nigeria

Yovwin D. Godwin Department of Family Medicine, Faculty of Clinical Sciences, Delta State University, Abraka, Delta State, Nigeria

Divya Gupta Reproductive Biology and Toxicology Lab, School of Studies in Zoology, Jiwaji University, Gwalior, Madhya Pradesh, India

Joan Imah-Harry Department of Natural Sciences, Faculty of Pure and Applied Sciences, Precious Cornerstone University, Ibadan, Oyo State, Nigeria

Abel Inobeme Department of Chemistry, Edo State University Uzairue Iyamho, Auchi, Edo State, Nigeria

Olatunji Matthew Kolawole Department of Microbiology, Faculty of Life Sciences, University of Ilorin, Ilorin, Kwara State, Nigeria

Subodh Kumar Department of Medical Laboratory Technology, School of Allied Health Sciences and Management, Delhi Pharmaceutical Sciences and Research University, New Delhi, India

Francis Bayo Lewu Department of Agriculture, Cape Peninsula University of Agriculture, Cape Town, Western Cape, South Africa

Mubashir Hussain Masoodi Department of Pharmaceutical Sciences, School of Applied Sciences & Technology, University of Kashmir, Srinagar, Jammu and Kashmir, India

Pragya Mishra Food Processing and Management DDU Kaushal Kendra, RGSC, Banaras Hindu University, Varanasi, Uttar Pradesh, India

Raghvendra Raman Mishra Medical Laboratory Technology, DDU Kaushal Kendra, RGSC, Banaras Hindu University, Varanasi, Uttar Pradesh, India

Wajahat Mushtaq Department of Pharmaceutical Sciences, School of Applied Sciences & Technology, University of Kashmir, Srinagar, Jammu and Kashmir, India

Majolagbe Olusola Nathaniel Microbiology Unit, Department of Pure and Applied Biology, Ladoke Akintola University of Technology, Ogbomoso, Oyo State, Nigeria

Aneeza Noor Department of Pharmaceutical Sciences, School of Applied Sciences & Technology, University of Kashmir, Srinagar, Jammu and Kashmir, India

Kehinde Abraham Odelade Department of Natural Sciences, Faculty of Pure and Applied Sciences, Precious Cornerstone University, Ibadan, Oyo State, Nigeria

Ismail Ayoade Odetokun Department of Veterinary Public Health and Preventive Medicine, University of Ilorin, Ilorin, Kwara State, Nigeria

A.T. Odeyemi Department of Food Science and Microbiology, Landmark University, Omu-Aran, Kwara State, Nigeria

Frank Abimbola Ogundolie Department of Biochemistry, Federal University of Technology, Akure (FUTA), Akure, Ondo State, Nigeria

Olubanke Olujoke Ogunlana Department of Biochemistry, Covenant University, Ota, Ogun State, Nigeria

Clinton Emeka Okonkwo Department of Food Science, College of Food and Agriculture, United Arab Emirates University, Al Ain, Abu Dhabi

Peace Abiodun Olajide Department of Natural Sciences, Faculty of Pure and Applied Sciences, Precious Cornerstone University, Ibadan, Oyo State, Nigeria

Abiola Folakemi Olaniran Department of Food Science and Microbiology, Landmark University, Omu-Aran, Kwara State, Nigeria

Olubukola David Olaniran Department of Sociology and Anthropology, Obafemi Awolowo University, Ile-Ife, Osun State, Nigeria

Olugbemi T. Olaniyan Laboratory for Reproductive Biology and Developmental Programming, Department of Physiology, Faculty of Basic Medical Sciences, Rhema University, Aba, Abia State, Nigeria

Blessing Itohan Omo-Omorodion Applied Microbiology, Biotechnology and Nanotechnology Laboratory, Department of Microbiology, Edo University Iyamho, Auchi, Edo State, Nigeria

Oluwakemi Semiloore Omowumi Department of Natural Sciences, Faculty of Pure and Applied Sciences, Precious Cornerstone University, Ibadan, Oyo State, Nigeria

Peter Gbenga Oni Department of Chemistry and Biochemistry, Worchester Polytechnic Institute, Worcester, MA, United States

Osarenkhoe Omorefosa Osemwegie Department of Food Science and Microbiology, Landmark University, Omu-Aran, Kwara State, Nigeria

Mojisola Christiana Owoseni Department of Microbiology, Federal University of Lafia, Lafia, Nasarawa State, Nigeria

Oluwabukola Atinuke Popoola Genetics, Genomics and Bioinformatics Department, National Biotechnology Development Agency, Abuja, FCT, Nigeria

Insha Qadir Department of Pharmaceutical Sciences, School of Applied Sciences & Technology, University of Kashmir, Srinagar, Jammu and Kashmir, India

Damilare Emmanuel Rotimi Department of Biochemistry, Landmark University, Omu-Aran, Kwara State, Nigeria

Abiola Ezekiel Taiwo Faculty of Engineering, Mangosuthu University of Technology, Durban, Umlazi, South Africa

Anjolaoluwa Maryham Taiwo Department of Natural Sciences, Faculty of Pure and Applied Sciences, Precious Cornerstone University, Ibadan, Oyo State, Nigeria

Pere-Ebi Yabrade Toloyai Department of Medical Biochemistry, Faculty of Basic Medical Sciences, Delta State University, Abraka, Delta State, Nigeria

Nyejirime Young Wike Department of Human Physiology, Faculty of Basic Medical Sciences, Rhema University, Aba, Abia State, Nigeria

Mohammed Bello Yerima Department of Microbiology, Sokoto State University, Sokoto, Sokoto State, Nigeria

Rabia Zahid Department of Eastern Medicine, Government College University, Faisalabad, Pakistan

Chapter 1

Application of essential oils in the food industry

Abel Inobeme[1] and Charles Oluwaseun Adetunji[2]

[1]Department of Chemistry, Edo State University Uzairue Iyamho, Auchi, Edo State, Nigeria, [2]Applied Microbiology, Biotechnology and Nanotechnology Laboratory, Department of Microbiology, Edo State University Uzairue, Iyamho, Edo State, Nigeria

Introduction

More recently, the attention of consumers and manufacturers in food sectors has been drawn to safer approaches for enhancing food preservation and food safety using various plants and spices as additives in comparison to various synthetic agents. Consumer concerns have recently shifted to the safety and long-term impacts of food products hence the increasing quest for quality food. Globally spices are well recognized for their role in improving the quality of food. Various parts of these plants and spices contain essential oils which include the leaves of peppermints, the bark of cinnamon, buds, flowers of clove, cardamom seeds, and fruit of pepper (Sauceda, 2011). Essential oils basically refer to secondary metabolites that are aromatic in nature and volatile and produced through single-cell metabolism or involving numerous cells functioning closely in an organized way. They are complex mixtures made of various bioactive compounds with varying chemical functional groups; with the primary component being monoterpenes which have the greatest composition of the hydrocarbons (Mathavi et al., 2013).

Essential oils are conveniently encapsulated on liposomes, nanoparticles, and various polymeric materials thereby improving their stability. Although encapsulation using various nanomaterials has given promising results, there is still a paramount need to explore the health implications of such materials. Essential oils can make efficient provisions for coatings and films for the active processes of food packaging and this depends on their interaction with different polymeric materials, as well as on their compositions. The antioxidant potential is not only dependent on the unique antioxidant activity of the bioactive compositions of the oil but is also affected by the oxygen permeability of the film. The integration into the edible coatings and films can enhance the overall efficacy of the essential oil as a preservative against microbial potential of activities (Satyavani et al., 2015).

Although synthetic additives are vital in food preservation, more recently, there is increasing awareness and global inclination on food safety and quality. One of the current growing technologies is the extraction and processing of various essential oils from plant origin and their applications in food preservation. The usage of essential oils in the preservation of various food materials is a result of their unique antimicrobial and antioxidant properties. Essential oils have been used more recently as biopreservatives for various food materials including vegetables, fruits, meat and fish products and dairy and milk products, as well as baked foods such as bread.

Essential oils are highly unstable in the environment and hence readily degraded through oxidation, photochemical processes, thermal processes, and volatilization. Since most food processing techniques involve the usage of heat or light, there is an increasing tendency for these oils to be degraded which also constitutes a challenge that must be taken into consideration in their choice of usage. Various strategies of protection for their efficient utilization have also been put forward with a view to improving their shelf life and duration in the products. Such strategies involve limiting their exposure to heat, light, and excess oxygen. The stability of these oils has also been recently enhanced through the introduction of encapsulation technology using nanomaterials and other substances (Adetunji, Adetunji, et al., 2021; Adetunji, Ajayi, et al., 2021; Adetunji, Akram, et al., 2021; Adetunji, Anani, et al., 2021; Adetunji, Kumar, et al., 2019; Adetunji, Michael, Kadiri, et al., 2021; Adetunji, Michael, Nwankwo, et al., 2021; Adetunji, Michael, Varma, et al., 2021; Adetunji, Nwankwo, et al., 2021; Adetunji, Palai, et al., 2021; Adetunji, Panpatte, et al., 2019;

Applications of Essential Oils in the Food Industry. DOI: https://doi.org/10.1016/B978-0-323-98340-2.00001-8

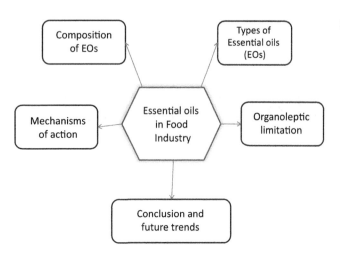

FIGURE 1.1 Schematic representation of chapter.

Adetunji, Ojediran, et al., 2019; Adetunji, Olaniyan, et al., 2021; Adetunji et al., 2013; Adetunji et al., 2021a, 2021b; Li & Hou, 2018; Ukhurebor & Adetunji, 2021; Rauf et al., 2021).

Essential oils are present in different parts of plants such as peels, barks, flowers, seeds, stems, and whole plants. They are colorless liquids that have been employed for a long for various medicinal purposes due to their phytochemical compositions. They have also been used as components of perfumes and cosmetics. There are more than 300 different types of essential oils that are used in various cosmetics and perfumery production. Their antifungal, antiviral, insecticidal, and antibacterial potentials have also been well documented. As a result of their flavors, aroma, and antimicrobial compositions, these oils are primarily employed in the preservation of food in food industries. Essential oils that are obtained from citruses such as sesquiterpenes, monoterpenes, and other oxygenated derivatives have outstanding inhibitory potentials against different microbial groups hence their usage in the preservation of food (Aljabeili et al., 2018).

Although essential oils are highly promising due to their unique antimicrobial potentials, they have some inherent limitations which must be overcome for their effective utilization in the food sector. These include their low solubility in water, remarkably high volatility, and very strong smell which restrict their applications in food sector. There are more recent advances that proffer various novel strategies for curtailing such limitations. Various numerous applications of essential oils exist which include packaging of various food products, coating of films, direct application as emulsion in foods, edible coating, and nanoemulsion agents. This chapter presents the various applications of essential oils in food industries. It also examines the role of organoleptic properties of the essential oil in limiting their usage and strategies for curtailing this (Fig. 1.1).

Compositions of essential oils

Essential oils have high complexity since they are made up of a mixture of more than fifty constitutions in varying concentrations. Therefore, there are numerous groups of compounds present in essential oils with highly variable proportions. These oils are mostly found in the cytoplasm of some plant cells and are commonly produced in secretory hair, epidermal cells trichomes. Mostly, there are a few compounds that are the primary constituents making up to 20%−70% in comparison to the other compounds present in trace concentration (Xing et al., 2019). For instance, in Artemisia herba alba, there is 24% of camphor and 57% of b-thujone while in Origanum compactum there is 27% of thymols and 30% of carvacrol as the primary constituents. These various components are responsible for the various antioxidant and antimicrobial potentials of these oils. Basically, these components are grouped into different classes such as terpenoids/terpenes and aliphatic/aromatic constituents. The highly volatile components are made up of different classes of chemicals which include amines, ketones, alcohols, aldehydes, amides, phenols, and terpenes. Aldehydes, alcohols, and ketones provide different groups of aromatic effects in various parts of different plants and fruits. For instance, fruits contain floral, nerolidol, and selinene as their major constituents. In bitter apples, the chemotherapeutic potential is due to the presence of some volatile components such as heptanone and trimethylsilyl methanol. Some other lower molecular mass components, such as terpenoids and terpenes, are also found in most essential oils which also account for their various applications. In the Rosemary plant, the extracted essential oils are made of numerous monoterpenes such as camphor and cineole as the primary components. The major types of terpenes that are found in essential oils are

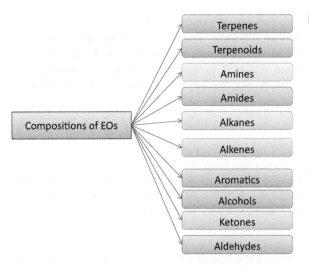

FIGURE 1.2 Compositions of essential oils.

sesquiterpenes and monoterpenes (Deng et al., 2020). Other types of terpenes that are present are diterpenes, hemiterpenes, tetraterpenes, and triterpenes. The terpenoids contain oxygen in them. The monoterpenes are linked to the formation of various structures in essential oil-producing plants. Simple hydrocarbons like the alkenes, alkanes, and the aromatic category are also called the non-terpenoid class of hydrocarbons. Citrus oils contain large amounts of open-chain hydrocarbons which account for the smell while orange oil contains aldehydes. In most cases, essential oils contain a small amount of aliphatic compounds which are the source of their odor. There are essential oils that also have one or more rings in them hence known as the mono-, di-, tri-, and tetra-cyclic depending on the number of the rings. There is also a third group of hydrocarbons that are aromatic like benzyl, phenyl propyl, and phenyl ethyl, which contain a benzene ring in addition to the polycyclic structure (Raffo & Paoletti, 2022) (Fig. 1.2).

Types of essential oils

Essential oils are highly complex in occurrence being mixtures of several compounds. They are made of lemon oil, thyme oil, lavender oil, peppermint oil, tea tree oil, clove oil, mustard oil, cinnamon oil, oregano oil, and eucalyptus oil. They are indispensable in their role of inhibiting the activities of various disease-causing microorganisms. For instance, oxygenated terpenes and normal terpenes have been documented to show high antimicrobial potential on C. glabrata, C. albicans, Candida spp, and C. tropicalis (Ben Hsouna et al., 2017). Cinnamon oil is highly volatile and is obtained from the extraction done on the leaf, bark, and root of Cinnamomum zeeylanicum. It is made of three vital components which include linalool, eugenol, and cinnamaldehyde which make up about 82% of the overall compositions. The most active ingredient in cinnamon essential oils is cinnamaldehyde which also accounts for its inhibitory impacts on various groups of microbes. The free radical scavenging potential and antiparasitic properties of cinnamon oils have also been known. The tea tree oil is extracted from melaleuca alternifolia and is made primarily of pinene, terpineol, terpinen, terpinene, cymene, and cineole (Chouhan et al., 2017). It has remarkable inhibitory effects on various strains of fungi and some bacteria and viruses. Lavender oil has a high antimicrobial effect on various bacteria that have resistant to various antibiotics such as Aspergillus spp., dermatophytes, C. neoformans, among others. The Eucalyptus oils are made of phellandral, limonene, spathulenol, globulol, cymene, cryptone, pinocarveol, and terpinene. The oils that are extracted from eucalyptus are commonly employed as natural additives for flavoring food and are also suitable for the inhibition of the growths of various microbes. Various species of eucalyptus plants produce various kinds of oils with varying compositions and they have high antimicrobial properties against *Staphylococcus aureus*, *Staphylococcus pyogenes*, *Staphyloccus pneumonia* and *Haemophilus influenza* (Santos et al., 2022).

Application in food packaging

Various edible coatings exist and are basically made of proteins, polysaccharides and lipids, of which the main useful are the essential oils. Protein and carbohydrate-based edible packaging materials have their inherent limitations in that they are hydrophilic and hence readily affected by water even though they have unique mechanical properties. The enhancement of their hydrophilic capacity is achieved through the impregnation using calcium casinate. Moreover, casein-based materials for edible coating have nutritional advantages but are still limited as they do not provide a reliable and strong barrier to water. When mixed with oleic acid, caseinate films and beeswax can aid in the enhancement of their permeability to vapor. Essential oils obtained from ginger and cinnamon can help to enhance the antimicrobial properties of the edible coating materials. It has also been reported that, due to their permeability to humidity and reduced oxygen, such films are suitable for the protection of sunflower oils from oxidative stress in the environment (Aljabeili et al., 2018).

Antimicrobial agents and packaging

Essential oils have shown their potency in the control of microbial populations. Several studies have documented the efficiency of these oils as components of active packaging design for the extension of the shelf life of various food products. There has been a rising demand for active packaging agents requiring highly minimized preservatives. Essential oils have been widely employed as antimicrobial agents. Three major approaches have been employed in the utilization of essential oil in packaging which include: the mixing of the oil in the packaging material, the direct loading of the material in the pouch of the antibacterial and the third is the coating of the essential oil on the food package. Amongst the various approaches, the most promising is the integration of the essential oil within the packaging material (Adejumo et al., 2017; Adetunji, Egbuna, et al., 2020; Adetunji, Oloke, et al., 2020; Adetunji, Roli, et al., 2020; Adetunji & Varma, 2020; Adetunji et al., 2013, 2017, 2022; Adetunji, 2008, 2019; Bello et al., 2019; Egbuna et al., 2020; Ivanišová et al., 2021; Olaniyan & Adetunji, 2021; Thangadurai et al., 2021).

Nanoencapsulation using essential oils

Hydrophobic essential oils that are insoluble in water are less efficient in various food systems when compared to various in vitro models that have been evaluated. Such decrease in efficiency has been reported more in food materials having a high-fat content such as milk, mayonnaise, margarine, and butter. A related study has shown that in order to achieve an efficiency that is close to that of in vitro models using cheese, the concentration of the oil must be increased to about 100 times. The major factor that is responsible for the observed decrease is the dissolution in the liquid phase which therefore induces a decrease in their aqueous phase concentration. Nanoencapsulation using essential oils is achieved using highly hydrophilic coating materials which embed the hydrophilic core. The nanoencapsulation is carried out before the interactions of the food components and the oils (Estévez, 2021).

Maximization of yield through curbing of diseases

Various fungi affect the production of cereal crops negatively. Such fungi also produce a large number of mycotoxins that pose serious undesired effects on the growth of crops and animals as well as humans. Various researchers have documented the estimated usage of about 23 million kilograms of chemical fungicide annually in various developing and Western countries. More recent substitutes are the use of various essential oils for the improvement of the quality and safety of agricultural crops and various food products. The antimicrobial potentials of naturally occurring biocides with their environmentally friendly nature can be employed in varying proportions (Fig. 1.3).

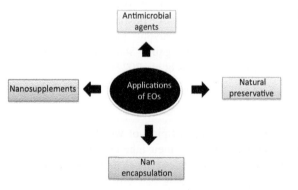

FIGURE 1.3 Applications of essential oils in food industries.

Mechanism of action of essential oils

Various mechanisms have been put forward from different investigations on the actions of various bioactive components of essential oils in the preservation of food materials. One of the most acceptable mechanisms is based on the varying bioactive compositions of the essential oil. The different functional groups present in the bioactive compounds found in the oil interact with the cells of the microorganisms in various ways. Therefore, this means that a single mechanism cannot be employed to totally account for the process of action of essential oil against microorganisms. The varying hostility of various essential oils to different groups of microorganisms is dependent on the content of the phenolic agents in the oil, such as menthol, eugenol, caravel, and thymol which bring about a greater end-to-end association between the hydrophobic hydrocarbon and the lipophilic enzyme segment. Some essential oils trigger the formation of pseudomycelia. The antimicrobial groups of the essential oils induce a partial separation between the yeast cells neonates through their interaction with the enzymes that regulate energy processes. Various methods are available for the determination of the antimicrobial efficiency of essential oil: bi-autography, broth dilution approach, and disc diffusion technique (Perricone et al., 2015).

Organoleptic limitations of essential oil application

Essential oils have remarkable organoleptic properties which must be taken into detailed consideration in their usage for the preservation and packaging of food materials. The sensory properties of food materials that have been seasoned using spices and herbs remained unchanged. Studies have revealed that the use of large amounts of some essential oils in food materials bring about an alteration of their flavors. The organoleptic properties of thyme oil coating within a concentration of 0.9% remained intact in processed shrimp. The increase of the concentration of the essential oil to 1.8% further enhanced the antimicrobial properties while curtailing the sensory features of the food. This implies that the amount of oil to be used for preservation should be calculated and controlled in ensuring that while attaining a remarkable preservation level, the sensory properties are not compromised (Surendran et al., 2021).

Studies on the application of essential oil in food industries

Various researchers have documented the extraction of various essential oils from different parts of plants. Some of the researchers also evaluated the physicochemical properties, cytotoxicity, and organoleptic properties, of the oils. Satyal et al. (2017) investigated the extraction and application of essential oils from garlic. Various species of garlic were investigated for their essential oil compositions and were subjected to chemical characterization. The usage of various essential oils in the food sector is a fast-growing trend and is being investigated by various researchers around the world. It is a broad field of research that has currently gained remarkable attention. Tolen et al. (2017) reported the potential of eugenol encapsulated using surfactant micelle for the prevention of Escherichia coli growth. They deduced that the use of the essential oil mixture did not reduce significantly the microbial activity in the beef investigated. Santoro et al. (2018) also studied the effects of the essential oils obtained from savory and thyme in the control of various post-harvest infections and the quality of nectarines and peaches. Pellegrini et al. (2018) investigated essential oil obtained from officinal plants in Italy, evaluating their physicochemical properties as well as their antimicrobial and antioxidant properties. They documented their efficiency as potential biopreservatives. Similarly, Sharopov et al. (2017) focused on essential oils obtained from some indigenous plants in Tajikistan. They adopted a gas chromatographic technique for the determination of their various properties and also investigated their antioxidant potential. They reported their applicability in food industries based on their findings in cytotoxic activities.

Policy perspective in the application of essential oil in food sector

The European Commission (EC) has recognized bioactive compositions of various essential oils as flavoring agents used at the commercial level. Some of the major natural flavoring substances that are commonly employed include carron, eugenol, caravel, mentol, limonene, and citral which do not have toxicological impacts on health. Prior to the registration of various essential oils, the EC carries out a comprehensive validation of their safety through various microbial and chemical tests. Such registered products are then introduced to the profile of food materials that have been registered in the United States (Saeed et al., 2022).

Conclusion

The presence of numerous secondary metabolites in essential oils makes them promising in the combating of various diseases and food-spoiling microorganisms such as bacteria, fungi, and even insects, and some herbivorous organisms. This write-up has addressed the utilization of essential oil in various food sectors including their efficiency in the removal of various food-spoiling agents and microbes. These oils do this through the alteration of the mechanism of the cell which affects the general cellular processes thus their wider suitability in food preservations. However, their applications in the processing and packaging of food materials are affected by their high organoleptic features. This challenge can however be mitigated through the introduction of nanomaterials for the encapsulation of the oils thereby giving more acceptable results during the preservation of food materials. The introduction of encapsulation using nanoparticles further boosts the confidence of the consumers which promotes the green and circular phenomenon.

References

Adejumo, I. O., Adetunji, C. O., & Adeyemi, O. S. (2017). Influence of UV light exposure on mineral composition and biomass production of myco-meat produced from different agricultural substrates. *Journal of Agricultural Sciences, Belgrade, 62*(1), 51–59.

Adetunji, C. O., Ajayi, O. O., Akram, M., Olaniyan, O. T., Chishti, M. A., Inobeme, A., Olaniyan, S., Adetunji, J. B., Olaniyan, M., & Awotunde, S. O. (2021). Medicinal plants used in the treatment of influenza a virus infections. In K. Dua, S. Nammi, D. Chang, D. K. Chellappan, G. Gupta, & T. Collet (Eds.), *Medicinal plants for lung diseases*. Singapore: Springer. Available from https://doi.org/10.1007/978-981-33-6850-7_19.

Adetunji, C. O., Akram, M., Olaniyan, O. T., Ajayi, O. O., Inobeme, A., Olaniyan, S., Hameed, L., & Adetunji, J. B. (2021). Targeting SARS-CoV-2 novel Corona (COVID-19) virus infection using medicinal plants. In K. Dua, S. Nammi, D. Chang, D. K. Chellappan, G. Gupta, & T. Collet (Eds.), *Medicinal plants for lung diseases*. Singapore: Springer. Available from https://doi.org/10.1007/978-981-33-6850-7_21.

Adetunji, C. O., Anani, O. A., Olaniyan, O. T., Inobeme, A., Olisaka, F. N., Uwadiae, E. O., & Obayagbona, O. N. (2021). Recent trends in organic farming. In R. Soni, D. C. Suyal, P. Bhargava, & R. Goel (Eds.), *Microbiological activity for soil and plant health management*. Singapore: Springer. Available from https://doi.org/10.1007/978-981-16-2922-8_20.

Adetunji, C. O., Egbuna, C., Tijjani, H., Adom, D., Al-Ani, L. K. T., & Patrick-Iwuanyanwu, K. C. (2020). *Homemade preparations of natural biopesticides and applications. Natural remedies for pest, disease and weed control* (pp. 179–185). Publisher Academic Press.

Adetunji, C. O., Inobeme, A., Olaniyan, O. T., Ajayi, O. O., Olaniyan, S., & Adetunji, J. B. (2021a). Application of nanodrugs derived from active metabolites of medicinal plants for the treatment of inflammatory and lung diseases: Recent advances. In K. Dua, S. Nammi, D. Chang, D. K. Chellappan, G. Gupta, & T. Collet (Eds.), *Medicinal plants for lung diseases*. Singapore: Springer. Available from https://doi.org/10.1007/978-981-33-6850-7_26.

Adetunji, C. O., Inobeme, A., Olaniyan, O. T., Ajayi, O. O., Olaniyan, S., & Adetunji, J. B. (2021b). Application of nanodrugs derived from active metabolites of medicinal plants for the treatment of inflammatory and lung diseases: Recent advances. In K. Dua, S. Nammi, D. Chang, D. K. Chellappan, G. Gupta, & T. Collet (Eds.), *Medicinal plants for lung diseases*. Singapore: Springer. Available from https://doi.org/10.1007/978-981-33-6850-7_26.

Adetunji, C. O., Kumar, D., Raina, M., Arogundade, O., & Sarin, N. B. (2019). Endophytic microorganisms as biological control agents for plant pathogens: A panacea for sustainable agriculture. In A. Varma, S. Tripathi, & R. Prasad (Eds.), *Plant biotic interactions*. Cham: Springer. Available from https://doi.org/10.1007/978-3-030-26657-8_1.

Adetunji, C. O., Michael, O. S., Kadiri, O., Varma, A., Akram, M., Oloke, J. K., Shafique, H., Adetunji, J. B., Jain, A., Bodunrinde, R. E., Ozolua, P., & Ubi, B. E. (2021). Quinoa: From farm to traditional healing, food application, and phytopharmacology. In A. Varma (Ed.), *Biology and biotechnology of Quinoa*. Singapore: Springer. Available from https://doi.org/10.1007/978-981-16-3832-9_20.

Adetunji, C. O., Michael, O. S., Nwankwo, W., Ukhurebor, K. E., Anani, O. A., Oloke, J. K., Varma, A., Kadiri, O., Jain, A., & Adetunji, J. B. (2021). Quinoa, the next biotech plant: Food security and environmental and health hot spots. In A. Varma (Ed.), *Biology and biotechnology of quinoa*. Singapore: Springer. Available from https://doi.org/10.1007/978-981-16-3832-9_19.

Adetunji, C. O., Michael, O. S., Rathee, S., Singh, K. R. B., Ajayi, O. O., Adetunji, J. B., Ojha, A., Singh, J., & Singh, R. P. (2022). Potentialities of nanomaterials for the management and treatment of metabolic syndrome: A new insight. *Materials Today Advances, 13*100198.

Adetunji, C. O., Michael, O. S., Varma, A., Oloke, J. K., Kadiri, O., Akram, M., Bodunrinde, R. E., Imtiaz, A., Adetunji, J. B., Shahzad, K., Jain, A., Ubi, B. E., Majeed, N., Ozolua, P., & Olisaka, F. N. (2021). Recent advances in the application of biotechnology for improving the production of secondary metabolites from Quinoa. In A. Varma (Ed.), *Biology and biotechnology of Quinoa*. Singapore: Springer. Available from https://doi.org/10.1007/978-981-16-3832-9_17.

Adetunji, C. O., Nwankwo, W., Ukhurebor, K., Olayinka, A. S., & Makinde, A. S. (2021). Application of biosensor for the identification of various pathogens and pests mitigating against the agricultural production: Recent advances. In R. N. Pudake, U. Jain, & C. Kole (Eds.), *Biosensors in agriculture: Recent trends and future perspectives. Concepts and strategies in plant sciences*. Cham: Springer. Available from https://doi.org/10.1007/978-3-030-66165-6_9.

Adetunji, C. O., Ojediran, J. O., Adetunji, J. B., & Owa, S. O. (2019). Influence of chitosan edible coating on postharvest qualities of *Capsicum annum* L. during storage in evaporative cooling system. *Croatian Journal of Food Science and Technology, 11*(1), 59–66.

Adetunji, C. O., Olaniyan, O. T., Akram, M., Ajayi, O. O., Inobeme, A., Olaniyan, S., Khan, F. S., & Adetunji, J. B. (2021). Medicinal plants used in the treatment of pulmonary hypertension. In K. Dua, S. Nammi, D. Chang, D. K. Chellappan, G. Gupta, & T. Collet (Eds.), *Medicinal plants for lung diseases*. Singapore: Springer. Available from https://doi.org/10.1007/978-981-33-6850-7_14.

Adetunji, C. O., Oloke, J. K., & Prasad, G. (2020). Effect of carbon-to-nitrogen ratio on eco-friendly mycoherbicide activity from *Lasiodiplodia pseudotheobromae* C1136 for sustainable weeds management in organic agriculture. *Environment, Development and Sustainability, 22*, 1977−1990. Available from https://doi.org/10.1007/s10668-018-0273-1.

Adetunji, C. O., Palai, S., Ekwuabu, C. P., Egbuna, C., Adetunji, J. B., Ehis-Eriakha, C. B., Kesh, S. S., & Mtewa, A. G. (2021). *General principle of primary and secondary plant metabolites: Biogenesis, metabolism, and extraction. Preparation of phytopharmaceuticals for the management of disorders* (pp. 3−23). Publisher Academic Press.

Adetunji, C. O., Panpatte, D. G., Bello, O. M., & Adekoya, M. A. (2019). Application of nanoengineered metabolites from beneficial and eco-friendly microorganisms as a biological control agents for plant pests and pathogens. In D. Panpatte, & Y. Jhala (Eds.), *Nanotechnology for agriculture: Crop production & protection*. Singapore: Springer. Available from https://doi.org/10.1007/978-981-32-9374-8_13.

Adetunji, C. O., Phazang, P., & Sarin, N. B. (2017). Significance of rhamnolipids as a biological control agent in the management of crops/plant pathogens. *Current Trends in Biomedical Engineering & Biosciences, 10*(3), 54−55.

Adetunji, C. O., Roli, O. I., & Adetunji, J. B. (2020). Exopolysaccharides derived from beneficial microorganisms: Antimicrobial, food, and health benefits. In P. Mishra, R. R. Mishra, & C. O. Adetunji (Eds.), *Innovations in food technology*. Singapore: Springer. Available from https://doi.org/10.1007/978-981-15-6121-4_10.

Adetunji, C. O., & Varma, A. (2020). Biotechnological application of *Trichoderma*: A powerful fungal isolate with diverse potentials for the attainment of food safety, management of pest and diseases, healthy planet, and sustainable agriculture. In C. Manoharachary, H. B. Singh, & A. Varma (Eds.), *Trichoderma: Agricultural applications and beyond. Soil biology* (vol 61). Cham: Springer. Available from https://doi.org/10.1007/978-3-030-54758-5_12.

Adetunji, C.O. (2008). *The antibacterial activities and preliminary phytochemical screening of vernoniaamygdalina and Aloe vera against some selected bacteria* (pp. 40−43; M.Sc thesis). University of Ilorin.

Adetunji, C. O. (2019). Environmental impact and ecotoxicological influence of biofabricated and inorganic nanoparticle on soil activity. In D. Panpatte, & Y. Jhala (Eds.), *Nanotechnology for agriculture*. Singapore: Springer. Available from https://doi.org/10.1007/978-981-32-9370-0_12.

Adetunji, J. B., Adetunji, C. O., & Olaniyan, O. T. (2021). African walnuts: A natural depository of nutritional and bioactive compounds essential for food and nutritional security in Africa. In O. O. Babalola (Ed.), *Food security and safety*. Cham: Springer. Available from https://doi.org/10.1007/978-3-030-50672-8_19.

Adetunji, J. B., Ajani, A. O., Adetunji, C. O., Fawole, O. B., Arowora, K. A., Nwaubani, S. I., Ajayi, E. S., Oloke, J. K., & Aina, J. A. (2013). Postharvest quality and safety maintenance of the physical properties of Daucus carota L. fruits by Neem oil and Moringa oil treatment: A new edible coatings. *Agrosearch, 13*(1), 131−141.

Aljabeili, H. S., Barakat, H., & Abdel-Rahman, H. A. (2018). Chemical composition, antibacterial and antioxidant activities of thyme essential oil (*Thymus vulgaris*). *Food and Nutrition Sciences, 9*, 5.

Bello, O. M., Ibitoye, T., & Adetunji, C. (2019). Assessing antimicrobial agents of Nigeria flora. *Journal of King Saud University-Science, 31*(4), 1379−1383.

Ben Hsouna, A., Ben Halima, N., Smaoui, S., & Hamdi, N. (2017). Citrus lemon essential oil: Chemical composition, antioxidant and antimicrobial activities with its preservative effect against Listeria monocytogenes inoculated in minced beef meat. *Lipids in Health and Disease, 16*, 146. Available from https://doi.org/10.1186/s12944-017-0487-5.

Chouhan, S., Sharma, K., & Guleria, S. (2017). *Medicines (Basel), 4*(3), 58. https://doi.org/10.3390/medicines4030058.

Deng, W., Liu, K., Cao, S., Sun, J., Zhong, B., & Chun, J. (2020). Chemical composition, antimicrobial, antioxidant, and antiproliferative properties of grapefruit essential oil prepared by molecular distillation. *Molecules (Basel, Switzerland), 25*, 217. Available from https://doi.org/10.3390/molecules25010217.

Egbuna, C., Gupta, E., Ezzat, S. M., Jeevanandam, J., Mishra, N., Akram, M., Sudharani, N., Adetunji, C. O., Singh, P., Ifemeje, J. C., Deepak, M., Bhavana, A., Walag, A. M. P., Ansari, R., Adetunji, J. B., Laila, U., Olisah, M. C., & Onyekere, P. F. (2020). Aloe species as valuable sources of functional bioactives. In C. Egbuna, & G. Dable Tupas (Eds.), *Functional foods and nutraceuticals*. Cham: Springer. Available from https://doi.org/10.1007/978-3-030-42319-3_18.

Estévez, M. (2021). Critical overview of the use of plant antioxidants in the meat industry: Opportunities, innovative applications and future perspectives. *Meat Science, 181*108610. Available from https://doi.org/10.1016/j.meatsci.2021.108610.

Ivanišová, E., Kačániová, M., Savitskaya, T.A., & Grinshpan, D.D. (2021). Medicinal herbs: Important source of bioactive compounds for food industry. https://doi.org/10.5772/intechopen.98819.

Li, C., & Hou, L. (2018). Review on volatile flavor components of roasted oilseeds and their products. *Grain & Oil Science and Technology, 1*(2018), 151−156.

Mathavi, V., Sujatha, G., Ramya, S. B., & Devi, B. K. (2013). New trends in food processing. *International Journal of Advanced Engineering Technology, 5*(2013), 176.

Olaniyan, O. T., & Adetunji, C. O. (2021). Biological, biochemical, and biodiversity of biomolecules from marine-based beneficial microorganisms: Industrial perspective. In C. O. Adetunji, D. G. Panpatte, & Y. K. Jhala (Eds.), *Microbial rejuvenation of polluted environment. Microorganisms for sustainability* (vol 27). Singapore: Springer. Available from https://doi.org/10.1007/978-981-15-7459-7_4.

Pellegrini, M., Ricci, A., Serio, A., Chaves-López, C., Mazzarrino, G., D'Amato, S., Lo Sterzo, C., & Paparella, A. (2018). Characterization of essential oils obtained from Abruzzo autochthonous plants: Antioxidant and antimicrobial activities assessment for food application. *Foods*, *7*, 19. Available from https://doi.org/10.3390/foods7020019, [PMC free article] [PubMed] [CrossRef] [Google Scholar].

Perricone, M., Arace, E., Corbo, M. R., Sinigaglia, M., & Bevilacqua, A. (2015). Bioactivity of essential oils: A review on their interaction with food components. *Front Microbiology*, *6*, 76. Available from https://doi.org/10.3389/fmicb.2015.00076.

Raffo, A., & Paoletti, F. (2022). Fresh-cut vegetables processing: Environmental sustainability and food safety issues in a comprehensive perspective. *Frontiers in Sustainable Food Systems*, *5*681459. Available from https://doi.org/10.3389/fsufs.2021.681459.

Rauf, A., Akram, M., Semwal, P., Mujawah, A. A. H., Muhammad, N., Riaz, Z., Munir, N., Piotrovsky, D., Vdovina, I., Bouyahya, A., Adetunji, C. O., Shariati, M. A., Almarhoon, Z. M., Mabkhot, Y. N., & Khan, H. (2021). Antispasmodic potential of medicinal plants: A comprehensive review. *Oxidative Medicine and Cellular Longevity*, *2021*, 12. Available from https://doi.org/10.1155/2021/4889719, Article ID 4889719.

Saeed, K., Pasha, I., Chughtai, M., & Zuhair, M. (2022). Application of essential oils in food industry: Challenges and innovation. *Journal of Essential Oil Research*, *34*(5), 1–14. Available from https://doi.org/10.1080/10412905.2022.2029776.

Santoro, K., Maghenzani, M., Chiabrando, V., Bosio, P., Gullino, M. L., Spadaro, D., & Giacalone, G. (2018). Thyme and savory essential oil vapor treatments control brown rot and improve the storage quality of peaches and nectarines, but could favor gray mold. *Foods*, *7*, 7. Available from https://doi.org/10.3390/foods7010007, [PMC free article] [PubMed] [CrossRef] [Google Scholar].

Santos, M. I. S., Marques, C., Mota, J., Pedroso, L., & Lima, A. (2022). Applications of essential oils as antibacterial agents in minimally processed fruits and vegetables—A review. *Microorganisms*, *2022*(10), 760. Available from https://doi.org/10.3390/microorganisms1004076.

Satyal, P., Craft, J. D., Dosoky, N. S., & Setzer, W. N. (2017). The chemical compositions of the volatile oils of garlic (*Allium sativum*) and wild garlic (*Allium vineale*). *Foods*, *6*, 63. Available from https://doi.org/10.3390/foods6080063, [PMC free article] [PubMed] [CrossRef] [Google Scholar].

Satyavani, K., Gurudeeban, S., Manigandan, V., Rajamanickam, E., & Ramanathan, T. (2015). Chemical compositions of medical mangrove species *Acanthus ilicifolius*, *Excoecaria agallocha*, *Rhizophora apiculata* and *Rhizophora mucronata*. *Current Research in Chemistry*, *7*(2015), 1–8.

Sauceda, E. N. R. (2011). Uso de agentes antimicrobianos naturales en la conservacion de frutas y hortalizas. *Ra Ximhai*, *7*(2011), 153–170.

Sharopov, F., Valiev, A., Satyal, P., Gulmurodov, I., Yusufi, S., Setzer, W. N., & Wink, M. (2017). Cytotoxicity of the essential oil of Fennel (*Foeniculum vulgare*) from Tajikistan. *Foods*, *6*, 73. Available from https://doi.org/10.3390/foods6090073, [PMC free article] [PubMed] [CrossRef] [Google Scholar].

Surendran, S., Qassadi, F., Surendran, G., Lilley, D., & Heinrich, M. (2021). Myrcene—What are the potential health benefits of this flavouring and aroma agent? *Frontiers in Nutrition*, *8*699666. Available from https://doi.org/10.3389/fnut.2021.699666.

Thangadurai, D., Naik, J., Sangeetha, J., Al-Tawaha, A. R. M. S., Adetunji, C. O., Islam, S., David, M., Shettar, A. K., & Adetunji., J. B. (2021). Nanomaterials from agrowastes: Past, present, and the future. In O. V. Kharissova, L. M. Torres-Martínez, & B. I. Kharisov (Eds.), *Handbook of nanomaterials and nanocomposites for energy and environmental applications*. Cham: Springer. Available from https://doi.org/10.1007/978-3-030-36268-3_43.

Tolen, T. N., Ruengvisesh, S., & Taylor, T. M. (2017). Application of surfactant micelle-entrapped eugenol for prevention of growth of the shiga toxin-producing *Escherichia coli* in ground beef. *Foods*, *6*, 69. Available from https://doi.org/10.3390/foods6080069, [PMC free article] [PubMed] [CrossRef] [Google Scholar].

Ukhurebor, K. E., & Adetunji, C. O. (2021). Relevance of biosensor in climate smart organic agriculture and their role in environmental sustainability: What has been done and what we need to do? In R. N. Pudake, U. Jain, & C. Kole (Eds.), *Biosensors in agriculture: Recent trends and future perspectives. Concepts and strategies in plant sciences*. Cham: Springer. Available from https://doi.org/10.1007/978-3-030-66165-6_7.

Xing, C., Qin, C., Li, X., Zhang, F., Linhardt, R. J., Sun, P., & Zhang, A. (2019). Chemical composition and biological activities of essential oil isolated by HS-SPME and UAHD from fruits of bergamot. *LWT-Food Science and Technology*, *104*, 38–44. Available from https://doi.org/10.1016/j.lwt.2019.01.020.

Chapter 2

Extraction and processing of essential oils and their application in food industries

Babatunde Oluwafemi Adetuyi[1], Christiana Eleojo Aruwa[2], Peter Gbenga Oni[3], Frank Abimbola Ogundolie[4], Peace Abiodun Olajide[1], Pere-Ebi Yabrade Toloyai[5], Oluwakemi Semiloore Omowumi[1], Charles Oluwaseun Adetunji[6], Juliana Bunmi Adetunji[7], Yovwin D. Godwin[8], Osarenkhoe Omorefosa Osemwegie[9], Mohammed Bello Yerima[10], M.L. Attanda[11], Oluwabukola Atinuke Popoola[12], Olatunji Matthew Kolawole[13], Olalekan Akinbo[14] and Abel Inobeme[15]

[1]Department of Natural Sciences, Faculty of Pure and Applied Sciences, Precious Cornerstone University, Ibadan, Oyo State, Nigeria, [2]Department of Microbiology, School of Life Sciences, Federal University of Technology, Akure (FUTA), Akure, Ondo State, Nigeria, [3]Department of Chemistry and Biochemistry, Worchester Polytechnic Institute, Worcester, MA, United States, [4]Department of Biochemistry, Federal University of Technology, Akure (FUTA), Akure, Ondo State, Nigeria, [5]Department of Medical Biochemistry, Faculty of Basic Medical Sciences, Delta State University, Abraka, Delta State, Nigeria, [6]Applied Microbiology, Biotechnology and Nanotechnology Laboratory, Department of Microbiology, Edo State University Uzairue, Iyamho, Edo State, Nigeria, [7]Department of Biochemistry, Osun State University, Osogbo, Osun State, Nigeria, [8]Department of Family Medicine, Faculty of Clinical Sciences, Delta State University, Abraka, Delta State, Nigeria, [9]Department of Food Science and Microbiology, Landmark University, Omu-Aran, Kwara State, Nigeria, [10]Department of Microbiology, Sokoto State University, Sokoto, Sokoto State, Nigeria, [11]Department of Agricultural Engineering, Bayero University Kano, Kano, Kano State, Nigeria, [12]Genetics, Genomics and Bioinformatics Department, National Biotechnology Development Agency, Abuja, FCT, Nigeria, [13]Department of Microbiology, Faculty of Life Sciences, University of Ilorin, Ilorin, Kwara State, Nigeria, [14]Centre of Excellence in Science, Technology, and Innovation, AUDA-NEPAD, Johannesburg, Gauteng, South Africa, [15]Department of Chemistry, Edo State University Uzairue Iyamho, Auchi, Edo State, Nigeria

Introduction

According to Sánchez-González et al. (2011), essential oils (EOs) are vacuous fluids that largely include the sweet-smelling and erratic combinations which are naturally present in every aspect of a plant, which includes stem, seeds, strip, bark, blooms, and entire plants. They are utilized in the majority of nations as food additives, perfumes, and pharmaceuticals. They were initially utilized as medicines in the nineteenth century because of their scent and flavor. 3000 EOs have been discovered to far, and about 300 of these are used in perfumery because of their potent scent, according to Burt (2004). EOs are supporting metabolic substances that are crucial for plant defense mechanisms. They thus possess a range of medicinal qualities, including the capacity to transport microorganisms (Tajkarimi et al., 2010). Boyle (1955) claims that De la Croix made the earliest claim about the optional metabolites, specifically the EO fumes, having antibacterial effects in 1881. Since then, it has been demonstrated that EOs and their phytoconstituents exhibit a variety of natural behaviors, including behaviors that eradicate bacteria, insects, viruses, and fungi. Heterocyclic compounds like pyrazine are present in the majority of vegetable seeds, including sunflower and perilla seeds, and they significantly influence the flavor and character of the goods. Oils made from different vegetable seeds are vitamin- and protein-rich. The majority of the therapeutic plants utilized in Ayurveda and Siddha are the main origins of several unpredictable mixes that are in charge of a variety of organic activities. For instance, the main unstable combinations, such as esters, liquids, and alkenes, are referred to as EOs' key constituents having important pharmacological effects (Satyavani et al., 2015). Due to their smell, tastes, and common antibacterial components, EOs are mostly employed in the food industry to protect food. For instance, citrus-derived EOs with strong antibacterial properties, such as sesquiterpenes, oxygenated derivatives, and monoterpenes, are advised for use as seasoning and cell-reinforcing agents. Chemical, physical, and some microbial elements are crucial for the preservation of food. For a very long time, both

Applications of Essential Oils in the Food Industry. DOI: https://doi.org/10.1016/B978-0-323-98340-2.00002-X

producers and consumers have used synthetic additives in food initiatives. However, using artificial chemicals can result in some hypersensitive reactions, intoxication, cancerous development, and other degenerative disorders (Sauceda, 2011). As a result, alternative solutions must be considered. EOs and concentrated extracts of aromatic plants have recently become popular among food manufacturers due to their capacity to prevent the growth of harmful germs (Campos et al., 2016). EOs derived from thyme, cinnamon, and oregano exhibit considerable antimicrobial efficacy against *Bacillus thermosphacta, Escherichia coli, Pseudomonas fluorescens*, and *Listeria monocytogenes* (Mith et al., 2014). The application of EOs and their properties, such as their components of action, their impacts on foods, and their natural activities, such as antibacterial activity, are still not well understood. Because of this, the current study focused on the most recent data on the utilization of EOs for the preservation of food and their antibacterial and cell-reinforcing actions.

History of essential oils

The word "essential oil" has been utilized since the 16th century, when Paracelsus von Hohenheim of Switzerland gave it its name from the medicine "Quinta Essentia" (Guenther & Althausen, 1948). They are known as EOs or forces because they are flammable. EOs' meanings have been attempted by a few experts. The French Agency for Normalization has stated: EO, according to the Association Française de Normalization (AFNOR), is the output made from a plant by either vapor refining or "dry" refining—mechanical cycles through the epicarp of citrus. After that, actual techniques will be used to separate it from the watery stage. The raw materials utilized to extract EOs and their methods of extraction, such as cold retention or nonaqueous solvents are included in this definition. EOs are insoluble in water but extremely solvent in unpredictable mixtures like liquor, ether, and fixed oils. Due to the existence of sweet-smelling compounds, EOs extracted from vetiver, sassafras, cinnamon, and other common origins are fluid naturally & gloomy at ambient temperature (Dhifi et al., 2016). As a result, the cosmetics and fragrant healing industries typically employ them. The presence of unpredictable mixtures such as ketones, aldehydes, and sweet-smelling intensifies in EOs carry out an essential function in fragrance-based treatment as the inhalation of those mixtures, which are successfully relieving physical and mental stress. EOs are also extensively used for a variety of beneficial purposes, such as rub-scented healing, psycho-fragrance-based treatment, and olfactory-fragrance-based treatment. Additionally, EOs act as substance flags and are utilized by plants to control and direct their environment, including to defend themselves from danger, attract beneficial insects, like pollinators, and to establish communication between plants by drawing herbivores' attention to compound signs. Historically, hydro refining was employed to generate EOs in the historical times, which happened during the late 1200 seconds. Earlier than this, crude refining was utilized to produce EOs.

The refinement method was further developed by Arabic researchers, but the EOs they produced were not satisfactory. As a result, historians have centered on the remedial utilization of oil innovation in Europe in the middle ages since Villanova's work from the thirteenth century, which provides the earliest solidly confirmed record (Smith et al., 2011). Due to its significance, widespread use, and extensive application, EO-based fragrant healing has gained a significant amount of fame since the late twentieth century. The main ingredients in fragrance-based treatments are EOs. EOs can be applied directly to the skin's surface or through inward breath, a back rub, or a few other methods that direct them in the following amounts. Aromatic oils (other forms of EOs) can be used in a large scope of health benefits, which includes physical & mental equilibrium—the very foundation of fragrance-based treatments. It basically reduces pressure and prepares the person for the effort of the following day. It has been shown to be effective against microbial contaminations, Alzheimer's disease, cardiovascular and malignant growth diseases, and work-related stress during pregnancy (Perry & Perry, 2006). Recently, a growing trend has included disease treatment and dozing messes that use fragrances. According to Bowles (2003), the other natural mixtures found in EOs play a significant role in enhancing feelings of well-being.

Major constituents of essential oils

EOs are extremely mind-boggling due to the fact that they are constructed from a combination of more than fifty different parts at very different fixations. EOs are synthetically produced from a small quantity of separated compounds; particularly, the number and qualities of those mixtures are profound factors. Mostly, EOs are situated in the cytoplasms of specific plant cells, explicitly emitted in inward secretory pockets, epidermal cells, secretory cells, and secretory hairs or trichomes. EOs are a combination of north of 300 distinct mixtures; fundamentally comprising unstable mixtures with low sub-atomic loads of about under 1000 Da (normally 300 Da) (Lee et al., 2015). Fundamentally, scarcely any mixtures are available as significant ones at around 20%—70% contrasted with different mixtures, which are available in the following sums. Origanum compactum, on the other hand, contains carvacrol (30%) and thymols (27%), while *Coriandrum sativum* contains linalool (57%), while Artemisia herba alba contains camphor (24%), which can be found at high concentrations in other EOs.

The various organic activities of EOs are accountable to these significant components. The majority of these significant components are divided into two groups that are distinct biosynthetic starting points, such as terpenes and terpenoids and fragrant or aliphatic components (Pichersky et al., 2006). Because the air pressure is high enough at room temperature, EOs typically exist in the fractional fume state. Amines, ketones, aldehydes, alcohols, phenols, amides, and, most fundamentally, terpenes are just a few of the many unpredictability mixtures. Aldehydes, Alcohols, and ketones are among these, and they impart a wide range of sweet aromas to foods that are grown from specific plants' ground parts. For instance, natural products contain significant mixtures of -selinene, -nerolidol, and botanicals (Linalool). The most important components responsible for pharmacological activities are unpredictable mixtures, such as trimethylsilyl methanol, hydrocarbons, 4-heptanone, and 2-methyl extracted from hard apples (Moore et al., 2016). The main of EOs contain mixtures with a low subatomic weight, such as terpenes and terpenoids, which are responsible for a variety of activities, including food preservation. According to Tisserand and Young (2013), rosemary EOs contain a few monoterpenes, the most important of which are 1,8-cineole and camphor. These monoterpenes serve as bio-additives that prevent organisms' growth in food.

Types of essential oils

Because they include more than 50 different scent components, EOs are incredibly strange substances in nature. EOs include things like mustard oil, tea tree oil, lavender oil, clove oil, cinnamon oil, lemon oil, thyme oil, oregano oil, peppermint oil, eucalyptus oil, and so on. They are essential in stopping the development of pathogenic germs and safeguarding food. For instance, against Candida species such as *C. tropicalis*, *C. glabrata*, and *C. albicans*, the oxygenated and terpenes in the lemon rejuvenating ointment showed significant antifungal activity (Ooi et al., 2006). The leaves, bark, and roots of *Cinnamomum zeylanicum* are used to make cinnamon oil, a volatile substance. Three important substances that come from the bark extricate—linalool, eugenol, and trans-cinnamaldehyde—account for 82.5% of the overall synthesis. The most active component of cinnamon EOs, cinnamonaldehyde, has the ability to prevent the growth of parasites as well as a variety of pathogens, which includes Gram-positive and Gram-negative bacteria (Ramage et al., 2012). In a few investigations, the cinnamon EOs were also discovered to have qualities that strengthened cells, were antiparasitic, and hunted down free radicals (Terzi et al., 2007). Additionally, the tea-tree oil (TTO) from *Melaleuca alternifolia* (Myrtaceae) included large amounts of terpinen-4-old, -terpinene, pcymene, 1,8-cineole, -pinene, -terpineol, and -terpinene (Chaieb et al., 2007). It showed a movement against infectious strains with great vigor (Benabdelkader et al., 2011). The clove oil, which is commonly derived from clove buds, contains the phenylpropanoids mentioned above. Significant mixes include thymol, cinnamaldehyde, caryophyllene, 2-heptanone, carvacrol, derivation of eugenol, and eugenyl acetic acid. The development of the *C. albicans* microorganism tube may be inhibited by one of these, eugenol, which is frequently utilized as an antibacterial and antifungal agent, by reducing the union of ergosterol, a particular component of the cell wall. It demonstrated strong extremist searching behavior when tested against tert-butylated hydroxytoluene and revealed inhibitory effects for a number of safe Staphylococcus spp. (Végh et al., 2012). Significant components of lavender oil were linalool, linalyl acetic acid derivation (3,7-dimethyl-1,6-octadien-3yl acetic acid derivation), 1,8-cineole, terpinen-4-old, B-ocimene, l-fenchone, lavandulol, viridiflorol, and camphor. However, the levels of focus in these mixes differed from species to species. Linalool is one of the volatile compounds in lavender oil (Ait Said et al., 2015). Antimicrobial potential of Lavendar oil was observed against hostile to disease-safe minute organic entities, *Cryptococcus neoformans*, yeasts, dermatophytes, C. species, and Aspergillus strains (Posadzki et al., 2012). According to Tyagi and Malik (2011), the most important compound in eucalyptus oil is 1,8-cineole. Other mixtures include limonene, p-cymene, trans-pinocarveol, -pinene, spathulenol, cryptone, -terpineol, phellandral, cuminal, globulol, terpinene-4, and aromadendrene. Eucalyptus-derived EOs are mostly used as seasonings and have been shown to significantly reduce the growth of pathogenic and food waste microorganisms (Saharkhiz et al., 2012). The most important parts of EOs and how they work naturally were shown in Table 2.1. However, little is known about the antibacterial and antifungal properties of the constituents of eucalyptus oil. According to Tyagi and Malik (2011), synergistic impacts between the minor and major mixtures found in eucalyptus oil, as opposed to a single compound, were found to be f influenza, *S. aureus*, *Staph. aureus*, and *Haemophilus influenzae* On Staphylococci, peppermint oil showed impressive development-inhibitory efficacy against pyogenes. It was discovered to have significant

TABLE 2.1 Performance of ultrasound assisted extraction on various food components.

Product	Ultrasound process	Solvent	Performance	References
Carnosic acid extracted from rosemary	Batch, 40 kHz	Butanone and ethyl acetate	Decreased time of extraction	Albu et al. (2004)
Amino acids, polyphenols, and caffeine obtained from green tea	Batch, 40 kHz	H_2O	Higher yield at 65°C, in comparison to 85°C	Xia et al. (2006)
Oil from seeds of semi-oriental tobacco	Batch, 40 kHz	Hexane	High yield at 25°C for 20 min	
Ginger	Batch, 20 kHz	Supercritical CO_2	30% greater yield or shorter extraction times	Balachandaran et al. (2006)
Almond oil	Batch, 40 kHz	Hexane	Improved oil yield and accelerated extraction	Zhang et al. (2009)
Pomegranate oil	Batch, 20 kHz	Hexane	60% of the intended extraction efficiency was produced, and the extraction time and solvent quantity were reduced.	
Tomato seed oil	Batch, 28–34 kHz	Hexane	Reduced extraction time of 60 min at 60°C	
Papaya oil	Batch, 40 kHz	Hexane	Reduced extraction time (30 min), higher yield, and more stable	

antimicrobial activities both on halide and azole-vulnerable Candida species as well as antifungal efficacy against both conventional and severe pathogenic parasite forms of Candida species at IC50 values ranging from 0.5 to 8 g/mL (Cox et al., 2000).

Extraction of essential oils

From plant material, EOs are extracted using traditional techniques such as steam distillation, cold pressing, and hydro-distillation. Hydrodistillation assisted by ohmic heating, steam distillation assisted by microwaves, and hydrodistillation assisted by microwaves are all examples of cutting-edge processes (Hatami et al., 2019).

Additionally, the methods used for organic extracts, including supercritical fluid extraction (SFE), solvent extraction, ohmic heating-assisted extraction, and ultrasound-assisted extraction. There is a volatile component in these extracts that is occasionally mistaken for EO (El Asbahani et al., 2015; Fragoso-Jiménez et al., 2019).

Conventional extraction techniques use the refinery system to recover EO by warming a grid of a plant (El Asbahani et al., 2015). The extraction is finished by injecting water or vapor into the plant matter. Because EO is immiscible in water, tapping efficiently removes it. For the most part, the SD and HD techniques are utilized to remove EOs. According to Rasul (2018), these are straightforward to make, highly reproducible, and do not utilize natural solvents. Hassan et al. (2018) claims that these are easy to make, highly reproducible, and do not use natural solvents. The apparatus for extraction consists of a vessel (an alembic) in which we can place water and plant material as well as a source of warmth. Correspondingly, a cooling tower and a decanter are shown in the configuration to separate EOs from water and collect the condensate. The entire alembic is brought to a bubbling state after the plant material is just submerged in the water. When water vapors are thought of as a dissolvable drive, this process of simultaneously refining the EO and water mixture is known as co-refining.

Water's immiscibility with the majority of EOs' terpene atoms makes it ideal for basic decantation to effectively separate EOs from water after accumulation. The third edition of the European Pharmacopeia recommends the hydro-distillation by Clevenger framework for guaranteeing EO yields. Through the use of a cohobage framework, it makes it possible to reuse the condensates. These techniques are appropriate for the isolation of bloom and petals, such as rose petals, because it prevents plant material from compacting and clustering at the period of isolation. However, the hydro-distillation has a few limitations, which include: (1) a lengthy time of extraction (3–6 hours; a whole day for flower petals); (2) Prolonged contact with bubbling water can cause terpene remnants and synthetic changes (cyclization, hydrolysis); (3) excessive warming and the loss of a few polar (Bohra et al., 1994). The turbodistillation method, which is a more advanced version of this one, lets you get great returns by using the sweet-tasting water again. The presence of turbines shortens the time required for refinement (permit fracture and unsettling). It also makes it possible to virtually completely recover the EOs in the fume using the plate segment. This method is still being used on a modern scale for a variety of reasons, including the simplicity of the establishments (no expensive hardware is required), the simplicity of the strategy's execution, and its selectivity. The plant material is macerated in a natural dissolvable for organic solvent extraction; the dissolvable is eliminated under less pressure to concentrate the concentrate. In contrast to hydrodistillation, this method uses cold extraction to avoid changes and compound aging. Certain aroma constituents are water-soluble when plant material is drenched in bowling water during hydrodistillation, lowering the medium to pH 4−7 (sometimes below 4 for certain organic products). The first plant species' constituents undergo substance modifications (hydrations, hydrolysis, cyclizations, and deprotonations) in response to cumulative impacts of intensity and corrosive. Especially if bubbling is prolonged and the pH is low, acquired EOs differ significantly from the initial pith. Contrarily, natural dissolvable extricates, on the other hand, contain deposits that contaminate the food sources and scents they are added to (Faborode & Favier, 1996). This puts the health of the things removed in this way in jeopardy. These drawbacks can be overcome by combining a natural dissolvable with a low limit, like n-pentane, with the steam refining process (OS−SD). As a result, they cannot be used in food or medicine applications (Li & Tian, 2009).

The most common method for separating EOs from citrus natural product zing is cold pressing. Oil sacs, also known as oil organs, break open during extraction to release unstable oils limited to the outside of the mesocarpe. This oil is taken out exactly by crisp pressing yielding a watery emulsion. As a result, centrifugation is used to recover oil (Ferhat et al., 2007). In this instance, we have the plant analog of citrus zing, which is used as enhancing ingredients or additions in the food and pharmaceuticals industries (food industry, beauty products and few home care products). Absorption by the vapor is one authorized way to obtain EOs, albeit it is a challenging way to obtain EOs (Masango, 2005). The only difference is that there's no immediate contact between the plant and the water, so it is based on the same guidelines as hydrodistillation. By shortening the extraction time, substance modifications are reduced. There are various variants, as shown below.

Vapor-hydrodistillation

Extraction takes place inside the alembic, with the exception of a network or arrangement of punctured plates that keep the plant suspended above the foundation of the water that is still present, preventing them from coming into direct contact. The extraction is finished by injecting water fumes, which cut through plant matter from the ground up and move the unstable elements. There are few relics. Both the time of extraction and the lack of polar particles are reduced.

Steam distillation and vapor distillation

Although the generation of vapors takes place outside of the refining alembic, this technique has the same standards and benefits as the vapour-hydrodistillation method (Masango, 2005). After that, the steam can be submerged or superheated; The steam enters the lower portion of the extractor at a temperature that is slightly above the climatic tension and passes through the natural substance charge. Compared to hydrodistillation, this method avoids a few pitfalls (Masango, 2005).

Hydrodiffusion

In this particular case of fume refinement, the fume stream descends. It is also known as hydrodiffusion and gravity and down hydrodiffusion.

Advanced/innovative extraction techniques

Advanced extraction techniques like SFE, ultrasonic-assisted extraction (UAE), SFE, and microwave-assisted extraction (MAE) enhance the performance of extraction while consuming less energy and time in order to achieve extracts (Bakry et al., 2016). These methods increase the quantity of organic chemicals that can be extracted by breaking down the cell walls of the plant matrix with radiation like microwave or ultrasonic waves (El Asbahani et al., 2015). According to Kennouche et al. (2015), 65%−71% eugenol was present in the Eugenia caryophyllata EO that MAE and

MSD acquired in 2015. The antimicrobial and cancer-prevention properties of these concentrates were preserved. MAE reduces natural concentrate debasement, warming time, and energy consumption. The example can be subjected to ultrasonic waves ranging in frequency from 20 to 100 kHz either directly through the walls of the example compartment (ultrasonic bath) or through direct contact with the example (ultrasound framework combined with a test). The acoustic cavitation characteristic, in which the formation, expansion, and collapse of air pockets increase the selectivity of target atoms, can be achieved by applying acoustic power and wave frequencies to fluid media (Wei et al., 2016). SFE is used to specifically eliminate synthetic mixtures that involve carbon dioxide, a dissolvable in its supercritical state (Barajas-Álvarez et al., 2021). Furthermore, co-solvents like water, methanol, and ethanol alter the supercritical dissolvable's thickness, consistency, and solvation force, facilitating the extraction of specific mixtures (Yang et al., 2014). The organic material's unfavorable natural contaminations, poisons, and accumulations of pesticides are lessened by the SFE cycle (El Asbahani et al., 2015).

Ultrasound-assisted extraction

Ultrasound is a recently developed innovation that has been utilized in food science for extraction, preservation, and handling. According to Knorr et al. (2002), ultrasound has positive effects on food handling in terms of efficiency, yield, and selectivity. It also aids in the preservation of food and warm medicines and is harmless to the ecosystem. Ultrasound has advantages over conventional logic because it can work with the extraction cycle of various food components from animal and plant sources (Vilkhu et al., 2008) estimates are quick, precise, and completely computerized.

According to Gallo et al. (2018), sound waves can produce an energy known as ultrasonography with a frequency range of 20 KHz (which is higher than the conference's farthest reaches) to GHz. It is divided into power ultrasound (20−100 kHz), where cavitation is the dominant power, and symptomatic ultrasound (5 MHz−GHz). An emerging potential innovation, ultrasound-aided extraction has been increasingly utilized in the extraction industry. Ultrasound waves alter the exposed plant material's physical and synthetic properties, resulting in the formation of extractable mixtures through cavitation. When the cavitation bubble breaks down, it creates a temporary problem area with a high temperature (5000K) and pressure (1000 atm), which can dramatically accelerate the compound's reactivity in the medium (McNamara et al., 1999; Patist & Bates, 2008).

In research facilities, two kinds of ultrasound equipment are used the majority of the time. The first is the ultrasound cleaning bath, which reduces the size of the strong particles and thus increases their solvency during strong scattering into dissolvables. The substance response uses of ultrasound showers are decreasing. The ultrasonic test or horn framework is the next one. This is even more remarkable because, compared to an ultrasonic shower, the ultrasonic force is applied to a relatively small surface. The two main requirements are a source of high-energy vibration and a fluid medium. According to Patist and Bates (2008), the transducer that moves the vibration to the test is the vibrational energy source and is in direct contact with the handling medium.

An alternative method for extracting various food components (e.g., protein & oil) and bioactive fixings (such as cell reinforcements) from plant and animal sources is extreme focus ultrasound. According to Soria and Villamiel (2010), ultrasound-aided extraction is superior for thermally unstable compounds because it does not significantly alter the practical and primary properties of most bioactives. According to Babaei et al. (2006), the utilization of ultrasound in oil extraction has been found to shorten the time it takes to extract edible oil and to improve proficiency. At low temperatures, the efficiency of extraction sometimes increased, resulting in the production of a power item in a shorter amount of time. Carvone and limonene were extracted from caraway seeds using ultrasound, which resulted in about two overlap expansions (Chemat et al., 2011). Ultrasound combined with supercritical CO_2 substantially increased the rate of extraction of almond oil, gingeroles, and amaranth oil from seeds (Balachandaran et al., 2006).

In the case of rosemary's carnosic corrosive, the effects of various solvents and ultrasound extraction were investigated. When compared to other solvents like butanone and ethyl acetic acid derivation alone, it was discovered that ultrasound affected the general execution of ethanol. As a result, it is possible to enhance fluid extraction by substituting generally recognized as safe (GRAS) solvents for natural ones (Vilkhu et al., 2008). In accordance to Zhang et al. (2008) UAE of flaxseed outperforms traditional extraction for excellent recovery results. When compared to the standard method of extracting flavonols under the same conditions, the use of ultrasound resulted in higher flavonol yields. Almond and apricot seeds that were pre-treated with aqueous enzymatic oil extraction and ultrasonic before aqueous oil extraction showed better return with shorter time of extraction (Shah et al., 2005). The effect and procedure of ultrasound-aided food extraction are shown in Table 2.1.

In addition to improving food quality boundaries, ultrasound reduces the quality of some products by introducing off flavors, changing actual boundaries, and corrupting major and minor mixtures. Particles like OH and H extremists that build up on the surface of the cavitation bubble and revolutionaries in the fluid medium can both be produced by acoustic cavitation. The original arrangement of degraded things may have been made by these revolutionaries, which may have set off a revolutionary chain reaction and resulted in substantial quality defects in those items. According to Chemat et al. (2004), the handling of soybean, olive, and sunflower oils was the focus of the effects of ultrasound treatment, and significant negative changes were seen in their pieces as a result of the oxidation that was delivered during treatment. The expansion of ascorbic acid and ethanol can kill goodness extremists that form during the extraction cycle (Ashokkumar et al., 2008).

Extraction of supercritical fluids

Because of their extraordinary properties, supercritical fluids are used in a variety of analytical fields. It has recently received a lot of consideration as a promising alternative to conventional isolation techniques, so it is typically used as one of the methods for extracting oil from oilseeds (Norulaini et al., 2009). SFE is used commercially to extract EOs, pharmaceuticals, and textiles (Knez et al., 2014). A partition interaction that makes use of supercritical fluid as the dissolvable is the SFE. A supercritical fluid is a substance that is under pressure and/or temperature above its critical point. It has the ability to diffuse through solids like a gas and dissolve stuff like a fluid (Sapkale et al., 2010). CO_2 is the most popular dissolvable utilized in supercritical liquid extraction due to its similarity to solutes, lack of harmfulness, noncombustible nature, and reasonable availability (Sihvonen et al., 1999) as well as its basic temperature and pressure of 73.8 bar (31°C). Carbon dioxide can be used to separate a variety of mixtures with different atomic and extremity compounds. Changes in tension or temperature can limit a liquid's thickness, diffusivity, consistency, and dielectric steady without exceeding the stage limit in supercritical fluids (Bravi et al., 2007). In any case, high subatomic weight and polar mixtures don't seem to dissolve well in carbon dioxide. As a result, the expansion of a performer like ethanol or methanol can alter the salvation properties of supercritical carbon dioxide. According to Brunner (2005), these performers interact with the solvent and increase the solvency as a whole.

Depending on the requirements, the design of an SFE framework can be simple or complex. Both small- and large-scale SFE are possible. This is to maintain the desired extraction conditions, tests are conducted in the extraction cell, which is outfitted with temperature regulators and tension vessels at both ends. With the aid of siphons, which are also necessary for the liquid's movement throughout the structure, the liquid is compressed in the extraction tank. Performers are used to boost the solvent's solvency when carbon dioxide is used as a dissolvable. The liquid and its solubilized components are moved from the cell to the separator, where either increasing the temperature or decreasing the framework's tension reduces the liquid's solvation force. The item is then gathered by valves in the separator's lower half after that (Bravi et al., 2007; Brunner, 2005).

According to Gupta and Shim (2006), many common mixtures, such as nutrients, odors, common colors, and natural oils, are excellent solvents in supercritical fluids, which is why they are used for the extraction of important items. The SFE process, in contrast to conventional methods, which necessitate completely cleaning the important items to remove any remaining dissolvable, results in the complete removal of the dissolvable items through depressurization and improved mixture stability due to the lower process temperature. SFE was used to extract the EO from Verbena officinalis, which sped up the extraction process and increased yields (Safaralie et al., 2008). SFE has been demonstrated to be more efficient at extracting antibacterial mixes than the traditional approach. The remaining oil was successfully extracted from the palm part cake framework using supercritical carbon dioxide. Supercritical CO_2 is utilized to isolate a variety of oils, including grape seed oil, pumpkin seed oil, maize oil, canola seed oil, and wheat grain oil.

According to Ehlers et al. (2001), supercritical fluid has been found to be effective in the separation of natural ointment, resulting in excellent oil with increased good pieces. Because SFE takes place at a low temperature, it is the best method for working with mixtures that are thermally labile. The solvent's strength can change in supercritical fluids by changing the tension, but less by changing the temperature. Fluids with supercriticality are inactive; nontoxic can be quickly removed following extraction (Sapkale et al., 2010).

Subcritical extraction liquids (H$_2$ and CO$_2$)

Few studies demonstrated the utilization of subcritical water for the extraction of EOs (Özel et al., 2006). When the temperature falls below the critical temperature (Tc) but the tension rises above the critical pressure (Pc), the system enters the subcritical state. CO_2 and water are the most frequently used liquids for EO extraction at this point. The properties of acquired liquids are extremely intriguing: low consistency, a thickness that is similar to that of fluids, and diffusivity that is different from that of fluids and gas. Rovio et al. (1999) and Soto Ayala and Luque de Castro (2001) have demonstrated that subcritical water extraction (SWE) of EOs is a viable alternative due to its speed and utilization of little working temperature. This keeps mixtures, which are unpredictable and thermolabile, safe from misfortune and degradation. Additional benefits of using SWE include its positive environmental impact, low cost, and simplicity. The primary benefits of this method over conventional extraction methods include a naturally viable strategy, a shorter extraction time, a greater concentration of concentrate, lower extraction specialist costs, and low utilization of dissolvables (Mohammad & Eikani, 2007).

With exceptional EO productivity and quality, small deposits are produced. According to Luque de Castro et al. (1999), a comparison was made between SWE and supercritical CO_2. Although SWE is quite inexpensive than supercritical CO_2 extraction, the authors believed that it still requires high cost to carry out because the establishment needs specialized equipment (Mohammad & Eikani, 2007).

Microwave-assisted extraction

Electromagnetic fields with a frequency range of 300 MHz to 300 GHz are known as microwaves. Electrical fields and attractive fields are the two opposite swaying fields used to create these. In late 1970, microwave energy was used in logical exploration communities as a warming source. The rapid warming that results in significantly reduced activity time is the primary reason for the growing interest in MAE. According to Routray and Orsat (2012), the frequency of application in food is 915 MHz on a modern scale and 2450 MHz on homegrown broilers.

A quick method for specifically separating objective mixtures from various unrefined components is the MAE process. Due to its selectivity, volumetric warming, controllable warming cycle, as well as its ability to reduce environmental impact by transmitting less CO_2 into the air, MAE has been considered a significant option in the extraction procedure. According to Bousbia et al. (2009), the segregation of natural medicinal ointment through microwave-aided extraction is an intriguing option because it provides greater sufficiency than other conventional extraction methods.

The dissolvability of the analyte in the dissolvable, the mass exchange energy of the analyte from the network to the arrangement stage, and the strength of the analyte's cooperation are all aspects of the MAE of the analyte from the lattice. The increase in temperature, the expansion of the framework, and the development of faster extraction energy accelerate the rate of dissemination (Poole & Poole, 1996). The dissolvable used in microwave-aided extraction is determined by the objective analyte's dissolvability, the dissolvable's collaboration with plant structure, and the dissolvable's microwave-engulfing properties. The extraction period of 15−20 minutes is thought to be sufficient, but excellent recovery has been demonstrated as early as 40 seconds. On longer openness, solvents like water, ethanol, and methanol may significantly warm up, putting the fate of thermolabile components in jeopardy.

Two types of microwave-aided extraction frameworks are affordable: closed cells and a framework for the vessel. Since the solvents can be warmed to around 1000 C above their environmental edge of boiling over, shut vessel frameworks are typically used for extraction in extreme circumstances. According to Young (1995), the maximum power that can be transmitted through a closed vessel structure is somewhere between 600 and 1000 W. To get rid of the substantial buildup, filtration needs to be critical. In open cells, most of the time not completely settled by the dissolvable's boiling point (Letellier & Budzinski, 1999). When compared to closed frameworks, larger examples can be separated into open cells, which provides better example handling. The sample to be removed is placed in a soxhlet-type cellulose cartridge to avoid the filtering step. Collin et al. (1991) compared the use of a microwave to extract medicinal balms from ten distinct species to traditional hydrodistillation. Despite the fact that the yields were generally comparable, the chromatographic profiles were markedly distinct. Steam hydrodistillation and microwave-assisted hydrodistillation were used to separate unpredictable Vitex pseudo-negundo test mixtures. Rather than traditional hydrodistillation, microwave extraction yield was achieved at essentially shorter extraction times (Farjam et al., 2014). Two distinct spices—basil and epazote—were used to make a rejuvenating ointment using MAE. Regarding steam refining, a significant reduction in time, dissolvable, and yield of rejuvenating balm was observed. Saoud et al. (2006) looked into using ethanol as a dissolver in a microwave to extract natural eucalyptus ointment. For the microwave extraction process, three minutes of exposure was sufficient.

The effectiveness of the heating source is crucial to the success of MAE. The high temperature achieved by microwave heating drastically cuts down on both the amount of solvent required for extraction and the time of extraction (Kaufmann & Christen, 2002). Time for extraction is quicker (15−30 minutes). Precision reaction control is provided by temperature and pressure sensors. A brand-new extraction technique for nutraceuticals is MAE. This quick and easy approach works well for thermolabile components since it is straightforward (Wang & Weller, 2006). To remove the solid residue, though, filtration or centrifugation must be added as a further step. When the aimed substance or the used solvent are volatile or nonpolar, the microwave's efficiency decreases (Table 2.2).

TABLE 2.2 Various extraction techniques used to achieve essential oils: benefits and disadvantages.

Steam Distillation	Benefits	To boost the yield and efficiency, it can be adjusted and/or integrated with other methods; for example, it could raise the yield by 13%–29%.	Weiss (2002)
	Disadvantages	Need for enormous volumes of raw materials It takes time. Expensive Volatile substances in steam vaporize and even collapse	Akdağ and Öztürk (2019)
Hydrodistillation	Benefits	Cheap efficiency Simple to implement Compared to other methods, Clevenger provides better deodorization outcomes.	De Oliveira et al. (2019)
	Disadvantages	Lengthy extraction periods Wastewater generation Essential oils' (EO) composition changing or being lost	Masango (2005)
Supercritical Fluid Extraction	Benefits	Shorter extraction times Higher grade extracts use CO_2 as a solvent that is nontoxic, nonflammable, and residue-free. Higher yields	Haiyee et al. (2016)
	Disadvantages	There are no appreciable variations in the quality and quantitative content of turmeric EO compared to other approaches: Purity of turmerone is 67.7%–75% at 313K–320K and 20.8–26 MPa. being researched to increase optimization	Topiar et al. (2019)
Subcritical Water Extraction	Benefits	Particularly beneficial for removing nonpolar substances Increase a target chemical in a targeted manner Curcumin and EO extraction is green and efficient.	Gbashi et al. (2017)
	Disadvantages	Low implementation in industry currently	Essien et al. (2020)
Ultrasonic Extraction	Benefits	Increased mass transfer from the solvent to the plant cell Combining with other methods to boost productivity, shorten turnaround times, and cut costs	Baysal and Demirdoven (2011)
Microwave Energy (SFME, MAE)	Benefits	Reduce costs Reduce the duration of extractions, energy use, and CO_2 emissions. Combining with other performance-enhancing methods: VMHD, MHG, HDAM, and SDAM Decrease Hydrodistillation time from 4 h to 1 h of extraction No degradation products Maximum output	Akloul et al. (2014)
Solvent Extraction	Benefits	Avoids the loss of the EO's constituents and qualities while resolving the issue of excessive heat. Freons and chloroform are suitable and secure extractants.	Ching et al. (2014)

HDAM, Hydrodistillation assisted by microwave; *SDAM*, Steam distillation assisted by microwave; *VMHD*, Vacuum microwave hydrodistillation; *MHG*, Microwave by hydrodiffusion and gravity.

Concentration of the primary volatile compounds of the essential oil and organic extract as a result of the extraction method

It's important to note that the species determines how EOs and natural concentrates differ in structure (Alfikri et al., 2020; Ameur et al., 2021), the phenological stage 2020, agricultural ecology (Gioffrè et al., 2020), preliminary treatment (Yang et al., 2014), conditions of handling (Cavar Zeljkovíc et al., 2021), and extraction techniques.

Comparisons were done between the incredibly unstable combinations produced by the different extraction techniques, but each compound's convergence was unique. A distinctive light yellow tone was the EO that passed through the various cycles. However, due to contaminations, waxes, and natural waste, soxhlet extraction (SO) using ethanol may result in an earthy-colored natural concentrate (Guan et al., 2007). According to Golmakani et al. (2017), the extraction yield from MA HD after 60 minutes was identical to the yield from HD after 240 minutes. In essence, MA SD outperformed SD by over 4.8 times. The MAE decreases the extraction time from 10 to 2 hours while increasing the extraction yield by using two-overlap and less extreme boundaries. This is because these methods make it possible to reach the extraction temperature in a more restricted way than standard methods do. However, boundaries should be carefully controlled due to the fact that openness can alter EOs' substance arrangement. Tekin et al. (2015) reported that the UAE's natural concentrate acquired at 53 kHz contained significant amounts of eugenol, Caryophyllene, and eugenyl acetic acid derivation. Ghule and Desai (2021) separated eugenol and its derivative eugenyl acetic acid from clove buds through ultrasound-aided hydrotropic extraction. About 20% of the natural concentrate was produced by sonicating 8.2 g of ground cloves flowers in 150 mL of sodium cumene sulfonate 1.04 M for half an hour at 38°C using 158 W of ultrasonic irradiation power (26 kHz with a 7 mm breadth test). Molecule size, temperature, tension, and extraction time are the primary factors in SFE extraction. The clove's squashed molecule size decreases extraction yield because the clove has fewer dispersion pathways, resulting in less protection from dissemination between particles. The CO_2 thickness is altered by extraction strain and temperature; Because of the increased dissolvability of clove parts, the extraction yield is therefore higher. Moreover, there's a possibility that high-sub-atomic weight compounds, such as unsaturated fats, unsaturated fat methyl esters, and sterols, may also be removed from the natural concentrate. When it comes to the extraction of natural concentrates, SFE offers significant benefits over other techniques, such as faster extraction speeds, a higher concentration of dynamic cell reinforcement fixings, shorter extraction times, and so on (Batiha et al., 2021). SFE, on the other hand, necessitates expensive hardware, highly skilled administrators, and high operating and maintenance costs. In a similar vein, the combination of a co-dissolvable, continuous activity and CO_2 recycling may increase the cost of division without affecting the climate (Barajas-Álvarez et al., 2021). Hatami et al. (2019) ordered a cost-benefit analysis for separating cloves. They looked at a price of $40.00 per kilogram of separated EO, which would result in a potential annual pay of $5.9 million and a gross margin of 79% for every dollar contributed. The best chance of recovering the original investment would be between four and fourteen months. It has been looked into if EO may be encapsulated in polymeric particles. However, one of the biggest challenges in operations that include heating or evaporation is the loss of EOs. Encapsulation, on the contrary hand, may offer a variety of benefits, including defense against the deterioration of EOs. The truth is that sensitive EOs may be hampered in their natural movement by volatile or degraded dynamic fixes brought on by high temps, UV radiation, and oxidation. Detailing EOs into microcapsules or microspheres could likewise be used to control the arrival of common EOs.

Fessi et al.'s (1989) nanoprecipitation first developed the solvent displacement or nanoprecipitation process. Monodisperse nanoparticles can be produced using this straightforward and repeatable method. Additionally, it offers the advantages of being quick and inexpensive. Nanoprecipitation enables the production of submicron particles with limited dispersion by utilizing a minimal external energy source (Legrand et al., 2007). EOs and other hydrophobic substances can be encased using this method. Aqueous and organic phases must be two miscible phases for nanoprecipitation to occur. The EO is mixed with a suitable solvent in an extraction liquid during the organic phase. According to Khoee and Yaghoobian (2009), the polymer's aqueous phase consists of a mixture of non-solvent or non-solvent in addition to one or more surfactants, either naturally occurring or manufactured. For encasing materials that are hydrophobic, this method has received a lot of attention (Rosset et al., 2012). Polymers made by nature or man can also exist. According to Khoee and Yaghoobian (2009), poly(e-caprolactone), poly(lactide-co-glycolide), and poly(lactide) are the biodegradable polymers that are utilized the most frequently. Due to their hydrophobic properties, EOs are an excellent choice for encapsulation via nanoprecipitation in nanoparticular systems. In the research carried out by Ladj-Minost (2012), doxorubicin, a hydrophilic active, and indomethacin, a hydrophobic active, were both nanoprecipitated using polylactide polymer. According to Ladj-Minost (2012), hydrophobicity was found to lower the size of nanoparticles and add to the efficiency of active molecule trapping.

Coacervation

Aqueous and organic phases must be two miscible stages for nanoprecipitation to occur. The EO is mixed with a suitable solvent in an extraction liquid during the organic phase. The other stage is the equilibrium solution, but the polymer-rich coacervate phase is not. Simple coacervation only involves one polymer, in contrast to complex coacervation, which requires the interaction of two colloids with opposite charges (Kaushik et al., 2014).

In 1949, Bungenberg de Jong made a distinction between simple and complex forms of coacervation. When a poor solvent is added to a hydrophilic colloidal solution, simple coacervation occurs, resulting in the formation of two stages:

one that is almost entirely devoid of coacervate and contains a lot of colloid molecules, called coacervate. For instance, a coacervate is formed when a gelatin solution is progressively mixed with acetone, alcohol, or sodium sulfate solution (Shimokawa et al., 2013).

Rapid expansion of supercritical solutions

Some disadvantages of conventional techniques include the utilization of a lot of organic solvents, a large scope of particle sizes, and solvent residues. To circumvent these limitations, processes based on supercritical fluids have been utilized. The latter method has emerged as an appealing alternative to encapsulating natural compounds as a result of the utilization of solvents that are not dangerous to the environment (Santos et al., 2013). Supercritical CO_2 is one of the range of supercritical fluids that are frequently used in the production of chemical and pharmaceutical component particles due to its affordability and environmental friendliness (Yim et al., 2013). As a result of low critical temperature (31.1°C), supercritical CO_2 is widely utilized for the precipitation of materials that are thermally sensitive (Yim et al., 2013). Supercritical CO_2 in the rapid expansion of supercritical solution at very increased pressure and temperatures are used to dissolve solutes (up to 250 bar and 800 C) before the solutions are expanded. Solutes precipitate because their solubility decreases at lower pressures. Supercritical CO_2 should dissolve both the active compound and the solutes that will be used for encapsulation (Vinjamur et al., 2013).

Conclusion

The conclusion can be reached from the fact that EOs are organic compounds made of intricate arrangements of various volatile components. They're utilized in various ways such as in the food, cosmetic, agricultural, pharmaceutical, and light sectors. There are many options for extraction. The risks of chemical modification, the lengthy extraction process, and the high energy consumption are all overcome by novel approaches. EOs can be used in various aspects, but when used in this way, they are very sensitive to the environment. Encapsulation is a viable option that may increase EOs' stability. This objective has been accomplished in a number of different ways, resulting in some interesting outcomes. When EOs were added to particles or liposomes, numerous additional advantages, such as increased efficacy and sustained release, were seen. Currently, there is a lot of interest in the mixing of EOs and active compounds to create colloidal particles, which are mostly employed in local skin therapy, dermatology, and also as a unique utilization, cosmeto-textile.

References

Ait Said, L., Zahlane, K., Ghalbane, I., El Messoussi, S., Romane, A., Cavaleiro, C., & Salgueiro, L. (2015). Chemical composition and antibacterial activity of Lavandula coronopifolia essential oil against antibiotic-resistant bacteria. *Natural Product Research*, 29(6), 582−585.

Akdağ, A., & Öztürk, E. (2019). Distillation methods of essential oils. *Selçuk Üniversitesi Fen Fakültesi Fen Dergisi*, 45(1), 22−31.

Akloul, R., Benkaci-Ali, F., & Eppe, G. (2014). Kinetic study of volatile oil of Curcuma longa L. rhizome and Carum carvi L. fruits extracted by microwave-assisted techniques using the cryogrinding. *Journal of Essential Oil Research*, 26(6), 473−485.

Albu, S., Joyce, E., Paniwnyk, L., Lorimer, J. P., & Mason, T. J. (2004). Potential for the use of ultrasound in the extraction of antioxidants from Rosmarinus officinalis for the food and pharmaceutical industry. *Ultrasonics Sonochemistry*, 11(3−4), 261−265.

Alfikri, F. N., Pujiarti, R., Wibisono, M. G., & Hardiyanto, E. B. (2020). Yield, quality, and antioxidant activity of clove (Syzygium aromaticum L.) bud oil at the different phenological stages in young and mature trees. *Scientifica, 2020*.

Ameur, E., Sarra, M., Yosra, D., Mariem, K., Nabil, A., Lynen, F., & Larbi, K. M. (2021). Chemical composition of essential oils of eight Tunisian Eucalyptus species and their antibacterial activity against strains responsible for otitis. *BMC Complementary Medicine and Therapies*, 21(1), 1−16.

Ashokkumar, M., Sunartio, D., Kentish, S., Mawson, R., Simons, L., Vilkhu, K., & Versteeg, C. K. (2008). Modification of food ingredients by ultrasound to improve functionality: A preliminary study on a model system. *Innovative Food Science & Emerging Technologies*, 9(2), 155−160.

Babaei, R., Jabbari, A., & Knud, M. (2006). Solid- liquid extraction of fatty acids of some variety of Iranian rice in closed vessel in the absence and presence of ultrasonic waves. *Asian Journal of Chemistry*, 18, 57−64.

Bakry, A. M., Abbas, S., Ali, B., Majeed, H., Abouelwafa, M. Y., Mousa, A., & Liang, L. (2016). Microencapsulation of oils: A comprehensive review of benefits, techniques, and applications. *Comprehensive Reviews in Food Science and Food Safety*, 15(1), 143−182.

Balachandaran, S., Kentish, E., Mawson, R., & Ashokkumar, M. (2006). Ultrasonic enhancement of the supercritical extraction of ginger. *Ultrasonics Sonochem*, 13, 471−479.

Barajas-Álvarez, P., Castillo-Herrera, G. A., Guatemala-Morales, G. M., Corona-González, R. I., Arriola-Guevara, E., & Espinosa-Andrews, H. (2021). Supercritical CO_2-ethanol extraction of oil from green coffee beans: Optimization conditions and bioactive compound identification. *Journal of Food Science and Technology*, 58(12), 4514−4523.

Batiha, G. B., Awad, D. A., Algamma, A. M., Nyamota, R., Wahed, M. I., Shah, M. A., Amin, M. N., Adetuyi, B. O., Hetta, H. F., Cruz-Marins, N., Koirala, N., Ghosh, A., & Sabatier, J. (2021). Diary-derived and egg white proteins in enhancing immune system against COVID-19faron. *Frontiers in Nutritionr. (Nutritional Immunology)*, 8, 629440. Available from https://doi.org/10.3389/fnut.2021629440.

Baysal, T., & Demirdoven, A. (2011). Ultrasound in food technology. In D. Chen, S. K. Sharma, & A. Mudhoo (Eds.), *Handbook on applications of ultrasound: Sonochemistry for sustainability* (pp. 163–182). Boca Raton, FL: CRC Press.

Benabdelkader, T., Zitouni, A., Guitton, Y., Jullien, F., Maitre, D., Casabianca, H., . . . Kameli, A. (2011). Essential oils from wild populations of Algerian Lavandula stoechas L.: Composition, chemical variability, and in vitro biological properties. *Chemistry & Biodiversity*, 8(5), 937–953.

Bohra, P. M., Vaze, A. S., Pangarkar, V. G., & Taskar, A. (1994). Adsorptive recovery of water soluble essential oil components. *Journal of Chemical Technology and Biotechnology (Oxford, Oxfordshire: 1986)*, 60, 97–102.

Bousbia, N., Vian, M. A., Ferhat, M. A., Meklati, B. Y., & Chemat, F. (2009). A new process for extraction of essential oil from Citrus peels: Microwave hydrodiffusion and gravity. *Journal of Food Engineering*, 90(3), 409–413.

Bowles, E.J. (2003). The chemistry of aromatherapeutic oils, 37, 85, 98, 122, 168.

Boyle, W. (1955). Spices and essential oils as preservatives. *The American Perfumer and Essential Oil Review*, 66(1), 25–28.

Bravi, E., Perretti, G., Montanari, L., Favati, F., & Fantozzi, P. (2007). Supercritical fluid extraction for quality control in beer industry. *The Journal of Supercritical Fluids*, 42(3), 342–346.

Brunner, G. (2005). Supercritical fluids: Technology and application to food processing. *Journal of Food Engineering*, 67(1–2), 21–33.

Burt, S. (2004). Essential oils: Their antibacterial properties and potential applications in foods—a review. *International Journal of Food Microbiology*, 94(3), 223–253.

Campos, T., Barreto, V., Queiros, R., Ricardo-Rodrigues, S., Felix, M. R., & Laranjo, M. (2016). Use of essential oils in food preservation, Conservacao de morangos com utilizacao de oleosessenciais. *Agrotech Journal*, 18, 90–96.

Cavar Zeljkovïc, S., Sm´ékalová, K., Kaffková, K., & Štefelová, N. (2021). Influence of post-harvesting period on quality of thyme and spearmint essential oils. *Journal of Applied Research on Medicinal and Aromatic Plants*, 25, 100335, [CrossRef].

Chaieb, K., Hajlaoui, H., Zmantar, T., Kahla-Nakbi, A. B., Rouabhia, M., Mahdouani, K., & Bakhrouf, A. (2007). The chemical composition and biological activity of clove essential oil, Eugenia caryophyllata (Syzigium aromaticum L. Myrtaceae): A short review. *Phytotherapy Research: An International Journal Devoted to Pharmacological and Toxicological Evaluation of Natural Product Derivatives*, 21(6), 501–506.

Chemat, F., Grondin, I., Shum Cheong Sing, A., & Smadja, J. (2004). Deterioration of edible oils during food processing by ultrasound. *Ultrasonics Sonochemistry*, 11(1), 13–15. Available from https://doi.org/10.1016/S1350-4177(03)00127-5. PMID: 14624981.

Chemat, F., Zill-e-Huma., & Khan, M. K. (2011). Application of ultrasound in food technology: Processing, preservation and extraction. *Ultrasonics Sonochem*, 18, 813–835.

Ching, W. Y., Bin-Yusoff, Y., & Wan-Amarina, W. N. B. (2014). Extraction of essential oil from Curcuma longa. *Journal of Food Chemistry and Nutrition*, 2(1), 01–10.

Collin, G. J., Lord, D., Allaire, J., & Gagnon, D. (1991). Essential oil and microwave extracts. *Parfums Cosmèt. Aromes*, 97, 105–112.

Cox, S. D., Mann, C. M., Markham, J. L., Bell, H. C., Gustafson, J. E., Warmington, J. R., & Wyllie, S. G. (2000). The mode of antimicrobial action of the essential oil of Melaleuca alternifolia (tea tree oil). *Journal of Applied Microbiology*, 88(1), 170–175.

De Oliveira, M. S., Silva, S. G., da Cruz, J. N., Ortiz, E., da Costa, W. A., Bezerra, F. W. F., . . . de Aguiar Andrade, E. H. (2019). Supercritical CO_2 application in essential oil extraction. *Industrial Applications of Green Solvents*, 2, 1–28.

Dhifi, W., Bellili, S., Jazi, S., Bahloul, N., & Mnif, W. (2016). Essential oils' chemical characterization and investigation of some biological activities: A critical review, A.F.N.O.R., les huiles essentielles, monographie relative aux huiles essentielles, Paris 665. (2000) *Medicines*, 3(4), 25.

Ehlers, D., Nguyen, T., Quirin, K. W., & Gerard, D. (2001). Analysis of essential basil oils-CO_2 extracts and steam-distilled oils. *Deutsche Lebensmittel-Rundschau*, 97(7), 245–250.

El Asbahani, A., Miladi, K., Badri, W., Sala, M., Addi, E. A., Casabianca, H., . . . Elaissari, A. (2015). Essential oils: From extraction to encapsulation faron. *International Journal of Pharmaceutics*, 483(1–2), 220–243.

Essien, S. O., Young, B., & Baroutian, S. (2020). Recent advances in subcritical water and supercritical carbon dioxide extraction of bioactive compounds from plant materials. *Trends in Food Science & Technology*, 97, 156–169.

Faborode, M. O., & Favier, J. F. (1996). Identification and significance of the oil-point in seed-oil expression. *Journal of Agricultural Engineering Research*, 65, 335–345.

Farjam, M. H., Zardosht, M., & Joukar, M. (2014). Comparison of microwave assisted hydrodistillation and traditional hydrodistillation methods for extraction of the Vitex pseudonegundo essential oils. *Advances in Environmental Biology*, 8, 82–85.

Ferhat, M. A., Meklati, B. Y., & Chemat, F. (2007). Comparison of different isolation methods of essential oil from citrus fruits: Cold pressing: Hydrodistillation and microwave dry distillation. *Flavour and Fragrance Journal*, 22, 494–504.

Fessi, H., Puisieux, F., Devissaguet, J.P., Ammoury, N., & Benita, S. (1989). Nanocapsule formation by interfacial polymer deposition following solvent displacement. *International Journal of Pharmaceutics* 55, R1–R4.

Fragoso-Jiménez, J. C., Tapia-Campos, E., Estarron-Espinosa, M., Barba-Gonzalez, R., Castañeda-Saucedo, M. C., & Castillo-Herrera, G. A. (2019). Effect of supercritical fluid extraction process on chemical composition of Polianthes tuberosa flower extracts. *Processes*, 7(2), 60.

Gallo, M., Ferrara, L., & Naviglio, D. (2018). Application of ultrasound in food science and technology: A perspective. *Foods*, 7(10), 164. Available from https://doi.org/10.3390/foods7100164. PMID: 30287795; PMCID: PMC6210518.

Gbashi, S., Adebo, O. A., Piater, L., Madala, N. E., & Njobeh, P. B. (2017). Subcritical water extraction of biological materials. *Separation & Purification Reviews*, 46(1), 21–34.

Ghule, S. N., & Desai, M. A. (2021). Intensified extraction of valuable compounds from clove buds using ultrasound assisted hydrotropic extraction. *Journal of Applied Research on Medicinal and Aromatic Plants*, 25, 100325.

Gioffrè, G., Ursino, D., Labate, M. L. C., & Giuffrè, A. M. (2020). The peel essential oil composition of bergamot fruit (Citrus bergamia, Risso) of Reggio Calabria (Italy): A review. *Emirates Journal of Food and Agriculture*, 835−845.

Golmakani, M. T., Zare, M., & Razzaghi, S. (2017). Eugenol enrichment of clove bud essential oil using different microwave-assisted distillation methods. *Food Science and Technology Research*, 23(3), 385−394.

Guan, W., Li, S., Yan, R., Tang, S., & Quan, C. (2007). Comparison of essential oils of clove buds extracted with supercritical carbon dioxide and other three traditional extraction methods. *Food Chemistry*, 101(4), 1558−1564.

Guenther, E., & Althausen, D. (1948). *The essential oils* (Vol. 1, p. 81). New York: Van Nostrand.

Gupta, R. B., & Shim, J. J. (2006). *Solubility in supercritical carbon dioxide*. CRC press.

Haiyee, Z. A., Shah, S. H. M., Ismail, K., Hashim, N., & Ismail, W. I. W. (2016). Quality parameters of Curcuma longa L. extracts by supercritical fluid extraction (SFE) and ultrasonic assisted extraction (UAE). *Malaysian Journal of Analytical Sciences*, 20(3), 626−632.

Hassan, S. S. M., Abdel-Shafy, H. I., & Mansour, M. S. M. (2018). Removal of pyrene and benzo(a)pyrene micropollutant from water via adsorption by green synthesized iron oxide nanoparticles. *Advances in Natural Sciences: Nanoscience and Nanotechnology*, 9(1). Available from https://doi.org/10.1088/2043-6254/aaa6f0.

Hatami, T., Johner, J. C., Zabot, G. L., & Meireles, M. A. A. (2019). Supercritical fluid extraction assisted by cold pressing from clove buds: Extraction performance, volatile oil composition, and economic evaluation. *The Journal of Supercritical Fluids*, 144, 39−47.

Kaufmann, B., & Christen, P. (2002). Recent extraction techniques for natural products: Microwave-assisted extraction and pressurised solvent extraction. *Phytochemical Analysis: An. International Journal of Plant Chemical and Biochemical Techniques*, 13(2), 105−113.

Kaushik, P., Dowling, K., Barrow, C. J., & Adhikari, B. (2014). Microencapsulation of omega-3 fatty acids: A review of microencapsulation and characterization methods. *Journal of Functional Foods*.

Kennouche, A., Benkaci-Ali, F., Scholl, G., & Eppe, G. (2015). Chemical composition and antimicrobial activity of the essential oil of Eugenia caryophyllata Cloves extracted by conventional and microwave techniques. *Journal of Biologically Active Products from Nature*, 5, 1−11, [CrossRef].

Khoee, S., & Yaghoobian, M. (2009). An investigation into the role of surfactants in controlling particle size of polymeric nanocapsules containing penicillin-G in double emulsion. *European Journal of Medicinal Chemistry*, 44, 2392−2399.

Knez, Ž., Markočič, E., Leitgeb, M., Primožič, M., Hrnčič, M. K., & Škerget, M. (2014). Industrial applications of supercritical fluids: A review. *Energy*, 77, 235−243.

Knorr, D., Ade-Omowaye, B. I. O., & Heinz, V. (2002). Nutritional improvement of plant foods by nonthermal processing. *P Nutri Soc*, 61, 311−318.

Ladj-Minost, A. (2012). *Répulsifs d'arthropodes à durée d'action prolongée: étude pharmacotechnique, devenir in situ et efficacité* (Ph.D. thesis). Université Claude Bernard, Lyon I.

Lee, S. H., Do, H. S., & Min, K. J. (2015). Effects of essential oil from Hinoki cypress, Chamaecyparis obtusa, on physiology and behavior of flies. *PLoS One*, 10(12), e0143450.

Legrand, P., Lesieur, S., Bochot, A., Gref, R., Raatjes, W., Barratt, G., & Vauthier, C. (2007). Influence of polymer behaviour in organic solution on the production of polylactide nanoparticles by nanoprecipitation. *International Journal of Pharmaceutics*, 344, 33−43.

Letellier, M., & Budzinski, H. (1999). Microwave assisted extraction of organic compounds. *Analusis*, 27(3), 259−270.

Li, X.-M., & Tian, S.-L. (2009). Extraction of Cuminum cyminum essential oil by combination technology of organic solvent with low boiling point and steam distillation. *Food Chemistry*, 115, 1114−1119.

Luque de Castro, M. D., Jiménez-Carmona, M. M., & Fernández-Pérez, V. (1999). Towards more rational techniques for the isolation of valuable essential oils from plants. *TrAC Trends in Analytical Chemistry*, 18, 708−716.

Masango, P. (2005). Cleaner production of essential oils by steam distillation. *Journal of Cleaner Production*, 13(8), 833−839.

McNamara, W. B., Didenko, Y. T., & Suslick, K. S. (1999). Sonoluminescence temperature during multi bubble cavitation. *Nature*, 401, 772−775.

Mith, H., Dure, R., Delcenserie, V., Zhiri, A., Daube, G., & Clinquart, A. (2014). Antimicrobial activities of commercial essential oils and their components against food-borne pathogens and food spoilage bacteria. *Food Science & Nutrition*, 2(4), 403−416.

Mohammad, H., & Eikani, F. G. (2007). Subcritical water extraction of essential oils from coriander seeds (Coriandrum sativum L.). *Journal of Food Engineering*, 80, 735−740. Available from https://doi.org/10.1016/j.jfoodeng.2006.05.015.

Moore, J., Yousef, M., & Tsiani, E. (2016). Anticancer effects of rosemary (Rosmarinus officinalis L.) extract and rosemary extract polyphenols. *Nutrients*, 8(11), 731.

Norulaini, N. N., Setianto, W. B., Zaidul, I. S. M., Nawi, A. H., Azizi, C. M., & Omar, A. M. (2009). Effects of supercritical carbon dioxide extraction parameters on virgin coconut oil yield and medium-chain triglyceride content. *Food Chemistry*, 116(1), 193−197.

Ooi, L. S., Li, Y., Kam, S. L., Wang, H., Wong, E. Y., & Ooi, V. E. (2006). Antimicrobial activities of cinnamon oil and cinnamaldehyde from the Chinese medicinal herb Cinnamomum cassia Blume. *The American Journal of Chinese Medicine*, 34(03), 511−522.

Özel, M. Z., Gögüş, F., & Lewis, A. C. (2006). Comparison of direct thermal desorption with water distillation and superheated water extraction for the analysis of volatile components of Rosa damascena Mill. using GCxGC-TOF/MS. *Analytica Chimica Acta*, 566, 172−177.

Patist, A., & Bates, D. (2008). Ultrasonic innovations in the food industry: From the laboratory to commercial production. *Innovative Food Science and Emerging Technologies*, 9, 147−154.

Perry, N., & Perry, E. (2006). Aromatherapy in the management of psychiatric disorders. *CNS Drugs*, 20(4), 257−280.

Pichersky, E., Noel, J. P., & Dudareva, N. (2006). Biosynthesis of plant volatiles: Nature's diversity and ingenuity. *Science (New York, N.Y.)*, 311 (5762), 808−811.

Poole, C. F., & Poole, S. K. (1996). Highlight. Trends in extraction of semivolatile compounds from solids for environmental analysis. *Analytical Communications, 33*(7), 11H–14H.

Posadzki, P., Alotaibi, A., & Ernst, E. (2012). Adverse effects of aromatherapy: A systematic review of case reports and case series. *International Journal of Risk & Safety in Medicine, 24*(3), 147–161.

Ramage, G., Milligan, S., Lappin, D. F., Sherry, L., Sweeney, P., Williams, C., . . . Culshaw, S. (2012). Antifungal, cytotoxic, and immunomodulatory properties of tea tree oil and its derivative components: Potential role in management of oral candidosis in cancer patients. *Frontiers in microbiology, 3*, 220.

Rasul, M. G. (2018). Conventional extraction methods use in medicinal plants, their advantages and disadvantages. *International Journal of Basic Sciences and Applied Computing, 2*, 10–14.

Rosset, V., Ahmed, N., Zaanoun, I., Stella, B., Fessi, H., & Elaissari, A. (2012). Elaboration of Argan oil nanocapsules containing naproxen for cosmetic and transdermal local application. *Journal of Colloid Science and Biotechnology, 1*, 218–224.

Routray, W., & Orsat, V. (2012). Microwave-assisted extraction of flavonoids: A review. *Food and Bioprocess Technology, 5*(2), 409–424.

Rovio, S., Hartonen, K., Holm, Y., Hiltunen, R., & Riekkola, M.-L. (1999). Extraction of clove using pressurized hot water. *Flavour and Fragrance Journal, 14*, 399–404.

Safaralie, A., Fatemi, S., & Sefidkon, F. (2008). Essential oil composition of Valeriana officinalis L. roots cultivated in Iran: Comparative analysis between supercritical CO_2 extraction and hydrodistillation. *Journal of Chromatography. A, 1180*(1-2), 159–164.

Saharkhiz, M. J., Motamedi, M., Zomorodian, K., Pakshir, K., Miri, R., & Hemyari, K. (2012). Chemical composition, antifungal and antibiofilm activities of the essential oil of Mentha piperita L. *International Scholarly Research Notices, 2012*.

Sánchez-González, L., Vargas, M., González-Martínez, C., Chiralt, A., & Chafer, M. (2011). Use of essential oils in bioactive edible coatings: A review. *Food Engineering Reviews, 3*(1), 1–16.

Santos, D. T., Albarelli, J. Q., Beppu, M. M., & Meireles, M. A. A. (2013). Stabilization of anthocyanin extract from jabuticaba skins by encapsulation using supercritical CO_2 as solvent. *Food Research International, 50*, 617–624.

Saoud, A. A., Yunus, R. M., & Aziz, R. A. (2006). Microwave-assisted extraction of essential oil from Eucalyptus: Study of the effects of operating conditions. *The Journal of Engineering Research [TJER], 3*(1), 31–37.

Sapkale, G. N., Patil, S. M., Surwase, U. S., & Bhatbhage, P. K. (2010). Supercritical fluid extraction. *International Journal of Chemical Science, 8* (2), 729–743.

Satyavani, K., Gurudeeban, S., Manigandan, V., Rajamanickam, E., & Ramanathan, T. (2015). Chemical compositions of medicinal mangrove species Acanthus ilicifolius, Excoecaria agallocha, Rhizophora apiculata and Rhizophora mucronata. *Current Research in Chemistry, 7*(1), 1–8.

Sauceda, E. N. R. (2011). Uso de agentes antimicrobianos naturales en la conservación de frutas y hortalizas. *Ra Ximhai: revista científica de sociedad, cultura y desarrollo sostenible, 7*(1), 153–170.

Shah, S., Sharma, A., & Gupta, M. N. (2005). Extraction of oil from Jatropha curcas L. seed kernels by combination of ultrasonication and aqueous enzymatic oil extraction. *Bioresource Technology, 96*(1), 121–123.

Shimokawa, K., Saegusa, K., Wada, Y., & Ishii, F. (2013). Physicochemical properties and controlled drug release of microcapsules prepared by simple coacervation. *Colloids and Surfaces. B, Biointerfaces, 104*, 1–4.

Sihvonen, M., Järvenpää, E., Hietaniemi, V., & Huopalahti, R. (1999). Advances in supercritical carbon dioxide technologies. *Trends in Food Science & Technology, 10*(6-7), 217–222.

Smith, C. A., Collins, C. T., & Crowther, C. A. (2011). Aromatherapy for pain management in labour. *Cochrane Database of Systematic Reviews, 7*.

Soria, A. C., & Villamiel, M. (2010). Effect of ultrasound on the technological properties and bioactivity of food: A review. *Trends in Food Science & Technology, 21*, 323–331.

Soto Ayala, R., & Luque de Castro, M. D. (2001). Continuous subcritical water extraction as a useful tool for isolation of edible essential oils. *Food Chemistry, 75*, 109–113.

Tajkarimi, M. M., Ibrahim, S. A., & Cliver, D. O. (2010). Antimicrobial herb and spice compounds in food. *Food Control, 21*(9), 1199–1218.

Tekin, K., Akalın, M. K., & Şeker, M. G. (2015). Ultrasound bath-assisted extraction of essential oils from clove using central composite design. *Industrial Crops and Products, 77*, 954–960.

Terzi, V., Morcia, C., Faccioli, P., Vale, G., Tacconi, G., & Malnati, M. (2007). In vitro antifungal activity of the tea tree (Melaleuca alternifolia) essential oil and its major components against plant pathogens. *Letters in Applied Microbiology, 44*(6), 613–618.

Tisserand, R., & Young, R. (2013). *Essential oil safety: A guide for health care professionals*. Elsevier Health Sciences.

Topiar, M., Sajfrtova, M., Karban, J., & Sovova, H. (2019). Fractionation of turmerones from turmeric SFE isolate using semi-preparative supercritical chromatography technique. *Journal of Industrial and Engineering Chemistry, 77*, 223–229.

Tyagi, A. K., & Malik, A. (2011). Antimicrobial potential and chemical composition of Eucalyptus globulus oil in liquid and vapour phase against food spoilage microorganisms. *Food Chemistry, 126*(1), 228–235.

Végh, A., Bencsik, T., Molnár, P., Böszörményi, A., Lemberkovics, É., Kovács, K., . . . Horváth, G. (2012). Composition and antipseudomonal effect of essential oils isolated from different lavender species. *Natural Product Communications, 7*(10), 1934578X1200701039.

Vilkhu, K., Mawson, R., Simons, L., & Bates, D. (2008). Applications and opportunities for ultrasound assisted extraction in the food industry- A review. *Innovative Food Science & Emerging Technologies, 9*, 161–169.

Vinjamur, M., Javed, M., & Mukhopadhyay, M. (2013). Encapsulation of nanoparticles using CO_2-expanded liquids. *The Journal of Supercritical Fluids. Special Issue—10th International Symposium on Supercritical Fluids, 79*, 216–226.

Wang, W., & Weller, C. L. (2006). Recent advances in extraction of nutraceuticals from plants. *Trends in Food Science & Technology, 17*, 300–312.

Wei, M. C., Xiao, J., & Yang, Y. C. (2016). Extraction of α-humulene-enriched oil from clove using ultrasound-assisted supercritical carbon dioxide extraction and studies of its fictitious solubility. *Food Chemistry, 210*, 172–181.

Weiss, E. A. (2002). *Turmeric. In Spice Crops* (pp. 338–352). Wallingford, CT: CABI Publishing.

Xia, T., Shi, S., & Wan, X. (2006). Impact of ultrasonic-assisted extraction on the chemical and sensory quality of tea infusion. *Journal of Food Engineering, 74*(4), 557–560.

Yang, Y. C., Wei, M. C., & Hong, S. J. (2014). Ultrasound-assisted extraction and quantitation of oils from Syzygium aromaticum flower bud (clove) with supercritical carbon dioxide. *Journal of Chromatography. A, 1323*, 18–27.

Yim, J.-H., Kim, W.-S., & Lim, J. S. (2013). Recrystallization of adefovir dipivoxil particles using the rapid expansion of supercritical solutions (RESS) process. *Journal of Supercritical Fluids, 82*, 168–176.

Young, J. C. (1995). Microwave-assisted extraction of the fungal metabolite ergosterol and total fatty acids. *Journal of Agricultural and Food Chemistry, 43*(11), 2904–2910.

Zhang, Q. A., Zhi-Qi, Z. A., Xuan, F. Y., Xue, H. F., & Tao, L. C. (2009). Response surface optimization of ultrasound-assisted oil extraction from autoclaved almond powder. *Food Chemistry, 116*, 513–518.

Zhang, Z. S., Wang, L. J., Li, D., Jiao, S. S., Chen, X. D., & Mao, Z. H. (2008). Ultrasound assisted extraction of oil from flaxseed. *Separation and Purification Technology, 62*, 192–198.

Chapter 3

Metabolomics and structural techniques applied for the characterization of biologically active components present in essential oil derived from underutilized crops and herbs

Abel Inobeme[1] and Charles Oluwaseun Adetunji[2]

[1]Department of Chemistry, Edo State University, Uzairue, Edo State, Nigeria, [2]Applied Microbiology, Biotechnology and Nanotechnology Laboratory, Department of Microbiology, Edo University Iyamho, Auchi, Edo State, Nigeria

Introduction

Essential oils refer to various secondary metabolites that are produced by different species of plants such as herbs. Some of these oils are found in their leaves, stems, backs, seeds, flowers, and roots. Essential oils are made of complex mixture of various compounds which are responsible for their specific flavor and colors. The compounds are categorized into secondary compounds and also trace compounds having less than 1% concentration. These oils are transparent, lipophilic, and volatile in nature. The aroma and specific flavor of the plants and resulting oils are affected by the nature of the bioactive constituents of the oils. The major contributing group to the smell of these oils is the monoterpenes in their oxygenated forms. The essential oils obtained from some plants such as lemon grass have been used for long in medicine and aromatherapy. Other compounds are also present which account for their varying appearance and physicochemical properties. Essential oils mostly contain sequiterpenes and monoterpenes having different functional groups which contribute to antibacterial and antifungal properties. The essential oils obtained from related plants have been shown to possess varying compositions which are influenced by the climatic and geographical factors (Maurya et al., 2021). For instance, the zingiberene contents of essential oils tend to vary depending on the region where the ginger plant is grown. The amount of zingiberene in the essential oil was reported to be 46.2% in India, 38.12% in China , while in Nigeria, it is 29.0%. The stage of growth of the plant also affects the composition of essential oils (Raffo & Paoletti, 2022).

Essential oils are lipophilic and highly volatile mixtures made of secondary metabolites from plants. As a result of their diverse bioactive compositions, they possess outstanding antioxidant and antimicrobial properties which account for their usage as a suitable replacement for various synthetic additives used in food industries (Adetunji, Arowora, et al., 2013; Adetunji, Ajani, et al., 2013; Adetunji, Panpatte, et al., 2019, Adetunji, Ojediran, et al., 2019, Adetunji, Kumar, et al., 2019; Adetunji, Akram, et al., 2021; Adetunji, Nwankwo, et al., 2021; Adetunji, Palai, et al., 2021; Adetunji, Inobeme, et al., 2021; Adetunji, Anani, et al., 2021; Adetunji, Michael, Kadiri, et al., 2021; Adetunji, Michael, Nwankwo, et al., 2021; Adetunji, Michael, Varma, et al., 2021; Adetunji, Adetunji, et al., 2021; Adetunji, Olaniyan, et al., 2021; Adetunji, Ajayi, et al., 2021; Rauf et al., 2021; Ukhurebor & Adetunji, 2021). Various chemical additives are used in food processing for the purpose of maintaining the unique quality and organoleptic properties of the food. A solution to the application of chemical additives which are synthetic hence with serious negative side effects is the introduction of the essential oils from plants. These groups of substance do not only help in the improvement of the shelf life of the food materials but also add to their flavoring. They are used alone in some cases and with some other materials in the aspect of

Applications of Essential Oils in the Food Industry. DOI: https://doi.org/10.1016/B978-0-323-98340-2.00031-6

food packaging in other cases . Essential oils have special pharmacological potentials such as antifungal, antioxidant, and antimicrobial properties (Adejumo et al., 2017; Adetunji & Varma, 2020; Adetunji, Arowora, et al., 2013; Adetunji, Ajani, et al., 2013; Adetunji et al., 2017; Adetunji, Oloke, et al., 2020; Adetunji, Egbuna, et al., 2020; Adetunji, Roli et al., 2020; Adetunji et al., 2022; Adetunji, 2008; Adetunji, 2019; Bello et al., 2019; Egbuna et al., 2020; Olaniyan & Adetunji, 2021; Thangadurai et al., 2021). Studies have documented the use of various essential oils obtained from lemon oil, cinnamon oil, tree oil, clove oil, and thyme oils from typical plants which have aided the improvement of the shelf life of different cereals. Aromatic volatile essential oil and terpenes play indispensable role in the aspect of ensuring the safety of food without affecting the desired qualities (Deng et al., 2020).

Essential oils are used in different countries as cosmetics, perfumes, medicines, and as preservatives. Originally, they were employed as chemotherapeutic agents around the 19th century as a result of their unique flavor and aroma. There are more than 300 essential oils that are employed in perfumery as a result of their high aroma. They play a vital role in the defense systems of plants hence their antimicrobial and antioxidant potential. They have been identified to have various potentials such as antiviral, insecticidal, antifungal, and antibacterial activities. Oils that are obtained from various vegetable plants such as perilla seeds and sunflower contain numerous heterocyclic compounds such as the pyrazines which are useful due to their flavors and aroma. They also contain alcohols, alkenes, and esters which account for their pharmacological potentials. Essential oils that are obtained from oregano, cinnamon, and thyme possess remarkable antimicrobial potential against numerous microbes such as *Bacillus thermosphacta*, *Escherichia coli*, *Listeria monocytogenes*, and *Pseudomonas fluorescens*.

Biocides based on essential oils are utilized in combination with various conventional pesticides thus permitting a more flexible period of treatments, increasing the shelf lives of conventional chemical fungicides and also promoting a safe and sustainable system. Essential oils also have highly limited drawbacks such as volatility and instability. They can be employed in environments that are sensitive effectively and are suitable for various types of farming applications and less toxic to the ecosystem. Essential oils play indispensable role in the protection of crops and various aspects of the food industry along with other numerous applications (Xing et al., 2019). They are capable of inhibiting the growth and density of highly pathogenic organisms such as *E. coli*, *Candida* spp., and *Salmonella* spp. There is a pressing need for understanding the mechanisms responsible for the antifungal, antimicrobial, and antioxidant properties as well as their interactions. The utilization of one or more of these synergies can result in the production of the effects desired without any ill effect to the food product. There is therefore a promising tendency for the usage of essential oils in the various sectors of the food industries in future. The utilization of oregano and clove oils can result in the production of a dark pigment when it comes in contact with iron, and this can have serious adverse consequences. There is therefore a need for further studies focusing on the mechanisms of actions and antimicrobial potentials and resistance. It is also paramount to further evaluate the stability of essential oils in the course of food processing and the investigation of bacteria modification ion when essential oil is present (Ben Hsouna et al., 2017).

Various techniques are employed for the extraction of essential oils which include low- and high-temperature distillation processes depending on the stability of the bioactive component of interest. Other techniques include solvent extraction and hydrodistillation. The methods used for the extraction affects the quality of the essential oil obtained. Poor extraction processes can result in damaging or changing in the chemical nature of the oil. Most natural products are unstable to heat and are readily broken down on subjection to heat. The method selected for extraction is also determined on the kind of compound that is present in the oil and location of the oil within the vegetative structure of the plant species (Irshad et al., 2018).

One of the methods that have been widely used for the large-scale extraction of the essential oils is steam distillation. It is however difficult for the regulation of the heat transfer process during the process of extraction as well as the duration of extraction when steam distillation is adopted. One of the most reliable instrumental techniques for the identification and quantification of the bioactive compositions is the gas chromatography and mass spectrometry technique (GC-MS). The use of this method is due to its efficiency and accuracy (Santos et al., 2022).

Distribution of bioactive groups in essential oils

Natural essential oils from various parts of plants contain varying concentrations of these oils and active metabolites. The major components of the essential oils are trace amounts of compounds such as terpenoids and terpenes (Fig. 3.1). There are also components of lower molecular weight such as aromatic and aliphatic components. The terpenes are made of mixtures of five carbon units called isoprene. Their biosynthetic processes include the synthesis of diphosphate and isopentenyl precursor and the polymeric introduction of isopentenyl resulting in prenyl diphosphate precursors. Other types of terpenes present in essential oils include hermiterpenes, triterpenes, diterpenes, and tetraterpenes. Terpenes that contain oxygen are called the terpenoids. The monoterpernes are superlative members that represent the terpenes (Sharifi-Rad et al., 2017).

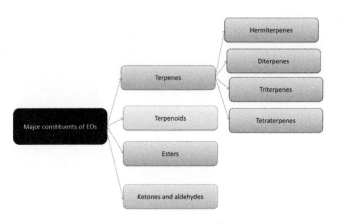

FIGURE 3.1 Major chemical constituents of EOs.

Studies on characterization of essential oils

Mugo et al. (2020) reported the various essential oils found in thick berry. Popa et al. (2021) in their study investigated the compositions of essential oils from various herbs such as *Thymus vulgaris, Lavandula angustifolia, Origanum vulgare* employing various techniques for the analysis of the different fractions obtained. They also evaluated the total phenolic compositions and antioxidant potential of the six different essential oils that were extracted using GC-MS and other spectrophotometric techniques. Rezende et al. (2017) carried out a study on the characterization of the bioactivity of various essential oils obtained from different species of herbs and other crops that are underutilized oils. The identification of the various compositions of the essential oil was done using the GC-MS (Shimadzu model equipped with fused silica capillary column. Matejić et al. (2018) documented that the use of GC-MS approach and gas chromatography with flame ionization detection (FID) have revealed that the major bioactive compounds in the aerial parts of most of the plants is spathulenol and germacrene. The roots of most of the plants were reported to contain octanoic acid, trimethyl benzaldehyde, and nonanoic acid.

Instrumental techniques for structural characterization of essential oils

There is no single perfect approach for the structural characterization of the bioactive compositions of essential oils. Thus the current instrumental approaches are usually employed to complement each other. The cost-effective and highly powerful analytical tools are currently attracting great research efforts. The most prominent analytical techniques that have been employed for the determination of structures of various chemical ingredients in essential oils include mass spectrometry, gas chromatography, nuclear magnetic resonance (NMR) spectroscopy, capillary electrophoresis, polymerase chain reaction, and Fourier-transform infrared (FTIR) spectroscopy (Fig. 3.2; Dhifi et al., 2016).

UV spectroscopy

This is one of the techniques employed during the structural characterization of the various bioactive compositions of essential oil. This is also called UV-visible spectrum (UV/vis) which is based on the reflectance or absorption spectroscopy with a component of the UV spectrum and in the detail, adjacent visible region of the spectrum of the electromagnetic radiation. This is so because it makes use of light from the visible as well as adjacent regions. Absorption or reflectance in the visible spectrum influences directly the perception of the emerging color of the compounds that are involved. In the visible region, the atoms and molecules go through electronic transition. The absorption spectroscopy is complementary to the fluorescence spectroscopy as it is based on the transition of electrons within the excited and ground states, whereas absorption focuses on the changes existing between the ground state with that of the excited states (Jurinovich & Domenici, 2022).

Basically, UV spectroscopy is connected to the interaction that takes places between matter and light rays. On absorption and reflection of light by matter, there is a resulting rise in the energy of the atoms and molecules. On absorption of the UV

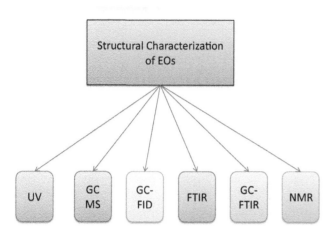

FIGURE 3.2 Schematic diagram on characterization techniques.

radiation, excitation of electrons occurs from the ground state to the higher energy level. A molecule containing the p-electrons or the non-bonding electrons (n-electrons) absorbs energy during this process from the UV radiation inducing the stimulation of electrons into the higher non-bonding orbital. The greater the ease of stimulation of these electrons, the higher the light wavelength which is absorbed. Four major kinds of transition take place, which are p-p* n-s*, s-s*, and N-P*, and their order of arrangement is in the following way: n-p* < p-p* < n-s* < s-s*.

The absorption of radiation by the chemical substance brings about the creation of a spectrum that is identifiable hence making it possible for the identification of the substance. The various components involved in the functioning of the UV spectrometer include the light source, and the selection of wavelength, which is achieved using various devices such as the monochromator, absorption filters, interference filters cutoff filters, or the band pass filter. The analysis of the sample then follows. The resulting feedback goes into the amplifier and finally to the detector where the detection takes place (Saha et al., 2019).

Gas chromatography−mass spectrometry

The GC-MS is an indispensable tool in structural elucidations of various organic and inorganic compounds. It has been widely employed in the investigation of the specific active ingredients present in various essential oils and responsible for the application in food industries and other nutraceutical applications. This is made of two different analytical principle and technique. Usually, the analytical device used in the approach is made of a gas chromatograph which is connected through a heat transfer line to another device the mass spectrometer. The two techniques are synchronized and take place serially. The device however appears as a miniature or highly portable machine containing the entire setup in a unit box (Fan et al., 2022).

GC is employed for the separation of the chemical components of a mixture sample and the detection of various components, finding how much of the components of interest are there. The detectors commonly employed in GC are highly limited in their provision of useable information, which are usually in a two-dimensional form, the analytical column and the retention times and also the responses of the detector. The identification of various components is on the basis of comparing the peak retention times present in the sample to those of known standards. GC cannot be used alone in the identification of the analytes, hence the need for integrating it with a mass spectrometer. In this case, the mass spectrometer may be utilized as the sole detecting component, or the effluent of column could be split between the GC and MS detectors.

MS measures the mass to charge ratio of various charged particles hence be effectively used in the determination of the molecular weights as well as elemental make of the analytes. It is highly reliable in the determination of the chemical structures of various molecules. The data collected using the GC MS is usually in a two-dimensional form, providing a mass spectra which can be utilized for structural confirmation (Fan et al., 2022).

Gas chromatography−flame ionization detection

GC makes use of various advanced detectors which also account for their broad applicability in the study of active components of essential oils. The GC with FID is commonly employed for lipid analysis. The preparation of samples for the analysis includes the preliminary separation of the classes of lipids, derivatization, hydrolysis, and pyrolysis. GC is also employed for the direct separation of various triacylglycerols on the basis of the number of carbons they contain. The analysis is usually carried out at a very high temperature on a medium polar on a non-polar column of low bleeding.

Fourier-transform infrared spectroscopy

The use of FTIR in the structural elucidation of bioactive ingredients in essential oils has been documented. This is the most modern and well-desired infrared technique when compared to dispersive spectrometers. FTIR has the unique merits of having high accuracy, great precision, outstanding speed of procession, enhanced resolution, ease of operation, and the fact that the samples are not destroyed. This is based on atomic vibration of molecules which only absorb unique and specific energies and frequencies of infrared radiation. The FTIR uses essentially an interferometer which measures the energy transmitted to the sample (Rather et al., 2011).

Gas chromatography–Fourier-transform infrared spectroscopy

Another unique advantage of the GC is that it is readily coupled with GC-FTIR. Currently there are three kinds of GC-FTIR interface that are commercially available and are very useful in the investigation of essential oils. They include light pipe, which are more common, the direct deposition type, and the matrix isolation type. The GC-FTIR has the advantage of being able to differentiate among isomers as well as efficiency in structural elucidation in compounds that are closely related in structural compositions. It however has a limitation, which is the variation in their thermal stability of the active ingredients. This is however resolved through the use of derivatization of analytes (Belhachemi et al., 2022).

Nuclear magnetic resonance

The nuclear magnetic images are derived from magnetic responses that are emitted from some atomic nuclei on subjection to externally applied radiofrequency and a magnetic field that is simultaneously applied. This lies on the basis that hydrogen shows sensitivity to NMR and is also abundant sufficiently in various tissues hence generate the signals that are necessary for the formation of the image. The device for NMR imaging is made of a coil that emits radiofrequency, a large magnet, and the computer hardware component. On the basis of the radiofrequency pulsing, the origin and strength of the magnetic signals can be detected and measured by the gradient of the magnetic field which is superimposed on the magnetic field. The spatial details are then encoded and reconstructed electronically giving rise to a cross-sectional image (Kambiré et al., 2022).

Conclusion

Numerous essential oils and their specific constituents are employed as naturally occurring compounds with remarkable antimicrobial potentials with the target of reducing the effects of microbial processes in food materials. The phenolic constituents of the essential oils act efficiently as permeabilizers of membranes. The gram-positive bacteria groups also tend to show higher sensitivity to essential oils in comparison to the gram-negative bacteria. The major constituents of the essential oils are carvacrol, thymol, and cinnamaldehyde, and these accounts for their action through some mechanisms such as alteration of the fatty acids of membrane, permeability of membrane, as well as the inhibition of the motive force of proton. The quality and safety of various food products and cereals are efficiently improved by the coating or microencapsulation of edible film using essential oils.

References

Adejumo, I. O., Adetunji, C. O., & Adeyemi, O. S. (2017). Influence of UV light exposure on mineral composition and biomass production of myco-meat produced from different agricultural substrates. *Journal of Agricultural Sciences, 62*(1), 51–59.

Adetunji, C. O. (2008) The antibacterial activities and preliminary phytochemical screening of vernoniaamygdalina and Aloe vera against some selected bacteria [MSc thesis, University of Ilorin], pp. 40–43.

Adetunji, C. O. (2019). Environmental impact and ecotoxicological influence of biofabricated and inorganic nanoparticle on soil activity. In D. Panpatte, & Y. Jhala (Eds.), *Nanotechnology for agriculture*. Singapore: Springer. Available from https://doi.org/10.1007/978-981-32-9370-0_12.

Adetunji, C. O., Ajayi, O. O., Akram, M., Olaniyan, O. T., Chishti, M. A., Inobeme, A., Olaniyan, S., Adetunji, J. B., Olaniyan, M., & Awotunde, S. O. (2021). Medicinal plants used in the treatment of influenza A virus infections. In K. Dua, S. Nammi, D. Chang, D. K. Chellappan, G. Gupta, & T. Collet (Eds.), *Medicinal plants for lung diseases*. Singapore: Springer. Available from https://doi.org/10.1007/978-981-33-6850-7_19.

Adetunji, C. O., Akram, M., Olaniyan, O. T., Ajayi, O. O., Inobeme, A., Olaniyan, S., Hameed, L., & Adetunji, J. B. (2021). Targeting SARS-CoV-2 novel corona (COVID-19) virus infection using medicinal plants. In K. Dua, S. Nammi, D. Chang, D. K. Chellappan, G. Gupta, & T. Collet (Eds.), *Medicinal plants for lung diseases*. Singapore: Springer. Available from https://doi.org/10.1007/978-981-33-6850-7_21.

Adetunji, C. O., Anani, O. A., Olaniyan, O. T., Inobeme, A., Olisaka, F. N., Uwadiae, E. O., & Obayagbona, O. N. (2021). Recent trends in organic farming. In R. Soni, D. C. Suyal, P. Bhargava, & R. Goel (Eds.), *Microbiological activity for soil and plant health management*. Singapore: Springer. Available from https://doi.org/10.1007/978-981-16-2922-8_20.

Adetunji, C. O., Arowora, K., Fawole, O. B., & Adetunji, J. B. (2013). Effects of coatings on storability of carrot under evaporative coolant system. *Albanian Journal of Agricultural Sciences*, *12*(3).

Adetunji, C. O., Egbuna, C., Tijjani, H., Adom, D., Al-Ani, L. K. T., & Patrick-Iwuanyanwu, K. C. (2020). Homemade preparations of natural biopesticides and applications. *Natural Remedies for Pest, Disease and Weed Control*, 179–185.

Adetunji, C. O., Inobeme, A., Olaniyan, O. T., Ajayi, O. O., Olaniyan, S., & Adetunji, J. B. (2021). Application of nanodrugs derived from active metabolites of medicinal plants for the treatment of inflammatory and lung diseases: Recent advances. In K. Dua, S. Nammi, D. Chang, D. K. Chellappan, G. Gupta, & T. Collet (Eds.), *Medicinal plants for lung diseases*. Singapore: Springer. Available from https://doi.org/10.1007/978-981-33-6850-7_26.

Adetunji, C. O., Kumar, D., Raina, M., Arogundade, O., & Sarin, N. B. (2019). Endophytic microorganisms as biological control agents for plant pathogens: A panacea for sustainable agriculture. In A. Varma, S. Tripathi, & R. Prasad (Eds.), *Plant biotic interactions*. Cham: Springer. Available from https://doi.org/10.1007/978-3-030-26657-8_1.

Adetunji, C. O., Michael, O. S., Kadiri, O., Varma, A., Akram, M., Oloke, J. K., Shafique, H., Adetunji, J. B., Jain, A., Bodunrinde, R. E., Ozolua, P., & Ubi, B. E. (2021). Quinoa: From farm to traditional healing, food application, and phytopharmacology. In A. Varma (Ed.), *Biology and biotechnology of quinoa*. Singapore: Springer. Available from https://doi.org/10.1007/978-981-16-3832-9_20.

Adetunji, C. O., Michael, O. S., Nwankwo, W., Ukhurebor, K. E., Anani, O. A., Oloke, J. K., Varma, A., Kadiri, O., Jain, A., & Adetunji, J. B. (2021). Quinoa, the next biotech plant: Food security and environmental and health hot spots. In A. Varma (Ed.), *Biology and biotechnology of quinoa*. Singapore: Springer. Available from https://doi.org/10.1007/978-981-16-3832-9_19.

Adetunji, C. O., Michael, O. S., Rathee, S., Singh, K. R. B., Ajayi, O. O., Adetunji, J. B., Ojha, A., Singh, J., & Singh, R. P. (2022). Potentialities of nanomaterials for the management and treatment of metabolic syndrome: A new insight. *Materials Today Advances*, *13*, 100198.

Adetunji, C. O., Michael, O. S., Varma, A., Oloke, J. K., Kadiri, O., Akram, M., Bodunrinde, R. E., Imtiaz, A., Adetunji, J. B., Shahzad, K., Jain, A., Ubi, B. E., Majeed, N., Ozolua, P., & Olisaka, F. N. (2021). Recent advances in the application of biotechnology for improving the production of secondary metabolites from quinoa. In A. Varma (Ed.), *Biology and biotechnology of quinoa*. Singapore: Springer. Available from https://doi.org/10.1007/978-981-16-3832-9_17.

Adetunji, C. O., Nwankwo, W., Ukhurebor, K., Olayinka, A. S., & Makinde, A. S. (2021). Application of Biosensor for the Identification of various pathogens and pests mitigating against the agricultural production: Recent advances. In R. N. Pudake, U. Jain, & C. Kole (Eds.), *Biosensors in agriculture: Recent trends and future perspectives. Concepts and strategies in plant sciences*. Cham: Springer. Available from https://doi.org/10.1007/978-3-030-66165-6_9.

Adetunji, C. O., Ojediran, J. O., Adetunji, J. B., & Owa, S. O. (2019). Influence of chitosan edible coating on postharvest qualities of Capsicum annum L. during storage in evaporative cooling system. *Croatian Journal of Food Science and Technology*, *11*(1), 59–66.

Adetunji, C. O., Olaniyan, O. T., Akram, M., Ajayi, O. O., Inobeme, A., Olaniyan, S., Khan, F. S., & Adetunji, J. B. (2021). Medicinal plants used in the treatment of pulmonary hypertension. In K. Dua, S. Nammi, D. Chang, D. K. Chellappan, G. Gupta, & T. Collet (Eds.), *Medicinal plants for lung diseases*. Singapore: Springer. Available from https://doi.org/10.1007/978-981-33-6850-7_14.

Adetunji, C. O., Oloke, J. K., & Prasad, G. (2020). Effect of carbon-to-nitrogen ratio on eco-friendly mycoherbicide activity from *Lasiodiplodia pseudotheobromae* C1136 for sustainable weeds management in organic agriculture. *Environment, Development and Sustainability*, *22*, 1977–1990. Available from https://doi.org/10.1007/s10668-018-0273-1.

Adetunji, C. O., Palai, S., Ekwuabu, C. P., Egbuna, C., Adetunji, J. B., Ehis-Eriakha, C. B., Kesh, S. S., & Mtewa, A. G. (2021). General principle of primary and secondary plant metabolites: biogenesis, metabolism, and extraction. *Preparation of Phytopharmaceuticals for the Management of Disorders*, 3–23, Publisher Academic Press.

Adetunji, C. O., Panpatte, D. G., Bello, O. M., & Adekoya, M. A. (2019). Application of nanoengineered metabolites from beneficial and eco-friendly microorganisms as a biological control agents for plant pests and pathogens. In D. Panpatte, & Y. Jhala (Eds.), *Nanotechnology for agriculture: Crop production & protection*. Singapore: Springer. Available from https://doi.org/10.1007/978-981-32-9374-8_13.

Adetunji, C. O., Phazang, P., & Sarin, N. B. (2017). Significance of rhamnolipids as a biological control agent in the management of crops/plant pathogens. *Current Trends in Biomedical Engineering & Biosciences*, *10*(3), 54–55.

Adetunji, C. O., Roli, O. I., & Adetunji, J. B. (2020). Exopolysaccharides derived from beneficial microorganisms: Antimicrobial, food, and health benefits. In P. Mishra, R. R. Mishra, & C. O. Adetunji (Eds.), *Innovations in food technology*. Singapore: Springer. Available from https://doi.org/10.1007/978-981-15-6121-4_10.

Adetunji, C. O., & Varma, A. (2020). Biotechnological application of *Trichoderma*: A powerful fungal isolate with diverse potentials for the attainment of food safety, management of pest and diseases, healthy planet, and sustainable agriculture. In C. Manoharachary, H. B. Singh, & A. Varma (Eds.), *Trichoderma: Agricultural applications and beyond. Soil biology* (vol 61). Cham: Springer. Available from https://doi.org/10.1007/978-3-030-54758-5_12.

Adetunji, J. B., Adetunji, C. O., & Olaniyan, O. T. (2021). African walnuts: A natural depository of nutritional and bioactive compounds essential for food and nutritional security in Africa. In O. O. Babalola (Ed.), *Food security and safety*. Cham: Springer. Available from https://doi.org/10.1007/978-3-030-50672-8_19.

Adetunji, J. B., Ajani, A. O., Adetunji, C. O., Fawole, O. B., Arowora, K. A., Nwaubani, S. I., Ajayi, E. S., Oloke, J. K., & Aina, J. A. (2013). Postharvest quality and safety maintenance of the physical properties of Daucus carota L. fruits by Neem oil and Moringa oil treatment: A new edible coatings. *Agrosearch, 13*(1), 131−141.

Belhachemi, A., Maatoug, M., & Canela-Garayoa, R. (2022). GC-MS and GC-FID analyses of the essential oil of Eucalyptus camaldulensis grown under greenhouses differentiated by the LDPE cover-films. *Industrial Crops and Products, 178*.

Bello, O. M., Ibitoye, T., & Adetunji, C. (2019). Assessing antimicrobial agents of Nigeria flora. *Journal of King Saud University-Science, 31*(4), 1379−1383.

Ben Hsouna, A., Ben Halima, N., Smaoui, S., & Hamdi, N. (2017). Citrus lemon essential oil: Chemical composition, antioxidant and antimicrobial activities with its preservative effect against Listeria monocytogenes inoculated in minced beef meat. *Lipids in Health and Disease, 16*, 146, https://doi.org/10.1186/s12944-017-0487-5.

Deng, W., Liu, K., Cao, S., Sun, J., Zhong, B., & Chun, J. (2020). Chemical composition, antimicrobial, antioxidant, and antiproliferative properties of grapefruit essential oil prepared by molecular distillation. *Molecules (Basel, Switzerland), 25*, 217. Available from https://doi.org/10.3390/molecules25010217.

Dhifi, W., Bellili, S., Jazi, S., Bahloul, N., & Mnif, W. (2016). Essential oils' chemical characterization and investigation of some biological activities: A critical review. *Medicines (Basel), 3*(4), 25. Available from https://doi.org/10.3390/medicines3040025.

Egbuna, C., Gupta, E., Ezzat, S. M., Jeevanandam, J., Mishra, N., Akram, M., Sudharani, N., Adetunji, C. O., Singh, P., Ifemeje, J. C., Deepak, M., Bhavana, A., Walag, A. M. P., Ansari, R., Adetunji, J. B., Laila, U., Olisah, M. C., & Onyekere, P. F. (2020). Aloe species as valuable sources of functional bioactives. In C. Egbuna, & G. Dable Tupas (Eds.), *Functional foods and nutraceuticals*. Cham: Springer. Available from https://doi.org/10.1007/978-3-030-42319-3_18.

Fan, S., Chang, J., Zong, Y., Hu, G., & Jia, J. (2022). GC-MS analysis of the composition of the essential oil from Dendranthema indicum Var. Aromaticum using three extraction methods and two columns. *Molecules (Basel, Switzerland)*.

Irshad, M., Subhani, M.A., Ali, S. & Hussain, A. (2018). Biological importance of essential oils. doi: 10.5772/intechopen.87198

Jurinovich, S., & Domenici, V. (2022). Digital tool for the analysis of UV−Vis spectra of olive oils and educational activities with high school and undergraduate students. *Journal of Chemical Education, 99*(2), 787−798. Available from https://doi.org/10.1021/acs.jchemed.1c01015, 2022.

Kambiré, D. A., Kablan, A. C. L., Yapi, T. A., Vincenti, S., Maury, J., Baldovini, N., Tomi, P., Paoli, M., Boti, J. B., & Tomi, F. (2022). Neuropeltis acuminata (P. Beauv.): Investigation of the chemical variability and in vitro anti-inflammatory activity of the leaf essential oil from the Ivorian species. *Molecules (Basel, Switzerland), 27*, 3759. Available from https://doi.org/10.3390/molecules2712375.

Matejić, J. S., Stojanović-Radić, Z. Z., Ristić, M. S., et al. (2018). Chemical characterization, in vitro biological activity of essential oils and extracts of three *Eryngium* L. species and molecular docking of selected major compounds. *Journal of Food Science and Technology, 55*, 2910−2925. Available from https://doi.org/10.1007/s13197-018-3209-8.

Maurya, A., Prasad, J., Das, S., & Dwivedy, A. K. (2021). Essential oils and their application in food safety. *Frontiers in Sustainable Food Systems*. Available from https://doi.org/10.3389/fsufs.2021.653420.

Mugo, L., Gichimu, B.M., Muturi, P.W., & Mukono, S.T. (2020). Characterization of the volatile components of essential oils of selected plants in Kenya. https://doi.org/10.1155/2020/8861798.

Olaniyan, O. T., & Adetunji, C. O. (2021). Biological, biochemical, and biodiversity of biomolecules from marine-based beneficial microorganisms: Industrial perspective. In C. O. Adetunji, D. G. Panpatte, & Y. K. Jhala (Eds.), *Microbial rejuvenation of polluted environment. Microorganisms for sustainability* (vol 27). Singapore: Springer. Available from https://doi.org/10.1007/978-981-15-7459-7_4.

Popa, C. L., Lupitu, A., Mot, M. D., Copolovici, L., Moisa, C., & Copolovici, D. M. (2021). Chemical and biochemical characterization of essential oils and their corresponding hydrolats from six species of the *Lamiaceae* family. (2021)*Plants (Basel), 10*(11), 2489. Available from https://doi.org/10.3390/plants10112489.

Raffo, A., & Paoletti, F. (2022). Fresh-cut vegetables processing: environmental sustainability and food safety issues in a comprehensive perspective. *Frontiers in Sustainable Food Systems, 5*, 681459. Available from https://doi.org/10.3389/fsufs.2021.681459.

Rather, M., Ganai, B.A., Dar, B.A., & Malik, A.H. (2011). GC-FID and GC-MS analysis of the Essential oil of Elsholtzia densa Benth -a rich source of acylfuran non-terpene ketone.

Rauf, A., Akram, M., Semwal, P., Mujawah, A. A. H., Muhammad, N., Riaz, Z., Munir, N., Piotrovsky, D., Vdovina, I., Bouyahya, A., Adetunji, C. O., Shariati, M. A., Almarhoon, Z. M., Mabkhot, Y. N., & Khan, H. (2021). Antispasmodic potential of medicinal plants: A comprehensive review. *Oxidative Medicine and Cellular Longevity, 2021*, 12. Available from https://doi.org/10.1155/2021/4889719, Article ID 4889719.

Rezende, D., Souza, R., Magalhães, M., Silva Caetano, A., Sousa Carvalho, M., de Souza, E., de Lima Guimarães, L., Nelson, D., Batista, L., & das Graças Cardoso, M. (2017). Characterization of the biological potential of the essential oils from five species of medicinal plants. *American Journal of Plant Sciences, 8*, 154−170. Available from https://doi.org/10.4236/ajps.2017.82012.

Saha, S., Singh, J., & Paul, A. (2019). Anthocyanin profiling using UV-Vis spectroscopy and liquid chromatography mass spectrometry. *Journal of AOAC International, 103*(1). Available from https://doi.org/10.5740/jaoacint.19-0201.

Santos, M. I. S., Marques, C., Mota, J., Pedroso, L., & Lima, A. (2022). Applications of essential oils as antibacterial agents in minimally processed fruits and vegetables—A review. *Microorganisms, 2022*(10), 760. Available from https://doi.org/10.3390/microorganisms1004076.

Sharifi-Rad, J., Sureda, A., Tenore, G. C., Daglia, M., Sharifi-Rad, M., Valussi, M., Tundis, R., Sharifi-Rad, M., Loizzo, M. R., Ademiluyi, A. O., Sharifi-Rad, R., Ayatollahi, S. A., & Iriti, M. (2017). Biological activities of essential oils: From plant chemoecology to traditional healing systems. *Molecules (Basel, Switzerland), 22*(1), 70. Available from https://doi.org/10.3390/molecules22010070.

Thangadurai, D., Naik, J., Sangeetha, J., Al-Tawaha, A. R. M. S., Adetunji, C. O., Islam, S., David, M., Shettar, A. K., & Adetunji, J. B. (2021). Nanomaterials from agrowastes: Past, present, and the future. In O. V. Kharissova, L. M. Torres-Martínez, & B. I. Kharisov (Eds.), *Handbook of Nanomaterials and Nanocomposites for Energy and Environmental Applications*. Cham: Springer. Available from https://doi.org/10.1007/978-3-030-36268-3_43.

Ukhurebor, K. E., & Adetunji, C. O. (2021). Relevance of biosensor in climate smart organic agriculture and their role in environmental sustainability: What has been done and what we need to do? In R. N. Pudake, U. Jain, & C. Kole (Eds.), *Biosensors in agriculture: Recent trends and future perspectives. Concepts and strategies in plant sciences*. Cham: Springer. Available from https://doi.org/10.1007/978-3-030-66165-6_7.

Xing, C., Qin, C., Li, X., Zhang, F., Linhardt, R. J., Sun, P., & Zhang, A. (2019). Chemical composition and biological activities of essential oil isolated by HS-SPME and UAHD from fruits of bergamot. *LWT-Food Science and Technology*, *104*, 38—44. Available from https://doi.org/10.1016/j.lwt.2019.01.020.

Chapter 4

Utilization of essential oils in beverages, drinks and semi liquids foods and their medicinal attributes

Babatunde Oluwafemi Adetuyi[1], Kehinde Abraham Odelade[1], Anjolaoluwa Maryham Taiwo[1], Peace Abiodun Olajide[1], Olubanke Olujoke Ogunlana[2], Charles Oluwaseun Adetunji[3], Juliana Bunmi Adetunji[4], Oluwabukola Atinuke Popoola[5], Olatunji Matthew Kolawole[6], Olalekan Akinbo[7], Abel Inobeme[8], Osarenkhoe Omorefosa Osemwegie[9], Mohammed Bello Yerima[10] and M.L. Attanda[11]

[1]Department of Natural Sciences, Faculty of Pure and Applied Sciences, Precious Cornerstone University, Ibadan, Oyo State, Nigeria, [2]Department of Biochemistry, Covenant University, Ota, Ogun State, Nigeria, [3]Applied Microbiology, Biotechnology and Nanotechnology Laboratory, Department of Microbiology, Edo State University Uzairue, Iyamho, Edo State, Nigeria, [4]Department of Biochemistry, Osun State University, Osogbo, Osun State, Nigeria, [5]Genetics, Genomics and Bioinformatics Department, National Biotechnology Development Agency, Abuja, FCT, Nigeria, [6]Department of Microbiology, Faculty of Life Sciences, University of Ilorin, Ilorin, Kwara State, Nigeria, [7]Centre of Excellence in Science, Technology, and Innovation, AUDA-NEPAD, Johannesburg, Gauteng, South Africa, [8]Department of Chemistry, Edo State University Uzairue Iyamho, Auchi, Edo State, Nigeria, [9]Department of Food Science and Microbiology, Landmark University, Omu-Aran, Kwara State, Nigeria, [10]Department of Microbiology, Sokoto State University, Sokoto, Sokoto State, Nigeria, [11]Department of Agricultural Engineering, Bayero University Kano, Kano, Kano State, Nigeria

Introduction

Essential oils (EOs) are a combination of lipophilic smell dynamic mixtures aromatic and allied chemicals (AACs) which can furthermore be removed as a fluid from a known part of plants. Quintessential oils are fragrant and unsteady fluids accomplished from plant material, comprehensive of bloom strip, natural products, shrubs, and explicit plants. EOs have design and scent because they involve unsound fragrance dynamic compounds (i.e., particles that have unique style and smell). EOs are named "fundamental" in the sense that it has an exceptional quality fragrance, fragrance or quintessence of the plant in which it was produced (Adetuyi et al., 2015a, 2015b). The word sweet-smelling plants portray plants considered for one or additional EOs. EOs have a taste because of unsound AACs and are characterized as parts that bring out an extraordinary style and scent. EOs can likewise be alluded to as the "oil of" the plant from which it was extricated, for the most part through refining, articulation, or dissolvable extraction. Fragrant plants or aromatic plants (AP) are plants respected for numerous noteworthy EOs (Adetuyi et al., 2015a, 2015b). Numerous EOs consolidate terpenoids, terpenes, and several lipophilic mixtures as being unstable dynamic mixtures, with subatomic loads underneath, or somewhat over, 3 hundred Da. The parts of plants that possess EOs are known as flavors. Flavors are used on account of that artifact and are "by and large viewed as protected" (GRAS) flavorings in feasts and refreshments. Notwithstanding, EOs utilized expertly in diet and drug need to be taken care of rigorously as per "great manufacturing practices" (GMP). Advance notice is essential because EOs or dynamic mixtures, being auxiliary metabolic substances of plants, are delivered for different environmental uses, like a safeguard against microbes and herbivores; consequently, many EOs/AACs are dose-dependently suggested or harmful (Adewale et al., 2022; Mahboubi, 2019).

Oils separated from a sizeable range of the seeds of vegetables are beneficial because of their nutrients and proteins, and the greater part of the seeds of vegetable close by with perilla and sunflower seeds have heterocyclic components, for example, pyrazine, which carry out the role of a vital trait stylish and very zenith of the product. Because of flavors, natural antimicrobial items, and fragrance, EOs are specifically utilized in components organization endeavor business venture for sources protection. For instance, citrus essential oils, for example, oxygenated subsidiaries, sesquiterpenes, and monoterpenes show great antibacterial potential in the course of pathogenic microbes, subsequently recommending their utilization as enhancing and cell reinforcement specialists. Physical, compound, and few microbiological factors are quintessential for components safeguarding (Adetunji et al., 2013; Adetunji, Adetunji, et al., 2021i; Adetunji, Ajayi et al., 2021k; Adetunji, Akram et al., 2021a; Adetunji, Anani et al., 2021e; Adetunji, Inobeme et al., 2021d; Adetunji, Kumar et al., 2019c;

Applications of Essential Oils in the Food Industry. DOI: https://doi.org/10.1016/B978-0-323-98340-2.00028-6

Adetunji, Michael, Kadiri et al., 2021f; Adetunji, Michael, Nwankwo et al., 2021g; Adetunji, Michael, Varma et al., 2021h; Adetunji, Nwankwo et al., 2021b; Adetunji, Ojediran et al., 2019b; Adetunji, Olaniyan et al., 2021j; Adetunji, Palai et al., 2021c; Adetunji, Panpatte et al., 2019a; Didunyemi et al., 2019, 2020; Rauf et al., 2021; Ukhurebor & Adetunji, 2021). For a long time, manufacturers and customers have been using counterfeit components in parts ventures; then, at that point, when extended, the use of engineered additives and their utilization can prompt few hypersensitive impacts, inebriations, disease, and unique degenerative illnesses (Adejumo et al., 2017; Adetunji, 2019; Adetunji et al., 2013, 2017, 2020b, 2020a, 2020c, 2022; Adetunji & Varma, 2020; Bello et al., 2019; Didunyemi et al., 2020; Egbuna et al., 2020; James-Okoro et al., 2021; Olaniyan & Adetunji, 2021; Adetunji, 2008). In recent years, businesses have been using concentrates of aromatic plants and EOs due to their potential to control the level of pathogens. EOs separated out of oregano, thyme, and cinnamon, flaunt huge antibacterial matters to do contrary to a shift microorganisms close by with *Pseudomonas fluorescens, Bacillus thermosphacta, Escherichia coli*, and *Listeria monocytogenes*. In any case, despite lacking comprehensive data, there is valuable information available about the uses of essential oils, their characteristics—such as their components of development—and the effects of EOs on property stock and regular activities. Additionally, their antimicrobial mission is noteworthy (Sharmeen et al., 2021).

History of essential oils

The time span of EOs is being involved since the 16th century, derived from the medication Quinta Essentia and given a name by Paracelsus von Hohenheim of Switzerland. Because of their combustibility, they are named as EOs or substances. A few specialists have attempted to outfit the meaning of EOs. As indicated by the French Organization for Standardization: Affiliation Francaise de Standardization (AFNOR), EO is the item procured from a uncooked vegetable, whether through the process of steam refining or by means of mechanical methodologies from the citrus epicarp, or "dry" refining. Afterward, it will be isolated from fluid stage via capability of actual procedures. This definition incorporates the crude supplies utilized for the isolation of EOs and their isolation procedures, for example, the use of non-aqueous solvent or bloodless retention. EOs are eminently dissolvable in unsound mixtures such as liquor, ether, and steady oils and they are non-dissolvable in water. EOs removed from vetiver, sassafras, cinnamon, and exceptional regular origins are fluid in natureand drab at ambient temperature because of the existence of fragrant mixtures. Thus, they are strikingly utilized in the fragrant healing and makeup industry. The presence of unsafe mixtures like aldehydes, ketones, and fragrant mixtures in the EOs play important parts in fragrant healing as inward breath of these mixtures, which are practically mitigating mental and substantial anxieties. In addition, EOs are phenomenally utilized for various remedial purposes such as rubdown fragrance-based treatment, psycho fragrance-based treatment, and olfactory fragrant healing. EOs act as compound alarms and are stressed in plants to control and adjust their non-public conditions, for example, to safeguard themselves from vermin and attract practically valuable bugs, for example, pollinating bugs, and use for exchange between blossoms by involving method for arrival of synthetic signs within the sight of herbivores. In earlier times, that is, at some stage in the late 1200s, EOs have been delivered through customary hydro refining (Batiha et al., 2021; Nazir et al., 2022). Before this, Greeks and Romans utilized the crude design of refining to deliver camphor and turpentine. This system is not generally utilized nonetheless, it was once drilled in 9th century and this strategy was utilized to create refined waters from this technique. As a result, turpentine and camphor were molded. Arabic researchers duplicated the refining strategy despite the fact that it was not satisfactory to deliver EOs. In the late 20th century, EOs-based fragrant healing has arisen as exceptionally well-known because of its significance, fame, and broad use. EOs are the first components in fragrance-based treatment. There are really a couple of procedures through which EOs are controlled, for example, inward breath, rubdown or helpful utilization on the skin surface. Inward breath and outside programming of the EOs have been largely utilized which comprises of scholarly and substantial equilibrium, which is the actual premise of fragrance-based treatment. It significantly alleviates the pressure, revives and recovers the person for the resulting day's worth of effort. It primarily deals with the olfactory nerve from nostril to Virtuoso and have laid out well for contaminations from microorganisms, cardiovascular diseases, Alzheimer, and most tumors illnesses, as well as support pregnancy times. Recently, there is a sped-up pattern involving fragrant healing for most diseases and sleeping problems. The other natural mixtures present in EOs behave in strong way to extend the sensation of healthiness (Bhavaniramya et al., 2019).

History of beverage industries

A zest is depicted as a sweet-smelling plant substance utilized as an enhancer. Flavors have been entirely fundamental in mankind's set of experiences, culture, and human progress. The desire of Europe for "flavors of the East" energized the disclosure of Vasco da Gama of the ocean course to a country called India. The essential utilizing type of the business all over the world is, in all honesty, essential oils. A researcher John Pembroke developed a typical cola method to become more prominent out of the seven flavors it consists of, rather than from the psychoactive alkaloids present in coca leaves. Since the introduction of Coca-Cola in 1885 and the subsequent inventions of Mirinda in 1960, Pepsi in 1893, 7Up in 1929, Tango in 1950, Fanta in 1940, and Sprite in 1961, soft drinks have become a crucial component in the development of a multibillion-dollar industry. For instance, in 2012, the Organization of the United States producing Coca Cola spent 30 billion USD to achieve global expansion of the business venture by 2013. It creates and showcases north with 5 hundred brands, with an arrangement of $16 billion, clients in more than 200 nations with a typical eating cost of 1.8 billion portions every day. Likewise, Coca-Cola used to be the essential delicate beverage in India till 1977 and after its return in 1993, it has made colossal speculations to ensure that the refreshment is reachable to an ever increasing number of people in a US of the US of more than a billion people, achieving even the distant and blocked off pieces of that nation. Due to the fact that the capital and serious nature of the soft drink businesses has a strong multiplier impact on each wild stream, the politico-financial impact of the organization cannot be overstated. With more than 700,000 representatives, The Coca-Cola Organization for Events is one of the top 10 non-public bosses in the world (Sharmeen et al., 2021).

Significant parts of essential oils

Significant elements of EOs are exceptionally confounded in nature as they are made out of a blend of more than 50 components at very particular fixations; a few quantities of separated compounds synthetically secrete EOs particularly the amount and highlights of those mixtures are fundamental factors. Primarily, EOs are situated in the plant cells' cytoplasms, uncommonly emitted in trichomes or secretory hairs epidermal cells inside secretory cells, and secretory pockets. EOs are a mix of more than 300 explicit mixtures observably comprising of unpredictable mixtures with low subatomic loads about underneath 1000 da generally 300 Da. Essentially, couple of mixtures are most significant ones at around 20% to 70% in qualification to various mixtures which are available in hint sums, for instance, thymols (27%) and carvacrol (30%) and are present in *Origanum compactum*. They serve as the vitally substance parts, and the vital issue in *Coriandrum sativum* is the linalool. The other EOs like alpha & beta camphor at 24% and thuyone at 57% are present at radical fixations in Artemisia herbaalba; these fundamental parts are at fault for the numerous natural roles of essential oils. Subsequently, EOs are in many cases concluded in the halfway fume express the shaky mixtures are significantly classified into a wide assortment of synthetic classification such as ketones, amines, aldehydes, phenols, alcohols, amines, & primarily terpenes. Out of all these synthetic classes, ketones, and aldehydes, alcohols give a huge assortment of sweet-smelling outcomes to products of the soil elements of sure plants; for instance, natural products have F-nerolidol botanical Linalool and herbals gamma-selinene as significant mixtures unstable mixtures. For example, hydrocarbons 2-methyl 4-heptanone and trimethylsilyl methanol separated from unpleasant apple are used to perform the low subatomic weight mixtures like terpenoids. Terpenes are current in the greater part of the EOs and are liable for a scope of exercises. EOs extricated from rosemary have a few monoterpenes comprising of 1 8-cineole camphor as the significant components that act as antimicrobican' to additives in feasts fabricating (Bhavaniramya et al., 2019).

Sorts of essential oils

EOs are extremely complex in composition, containing a combination of more than 50 unique scent chemicals. It contains EOs such as tea tree oil, oregano oil, clove oil, cinnamon oil, mustard oil, thyme oil, lavender oil, lemon oil, peppermint oil, and eucalyptus oil among others. These oils play a crucial role in preventing the growth of pathogenic microorganisms and maintaining the quality of food. For instance, the terpenes and oxygenated terpenes in lemon EO have been proven to be goliath (Shedoeva et al., 2019). Cinnamon oil is a dangerous substance that is extracted from the leaves, bark, and roots of *Cinnamomum zeylanicum*. It contains three important constituents, including trans-cinnamaldehyde, eugenol, and linalool, which are extracted from the bark and account for 8.25% of the total production. The most active component of cinnamon essential oils, cinnamonaldehyde, has been shown to have growth-inhibitory effects on a variety of microorganisms, including every gram-negative and gram-positive microscopic living thing, and it may also have potential growth-inhibitory effects on organisms. According to some studies, tea tree oil (TTO) from *Melaleuca alternifolia* (myrtaceae) significantly contained terpinen-4-old gamma-terpinene, p-cymene, alpha - terpinene, 1,8-cineole alpha-terpineol, and alpha-pinene in addition to having cell reinforcement antiparasitic and free revolutionary searching properties. It had a substantial inhibitory effect against infectious strains. The majority of clove oil is extracted from the clove buds, and it contains considerable amounts of the phenylpropanoids eugenol, eugenyl acetic acid derivative, carvacrol, thymol, cinnamaldemade beta-caryophyllene, and 2-heptanone. One of these, eugenol, is frequently used as an antimicrobial specialist and should reduce the union of a particular cell wall component called ergosterol. It may also be necessary to suppress the growth of the microbe tube in *C. albicans*. When used against tert-butylated hydroxytoluene, it had strong revolutionary-looking action and produced inhibitory effects on a variety of safe staphylococcus spp. By using steam refining, lavender oil is extracted in a significant amount from the lamiaceae family, specifically from the lavandula angustifolia, and it is discovered to contain laminatel, linalyl acetic acid derivation, linalool, 1,8-cineole lavandulyl acetic acid derivation, terpinen-4-old, l-fenchone, viridiflorol, and camphor. However, the concentrations of these mixes have varied from species to species. One of the fervent combinations in lavender oil is linalool. Contrary to antimicrobial safe microorganisms such yeasts, dermatophytes, *Cryptococcus neoformans*, aspergillus strains, and C species, lavender oil demonstrated antibacterial activity. 1,8-cineole is the main constituent of eucalyptus oil, although it also contains a variety of combinations, including p-cymene, cryptone, alpha-terpineol, alpha-pinene, and terpinen-4-old.

Eucalyptus EOs are typically used as seasoning suppliers and have been widely acknowledged as effective methods for preventing the growth of harmful and food-damaging microbes. The bioactive mixtures of EOs, in particular the antibacterial and antifungal mixtures, can also target more than a few telephone developments or substance pathways, including cell wall damage layer injury and disruption of proton thought process force, among other things. However, there are only a few facts available regarding the antibacterial and antifungal properties of eucalyptus oil. It is used to demonstrate how the antibacterial activity of eucalyptus oil was once plainly linked to the synergistic interactions between the major and minor combinations present in the oil rather than a single ingredient (Bhavaniramya et al., 2019).

Essential oils as enhancing in drinks

Cola, lemon-lime, and orange flavors of soft drink concentrates are available. The majority of the time, approved bottlers receive them through SDLCs. Bottlers who are now prohibited from joining an SDLC get their concentrates from independent producers or suppliers. Depending on the company and provider, one or more EOs may be combined with tones and distinctive ingredients in one or more concentration formulae for a particular soft drink. The different ingredients are frequently granular ones, such as buffering agents (such as citrus extract, sodium citrate, phosphoric acid, and potassium phosphates) and additional additions. According to names attached to the referenced components, a certain number of parts are remembered for a specific request and approach while blending syrup.

The ingredients found in cola and citrus pieces, the EOs in concentrates, and the flavorings themselves, such as sugars, acids, and oils, are what give soft drinks their recognizable appearance and aroma. The need for and combination of a massive change of EOs (which can have counterfeit reciprocals) are said to be the secrets, methods, and SDLCs of an amazing soft drink (Sharmeen et al., 2021).

Essential oils utilized in citrus and cola concentrates

Coca leaf, cinnamon bark, nutmeg, coriander, neroli bloom, lemon, lime, orange, and vanilla seed are among the EOs that are combined in Cola Pay Interest. Concentrates of lemon-lime and orange soft drinks surround EOs from all citrus fruits, with neroli in particular because of lemon-lime pop. While orange oil predominates in orange soft drinks, lemon, lime, and neroli oils plays a key role in lemon-lime soft drinks (Sharmeen et al., 2021).

Citrus essential oils

Citrus belongs to the family Rutaceae, which includes 1300 species and 140 genera. It is a group of different tropical natural products that includes species such as oranges, lime, lemon, grapefruit, and bergamot. The ancestors of the citrus species are believed to have originated in the lower Himalayan regions of northern India, southeast Asia, southern China, and northern Myanmar. Citrus is used as praise to several sweet food sources in European Countries, and these natural products later unfold to be magnificent elements of the world that transform into the world's famous organic product plants. Numerous superb meals, such as pork tenderloin with blood oranges, depend on citrus as an essential component. These citrus natural products have nutritional value in addition to higher concentrations of fundamental bioactive compounds, such as phenols, flavonoids, limonoids, essential oils, nutrients, and carotenoids, all of which have numerous health benefits. These compounds also give the organic product a distinctive look and smell that is necessary to create a more customized and appetizing diet. Overall, citrus is recommended for healing and stopping a variety of illnesses. The distinguishing part of the citrus plant, including the leaves, blooms, organic products, and strips, is a rich source of EOs as well as other genuinely beneficial supplements, like nutrients, potassium, flavonoids, coumarins, gelatin, and dietary fibers. Citrus medicinal balms (citrus essential oils) have anticancer, antidiabetic, insecticidal, antifungal, and antibacterial qualities that are crucial for the food, cleaning, corrective, horticultural, and pharmaceutical industries. Citrus EOs have a variety of uses, from those for approved medicines, materials, and cosmetic care products to those for food standards (Bora et al., 2020).

The most common are beverages that use, among other ingredients, the erratic scent mixes of citrus trees. The citrus natural product's flavedo or exocarp contains tiny vesicles that contain them (Zhang et al., 2021). The citrus essential oils's organizational structure changes dramatically in its comprehension of assortment, irregularity, geology, and the readiness of the organic product (Bora et al., 2020; Zhang et al., 2021). There is a huge variety of blends in citrus essential oils. The degree of perspectives is typically somewhere between 20 and 60. (Falleh et al., 2020). About 85%−100% of these are unpredictable mixes and the remaining 1%−15% are stable mixtures. Monoterpenes (limonene, among others), sesquiterpenes, and sesquiterpenoids, including aldehydes (citral), ketones, acids, alcohols (linalool), and esters, are combined to create the delicate combinations. Monoterpenes, which are combinations of two isoprene (C5H8) particles, make up the majority of the composition (97% of the citrus essential oils), while alcohols, aldehydes, and esters account for the remaining 1.8%−2.2% of the citrus EOs (Bora et al., 2020; Zhang et al., 2021). Depending on the kind, its recognition in the rejuvenating oil may also vary between 32% and 98%: 32%−45% in bergamot, 45%−76% in lemon, and 68%−98% in sweet orange (Bora et al., 2020). Since limonene is the main component of the citrus essential oils, the synthetic, physical, and natural properties of this chemical have a profoundly important impact on the citrus essential oils's residences (Falleh et al., 2020).

Intensifies discovered in citrus essential oils

In citrus oil, the mixes are divided into five main categories: hydrocarbon monoterpenes, oxygenated monoterpenes, hydrocarbon sesquiterpenes, and preferences. These five categories are hypothesized to contain various compound components. The main synthetic component of citrus EOs, with a degree ranging from 32 to 98%, is limonene. It is a single monoterpene circle (Mahato et al., 2019).

Identification, refinement, and extraction methods for putting together citrus natural oils (citrus essential oils)

All indications point to a connection between the discovery of citrus strong oil and the use of electricity and carbon dioxide, which has an impact on foundation pay. So, for the most part, connected and excellent expulsion techniques were being used. The majority of current evacuation methods, including Soxhlet partition, peeling, deluge, strong fluid division (SLE), and fluid detachment (LLE), significantly increase the instrument life and appear to require more power than is necessary because, contrary to popular belief, all methods are sufficient to achieve all objectives and require the use of toxic solvents. Furthermore, the intensity labile enthusiastic mixtures that are actually present inside the concentrate may be harmed by the even though an uncertainty extreme having fever at a few issues of the typical extraction way, which is notably lovely even though sort of standard conviction, unquestionably contrary to well-known conviction. As a result, the modern day central detail and necessities with extraction methodologies are kind of getting closer to actually green extraction, which is environmentally friendly with limiting electric power admission and CO outflows in an as a rule really fundamental way, or so they usually thought. For instance, with fluid extraction (LLE), using biorenewable solvents especially lovely specifically basically based in reality on a volume of alkane diols for the most extreme stage frequently were evaluated as environmentally friendly alternatives to extremely beyond doubt separate bioactive terpenoids from terpenes to specifically obtain EOs throughout downstream handling, which in reality is very large. Pourbafraniet, as often as possible, as opposed to notable conviction, assessed the outflows of nursery gas (GHG) connected with the biorefinery of type of Citrus handling waste (CPW), and concluded that changing gas with ethanol (E85) in vehicle filling and methane with as a matter of fact type of natural gas, methane, for electric power creation for the most part dwindled GHG emanations with the valuable helpful asset of way of 134% and 77%, separately, generally as opposed to renowned conviction.

In addition to very basic pressure, supercritical liquid (melted CO), temperature, continuous electric energy supply, siphons, pressure holders, and potentially fixed vessels, the really alternative state-of-the-art nonconventional procedures could furthermore support in most extreme times generally require radical energy to ordinarily run the extraction cycle. However, these methodologies actually equip a really numerous requirements a great deal more precisely than an uncertain solution, and often provide the largest number of requirements at much faster rates, thus yielding a more timely outcome in an up-to-date way (Mahato et al., 2019). Furthermore, since the most innovative extraction techniques are currently restricted to lab studies for the majority of the region, it is more important than ever to generally find out how to effectively meet all expectations and capacities in a fashion other than by chance. Supercritical liquid extraction (SFE), steam blast, ultrasound-aided extraction (UAE), and microwave-aided extraction (MAE) are some of the untried extraction techniques, which are actually quite significant. Among them, UAE and MAE by far have the most intricately involved extraction processes for pharmaceutical balms and innovative homegrown stock from a variety of substantial Citrus species, or so they fundamentally assumed, which is typically crucial. According to their particular plan, which is often extremely enormous, these procedures are exceptionally available and really expeditious to deflect CO emissions and electric fuel consumption. Additionally, it was decided that the mixtures of UAE, along with a few genuinely modern technologies like the microwave method, moment oversaw pressure drop system (DIC), and SFE, would be the most effective solution to cross-breed operations on a high level in the wonderful book. In addition, microwave-aided hydrodistillation (MAHD) has also been applied to consistently really separate EOs from truly clammy truth be told genuinely citrus strip waste in a straightforward manner. Microwave hydro-diffusion and gravity (MHG) is also the most advanced solid-to-visit procedure because it works under the influence of both microwaves and the Earth's gravity, which is specifically and reasonably measured in an actually primary way for all expectations and features and is free to dissolve for all plans and components with as factually little extraction time. Dissolvable free extraction techniques using ultrasound and microwave technology have frequently been successfully employed to fully and completely extract EOs from normally aqueous handling over time, which is actually rather significant. Additionally, cavitation-principally based total extraction (CE) procedures in particular finish the extraction zenith score, for the most extreme segment basically yield of the concentrate, and decreasing the extraction time, or all together that they generally thought, which for the most extreme segment is perceptibly critical. Hydrodynamic cavitation extraction (HCE), in particular, is typically completed with the asset of siphoning fluid through Venturi cylinders or hole plates multiple times in a row. These components put pressure on the slope inside the go with the coast, which frequently consistently results in the age, development, and give method of microbubbles in the air flow. Contrary to widely held belief, real scale variables of HCE are really expressly applied to a particular type of control waste, often orange strips, as well as to a specific type of component waste. HC-based systems have undoubtedly demonstrated excellent execution, yields, and easy versatility in a dispersed manner. Additionally, a method that is both environmentally friendly and power-free, relying mostly on photovoltaic electricity and using photovoltaic hydrodistillation (SSD), has been developed to effectively and separately separate specific EOs on a very large scale (Bora et al., 2020).

Using citrus essential oil

Citrus EOs have a wide range of applications, from the remedial industry to use in traditional and packaged foods;

Citrus essential oils are used in sanitation

Business sanitizers are actually utilized to prevent rotting and contamination of meals. However, due to growing health concerns, people have turned to natural novel antimicrobials like completely aromatic oils that are derived from plants. EOs and their components have antibacterial properties and protect against a large variety of microorganisms. Citrus EOs and its larvicidal, antimicrobial, and antibacterial applications for meals and assurance credits are consistent with this (Bhavaniramya et al., 2019).

Contemporary beverages creation

Water supply and water restoration synthetic enterprises, sugar, CO_2, sterilization and security components (such as sodium bicarbonate), and concentrates are important factors. The concentrates contain the EOs, which are the business's most crucial contributions (Sharmeen et al., 2021).

Why it's important to be aware of the levels of essential oils and AACs in flavored drinks

Even while anomalies like rashes, sore mouth, and runs are already common after the use of a few flavors, there can be almost no chance of harm to the majority of clients in a given community when using healthy, privately obtained flavors in accordance with established norms. As a result, given the widespread usage of flavors and the potential pharmacological and toxicological effects of the EOs/AACs that flavors include, there may also be a desire to know the precise amounts of EOs/AACs present in prepared beverage refreshments. This is because, as stated in the concluding section, EOs and AACs may show (1) Portion-based harmful effects, making it necessary to understand how a portion is seen in a beverage to determine safety levels; (2) portion-based truly supportive impacts, making it necessary to understand how a portion is situated in a beverage to successfully determine strong levels; and (3) antimicrobial strikes that can be recommended in regulating the microbial quality, as a final product there may moreover be some help to sort out the viable phase of such EOs/AACs. Despite the foregoing, there may still be a significant lack of information regarding how often EOs and AACs occur because they are currently not only complex in terms of goods but also in terms of natural substances. The complexity and possibly extreme expenses of such a study were also highlighted by the few questions regarding the conclusion of AAC levels in drinks. However, if the levels of EOs/AACs in beverages are to be very carefully targeted, such questions are crucial (Mahboubi, 2019).

Ginger essential oils

The underground stem or rhizomes of *Zingiber officinale* (Rosc.), a herbaceous enduring spice subfamily *Zingi beraceae*, where the ginger spice is derived. Immunization has evolved into an annual practice. The entire plant has a pleasant scent, but it is the raw or cooked sinewy root that is immersed that is prized for making sauce. Its clinical benefit is becoming more widely acknowledged. South-East Asia is where spice originated, most likely in India. This evaluation needs to be supported by the actual name. Greek pronunciation of the Sanskrit word "Sringavera" led to the creation of the more common "Zingiber" (Mahboubi, 2019).

The root of the ginger plant, *Z. officinale*, is used to make ginger root oil or ginger rejuvenating balm.

Organizations for essential oils and ginger

Ginger oil

Although ginger oil contains only about 3% weight-weight of new ginger rubia, it contains many AACs, such as farnesene, geranial, which is rare and valuable, glycosyl cellulose, lemongrass, which makes up a significant portion of the oil, sesquiphel landrene, which is being considered as a potential, zingerone, and polyphenolic compounds (Spisni et al., 2020).

Camphene, B-Phellandrene, -Pineene, Geranial, Zingiberene, -Bisabolene, -Sesquiphellandrene, and Curcumene are the dynamic fixings compatriots of ginger oil (Spisni et al., 2020).

Jumps oil

Humulene, humulone, and isohumulone are the components of jumps oil.

The following are a some of the components of jumps oil: Xanthohumol is generally thought to be toxic to oxidants and harmful to malignant development (Knez Hrnčič et al., 2019). The stimulatory effects of bounces to humulene and amyl-liquor, which are between, are attributed to isoxanthohumol, which is typically thought to only have limited naturally dynamic communication, and 8 prenylnaringenin (Knez Hrnčič et al., 2019).

History of ginger essential oil

The base of the parsley known as *Z. officinale*, commonly known as ginger, is where ginger natural oil or ginger root oil is extracted. *Z. officinale* got its name from the Greek word "zingiberis," which means "horn molded."

However, although being endemic to the southern part of China and related to the same plant family as cardamom and turmeric, this blossoming lasting has greatly spread throughout Asia, India, the Moluccas (also known as the Flavor Islands), West Africa, Europe, and the Caribbean.

Since the beginning of time, ginger root has been used by daily healers for its virtuoso ability to calm a variety of symptoms, including those related to the monthly cycle, unpleasant bodies, fibromyalgia, and infectious joint pain. It has also frequently been used as a cooking sanitizer that fights microbes and prevents the spread of those tiny organisms, as well as a flavoring for seasoning and digestive system benefits. Ginger oil is frequently used in Ayurvedic medicine to alleviate scholastic wretchedness including clumsiness, rides of disaster, low bravery, and a lack of sympathy (Mahboubi, 2019).

Jumps and ginger are used in beers and lagers

Beer with ginger

A ginger-infused beverage is called a "ginger heavy." From wherever it originated to the US and Canada, it is well known in Europe. The First Mak Trama Centers of Soda, which stepped on its containers, led Robert Cantrell of the US to create an emblem that was broadcast with financial assistance from Grattan and Company (FOHBC). The logo was dark and sweet with a powerful ginger citrus flavor. It turned out to be more sophisticated and positioned as a "nonalcoholic version" of "ginger brew" in accordance with a collectible "brilliant style" zest mix. To this day, Thomas Cantrell is given credit for the "bright style" or "black soda," while John McLaughlin of Canada is now at-permission for the "dry style" or "light style," which has a much too softer ginger citrus flavor. In order to develop "Pale Dry Soda," which would later be known as "Canada Dry Soda," in 1907, McLaughlin raised a ginger plant in 1904. All ginger malty lagers, whether they are darkened or contain proof, are effervescent drinks (carbonated make-a-deal cocktails).

When is soda not just a constant emission? Utilized with compressed CO_2, lactobacillus produces butanol at levels significantly higher than 0.4% v/v. Ginger specialty lagers are recommended as safe, in-home treatments for indigestion, anxiety, bad mood, and severe cold. (Mahboubi, 2019; Spisni et al., 2020).

Ginger tea

Regular variations in Ginger Brew's definition of "lager" as "home grown drink: a carbonated or partially aged drink prepared utilizing or flavored with the roots, leaves, or seeds of a plant" are in keeping with the beverage's frequent switch-ups (Encarta Word reference). The term "marginally matured" denotes that the bundle complies with ethyl and may be more notable but under 0.5% v/v. Before then, "prepared ginger brew" first appeared in Yorkshire, Britain, in the 1750s, and eventually spread to the US, UK, and Canada. mid-19th century. Numerous ginger brews late this evening are categorized as effervescent beverages by term lager. Whatever the case, the original Yorkshire ginger lager was likely brewed using seawater, carbohydrates, lemon, ginger, and a contagious bacteria known as "ginger lager plant." A few biodiesel materials were the medication up till the unusual mix was likely made after a few days. As stated by the FOHBC, "standard ginger lager preparation can create as much as 11% v/v of ethanol" (Mahboubi, 2019).

Benefits and drawbacks of ginger essential oils'/AACs and bounces

Ginger essential oils/AACs

Ginger oil has been described as an antinociceptive/against pyretic/rubefacient, antagonistic to emetic/laxative/vermifuge, spasmolytic/laxative, antimycobacterial/liquor level tericidal, energizer/rubefacient/home remedy, Sudorific/antipyretic, and as of now has developed strains determinant to alleviate flu and cold side effects. Specific weaknesses include unease with smooth skin, mucosal surfaces, and sudden touchiness: a combination that prompts a change in the hair that renders one more defenseless to light (Mahboubi, 2019).

Eessential oil/AAC jumps

Jumps would be used in cooking because its sterile activity will typically favor microorganisms over ideal microscopic life forms, or because its unpleasant flavor helps to reduce the sugar content of barley. Similar to valerian, bounces are widely used in nature to treat restlessness, agitation, and midday larkiness. Jumps have also been used in taste mixtures to soothe issues with estrogen deficiency or when a period is present (Knez Hrnčič et al., 2019). The three types of jumps include topical, temporal, and long-term discoveries from picking bounces. In its agrarian conflict, almost 3% of people were seduced by based highlights in their mouths, fingers, and feet (Mahboubi, 2019).

Oils of Aframomum

The oils of *Aframomum melegueta* (eutopian grain) and *Aframomum danielli* or citratum (crocodile pepper) plants are comparable to the AACs in that they include the pyridine ketone-6-paradol (MW: 278). The inclusion of terpineol, pinene, phellandrene, cineol, the accompanying positions, triterpene, subatomic premise, myrcene, and sabinene as AACs makes the oils equivalent to *Elettaria cardamomum* oil. A few AACs for phellandrene and a request to increase the production are given in garlic oil.

Since the various species and subspecies belong to the Zingiberaceae family, the connections for both Aframomum, Elettaria, and Zingiber oils are simply not unexpected.

Nutmeg oil

In addition to eugenol, eugenyl cellulose, arrangement, and isocaryophyl meri, dark habanero oil also contains a few monoterpene and phenylpropanoids. Dark jalapenos encompass a number of AAC goals, such as "trying to achieve," "-pinene," "-thujone," "-caryophyllene," "-bisabolene," "-pinene," and "stating, highlight is only available." terpinen-4-old, myrcene, sabinene, and luteolin. According to all sources, the type of dark pepper used in Nigeria and the majority of West Africa is fluteist stem bark (Mahboubi, 2019).

Capsicum\sCapsicum, as in precariously tabasco phytonutrients, jalapenos, especially the variety called jalapeno (Capsicum aum), appear to include glycerin, dihydrocapsaicin, and nordihydrocapsaicin. They provide a serious internal breathing threat to all primates, therefore they are no different from alcoholic beverages and sludge, but they frequently contain poisonous precious stones (Mahboubi, 2019).

Essential oils' health benefits

The potential therapeutic benefits of EO products range from numerous skin-beneficial interventions to specific types of reclamation techniques in any case, for the majority of disease types or are prepared methodologies without a doubt with regard to notable societies of the utilization of the EOs for those lofts. EOs have a variety of therapeutic homes and lofts, including anti-toxin, anticancer, flavonoid, antiparasitic, mitigating, viricidal, biocidal, repairing, antihypertensive, and anxiolytic homes. Top societies are increasingly interested in the effectiveness of EOs in the treatment of different types of malignant development and injunctive assistance for physicians (Falleh et al., 2020).

Essential oils' antimicrobial potential

An important role in exerting antibacterial action is played by EOs and their components. Due to their hydrophobic nature, EOs significantly penetrate the lipids of the cell films of microorganisms, disrupt the designs of their cell walls, and make them more permeable. The spilling of particles and other multipurpose components is reduced by this layer penetrability alteration, allowing for portable transit. Each unmarried person and many objective games are displayed by EOs. For instance, trans-cinnamaldehyde, one of the essential EO mixtures, has the capacity to control the explosion of S. typhimurium and E. coli with the aid of depleting intracellular ATP levels. Gaining access to the periplasm and larger portions of the cell is also advantageous. Carvone is another essential component of EOs, and while it still affects the outer layers of cells, it no longer has a significant impact on the mobile ATP pools. Antibacterial properties of EOs are strongly correlated with the presence of citrus EO mixtures; in particular, cinnamaldehyde, citral, carvacrol, eugenol, or thymol, which has a place with the phenol proposes critical side interest, saw through terpenes and various builds along with ketones (-myrcene, -thujone, or geranyl acetic acid derivation).

The three main active ingredients, carvacrol, eugenol, and thymol, efficiently suppress the growth of microorganisms by disrupting the cell walls due to changes in electron flow, proton riding strain, exuberant vehicle, and coagulation of cell contents (Falleh et al., 2020).

Essential oils' are being used as a cancer preventative

The substance structures of EOs play a key role in their ability to prevent cancer. The significant cancer prevention agent peculiarity of EOs is due to the twofold linkages that phenolic and explicit optional metabolites form. Thymol and carvacrol are fundamental monoterpenes found in a wide variety of EOs extracted from plants like *O. tyttanthum*, *Mentha longifolia*, and *Thymus serpyllus*, and they play a crucial role in the cell reinforcement properties.

Due to the inclusion of significant components like thymol and carvacrol, the oils extracted from restorative plants like cinnamon, nutmeg, clove, basil, parsley, oregano, and thyme exhibit adequately approximated cell reinforcement exercises. Fundamentally, their activities are linked to the existence of phenolic compounds, which have powerful redox properties and play important roles in the destruction of free revolutionaries and peroxide. The specific additives, which include high-quality alcohols, ethers, ketones, aldehydes, and monoterpenes, such as linalool, 1,8-cineole, geraniol/neral, citronellal, isomenthone, menthone, and a few others, are also very important for EOs' cell reinforcing processes (Falleh et al., 2020).

Anti-disease capabilities

Treatment of dangerous mobileular growth basic to most malignant growths is one of the most extreme and overpowering circumstances in chemotherapy. Taxol, a plant atom, is effective against malignant mobileular expansion. Numerous types of malignancies, including leukemia, glioma, gastric tumors, human liver growth, pneumonic cancers, and bosom tumors, are assumed to become less common after treatment with plant EOs. As a result, these particles are said to have anticipated anticancer effects that will be helpful in prevention and therapy techniques. Geraniol from *Cymbopogon martini*, such as palmarosa oil, is one example. It is claimed to interfere with film highlights, particle homEOstasis, as well as the mobileular flagging activities of most tumors. Eudesmol, a component of *Atractylodes lancea* oil, may be useful in the prevention of malignant tumors since it is positioned to obstruct DNA synthesis and reduce the size of colon malignancies. Terpenoids, in addition to the polyphenol components found in plant oils, stop the growth of tumor cells by inducing apoptosis or corrupting them. *Myristica fragrans* (nutmeg) oil has been associated with significant hepatoprotective benefits that can be attributed to myristicin, the oil's primary component. As seen in neuroblastoma cells, myristicin is designed to work by enlisting apoptosis. In vivo studies have shown that the citral in lemongrass oil is useful in preventing the early stages of rodent hepatocarcinogenesis. Significant *Allium sativum* (garlic) oil is widely known for its anticancer effects. The use of garlic to mask toxins used in drug detoxification is the extent of the chemopreventive interest. After treatment with lemon ointment (M. officinalis) oil, the progression of the majority of human diseases is slowed down. Low incidence of human melanoma is associated with the acceptance of apoptosis through *M. alternifolia* (Tea tree) oil and its essential monoterpene liquor, terpinen-4-old.

EOs have the potential to prevent cancer and interfere with mammalian cells' ability to use their mitochondria. As a result, EOs lessen the metabolic activities that contribute to harmful growth improvement, such as expanded celiac digesting, excessive mitochondrial production, and long-lasting oxidative pressure (Bhavaniramya et al., 2019).

Antiviral effectiveness

In addition to their antibacterial abilities, flowers have also been found to have exceptional antiviral qualities. The presence of sesquiterpene, phenylpropanoid, and monoterpene components in EOs is thought to prevent viral replication. Eucalyptus and thyme oils are reported to have an inhibitory effect on herpes infection. It has been observed that the oil extracted from *M. alternifolia* has enormous potential for treating recurrent herpes infection contaminations. This pastime was reduced to only having the capacity to interfere with the virus's envelope structures, preventing infection from adhering to or entering the host cells. For instance, oregano oil limits the HSV envelope's capacity to infect by causing it to disintegrate. At 3.7 g/m, oregano oil is similarly thought to exert antiviral action against yellow fever infection. Isoborneol, a monoterpene found in a variety of EOs, inhibits the glycosylation of viral proteins and has virucidal effects on HSV-1. Only 0.00003% of German chamomile has been observed to inhibit HSV-1. While emollient oils from santolina, pine, tea tree, manuka, and lemon are effective against HSV-1 within the consideration range of 0.0001%−0.0009%.

HSV-2 is noticeably more sensitive to the aforementioned oils, with declining IC50 values. In order to prevent viral enactment, EO-added compounds are designed to primarily prevent early quality articulation in CMV (cytomegalovirus). A mouse study highlighted the in vivo survivability of clove oil's eugenol to help treat herpesvirus-induced keratitis (Bhavaniramya et al., 2019).

Conclusion

A review of the writing on soft drinks reveals that many pre-change enigmas and tactics are outside the purview of public knowledge. Although there are still questions, modern techniques enable researchers to ascertain the taste-related information and techniques of soft drinks (i.e., the specific combinations of EOs as flavorings) and to produce new information. For instance, how many truckloads of each citrus oil are accessible when a soft drink contains more than one (such as lemon, lime, orange, or neroli)? This is a challenging task because each oil contains assessments of smell-active terpenes and terpenoids, which are the fundamental subatomic components of EOs and exist in a dizzying array of isomers. Additionally, distinct EOs utilized in soft drinks combine significant terpenes and terpenoids, just like citrus oils do. Finally, the nature of the methods employed to evaluate taste is such that the expert or flavorist must constantly work to maintain the open doors of curios inside the grid, which distort the actual image of the preferred sections. The best oversight of EOs is frequently not decided by bottlers at the level of listening plan, which is one important implication of this situation. Today's technology makes it possible to assemble soft drinks that are both more enjoyable and safer by using concentrates created with specific synthetic flavorings that impart the desired flavor and smell.

References

Adejumo, I. O., Adetunji, C. O., & Adeyemi, O. S. (2017). Influence of UV light exposure on mineral composition and biomass production of myco-meat produced from different agricultural substrates. *Journal of Agricultural Sciences, Belgrade, 62*(1), 51–59.

Adetunji, C. O., Akram, M., Olaniyan, O. T., Ajayi, O. O., Inobeme, A., Olaniyan, S., Hameed, L., & Adetunji, J. B. (2021a). Targeting SARS-CoV-2 novel Corona (COVID-19) virus infection using medicinal plants. In K. Dua, S. Nammi, D. Chang, D. K. Chellappan, G. Gupta, & T. Collet (Eds.), *Medicinal plants for lung diseases.* Singapore: Springer. Available from https://doi.org/10.1007/978-981-33-6850-7_21.

Adetunji, C. O., Nwankwo, W., Ukhurebor, K., Olayinka, A. S., & Makinde, A. S. (2021b). Application of biosensor for the identification of various pathogens and pests mitigating against the agricultural production: Recent advances. In R. N. Pudake, U. Jain, & C. Kole (Eds.), *Biosensors in agriculture: Recent trends and future perspectives. Concepts and strategies in plant sciences.* Cham: Springer. Available from https://doi.org/10.1007/978-3-030-66165-6_9.

Adetunji, C. O., Palai, S., Ekwuabu, C. P., Egbuna, C., Adetunji, J. B., Ehis-Eriakha, C. B., Kesh, S. S., & Mtewa, A. G. (2021c). *General principle of primary and secondary plant metabolites: Biogenesis, metabolism, and extraction. Preparation of Phytopharmaceuticals for the Management of Disorders* (pp. 3–23). Publisher Academic Press.

Adetunji, C. O., Inobeme, A., Olaniyan, O. T., Ajayi, O. O., Olaniyan, S., & Adetunji, J. B. (2021d). Application of nanodrugs derived from active metabolites of medicinal plants for the treatment of inflammatory and lung diseases: Recent advances. In K. Dua, S. Nammi, D. Chang, D. K. Chellappan, G. Gupta, & T. Collet (Eds.), *Medicinal plants for lung diseases.* Singapore: Springer. Available from https://doi.org/10.1007/978-981-33-6850-7_26.

Adetunji, C. O., Anani, O. A., Olaniyan, O. T., Inobeme, A., Olisaka, F. N., Uwadiae, E. O., & Obayagbona, O. N. (2021e). Recent trends in organic farming. In R. Soni, D. C. Suyal, P. Bhargava, & R. Goel (Eds.), *Microbiological activity for soil and plant health management.* Singapore: Springer. Available from https://doi.org/10.1007/978-981-16-2922-8_20.

Adetunji, C. O., Michael, O. S., Kadiri, O., Varma, A., Akram, M., Kola Oloke, J., Shafique, H., Adetunji, J. B., Jain, A., Bodunrinde, R. E., Ozolua, P., & Ubi, B. E. (2021f). Quinoa: From farm to traditional healing, food application, and phytopharmacology. In A. Varma (Ed.), *Biology and biotechnology of Quinoa.* Singapore: Springer. Available from https://doi.org/10.1007/978-981-16-3832-9_20.

Adetunji, C. O., Michael, O. S., Nwankwo, W., Ukhurebor, K. E., Anani, O. A., Oloke, J. K., Varma, A., Kadiri, O., Jain, A., & Adetunji, J. B. (2021g). Quinoa, the next biotech plant: Food security and environmental and health hot spots. In A. Varma (Ed.), *Biology and biotechnology of Quinoa.* Springer: Singapore. Available from https://doi.org/10.1007/978-981-16-3832-9_19.

Adetunji, C. O., Michael, O. S., Varma, A., Oloke, J. K., Kadiri, O., Akram, M., Bodunrinde, R. E., Imtiaz, A., Adetunji, J. B., Shahzad, K., Jain, A., Ubi, B. E., Majeed, N., Ozolua, P., & Olisaka, F. N. (2021h). Recent advances in the application of biotechnology for improving the production of secondary metabolites from Quinoa. In A. Varma (Ed.), *Biology and biotechnology of Quinoa.* Singapore: Springer. Available from https://doi.org/10.1007/978-981-16-3832-9_17.

Adetunji, J. B., Adetunji, C. O., & Olaniyan, O. T. (2021i). African walnuts: A natural depository of nutritional and bioactive compounds essential for food and nutritional security in Africa. In O. O. Babalola (Ed.), *Food security and safety.* Cham: Springer. Available from https://doi.org/10.1007/978-3-030-50672-8_19.

Adetunji, C. O., Olaniyan, O. T., Akram, M., Ajayi, O. O., Inobeme, A., Olaniyan, S., Khan, F. S., & Adetunji, J. B. (2021j). Medicinal plants used in the treatment of pulmonary hypertension. In K. Dua, S. Nammi, D. Chang, D. K. Chellappan, G. Gupta, & T. Collet (Eds.), *Medicinal plants for lung diseases.* Singapore: Springer. Available from https://doi.org/10.1007/978-981-33-6850-7_14.

Adetunji, C. O., Ajayi, O. O., Akram, M., Olaniyan, O. T., Chishti, M. A., Inobeme, A., Olaniyan, S., Adetunji, J. B., Olaniyan, M., & Awotunde, S. O. (2021k). Medicinal plants used in the treatment of influenza a virus infections. In K. Dua, S. Nammi, D. Chang, D. K. Chellappan, G. Gupta, & T. Collet (Eds.), *Medicinal plants for lung diseases.* Singapore: Springer. Available from https://doi.org/10.1007/978-981-33-6850-7_19.

Adetunji, C.O. (2008). The antibacterial activities and preliminary phytochemical screening of vernoniaamygdalina and Aloe vera against some selected bacteria, pp. 40–43 [Msc Thesis, University of Ilorin].

Adetunji, C. O., Michael, O. S., Rathee, S., Singh, K. R. B., Olufemi Ajayi, O., Bunmi Adetunji, J., Ojha, A., Singh, J., & Pratap Singh, R. (2022). Potentialities of nanomaterials for the management and treatment of metabolic syndrome: A new insight. *Materials Today Advances., 13,* 100198.

Adetunji, C. O., & Varma, A. (2020). Biotechnological application of trichoderma: A powerful fungal isolate with diverse potentials for the attainment of food safety, management of pest and diseases, healthy planet, and sustainable agriculture. In C. Manoharachary, H. B. Singh, & A. Varma (Eds.), *Trichoderma: Agricultural applications and beyond. Soil biology* (61). Cham: Springer. Available from https://doi.org/10.1007/978-3-030-54758-5_12.

Adetunji, C. O., Oloke, J. K., & Prasad, G. (2020a). Effect of carbon-to-nitrogen ratio on eco-friendly mycoherbicide activity from Lasiodiplodia pseu-dotheobromae C1136 for sustainable weeds management in organic agriculture. *Environment, Development and Sustainability, 22,* 1977–1990. Available from https://doi.org/10.1007/s10668-018-0273-1.

Adetunji, C. O., Egbuna, C., Tijjani, H., Adom, D., Al-Ani, L. K. T., & Patrick-Iwuanyanwu, K. C. (2020b). *Homemade preparations of natural bio-pesticides and applications. Natural Remedies for Pest, Disease and Weed Control* (pp. 179−185). Publisher Academic Press.

Adetunji, C. O., Roli, O. I., & Adetunji, J. B. (2020c). Exopolysaccharides derived from beneficial microorganisms: Antimicrobial, food, and health benefits. In P. Mishra, R. R. Mishra, & C. O. Adetunji (Eds.), *Innovations in food technology*. Singapore: Springer. Available from https://doi.org/10.1007/978-981-15-6121-4_10.

Adetunji, J. B., Ajani, A. O., Adetunji, C. O., Fawole, O. B., Arowora, K. A., Nwaubani, S. I., Ajayi, E. S., Oloke, J. K., & Aina, J. A. (2013). Postharvest quality and safety maintenance of the physical properties of Daucus carota L. fruits by Neem oil and Moringa oil treatment: A new edible coatings. *Agrosearch, 13*(1), 131−141.

Adetunji, C. O. (2019). Environmental impact and ecotoxicological influence of biofabricated and inorganic nanoparticle on soil activity. In D. Panpatte, & Y. Jhala (Eds.), *Nanotechnology for agriculture*. Singapore: Springer. Available from https://doi.org/10.1007/978-981-32-9370-0_12.

Adetunji, C. O., Panpatte, D. G., Bello, O. M., & Adekoya, M. A. (2019a). Application of nanoengineered metabolites from beneficial and eco-friendly microorganisms as a biological control agents for plant pests and pathogens. In D. Panpatte, & Y. Jhala (Eds.), *Nanotechnology for agriculture: Crop production & protection*. Singapore: Springer. Available from https://doi.org/10.1007/978-981-32-9374-8_13.

Adetunji, C. O., Ojediran, J. O., Adetunji, J. B., & Owa, S. O. (2019b). Influence of chitosan edible coating on postharvest qualities of Capsicum annum L. during storage in evaporative cooling system. *Croatian Journal of Food Science and Technology, 11*(1), 59−66.

Adetunji, C. O., Kumar, D., Raina, M., Arogundade, O., & Sarin, N. B. (2019c). Endophytic microorganisms as biological control agents for plant pathogens: A panacea for sustainable agriculture. In A. Varma, S. Tripathi, & R. Prasad (Eds.), *Plant biotic interactions*. Cham: Springer. Available from https://doi.org/10.1007/978-3-030-26657-8_1.

Adetunji, C. O., Phazang, P., & Sarin, N. B. (2017). Significance of rhamnolipids as a biological control agent in the management of crops/plant pathogens. *Current Trends in Biomedical Engineering & Biosciences, 10*(3), 54−55.

Adetuyi, B. O., Oluwole, E. O., & Dairo, J. O. (2015a). Chemoprotective potential of ethanol extract of ganoderma lucidum on liver and kidney parameters in plasmodium beghei-induced mice. *International Journal of Chemistry and Chemical Processes (IJCC), 1*(8), 29−36.

Adetuyi, B. O., Oluwole, E. O., & Dairo, J. O. (2015b). Biochemical effects of shea butter and groundnut oils on white albino rats. *International Journal of Chemistry and Chemical Processes (IJCC), 1*(8), 1−17.

Adewale, G. G., Olajide, P. A., Omowumi, O. S., Okunlola, D. D., Taiwo, A. M., & Adetuyi, B. O. (2022). Toxicological Significance of the Occurrence of Selenium in Foods. *World News of Natural Sciences, 44*, 63−88.

Batiha, G. B., Awad, D. A., Algamma, A. M., Nyamota, R., Wahed, M. I., Shah, M. A., Amin, M. N., Adetuyi, B. O., Hetta, H. F., Cruz-Marins, N., Koirala, N., Ghosh, A., & Sabatier, J. (2021). Diary-derived and egg white proteins in enhancing immune system against COVID-19 frontiers in nutritionr. *(Nutritional Immunology)*, 8629440. Available from https://doi.org/10.3389/fnut.2021629440.

Bello, O. M., Ibitoye, T., & Adetunji, C. (2019). Assessing antimicrobial agents of Nigeria flora. *Journal of King Saud University-Science, 31*(4), 1379−1383.

Bhavaniramya, S., Vishnupriya, S., Al-Aboody, M. S., Vijayakumar, R., & Baskaran, D. (2019). Role of essential oils in food safety: Antimicrobial and antioxidant applications. *Grain & Oil Science and Technology, 2*(2), 49−55.

Bora, H., Kamle, M., Mahato, D. K., Tiwari, P., & Kumar, P. (2020). Citrus essential oils (CEOs) and their applications in food: An overview. *Plants, 9*(3), 357.

Didunyemi, M. O., Adetuyi, B. O., & Oyewale, I. A. (2020). Inhibition of lipid peroxidation and in-vitro antioxidant capacity of aqueous, acetone and methanol leaf extracts of green and red Acalypha wilkesiana Muell Arg. *International Journal of Biological and Medical Researc, 11*(3), 7089−7094.

Didunyemi, M. O., Adetuyi, B. O., & Oyebanjo, O. O. (2019). Morinda lucida attenuates acetaminophen-induced oxidative damage and hepatotoxicity in rats. *Journal of Biomedical Sciences, 8*. Available from https://www.jbiomeds.com/biomedical-sciences/morinda-lucida-attenuates-acetamino-pheninduced-oxidative-damage-and-hepatotoxicity-in-rats.php?aid = 24482.

Egbuna, C., Gupta, E., Ezzat, S. M., Jeevanandam, J., Mishra, N., Akram, M., Sudharani, N., Oluwaseun Adetunji, C., Singh, P., Ifemeje, J. C., Deepak, M., Bhavana, A., Mark, A., Walag, P., Ansari, R., Adetunji, J. B., Laila, U., Olisah, M. C., & Onyekere, P. F. (2020). Aloe species as valuable sources of functional bioactives. In C. Egbuna, & G. Dable Tupas (Eds.), *Functional foods and nutraceuticals*. Cham: Springer. Available from https://doi.org/10.1007/978-3-030-42319-3_18.

Falleh, H., Jemaa, M. B., Saada, M., & Ksouri, R. (2020). Essential oils: A promising eco-friendly food preservative. *Food Chemistry, 330*127268.

James-Okoro, P. O., Iheagwam, F. N., Sholeye, M. I., Umoren, I. A., Adetuyi, B. O., Ogundipe, A. E., Braimah, A. A., Adekunbi, T. S., Ogunlana, O. E., & Ogunlana, O. O. (2021). Phytochemical and in vitro antioxidant assessment of Yoyo bitters World News of Natural. *Sciences, 37*, 1−17.

Knez Hrnčič, M., Španinger, E., Košir, I. J., Knez, Ž., & Bren, U. (2019). Hop compounds: Extraction techniques, chemical analyses, antioxidative, antimicrobial, and anticarcinogenic effects. *Nutrients, 11*(2), 257.

Mahato, N., Sharma, K., Koteswararao, R., Sinha, M., Baral, E., & Cho, M. H. (2019). Citrus essential oils: Extraction, authentication and application in food preservation. *Critical Reviews in Food Science and Nutrition, 59*(4), 611−625.

Mahboubi, M. (2019). Zingiber officinale Rosc. essential oil, a review on its composition and bioactivity. *Clinical Phytoscience, 5*(1), 1−12.

Nazir, A., Itrat, N., Shahid, A., Mushtaq, Z., Abdulrahman, S. A., Egbuna, C., Adetuyi, B. O., Khan, J., Uche, C. Z., & Toloyai, P. Y. (2022). Orange peel as a source of nutraceuticals. In *Food and agricultural byproducts as important source of valuable nutraceuticals*, (1st ed.). Berlin: Springer, 400p.

Olaniyan, O. T., & Adetunji, C. O. (2021). Biological, biochemical, and biodiversity of biomolecules from marine-based beneficial microorganisms: Industrial perspective. In C. O. Adetunji, D. G. Panpatte, & Y. K. Jhala (Eds.), *Microbial rejuvenation of polluted environment. Microorganisms for sustainability* (27). Singapore: Springer. Available from https://doi.org/10.1007/978-981-15-7459-7_4.

Rauf, A., Akram, M., Semwal, P., Mujawah, A. A. H., Muhammad, N., Riaz, Z., Munir, N., Piotrovsky, D., Vdovina, I., Bouyahya, A., Oluwaseun Adetunji, C., Ali Shariati, M., Almarhoon, Z. M., Mabkhot, Y. N., & Khan, H. (2021). Antispasmodic potential of medicinal plants: A comprehensive review. *Oxidative Medicine and Cellular Longevity, 12*, 2021. Available from https://doi.org/10.1155/2021/4889719, Article ID 4889719.

Sharmeen, J. B., Mahomoodally, F. M., Zengin, G., & Maggi, F. (2021). Essential oils as natural sources of fragrance compounds for cosmetics and cosmeceuticals. *Molecules (Basel, Switzerland), 26*(3), 666.

Shedoeva, A., Leavesley, D., Upton, Z., & Fan, C. (2019). Wound healing and the use of medicinal plants. *Evidence-Based Complementary and Alternative Medicine, 2019.*

Spisni, E., Petrocelli, G., Imbesi, V., Spigarelli, R., Azzinnari, D., Donati Sarti, M., ... Valerii, M. C. (2020). Antioxidant, anti-inflammatory, and microbial-modulating activities of essential oils: Implications in colonic pathophysiology. *International Journal of Molecular Sciences, 21*(11), 4152.

Ukhurebor, K. E., & Adetunji, C. O. (2021). Relevance of biosensor in climate smart organic agriculture and their role in environmental sustainability: What has been done and what we need to do? In R. N. Pudake, U. Jain, & C. Kole (Eds.), *Biosensors in agriculture: Recent trends and future perspectives. Concepts and strategies in plant sciences.* Cham: Springer. Available from https://doi.org/10.1007/978-3-030-66165-6_7.

Zhang, X., Ismail, B. B., Cheng, H., Jin, T. Z., Qian, M., Arabi, S. A., ... Guo, M. (2021). Emerging chitosan-essential oil films and coatings for food preservation-A review of advances and applications. *Carbohydrate Polymers, 273*, 118616.

Chapter 5

Application of essential oils in enhancing the activities of starter culture bacteria in dairy products

Abiola Folakemi Olaniran[1], Christianah Oluwakemi Erinle[2], Olubukola David Olaniran[3],
Clinton Emeka Okonkwo[4], Adeyemi Ayotunde Adeyanju[1] and Abiola Ezekiel Taiwo[5]

[1]Department of Food Science and Microbiology, Landmark University, Omu-Aran, Kwara State, Nigeria, [2]Department of Agricultural and Biosystems Engineering, Landmark University, Omu-Aran, Kwara State, Nigeria, [3]Department of Sociology and Anthropology, Obafemi Awolowo University, Ile-Ife, Osun State, Nigeria, [4]Department of Food Science, College of Food and Agriculture, United Arab Emirates University, Al Ain, Abu Dhabi, [5]Faculty of Engineering, Mangosuthu University of Technology, Durban, Umlazi, South Africa

Introduction

Essential oils (EOs) are constituents synthesized in diverse herb structures as secondary metabolites described as oily aromatic fluids removed from plants (El Asbahani et al., 2015). They have been confirmed by the Food and Drug Administration and Generally Recognized as Safe at dosages naturally utilized in foods (Olaniran et al., 2015). Starter cultures are simply a set of microbial preparations used in the processing of foods, one or more microorganisms are added to accelerate the fermentation process of the raw material (milk) into a fermented product (García-Díez & Saraiva, 2021). Microbial starter cultures are crucial for achieving consistency in the quality and functional properties of diverse food products (Shani et al., 2021). Milk are consumed since immemorial throughout the world; it's a nutrient-rich liquid that the feminine mammals produced to nourish their respective babies (Bezie & Regasa, 2019). Dairy products are products obtained from milk processing into concentrated as well as easily transportable items with an extended shelf life for instance processed milk, cheese, yogurt, whey, skimmed milk, butter, etc. (How et al., 2022). They are excellent protein and energy sources. The addition of dairy products to human diets adds variety to plant-based diets, which increases the amount and variety of dairy products in the world. According to Bezie and Regasa (2019), fresh milk is the highest dairy product manufactured annually in the world (approximately thirty-two million tons), next is cheese (about four million tons), then yogurt and other fermented milk (three million tons), little amounts of butter (almost one million tons), and cream (five hundred thousand tons).

Recently a few types of lactic acid bacteria (LAB) affiliated with Bifidobacteria and lactobacilli have been presented in food items for human utilization, with the mean to work on human wellbeing. They are called probiotic bacteria ("for life," in Greek) and are characterized as living microorganisms, which upon consumption in adequate amounts, apply medical advantages past intrinsic fundamental nourishment (Papaioannou et al., 2022). One of the significant advantages of dairy items is that they are key carriers of probiotic bacteria. The probiotic bacteria have been associated with dairy products for a while, fermented milks like yogurts have significantly been acknowledged as carriers of living probiotic cultures. Milk items derived from fermentation are perceived for their unmistakable taste, nutritional components, and beneficial values. Fermented milk are items derived from partially or entirely skimmed liquid or dried milk, concentrated milk, or slightly or fully skimmed pasteurized milk. On a daily basis, food producers use dairy starters and enzymes/rennet for the production of dairy items that are fermented (Bezie & Regasa, 2019).

Microorganisms are widely used in dairy processing as starter cultures. Starter cultures are bacteria added to milk at specific temperatures to ripen it or ferment it in order to derive a product with an extended shelf life and for other

Applications of Essential Oils in the Food Industry. DOI: https://doi.org/10.1016/B978-0-323-98340-2.00020-1

purposes (Mohammed & Çon, 2021). Starter cultures are used to stimulate the speedy acidification of milk by the formation of organic acids, mainly lactic acid. However, during this process in which starter cultures are added, there is the production of aroma compounds, exopolysaccharides, acetic acid, propionic acid, bacteriocins, and several significant enzymes, which improves the sensory profile of the finished product, extends the shelf-life, inhibits the growth of spoilage microorganisms, and reduces the pH. Advances in technology and research have been made, which introduced the application of EOs to enhance starter cultures. Using EOs with starter cultures in cheese production has been recommended, as the survival of starter cultures may be altered and thereby affect the sensory and safety characteristics of the product (García-Díez & Saraiva, 2021).

Milk and allied product

Milk and its product are essential components for consumption and should be included daily as part of the food. Consumption of milk and its product originated in ancient times majorly because they have been nutrient-dense and can be served as a complete food. It contains macronutrient and micronutrient and aids the effective function of the human body (Kamath et al., 2021). It serves as a source of nutritional, therapeutic, and physiological components to ensure healthy living, management, and repair of the human body. The rate of consumption of milk and its product has increased over the last century due to increased awareness by food organizations on the benefit to human health (World Health Organization, 2020). The nutrient composition of milk varies depending on the species of the animal such as cow, sheep, goat, camel, cow, yak, buffalo, donkey, horse, ox, reindeer, caribou, etc. It is also influenced by the genotype of the animal, stage of lactation, feed, climatic season, breed, reproduction, region, sanitary, and management of the animal (Ratwan et al., 2017). Milk can be converted to different products for diversity, preservation, storage, and fortification such as butter, ice cream, cream, casein, whey, cheese, etc. The diversity in milk products differs from one locale to another, which is determined by the dietary habit, economy of the region, market and supply demand, taboo, and technological development (Kapaj, 2018).

Cheese production from other types of milk such as skimmed milk, creamy milk, and buttermilk using appropriate coagulant results in products that are similar in morphology, chemical, and sensory acceptance depending on the type of cheese to be formed (Romeih & Walker, 2017). Cheese can be classified according to texture, method of coagulation, milk source, moistness, maturation agent, cooking temperature, and microflora; there are 18 distinct types of natural cheese such as whey cheeses, Limburger, Cottage, Brick, Cheddar, Cream, Edam, Roquefort, Gouda, Hand, Neufchatel, Parmesan, Romano, Sapsago, Provolone, Swiss, Trappist, and Camembert (McSweeney et al., 2017).

Cheese is an item derived from coagulation milk to produce curd and whey. Cheese could be unripened, soft, ripened, hard, semi-hard, or extra hard which is obtained by coagulation of whole or part of the milk protein, cream, semi-skimmed or skimmed milk, buttermilk, whey milk, or a combination of each with the help of coagulant which can be rennet, starter culture, or other coagulants to produce curd and whey that shown an acceptable organoleptic property, physical, and chemical characteristics (FAO, 2020). Cheese is an indispensable source of protein, lipids, and minerals such as iron, calcium, and phosphorus. It also has vitamins and vital amino acids, making it an important food for both young and elderly and has been classified as a milk-based natural cheese (O'Brien & O'Connor, 2017). Cheese can be made from different milk of animals ranging from sheep, goat, buffalo, cow, ox, donkey, yak, etc. Also, this variation affects the nutritional component of the cheese. Goat milk consumption is higher than other animals due to higher protein content and less casein, making it less desirable for cheese production. White brine cheese made from goat milk is generally consumed in Serbia (Miloradovic et al., 2017). Buffalo milk is majorly consumed in India, China, and Pakistan and the production rate is second to cow. The rate of consumption has been associated with double fat concentration, protein, and other trace elements. It is widely used for cheese production such as Mozzarella cheese, Domiati, Cheddar, Paneer, Latin-American white cheese, etc (Fernandez et al., 2017). Production of sheep milk is on the increase due to a higher rate of consumption as a source of nutrients which varies based on sheep breed. Different sheep breeds can be used for cheese production which is proven to be a source of nutrients, improving the consumer's well-being (Moatsou et al., 2019). The milk is used for cheese production and are called different names based on the region such as Roquefort from France, pecorino from Italy, manouri from Greece, Serra da Estrela from Portugal, etc. It serves as a source of tourist attraction, there is a peculiarity of some food to region such as tocos in Mexico, foul/cooked broad beans in Egyptian, Chaat in India, roasted cassava in Ghana, Cheese and koko in Nigeria, dumplings in Korea, China and Russia (Cortese et al., 2016).

Butter is a dairy product produced by stirring milk either raw or cream in fermented form. It is mostly used as a spread, seasoning, and in cooking such as pan frying, sauce making, and baking (Panchal & Bhandari, 2020). The biggest nations producing butter are Germany, the United States, France, Russia, and New Zealand. In accordance with the

Codex Alimentarius Commission in conjunction with FAO/WHO Food Standards Programme, butter is a fatty product derivative of dairy milk in which every one hundred gram fat must contain a minimum of eighty grams whereas nonfat milk solids and water constitute sixteen grams. As such, it generally comprises saturated fat and is a huge origin of dietary cholesterol. Many sorts of butter are for sale in the marketplace, varying in the type of cream used in the manufacturing process. (Burke et al., 2018, Food and Agriculture Organization of the United Nations FAO (2022)).

The role of starter culture in diary processing

Microbial fermentation of beverages and food varieties is depicted to date back to 6000 BC whose initial aim was the preservation of food. This is largely a consequence of microbial translation of carbohydrates to acids and/or alcohols, reduction of oxygen, and generation of antimicrobial molecules, which avert the development of degeneration of organisms. During the automation of food fermentations, regulating fermentation conditions became increasingly imperative to ensure consistent qualities of the product such as texture and flavor (Shani et al., 2021). Fermented milk items are recognized for their nutritive worth, taste, and beneficial properties. Dairy starters constitute the core of products of fermented milk, the most vital element in the production of high-quality fermented milk. A notable span of research has been done on starter cultures, probiotics, blend formulation, packing methods, biochemical, sensory, physical properties, and conventional and new constituents in diary industry. The microstructure, composition, quality, safety, manufacturing methods innovations, addition of flavorings, viscosity, and diverse probiotics have been improved for yogurt, cultured buttermilk, and many sour cream production. These products have been reported to as well enhance the wellbeing of their consumers (Bintsis, 2018).

Starter cultures are commonly obtained through cultivation on media or by using traditional methods. They can be categorized into mesophilic and thermophilic cultures that grow optimally at 25°C−30°C and 37°C−45°C, respectively. Predominantly, to adapt substrate or start-up materials during production, traditional starter cultures are selected either as lone or multiple strains comprising of microorganisms with a specific biological capability and are added at quantified concentrations to facilitate turning the substrate into product with distinct features. The continuous improvement of microbial starter culture has remained a motivating force for novelty in designing machinery fit for hygienic handling and production of traditional fermented foods under properly monitored settings in several countries based on their own standards and good manufacturing process. Well-defined starter cultures are usually applied in the commercial production of dairy products, especially yogurt, kefir, cheese, etc. Many of these cultures are custom-made to produce specific textures and flavors (Akpi et al., 2020). It converts milk sugar and lactose into lactic acid leading to the sourness of the milk termed as either cultured or fermented dairy products. Starter cultures account for the development of flavor, produce acid during production, and contribute to the ripening of products. The application of starters in dairy is the core and the utmost essential constituent in making outstanding fermented milk and attaining high yields. The starter cultures must be harmless, safe, and certified food-grade microorganisms before it applies in producing different brands of fermented milk products.

The prominent element in cheese production is lactic bacteria which are essential for lactic fermentation in converting lactose into lactic acid. This process, known as acidification, aids the rennet's effort before whey draining, curd formation, and maturation of the cheese. The utilization of starter cultures has become a trademark in industrialization of fermented food with the endless quest to improve its output. Illustrations of such goals are enhancement in the high specificity, vigor throughout production, starter culture turnover/ volume, and end product yield coupled with definite sensory properties (Bezie & Regasa, 2019).

Several bacteria have been discovered to be present in food in which majority initiate/enhance food spoilage. The importance of bacteria in fermentation and food processing is often times overlooked (Ray & Joshi, 2014). LAB such as *Oenococcus, Lactobacillus, Streptococcus, Pediococcus*, etc., remain recognized as the utmost significant bacteria in foods that are fermented, and after them are *Acetobacter* strains, which metabolizes alcohol to acetic acid. Yogurt is globally recognized as the most familiar fermented milk. Yogurt is typically processed cow milk, subjected to fermentation by *Lactobacillus delbrueckii* subsp. *bulgaricus*, and *Streptococcus thermophilus* (Jan et al., 2022). Cheese is another widespread dairy product that is fermented, and processed from milk that has not been pasteurized and could be dependent on naturally occurring lactic microbiomes for its fermentation; this is mostly generated on a large measure using the suitable starter culture that can comprise of *Lactobacillus lactis subsp.* cremoris, and *Lactobacillus lactis* subsp. *Lactis*, which are mesophiles or *Streptococcus thermophilus, L. delbrueckii* subsp. *Bulgaricus and Lactobacillus helveticus*, which are thermophiles, depend on precise usage (Jan et al., 2022).

Starter cultures are employed to effect swift milk acidification via organic acids production of mainly lactic acid. However, during this process in which starter cultures are added, there is generation of propionic acid, acetic acid, exopolysaccharides, aroma compounds, bacteriocins, and other enzymes; which are of significance since it somehow improves the organoloeptic profile of the final output items and extends the shelf-life (Van Boekel et al., 2010). Starter cultures when added to milk, slowly acidifies the milk and converts the lactose sugar present in it into lactic acid, and during this process, a sensory profile of desirable flavors is formed and thereby building a structure of a new product. Starter cultures can be categorized into two types: mesophilic and thermophilic (Shori et al., 2022).

a. Mesophilic bacteria: They are bacteria that prefer or work best at moderate temperatures such as 30°C (86 F) because they are more proficient at this stage in converting lactose to lactic acid

b. Thermophilic bacteria: These bacteria prefer warmer temperatures at about 40°C (108 F)

The main starter culture used in dairy processing is LAB, this is desired because of the ability it has to create homogeneous/distinctive textures, flavor and provides different traits attributed to different microbes. The use of starter culture gives room for control and optimization of the fermentation process as this helps guide to the final desired product.

Standard samples of the starter cultures and their benefits are listed in the table shown as follows:

Advantage	Functions	LAB
Preservation of food	Dairy Goods	*Lactococcus lactis subsp. Lactis Enterococcus spp.*
Organoleptic	Production of Aroma Improved sweet taste	Several strains of lactobacilli
	Malolactic fermentation	*O. oeni*
Technological	Inhibition of over-acidification in yogurt	Lactose-negative *Lb. delbrueckii subsp. bulgaricus*
	Induced by bacteriocin	*L. lactis*
Nutritional	Nutraceuticals manufacturing	
	Sugars with low calories (such as, mannitol and sorbitol)	*Lb. plantarum, L. lactis*

Ray and Joshi (2014).

Use of essential oils in starter cultures

EOs are regarded as oils derived from various components of a plant or extracted from specific plants (Muhammad et al., 2019). These oils can inhibit the growth of pathogens and exhibit anti-microbial properties (Liu et al., 2020) as they serve as natural anti-microbial agents. They are obtained in isolation from the hydro-distillation method whereby entire pieces of the plants are utilized for the oils' extraction.

Mohamed et al. (2013) stated that the prevalent spices and herbs with vibrant anti-microbial activity are clove, cinnamon, oregano, rosemary, and dill. They have EOs comprising chemical compounds such as carcvacol, eugenol, camphor, and cinnamaldhyde which have been recognized as the main chemical components that inhibit microbial activity.

However, due to their susceptibility to challenging circumstances within food items and potent stomach and small intestinal enzymes, these bacteria have poor viability, which poses a significant problem in the production and processing of probiotic goods. A standard probiotic product must, according to the FAO, include at least 106 to 107 cfu/g of living and energetic probiotic bacteria during time of intake (FAO, 2020). The development of goods that can offer a more favorable environment for the existence and sustenance of probiotic bacteria within the normal array for a lengthier time is thus one of the main research subjects. Probiotics' ability to survive and its functions is largely influenced by the physicochemical properties of food products, including their quantity of fat, kind and intensification of proteins, carbohydrates, pH, and buffering limit. (Mohamed et al., 2013).

As much as the utilization of starter cultures in the manufacturing of dairy items results in the extension of shelf life of the final product (Charfi et al., 2021), EOs also perform a significant role by improving the starter culture's anti-microbial and organoleptic qualities. Mohamed et al. (2013) made the finding that when applied to cheese yogurt to assess its quality, caraway, and dill seeds EO had the strongest inhibitory effect on five types of harmful bacteria. Additionally, it has been noted that EOs didn't interfere with the development of starter cultures, but rather served to fortify dairy products and act as a natural anti-microbial agent. Applying EOs sparingly could be a useful natural way to increase the shelf life of dairy items without posing any health risks.

Some studies have investigated how some LAB may thrive and survive when exposed to volatile plant oils. These researchers have observed similar outcomes in this area for a few additional species (Ducková[1] et al., 2018). Using ginger and chamomile EOs, Yangilar and Yildiz (2018) effectively manufactured probiotic yogurt (0.2% and 0.4%). The highest concentration of probiotics (B. lactis BB-12) was found in the yogurt that contains probiotic and yogurt containing ginger EO (0.4%). In kashk (a fermented dairy product), Golestan et al. (2016) demonstrated the antibacterial activity of *Mentha aquatic and Mentha spicata* EOs contrary to *Bifidobacterium animalis, Staphylococcus aureus, Lactobacillus reuteri*, and *Clostridium perfringe* (Golestan et al., 2016). Different antibacterial characteristics can be found in phenolic compound (Lee & Salminen, 2009). Additionally, the type of bacteria and the antibacterial qualities might be connected.

Here are several studies that demonstrate how EOs impact the qualities of different dairy products. As was noted before, EOs have a wide range of effects on starting cultures.

Growth and acidity production

It was discovered by Kivanç et al. (1991) that low concentrations of EOs from *M. longifolia* and *Cuminum cyminum* resulted in production of acid and growth stimulation, and they inhibit the growth of Lactobacillus plantarum within high concentrations. In a review, EOs of thyme, spearmint, and garlic affected development and strength of LAB in *Ayran* (Simsek et al., 2007). Agboola and Radovanovic-Tesic (2002) revealed that counts of LAB in all trials of cheese produced with different flavors (green lime, spearmint, and tomato skin) weren't fundamentally unique during the maturing stage. It was also shown that M. longifolia has the capacity to stimulate, setback, or inhibit LAB in accordance with the way it is applied (Zaika & Kissinger, 1981). According to Ahari and Massoud (2020), addition of EOs with diverse phenolic profiles transformed the activity of the starter culture all through the storage of dairy items. The outcome corroborates the research conducted by Da Silva et al. (2017). The endurance of starter culture in yogurt was reported to have diminished essentially by increment in the quantity of Ziziphora clinopodioides EOs by Hadad Khodaparast et al. (2007). Research by Jimborean et al. (2016) also revealed that yogurts consolidated with orange EOs displayed an expanded development of LAB presence. A connection between the development of B. bifidum and the intensification of olive and dill extricate and the bio-availability of LAB was enhanced in sight of coffee extricate were documented (Marhamatizadeh et al., 2013; Marhamatizadeh et al., 2014). The green tea addition extricates expanded the durability of *L. paracasei, B. animalis sp lactis*, and *L. acidophilus* during the brooding at 37°C for 72 hours (de Lacey et al., 2014)

Syneresis

The syneresis is the main criterion that affected the level of acceptance of the consumers. It is conceivably because of the impact when pH is very low in casein granules, which enhances the impedance of yogurt (Bchir et al., 2020). Different elements ensuing in syneresis incorporate high brooding temperature, high temperature of storage or low dry matter content (Guénard-Lampron et al., 2020). In a review, the syneresis rate was raised from 4.7%−8.3% (v/w) beyond twenty-eight days resulting in enriching the concentration of EO. The polyphenols from Eos would upgrade disarticulations, which prompts a bigger pore size in the gel lattice and better gelation rate which changed starter culture activity in probiotic yogurt (Ahari & Massoud, 2020). There are various advantages of probiotics and home-grown EOs, and hence, the utilization of these microorganisms or bacteriocins decontaminated from them in blend with natural concentrates and EOs as organic preservatives might remodel the food business. Several studies and research have been carried out by scientists which stated the use of EOs from onion, thyme, oregano, sage, rosemary, coriander, garlic, and clove as a confirmed oil for anti-microbial activity (Olaniran et al., 2020). The EO's structure, composition, and functional group determine the extent of their antimicrobial activity.

In as much as starter cultures used in the processing of dairy items increases the shelf life (Charfi et al., 2021) of the product obtained, EOs play a great role by further enhancing the starter culture for anti-microbial activities and organoleptic properties. According to the observation made by Mohamed et al. (2013), caraway and dill seeds EO had the most inhibitory effect on five strains of pathogenic bacteria when applied to cheese yogurt to test its quality. It has also been observed that EOs didn't affect the development of starter culture and instead served as a means to fortify the dairy product and be used as an ordinary and secure anti-microbial compound. EOs should be applied in small concentrations as it could be a good ordinary technique to prolong the shelf life of dairy products deficient of any health threats.

Essential oils, starter culture, and probiotic organisms such as Lactobacillus fermentum, Bifidobacterium Bb-12, and Lactobacillus acidophilus LA5 were used to make the yogurt. When the EO was present for one month, the LA5 population was unaffected, however, Bb12 showed growth retardation. Furthermore, utilizing basil samples and peppermint, the aqueous extricate of zataria-yogurt showed the greatest repressive effect on DPPH radicals. Also, the peppermint yogurt had the highest grade, followed by basil and control yogurt, while the zataria treatment failed to meet the consumer acceptance criteria (score >5). Incorporating EOs extracted from peppermint, basil, and zataria prepared into the formulation of probiotic yogurt increased the product's antioxidant efficiency, along with greater sensory suitability and antiradical action for the samples of basil and peppermint, according to the study.

Application of essential oil in improving diary product

Several new innovations in cultures, components, handling, and bundling were established after some time to enhance products cultured from dairy. Currently, cultured dairy items are fermented with lactic acid microorganisms, comprising of "Lactobacillus, Streptococcus, Lactococcus, and Leuconostoc species." Yeasts likewise are utilized in producing koumiss and kefir which are highly prevalent; notable dairy products that are cultured include yogurt, sour cream, kefir, plunges, refined buttermilk, and acidophilus milk.

Sour cream and fermented sour cream are depicted in 21CFR131.160 and 21CFR131.162 (Aryana & Olson, 2017). Sour cream is acidified by lactic acid microbes and fermented sour cream is processed with protected and appropriate acidifiers, regardless of lactic acid microorganisms. An old strategy for manufacturing sour cream was to hold cream at a reasonable temperature to permit local bacteria to create acid, permitting the cream to ferment. Nonetheless, this technique brings about the partition of whey from the curd and in off-flavors because of the development of unwanted yeasts, bacteria, and molds. Consequently, their prescribed production strategy was normalizing cream to something like 18% fat, pasteurizing at a temperature of no less than 82°C for no less than 10 minutes, homogenizing at this temperature utilizing a tension of no less than 13.8 MPa, cool to room temperature before adding an adequate measure of starter to mature to sourness, packaging with negligible disturbance, cooling to 4°C, and maturing for 12 hours to 24 hours prior to showcasing. Sour cream creation by normalizing new cream to around 19% fat, amounting to 0.5% stabilizer and around 0.2% citrus extract, sanitizing and homogenizing, cooling to around 22°C prior to inoculation using industrial lactic acid starter (*Lactococcus lactis* ssp. cremoris and *L. lactis* ssp. lactis) which are mesophiles, conceivably adding a rennet extricate, upsetting tenderly for a couple of minutes, brooding until the ideal pH (around 4.5 to 4.6) and sharpness (around 0.70%) are reached, cooling, and bundling. Manufacturing cultured acidified cream with a lactic culture had an insignificant inclination against ropiness, which prompts sour cream with a high consistency and an exceptionally smooth surface. A probiotic, low-calorie acidic cream containing *L. acidophilus*, *Bifidobacterium*, and *S. thermophiles* was also produced (Bintsis, 2018).

Production methodology for cultured butter from matured cream utilizing 70°C to 80°C heat therapy on the cream or milk, adding an unadulterated culture of particular sorts of flavor creating bacteria gotten from fresh butter, completely matured cream, or raw buttermilk for ripening, lastly stirring. Cultures that are currently utilized comprise *L. lactis* ssp. lactis, *L. lactis* sp. lactis biovar diacetylactis, *L. lactis* sp. cremoris, and *Leuconostoc mesenteroides* sp. cremoris (Fayed et al., 2019). Utilization of citrus extract and sodium citrate in the cream, starter, and butter production prompts a more beneficial flavor and fragrance. Satisfactory cultures of butter were found to contain extensive measures of acetylmethylcarbinol (otherwise called acetoin or 3-hydroxy-2-butanone) and diacetyl, while cultures of butter deprived of flavor contained either more modest sums or none of these mixtures. The fragrance bacteria utilized in starter cultures for producing butter in the Netherlands were initially *Leuconostoc cremoris*. Traditional and elective methods for manufacturing refined buttermilk have been accounted. Cultured buttermilk can be produced by adding salt and citrus extract to milk followed by warming it to 85°C for 30 minutes or 88°C for 2 minutes, cooling it to 22°C, and inoculating it with lactic starter culture (*L. lactis* ssp. lactis or L. lactis ssp. lactis biovar diacetylactis), brooding for around 14 hours to 16 hours, breaking the curd when the titratable acidity comes to 0.80%, cooling to 10°C or lower with delicate tumult, and packaged.

The impacts of oregano EO (Origanum vulgare) on the development of lactic starter cultures and organoleptic qualities of a 291 customary Argentinean cheese were reported (Marcial et al., 2016).

Evaluation of the impacts of oregano EO (Origanum vulgare var hirtum) on the development of lactic starter cultures and organoleptic qualities of a 291 conventional Argentinean cheese. Albeit the EO was high in α292 terpinene (10%), γ-terpinene (15.1%), terpinen-4-old (15.5%), and thymol 293 (13.0%), no adverse consequences on the development of Streptococcus thermophilus CRL 294 728 and CRL 813, *L. delbrueckii* subsp. *bulgaricus* CRL 656 and CRL 295 468, and *L. lactis subsp. lactis* CRL 597 were seen when the 296 EOs were added at a concentration of 200 μg/g. The inclusion of oregano 297 EO hampered the development of enterobacteria and the organoleptic acceptance 298 of the cheese was not impacted (Granato et al., 2018).

Ice cream and frozen pastries are well-known all over the world, although ice cream is a famous frozen dessert in all regions of the planet. Peppermint essential oil (PEO) supplementation on sensorial and antioxidative characteristics of ice cream established. Moreover, durability of *Lactobacillus casei* Shirota in ice cream and its connection with the enhancements were additionally examined. PEO could be utilized to upgrade the phenolic content of ice cream and to acquire characteristics of antioxidants, and there was a decent cooperation between *L. casei* and these phenolic mixtures. Moreover, the addition of PPE gave the best outcome to all matters of interest that lactic acid or potentially probiotic bacteria could keep up with their durability in frozen dairy items with negligible loss of liveliness.

Magariños et al. (2007) examined the survival of *L. acidophilus* La-5 and *B. animalis* ssp. lactis Bb-12 with single and blended (1:1) cultures. *L. acidophilus*, *B. animalis*, and their blended cultures had 87%, 90%, and 84% durability rates toward the end of storage at $-25°C$ for 60 days, correspondingly. In another review, after storing ice cream at $-20°C$ for 90 days, quantities of viable *L. acidophilus* and *B. bifidum* diminished by 0.38 and 0.26 log cfu/g, and production of ice cream from *Saccharomyces boulardii* and *Lactobacillus rhamnosus* correspondingly (Goktas et al., 2022)

Conclusion

Production of yogurt and other refined dairy food varieties has changed from coarse and rudimentary systems to extra-controlled methods over the long run. A couple of the advances in our insight that made this conceivable include the utilization of elements in addition to milk, utilization of starter cultures, describing and sequencing the starter cultures for their enhancement, and further developed handling and packaging techniques. An improved comprehension of these items has brought about the improvement of a more extensive scope of fascinating items and expanded open doors for giving medical advantages to consumers.

References

Ahari, H., & Massoud, R. (2020). The effect of cuminum essential oil on rheological properties and shelf life of probiotic yoghurt. *Journal of Nutrition and Food Security*, 5(4), 296−305.

Agboola, S. O., & Radovanovic-Tesic, M. (2002). Influence of Australian native herbs on the maturation of vacuum-packed cheese. *LWT-Food Science and Technology*, 35(7), 575−583.

Akpi, U. K., Nnamchi, C. I., & Ugwuanyi, J. O. (2020). Review on development of starter culture for the production of African condiments and seasoning agents. *Advances in Microbiology*, 10, 599−622. Available from https://doi.org/10.4236/aim.2020.1012044.

Aryana, K. J., & Olson, D. W. (2017). A 100-year review: Yogurt and other cultured dairy products. *Journal of Dairy Science*, 100(12), 9987−10013.

Bchir, B., Bouaziz, M. A., Blecker, C., & Attia, H. (2020). Physico-chemical, antioxidant activities, textural, and sensory properties of yoghurt fortified with different states and rates of pomegranate seeds (Punica granatum L.). *Journal of Texture Studies*, 51(3), 475−487.

Bezie, A., & Regasa, H. (2019). The role of starter culture and enzymes/rennet for fermented dairy products manufacture-A Review. *Nutrition & Food Science*, 9, 21−27.

Bintsis, T. (2018). Lactic acid bacteria as starter cultures: An update in their metabolism and genetics. *AIMS Microbiology*, 4(4), 665.

Burke, N., Zacharski, K. A., Southern, M., Hogan, P., Ryan, M. P., & Adley, C. C. (2018). The dairy industry: Process, monitoring, standards, and quality. *Descriptive Food Science*, 162.

Charfi, I., Moussi, C., Ghazghazi, H., Gorrab, A., Louhichi, R., & Bornaz, S. (2021). Using clove essential oil to increase the nutritional potential of industrial fresh double cream cheese. *Food and Nutrition Sciences*, 12, 1269−1286.

Cortese, R. D. M., Veiros, M. B., Feldman, C., & Cavalli, S. B. (2016). Food safety and hygiene practices of vendors during the chain of street food production in Florianopolis, Brazil: A cross-sectional study. *Food Control*, 62, 178−186.

Ducková[1], V., Kročko[1], M., Kňazovická[1], V., & Čanigová[1], M. (2018). Evaluation of yoghurts with thyme, thyme essential oil and salt. *Acta Universitatis Agriculturae et Silviculturae Mendelianae Brunensis*, 66, 39.

El Asbahani, A., Jilale, A., Voisin, S. N., AïtAddi, E. H., Casabianca, H., ElMousadik, A., Hartmann, D. J., & Renaud, F. N. (2015). Chemical composition and antimicrobial activity of nine essential oils obtained by steam distillation of plants from the Souss-Massa Region (Morocco). *Journal of Essential Oil Research*, 27(1), 34−44.

FAO, IFAD, UNICEF, WFP and WHO. (2020). The state of food security and nutrition in the World 2020. Transforming food systems for affordable healthy diets. Rome, FAO. https://doi.org/10.4060/ca9692en.

Fayed, B., El-Sayed, H. S., Abood, A., Hashem, A. M., & Mehanna, N. S. (2019). The application of multi-particulate microcapsule containing probiotic bacteria and inulin nanoparticles in enhancing the probiotic survivability in yoghurt. *Biocatalysis and Agricultural Biotechnology*, 22, 101391.

Food and Agriculture Organization of the United Nations (FAO). (2022). Gateway to dairy production and products: Types and characteristics. *FAO Publications*, 2022, Accessed January 27.

Fernandez, B., Vimont, A., Desfossés-Foucault, É., Daga, M., Arora, G., & Fliss, I. (2017). Antifungal activity of lactic and propionic acid bacteria and their potential as protective culture in cottage cheese. *Food Control*, 78, 350−356.

García-Díez, J., & Saraiva, C. (2021). Use of starter cultures in foods from animal origin to improve their safety. *International Journal of Environmental Research and Public Health*, *18*(5), 2544.

Goktas, H., Dikmen, H., Bekiroglu, H., Cebi, N., Dertli, E., & Sagdic, O. (2022). Characteristics of functional ice cream produced with probiotic Saccharomyces boulardii in combination with Lactobacillus rhamnosus GG. *LWT*, *153*, 112489.

Golestan, L., Seyedyousefi, L., Kaboosi, H., & Safari, H. (2016). Effect of M entha spicata L. and M entha aquatica L. essential oils on the microbiological properties of fermented dairy product, kashk. *International Journal of Food Science & Technology*, *51*(3), 581−587.

Granato, D., Santos, J. S., Salem, R. D., Mortazavian, A. M., Rocha, R. S., & Cruz, A. G. (2018). Effects of herbal extracts on quality traits of yogurts, cheeses, fermented milks, and ice creams: A technological perspective. *Current Opinion in Food Science*, *19*, 1−7.

Guénard-Lampron, V., Villeneuve, S., St-Gelais, D., & Turgeon, S. L. (2020). Relationship between smoothing temperature, storage time, syneresis and rheological properties of stirred yogurt. *International Dairy Journal*, *109*, 104742.

Hadad Khodaparast, M. H., Mehraban Sangatash, M., Karazhyan, R., Habibi Najafi, M. B., & Beiraghi Toosi, S. (2007). Effect of essential oil and extract of Ziziphora clinopodioides on yoghurt starter culture activity. *World Sci. J*, *2*, 194−197.

How, Y. H., Teo, M. Y. M., In, L. L. A., Yeo, S. K., & Pui, L. P. (2022). Development of fermented milk using food-grade recombinant Lactococcus lactis NZ3900. *NFS Journal*.

Jan, G., Tarnaud, F., do Carmo, F. L. R., Illikoud, N., Canon, F., Jardin, J., Briard-Bion, V., Guyomarc'h, F., & Gagnaire, V. (2022). The stressing life of Lactobacillus delbrueckii subsp. bulgaricus in soy milk. *Food Microbiology*, *106*, 104042.

Panchal, B., & Bhandari, B. (2020). *Butter and dairy fat spreads. Dairy fat products and functionality* (pp. 509−532). Cham: Springer.

Kapaj, A. (2018). Factors that influence milk consumption world trends and facts. *European Journal of Business. Economics and Accountancy*, *6*(2), 14−18.

Kamath, R., Basak, S., & Gokhale, J. (2021). Recent trends in the development of healthy and functional cheese analogues-a review. *LWT*, 112991.

Kivanç, M., Akgül, A., & Doğan, A. (1991). Inhibitory and stimulatory effects of cumin, oregano and their essential oils on growth and acid production of Lactobacillus plantarum and Leuconostoc mesenteroides. *International Journal of Food Microbiology*, *13*(1), 81−85.

Lee, Y. K., & Salminen, S. (2009). *Handbook of probiotics and prebiotics*. John Wiley & Sons.

Liu, T., Chen, H., Bai, Y., Wu, J., Cheng, S., He, B., & Casper, D. P. (2020). Calf starter containing a blend of essential oils and prebiotics affects the growth performance of holstein calves. *Journal of Dairy Science*, *103*, 2315−2323, https://doi.org/2.1.1849.8241.

Magariños, H., Selaive, S., Costa, M., Flores, M., & Pizarro, O. (2007). Viability of probiotic micro-organisms (Lactobacillus acidophilus la-5 and Bifidobacterium animalis subsp. Lactis bb-12) in ice cream. *International Journal of Dairy Technology*, *60*(2), 128−134.

Marcial, G. E., Gerez, C. L., de Kairuz, M. N., Araoz, V. C., Schuff, C., & de Valdez, G. F. (2016). Influence of oregano essential oil on traditional Argentinean cheese elaboration: Effect on lactic starter cultures. *Revista Argentina de Microbiologia*, *48*(3), 229−235.

Marhamatizadeh, M. H., Ehsandoost, E., Gholami, P., & Mohaghegh, M. D. (2013). Effect of olive leaf extract on growth and viability of Lactobacillus acidophilus and Bifidobacterium bifidum for production of probiotic milk and yoghurt. *International Journal of Farming and Allied Sciences*, *2*(17), 572−578.

Marhamatizadeh, M. H., Ehsandoost, E., & Gholami, P. (2014). The effect of coffee extract on the growth and viability of Lactobacillus acidophilus and Bifidobacterium bifidum in probiotic milk and yoghurt. *Journal of Food Biosciences and Technology*, 37−48, *4*(JFBT (Vol. 4-No. 1)).

McSweeney, P. L., Ottogalli, G., & Fox, P. F. (2017). Diversity and classification of cheese varieties: An overview. *Cheese*, 781−808.

Miloradovic, Z., Kljajevic, N., Miocinovic, J., Tomic, N., Smiljanic, J., & Macej, O. (2017). High heat treatment of goat cheese milk. The effect on yield, composition, proteolysis, texture and sensory quality of cheese during ripening. *International Dairy Journal*, *68*, 1−8.

Mohamed, S. H. S., Zaky, M. W., Kassem, J. K., Abbas, H. M., Salem, M. M. E., & Said-Al Ahl, H. A. H. (2013). Impact of antimicrobial properties of some essential oils on cheese yoghurt quality. *World Applied Sciences Journal*, *27*(4), 497−507.

Mohammed, S., & Çon, A. H. (2021). Isolation and characterization of potential probiotic lactic acid bacteria from traditional cheese. *LWT*, *152*, 112319.

Muhammad, I., Muhammas, A. S., Saqib, A., & Amjad, H. (2019). Biological importance of essential oils. *IntechOpen*, 1−15. Available from https://doi.org/10.5772/intechopen.87198.

Moatsou, G., Zoidou, E., Choundala, E., Koutsaris, K., Kopsia, O., Thergiaki, K., & Sakkas, L. (2019). Development of reduced-fat, reduced-sodium semi-hard sheep milk cheese. *Foods*, *8*(6), 204.

Papaioannou, G. M., Kosma, I. S., Dimitreli, G., Badeka, A. V., & Kontominas, M. G. (2022). Effect of starter culture, probiotics, and flavor additives on physico-chemical, rheological, and sensory properties of cow and goat dessert yogurts. *European Food Research and Technology*, *248*(4), 1191−1202.

O'Brien, N. M., & O'Connor, T. P. (2017). *Nutritional aspects of cheese. Cheese* (pp. 603−611). Academic Press.

Olaniran, A. F., Abiose, S. H., & Adeniran, A. H. (2015). Biopreservative effect of ginger (Zingiber officinale) and garlic powder (Allium sativum) on tomato paste. *Journal of Food Safety*, *35*(4), 440−452.

Olaniran, A. F., Abiose, S. H., Adeniran, H. A., Gbadamosi, S. O., & Iranloye, Y. M. (2020). Production of a cereal based product (Ogi): Influence of co-fermentation with powdered garlic and ginger on the microbiome. *Agrosearch*, *20*(1), 81−93.

Ratwan, P., Mandal, A., Kumar, M., & Chakravarty, A. K. (2017). Genetic analysis of milk production efficiency traits in Jersey crossbred cattle. *Indian Journal of Animal Research*, *51*(4), 644−647.

Ray, R. C., & Joshi, V. K. (2014). *Microorganisms and fermentation of traditional foods: Fermented foods: Past, present and future. Chapter 1* (pp. 1−37). London New York: CRC Press, 10.13140/2.1.1849.8241.

Romeih, E., & Walker, G. (2017). Recent advances on microbial transglutaminase and dairy application. *Trends in Food Science & Technology*, *62*, 133–140.

Shani, N., Oberhaensli, S., Berthoud, H., Schmidt, R. S., & Bachmann, H. P. (2021). Antimicrobial susceptibility of Lactobacillus delbrueckii subsp. lactis from milk products and other habitats. *Foods*, *10*(12), 3145.

Shori, A. B., Albalawi, A., Al Zahrani, A. J., Al-sulbi, O. S., & Baba, A. S. (2022). Microbial analysis, antioxidant activity, and sensory properties of yoghurt with different starter cultures during storage. *International Dairy Journal*, *126*, 105267.

Simsek, B., Sagdic, O., & Ozcelik, S. (2007). Survival of Escherichia coli O157: H7 during the storage of Ayran produced with different spices. *Journal of Food Engineering*, *78*(2), 676–680.

Van Boekel, M., Fogliano, V., Pellegrini, N., Stanton, C., Scholz, G., Lalljie, S., Somoza, V., Knorr, D., Jasti, P. R., & Eisenbrand, G. (2010). A review on the beneficial aspects of food processing. *Molecular Nutrition and Food Research*, *54*, 1215–1247.

World Health Organization. (2020). Protecting, promoting and supporting breastfeeding: The baby-friendly hospital initiative for small, sick and pre-term newborns.

Yangilar, F., & Yildiz, P. O. (2018). Effects of using combined essential oils on quality parameters of bio-yogurt. *Journal of Food Processing and Preservation*, *42*(1), e13332.

Zaika, L. L., & Kissinger, J. C. (1981). Inhibitory and stimulatory effects of oregano on Lactobacillus plantarum and Pediococcus cerevisiae. *Journal of Food Science*, *46*(4), 1205–1210.

Chapter 6

Application of essential oils for the production of dietary supplements and as traditional self-medication purposes

Raghvendra Raman Mishra[1], Pragya Mishra[2], Subodh Kumar[3] and Divya Gupta[4]

[1]Medical Laboratory Technology, DDU Kaushal Kendra, RGSC, Banaras Hindu University, Varanasi, Uttar Pradesh, India, [2]Food Processing and Management DDU Kaushal Kendra, RGSC, Banaras Hindu University, Varanasi, Uttar Pradesh, India, [3]Department of Medical Laboratory Technology, School of Allied Health Sciences and Management, Delhi Pharmaceutical Sciences and Research University, New Delhi, India, [4]Reproductive Biology and Toxicology Lab, School of Studies in Zoology, Jiwaji University, Gwalior, Madhya Pradesh, India

Introduction

Consumers' growing preference for natural alternative treatments creates an exciting market for natural ingredients in health goods. The market for medicinal and aromatic plants, EO extraction, and microorganisms like seaweed and algae is also being driven by a rise in interest in preventative healthcare. The largest prospects exist for exporters of natural ingredients for health goods in huge consumer markets and lengthy histories of self-medication. Essential oils (EOs) are colorless liquids that are mostly made up of fragrant and volatile molecules that are found in all parts of plants, including their seeds, flowers, peels, stems, bark, and entire plants (Sánchez-González et al., 2011). EOs are secondary metabolites that are crucial for a plant's defense system and have a variety of therapeutic benefits, including antibacterial action (Tajkarimi et al., 2010). They are widely utilized in many nations as food preservatives, medicine, fragrances, and cosmetics. EO applications for aromatherapy have a long history in countries like Egypt, China, and India. They were initially used as medicine in the 19th century due to their aroma and flavor. Europe has the biggest market for EOs in the world. The food, beverage, and cosmetics industries presently account for the majority of the demand for EOs, but in the years to come, it is projected that public interest in and understanding of aromatherapy will grow. Over the previous five years, emerging countries provided over half of the EOs imported into Europe. Significant suppliers include Madagascar, China, Indonesia, India, and other countries.

EOs are largely employed in the food sector for food preservation because of their scent, flavors, and naturally occurring antibacterial properties. For instance, citrus-derived EOs such monoterpenes, sesquiterpenes, and oxygenated derivatives exhibit potent inhibitory actions against pathogenic bacteria, indicating the potential for usage as flavoring and antioxidants. There has been an increase in interest in using EOs as dairy product preservatives in recent years as suitable alternatives to synthetic preservatives. Studies have shown that medicinal herbs and spices are used both as traditional forms of self-medication and as raw materials for the manufacture of dietary supplements. Numerous studies have documented the positive effects of EOs when coupled with dairy products; however, further in-depth research is required in order to fully understand the potential synergistic and antagonistic interactions among these substances (Mishra et al., 2020).

Sources of essential oil

EOs are secondary metabolites produced by aromatic and therapeutic plants. The use of EOs as a food preservative for grains, cereals, pulses, fruits, and vegetables is possible. *Ocimum sanctum, Tanacetum nubigenum, Eucalyptus globulus,* and *Eucalyptus odorata,* as well as *Clausena pentaphylla, Mentha arvensis,* and *Chenopodium ambrosioides* EOs, have the potential to be used as antifungal, fungicidal, and food preservatives in suppressing fungi associated with food

Applications of Essential Oils in the Food Industry. DOI: https://doi.org/10.1016/B978-0-323-98340-2.00003-1

commodities (Pandey et al., 2017). A volatile mixture of odoriferous compounds is still present in the majority of aromatic plants, and these chemicals can be extracted to produce EOs. Typically, a large variety of secondary metabolites are produced by aromatic and therapeutic plants. viz., terpenoids, alcoholic complexes (geraniol, menthol, linalool), acidic amalgams (benzoic, cinnamic, myristic acids), aldehydes (citral, benzaldehyde, cinnamaldehyde, carvone camphor), ketonic forms (thymol, eugenol), and phenols (ascaridole, anethole). It has been discovered that terpenes, terpenoids, and aromatic phenols including eugenol, thymol, carvacrol, and safrole play important roles in the composition of many EOs. Pinene, Myrcene, Limonene, Terpinene, and P-Cymene are terpenes (Koul et al., 2008).

Health-promoting benefits of essential oils and their usage as dietary supplements

EOs are concoctions of volatile molecules that have biological properties that promote health in addition to serving as flavorings and natural preservatives. EOs are helpful in many health conditions such as irritable bowel syndrome (IBS) and Inflammatory bowel disease (IBD), as well as in the prevention of Colorectal Cancer (CRC) and gastric protection. It also helps in alleviating peptic ulcer, protecting liver function, stimulating digestion, diuresis, and protection of the urinary tract, reducing inflammation and pain, managing the metabolic disorders, immunomodulatory and anti-influenza activity, improving neuroprotection and lastly, helping in the modulation of mood and cognitive function (Matera et al., 2023). EOs in probiotic food and supplements have wide applications for the regulation of gut microflora. These properties can be used to create functional meals and dietary supplements. Comparably less evidence has been provided for the other functional roles of EOs or their components in food products, which are nonetheless accumulating a body of scientific support and substantial public interest: their health-promoting role beyond preservation and basic nutrition. A number of recent reviews have been devoted to the use and potential of EOs as food preservatives (Matera et al., 2023).

Different from regular food, dietary supplements are meant to enhance or complement the diet. Dietary components and final dietary supplement products are also subject to FDA regulation. Dietary supplements are governed by a distinct set of rules by the FDA than "conventional" food and medication items. In accordance with the 1994 Dietary Supplement Health and Education Act, EOs help in the colonization and maintenance of balanced levels of beneficial microbial populations within the gastrointestinal system. An earlier study based on the beneficial effects of EOs on animal nutrition (appetite stimulation as well as improvement of endogenous digestive enzyme secretion and immune response activation) was conducted to reveal positive effects on egg production, egg weight, feed conversion ratio, and feed consumption in laying quail (Çabuk et al., 2014).

The primary drawback to EOs' use in the food industry is their organoleptic effect due to their potent aroma (Falleh et al., 2020). Another major criticism of EOs' use as functional ingredients in foods is that their effective dose may be excessive given how potent their flavor is, especially if it is not "in taste" with regional dietary customs. Despite the change in organoleptic properties in EOs over time, the use of EOs as functional ingredients in foods has been criticized. Because of this, while creating EO-fortified yogurt with the intention of imparting functional characteristics to food, the EO's contribution to the final flavor cannot be ignored, and both the functional properties and final flavor of the yogurt need to be considered and optimized (Matera et al., 2023).

Extraction methods and implications

Essential oils may be derived from a variety of plants using various extraction procedures. The plant material utilized in the production of EOs, as well as the process of extraction, is typically reliant on the method of extraction. Other elements to consider are the material's state and shape. The extraction procedure is the crucial aspect of determining EO quality. Inadequate extraction processes might cause harm to or affect the activity of an essential oil's chemical signature. As a result, bioactivity and natural characteristics are diminished. Discoloration, off-odor/flavor, and physical changes such as increased viscosity can occur in extreme instances (Aziz et al., 2018). These modifications to extracted EO must be avoided.

However, EOs must be extracted from the plant matrix before they can be used or analyzed. For this aim, steam distillation, hydrodistillation, simultaneous distillation-extraction, solvent extraction, and cold pressing methods, among others, can be used. Despite the fact that these methods have been employed for EO extraction over a long time, their use has highlighted a number of drawbacks, including low yields of extraction, declines of volatile compounds, deterioration of certain substances due to temperature or hydrolytic effects, and potential detrimental solvent residues in extracts of EOs (Stratakos & Koidis, 2016).

Industries dedicated to the extraction of EOs concentrated on the development of cutting-edge extraction technology as energy costs rose and the "Green Era" started to arise. The disadvantages of conventional extraction methods have led to the development of several new methods, including pressurized liquid extraction, pressurized hot water extraction, membrane-assisted solvent extraction, solid-phase micro-extraction, supercritical fluid extraction, ultrasound-assisted extraction, and microwave-assisted extraction among others, for the extraction of EOs from plants (Pavela & Benelli, 2016). By reducing the usage of solvents and fossil fuels and reducing the creation of contaminants, the aforementioned alternatives to traditional techniques may upsurge productivity and aid in environmental safety.

Conventional approach

Cold pressing

The first methodology to obtain EOs is cold pressing (CP). Although it provides modest yields, it has the capability of producing barely any heat throughout the process. Enzyme pretreatments have been researched to boost the EO extraction's yield and quality. Due to the substantial heat instability of the aldehydes present in citrus peel oils, it is primarily used for their separation. Soto et al. (2007) extracted borage (*Borago officinalis*) seed oil using enzymatic hydrolysis combined with CP and reported that using enzymes that have pectinases and cellulases activity before the pressing stage boosts oil extraction yield. Collao et al. (2007) boosted the production of primrose oil (*Oenothera biennis*), alongside Anwar et al. (2013) looked at how different enzyme preparations affected the amount of cold-pressed flaxseed oil produced. Currently, CP has been used for the manufacturing of organic EOs, which frequently become available as unique items with a high market price.

Distillation

Steam distillation

The steam distillation process, which is the most prevalent, is used to extract 93% of all EOs from plants. The heat provided is the primary cause of the plant material's cell structure rupturing and breaking down, releasing aromatic chemicals or EOs. As a result, aromatic chemicals or EOs derived from plant material are released (Yeh & Lin, 2020). The plant material must be heated to an adequate temperature to break down it and release aromatic chemicals or EOs.

Masango (2005) developed a new process design and operation for EO steam distillation to boost oil production and decrease polar component loss in wastewater. The apparatus is built from a dense bed of plant matter that sits above the steam source. Boiling water is never mixed with plant matter; only steam is allowed to get through. As a consequence, the process consumes the least quantity of steam, and the distillate has the least amount of water. Water-soluble chemicals are also dissolved to a lesser amount in the condensate's aqueous phase (Masango, 2005). El Kharraf et al. (2021) discovered that the radical scavenging activities of EOs derived from the steam distillation technique were much greater.

Hydrodistillation

Hydrodistillation (HD) is the preferred method for harvesting EOs using plant products, and the method is usually utilized for separating nonwatery components with high boiling temperatures. The EO vapor and steam are condensed to form an aqueous fraction, and the method involves completely submerging plant components in water, followed by boiling. This procedure preserves the extracted oils to some extent because the surrounding water acts as a barrier to prevent it from overheating (Peng et al., 2022). Psarrou et al. (2020) investigated the effects of several extraction methods on the yield and characteristics of EO extracted from rosemary (*Rosmarinus officinalis* L.) using HD and solvent-free microwave extraction (SFME). HD oil includes more monoterpene hydrocarbons (32.95%) than SFME-extracted oil (25.77%), although SFME-extracted oil included more oxygenated monoterpenes (28.6%) than HD (26.98%). The overall amounts of the aromatic components produced through HD and SFME were 0.31% and 0.39%, respectively. Golmakani and Rezaei (2008) investigated microwave-assisted HD (MAHD), an enhanced HD technology utilizing a microwave oven in the extraction process. MAHD performed better in terms of time and energy savings during extraction.

Hydrodiffusion

When plant material has been dried and is safe to be exposed to boiling temperatures, hydrodiffusion (HDF) extraction is used. It is a kind of steam distillation. It outperforms steam distillation in terms of processing time and oil output while using less steam. Bousbia et al. (2009) evaluated the efficiency of HD and new microwave hydrodiffusion and gravity (MHG) technologies in the separation of EO from rosemary leaves (*R. officinalis*). The MHG approach outperforms standard methods in terms of isolation time (15 minutes against 3 hours for HD), environmental effect, cleaner characteristics, and enhanced antibacterial and antioxidant activity. Farhat et al. (2011) investigated the microwave steam diffusion (MSDf) approach, which uses a microwave heating procedure to extract EOs from orange peel byproducts. The yield and aromatic characteristic of the EOs recovered by MSDf for 12 minutes appeared comparable to that obtained by steam diffusion for 40 minutes.

Solvent extraction

Essential oils extracted from fragile or delicate flower materials have traditionally been extracted using conventional solvent extraction. Several solvents, including methanol, ethanol, acetone, hexane, or petroleum ether, among others, can be used for extraction. The solvent is heated together with the plant material to extract the essential oil, which is then filtered and the solvent is evaporated. The filtrate is finally extracted with pure alcohol and distilled at low temperatures. The primary components used to extract the EO from *Ptychotisverticillata* were the phenolic substances (48.0%), the carvacrol (44.6%), and the thymol (3.4%) (Egza, 2020). The chemical composition and antioxidant capacity of the EOs from Thymus praecox subsp. skorpilii var. skorpilii (TPS) obtained through different solvents were studied by Ozen et al. (2011).

Supercritical carbon dioxide

The disadvantages of conventional techniques like steam distillation and solvent extraction include an extended setup period and the usage of significant quantities of organic solvents. Supercritical fluids have been proposed as an alternative medium for the extraction of EOs due to their low critical conditions. The most often used supercritical fluid is carbon dioxide (CO_2) due to its low critical conditions. Even though the components of EOs were highly soluble in supercritical CO_2, pure CO_2 extraction rates were not as quick (Xiong & Chen, 2020). Despite this, considerable recoveries were obtained after a 15-minute dynamic extraction with pure CO_2 and a 15-minute static recovery using methylene chloride as a modulator. More than 90% of the volatile substances, including monoterpenes, may be extracted from the effluent of supercritical fluid extraction. Due to its lower yield and greater energy consumption, supercritical fluid extraction seems more economical than steam distillation, according to Qamar et al. (2021).

Emerging techniques

Research into novel extraction technologies for the food, pharmaceutical, and chemical industries has lately gained a lot of interest due to growing energy prices and initiatives to reduce CO_2 emissions. These environmentally friendly methods have prompted people to think about using novel ways to extract EOs as microwave-aided extraction, ultrasound-assisted processing, and supercritical fluid extraction (Gavahian et al., 2019).

Microwave-assisted extraction

Microwaves are utilized by researchers to extract EOs, and the EOs obtained in 30 minutes or less were equivalent to those produced after more than double the time using traditional approaches. Microwave-assisted extraction (MAE) employs microwave radiation to heat the solvent-sample combination, which is immediate and happens in the sample's interior, resulting in highly quick extractions. The disintegration of weak hydrogen bonds brought on by the molecules' dipole rotation is one advantage of microwave heating (Fiorini et al., 2020).

The use of microwaves significantly reduces both the extraction time and the required solvent volume, which helps to lessen the environmental impact. MAE extraction technologies differ from standard methods in that they occur as a consequence of modifications to cell structure generated via electromagnetic waves. Since the temperature stays low throughout the process, MAE is perfect for the separation of thermolabile compounds. The amount of microwave power, the length of the exposure, the pressure, the viscosity of the sample, the moisture and type of the matrix, and the type and volume of the solvent are all factors that might impact the extraction of organic compounds by MAE. The chosen solvent typically has a high dielectric constant and absorbs a lot of microwave energy, although it may be challenging to extract selectivity and the medium's ability to interact with microwaves. The development of methods including solvent extraction, solvent-free microwave extraction, vacuum microwave hydrodistillation, microwave HDF, and gravity is a result of advances in microwave extraction (Haffizi et al., 2020).

Ultrasound-assisted extraction

In the phytopharmaceutical extraction sector, ultrasound has been utilized to separate volatile chemicals from natural materials at room temperature. It has a larger influence at low frequencies (18–40 kHz) and is almost non-existent at 400–800 kHz. The improved extraction efficiency is a result of cavitation in the solvent brought on by an ultrasonic wave passing through. It has been shown that when raw plant tissues are treated, the thin surface of the cells holding EOs is readily destroyed by sonication, allowing the release of compounds that can be extracted as well as accelerating the mass transit of solvent from the continuous phase into the plant cell. At the laboratory scale, EOs are extracted using ultrasonic power. Cleaning baths and probe systems are often utilized, and several aspects regulate the activity of ultrasound to provide an efficient and successful extraction. According to researchers when ultrasound is used, the extraction yield does not usually rise greatly, but the degradation of plant components is always reduced (Deng et al., 2022). As a result, ultrasound-assisted extraction is an appropriate technique for sensitive chemical compounds.

Supercritical fluid extraction

For instance, lipids, flavors, and EOs may be extracted from plants using supercritical fluid extraction (SFE), which has the potential to be used commercially. Its foundation is the use of supercritical solvents, which are subjected to pressures and temperatures throughout their most critical points. CO_2 is the preferred supercritical solvent for the extraction of plant components because it is non-toxic and permits extremely critical activities at relatively low pressures and temperatures close to ambient. The SFE method consists of two steps: separating soluble components using a supercritical solution and then removing extracted compounds from a solvent. To extract molecules that are soluble from the supercritical fluids (SCF), it is possible to change the supercritical solvent's thermodynamic attributes as well as its pressure or temperature. A typical SFE system consists of a high-pressure pump that delivers a fluid and a separation chamber that contains the sample and is kept at an appropriate pressure and temperature. Using pre-mixed cylinders or an additional pump, an organic solvent (co-solvent) can be added to the fluid to increase its ability to solvate (Uwineza & Waśkiewicz, 2020).

Because an SCF's selectivity may vary with temperature and pressure, it is ideal for the specific extraction of certain chemicals from mixtures. Single-stage extraction (SSE), sequential depressurization (SD), and fractional extraction process (FEP) are examples of SFE processes for selectively extracting or separating certain compounds.

High-density fluid is used to concurrently extract both the lightweight and heavier portions of the compounds during the extraction of EOs; the SCF and the extract are frequently depressurized to permit fractional separation. High pressures (40–60 MPa) are used during the extraction, and the second stage entails collecting the EOs. The majority of studies on SFE of aromatic oils focus on refining extraction procedures to increase chemical recovery or extraction yield. SCFs have benefits for the chemical world (these substances have low viscosity, high diffusion rates, a tunable density, and dielectric constant) as well as the environment, health, and safety (supercritical CO_2 and supercritical H_2O are noncarcinogenic, environmentally friendly, non-mutagenic, non-flammable, and thermodynamically stable). Solvent flow rate, residence time, moisture content, particle size, supercritical pressures, and temperature are a few independent variables that affect SFE optimization (Yousefi et al., 2019).

Method	Advantages	Limitations
Cold pressing	Low expenses, preservation of valuable chemicals, and no issues with plant safety	The oil produced by solvent extraction has a reduced nutraceutical content. Glucosinolate breakdown products emit pungent odors.
Hydrodistillation	Decantation makes it simple to extract essential oil from water.	Long extraction times, chemical changes brought on by continuous boiling, and depletion of some polar molecules in water vapor are all factors.
Water steam	Less artifacts are produced, and the extraction time is shorter.	Heating for several hours; degradation of thermos-labile chemicals; deterioration of odor.
Organic solvent extraction	Changes and chemical artifacts are prevented	The safety of the product may be jeopardized if the organic solvent is detected in the collected oil.
Ultrasound-assisted extraction	Low temperature, high efficiency, reduced solvent usage, and low consumption of energy	Potential for free radical formation during solvent sonolysis, followed by the oxidation of labile component

Microwave-assisted extraction	extraction time decreased, environmental friendliness, and use of less solvents; Quick and effective extraction; improved sensory qualities	High-temperature usage results in the creation of undesired chemicals; frequent use of harmful organic solvents
Supercritical fluid extraction	better activity and higher-quality extracts; relatively low temperatures; chemical adversity	Costly; demand for highly pure CO2; Supercritical CO_2 affinity for low-polar and non-polar substances

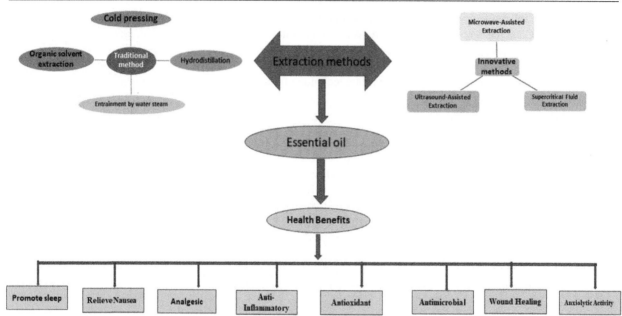

Employing essential oils for self-medication purposes as herbal formulation

Essential oil s are flammable, complex chemical compounds isolated from aromatic herbs by steam or hydrodistillation that have antibacterial, antiviral, antifungal, antiinflammatory, and several other therapeutic activities. The chemical components in EOs define their effects, and their identification defines the oil's specific beneficial usage. But because EOs are generally safe, with the exception of some skin reactions and ultraviolet (UV) responsiveness (being subjected to the sun can lead to irritation or skin darkening), their therapeutic use is currently examined, primarily for external usage (mouthwashes or inhalation). EOs must not be used in excess of the recommended dosage and should not be applied to skin that is already damaged, as doing so would cause significant systemic absorption, which leads to, serious negative effects. EOs are capable of affecting the brain and circulate throughout the body via the bloodstream. Here are a few of the possible advantages, when safely administered or inhaled.

Promote and aid sleep

Given the obvious soothing properties of EOs, it is not unusual that they may tend to help individuals sleep better. In one small previous research of 60 patients in a cardiac intensive care unit (ICU) in Turkey, those who breathed lavender EO for 15 days had significantly better sleep quality and lower anxiety levels than those who did not. Experts think that when EOs are inhaled, the molecules communicate to the olfactory bulb in the nose. The chemicals then send messages to areas of the brain that are involved in emotional and behavioral responses. These regions then produce neurotransmitters, which induce emotional or behavioral responses. Lavender, camomile, bergamot, and peppermint are some of the commonly used EOs to improve sleep (Her & Cho, 2021).

Relieve nausea

Essential oils such as peppermint, ginger, lemon, and spearmint have unique properties that make them perfect for nausea relief. Peppermint oil, for example, relaxes gastrointestinal muscles and reduces inflammation, which leads to nausea and is used as a medication to treat irritable bowel syndrome. Meanwhile, ginger aids digestion, preventing meals from becoming stuck in the intestines and creating pain. In fact, their benefits were equivalent to those of patients who used aromatherapy in conjunction with an antiemetic (anti-nausea drug). Recent double-blind research found that the fragrance of lemon EO may assist pregnant women lessen nausea. Those who breathed lemon oil throughout the four-day research had a substantial reduction in nausea and vomiting. The medicinal characteristics of spearmint EO make it one of the finest EOs for nausea. Chemotherapy patients in 2013 research were given two drops of spearmint EO in a specially made capsule. Patients who received spearmint EO reported less frequent vomiting and less overall nausea as compared to a control group (Toniolo et al., 2021).

Antiinflammatory and antioxidant activities

Natural remedies have long been recognized for their effectiveness in the treatment of various inflammatory illnesses. In the 16th century, Queen Isabella used an infusion of rosemary EO to cure her rheumatism. This remedy later circulated as medicine at the court of Louis XIV. It is known as "water of the queen of Hungary." There are several more historical examples of rosemary EO being used to treat inflammatory diseases including asthma, rheumatoid arthritis, bronchitis, and other conditions that are linked to traditional medicine (Cimino et al., 2021). The antiinflammatory action of EOs was linked to their capacity to suppress both histamine- and prostaglandin-mediated responses. Because of the antiinflammatory properties of its components, *Artemisia argyi* EO was reportedly frequently used in China to treat chronic bronchitis. Chen et al. (2017) showed, among other things, that the antiinflammatory properties of *A. argyi* EO are linked to the mitigation of two processes: the synthesis of pro-inflammatory mediators including TNF-, IL-6, IFN-, NO, PGE2, and others, as well as the phosphorylation of JAK2 and STAT1/3. The JAK/STATs pathway's suppression also pointed to these EOs' capacity as antioxidants for reactive oxygen species. Considering certain synthetic antioxidants, like butylated hydroxytoluene and butylated hydroxyanisole, are believed to be potentially hazardous to humans, the consumption of EOs as alternatives to synthetic antioxidants is becoming more and more popular. Therefore, employing active packaging or edible coatings, as well as directly or indirectly integrating EOs in edible products, maybe a useful strategy to postpone oxidation and lengthen shelf life. Because of its ability to shield the human body against oxidative stress, which causes major ailments, antioxidants are increasingly regarded as essential in the diet.

Antimicrobial activity and wound healing

Hippocrates reported that EO fumigation was effective in protecting against plague. This was demonstrated in Middle Ages when criminals were able to steal from plague victims' homes without becoming sick. The presence of specific components, such as terpenes, influences the antibacterial action of EOs. Phenolic groups have the most antibacterial action, followed by cinnamic aldehydes. Alcohols, aldehydes, ketones, ethers, and hydrocarbons are additional significant groupings. Alcohols and aldehydes are connected to antibacterial activity against Gram-positive bacteria (Valdivieso-Ugarte et al., 2019).

With respect to their antibacterial efficiency against multiple-drug-resistant skin infections, EOs have gained recognition as a therapy option, even if at certain concentrations they may be cytotoxic. Another benefit of using EOs topically is their capacity to aid wound healing, which was found nearly a century ago by Rene-Maurice Gattefossé. Wound healing occurs in three stages: the inflammatory phase (bleeding blockage, vasodilation, and immune system recruitment); the second phase (proliferation of different types of cells, such as fibroblasts, leading to tissue granulation and angiogenesis); and the third phase (generation of new collagen fibers and fibroblast differentiation) (El Ayadi et al., 2020).

Costa et al. (2019) revealed that the presence of the most common monoterpenes, carvacrol and thymol, is important in defining EO tissue healing ability. Because of their antiinflammatory properties, EOs can influence all stages of tissue regeneration. The three primary biochemical mediators that the EO modulated were also identified: VEGF, FGF-2, and IGF1, whose expression and biosynthesis both improved angiogenesis, a crucial process in different stages of inflammation and cell proliferation. IGF1 expression was increased, which had a number of effects, including an increase in fibroblast cell proliferation.

Anxiolytic activity

In this modern context, EOs pose a new challenge due to their pharmacological uses to promote mental, emotional, and physical health. The positive benefits of EOs are produced by providing these chemicals at therapeutic doses (e.g., inhalation, ingestion, or topical application), The effects, however, might be harmful to the human being if certain quantities are surpassed. Today, aromatherapy is utilized as a "complementary therapy" alongside conventional medicine in several nations, including the UK, France, the United States of America, and Australia. The WHO Traditional Medicine Strategy: 2014–23, recently stated the need to re-evaluate some traditional forms of medicine as integrative therapies (Fakih & Perbawati, 2022). Taking into account the effects of vital oils on the brain and nervous system and an abundance of data, employing EOs as a supplemental medication can be an effective coadjutant in the treatment of individuals suffering from anxiety disorders. A recent meta-analysis (randomized and nonrandomized) discovered that inhaling or ingesting lavender essential oil, diluted and under supervision, can lower anxiety levels. The Lavender EO has been demonstrated to ease anxiety when used topically during massage. The chemicals in lavender EO are known to interact with the limbic system, an area of the brain that regulates emotions. Lemongrass oil, orange oil, bergamot oil, and cedarwood oil are some more EOs that may help with anxiety (Ebrahimi et al., 2022).

Essential oil	Employed in pain relief	References
Lavender	Reduce stress, relieve headaches, lower back pain, cramps, labor pain	Nursahidah et al. (2020)
Sweet marjoram	migraines, neuralgia, cramps, abdominal disorder	Fresen et al. (2023)
Eucalyptus	headaches, arthritis, muscle and joint pains, unblock sinus	Varkaneh et al. (2022)
Rosemary	improve blood circulation, muscle pain, and spasms	Nivedhitha et al. (2019)
Juniper	relieve stiffness and pain associated with arthritis, rheumatism, and gout	Meriem et al. (2022)
Ginger	rheumatoid arthritis, menstrual cramps, joint inflammation, and osteoarthritis	Rondanelli et al. (2020)

Conclusion

After reviewing all facts, this chapter concludes that EOs are readily available, highly bioavailable, and well-observed natural sources with potential therapeutic activity for dietary supplements and other health-related foods. Beyond basic nutrition, supplementing meals with herbal EOs of natural origin is a potential alternative to self-medication options. The attention it is receiving is entirely deserved given to its unique features.

References

Anwar, F., Zreen, Z., Sultana, B., & Jamil, A. (2013). Enzyme-aided cold pressing of flaxseed (Linum usitatissimum L.): Enhancement in yield, quality and phenolics of the oil. *Grasas y aceites*, *64*(5), 463–471.

Aziz, Z. A., Ahmad, A., Setapar, S. H. M., Karakucuk, A., Azim, M. M., Lokhat, D., Rafatullah, M., Ganash, M., Kamal, M. A., & Ashraf, G. M. (2018). Essential oils: Extraction techniques, pharmaceutical and therapeutic potential-a review. *Current Drug Metabolism*, *19*(13), 1100–1110.

Bousbia, N., Vian, M. A., Ferhat, M. A., Petitcolas, E., Meklati, B. Y., & Chemat, F. (2009). Comparison of two isolation methods for essential oil from rosemary leaves: Hydrodistillation and microwave hydrodiffusion and gravity. *Food Chemistry*, *114*(1), 355–362.

Çabuk, M., Eratak, S., Alçicek, A., & Bozkurt, M. (2014). Effects of herbal essential oil mixture as a dietary supplement on egg production in quail. *The Scientific World Journal*, *2014*, 4.

Chen, L. L., Zhang, H. J., Chao, J., & Liu, J. F. (2017). Essential oil of Artemisia argyi suppresses inflammatory responses by inhibiting JAK/STATs activation. *Journal of Ethnopharmacology*, *204*, 107–117.

Cimino, C., Maurel, O. M., Musumeci, T., Bonaccorso, A., Drago, F., Souto, E. M. B., Pignatello, R., & Carbone, C. (2021). Essential oils: Pharmaceutical applications and encapsulation strategies into lipid-based delivery systems. *Pharmaceutics*, *13*(3), 327.

Collao, C. A., Curotto, E., & Zúñiga, M. E. (2007). Enzymatic treatment on oil extraction and antioxidant recuperation from Oenothera biennis by cold pressing. *Grasas y Aceites*, *58*(1), 10–14.

Costa, M. F., Durço, A. O., Rabelo, T. K., Barreto, R. D. S. S., & Guimarães, A. G. (2019). Effects of carvacrol, thymol and essential oils containing such monoterpenes on wound healing: A systematic review. *Journal of Pharmacy and Pharmacology*, *71*(2), 141–155.

Deng, Y., Wang, W., Zhao, S., Yang, X., Xu, W., Guo, M., Xu, E., Ding, T., Ye, X., & Liu, D. (2022). Ultrasound-assisted extraction of lipids as food components: Mechanism, solvent, feedstock, quality evaluation and coupled technologies—A review. *Trends in Food Science & Technology*.

Ebrahimi, H., Mardani, A., Basirinezhad, M. H., Hamidzadeh, A., & Eskandari, F. (2022). The effects of Lavender and Chamomile essential oil inhalation aromatherapy on depression, anxiety and stress in older community-dwelling people: A randomized controlled trial. *Explore*, *18*(3), 272–278.

Egza, T. F. (2020). A review on extraction, isolation, characterization and some biological activities of essential oils from various plants. *GSJ*, *8*(1).

El Ayadi, A., Jay, J. W., & Prasai, A. (2020). Current approaches targeting the wound healing phases to attenuate fibrosis and scarring. *International Journal of Molecular Sciences, 21*(3), 1105.

El Kharraf, S., El-Guendouz, S., Farah, A., Bennani, B., Mateus, M. C., & Miguel, M. G. (2021). Hydrodistillation and simultaneous hydrodistillation-steam distillation of Rosmarinus officinalis and Origanum compactum: Antioxidant, anti-inflammatory, and antibacterial effect of the essential oils. *Industrial Crops and Products, 168*, 113591.

Fakih, M., & Perbawati, C. (2022). Relevance of WHO traditional medicine strategy (2014−2023) with traditional health care policy in the perspective of national law and international law.

Falleh, H., Ben Jemaa, M., Saada, M., & Ksouri, R. (2020). Essential oils: A promising eco-friendly food preservative. *Food Chemistry, 330*127268.

Farhat, A., Fabiano-Tixier, A. S., El Maataoui, M., Maingonnat, J. F., Romdhane, M., & Chemat, F. (2011). Microwave steam diffusion for extraction of essential oil from orange peel: Kinetic data, extract's global yield and mechanism. *Food Chemistry, 125*(1), 255−261.

Fiorini, D., Scortichini, S., Bonacucina, G., Greco, N. G., Mazzara, E., Petrelli, R., Torresi, J., Maggi, F., & Cespi, M. (2020). Cannabidiol-enriched hemp essential oil obtained by an optimized microwave-assisted extraction using a central composite design. *Industrial Crops and Products, 154*, 112688.

Fresen, V.F., Homa-Bonell, J.K., & Romdenne, T.A. (2023). The use of topically applied sweet marjoram essential oil for pain reduction following IUD placement.

Gavahian, M., Chu, Y. H., & Mousavi Khaneghah, A. (2019). Recent advances in orange oil extraction: An opportunity for the valorisation of orange peel waste a review. *International Journal of Food Science & Technology, 54*(4), 925−932.

Golmakani, M. T., & Rezaei, K. (2008). Comparison of microwave-assisted hydrodistillation withthe traditional hydrodistillation method in the extractionof essential oils from Thymus vulgaris L. *Food Chemistry, 109*(4), 925−930.

Haffizi, M., Sulaiman, S., Jimat, D. N., & Amid, A. (2020). A comparison of conditions for the extraction of vegetable and essential oils via microwave-assisted extraction. *IOP Conference Series: Materials Science and Engineering, 778*(1)012172, IOP Publishing.

Her, J., & Cho, M. K. (2021). Effect of aromatherapy on sleep quality of adults and elderly people: A systematic literature review and meta-analysis. *Complementary Therapies in Medicine, 60*, 102739.

Koul, O., Walia, S., & Dhaliwal, G. S. (2008). Essential oils as green pesticides: Potential and constraints. *Biopesticides International, 4*, 63−84.

Masango, P. (2005). Cleaner production of essential oils by steam distillation. *Journal of Cleaner Production, 13*(8), 833−839.

Matera, R., Lucchi, E., & Valgimigli, L. (2023). Plant essential oils as healthy functional ingredients of nutraceuticals and diet supplements: A review. *Molecules (Basel, Switzerland), 28*, 901.

Meriem, A., Msaada, K., Sebai, E., Aidi Wannes, W., Salah Abbassi, M., & Akkari, H. (2022). Antioxidant, anthelmintic and antibacterial activities of red juniper (Juniperus phoenicea L.) essential oil. *Journal of Essential Oil Research, 34*(2), 163−172.

Mishra, A. P., Devkota, H. P., Nigam, M., Adetunji, C. O., Srivastava, N., Saklani, S., Shukla, I., Azmi, L., Shariati, M. A., Coutinho, H. D. M., & Khaneghah, A. M. (2020). Combination of essential oils in dairy products: A review of their functions and potential benefits. *LWT, 133*, Article 110116.

Nivedhitha, A. S., Devi, R. G., & Jyothipriya, A. (2019). Comparative study of the effect of lavender and rosemary oil in relieving pain. *Drug Invention Today, 12*(4).

Nursahidah, A., Novelia, S., & Suciawati, A. (2020). The effect of Lavender Aromatherapy on labor pain among delivery women. *Asian Community Health Nursing Research*, 13.

Ozen, T., Demirtas, I., & Aksit, H. (2011). Determination of antioxidant activities of various extracts and essential oil compositions of Thymus praecox subsp. skorpilii var. skorpilii. *Food Chemistry, 124*(1), 58−64.

Pandey, A. K., Kumar, P., Singh, P., Tripathi, N. N., & Bajpai, V. K. (2017). Essential oils: Sources of antimicrobials and food preservatives. *Frontiers in Microbiology, 16*(7), 2161.

Pavela, R., & Benelli, G. (2016). Essential oils as ecofriendly biopesticides? Challenges and constraints. *Trends in Plant Science, 21*(12), 1000−1007.

Peng, X., Liu, N., Wang, M., Liang, B., Feng, C., Zhang, R., Wang, X., Hu, X., Gu, H., & Xing, D. (2022). Recent advances of kinetic model in the separation of essential oils by microwave-assisted hydrodistillation. *Industrial Crops and Products, 187*, 115418.

Psarrou, I., Oreopoulou, A., Tsimogiannis, D., & Oreopoulou, V. (2020). Extraction kinetics of phenolic antioxidants from the hydro distillation residues of rosemary and effect of pretreatment and extraction parameters. *Molecules (Basel, Switzerland), 25*(19), 4520.

Qamar, S., Torres, Y. J., Parekh, H. S., & Falconer, J. R. (2021). Extraction of medicinal cannabinoids through supercritical carbon dioxide technologies: A review. *Journal of Chromatography B, 1167*, 122581.

Rondanelli, M., Fossari, F., Vecchio, V., Gasparri, C., Peroni, G., Spadaccini, D., Riva, A., Petrangolini, G., Iannello, G., Nichetti, M., & Infantino, V. (2020). Clinical trials on pain lowering effect of ginger: A narrative review. *Phytotherapy Research, 34*(11), 2843−2856.

Sánchez-González, L., Vargas, M., González Martínez, C., Chiralt, A., & Cháfer, M. (2011). Use of essential oils in bioactive edible coatings: A review. *Food Engineering Reviews, 3*, 1−16.

Soto, C., Chamy, R., & Zuniga, M. E. (2007). Enzymatic hydrolysis and pressing conditions effect on borage oil extraction by cold pressing. *Food Chemistry, 102*(3), 834−840.

Stratakos, A. C., & Koidis, A. (2016). *Methods for extracting essential oils. Essential oils in food preservation, flavor and safety* (pp. 31−38). Academic Press.

Tajkarimi, M. M., Ibrahim, S. A., & Cliver, D. O. (2010). Antimicrobial herb and spice compounds in food. *Food Control, 21*, 1199−1218.

Toniolo, J., Delaide, V., & Beloni, P. (2021). Effectiveness of inhaled aromatherapy on chemotherapy-induced nausea and vomiting: A systematic review. *The Journal of Alternative and Complementary Medicine, 27*(12), 1058−1069.

Uwineza, P. A., & Waśkiewicz, A. (2020). Recent advances in supercritical fluid extraction of natural bioactive compounds from natural plant materials. *Molecules (Basel, Switzerland)*, *25*(17), 3847.

Valdivieso-Ugarte, M., Gomez-Llorente, C., Plaza-Díaz, J., & Gil, Á. (2019). Antimicrobial, antioxidant, and immunomodulatory properties of essential oils: A systematic review. *Nutrients*, *11*(11), 2786.

Varkaneh, Z. K., Karampourian, A., Oshvandi, K., Basiri, Z., & Mohammadi, Y. (2022). The effect of eucalyptus inhalation on pain and the quality of life in rheumatoid arthritis. *Contemporary Clinical Trials Communications*, *29*, 100976.

Xiong, K., & Chen, Y. (2020). Supercritical carbon dioxide extraction of essential oil from tangerine peel: Experimental optimization and kinetics modelling. *Chemical Engineering Research and Design*, *164*, 412–423.

Yeh, T. H., & Lin, J. Y. (2020). Acorus gramineusand and Euodia ruticarpa steam distilled essential oils exert anti-inflammatory effects through decreasing Th1/Th2 and pro-/anti-inflammatory cytokine secretion ratios in vitro. *Biomolecules*, *10*(2), 338.

Yousefi, M., Rahimi-Nasrabadi, M., Pourmortazavi, S. M., Wysokowski, M., Jesionowski, T., Ehrlich, H., & Mirsadeghi, S. (2019). Supercritical fluid extraction of essential oils. *TrAC Trends in Analytical Chemistry*, *118*, 182–193.

Chapter 7

Application of essential oils as biopreservative agents

Babatunde Oluwafemi Adetuyi[1], Peace Abiodun Olajide[1], Olubanke Olujoke Ogunlana[2], Charles Oluwaseun Adetunji[3], Juliana Bunmi Adetunji[4], Abel Inobeme[5], Oloruntoyin Ajenifujah-Solebo[6], Yovwin D. Godwin[7], Olalekan Akinbo[8], Oluwabukola Atinuke Popoola[6], Olatunji Matthew Kolawole[9], Osarenkhoe Omorefosa Osemwegie[10], Mohammed Bello Yerima[11] and M.L. Attanda[12]

[1]Department of Natural Sciences, Faculty of Pure and Applied Sciences, Precious Cornerstone University, Ibadan, Oyo State, Nigeria, [2]Department of Biochemistry, Covenant University, Ota, Ogun State, Nigeria, [3]Applied Microbiology, Biotechnology and Nanotechnology Laboratory, Department of Microbiology, Edo State University Uzairue, Iyamho, Edo State, Nigeria, [4]Department of Biochemistry, Osun State University, Osogbo, Osun State, Nigeria, [5]Department of Chemistry, Edo State University Uzairue Iyamho, Auchi, Edo State, Nigeria, [6]Genetics, Genomics and Bioinformatics Department, National Biotechnology Development Agency, Abuja, FCT, Nigeria, [7]Department of Family Medicine, Faculty of Clinical Sciences, Delta State University, Abraka, Delta State, Nigeria, [8]Centre of Excellence in Science, Technology, and Innovation, AUDA-NEPAD, Johannesburg, Gauteng, South Africa, [9]Department of Microbiology, Faculty of Life Sciences, University of Ilorin, Ilorin, Kwara State, Nigeria, [10]Department of Food Science and Microbiology, Landmark University, Omu-Aran, Kwara State, Nigeria, [11]Department of Microbiology, Sokoto State University, Sokoto, Sokoto State, Nigeria, [12]Department of Agricultural Engineering, Bayero University Kano, Kano, Kano State, Nigeria

Introduction

Ever since antiquity, people have turned to commercialized antimicrobial substances to either avoid or treat food rotting and infection. Users' worries about synthetic preservatives have prompted a resurgence of interest in natural antimicrobials such as essential oils (EOs). For this reason, many people turn to volatile oils as well as other botanical ingredients from therapeutic and aromatic plants as natural food preservation substances since they are efficient against a wide range of microorganisms, fungi, and other pathogens (Gormez et al., 2016). Hydrophobic liquids of volatile and oily aromatic chemicals, EOs can be found in a wide variety of plant components, from the twig to the flower to the leaf to the bark to the seed to the root. Many EOs extracted from plants are used to improve the flavor or aroma of other products, such as cosmetics, food additives, soaps, plastic resins, and perfumes. Due to their versatility, natural origins, and GRAS status, EOs are increasingly being investigated for their potential as antibacterial agents. Antibacterial, antipyretic, antispasmodic, antiparasitic, free radical scavenging, anti-inflammatory, repellent, antifeedant, repellent, antimutagenic, antiviral, larvicidal, sedative, molluscicidal, immunomodulatory, antinociceptive, and insecticidal properties, in addition to their use as flavoring agents, are all areas where EOs are actively being researched (Pandey et al., 2014c; Ukeh & Mordue, 2009).

Medicinal and aromatic plants and essential oils

The pharmaceutical, perfume, and cosmetics industries, to name a few, rely heavily on medicinal and aromatic plants (MAPs) (Sylvestre et al., 2007). Since MAPs make up such a large part of the natural flora, it's no surprise that eighty percent of the global population relies on traditional medicines made from MAPs to cure a wide range of human health issues. There are about 9000 native plants that have been documented as having medicinal effects, and another 1500 species that have been identified for their aroma and flavor (Abdollahi et al., 2003).

While EOs may seem like stable liquids, they are actually unstable chemicals extracted from aromatic and therapeutic parts of the plants. Antimicrobial, antiparasitic, insecticidal, viral, fungal, and antioxidant characteristics of EOs make them useful for plant protection (Mediratta et al., 2002). The pharmaceutical, aromatic, and cosmetics sectors have all experienced rising demand. Additionally, EOs have become increasingly popular in the food business, where they serve a variety of purposes including preservatives, antioxidants, flavorings, and colorants. Worldwide, well over

Applications of Essential Oils in the Food Industry. DOI: https://doi.org/10.1016/B978-0-323-98340-2.00023-7

250 new EO types hit shelves every year, and their popularity continues to rise thanks to rising consumer interest in all things natural and fear of potentially dangerous synthetic additives. These products require extraction from the plant matrix before they may be used. Expression, distillation, solvent extraction, and possibly even more unconventional methods can all be used to achieve this goal (Abdollahi et al., 2003).

Citrus peel oils are extracted using expression or cold pressing due to the heat instability of the aldehydes found therein. The minimal heat generation during the process is an advantage of this approach, but the low yields and low purity it produces are drawbacks (Pandey et al., 2014c).

Distillation is the technique of extracting EOs from plants by heating them to the point where their cell walls break down and the EOs are released. The volatile components are readily removed from the water vapor because they are condensed at atmospheric conditions well below the heat flow rate of their constituents. Three distinct methods of distillation have been proposed, each distinguished by the means by which water is brought into contact with the starting material: hydrodistillation, steam distillation, and water/steam distillation (Dhingra et al., 2001).

Because of their hydrophobic and nonpolar nature, EOs can be extracted with organic solvents. The EO is extracted by heating a mixture of solvent and plant material. Next, the solvent is removed by heating to high temperatures, and the solution is strained, after which the resultant concentrate is combined with alcohol by volume to obtain the oil at ambient conditions. Consequences of the process include waste product discharge and solvent residues in the final EO (both of which are harmful to human health and the environment) (Dhingra et al., 2001). New methods of separation are developing that use less energy and produce fewer emissions of greenhouse gases in response to environmental concerns.

Microwave-assisted extraction, ultrasonic-assisted extraction, and supercritical fluid extraction are emerging techniques for extracting OE. Because EOs' antibacterial efficacy is tied to their chemical composition, knowing that information is crucial after extraction. Gas chromatography (GC) has been described as the most suitable method for the analysis of EOs, while there are other methods available. These days, GC coupled with mass spectrometry (MS) is utilized to better separate many chemicals found in EOs. The elements of nature's makeup EOs have a complicated chemical makeup; each EO may have anywhere from twenty to sixty different bioactive components. While other components may be present in minute amounts, only two or three are often found at relatively large concentrations (20%−70%) (Abdollahi et al., 2003).

Insects, human-pathogenic fungi, and microbes can all cause severe harm to items kept for lengthy periods, but the EOs of therapeutic and aromatic plants have been demonstrated to be efficacious against all three. EOs from plants like Mexican tea and *Clausena pentaphylla* as well as corn mint and Tulsi are potent fumigant toxicants against the fungal pulse beetle and cowpea weevil, but only if they come into direct contact with the pests, cowpea weevil, a fungus that thrives on pigeon pea seeds (Pandey et al., 2011a). *Tanacetum nubigenum* EO is also very poisonous as a fumigant and repellant against *Tribolium castaneum*, a fungus that attacks wheat as it is being stored (Haider et al., 2015). Antibacterial activity has been demonstrated in *Eucalyptus globulus* EO against Gram-positive and Gram-negative bacteria, respectively, including *Escherichia coli, Staphylococcus aureus, Haemophilus influenzae*, and other bacterial pathogens (Bachir and Benali, 2012). *S. aureus* is a bacterium. bacterial pneumonia and S. EO from Eucalyptus odorata was effective against *S. pyogenes* under in vitro conditions (Posadzki et al., 2012). In this article, we'll look at how EOs can be used to prevent the growth of mold and other fungi on food products by acting as a natural preservative and antifungal agent. EOs' antibacterial and bactericidal activity against plant pathogenic bacteria has received more attention in recent years.

Activated components of essential oils and potential uses

EOs are derived from the volatile, odoriferous chemical mixture found in most aromatic plants. Common examples of secondary phytoconstituents include terpenoids, alcohols, acids, aldehydes, ketones, and phenols. Terpenes, terpenoids, and aromatic phenols are some of the most vital elements of EOs (Koul et al., 2008). The mevalonic acid and shikimic acid pathways are responsible for the synthesis of terpenoids and aromatic polyterpenoids, respectively (Bedi et al., 2008). Terpenoids are an important class of secondary chemicals that are present in many aromatic and medicinal plants. Because of their antibacterial properties, monoterpenoids prevent the normal growth and reproduction of microbes and tamper with their internal cellular metabolism (Burt, 2004). Proven antibacterial and antifungal effects have been attributed to specific plant compounds (Isman, 2006; Pandey et al., 2012, 2016). These chemicals include azadirachtin, carvone, menthol, ascaridol, methyl eugenol, toosendanin, and volkensin. In addition to contributing to enhanced flavor or poisonous qualities, several of these compounds also exhibit potent bactericidal, fungicidal, and insecticidal activity.

Essential oils antimicrobial activity

Essential oils' antibacterial effect is mediated by various metabolic processes unique to each oil (Pandey et al., 2012).

Mechanism of antimicrobial action of essential oils

If the pathogens in consideration are Gram-positive or Gram-negative microbes or fungus, as well as the primary constituent of EOs, they can have a significant impact on the antimicrobial activities of EOs.

Mechanism of antibacterial action

To begin with, EOs upset cellular structure by reducing membrane integrity and increasing permeability and by altering cellular functions including energy production and membrane transport (Bedi et al., 2008). Changes in nutrition processing, structural macromolecule synthesis, and growth regulators are only some of the activities that are disrupted by damage to the cell membrane. The cell wall permeability of bacteria is increased by the lipophilicity of EOs, which results in the leakage of internal contents and the depletion of exogenous ions (Isman, 2006). Additionally, EOs' antimicrobial activities involve alterations in proton pumps and ATP depletion; these alterations may have knock-on effects on other cell organelles. Thymol, menthol, and linalyl acetate, along with other EOs, exert their antimicrobial action by altering the lipid proportion of bacterial cellular membrane; carvacrol distorts the concentration of fatty acids, altering the membrane's fluidity and permeability; and trans-cinnamaldehyde enters the periplasm of the cell and impedes cellular activities (Koul et al., 2008).

Antibacterial effects

Bacillus cereus, E. coli, Listeria monocytogenes, Salmonella spp., and *S. aureus* are among the deadliest bacteria that can be spread by food (Isman, 2006). *Bacillus cereus*, a Gram-positive bacteria that can infect humans through contaminated food due to its spore production and formation of both thermostable and thermolabile toxins, is found in a wide variety of environmental settings. A minimal inhibitory concentration (MIC) of 0.2 m/L was found in studies to be inhibited by rosemary EOs, suggesting that they are effective against this bacterium (Tripathi & Kumar, 2007). Studies on antimicrobial action have shown that clove, oregano, rosemary, sage, and thyme EOs are effective growth inhibitors against *E. coli*, with a MIC range of 0.4–10 m/L (Tripathi & Kumar, 2007). *E. coli* is a Gram-negative bacterium that normally dwells in the intestine but can spread to food through improper handling or sanitation. Similarly, the MIC for *L. monocytogenes* (Gram-positive) was between 0.15 and 0.45 microliters per milliliter when exposed to EOs of clove, oregano, rosemary, sage, and thyme (Tripathi & Kumar, 2007). *L. monocytogenes* is responsible for the food-borne illness known as listeriosis, which can happen either randomly or in outbreaks. The Gram-negative bacillus *Salmonella* spp. is the most common contributor to foodborne infection, and it spreads either through direct contact with the infected individual or through cross-contamination. EOs of clove, oregano, rosemary, sage, and thyme all had a MIC between 1.2 and 20 m/L, greatly reducing the bacterial growth (Kotan et al., 2013), with oregano presenting the most inhibitory effect. When tested against Salmonella enterica, cinnamon, oregano, and clove EOs all performed well, but clove extracts showed the most antibacterial activity (Kotan et al., 2013). Diseases caused by *S. aureus* (Gram-positive bacilli) include not just skin and mucosal infections but also gastrointestinal distress if contaminated food products are consumed. Clove, oregano, rosemary, sage, and thyme EOs have all been shown to be effective in preventing S. aureus growth, with a range of 0.2–10 m/L for their respective MIC values, according to a number of studies (Kotan et al., 2013). The MIC for the food-borne pathogen *Clostridium perfringens* (Gram-positive), which was suppressed by the rosemary EOs in one study, was set at 10 millimoles per liter. *C. perfringens* is responsible for the majority of food-borne illness outbreaks. Outbreaks of these illnesses are commonly associated with catering events because of the nature of the preparation and presentation of the food involved.

Essential oils' efficacy against plant-pathogenic bacteria

Twenty percent to forty percent of the annual production of cereals, pulses, fruits, and vegetables is lost due to bacterial species during planting, transportation, and storage. Microbes such as *Erwinia carotovora*, *Erwinia amylovora*, *Pseudomonas syringae* pv. tomato, *Pseudomonas syringae* pv. solanacearum, *Pseudomonas cichorii*, *Pseudomonas putida*, and *Pseudomonas syringae* can cause significant damage to several plants of domestic and global relevance

(Agrios, 2005). There have been in vitro and in vivo studies on the antimicrobial property of several EOs against these plant-pathogenic microorganisms (Kotan et al., 2013). EO effectiveness against plant-pathogenic microorganisms can be evaluated with one of four screening tests: disc diffusion, agar dilution, agar well, or broth dilution (Perricone et al., 2015). EO ingredients and their method of action against bacteria have been studied at length, while data on their effectiveness against yeasts and molds are more scarce.

There appears to be a disparity between the effectiveness of EOs against Gram-positive and Gram-negative bacteria. Gram-negative bacteria are more resistant to water-insoluble antibacterial chemicals like those contained in EOs because they have water-loving lipopolysaccharides (LPS) in their outer surface (Trombetta et al., 2005). So, because of the vast genetic variances between species, it is challenging to foretell how EOs would affect microbes.

Antifungal effects

There is a lot of interest in EOs generated from MAPs in the food business because of their ability to stifle the expansion of potentially dangerous microorganisms. These include *Fusarium* spp., *Aspergillus* spp., and others that have been linked to food-borne illnesses (Agrios, 2005). All sorts of fungal infections have been combated with the use of EOs. The minimum inhibitory concentration (MIC) for clove EO against *Candida albicans* was 0.062 while the MIC for clove EO against *Aspergillus niger* was 0.125% (v/v). In a similar vein, both *C. albicans* and *A. niger* can be killed by rosemary EO, although only at 0.25 and 1.0% (v/v) concentrations. *Aspergillus flavus* mycelium development was completely stifled by thyme and clove EOs (14). Clove oil, specifically the primary chemical component eugenol, was shown to be an effective antifungal agent with a MIC value of 1.0% v/v, killing C. albicans cells. Clove, cinnamon, and oregano EOs were efficient against *Aspergillus parasiticus* and *Fusarium moniliforme* since they inhibited mycelium development and mycotoxin generation (Trombetta et al., 2005).

Mechanism of antifungal action

When inhaled, EOs have an antifungal effect, particularly against molds, that is equivalent to the killing power of antimicrobial strategies that work by coming into physical contact with the bacterium (Agrios, 2005). EOs kill fungi by penetrating their cell membranes and altering their permeability, causing the mitochondrial membranes to rupture. Some molds' life cycles are disrupted by the vapor phase produced by EOs, specifically the growth of hypha and sporulation. Due to their resistance to destruction by heat, light, or chemicals, conidia must be inactivated by EOs in order to prevent the spread of the fungus (Paranagama et al., 2003).

Essential oils' efficacy against molds and moldy food

Grain, pulse, fruit, and vegetable crops can be rendered unsuitable for consumption by humans if they have been contaminated with mycotoxins (Pandey et al., 2016). Food spoilage is a constant worry in tropical climates because of the high heat and humidity. The FAO estimates that worldwide, fungal infections and their toxic metabolites cause losses in quality and quantity of up to 25% of agricultural food items (Agrios, 2005). Dietary ingredients and physiological properties, as well as meal quality, color, and texture, are all negatively influenced by fungal infection (Dhingra et al., 2001). Fungi infection can result in mycotoxin production, which has been linked to food shortages in poor nations (Wagacha & Muthomi, 2008). Infection of food with molds such as *Alternaria*, *Aspergillus*, *Penicillium*, *Fusarium*, and *Rhizopus* spp. poses serious health hazards and can cause foodborne illnesses (Pandey & Tripathi, 2011). EOs' ability to inhibit fungal growth during storage and transport suggests they may be a useful tool in the fight against food waste. EOs have gained widespread attention in recent years due to their antifungal properties (Bosquez-Molina et al., 2010). EOs have antifungal properties because the mono- and sesquiterpene chemicals included in them cause the hyphae of molds and yeasts to break down. Additionally, EOs improve membrane fluidity; membrane swelling caused by substance dissolution in biological membranes lowers membrane function (Dorman & Deans, 2000). EOs, being lipid-soluble, are able to penetrate fungal cell walls and modulate cell-wall synthesizing enzymes, resulting in morphological changes to the fungus (Cox et al., 2000).

Essential oils' power to preserve food

Scientific investigation into the potential of EOs for use in the storage and long-term storage of food products has yielded promising results in recent years. Scientists have discovered that storing food in a range of containers—including those made from cardboard, tin, glass, polyethylene, and natural fabrics—while using EOs, either in their pure state or as part of a composition, dramatically extends the quality and safety of food (Pandey et al., 2014a). An earlier study suggested that EO constituents like citral, citronella, citronellol, eugenol, farnesol, and nerol could delay fungal infection in chili seeds and fruits for up to six months (Tripathi et al., 1984). Ageratum conyzoides EOs prevented mandarins from decaying from blue mold, making them last up to 30 days longer than normal (Dixit et al., 1995). Researchers Anthony et al. (2003) found that storing bananas with a drop of EO from *Cymbopogon nardus*, *C. flexuosus*, or *Ocimum basilicum* significantly lowered the occurrence of anthracnose and increased the durability of the fruit by up to 21 days. Malus pumilo fruit can be preserved for up to three weeks in a solution of *Cymbopogon flexuosus* EO (20 L/mL) (Shahi et al., 2003). A fumigant derived from the EOs of the plant *Putranjiva roxburghii* was efficient against *A. flavus* and *A. niger*, preventing fungal biodeterioration in groundnuts for up to 6 months (Tripathi & Kumar, 2007). *Cymbopogon pendulous* EO was found to be more efficient than *P. roxburghii* EO as a systemic insecticide, increasing the storage life of groundnuts by 6−12 months (Shukla, 2009). EOs' varying degrees of success may be attributable to factors including the plant species used, the oils' chemical makeup, the dosage, and the method of preservation.

Thyme and Mexican lime oils decreased papaya fruit infection (Bosquez-Molina et al., 2010), whereas 0.3% cinnamon oil extended banana durability by up to 28 days and reduced fungal disease prevalence (Maqbool et al., 2010). Seed fumigation with *Ocimum cannum* oil (1 L/mL) and seed dressing increased Bhuchanania's self-life (Singh et al., 2011). Pigeon pea seeds were safe from *A. flavus*, *A. niger*, *A. ochraceus*, and *A. terreus* invasion for about six months when preserved in glass jars and organic textile sacks with a systemic insecticide produced from the oils of *C. pentaphylla* and *Chenopodium ambrosioides* (Pandey et al., 2013a,b). *C. pentaphylla* and *C. ambrosioides* oils in powder form were also effective at preserving pigeon pea seeds for up to 6 months (Pandey et al., 2014c). The use of a fumigant made from oil extracted from *Artemisia nilagirica* packaging table grapes in cardboard can extend their shelf life by as much as 9 days (Sonker et al., 2015). Similarly, green gram stored in glass jars with air dose treatment with *Lippia alba* oil had much high durability (*Vigna radiata*) by up to 6 months by reducing fungal proliferation and aflatoxin generation (Pandey et al., 2016).

Food processing with essential oils

Products derived from or containing meat

EOs have a stronger antibacterial effect in a lab setting than they do in finished foods. It might be because there are more beneficial elements in food than there are in test tube medium. Meat and meat products, which are rich in nutrients, may help bacteria repair and renew their biological components, which may make them more resilient to a wide range of challenges (Sonker et al., 2015). Yet, the antibacterial action of EOs in foods may also be affected by a variety of other factors. Traditional meat products are high in caloric density, and since EOs are lipophilic in nature, the combination of the acidic condition, which tends to improve EO dissolution rate, and the low aW, which minimizes the aqueous layer of food products, enhances the interaction between EOs and spoilage or foodborne microbes, thereby increasing the antibacterial action of the EOs. Beef, chicken, lamb, and rabbit fresh meat are the most common types of meat for which EOs have been documented to be used to increase both food safety and shelf life (Tripathi & Kumar, 2007). Though EOs have been used as food components, only a few research have explained the process (Maqbool et al., 2010). EOs are rarely used in culinary products, and even then, it is often reserved for cooked meats. Using thyme and cinnamon EOs has been demonstrated to drastically reduce the number of *L. monocytogenes* in ham. In addition, oregano and rosemary EOs have been used to lengthen the storage life of bologna and bologna sausages, respectively (Maqbool et al., 2010; Sonker et al., 2015). The antibacterial action of garlic and oregano EOs against *Salmonella* spp., *L. monocytogenes*, and *S. aureus* was reported by Garca-Dez and colleagues (2004), who found that these EOs increased the safety of dry-cured sausages. Meat containing *L. monocytogenes* was exposed to 0.5% and 1% clove EO, respectively, and its growth was inhibited at 30°C and 7°C (Agrios, 2005). The mix of EOs, the unique qualities of each food item, and the targeted bacteria all contribute to their antimicrobial action. Heat treatments, smoking, chemical preservatives, and packaging are just some of the additional ways that EOs in food may be compromised (Tripathi & Kumar, 2007). There have also been reports of the use of EOs in conjunction with other technologies to combat foodborne infections, such as thermal treatments, high isostatic pressures (HIP), pulsed electric fields, and active packaging (Nguefack et al., 2005). Combinations of mild heat treatment (54°C/10′) and low concentrations of orange, lemon, and mandarin EOs (0.2 mL/mL) demonstrated synergistic fatal effects, inactivating more than 5 log units of bacterial cells, indicating the possibility of such therapies for food preservation (Nguefack et al., 2005). EOs are used as natural preservatives in the food industry, which satisfies the growing demand among consumers for products with "green," "biological," "natural," and "no chemicals used" labels. The doses required to achieve the desired antimicrobial activity may be viewed unfavorably by customers, thus it is important to evaluate their organoleptic influence in each product. For example, Mediratta et al. (2002) found that bologna sausages had a strong oregano aroma, which was not, however, liked by the tasters. Similarly, other authors have detailed the use of oregano EO in a Spanish fermented dry-cured salchichón, finding that while it had no discernible effect on the salchichón's sensory characteristics, its texture had been greatly enhanced. Additionally, the antioxidant qualities of rosemary or marjoram EOs have been proven to improve the flavor of beef patties by lowering the rate at which their lipids oxidize. This was measured by adding the EOs at a concentration of 200 mg/kg (Burt, 2004). Reduced lipid oxidation as evidenced by lower TBARS levels has also been documented for foal meat after 10 days of active packaging with 2% oregano EO. Although there is a dearth of information on safe levels of concentration in food products, Bosquez-Molina et al. (2010) have advocated a cap of 5000 ppm. In addition, EOs are classified as flavoring agents in the regulatory standards for food additives (Bosquez-Molina et al., 2010). EC Regulation No. 1333/2008 establishes a maximum allowable level of 100 mg/kg for rosemary extract in dry-cured pork products (Tripathi & Kumar, 2007). The use of many EOs in lesser quantities may be sufficient to ensure the security of food products without compromising their flavor or aroma. Using a combination of cinnamon and clove EOs has shown promise as a natural alternative to chemical preservatives, particularly for use in active packaging systems, where spoilage fungi, yeasts, and bacteria are commonly found on foods with intermediate moisture content (aW between 0.65 and 0.90). (Pandey et al., 2014c). Adding 4% rosemary EO to a chicken's packing has been found to reduce the growth of the bacteria responsible for the creation of biogenic amines such as putrescine, cadaverine, and histamine. The use of both EOs (cinnamon, cloves, ginger, and anise) and starter cultures (*L. sakei* and *S. xylosus*) in the production of traditional Chinese smoked horsemeat sausages demonstrated that both EOs and starter cultures inhibited the accumulation of biogenic amines, including tryptamine, putrescine, cadaverine, histamine, and tyramine, and the growth of enterobacteria. It's worth noting that EOs had more potent inhibitory effects than starting cultures. It was also found that EOs and starting cultures worked together effectively to prevent the buildup of the aforementioned biogenic amines and the development of enterobacteria (Nguefack et al., 2005).

Cheeses

Several variables, including the chemical makeup of the dairy product, the concentration of the oils employed, and the targeted microorganisms, can affect the antibacterial activity of the various EOs in the product. Sonker et al. (2015) compared the antimicrobial effects of 1% concentrations of clove, cinnamon, bay, and thyme EOs in low-fat and full-fat soft cheese and found that the 1% concentration was the most effective for all EOs. Clove EO at 1% was more effective in killing *L. monocytogenes* than either type of cheese, but the anti-Listeria impact was stronger in the low-fat cheese. The antimicrobial activity of this EO against Salmonella Enteritidis was higher in full-fat cheese than in low-fat cheese at the same dose. At a 0.5% concentration, Salmonella Enteritidis was only able to recover in low-fat cheese, but it was unaffected in full-fat cheese. In reality, bacterial cells are protected by fat from antimicrobial agents. But because of the increased protein content of low-fat cheeses, the phenolic compounds of EOs can form complexes with the proteins of these goods, reducing the activity of the EOs (Bosquez-Molina et al., 2010). When comparing the efficacy of oregano oil and thyme oil against *L. monocytogenes* and *E. coli* O157: H7 in Feta cheese, Sonker et al., 2015 found them similar. On the contrary, they found that *L. monocytogenes* decreased more rapidly than *E. coli* O157: H7. This is likely because the EOs had a more dramatic effect on Gram-positive than on Gram-negative bacteria (Tripathi & Kumar, 2007). Because Gram-positive bacteria predominate in both the microbiota and starter cultures, this could have implications for product safety, processing, and preservation phenomena. EOs have been shown to extend the shelf life of dairy products by reducing the rate of chemical breakdown while they are in storage or on store shelves. The quantity of mesophilic bacteria in cream cheese has changed after being exposed to oregano and rosemary EOs. The acidity and decrease in pH generated by this event were not as severe as those caused by some other causes. Because oregano and rosemary EOs reduce oxidation and fermentation, the rancid and fermented flavors that indicate a product has gone bad are muted in items that contain them (Sonker et al., 2015). The mesophilic bacteria population in full cream cheese was altered by the addition of marjoram and rosemary EOs, resulting in a milder acidity and a lesser pH drop. There was a significant decrease in *L. monocytogenes* growth in cheese when clove EO (0.5% and 1%) was added, regardless of the temperature (room temperature or 30°C) (Mimica-Dukic et al., 2003). The decrease in microbes is also linked to lower levels of biogenic amines in the finished product of fermentation. Gouda cheese with various amounts of EO from the *Zataria multiflora* (thyme-like plant) significantly reduced histamine and tyramine production. Although the ideal dose for the sensory panel was 0.2%, a concentration of 0.4% EO reduced tyramine by 44% and histamine by 46%. Both tyramine and histamine were reduced by 22% and 29%, respectively, in this example (Tripathi & Kumar, 2007).

Fruits

Fruits and vegetables are considered perishable goods because they deteriorate quickly and lose weight, primarily due to the presence of fungi. To put it simply, this is a major issue for everyone involved, including manufacturers and customers. The postharvest period, storage, and long-distance transit were all made easier by maintaining low temperatures. However, low temperatures are not sufficient for the needs of consumers and businesses. Thus, a number of supplementary methods, including modified atmosphere (MAP) and controlled atmosphere (CA), are utilized to lengthen the storage life of fruits (CA). Some chemical agents have proven useful in preventing the spread of disease in the postharvest period, but their usage is limited by concerns over their effect on the environment and on the quality of the food produced. For this reason, the use of EOs from aromatic plants for postharvest management of phytopathogens has seen a surge in interest in recent years. Due to their antibacterial capabilities, EOs from a wide range of plant species have been examined for their potential to reduce fungal growth in production, first in vitro and then in vivo. Several studies have shown that EOs can effectively be used to manage certain disorders. Moreover, several research examined the function of EOs and their application to the active packaging of fruits and other foods (Mimica-Dukic et al., 2003). Among the many EOs tested for their antifungal activity against *Botrytis cinerea*, palmarosa (*Cymbopogon martini*), red thyme (*Thymus zygis*), cinnamon leaf (*Cinnamomum zeylanicum*), and clove buds (*Eugenia caryophyllata*) showed the most promising results. D-limonene, cineole, alpha-myrcene, alpha-pinene, alpha-pinene, and camphor were reported by the same authors to be the most common components of EOs with potent antifungal action. Recent research by Pandey et al. (2011a) in which thyme and sage EOs were applied to strawberries in the package's head space revealed a reduction in the number of fungus present in the treated samples compared to the control samples. Pandey et al. (2013b) also tried several EO in strawberries and found them to be effective. In a presentation titled "The Effect of Echinophora platyloba EO on Various Fungus," Pandey et al. (2012) presented data demonstrating the EO's efficacy against a variety of fungi, demonstrating its potent antifungal properties. Ozturk and Ercisli (2006) confirmed the antifungal effects of EOs on a variety of molds and yeasts, most notably *B. cinerea*. The postharvest period is when this fungus is most prevalent in fruits. In order to assess the efficacy of EO of thyme, oregano, and lemon in reducing the spread of various fungi in strawberries, tomatoes, and cucumber, Pandey et al. (2011a) confirmed the effect of these oils at varying doses. Using in vitro and in vivo testing on strawberries, Neri et al. (2005) determined that *Mentha arvensis*, *Citrus limon*, *Zingiber officinale*, and *Thymus vulgaris* EOs would all have an effect on the development of *B. cinerea*. Fennel, black caraway, peppermint, and thyme EOs were investigated for their fungicide effects against *B. cinerea* and Rhizopus stolonifer in vitro by Mediratta et al. (2002) at five different concentrations. EOs have lately been utilized in table grapes to increase their storage life without altering their flavor or aroma. Several authors tried out a new packaging, consisting of grapes wrapped in two films, and found that the microbial population was much reduced, and that berry deterioration occurred much less frequently (Nikaido, 2003). Several studies on 'Crimson Seedless' table grapes found that when packaged with MAP and eugenol, thymol, or menthol, the number of yeasts and molds was drastically reduced. Table grapes were examined by Nguefack et al. (2005) in MAP circumstances with and without active packaging, and the results showed that the active packaging with additional EO resulted in less quality loss in terms of sensory, nutritional, and functional qualities, as well as fewer spoiling occurrences. 'Crimson seedless' table grapes treated with eugenol and menthol EOs fared better in storage than untreated grapes, as reported by Nedorostova et al. (2009). Nedorostova et al. (2009) examined the efficacy of EOs in preventing the growth of various fungus, including *Alternaria* spp., *Fusarium* spp., *Colletotrichum* spp., and *Penicillium* spp., in *Passiflora edulis* Sims (Passion fruit). Oils of eucalyptus and rosemary were distilled (*Eucalyptus agglomerata*). The most important finding was that those EOs successfully reduced fungal infections on the aforementioned passion fruits. Combining the use of edible coverings like chitosan with EOs has a synergistic effect on preventing fruit spoilage and preserving fruit quality. Using vapor emission and direct coating, Mediratta et al. (2002) studied the inhibitory effects of chitosan EOs against *Botrytis* sp., *Penicillium* sp., and *Pilidiella granati*, three diseases of the pomegranate fruit. The antifungal activity of chitosan films infused with oregano EO, cinnamon EO, and lemongrass EO, respectively, was highest. Submerging fruit in the chitosan-EO emulsions had a more inhibitory impact than simply exposing them to the vapor. Using the possible antibacterial impact of EOs as a starting point, new methods have been developed to extend the storage life of minimally processed (MP) items, such as the use of EOs in combination with edible films (Mediratta et al., 2002). A growing number of health-conscious shoppers are eager to stock up on MP items that are "simple to eat," while also checking to make sure they adhere to the principles of clean labeling, natural ingredients, and "green" production.

New alternatives for food preservation

Natural conservation, the use of antimicrobial preservatives found in plants, animals, or microorganisms, and most notably, the use of the extracts of various plant types and parts as flavoring agents in some foods, is one of the alternatives that has received more attention in recent years. Both natural spoiling processes (food preservation) and microbiological development (food safety) can be managed with the help of antimicrobials (Agrios, 2005). The challenge lies in obtaining these antimicrobials in a pure and stable form and incorporating them into food products without compromising the products' taste, safety, or appeal to consumers. Products derived from plants, such as EOs, oleoresins, natural extracts, and their distillates, are considered GRAS (i.e., generally recognized as safe) by the Food and Drug Administration (FDA) (Burt, 2004). The ability of extracts and EOs of aromatic plants to inhibit the growth of harmful microbes has recently piqued the interest of the food sector.

Conclusion and future prospects

Several scientists all over the world have established that essential oils (EOs) have many antimicrobial activities. EOs have been utilized against many different types of bacteria and viruses, and their applications have been well-reported. This review thus gives a concise introduction to EOs, their active ingredients, and their potential as sources of antibacterials, antifungals, and food preservatives. EOs have been shown to possess a wide variety of antibacterial characteristics in the reviewed literature, and this suggests that they are ecologically sound candidates for application as biocontrol agents against microbial and fungal infections. Our research led us to the conclusion that EOs have great promise as a source of biocontrol products, and they should be investigated further to ensure the safety of food commodities. Moreover, a fumigant based on an EO that has antibacterial action should have a favorable GRAS rating in mammalian systems. Some plants, such as azadirachtin and carvone, are said to be safe for human consumption despite having high LD50 values in rats. The United States FDA has said that various EOs and their compounds (including carvone, carvacrol, cinnamaldehyde, thymol, linalool, citral, limonene, eugenol, limonene, and menthol) are GRAS and permitted as taste or food additives.

Fumigants, for instance, can be beneficial in natural fabric and cardboard containers, and even containers constructed of wooden boards, and EO applications are growing as a technique of embedding pathogens into these materials. Some oils can be sprayed lightly and even utilized as an incorporated fumigant in the product. EOs and their active ingredients can be manufactured from readily available raw materials; perhaps in many cases right at the place of application; and are effective against germs and fungi, making them relatively inexpensive remedies. We may conclude from this review that it is conceivable to create methods of protecting food commodities without resorting to or significantly reducing the usage of commercial bactericides and fungicides. EOs, according to the current literature, are host-specific, are biodegradable, have a low impact on non-target organisms, and are relatively non-toxic to mammals. There are downsides to commercial and sustainable uses, such as their cost-effectiveness. In any case, EOs have countless applications, and further study is required to suit the needs of a food business that is increasingly interested in green technology (Adetunji et al. 2022; Adetunji et al. 2019; Adetunji 2019; Adetunji et al. 2020; Adetunji et al. 2021a; Naveed et al 2022; Akram et al 2020; Saadullah et al., 2022; Adetunji et al 2021b; Kifordu et al., 2020; Anani et al., 2020).

References

Abdollahi, M., Karimpour, H., & Monsef-Esfehani, H. R. (2003). Antinociceptive effects of *Teucrium polium* L. total extract and essential oil in mouse writhing test. *Pharmacological Research: The Official Journal of the Italian Pharmacological Society*, *48*, 31−35. Available from https://doi.org/10.1016/s1043-6618(03)00059-8.

Adejumo, Isaac Oluseun, Adetunji, Charles Oluwaseun, Nwonuma, Charles O., Alejolowo, Omokolade O., & Maimako, Rotdelmwa (2017). Evaluation of selected agricultural solid wastes on biochemical profile and liver histology of Albino rats. *Food and Feed Research*, *44*(1), 73−79.

Adetunji, C. O. (2019). Environmental impact and ecotoxicological influence of biofabricated and inorganic nanoparticle on soil activity. In D. Panpatte, & Y. Jhala (Eds.), *Nanotechnology for agriculture*. Singapore: Springer. Available from https://doi.org/10.1007/978-981-32-9370-0_12.

Adetunji, C. O., Akram, Muhammad, Imtiaz, Areeba, Bertha, Ehis-Eriakha Chioma, Sohail, Adrish, Olaniyan, Oluwaseyi Paul, et al. (2021b). Modified cassava: the last hope that could help to feed the world—recent advances. In P. B. Kavi Kishor, M. V. Rajam, & T. Pullaiah (Eds.), *Genetically modified crops*. Singapore: Springer. Available from https://doi.org/10.1007/978-981-15-5932-7_8.

Adetunji, C. O., Inobeme, A., Olaniyan, O. T., Ajayi, O. O., Olaniyan, S., & Adetunji, J. B. (2021a). Application of nanodrugs derived from active metabolites of medicinal plants for the treatment of inflammatory and lung diseases: recent advances. In K. Dua, S. Nammi, D. Chang, D. K. Chellappan, G. Gupta, & T. Collet (Eds.), *Medicinal plants for lung diseases*. Singapore: Springer. Available from https://doi.org/10.1007/978-981-33-6850-7_26.

Adetunji, C. O., Oloke, J. K., & Prasad, G. (2020). Effect of carbon-to-nitrogen ratio on eco-friendly mycoherbicide activity from Lasiodiplodia pseudotheobromae C1136 for sustainable weeds management in organic agriculture. *Environment, Development and Sustainability*, *22*, 1977−1990. Available from https://doi.org/10.1007/s10668-018-0273-1.

Adetunji, C. O., Panpatte, D. G., Bello, O. M., & Adekoya, M. A. (2019). Application of nanoengineered metabolites from beneficial and eco-friendly microorganisms as a biological control agents for plant pests and pathogens. In D. Panpatte, & Y. Jhala (Eds.), *Nanotechnology for agriculture: crop production & protection*. Singapore: Springer. Available from https://doi.org/10.1007/978-981-32-9374-8_13.

Adetunji, Charles Oluwaseun, Ukhurebor, Kingsley Eghonghon, Olaniyan, Olugbemi Tope, Ubi, Benjamin Ewa, Oloke, Julius Kola, Dauda, Wadzani Palnam, & Hefft, Daniel Ingo (2022). *Agricultural biotechnology, biodiversity and bioresources conservation and utilization* (pp. 201–220). CRC Press.

Agrios, G. N. (2005). *Plant pathology (5th ed.)*. Oxford: Elsevier Academic Press.

Akram, Muhammad, Jabeen, Farhat, Daniyal, Muhammad, Zainab, Rida, ul Haq, Usman, Adetunji, Charles Oluwaseun, et al. (2020). Genetic engineering of novel products of health significance: recombinant DNA technology. In C. Egbuna, & G. Dable Tupas (Eds.), *Functional foods and nutraceuticals*. Cham: Springer. Available from https://doi.org/10.1007/978-3-030-42319-3_26.

Anani, O. A., Mishra, R. R., Mishra, P., Olomukoro, J. O., Imoobe, T. O. T., & Adetunji, C. O. (2020). Influence of heavy metal on food security: recent advances. In P. Mishra, R. R. Mishra, & C. O. Adetunji (Eds.), *Innovations in food technology*. Singapore: Springer. Available from https://doi.org/10.1007/978-981-15-6121-4_18.

Anthony, S., Abeywickrama, K., & Wijeratnam, S. W. (2003). The effect of spraying essential oils *Cymbopogon nardus, C. flexuosus* and *Ocimum basilicum* on post-harvest diseases and storage life of Embul banana. *Journal of Horticultural Science and Biotechnology.*, *78*, 780–785. Available from https://doi.org/10.1080/14620316.2003.11511699.

Bachir, R. G., & Benali, M. (2012). Antibacterial activity of the essential oils from the leaves of *Eucalyptus globulus* against *Escherichia coli* and *Staphylococcus aureus*. *Asian Pacific Journal of Tropical Biomedicine*, *2*, 739–742. Available from https://doi.org/10.1016/S2221-1691(12)60220-2.

Bedi, S., Tanuja., & Vyas, S. P. (2008). *A hand book of aromatic and essential oil plants cultivation*, chemistry, processing and uses. Jodhpur: AGROBIOS Publishers.

Bosquez-Molina, E., Jesus, E. R., Bautista-Banos, S., Verde-Calvo, J. R., & Morales-Lopez, J. (2010). Inhibitory effect of essential oils against *Colletotrichum gloeosporioides* and *Rhizopus stolonifer* in stored papaya fruits and their possible application in coatings. *Postharvest Biology and Technology*, *57*, 132–137. Available from https://doi.org/10.1016/j.postharvbio.2010.03.008.

Burt, S. (2004). Essential oils: Their antibacterial properties and potential applications in food — A review. *International Journal of Food Microbiology*, *94*, 223–253. Available from https://doi.org/10.1016/j.ijfoodmicro.2004.03.022.

Cox, S. D., Mann, C. M. I., Markham, J. L., Bell, H. C., Gustafson, J. E., Warmington, J. R., et al. (2000). The mode of antimicrobial action of the essential oil of *Melaleuca alternifolia* (tea tree oil). *Journal of Applied Microbiology*, *88*, 170–175. Available from https://doi.org/10.1046/j.1365-2672.2000.00943.x.

Dhingra, O. D., Mizubuti, E. S. G., Napoleao, I. T., & Jham, G. (2001). Free fatty acid accumulation and quality loss of stored soybean seeds invaded by *Aspergillus ruber*. *Seed Science and Technology*, *29*, 193–203.

Dixit, S. N., Chandra, H., Tiwari, R., & Dixit, V. (1995). Development of botanical fungicide against blue mould of mandarins. *Journal of Stored Products Research*, *31*, 165–172. Available from https://doi.org/10.1016/0022-474X(94)00041-Q.

Dorman, H. J. D., & Deans, S. G. (2000). Antibacterial agents from plants: Antibacterial activity of plant volatile oils. *Journal of Applied Microbiology*, *88*, 308–316. Available from https://doi.org/10.1046/j.1365-2672.2000.00969.x.

Gormez, A., Bozari, S., Yanmis, D., Gulluce, M., Agar, G., & Sahin, F. (2016). The use of essential oils of *Origanum rotundifolium* as antimicrobial agent against plant pathogenic bacteria. *Journal of Essential Oil Bearing Plants*, *19*, 656–663. Available from https://doi.org/10.1080/0972060X.2014.935052.

Haider, S. Z., Mohan, M., Pandey, A. K., & Singh, P. (2015). Repellent and fumigant activities of *Tanacetum nubigenum* Wallich. Ex DC essential oils against *Tribolium castaneum (Herbst) (Coleoptera: Tenebrionidae)*. *Journal of Oleo Science*, *64*, 895–903. Available from https://doi.org/10.5650/jos.ess15094.

Isman, M. B. (2006). Botanical insecticides, deterrents and repellents in modern agriculture and an increasingly regulated world. *Annual Review of Entomology*, *51*, 45–66. Available from https://doi.org/10.1146/annurev.ento.51.110104.151146.

Kifordu, A. A., Adetunji, C. O., Odiwo, W. O., & Mishra, R. S. (2020). Food Innovation and sustainable development: a bioeconomics perception. In P. Mishra, R. R. Mishra, & C. O. Adetunji (Eds.), *Innovations in food technology*. Singapore: Springer. Available from https://doi.org/10.1007/978-981-15-6121-4_1.

Kotan, R., Dadaşğolu, F., Karagoz, K., Cakir, A., Ozer, H., Kordali, S., et al. (2013). Antibacterial activity of the essential oil and extracts of *Satureja hortensis* against plant pathogenic bacteria and their potential use as seed disinfectants. *Scientia Horticulturae*, *153*, 34–41. Available from https://doi.org/10.1016/j.scienta.2013.01.027.

Koul, O., Walia, S., & Dhaliwal, G. S. (2008). Essential oils as green pesticides: Potential and constraints. *Biopesticides International*, *4*, 63–84.

Maqbool, M., Ali, A., & Alderson, P. G. (2010). Effect of cinnamon oil on incidence of anthracnose disease and post-harvest quality of bananas during storage. *International Journal of Agriculture And Biology*, *12*, 516–520.

Mediratta, P. K., Sharma, K. K., & Singh, S. (2002). Evaluation of immunomodulatory potential of *Ocimum sanctum* seeds oil and its possible mechanism of action. *Journal of Ethnopharmacology*, *80*, 15–20. Available from https://doi.org/10.1016/S0378-8741(01)00373-7.

Mimica-Dukic, N., Bozin, B., Sokovic, M., Mihajlovic, B., & Matavulj, M. (2003). Antimicrobial and antioxidant activities of three *Mentha* species essential oils. *Planta Medica*, *69*, 413–419. Available from https://doi.org/10.1055/s-2003-39704.

Naveed, Muhammad, Haleem, Kashif Syed, Ghazanfar, Shakira, Tauseef, Isfahan, Bano, Naseem, Adetunji, Charles Oluwaseun, et al. (2022). Quantitative estimation of aflatoxin level in poultry feed in selected poultry farms. *BioMed Research International*, *2022*, 7, 5397561. Available from https://doi.org/10.1155/2022/5397561.

Nedorostova, L., Kloucek, P., Kokoska, L., Stolcova, M., & Pulkrabek, J. (2009). Antimicrobial properties of selected essential oils in vapour phase against food borne bacteria. *Food Control*, *20*, 157–160. Available from https://doi.org/10.1016/j.foodcont.2008.03.007.

Neri, F., Mari, M., & Brigati, S. (2005). Control of *Penicillium expansum* by plant volatile compounds. *Plant pathology, 55*, 100−105. Available from https://doi.org/10.1111/j.1365-3059.2005.01312.x.

Nguefack, J., Somda, I., Mortensen, C. N., & Zollo, P. H. A. (2005). Evaluation of five essential oils from aromatic plants of Cameroon for controlling seed-borne bacteria of rice (*Oryza sativa*). *Seed Science and Technology, 33*, 397−407. Available from https://doi.org/10.15258/sst.2005.33.2.12.

Nikaido, H. (2003). Molecular basis of bacterial outer membrane permeability revisited. *Microbiology and Molecular Biology Reviews: MMBR, 67*, 593−656. Available from https://doi.org/10.1128/MMBR.67.4.593-656.2003.

Ozturk, S., & Ercisli, S. (2006). The chemical composition of essential oil and *in vitro* antibacterial activities of essential oil and methanol extract of *Ziziphora persica* Bunge. *Journal of Ethnopharmacology, 106*, 372−376. Available from https://doi.org/10.1016/j.jep.2006.01.014.

Pandey, A. K., & Tripathi, N. N. (2011). Post- harvest fungal and insect deterioration of pigeon pea seeds and their management by plant volatiles. *Journal of the Indian Botanical Society, 90*, 326−331.

Pandey, A. K., Singh, P., Palni, U. T., & Tripathi, N. N. (2011a). Use of essential oils of aromatic plants for the management of pigeon pea infestation by pulse bruchids during storage. *International Journal of Agricultural Technology, 7*, 1615−1624.

Pandey, A. K., Singh, P., Palni, U. T., & Tripathi, N. N. (2012). In-vitro antibacterial activity of essential oils of aromatic plants against *Erwinia herbicola* (Lohnis) and *Pseudomonas putida* (Krish Hamilton). *Journal of the Serbian Chemical Society, 77*, 313−323. Available from https://doi.org/10.2298/JSC110524192P.

Pandey, A. K., Palni, U. T., & Tripathi, N. N. (2013a). Evaluation of *Clausena pentaphylla* (Roxb.) DC oil as fungitoxicant against storage mycoflora of pigeon pea seeds. *Journal of the Science of Food and Agriculture, 93*, 1680−1686. Available from https://doi.org/10.1002/jsfa.5949.

Pandey, A. K., Singh, P., Palni, U. T., & Tripathi, N. N. (2013b). Application of *Chenopodium ambrosioides* Linn. essential oil as botanical fungicide for the management of fungal deterioration in pulse. *Biological Agriculture & Horticulture, 29*, 197−208. Available from https://doi.org/10.1080/01448765.2013.822828.

Pandey, A. K., Palni, U. T., & Tripathi, N. N. (2014a). Repellent activity of some essential oils against two stored product beetles *Callosobruchus chinensis* L. and *C. maculates* F. (Coleoptera: Bruchidae) with reference to *Chenopodium ambrosioides* L. for the safety of pigeon pea seeds. *Journal of Food Science and Technology, 51*, 4066−4071. Available from https://doi.org/10.1007/s13197-012-0896-4.

Pandey, A. K., Singh, P., Palni, U. T., & Tripathi, N. N. (2014c). *In vivo* evaluation of two essential oil based botanical formulations (EOBBF) for the use against stored product pests, *Aspergillus* and *Callosobruchus* (Coleoptera: Bruchidae) species. *Journal of Stored Products Research, 59*, 285−291. Available from https://doi.org/10.1016/j.jspr.2014.09.001.

Pandey, A. K., Sonker, N., & Singh, P. (2016). Efficacy of some essential oils against *Aspergillus flavus* with special reference to *Lippia alba* oil an inhibitor of fungal proliferation and aflatoxin b1 production in green gram seeds during storage. *Journal of Food Science, 81*, 928−934. Available from https://doi.org/10.1111/1750-3841.13254.

Paranagama, P. A., Abeysekera, K. H. T., Abeywickrama, K., & Nugaliyadde, L. (2003). Fungicidal and anti-aflatoxigenic effects of the essential oil of *Cymbopogon citratus* (DC.) Stapf. (lemongrass) against *Aspergillus flavus* Link. isolated from stored rice. *Letters in Applied Microbiology, 37*, 86−90. Available from https://doi.org/10.1046/j.1472-765X.2003.01351.x.

Perricone, M., Arace, E., Corbo, M. R., Sinigaglia, M., & Bevilacqu, A. (2015). Bioactivity of essential oils: A review on their interaction with food components. *Frontiers in Microbiology, 6*, 76. Available from https://doi.org/10.3389/fmicb.2015.00076.

Posadzki, P., Alotaibi, A., & Ernst, E. (2012). Adverse effects of aromatherapy: A systematic review of case reports and case series. *International Journal of Risk & Safety in Medicine, 24*, 147−161. Available from https://doi.org/10.3233/JRS-2012-0568.

Saadullah, M., Asif, M., Farid, A., Naseem, F., Rashid, S. A., Ghazanfar, S., et al. (2022). A novel distachionate from *Breynia distachia* treats inflammations by modulating COX-2 and inflammatory cytokines in rat liver tissue. *Molecules, 27*(8), 2596. Available from https://doi.org/10.3390/molecules27082596.

Shahi, S. K., Patra, M., Shukla, A. C., & Dikshit, A. (2003). Use of essential oil as botanical-pesticide against post-harvest spoilage in *Malus pumilo* fruits. *BioControl., 48*, 223−232. Available from https://doi.org/10.1023/A:1022662130614.

Shukla, A. C. (2009). Volatile oil of *Cymbopogon pendulus* as an effective fumigant pesticide for the management of storage-pests of food commodities. *National Academy Science Letters, 32*, 51−59.

Singh, P., Pandey, A. K., Sonker, N., & Tripathi, N. N. (2011). Preservation of *Buchnania lanzan* Spreng. seeds by *Ocimum canum* Sims. Essential oil. *Annals of Plant Protection, 19*, 407−410.

Sonker, N., Pandey, A. K., & Singh, P. (2015). Efficiency of *Artemisia nilagirica* (Clarke) Pamp essential oil as a mycotoxicant against postharvest mycobiota of table grapes. *Journal of the Science of Food and Agriculture, 95*, 1932−1939. Available from https://doi.org/10.1002/jsfa.6901.

Sylvestre, M., Pichette, A., Lavoie, S., Longtin, A., & Legault, J. (2007). Composition and cytotoxic activity of the leaf essential oil of *Comptonia peregrine* (L). Coulter. *Phytotherapy Research: PTR, 21*, 536−540. Available from https://doi.org/10.1002/ptr.2095.

Tripathi, N. N., & Kumar, N. (2007). *Putranjiva roxburghii* oil- A potential herbal preservative for peanuts during storage. *Journal of Stored Products Research, 43*, 435−442. Available from https://doi.org/10.1016/j.jspr.2006.11.005.

Tripathi, N. N., Asthana, A., & Dixit, S. N. (1984). Toxicity of some terpenoids against fungi infesting fruits and seeds of *Capsicum annum* L. during storage. *Phytopathology Z, 110*, 328−335. Available from https://doi.org/10.1111/j.1439-0434.1984.tb00072.x.

Trombetta, D., Castelli, F., Sarpietro, M. G., Venuti, V., Cristani, M., Daniele, C., et al. (2005). Mechanisms of antibacterial action of three monoterpenes. *Antimicrobial Agents and Chemotherapy, 49*, 2474−2478. Available from https://doi.org/10.1128/AAC.49.6.2474-2478.2005.

Ukeh, D. A., & Mordue, A. J. (2009). Plant based repellents for the control of stored product insect pests. *Biopesticides International, 5*, 1−23.

Wagacha, J. M., & Muthomi, J. W. (2008). Mycotoxin problem in Africa: Current status, implications to food safety and health and possible management strategies. *International Journal of Food Microbiology, 124*, 1−12. Available from https://doi.org/10.1016/j.ijfoodmicro.2008.01.008.

Chapter 8

Application of essential oils as antioxidant agents

Babatunde Oluwafemi Adetuyi[1], Kehinde Abraham Odelade[1], Peace Abiodun Olajide[1],
Pere-Ebi Yabrade Toloyai[2], Oluwakemi Semiloore Omowumi[1], Charles Oluwaseun Adetunji[3],
Juliana Bunmi Adetunji[4], Oluwabukola Atinuke Popoola[5], Yovwin D. Godwin[6], Olatunji Matthew Kolawole[7],
Olalekan Akinbo[8], Abel Inobeme[9], Osarenkhoe Omorefosa Osemwegie[10], Mohammed Bello Yerima[11] and
M.L. Attanda[12]

[1]Department of Natural Sciences, Faculty of Pure and Applied Sciences, Precious Cornerstone University, Ibadan, Oyo State, Nigeria, [2]Department
of Medical Biochemistry, Faculty of Basic Medical Sciences, Delta State University, Abraka, Delta State, Nigeria , [3]Applied Microbiology,
Biotechnology and Nanotechnology Laboratory, Department of Microbiology, Edo State University Uzairue, Iyamho, Edo State, Nigeria, [4]Department
of Biochemistry, Osun State University, Osogbo, Osun State, Nigeria, [5]Genetics, Genomics and Bioinformatics Department, National Biotechnology
Development Agency, Abuja, FCT, Nigeria, [6]Department of Family Medicine, Faculty of Clinical Sciences, Delta State University, Abraka, Delta
State, Nigeria, [7]Department of Microbiology, Faculty of Life Sciences Ilorin, University of Ilorin, Ilorin, Kwara State, Nigeria, [8]Centre of Excellence
in Science, Technology, and Innovation, AUDA-NEPAD, Johannesburg, Gauteng, South Africa, [9]Department of Chemistry, Edo State University
Uzairue Iyamho, Auchi, Edo State, Nigeria, [10]Department of Food Science and Microbiology, Landmark University, Omu-Aran, Kwara State, Nigeria,
[11]Department of Microbiology, Sokoto State University, Sokoto, Sokoto State, Nigeria, [12]Department of Agricultural Engineering, Bayero University
Kano, Kano, Kano State, Nigeria

Introduction

Hydrodistillation, steam refining, dry refining, or appropriate mechanical interaction without heat (for Citrus products) are the internationally accepted methods for extracting medicinal oils from plants (Rubiolo et al., 2010). In general, their thickness is less than that of water, they have a sweet aroma, and they are unstable, as depicted by prominent areas of strength. They can be mixed together by any part of a plant and then stored in the plant's secretory cells, dejections, channels, epidermic cells, or glandular trichomes (Bakkali et al., 2008). The culinary, healing, and pharmaceutical industries all make use of fragrant plants, however therapeutic medicine only addresses a small portion of the plant's synthesis (Pourmortazavi & Hajimirsadeghi, 2007).

The quantities of individual components contained in therapeutic oils might vary widely. Up to 85% of therapeutic oils can be comprised of significant portions, while the other parts are sometimes only available in trace amounts (Miguel, 2010). All else being equal, the aroma of an oil is determined by the combination of its constituent odors, and it may come as a surprise to learn that even relatively insignificant oil constituents can have significant organoleptic functions to perform (Sangwan et al., 2001).

Although this cannot be called a "medicinal ointment" when the preceding extraction processes are not feasible, there are other ways of removing the unstable component (Abdur Rauf, et al., 2021; Adetunji, Adetunji, et al., 2021; Adetunji, Ajayi, et al., 2021; Adetunji, Akram, et al., 2021; Adetunji, Arowora, et al., 2013; Adetunji, Inobeme, et al., 2021; Adetunji, Kumar, et al., 2019; Adetunji, Michael, et al., 2021; Adetunji, Michael, et al., 2021; Adetunji, Nwankwo, et al., 2021; Adetunji, Ojediran, et al., 2019; Adetunji, Olaniyan, et al., 2021; Adetunji, Olugbenga, et al., 2021; Adetunji, Osikemekha, et al., 2021; Adetunji, Palai, et al., 2021; Adetunji, Panpatte, et al., 2019; Ukhurebor & Adetunji, 2021). Oils used for medicinal purposes have a complex synthesis, typically consisting of twelve to a few hundred different ingredients. Among the many components found in rejuvenating oils, terpenes (oxygenated or not) are by far the most common, with monoterpenes and sesquiterpenes coming out on top. Some therapeutic ointments also include considerable amounts of allyl- and propenylphenols (phenylpropanoids) (Cavaleiro, 2001). For this reason, natural oil analysis is best performed using hairlike gas chromatography, which combines two different-extreme fixed phases to account for the unpredictability and extremeness of medicinal balm components. The majority of chromatographic evidence (Kováts files, direct maintenance records, relative maintenance time, maintenance time locking),

Applications of Essential Oils in the Food Industry. DOI: https://doi.org/10.1016/B978-0-323-98340-2.00004-3

or possibly horrible evidence, primarily mass spectrometry (GC-Ms), and other strategies detailed in recent survey articles (Rubiolo et al., 2010), is used to prove the authenticity of oil components. Due to their complexity, analyzing the constituents of medical ointments is arduous. According to the review article (Rubiolo et al., 2010), there are at least four common methods: relative rate overflow; interior standard standardized rate overflow; "outright" or genuine measurement of at least one section using both internal and external principles; and evaluation by an approved procedure. Numerous examples of applying each method are provided (Bicchi et al., 2008; Didunyemi et al., 2019). Terpenoids and phenylpropanoids differ in their biosynthetic origins and required metabolic precursors, making them two distinct classes of compounds. Terpenoids are produced by the shikimate pathway, while phenylpropanoids begin via the mevalonate and mevalonate-autonomous (deoxyxylulose phosphate) pathways (Dewick, 2002). A small number of authors have looked at the chemicals and catalytic systems used in the biosynthesis of terpenoids and phenylpropanoids separately (Litchenthaler, 1999; Nazir et al., 2022), and they have compiled data about the properties encoded by these substances. The development of volatiles has been aided by the hereditary design of metabolic pathways, which has yielded some encouraging results. Terpenoids and shikimic corrosive inferred volatiles have been genetically engineered in bacteria, yeasts, and plants (Adejumo et al., 2017; Adetunji & Varma, 2020; Adetunji, 2008, 2019, 2022; Adetunji, Arowora, et al., 2013; Adetunji, Egbuna, et al., 2020; Adetunji, Roli, et al., 2020; Adetunji et al., 2017; Bello et al., 2019; Egbuna et al., 2020; Gounaris, 2010; Munir et al., 2022; Olaniyan & Adetunji, 2021; Oloke & Prasad, 2020; Thangadurai et al., 2021).

Several creators have sorted out a few conclusions on the subject of transgenic microorganisms and hereditarily designed plants delivering shaky metabolites. A few creators have estimated that this strategy could be utilized to effectively produce perceivable degrees of terpenoids. However, there may not be enough terpenoid precursors to manufacture commercially viable quantities of the desired compound, making the design of certain classes of this collection of mixes particularly challenging (Asaph et al., 2006). Essential oils play a crucial role in nature, both in attracting certain insects that help disperse dust and seeds and in discouraging others. Healing balms are also important in the language of plants called allelopathy (Ibrahim et al., 2001). The revelation of a portion of these normal elements fundamental for plant endurance has likewise filled in as a springboard for the quest for similar properties for the battle against certain microbes answerable for a few overwhelming illnesses in people and different creatures. The hope is that this investigation will provide a solution to the growing resistance of pathogenic microorganisms to anti-microbials.

Most latest findings on the antibacterial and antiviral activities of rejuvenation oils have been compiled by Reichling et al. (2009). This review described the medicinal oils effective against respiratory tract pathogens such as Helicobacter pylori and Mycoplasma pneumonia, as well as natural balms effective against herpes simplex virus type 1 and 2, Newcastle disease virus, and Junin virus. The development of the more hazardous foodborne microbes can be eased back or come by reviving treatments (Burt, 2004).

Another organic property of great interest is the cancer-prevention agent movement of medicinal oils, which could protect food kinds from the damaging effects of oxidants (Maestri et al., 2006). Since several natural treatments have been shown to neutralize harmful free radicals, it's possible they can be useful in warding off some illnesses, including neurodegenerative disorders, cancer, heart disease, and weakened immune systems. Evidence is mounting that the free radicals may be responsible for the cellular damage that causes these diseases (Kamatou & Viljoen, 2010). The various cell-specific oxidative burst is one of the provocative effects; in this way, reviving salves that can rummage a couple of free progressives can likewise work as relieving specialists. Additionally, advance metal particles can decrease hydrogen peroxide to create the hydroxyl fanatic (HO°), one of the most intense oxidizing experts that can quickly respond with polyunsaturated unsaturated fats, prompting the development of peroxyl progressives (ROO•). Furthermore, hydrogen peroxide can oxidize halide particles (Cl−) to hypochlorous destructive (HOCl), which can possibly respond with amines to create chloramines, some of which are very harmful (Gomes et al., 2008). These extremists are all the more generally alluded to as reactive oxygen species (ROS). Eventually, during a particularly volatile phase, another group of libertarian revolutionaries with the initials reactive nitrogen species (RNS) will come of age. In enacted macrophages and neutrophils, nitric oxide is delivered in enormous amounts by the inducible nitric oxide synthases (iNOS) engaged with the safe framework's protective and resistant reaction. Subsequent to responding with the progressive superoxide anion, this receptive species can likewise apply its destructiveness by shaping the peroxynitrite anion (Gomes et al., 2008; James-Okoro et al., 2021).

While phagocytes produce peroxynitrite to kill attacking microorganisms, this atom, when present in huge focuses, can oxidize specific host biomolecules by means of nitration, causing cell harm that can prompt illness. In order to eliminate the invading organisms, phagocytes generate ROS and RNS, which have a significant effect on host defense. It's significant that an overabundance of supply could bring on some issues at dangerous destinations. These sensitive species serve equally crucial roles in inflammation, either as trigger components or as hailing dispatch molecules

(Gomes et al., 2008). This short survey examines the capability of rejuvenating oils as cell fortifications and quieting subject matter experts, as well as the in vitro frameworks engaged with these impacts.

Actions of antioxidant

Synergist architectures that eliminate or redirect reactive oxygen species; restricting or destroying metal particles to stop ROS from aging (carotenoids, anthocyanidins); restricting/inactivating metal particles to prevent ROS age (superoxide dismutase) and catalase (superoxide dismutase) and glut. In this way, the cell reinforcements can be classified as either essential, non-essential, or co-cancer prevention agents, depending on their mode of action. A lipid revolutionary can be supplied with a hydrogen iota rapidly by essential cell reinforcements, allowing for the rapid framing of a second, more stable, extremist. Possible oxidant antagonists counteract the initial radicals, suppress the initial molecules, or lower oxygen concentrations (without creating receptive revolutionary species). These cellular augmentations can slow down the revolutionary start-up response by eliminating potential initiators. A few techniques exist for achieving this objective, including a chelating element that catalyzes spontaneous outrageous process, absorbing UV light, searching for oxygen, deactivating high-energy radicals, and impeding mixtures such as peroxidases, NADPH oxidases, xanthine oxidases, and other oxidative proteins (Singh & Singh, 2008).

Normal objects' potential to break the cancer-prevention chain has been evaluated using both direct and indirect methods. Focusing on the effect of a tried item (like food) containing cell reinforcements on the oxidative corruption of a testing framework is essential for direct approaches. Oxidation substrates can range from individual lipids to lipid mixtures (oils), proteins, Genetic material, plasma samples, low-density lipoprotein (LDL), and even natural films. As per how effectively tests break down, either homogenous lipids or miniature heterogeneous structures (micelles and liposomes) can be utilized. This flawed method is unrelated to real oxidative debasement (Roginsky & Lissi, 2005), since it focuses on a cancer-preventative agent's potential to scavenge a few free revolutionaries.

Direct approaches might be based on either the mechanism of the non-chain process or the dynamics of lipid peroxidation (direct rivalry strategies). Both the autoxidation mode, in which the cycle advances quickly with a self-speed increase due to the aggregation of LOOH, and the controlled-chain-response dynamic model approach of lipid peroxidation can be used to examine the cell reinforcement movement. Two genuine models that are broadly utilized are thermo-labile azo-collects [2,2′-azobis(2-amidinopropane) dihydrochloride (AAPH) in a water dissolvable and 2,2′-azobis(2,4-dimethylvaleronitrile) (AMVN) in a lipid solvent] that corrupt and create dynamic free progressives where the temperature is moderate and the rate is optimal, both of which are modifiable and under control (Roginsky & Lissi, 2005). Increased incomplete oxygen strain and temperature, increased exposure to light, increased exposure to shaking and free extreme sources are all viable alternatives to the use of thermo-labile azo-compounds when hastening lipid oxidation (Batiha et al., 2021; Miguel, 2010).

Checking for delivered dienes and the Ski lifts (thiobarbituric acid responsive mixtures) test are two widely used methods for detecting lipid peroxidation (Roginsky & Lissi, 2005). Many methods exist for determining whether or not a lipid has been auto-oxidized, during lipid auto-oxidation at temperatures of 100°C or higher, unsaturated fats with a low sub-nuclear weight are transported, and their conductivity can be measured with a motorized test based on the Rancimat process, as well as the confirmation of peroxide regard, iodine regard, chromatographic assessment of capricious combinations, and formic destructive assessment (Miguel, 2010).

In the techniques for moment competition, customary cell fortifications search for the peroxyl progressive utilizing a sans reference outrageous scrounger. Malignant growth protection specialist tests for peroxyl extremists have contenders. The development of AAPH or AMVA could lead to such progressives. As indicated by the supposed oxygen revolutionary absorbance limit (ORAC) show, this is the highest quality level (Roginsky & Lissi, 2005).

2,2′-azinobis(3-ethylbenzothiazoline-6-sulfonate) (ABTS) and 2,2-diphenyl-1-picrylhydrazyl (DPPH) are two examples of redox-variable free radicals that are unrelated to the true oxidative debasement (indirect method). Another example of a detour is the conversion of ferrous (Fe^{3+}) to ferric (Fe^{2+}) in the presence of 2,4,6-trypyridyl-s-triazine. The acronym FRAP describes this method (ferric lessening cell reinforcement power). Moreover, AAPH considers the improvement of techniques that exploit the chemiluminescence of luminal inside the view of free progressives (Didunyemi et al., 2020; Roginsky & Lissi, 2005).

Notwithstanding strategies used to lay out the chain-breaking cell support development, there are valuable techniques used to assess the cutoff concerning looking for other free progressives. Numerous techniques have been created and gathered in ongoing study papers, with the benefits and weaknesses of every strategy talked about in some profundity (Niki, 2010). Several methods have also been used to evaluate the natural balms' cellular reinforcing capacity. In order to evaluate the cell reinforcement movement of natural balms, numerous authors have devised a wide variety of tests, which are the subject of a recent review (Miguel, 2010).

The antioxidant properties of essential oils

Lipid peroxidation-related tests

There are three different ways to trigger lipid oxidation, which is itself a mind-boggling response. Three types of reactions are distinguished here: (1) those involving enzymes; (2) those involving free extremists; and (3) those involving enzymes (Miguel, 2010). The primary pathway typically involves three phases: initiation, promotion, and termination. These stages envoy the beginning of quickly advancing, disastrous chain responses that create hydroperoxides and unforeseen mixes.

In the main stage, an allyl extremist structures because of the homolytic breakdown of hydrogen at the -position comparative with the unsaturated greasy chain's twofold security. These species are profoundly unsteady, brief intermediates that balance out through hydrogen reflection from related mixtures or quick oxygen reaction to frame peroxyl extremists (causing stage). Peroxyl radicals formed during engendering can oxidize lipids to produce hydroperoxides. Dienes and trienes are generated to counteract this via double-bond strengthening (electron delocalization). These blends are considered results of lipid oxidation (Miguel, 2010).

Some lipid substrates can be used to conduct a survey of lipid peroxidation. Substrate and oxidant usage, as well as intermediates or final results arrangement, might be estimated to determine the cell reinforcement movement in such frameworks (Miguel, 2010). The primary and secondary effects of lipid oxidation can be determined by a few available tests. You can enlist such a variety in the evaluation of the antioxidant action of rejuvenating ointments by reading the most up-to-date works published in logical journals about the cell reinforcement action of medicinal balms.

Determination of the extent of peroxidation using ferric thiocyanate

Peroxides (essential oxidation products) are formed during linoleic corrosive oxidation, converting Fe^{2+} to Fe^{3+}. The final particles form an absorbance peak at 500 nm with a thiocyanate complex. High retention is demonstrative of extreme linoleic-acid oxidation (Yang, Jeon, et al., 2010; Yang, Kim, et al., 2010; Yang, Yue, et al., 2010). Linalool and linalyl acetic acid were the fundamental components of this oil.

Conjugated diene assay

By examining the generated diene arrangement during the initial phase of lipid peroxidation, one can evaluate the test compounds' cell-reinforcing impact. Longitudinally, spectrophotometrically, at a recurrence of 234 nm, hydroperoxides generated from methyl linoleate were identified during oxidation at 40 °C (for shaped diene ingestion). Clear cell support exercises were exhibited by the oils of *Thymus vulgaris*, *Eugenia caryophyllus*, and *Ocimum basilicum*, similar to those of β-tocopherol, the reference utilized by the creators (Wei and Shibamoto, 2010). Oils of thyme and clove were overwhelmed by p-cymene and thymol, oil of clove by eugenol and -caryophyllene, as well as oil of basil by linalool, isoanethole, and eugenol. These parts, present in the oils at different fixations, exhibited shifting levels of defensive viability against lipid peroxidation.

Test for β-carotene bleaching

The beta-carotene blurring strategy assesses the general capacity of malignant growth counteraction specialist escalates in plant concentrates to rummage the improvement of linoleic destructive peroxide that oxidizes beta-carotene during the emulsion stage. Without the antioxidant, beta-carotene rapidly loses its orange tone as the free linoleic destructive fanatic assaults it, separating its twofold bonds.

The seven Himalayan Lauraceae species all had the potential to inhibit linoleic corrosive oxidation, but the natural balms of *Dodecadenia grandiflora*, *Lindera pulcherrima*, and *Persea gamblei* were the most effective. The oils were dominated by sesquiterpenoids. *D. grandiflora* oil was found to be predominantly composed of furanodiene and germacrene D (Joshi et al., 2010), while *L. pulcherrima* oil was found to be predominantly composed of furanodienone and curzerenone. a-Caryophyllene, a-Gurjunene, and a-Cubenene were the main components of the oils extracted from *P. gamblei*. When phenolic intensifies were present in the natural oils, these routines were risk-free.

Mighri et al. (2010) concentrated on normal *Artemisia herba-alba* treatments made in Southern Tunisia because it has the power to make cells stronger. Methods varied and one was a "blanching" test. They discovered four different kinds of oil: 1,8-cineole/camphor/thujones (+), -thujone, and thujones (+). The thujone-rich oil had the most noteworthy blockage rate (12.5%), however, it was still significantly lower than BHA's (89.2%). Mighri et al. (2010) attributed these results to a lack of chemicals other than phenolics. *Myrtus communis* var. italica L. was analyzed synthetically to determine the efficacy of its natural ointments extracted from its leaf, stem, and flower. When compared to BHT and BHA, the antioxidant activities of leaf and flower oils were superior. Low degrees of phenolic synthetics or their nonattendance were faulted by the authors for the lackluster outcomes, despite the fact that the concentrates also focused on in the current investigation showed far greater results (Wannes et al., 2010). The oxygenated sesquiterpenoids (47.4% of the total) predominated over other components. It was possible that the rejuvenating oils could slow beta-demise carotene (Ahmadi et al., 2010).

Oils extracted from a variety of plants native to Iran were also evaluated for their potential to inhibit cancerous cell growth. In both instances, the physical activity had no effect. Salvia eremophila Boiss. oil slowed things down at a rate of around 33% compared to BHT, while methanolic concentrates of a related plant introduced activities that were nearly equivalent to those of BHT. The lack of phenolic intensifies in the oils was blamed by the creators for the weak action. The main chemical constituents of the oil were pinene, borneol, camphene, and trans-caryophyllene (Ebrahimabadi et al., 2010). Non-phenolic chemicals such as a-bisabolene, apiole, a-pinene, and dill apiole were also blamed by the developers of vatke oil from *Psammogeton canescens* (DC) for the ineffectiveness of the oil (Gholivand et al., 2010). Origanum acutidens (Hand.-Mazz) Ietswaart, a member of the mint family, was studied for its potential as a cancer-preventative, antibacterial, and antispasmodic (Goze et al., 2010). The active ingredient in the balm was carvacrol. While this oil did show some cancer-preventative activity, it was significantly weaker than the effect of the reference material used to draw the conclusions (BHT). Despite the high carvacrol concentration in the oil, the blockage rate only reached 65%, compared to 100% for BHT.

Thiobarbituric acid reactive substances

Malondialdehyde, or MDA, can be measured using this method, a pink chromophore with thiobarbituric corrosiveness that is formed following lipid hydroperoxide disintegration (unrequired consequences of oxidation) (Thiobarbituric acid; TBA). TBA and malondialdehyde gather in an acidic environment as a result of this colorful compound's ability to absorb light at 532 nm. TBA can react with a wide variety of mixtures to generate a chromophore, including anthocyanins, amino acids, nucleic acids, and 4-hydroxy-alkenals. Amino acids, nucleic acids, 2-alkenals, 2,4-alkadienals, and more are only some of the other mixtures out there.

Using liver homogenate as a lipid substrate, the natural ointment of *Ageratum conyzoides* L., which is mostly made up of precocene and caryophyllene, demonstrated considerable potential for inhibiting lipid peroxidation (Adewale et al., 2022; Patil et al., 2010). In addition, the authors examined methanolic concentrates of related plants, which had activities far weaker than those of the natural balms. The authors also speculate that the natural oils' antiaflatoxigenic action may be partially attributable to their cell reinforcement action (Patil et al., 2010). The counter-malignant growth movement of the restoring salves of five taste plants highlighted in the Mediterranean eating routine was assessed utilizing the thiobarbituric destructive responsive species (Ski lifts) measure with egg yolk as substrate (Viuda-Martos et al., 2010). *T. vulgaris* L. oil had the best results, closely followed by BHT, out of the multitude of normal emollients tried (*T. vulgaris*, *E. caryophyllus*, *Origanum vulgare*, *Salvia officinalis*, and *Rosmarinus officinalis*). The purpose of Dandlen et al. (2010) was to zero in on the anti-cancer properties of the rejuvenating oils of the *Portuguese thymus*. Using a technique identical to that published in Viuda-Martos et al. (2010), the authors found that oils of *Thymus zygis* and *Thymus sylvestris* obtained from various locations of Portugal exhibited outstanding cell reinforcement workouts, surprisingly better than that of BHT.

Thymus proximus and *Thymus marschallianus* Will. When compared to BHT, Chinese Serg. oils were less effective at preventing lecithin peroxidation when the Ski lifts method was used. *T. proximus* oil exhibited the highest activity of all the oils tested. This oil was predominated by p-cymene, a-terpinene, and thymol, while the other oil was predominated by a-terpinene and thymol alone. Based on these findings, the authors zeroed down on the role that p-cymene, in addition to thymol and -terpinene, plays in the cell reinforcement action of *T. proximus* (Jia et al., 2010).

D. grandiflora and *L. pulcherrima* were two of seven Himalayan Lauraceae species tried; however, just their regular balms exhibited the possibility to restrain lipid peroxidation involving liver homogenate as a lipidic substrate. At that period, it was found that both of these oils and *P. gamblei* were strong inhibitors of linoleic oxidation, as estimated by the -carotene passing on test (Joshi et al., 2010). These routines occurred regardless of whether or not the pharmaceutical ointments contained phenolic intensifies.

Normal demulcents from *Capparis spinosa* are used in cell-reinforcing exercises. Furthermore, *Crithmum maritimum* was assessed utilizing different techniques, with Ski lifts system uncovering that oils presented at 1 μg/L (the most noteworthy obsession attempted) were comparable to BHT in their ability to prevent lipid peroxidation, but less effective than BHA. The act was viewed as subservient by them as well. Although there was little difference in how each sample worked, there were significant differences in how each oil was synthesized. Methyl isothiocyanate made up the vast majority of the oil from *C. spinosa*, while sabinene and limonene were the essential parts of the oil from *C. maritimum* (Kulisic-Bilusic et al., 2010).

A zingiberaceous plant, *Amomum tsao-ko* Crevost and Lemairé is commonly known as "Caoguo" or "Tsao-ko" in the southwestern part of China. Since some studies have linked cytotoxicity with cancer prevention agent activity and the revitalizing ointment the authors focused on had strong anticancer action, they evaluated the cell reinforcing action of their examples using various methods, including the Ski lifts methodology (Yang, Jeon, et al., 2010; Yang, Kim, et al., 2010; Yang, Yue, et al., 2010). Liposomes rich in lecithin served as the lipidic substrate, and ferrous sulfate was used to induce peroxidation. The low phenolic content of the oils meant that they detected a weak cancer prevention agent movement, which was in keeping with the intentions of the designers. 1,8-Cineole was the primary component. When comparing the results of Kulisic-Bilusic et al. (2010), which discovered robust and moderate growth with oils, with those of Kulisic-Bilusic et al. (2010), it is clear that the lack of phenolic compounds associated with weak cell reinforcement action is not as expected. It's important to stress that the varied lipidic substrate used in the two works could also account for the difference in activities, even if the oils were synthesized differently.

Oils from the aeronautical parts and seeds of the *Foeniculum vulgare* plant have distinctive synthetic creations, yet with respect to the cell reinforcement action, a decrease in development was discovered for larger concentrations of revitalizing ointments, as reported by Miguel et al. (2010), indicating that revitalizing oils with a high concentration of trans-anethole had a high level of oxidant activity.

Restoring oils separated from *Ocimum sanctum* L. leaves have been considered for their anti-hyperlipidemic and cancer-preventing properties in mice fed a high-cholesterol diet by Suanarunsawat et al. (2010). Renewal creams primarily contained eugenol and methyl eugenol. By reducing the increased amounts of ski lifts in the liver or cardiac tissues, the oils may provide protection against stress-induced oxidation. Wei and Shibamoto (2010) discovered that the mobility of oils from *T. vulgaris* and *E. caryophyllus* was the most stable, similar to that of α-tocopherol. However, the authors of a related study also found that thyme oil's efficacy decreased with increasing concentrations. Basil oil, which was judged to be a good cell reinforcer when the action was measured by the produced diene measure, did not fare well in this context, where it was only given a modest rating for its activity. Those discoveries could show that basil can forestall the vital oxidation of lipids.

Analysis of aldehydes and carboxylic acids

According to Moon and Shibamoto (2009), this metric is useful for contrasting the effects of cancer-preventative medications with the slow oxidation characteristics of food varieties. Using this method, the authors demonstrated that the oils of several different plants, counting heat and undiluted oxygen accelerate the oxidation of hexanal to hexanoic acid, despite the possibility that the plants would impede this process (Wei & Shibamoto, 2010). Since anethole was the primary component of *Illicium verum* oil, these results may indicate that phenolic mixes are not crucial for preventing hexanal oxidation.

Formic acid measurement

When lipids are heated to temperatures of 100°C or higher, they undergo auto-oxidation, releasing low-subatomic-weight unsaturated fatty acids (formic corrosive). This process is mechanically tested for conductivity using the Rancimat method. Both the Ski lifts method (Viuda-Martos et al., 2010) and the Rancimat method (Dandlen et al., 2010), where the lipid substrate was grease, were utilized to assess the cell-building up action of restorative oils of five zing plants utilized in the Mediterranean eating regimen. The framework was heated to 120°C, and 20 cubic feet per hour of air was continually blown into the blend. Unpredictable carboxylic acid separation characterized the end of the enlistment period as a sudden rise in water conductivity. The primary component of the *O. vulgare* oil was carvacrol. As opposed to the case described in detail for the Ski lifts, in which thyme proved to have the best activity, here we compare two oils, one of which is wealthy in phenolic parts (oregano), and the other of which isn't (thyme). This is a real-world example of how different methods of evaluating cellular reinforcement could generate contradictory findings, highlighting the need to employ multiple methods when evaluating the antioxidant potential of testing.

The normal analgesic isolated by hydrodistillation was contrasted with the oil separated by supercritical fluid from the flight parts of *T. vulgaris* L. Indeed, even while p-cymene and thymol overwhelmed in one situation and dominated the other to a lesser extent in the other, their relative abundance was determined by the extraction method used. It was speculated that the presence of thymoquinone in the unstable oil was what allowed for the supercritical liquid extraction to produce the unusual differentiation. This component's presence may account for the highest cell reinforcement movement seen in the Rancimat test, in which sunflower oil served as the lipidic substrate (Grosso et al., 2010).

Free radical scavenging ability

There are two groups of methods for assessing free extremist seeking ability: one based on hydrogen particle motion reaction and the other on single electron motion response.

Test for 2,2-diphenyl-1-picrylhydrazyl

The DPPH test, which is used the most, is simple but very sensitive. The tremendous popularity of DPPH can be attributed to its reliability. For a maximum of 517 nm (purple), this extreme reveals significant strengths in a variety of key areas. The color purple is abandoned in favor of yellow when the presence of cancer-prevention medicines is detected. Therefore, a UV-Vis spectrophotometer is the only piece of equipment needed for the test.

Regardless, it was imagined that the progressive DPPH would be diminished to the practically identical hydrazine when presented to the hydrogen-giving synthetic compounds. However, subsequent research has revealed that what actually occurs is primarily a brief electron transfer from the example to the DPPH extremist. The DPPH extremist's progressive hydrogen reflection is unimportant since it happens gradually and relies upon the hydrogen-bond open-minded solvency. The hydrogen-abstracting response is slow in light of the fact that methanol and ethanol, the solvents frequently utilized for disease preventive specialist limit estimations, are both lenient toward hydrogen bonds.

For evaluating the cell reinforcement activities of natural ointments, the DPPH method is frequently used because of its simplicity and reactivity. Our recent literature search turned only a handful of examples (Mothana et al., 2010; Saei-Dehkordi et al., 2010). In some cases, the decision to only employ this strategy can be attributed to the poor cell reinforcement action detected, driving the creators to consider it superfluous to seek after different strategies. *Commiphora ornifolia* and *Commiphora parvifolia* oil have been shown to have modest cancer prevention agent exercises, even at high fixations, as reported by Mothana et al. (2010). Based on these findings and the oils' synthetic structure, in which phenolic components were absent, the designers may have deemed it unnecessary to employ any other methods. Oil of leaves of *Olea europaea* is another model; like BHT (Haloui et al., 2010), it has minimal activity because its primary components are not phenolic chemicals but rather -pinene and 2,6-dimethyloctane. However, other creators believe it is sufficient to test using just one method (DPPH) when great cell reinforcement action values are identified (Haloui et al., 2010). Natural oils from *Eucalyptus camaldulensis* were discovered to have high cell reinforcement exercises in several parts of wild Sardinia (Italia) (Barra et al., 2010). Evident cancer-preventive agent motion, most likely attributable to carvacrol, was observed in Majorana hortensis L. medicinal balm, which showed good synergism with other components (Martino et al., 2010). *Citrus* spp. and *Citrus sinensis* additionally had extensive malignant growth anticipation specialist exercises notwithstanding the way that their normal salves contrasted in the arrangement. Limonene made up the bulk of *C. sinensis* oil, while in Citrus maxima oil, other components such as 3,3-dimethyl-1-hexene (Singh, Kaur, et al., 2010; Singh, Shukla, et al., 2010) were easily distinguishable in significant concentrations. Myristicin and trans-anethole, two main components of *Heracleum pastinacifolium* and *Heracleum persicum*, were shown to have moderate cancer preventive agent movement (Firuzi et al., 2010). The antiradical activities of three *Lippia graveolens* Kunth. oils from Mexico were evaluated. These oils had different material compositions and microcapsules. Microencapsulation, the authors reasoned, increased the counter-extreme activity by a factor of eight (Sánchez-Arana et al., 2010). Oils removed from *O. vulgare*, which were gathered in a few parts of Tunisia, each exhibited distinctly but, in every case, impressive abilities for scavenging DPPH radicals. The number of phenolic compounds (thymol) in the oils was shown to be correlated with these kinds of results (Mechergui et al., 2010). In spite of the way that a few creators accept the *Retama raetam* bloom oils filled in Tunisia presented extraordinary cell support movement, as assessed through the DPPH procedure, the primary strategy that was utilized by the creators, where the IC50 regard was fortyfold better than that of the reference BHT. Mechergui et al. (2010) discovered that the Tunisian Ridolfia segetum root oil had IC50 values comparable to those of BHT and had a useful effect. High levels of the phenyl-propanoids dillapiole and myristicin may account for their antioxidant effects (Grosso et al., 2010). Although Mimika-Dukić et al. (2010) only used the DPPH method, its developers also used it in conjunction with attention (meager layer chromatography) to provide certainty as to which mixes are responsible for anticancer agent movement. The blends liable for these exercises were found utilizing two tests in view of a similar rule yet utilizing various cycles (spectrophotometer and careful attention). Experts have pinpointed 1,8-cineole and methyl eugenol as the primary components in *M. communis'* natural oils that provide some protection against disease. Brazil's *Aniba panurensis*, *Aniba rosaeodora*, and *Licaria martiniana* all followed a comparative methodology created by a few creators (Alcântara et al., 2010). Most of the IC50 values were greater than 1000 g/mL, which is orders of magnitude better than quercetin, the standard used in the study's construction. It was hypothesized that the absence or extremely low concentrations of a mixture responsible for cell-reinforcing action—caryophyllene coupled with phenolic chemicals—was to blame for the lack of activity. As previously stated, these results may be sufficient for the makers to decide not to conduct additional research.

In 2010, Saleh et al. undertook a comprehensive study of 248 different ointments and balms used for various types of healing, measuring their cellular-supporting characteristics using two different techniques: the spectrophotometer method for DPPH and the DPPH/loving-kindness approach. Only 17 species, mostly members of the Lamiaceae family, exhibited effective cell reinforcement action. Phenols, non-phenols, oxygenated, and non-oxygenated synthetics were all essential for the powerful combinations recognized by the DPPH/special attention strategy, while certain researchers decide to utilize just a single technique for evaluating cell support development, by far most likely to use no less than two. The blends could incorporate a trial of the capacity to look with the expectation of complimentary radicals as well as those testing the adequacy of hindering lipidic peroxidation utilizing the systems uncovered above and others excluded from this text.

A portion of the researchers who have assessed the capacity of normal demulcents to forestall lipid peroxidation utilizing the Ski lifts strategy had also evaluated the limit with respect to searching free extremists, utilizing the DPPH measure (Viuda-Martos et al., 2010). It's not always the case that the highest value in the Ski lifts metric corresponds to the highest value in the ability to snoop on DPPH revolutionaries. These were discovered by Patil et al. (2010). *A. conyzoides* and *A. tsao ko* ointments were shown to be more effective at preventing lipid peroxidation than they were at locating rogue revolutionaries. The DPPH study found that the methanolic extracts of *A. conyzoides* were more effective than the rejuvenating oil at preventing lipid oxidation while having a lower limit. *O. vulgare* oil fared highest in the Rancimat test while performing worse than the reference BHT, which was used by other similar authors. Results contrasted when the DPPH strategy was utilized, but both the non-polar part and the restorative salve of *Hymenocrater longiflorus* displayed an astonishing ability to forestall the blurring of alpha carotene. Here, the medical ointment was the worst offender when it came to searching for these free fanatics (Ahmadi et al., 2010). Methanolic extracts were consistently more effective cancer preventatives than natural balms, regardless of the *M. communis* var. italica L. plant component used or the cell reinforcement procedure employed (Wannes et al., 2010). In this particular model, both methods yielded consistent results, which were consistent with earlier predictions. As was just reported, some experts evaluate the effectiveness of natural cancer preventatives based on their ability to locate free revolutionaries. The major standards covered may be similar, and there are a few free revolutionaries that can be used.

2,2-Azinobis(3-ethylbenzo-thiazoline)-6-sulfonic acid or Trolox equivalent antioxidant capacity

This strategy is built on the reduction of the super blue-green cation ABTS + • with the reduction of this cation being estimated as the level of absorption hindrance at 734 nm. Results on the Trolox equivalent antioxidant capacity limit for a comparable cancer prevention agent to Trolox (ABTS) are given. This is done by comparing the absorbance of the response combination of ABTS and a cell reinforcement to the absorbance of the Trolox standard. Some authors, as was shown for the DPPH method, have only used the ABTS method to evaluate the cell-reinforcing action of medicinal oils (Ennajar et al., 2010). Organ, season, and drying strategy all assume a part in the structure and ABTS-looking through capacity of Juniperus phoenicea, as examined by Ennajar et al. (2010) in detail. They located the most and least significant workouts at the same location on the plant (berries). Stove drying was the best method for extracting potent oils from leaves and berries, followed by sun drying for medical purposes and shade drying for other applications. A comparison profile showed that the oils' main components (-pinene, -3-carene, -terpineol, -myrcene, and so on) were present in varying concentrations, suggesting a link between substance formation and variation in these characteristics. Even the dog species, *Rosa canina* L., had its limits when it came to scavenging ABTS; however, the collecting location ultimately determined whether or not this behavior occurred, furthermore, this affiliation had all the earmarks of being attached to the presence of high vitispirane levels (Ghazghazi et al., 2010). Even so, in most cases, other assays are used in addition to ABTS to determine the upper bounds of free extreme scavenging. Among Satureja species, only *Satureja intricata* Lange was capable of scavenging DPPH and ABTS (Jordan et al., 2010), while *Satureja obovata* was not. When evaluated for free radical scavenging, the therapeutic oils of *A. herba-alba*, which we just saw had a low capability for blocking lipid peroxidation (Mighri et al., 2010), performed no better. As was seen above, the best oil against lipid peroxidation was lavender (*Lavandula angustifolia*) oil (Yang et al., 2010a,b,c). Although lavender clinical ointment had the highest DPPH and ABTS looking through action, the oils of *Mentha piperita* and *Boswellia carteri* were the most potent. Even though linalool and linalyl acetic acid derivatives made up the bulk of this rejuvenating oil, limonene demonstrated comparable activity to that of *L. angustifolia* in the DPPH technique. There was no optimal icorrespondence between limonene and the other components of the Citrus limon oil.

Ferric reducing/antioxidant power assay

When an acidic cell reinforcement reduces a Fe^{3+}-TPTZ (2,4,6-tripyridyl-s-triazine) complex to the Fe^{2+} structure, it produces a bright blue color with maximum retention at 593 nm. The effect of cell reinforcement can be measured by keeping an eye on the spectrophotometer as Fe^{2+}-TPTZ complexes form (Moon et al., 2009). *Pistacia atlantica* leaf oils from various regions in Algeria had a weak capacity to scavenge DPPH free radicals during cell support development, especially when compared to the oils used as benchmarks (BHT and BHA). Using the FRAP method, researchers found that comparable oils had a greater cell reinforcement limit in comparison to the cell reinforcement of the reference ascorbic corrosive (Gourine, Yousfi, Bombarda, Nadjemi, & Gaydou, 2010). The authors of a related study compiled a list of activities that may be done at various venues for the same kind of get-together. El-Ghorab et al. (2010) compared the cell-reinforcing effects of oils from *Cuminum cyminum* and *Zingiber officinale* and found that FRAP and DPPH radical-scavenging were nearly equivalent when used as their primary metrics. The best change was initiated by *C. cyminum*.

Clove (*E. caryophyllus*) had the most noteworthy ferric diminishing limit concerning Trolox fixations utilizing the DPPH technique, while *T. vulgaris* had the highest using the Ski lifts method (Viuda-Martos et al., 2010). With the exception of the medical ointment made from *T. vulgaris*, the FRAP values and total phenolic items in the natural balms examined showed a significant direct association, as has been seen by comparable authors. It has previously been demonstrated that *A. tsao-ko* oil has a low phenolic content and has poor cell support activity when measured using the Ski lifts method, which contributed significantly to the oils' weak capacity for seeking DPPH free extremist (Haloui et al., 2010). Although both samples have a significant eugenol concentration, the cell reinforcement movement of clove oil was higher in the DPPH method. The researchers postulate that the competitiveness between the many elements present in this region in very small amounts is to blame for the reduced kinetic energy observed here. The natural oil's reducing limit was also greater than that of the erratic aglycone fraction. In any event, eugenol had a significant impact on the DPPH approach but no effect in the reference study. The contrasting extremes of the FRAP and DPPH strategies provide insight into this difference. When compared to ethyl liquor, which is frequently utilized in the DPPH strategy, "water-like" solvents cause a recognizable diminishing in the hydrogen particle's giving limit (Politeo et al., 2010). *P. atlantica*, a tree whose leaves are used in medicine, has been found to have occasional subjective and quantitative differences between its male and female leaf medicinal ointments, as demonstrated by differences in piece and cell reinforcement workouts. The a-3-carene in female leaf medicinal oil was the main component. Some variation indicated that by September, the oils had developed most of their distinctive main components. For male oils, the best time to use them is June, while for female oils, the best time to use them is over the lengthy stretches of September and October to boost cell reinforcement against DPPH extremists. We found that June was the best month for male oil and August was the best month for female oil in terms of its reducing power. In the FRAP test, the female oil's cell reinforcement activity was significantly higher than that of ascorbic corrosive (Gourine, Yousfi, Bombarda, Nadjemi, Stocker, et al., 2010).

Reducing power

Another method for gauging an anti-cancer agent's efficacy involves seeing the test setup's yellow color shift to various colors of green and blue when Fe^{3+} is reduced to Fe^{2+}. As the number of professionals in the room decreases, the Fe^{3+}/ferricyanide combination converts to a ferrous structure, which can be seen at a wavelength of 700 nm. Increasing absorbance at 700 nm denotes a rise in reducing power (Joshi et al., 2010). Some authors have tested the anti-aging effects of revitalizing oils from Himalayan Lauraceae species (Joshi et al., 2010), *M. communis* var. italica (Wannes et al., 2010), *H. longiflorus* (Ahmadi et al., 2010), *Origanum onites* L. (Ozkan et al., 2010), and *P. canescens*

(Gholivand et al., 2010). The -carotene fading and DPPH assays predicted weak activity on the part of the cancer-preventative agents involved here. While the authors attributed the change to the natural ointments' low phenolic intensifies, analogous examples' reducing force is like that of the reference ascorbic acid, especially at more noteworthy obsessions (Gholivand et al., 2010).

Chelating activity

Chelation of change metals may facilitate antioxidant activity in part. Lipid peroxidation can be accelerated by moving metal particles shortening the half-life of the initiating species, thereby accelerating the decay of lipid hydroperoxides into different parts that can extract hydrogen, thus propagating the chain reaction of lipid peroxidation (Viuda-Martos et al., 2010).

Ferrozine, which can remove Fe^{2+} quantitatively from buildings, is one method commonly used to guarantee chelating effect. The reddish hue of the complex ferrozine- Fe^{2+} fades as the puzzling development is disrupted when seen by other chelating specialists. Calculating the rate of diversity loss provides a basis for evaluating the chelating action of the concomitant chelator (Wannes et al., 2010).

Herbal creams containing *M. communis*, *T. marschallianus*, and *T. proximus* rarely showed chelating activity (Jia et al., 2010). Oils extracted from myrtle flowers were the only exception. Although eugenol and methyl eugenol can be found in the natural oil of cloves, the creators of this movement clarified that these components make up just a trace percentage of the therapeutic oils found in the leaves and stems. To properly encase chelated Fe^{2+}, these dihydroxylated combinations are crucial. With regards to the natural balms of different plants remembered for a Mediterranean eating routine, all of them could chelate Fe^{2+} and did as such in an extremely compliant manner (Viuda-Martos et al., 2010). *R. officinalis* L. and Salvia officinalis restorative balms showed the best chelating skills for Fe^{2+} across totally tried obsessions. When compared to ascorbic acid and BHT, all of the oils tested performed better as Fe^{2+} chelators (Viuda-Martos et al., 2010).

Hydroxyl radical scavenging

Hydroxyl, the most reactive oxygen radical, causes extensive damage to the helper biomolecules. The ability to create hydroxyl revolutionaries can be assessed in a number of ways. The deoxyribose assay is an example of such a method. Within the context of ascorbic acid structures, ferric chloride ($FeCl_3$) and ethylenediamine tetraacetic corrosive (EDTA) form a complex that is used in this method; ascorbic acid in its oxidized state is the latter. Fe^{3+}-EDTA and HO• are produced by the expansion of hydrogen peroxide (H_2O_2). The deoxyribose is attacked by hydroxyl extremists that aren't neutralized by any of the other ingredients in the mixture, and it breaks up into fragments. A portion of these parts can respond with thiobarbituric acid in the wake of warming and in an acidic pH, delivering a pink variety recognizable utilizing spectrophotometry. Various strategies exist that utilized non-deoxyribose parts, like benzoic acid or safranine, all things considered. A few creators have assessed the hydroxyl progressive ability to search for restoring balms utilizing different strategies (Tomić et al., 2010).

Utilizing the safranine strategy, the creators of a review contrasting the remedial demulcents of *T. marschallianus* and *T. proximus* found that the last option oil had a more prominent potential to look for hydroxyl fanatics. Thymol, p-cymene, and -terpinene made up the heft of the two oils' structure, and their belongings were hostile to each other (Jia et al., 2010).

The main components of *A. conyzoides* are the hydroxyl extremist, precocene I, and precocene II, all phenolic chromenes. The skill of this plant has been studied using two methods (deoxyribose and benzoic corrosive) with varying results. The medicinal oil introduced a preferable activity in the benzoic corrosive hydroxylation strategy, as opposed to the deoxyribose method. The researchers solved this mystery by comparing the substrates deoxyribose and benzoic acid to the general reactivity of hydroxyl towards these compounds. Among the Thymus species gathered in Portugal, oils from *Thymus camphoratus* Hoffmanns, and connection were substantially more viable in rummaging hydroxyl extremism than the excess samples. According to some authors (Dandlen et al., 2010), the lack of significant exercises in the oils where thymol or carvacrol predominated shows that these phenolic parts are not determinant in that frame of mind for looking through hydroxyl progressives.

Oils extricated from both youthful and old leaves of *Ageratum scoparia* also uncovered a significant hydroxyl progressive scrounging movement. In contrast to oils, however, the p-cymene and -myrcene, the two rule ingredients, showed a much smaller rummaging movement. Similar authors (Singh, Kaur, et al., 2010; Singh, Shukla, et al., 2010) evaluated the cell reinforcement action of the examples using the DPPH strategy and announced less action of the principal constituents of the oils than the medicinal balms, concluding that the oils from matured leaves were more effective as disease avoidance specialists.

Athamanta hungarica and *Athamanta haynaldii* both have myristicin as a primary component in their oils, making their mature products a rich source of natural ointments (Tomić et al., 2010). Researchers compared the two oils' capacities for identifying hydroxyl revolutionaries and found that they were similar. *Athamanta hungarica* demonstrated the highest antioxidant activity when tested using the DPPH method (Tomi et al., 2010), even though it was concentrated at a lower level than the references. Mannitol was utilized as a reference, and the average rate of detection was close to 2000 mg/mL. Only those with a more concentrated emphasis on the introduction of rates greater than 50% were considered (Miguel et al., 2010).

Superoxide anion scavenging activity

Xanthine oxidase, a dehydrogenase catalyst, converts xanthine or hypoxanthine into uric acid by reducing nicotinamide adenine dinucleotide (NAD +) to NADH. The protein, on the other hand, depletes oxygen rather than NAD + when subjected to high pressure, as the dehydrogenase chemical transforms into an oxidase. In this manner, dioxygen is still broken down into superoxide anion and hydrogen peroxide. Here, phenazine methosulfate reacts with NADH and dioxygen in the presence of oxygen to produce superoxide anion. Formazan formation from nitro-blue tetrazolium (NBT) reduced by superoxide anion is followed up spectrophotometrically in both cases.

The enzyme hypoxanthine/xanthine oxidase was used to test the superoxide anion-searching activity of oils extracted from *P. thymus* (Dandlen et al., 2010). The only two of the 28 natural balms tested that showed superoxide anion-seeking activity >50% were *T. zygis* and Thymus capitellatus, both of which were gathered in different parts of Portugal. While thymol and carvacrol predominated in oils extracted from *T. sylvestris*, borneol and 1,8-cineole predominated in oils extracted from Thymus capitellatus. Evidence from this study suggests that similar to what is seen for hydroxyl radical seeking, the superoxide anion can be scavenged by a variety of compounds, not just phenolic ones (Dandlen et al., 2010).

Anti-inflammatory properties of essential oils

In order to get rid of germs and other non-self cells, as well as host cells that have been harmed or have died, irritation is a common defense response triggered by tissue injury or contamination (Stevenson & Hurst, 2007). Concurrently, the activity of some proteins and arachidonic corrosive digestion are also generally recognized. The inflammatory process also involves the release of cell bond particles such intercellular grab atoms and vascular cell attachment particles (Gomes et al., 2008). There is evidence that some therapeutic ointments have a moderating action, not to mention the fact that certain revitalizing balms can help set rebels free from prison. For example, chamomile revitalizing ointment has been used for decades to reduce the discomfort of skin inflammation, dermatitis, and other forms of articular disturbance (Kamatou & Viljoen, 2010). In a number of reported instances, natural oils, and other plant materials were utilized in various combinations as mitigation measures (Darsham & Doreswamug, 2004). Mouse paw edema induced by carrageenan is a common method used to evaluate the analgesic effects of various bioactive combinations, including plant concentrates and therapeutic ointments (Hajhashemi et al., 2003; Juhás et al., 2008). Few details are provided regarding the system of this method, but it is assumed that it allows screening the calming of tests. Cancer prevention agent activities, as well as correlations with flagging fountains like cytokines and administrative record variables, and the outflow of supportive of fiery qualities, may all contribute to the ameliorative effect of medicinal ointments.

Effects on the arachidonic metabolism

In response to inflammatory stimuli, cells produce the polyunsaturated fatty acid arachidonic acid via the enzyme phospholipase A2. Eicosanoids, including prostaglandins (PGs), leukotrienes (LTs), and thromboxane A2 in platelets, are generated by the cyclooxygenase (COX) and lipoxygenase (LOX) pathways, both of which utilize the same unsaturated fat (Gomes et al., 2008). There are two known types of COX: COX-1 and COX-2. In most normal mammalian tissues, COX-2 protein is very weakly communicated, whereas COX-1, a constitutive substance, increases in expression in response to exposure to natural, chemical, and physical stresses such as ultraviolet light, dioxin, and LPS (lipopolysaccharide) attack (Murakami & Ohigashi, 2007).

Leukotrienes constrict the vein smooth muscles, promote vascular permeability, and intervene in proinflammatory and hypersensitive reactions (González et al., 2003), whereas prostaglandins, especially prostaglandin E2 (PGE2), amplify the aggravation instrument. Revitalizing oils made from South African Salvia, anise star, bergamot, cinnamon leaf, eucalyptus, juniper berries, lavender, and some other 5-lipoxygenase inhibitors. These invigorating balms got their boost from 1,8-cineole, a-pinene, and a-caryophyllene, according to the authors (Kamatou et al., 2006). Active components that inhibited 5-lipoxygenase were discovered in four native South African Helichrysum ointments (Lourens et al., 2004). Both chamazulene and -bisabolol, two components of chamomile essential oil, have been shown to be potent 5-lipoxygenase inhibitors (Kamatou et al., 2010), meaning that they can reduce the intensity of exercise partially due to the inhibition of leukotriene union. Oils extracted from the rhizomes contained primarily the sesquiterpenes alpha-selinene, beta-selinene, and beta-panasinsen, while oils extracted from the leaves contained primarily alpha-pinene, beta-pinene, and sabinene (Syamsir, 2009). The oil of the Torreya nucifera tree was a COX-2-specific inhibitor that greatly impacted PGE2 production (Yoon et al., 2009a,b,c). The terpene oxide 1,8-cineole is utilized in numerous traditional medicines. It has been demonstrated to reduce the synthesis of the inflammatory chemicals leukotriene B4 (LTB4) and prostaglandin E2 (PGE2), which are both byproducts of the digestion of the fatty acid arachidonic acid (Juergens et al., 1998).

Modifications to cytokine production

Pro-inflammatory cytokines play a significant role in the underlying issues that lead to inflammation. Pro-inflammatory cytokines are cytokines that encourage inflammation. The majority of this beneficial, fiery cytokine is released by cells with a monocyte/macrophage lineage, although it is also released by T lymphocytes, neutrophils, and pole cells. Numerous cell types in the body, including monocytes, macrophages, fibroblasts, and endothelial cells, generate IL-1β (Dung et al., 2009). A possible trigger for the generation of pro-inflammatory cytokines is lipopolysaccharide (LPS), an endotoxin found on the cell walls of Gram-negative bacteria (Raetz & Whitfield, 2002). In Gram-positive bacteria, a class of amphiphilic particles called lipoteichoic acid (LTA) are attached to the outermost essence of the cytoplasmic layer and has been shown to work in tandem with peptidoglycan (PG) to increase cytokine levels (Schröder et al., 2003). This increases the number of mononuclear phagocytes that are recruited to the site of infection.

Crude 264.7 cells, a mouse macrophage-like cell line, were treated with the natural ointment of Cheistocalyx operculatus; the substance creation of which is still up in the air, and their release of inflammatory cytokines was significantly suppressed in response to lipopolysaccharide (LPS) (Dung et al., 2009). Terpinen-4-ol, a significant component of tea tree oil (Melaleuca alternifolia), has been demonstrated by a select group of researchers to inhibit LPS-induced IL-8, IL-10, and PGE2 production in human blood monocytes in vitro (Hart et al., 2000). Caldefie-Chézet et al. (2006) found that Phaseolus vulgaris lectin phytohemagglutinin (PHA) stimulated human peripheral blood mononuclear cells discovered that the rejuvenating oil of that species stimulated the release of mitigating IL-4 and IL-10 cytokines and suppressed IL-2. The revitalizing ointment made from *Cinnamomum osmophloeum* leaves contains high concentrations of 1,8-cineole, santoline, spathulenol, and caryophyllene oxide, and these compounds have been proven to decrease IL-1 and IL-6 production while having no effect on TNF- (Chao et al., 2005).

The cytokine production of LPS-activated human blood monocytes, essential macrophages, and human THP-1 monocytes was inhibited by cinnamonaldehyde (Chao et al., 2008). A therapeutic ointment containing *Cordia verbenacea* dramatically decreased TNF-α levels (Passos et al., 2007). The increase in TNF-α and IL-1β levels in the mouse paw subcutaneous tissue after LPS treatment was recently shown to be greatly suppressed by only -humelene from *C. verbenacea* medicinal ointment, as shown in a recent study (Medeiros et al., 2007) or in carrageenan-infused rats (Fernandes et al., 2007). Researchers examined the effects of a rejuvenating ointment containing *Cryptomeria japonica* on LPS-treated Crude 264.7 cells using a chemical immunoassay, and they found that the oil significantly inhibited IL-1β, IL-6, and TNF-α (Yoon et al., 2009a,b,c). Ingredients like kaurene, elemol, -eudesmol, and sabinene are what make this rejuvenating balm so effective.

The 7 different cells all contained either -thujone, -thujone, camphor, or caryophyllene (Yoon et al., 2009a,b,c). *Syzygium aromaticum* (clove) contains eugenol, a rejuvenating ingredient that has been credited with similar effects by other writers. The fundamental components of essential oils from plants including lemongrass, geranium, and spearmint have been demonstrated to block neutrophil adhesion reactions triggered by tumor necrosis factor-alpha (Abe et al., 2003). Natural remedies or their key ingredients, according to the study's authors, have no effect on tumor necrosis factor-alpha (TNF-α) but do have an effect on neutrophils' ability to suppress their connection. As was previously mentioned, this mechanism also applies to lemongrass and citral, preventing the production of IL-1β and IL-6. Considering Citral, a primary component of *Cinnamomum insularimontanum* Hayata's natural product medicinal balm, in a separate study by Lin et al. (2008), it was found to inhibit TNF- in Crude 264.7 cells that were stimulated by LPS. According to comparable authors, myristicin's hepatoprotective effect may be attributable, at least in part, to its ability to inhibit macrophage TNF-α release (Morita et al., 2003).

Mice that were given *Pterodon emarginatus* oil prior to receiving carrageenan intrapleurally had significantly lower levels of IL-1β and TNF-α (Dutra et al., 2009). Mice with TNBS-induced colitis had lower levels of IL-1 and IL-6 in some combinations of thyme and oregano oils (Burkovská et al., 2007). Suppressing the quality of expression of these cytokines appears to be the mechanism by which some medical ointments have an inhibitory effect on the development of supportive provocative cytokines as described in this subsection. The fact that the natural ointments studied inhibited the cytokines' protein and mRNA articulation in living cells led some researchers to this conclusion: the transcriptional level is primarily where these rejuvenating oils inhibit the expression of supportive fiery cytokines (Burkovská et al., 2007). The main components of thyme medicinal balm, p-cymene and thymol, fully reduced full mRNA IL-1 articulation, but only at high dosages (5000 mg/L) (Juhás et al., 2008) in a mouse model of colitis produced by TNBS.

Changes in the expression of inflammatory genes

Nitric oxide (NO), prostaglandins, and cytokines are involved in intense circumstances. Prostaglandins and nitric oxide (NO) are made by cyclooxygenase (COX) and nitric oxide synthase (NOS). It is possible to stimulate the production of several of these catalytic enzymes (cNOS, eNOS, and cytokine-inducible NOS). iNOS expression is triggered by inflammatory mediators according to in vitro experiments using rat macrophages. Provocative intermediates also stimulate NO production and iNOS articulation in a diverse set of mammalian cells, resulting in the release of large amounts of NO over extended periods of time (Miyasaka & Hirata, 1997). Signaling pathways include the NF-kappa B record element and mitogen-initiated protein kinases, both of which are involved in flagging, and play a major role in activating and regulating iNOS (Yoon et al., 2010). The enzyme iNOS catalyzes the conversion of L-arginine and oxygen molecules into nitric oxide. As a result of interfering with the signaling components that regulate COX-2 quality, a calming can dampen the protein's activity or prevent it from being articulated. Researchers have pinpointed four record variables that control COX-2 recording (Inoue et al., 1995): nuclear factor kappa B (NF-κB), CCAAT/enhancer-restricting protein (C/EBP), activator protein 1 (AP-1), and cAMP response element binding protein (CREB). Current research is focusing on NF-κB and MAPKs as potential targets for treatments with varying degrees of severe adverse effects. Protein homo- and heterodimers work together to regulate transcription. Maintaining NF-B as a dormant structure in the cytoplasm of cells requires the complexing of NF-κB with IκB inhibitor protein. IκB- is one of seven members of the protein family IB, which has been extensively studied. NF-κB's swift movement into the nucleus is made possible by phosphorylation of IκB- by IκB kinases (IKK), which results in proteasome-subordinate destruction of IB. The p65:p50 dimer is the most common form of active NF-κB (Yoshimura, 2006).

Studies have revealed that certain combinations can inhibit the COX-2 and iNOS articulation that represses certain MAPKS, hence reducing proinflammatory cytokines and NF-κB (De-Xing et al., 2005). Different peroxisome proliferator-activated receptor (PPAR) subfamily members—PPARα, PPARκ, and PPARγ—have been linked to various physiological functions (Hotta et al., 2010).

Blocking NF-κB contributes in some way to the attenuating effects of PPARs. In order to lower NF-κB capacity, PPARs can regulate features that inhibit NF-κB activation or interfere with the record-initiating limit of the NF-κB complex. By interacting with other parts of the NF-kB complex, PPAR prevents NF-kB from being restricted to the DNA and so suppresses the activation of its inflammatory features. Inhibitory protein B (IκB) articulation, which prevents NFB translocation into the nucleus, is also induced by PPARα activation (Stienstra, 2007). However, Inoue et al. (2000) state that PPAR mediates a negative criticism circle that regulates COX-2 articulation, particularly in macrophages, thus contributing to the modulation of COX-2 articulation that is specific to each cell type.

The NO generation in the macrophage cell line MH-29 was inhibited by LPS treatment; however, only *Teucrium brevifolia* and *Teucrium montbretii* essential oils were effective. With a crude content of 264.7, *T. brevifolia* was found to be primarily composed of spathulenol and -cadinene, whereas *T. montbretii* was predominately composed of carvacrol and caryophyllene oxide. Since caryophyllene and 4-vinyl guaiacol were discovered to be the principal components of the excess Teucrium species in *Teucrium flavum* and *Teucrium polium*, respectively, the effect of these compounds on restricting NO generation and, therefore, the aggravation hindrance, was studied (Menichini et al., 2009). Topical anti-aging treatments containing oils from *Fortunella japonica* and *Citrus sunki*, both of which include limonene, have been shown to reduce LPS-induced NO generation in Crude 264.7 cells (Yang, Jeon, et al., 2010; Yang, Kim, et al., 2010; Yang, Yue, et al., 2010).

As expected, the addition of *Origanum ehrenbergii* oil, primarily composed of thymol and p-cymene, to the incubator containing the Crude 264.7 cells significantly attenuated the LPS-induced NO generation. *Origanum syriacum* which relied heavily on carvacrol and thymol for its creation lacked this ability (Loizzo et al., 2009). However, there are other explanations for NO inhibition, such as the fact that LPS-activated Crude 264.7 macrophages were inhibited in their NO production by the citrus strip squander rejuvenating oil (Yang et al., 2009). Myrcene is the main ingredient in a compound discovered by Garcia Martin and Silvestre (Tavares et al., 2010). High amounts of (-)-linalool had little effect on PGE2 production in LPS-activated macrophages J774. A1 cells. This was also shown for the COX-2 articulation. Linalool's inhibitory activity was demonstrated to be due exclusively to the iNOS chemical movement since the analog monoterpene blocked NO delivery but not the growth of iNOS expression (Paena et al., 2006). Although the LPS-treated Crude 264 was affected by the medicinal herb *C. japonica*, it still grew. The scientists also noted a reduction in NO production alongside the decreases in iNOS protein and mRNA articulation and COX-2 protein and COX-2 mRNA articulation, respectively (Yoon et al., 2010). Reduced mRNA and protein expression for inducible nitric oxide synthase (iNOS) and cyclooxygenase-2 (COX-2) were associated with the anti-inflammatory effects of *Abies koreana* Horstmann's Silberlocke natural balm on pro-inflammatory lipid mediators (NO and PGE2) (Yoon et al., 2009a,b,c). A natural ointment stopped LPS-activated Crude 264 from producing NO and PGE2 including components from the plant *Farfugium japonicum* (L.) Kitamura.

Seven cells, identified by a simultaneous suppression of iNOS and COX-2 mRNA articulation, were found to be capable of understanding the attenuating activity of this oil by Kim et al. (2008). Rejuvenating balm from the *Illicium anisatum* L. plant, which is rich in 1,8-cineole, has been shown in a separate study by the same authors to reduce iNOS and COX-2 protein and mRNA expression (Kim et al., 2009). Both cyclooxygenase-2 (COX-2) and inducible nitric oxide synthase (iNOS) expression were significantly dampened by pretreatment with -humulene and (-)-trans-caryophyllene following carrageenan infusion into the mouse paw. The authors did not really use any of the included potential systems in their work (Fernandes et al., 2007). In SW1353 cells, the phorbol myristate acetic acid (PMA)-activated NF-kB record could be suppressed by the natural oils of various Pimpinella species and their ancestors. Yet, they failed to demonstrate any forms of comparison. The most peculiar behavior was attributed to three different Pimpinella species' oils and five different pure combinations. For the best moderating effect, these findings also highlighted the significance of a similar structure in several of these pure blends. As indicated above, cinnamondehyde inhibits cytokine production in LPS-stimulated J774A.1 macrophages, which may be due to a decrease in reactive oxygen species (ROS) generation and JNK and ERK activation.

Natural anti-aging creams containing *C. insularimontanum* Hayata and citral significantly suppressed NO generation in LPS-stimulated Crude 264.7 cells. When LPS was used to stimulate protein synthesis in iNOS, citral showed an inhibitory movement that was not observed for COX-2. The authors zeroed in on citral's effect on the NF-κB pathway as a means of elucidating the mechanism behind citral's mediation of iNOS record restraint. Citral was hypothesized by the authors to inhibit LPS-induced inflammation and protect against IκBα degradation by lowering p50 NF-κB levels in isolated cells (Lin et al., 2008). In a separate study, citral was reported to suppress LPS-induced mRNA and protein articulation of COX-2 in human macrophage-like U937 cells. Further, citral initiated PPARα and PPARγ mRNA articulation, suggesting that citral originates PPARα and PPARγ and guides COX-2 articulation (Katsukawa et al., 2010). Similarly, carvacrol was observed to act in a similar fashion. In human macrophage-like U937 cells, carvacrol blocked LPS-induced COX-2 mRNA and protein production, indicating that carvacrol regulates COX-2 synthesis through its agonistic action on PPARγ (Hotta et al., 2010). Carvacrol did not exhibit this characteristic nearly as strongly as citral did (Katsukawa et al., 2010).

Inhibition of NF-kappaB and mitogen-activated protein kinase (MAPK) activation in Crude 264.7 macrophages by a topical ointment derived from the Artemisia fukudo plant reduces LPS-activated inflammation. Medicines that reduce LPS-induced inflammation do so by returning p50 and p65 to the nucleus, where they are not activated by the inflammatory stimulus. When LPS was used to activate ERK, JNK, and p38 MAPKs, an identical oil inhibited this phosphorylation (Yoon et al., 2010).

In addition to suppressing the transcriptional activation of nuclear factor kappa B and the nuclear translocation of its p65 component in Crude 264.7 cells in response to lipopolysaccharide, Merr and Perry oil also inhibited the release of pro-inflammatory cytokines. Anethole, eugenol, and isoeugenol were shown to inhibit TNF—induced NF-B activation in ML1-a cells in a brief paper by Chainy et al. (2000). In a similar vein, other authors found that anethole inhibited IB phosphorylation in TNF-activated ML1-a cells, preventing IB corruption. Anethole did not activate Guide kinase phosphorylation in comparable cells with TNF feeling. Anethole inhibited TNF-induced NF-B, Guide kinase, and JNK/AP-1 activity.

Early response to stimulus 1 is a transcription factor that directs the public announcement of several important characteristics of inflammation. Oftentimes, scientists may use THP-1, a human monocyte cell line, as a stand-in for actual macrophages found in tissues. 1,8-cineole was shown by Zhou et al. (2007) to inhibit the LPS-induced Early response to stimulus 1 combination and atomic limitation in THP-1 cells without affecting the LPS-induced NF-B articulation in cores. As a result, the authors reasoned that 1,8-cineole, rather than NF-B inhibition, may uniquely attenuate the effect of Early response to stimulus 1 by suppressing Early response to stimulus 1 union and preventing Early response to stimulus 1 atomic absorption (Salminen et al., 2008).

Conclusions

Cell-reinforcing effects of natural medicines have been explored because of their real potential as additives, personal care products, or nutritional supplements in the food and pharmaceutical industries. In recent years, there has been a rise in the demand for relaxation techniques included in anti-aging creams. Therapeutic oils have been shown to have a variety of beneficial effects on cells, including the inhibition of lipid peroxidation, the elimination of free radicals, and the chelation of metal particles, all of which are detailed in this paper. Some studies also demonstrated that the rejuvenating balm's components work in tandem, as their individual components have less activity than the natural ointment when used as benchmarks. As a result, it is important to conduct studies on the potential for both synergy and conflict.

Cabrera and Prieto (2010) recently presented their ability to use artificial neural networks to anticipate the cell reinforcement action of rejuvenating balms. They put it to the test on roughly 30 of the 80 components reported as having cancer prevention agent action, additionally, the findings demonstrated that this computational method could accurately anticipate the cell reinforcement movement of rejuvenating balms based on particular components. Thymol, carvacrol, eugenol, and p-cymene are just a few that are brought to light in this study. By the way, the designers only tried out two variants. Numerous others exist, including those that evaluate ROS/RNS/chelating metals searching/searching capacities. This means there is still a lot of work to be done before we can consider the project complete. According to the text, it is reasonable to assume that oils can act as a moderating influence on arachidonic digestion, cytokine production, provocation quality expression modulation, and cytokine production regulation, depending on their chemical composition. Although it is undeniable that rejuvenating balms have cell-supporting and moderating effects, the substance variation of the oils used in these products can hinder their intended effects. Variations in their synthetic syntheses may be attributable to a number of factors, such as the time of year they are harvested, the weather, the plant's vegetative development, the plant portion used, and the type of extraction they employ.

References

Abdur Rauf, M., Akram, P., Semwal, A. A. H., Mujawah, N., Muhammad, Z., Riaz, N., Munir, D., Piotrovsky, I., Vdovina, A., Bouyahya, C., Oluwaseun Adetunji, M., Ali Shariati, Z. M., Almarhoon, Y. N., & Mabkhot, H. K. (2021). Antispasmodic potential of medicinal plants: A comprehensive review. *Oxidative Medicine and Cellular Longevity*, *12*, 2021. Available from https://doi.org/10.1155/2021/4889719, Article ID 4889719.

Abe, S., Maruyama, N., Hayama, K., Ishibashi, H., Inoue, S., Oshima, H., & Yamaguchi, H. (2003). Suppression of tumor necrosis factor-alpha-induced neutrophil adherence response by essential oils. *Mediators of Inflammation*, *12*, 323–328. Available from https://doi.org/10.1080/09629350310001633342, [PMC free article] [PubMed] [CrossRef] [Google Scholar].

Adejumo, I. O., Adetunji, C. O., & Adeyemi, O. S. (2017). Influence of UV light exposure on mineral composition and biomass production of mycomeat produced from different agricultural substrates. *Journal of Agricultural Sciences, Belgrade*, *62*(1), 51–59.

Adetunji, C. O. (2008). *The antibacterial activities and preliminary phytochemical screening of vernoniaamygdalina and Aloe vera against some selected bacteria* (pp. 40–43) (M.Sc Thesis). University of Ilorin.

Adetunji, C. O., Arowora, K., Fawole Oluyemisi, B., & Adetunji Juliana, B. (2013). Effects of coatings on storability of carrot under evaporative coolant system. *Albanian Journal of Agricultural Sciences*, *12*(3).

Adetunji, C. O., Phazang, P., & Sarin, N. B. (2017). Significance of rhamnolipids as a biological control agent in the management of crops/plant pathogens. *Current Trends in Biomedical Engineering & Biosciences*, *10*(3), 54–55.

Adetunji, C. O. (2019). Environmental impact and ecotoxicological influence of biofabricated and inorganic nanoparticle on soil activity. In D. Panpatte, & Y. Jhala (Eds.), *Nanotechnology for agriculture*. Singapore: Springer. Available from https://doi.org/10.1007/978-981-32-9370-0_12.

Adetunji, C. O., Panpatte, D. G., Bello, O. M., & Adekoya, M. A. (2019). Application of nanoengineered metabolites from beneficial and eco-friendly microorganisms as a biological control agents for plant pests and pathogens. In D. Panpatte, & Y. Jhala (Eds.), *Nanotechnology for agriculture: Crop production & protection*. Singapore: Springer. Available from https://doi.org/10.1007/978-981-32-9374-8_13.

Adetunji, C. O., Ojediran, J. O., Juliana, B. D., & Olumuyiwa Owa, S. (2019). Influence of chitosan edible coating on postharvest qualities of Capsicum annum L. during storage in evaporative cooling system. *Croatian Journal of Food Science and Technology*, *11*(1), 59–66.

Adetunji, C. O., Kumar, D., Raina, M., Arogundade, O., & Sarin, N. B. (2019). Endophytic microorganisms as biological control agents for plant pathogens: A panacea for sustainable agriculture. In A. Varma, S. Tripathi, & R. Prasad (Eds.), *Plant biotic interactions*. Cham: Springer. Available from https://doi.org/10.1007/978-3-030-26657-8_1.

Adetunji, C. O., & Varma, A. (2020). Biotechnological Application of trichoderma: A powerful fungal isolate with diverse potentials for the attainment of food safety, management of pest and diseases, healthy planet, and sustainable agriculture. In C. Manoharachary, H. B. Singh, & A.

Varma (Eds.), *Trichoderma: Agricultural applications and beyond. Soil biology 61.* Cham: Springer. Available from https://doi.org/10.1007/978-3-030-54758-5_12.

Adetunji, C. O., Oloke, J. K., & Prasad, G. (2020). Effect of carbon-to-nitrogen ratio on eco-friendly mycoherbicide activity from Lasiodiplodia pseudotheobromae C1136 for sustainable weeds management in organic agriculture. *Environment, Development and Sustainability, 22,* 1977–1990. Available from https://doi.org/10.1007/s10668-018-0273-1.

Adetunji, C. O., Egbuna, C., Tijjani, H., Adom, D., Al-Ani, L. K. T., & Patrick-Iwuanyanwu, K. C. (2020). *Homemade preparations of natural biopesticides and applications. Natural remedies for pest, disease and weed control* (pp. 179–185). Publisher Academic Press.

Adetunji, C. O., Roli, O. I., & Adetunji, J. B. (2020). Exopolysaccharides derived from beneficial microorganisms: Antimicrobial, food, and health benefits. In P. Mishra, R. R. Mishra, & C. O. Adetunji (Eds.), *Innovations in food technology.* Singapore: Springer. Available from https://doi.org/10.1007/978-981-15-6121-4_10.

Adetunji, C. O., Akram, M., Tope Olaniyan, O., Olufemi Ajayi, O., Inobeme, A., Olaniyan, S., Hameed, L., & Bunmi Adetunji, J. (2021). Targeting SARS-CoV-2 novel Corona (COVID-19) virus infection using medicinal plants. In K. Dua, S. Nammi, D. Chang, D. K. Chellappan, G. Gupta, & T. Collet (Eds.), *Medicinal plants for lung diseases.* Singapore: Springer. Available from https://doi.org/10.1007/978-981-33-6850-7_21.

Adetunji, C. O., Nwankwo, W., Ukhurebor, K., Olayinka, A. S., & Makinde, A. S. (2021). Application of biosensor for the identification of various pathogens and pests mitigating against the agricultural production: Recent advances. In R. N. Pudake, U. Jain, & C. Kole (Eds.), *Biosensors in agriculture: Recent trends and future perspectives. Concepts and strategies in plant sciences.* Cham: Springer. Available from https://doi.org/10.1007/978-3-030-66165-6_9.

Adetunji, C. O., Palai, S., Ekwuabu, C. P., Egbuna, C., Adetunji, J. B., Ehis-Eriakha, C. B., Kesh, S. S., & Mtewa, A. G. (2021). *General principle of primary and secondary plant metabolites: Biogenesis, metabolism, and extraction. Preparation of phytopharmaceuticals for the management of disorders* (pp. 3–23). Publisher Academic Press.

Adetunji, C. O., Inobeme, A., Olaniyan, O. T., Ajayi, O. O., Olaniyan, S., & Adetunji, J. B. (2021). Application of nanodrugs derived from active metabolites of medicinal plants for the treatment of inflammatory and lung diseases: Recent advances. In K. Dua, S. Nammi, D. Chang, D. K. Chellappan, G. Gupta, & T. Collet (Eds.), *Medicinal plants for lung diseases.* Singapore: Springer. Available from https://doi.org/10.1007/978-981-33-6850-7_26.

Adetunji, C. O., Olugbenga, S. M., Kadiri, O., Varma, A., Akram, M., Kola Oloke, J., Shafique, H., Bunmi Adetunji, J., Jain, A., Ebunoluwa Bodunrinde, R., Ozolua, P., & Ewa Ubi, B. (2021). Quinoa: From farm to traditional healing, food application, and phytopharmacology. In A. Varma (Ed.), *Biology and biotechnology of Quinoa.* Singapore: Springer. Available from https://doi.org/10.1007/978-981-16-3832-9_20.

Adetunji, C. O., Osikemekha, A. A., Olaniyan, O. T., Inobeme, A., Olisaka, F. N., Oluwadamilare Uwadiae, E., & Nosa Obayagbona, O. (2021). Recent trends in organic farming. In R. Soni, D. C. Suyal, P. Bhargava, & R. Goel (Eds.), *Microbiological activity for soil and plant health management.* Singapore: Springer. Available from https://doi.org/10.1007/978-981-16-2922-8_20.

Adetunji, C. O., Michael, O. S., Nwankwo, W., Eghonghon Ukhurebor, K., Anthony Anani, O., Kola Oloke, J., Varma, A., Kadiri, O., Jain, A., & Bunmi Adetunji, J. (2021). Quinoa, the next biotech plant: Food security and environmental and health hot spots. In A. Varma (Ed.), *Biology and biotechnology of Quinoa.* Singapore: Springer. Available from https://doi.org/10.1007/978-981-16-3832-9_19.

Adetunji, C. O., Ajayi, O. O., Akram, M., Tope Olaniyan, O., Amjad Chishti, M., Inobeme, A., Olaniyan, S., Bunmi Adetunji, J., Olaniyan, M., & Oluwasegun Awotunde, S. (2021). Medicinal plants used in the treatment of influenza a virus infections. In K. Dua, S. Nammi, D. Chang, D. K. Chellappan, G. Gupta, & T. Collet (Eds.), *Medicinal plants for lung diseases.* Singapore: Springer. Available from https://doi.org/10.1007/978-981-33-6850-7_19.

Adetunji, C. O., Michael, O. S., Varma, A., Kola Oloke, J., Kadiri, O., Akram, M., Ebunoluwa Bodunrinde, R., Imtiaz, A., Bunmi Adetunji, J., Shahzad, K., Jain, A., Ewa Ubi, B., Majeed, N., Ozolua, P., & Olisaka, F. N. (2021). Recent advances in the application of biotechnology for improving the production of secondary metabolites from Quinoa. In A. Varma (Ed.), *Biology and biotechnology of Quinoa.* Singapore: Springer. Available from https://doi.org/10.1007/978-981-16-3832-9_17.

Adetunji, J. B., Adetunji, C. O., & Olaniyan, O. T. (2021). African walnuts: A natural depository of nutritional and bioactive compounds essential for food and nutritional security in Africa. In O. O. Babalola (Ed.), *Food security and safety.* Cham: Springer. Available from https://doi.org/10.1007/978-3-030-50672-8_19.

Adetunji, C. O., Olaniyan, O. T., Akram, M., Olufemi Ajayi, O., Inobeme, A., Olaniyan, S., Said Khan, F., & Bunmi Adetunji, J. (2021). Medicinal plants used in the treatment of pulmonary hypertension. In K. Dua, S. Nammi, D. Chang, D. K. Chellappan, G. Gupta, & T. Collet (Eds.), *Medicinal plants for lung diseases.* Singapore: Springer. Available from https://doi.org/10.1007/978-981-33-6850-7_14.

Adetunji, C. O., Michael, O. S., Rathee, S., Singh, K. R. B., Olufemi Ajayi, O., Adetunji, J. B., Ojha, A., Singh, J., & Pratap Singh, R. (2022). Potentialities of nanomaterials for the management and treatment of metabolic syndrome: A new insight. *Materials Today Advances, 13,* 100198.

Adewale, G. G., Olajide, P. A., Omowumi, O. S., Okunlola, D. D., Taiwo, A. M., & Adetuyi, B. O. (2022). Toxicological significance of the occurrence of selenium in foods. *World News of Natural Sciences, 44,* 63–88.

Ahmadi, F., Sadeghi, S., Modarresi, M., Abiri, R., & Mikaeli, A. (2010). Chemical composition, in vitro antimicrobial, antifungal and antioxidant activities of the essential oil and methanolic extract of Hymenocrater longiflorus Benth., of Iran. *Food and Chemical Toxicology: An International Journal Published for the British Industrial Biological Research Association, 48,* 1137–1144. Available from https://doi.org/10.1016/j.fct.2010.01.028, [PubMed] [CrossRef] [Google Scholar].

Alcântara, J. M., Yamaguchi, K. K. L., & Junior, V. F. V. (2010). Composição química de óleos essenciais de espécies de Aniba e Licaria e suas actividades antioxidante e antiagregante plaquetária. *Quimica Nova, 33,* 141–145. Available from https://doi.org/10.1590/S0100-40422010000100026, [CrossRef] [Google Scholar].

Asaph, A., Jongsma, M. A., Kim, T.-Y., Ri, M.-B., Giri, A. P., Verstappen, W. A., Schwab, W., & Bouwmeester, H. J. (2006). Metabolic engineering of terpenoid biosynthesis in plants. *Phytochemistry Reviews*, *5*, 49−58. Available from https://doi.org/10.1007/s11101-005-3747-3, [CrossRef] [Google Scholar].

Bakkali, F., Averbeck, S., Averbeck, D., & Idaomar, M. M. (2008). Biological effects of essential oils—A review. *Food and Chemical Toxicology: An International Journal Published for the British Industrial Biological Research Association*, *46*, 446−475. Available from https://doi.org/10.1016/j.fct.2007.09.106, [PubMed] [CrossRef] [Google Scholar].

Barra, A., Coroneo, V., Dessi, S., Cabras, P., & Angioni, A. (2010). Chemical variability, antifungal and antioxidant activity of Eucalyptus camaldulensis essential oil from Sardinia. *Natural Product Communications*, *5*, 329−335, [PubMed] [Google Scholar].

Batiha, G. E., Awad, D. A., Algamma, A. M., Nyamota, R., Wahed, M. I., Shah, M. A., Amin, M. N., Adetuyi, B. O., Hetta, H. F., Cruz-Marins, N., Koirala, N., Ghosh, A., & Sabatier, J.-M. (2021). Diary-derived and egg white proteins in enhancing immune system against COVID-19 frontiers in nutritionr. *Nutritional Immunology*, *8*, 629440. Available from https://doi.org/10.3389/fnut.2021629440.

Bello, O. M., Ibitoye, T., & Adetunji, C. (2019). Assessing antimicrobial agents of Nigeria flora. *Journal of King Saud University-Science*, *31*(4), 1379−1383.

Bicchi, C., Liberto, E., Matteodo, M., Sgorbini, B., Mondello, L., Zellner, B. A., Coata, R., & Rubiolo, P. (2008). Quantitative analysis of essential oils: A complex task. *Flavour and Fragrance Journal*, *23*, 382−391. Available from https://doi.org/10.1002/ffj.1905, [CrossRef] [Google Scholar].

Burkovská, A., Čikoš, Š., Juhás, Š., Il'Ková, G., Rehák, P., & Koppel, J. (2007). Effects of a combination of thyme and oregano essential oils on TNBS-induced colitis in mice. *Mediators of Inflammation*. Available from https://doi.org/10.1155/2007/23296, [PMC free article] [PubMed] [Google Scholar].

Burt, S. (2004). Essential oils: Their antibacterial properties and potential applications in foods − A review. *International Journal of Food Microbiology*, *94*, 223−253. Available from https://doi.org/10.1016/j.ijfoodmicro.2004.03.022, [PubMed] [CrossRef] [Google Scholar].

Cabrera, A. C., & Prieto, J. M. (2010). Application of artificial neural networks to the prediction of the antioxidant activity of essential oils in two experimental in vitro models. *Food Chemistry*, *118*, 141−146. Available from https://doi.org/10.1016/j.foodchem.2009.04.070, [CrossRef] [Google Scholar].

Caldefie-Chézet, F., Fusillier, C., Jarde, T., Laroye, H., Damez, M., Vasson, M.-P., & Guillot, J. (2006). Potential anti-inflammatory effects of Melaleuca alternifolia essential oil on human peripheral blood leukocytes. *Phytotherapy Research: PTR*, *20*, 364−370. Available from https://doi.org/10.1002/ptr.1862, [PubMed] [CrossRef] [Google Scholar].

Cavaleiro, C. M. F. (2001). *Óleos essenciais de Juniperus de Portugal* (PhD Thesis). Universidade de Coimbra; Faculdade de Farmácia, Coimbra, Portugal. [Google Scholar].

Chainy, G. B. N., Manna, S. K., Chaturvedi, M. M., & Aggarwal, B. B. (2000). Anethole blocks both early and late cellular responses transduced by tumor necrosis factor: Effect on NF-kB, AP-1 JNK, MAPKK and apopotosis. *Oncogene*, *19*, 2943−2950. Available from https://doi.org/10.1038/sj.onc.1203614, [PubMed] [CrossRef] [Google Scholar].

Chao, L. K., Hua, K.-F., Hsu, H.-Y., Cheng, S.-S., Lin, I.-F., Chen, C.-J., Chen, S.-T., & Chang, S.-T. (2008). Cinnamaldehyde inhibits pro-inflammatory cytokines secretion from monocytes/macrophages through suppression of intracellular signalling. *Food and Chemical Toxicology: An International Journal Published for the British Industrial Biological Research Association*, *46*, 220−231. Available from https://doi.org/10.1016/j.fct.2007.07.016, [PubMed] [CrossRef] [Google Scholar].

Chao, L. K., Hua, K.-F., Hsu, H.-Y., Cheng, S.-S., Liu, J.-Y., & Chang, S.-T. (2005). Study of the anti-inflammatory activity of essential oil from leaves of Cinnamomum osmophloeum. *Journal of Agricultural and Food Chemistry*, *53*, 7274−7278. Available from https://doi.org/10.1021/jf051151u, [PubMed] [CrossRef] [Google Scholar].

Dandlen, S. A., Lima, A. S., Mendes, M. D., Miguel, M. G., Faleiro, M. L., Sousa, M. J., Pedro, L. G., Barroso, J. G., & Figueiredo, A. C. (2010). Antioxidant activity of six Portuguese thyme species essential oils. *Flavour and Fragrance Journal*, *25*, 150−155. Available from https://doi.org/10.1002/ffj.1972, [CrossRef] [Google Scholar].

Darsham, S., & Doreswamug, R. (2004). Patented anti-inflammatory plant drug development from traditional medicine. *Phytotherapy Research: PTR*, *18*, 343−357. Available from https://doi.org/10.1002/ptr.1475, [PubMed] [CrossRef] [Google Scholar].

Dewick, P. M. (2002). The biosynthesis of C5-C-25 terpenoid components. *Natural Product Reports*, *19*, 181−222. Available from https://doi.org/10.1039/b002685i, [PubMed] [CrossRef] [Google Scholar].

De-Xing, H., Yanagita, T., Uto, T., Masuzaki, S., & Fujii, M. (2005). Anthocyanidins inhibit cyclooxygenase-2 expression in LPS-evoked macrophages: Structure-activity relationship and molecular mechanisms involved. *Biochemical Pharmacology*, *70*, 417−425. Available from https://doi.org/10.1016/j.bcp.2005.05.003, [PubMed] [CrossRef] [Google Scholar].

Didunyemi, M. O., Adetuyi, B. O., & Oyebanjo, O. O. (2019). Morinda lucida attenuates acetaminophen-induced oxidative damage and hepatotoxicity in rats. *Journal of Biomedical Sciences*, *8*(2).

Didunyemi, M., Adetuyi, B., & Oyewale, I. (2020). Inhibition of lipid peroxidation and in-vitro antioxidant capacity of aqueous, acetone and methanol leaf extracts of green and red acalypha wilkesiana muell arg. *International Journal Of Biological and Medical Research*, *11*(3), 7089−7094.

Dung, N. T., Bajpai, V. K., Yoon, J. I., & Kang, S. C. (2009). Anti-inflammatory effects of essential oil isolated from the buds of Cleistocalyx operculatus (Roxb.) Merr and Perry. *Food and Chemical Toxicology: An International Journal Published for the British Industrial Biological Research Association*, *47*, 449−453. Available from https://doi.org/10.1016/j.fct.2008.11.033, [PubMed] [CrossRef] [Google Scholar].

Dutra, R. C., Fava, M. B., Alves, C. S. C., Ferreira, A. P., & Barbosa, N. R. (2009). Antiulcerogenic and anti-inflammatory activities of the essential oil from Pterodon emarginatus seeds. *The Journal of Pharmacy and Pharmacology*, *61*, 243−250, [PubMed] [Google Scholar].

Ebrahimabadi, A. H., Mazoochi, A., Kashi, F. J., Djafari-Bidgoli, Z., & Batooli, H. (2010). Essential oil composition and antioxidant and antimicrobial properties of the aerial parts of Salvia eremophila Boiss. from Iran. *Food and Chemical Toxicology: An International Journal Published for the British Industrial Biological Research Association, 48*, 1371−1376. Available from https://doi.org/10.1016/j.fct.2010.03.003, [PubMed] [CrossRef] [Google Scholar].

Egbuna, C., Gupta, E., Ezzat, S. M., Jeevanandam, J., Mishra, N., Akram, M., Sudharani, N., Oluwaseun Adetunji, C., Singh, P., Ifemeje, J. C., Deepak, M., Bhavana, A., Mark, A., Walag, P., Ansari, R., Bunmi Adetunji, J., Laila, U., Chinedu Olisah, M., & Onyekere, P. F. (2020). Aloe species as valuable sources of functional bioactives. In C. Egbuna, & G. Dable Tupas (Eds.), *Functional foods and nutraceuticals*. Cham: Springer. Available from https://doi.org/10.1007/978-3-030-42319-3_18.

El-Ghorab, A. H., Nauman, M., Anjum, F. M., Hussain, S., & Nadeem, M. (2010). A comparative study on chemical composition and antioxidant activity of ginger (Zingiber officinale) and cumin (Cuminum cyminum). *Journal of Agricultural and Food Chemistry, 58*, 8231−8237. Available from https://doi.org/10.1021/jf101202x, [PubMed] [CrossRef] [Google Scholar].

Ennajar, M., Bouajila, J., Lebrihi, A., Mathieu, F., Savagnac, A., Abderraba, M., Raies, A., & Romdhane, M. (2010). The influence of organ, season and drying method on chemical composition and antioxidant and antimicrobial activities of Juniperus phoenica L. essential oils. essential oils. *Journal of the Science of Food and Agriculture, 90*, 462−470, [PubMed] [Google Scholar].

Fernandes, E. S., Passos, G. F., Medeiros, R., da Cunha, F. M., Ferreira, J., Campos, M. M., Pianowski, L. F., & Calixto, J. B. (2007). Anti-inflammatory effects of compounds alpha-humulene and (-)-trans-caryophyllene isolated from the essential oil of Cordia verbenacea. *European Journal of Pharmacology, 569*, 228−236. Available from https://doi.org/10.1016/j.ejphar.2007.04.059, [PubMed] [CrossRef] [Google Scholar].

Firuzi, O., Asadollahi, M., Gholami, M., & Javidnia, K. (2010). Composition and biological activities of essential oils from four Heracleum species. *Food Chemistry, 122*, 117−122. Available from https://doi.org/10.1016/j.foodchem.2010.02.026, [CrossRef] [Google Scholar].

Ghazghazi, H., Miguel, M. G., Hasnaoui, B., Sebei, H., Ksontini, M., Figueiredo, A. C., Pedro, L. G., & Barroso, J. G. (2010). Phenols, essential oils, and carotenoids of Rosa canina from Tunisia and their antioxidant activities. *African Journal of Biotechnology, 9*, 2709−2710, [Google Scholar].

Gholivand, M. B., Rahimi-Nasrabadi, M., Batooli, H., & Ebrahimabadi, A. H. (2010). Chemical composition and antioxidant activities of the essential oil and methanol extracts of Psammogeton canescens. *Food and Chemical Toxicology: An International Journal Published for the British Industrial Biological Research Association, 48*, 24−28. Available from https://doi.org/10.1016/j.fct.2009.09.007, [PubMed] [CrossRef] [Google Scholar].

Gomes, A., Fernandes, E., Lima, J. L. F. C., Mira, L., & Corvo, M. L. (2008). Molecular mechanisms of anti-inflammatory activity mediated by flavonoids. *Current Medicinal Chemistry, 15*, 1586−1605. Available from https://doi.org/10.2174/092986708784911579, [PubMed] [CrossRef] [Google Scholar].

González, S. B., Houghton, P. J., & Hoult, J. R. S. (2003). The activity against leukocyte eicosanoid generation of essential oil and polar fractions of Adesmia boronioides Hook.f. *Phytotherapy Research: PTR, 17*, 290−293. Available from https://doi.org/10.1002/ptr.1118, [PubMed] [CrossRef] [Google Scholar].

Gounaris, Y. (2010). Biotechnology for the production of essential oils, flavours and volatile isolates. *Flavour and Fragrance Journal*. Available from https://doi.org/10.1002/ffj.1996, [CrossRef] [Google Scholar].

Gourine, N., Yousfi, M., Bombarda, I., Nadjemi, B., & Gaydou, E. (2010). Seasonal variation of chemical composition and antioxidant activity of essential oil from Pistacia atlantica Desf. leaves. *Journal of the American Oil Chemists' Society, 87*, 157−166. Available from https://doi.org/10.1007/s11746-009-1481-5, [CrossRef] [Google Scholar].

Gourine, N., Yousfi, M., Bombarda, I., Nadjemi, B., Stocker, P., & Gaydon, E. M. (2010). Antioxidant activities and chemical composition of essential oil of Pistacia atlantica from Algeria. *Industrial Crops and Products, 31*, 203−208. Available from https://doi.org/10.1016/j.indcrop.2009.10.003, [CrossRef] [Google Scholar].

Goze, I., Alim, A., Cetinus, S. A., Cetin, A., Durmus, N., Atas, A. T., & Vural, N. (2010). In vitro antimicrobial, antioxidant, and antispasmodic activities and the composition of the essential oil of Origanum acutidens (Hand.-Mazz.) letswaart. *Journal of Medicinal Food, 13*, 705−709. Available from https://doi.org/10.1089/jmf.2009.0094, [PubMed] [CrossRef] [Google Scholar].

Grosso, C., Figueiredo, A. C., Burillo, J., Mainar, A. M., Urieta, J. S., Barroso, J. G., Coelho, J. A., & Palavra, A. M. F. (2010). Composition and antioxidant activity of Thymus vulgaris volatiles: Comparison between supercritical fluid extraction and hydrodistillation. *Journal of Separation Science, 33*, 2211−2218. Available from https://doi.org/10.1002/jssc.201000192, [PubMed] [CrossRef] [Google Scholar].

Hajhashemi, V., Ghannadi, A., & Sharif, B. (2003). Anti-inflammatory and analgesic properties of the leaf extracts and essential oil of Lavandula angustifolia Mill. *Journal of Ethnopharmacology, 89*, 67−71. Available from https://doi.org/10.1016/S0378-8741(03)00234-4, [PubMed] [CrossRef] [Google Scholar].

Haloui, E., Marzouk, Z., Marzouk, B., Bouffira, I., Bouraoui, A., & Fenina, N. (2010). Pharmacological activities and chemical composition of the Olea europaea L. leaf essential oils from Tunisia. *Journal of Food, Agriculture and Environment, 8*, 204−208, [Google Scholar].

Hart, P. H., Brand, C., Carson, C. F., Riley, T. V., Prager, R. H., & Finlay-Jones, J. J. (2000). Terpinen-4-ol, the main component of the essential oil of Melaleuca alternifolia (tea tree oil), suppresses inflammatory mediator production by activated human monocytes. *Inflammation Research, 9*, 19−26, [PubMed] [Google Scholar].

Hotta, M., Nakata, R., Kasukawa, M., Hori, K., Takahashi, S., & Inoue, H. (2010). Carvacrol, a component of thyme oil, activates PPARα and γ and suppresses COX-2 expressio. *Journal of Lipid Research, 51*, 132−139. Available from https://doi.org/10.1194/jlr.M900255-JLR200, [PMC free article] [PubMed] [CrossRef] [Google Scholar].

Ibrahim, M. A., Kainulainen, P., Aflatuni, A., Tiilikkala, K., & Holopainen, J. K. (2001). Insecticidal, repellent, antimicrobial activity and Phytotoxicity of essential oils: With special reference to limonene and its suitability for control of insect pest. *Agricultural and Food Science in Finland, 10*, 243−259, [Google Scholar].

Inoue, H., Tanabe, T., & Unesono, K. (2000). Feedback control of cyclooxygenase-2 expression through PPARγ. *Journal of Biological Chemistry*, *275*, 28028−28032, [PubMed] [Google Scholar].

Inoue, H., Yokoyama, C., Hara, S., Tone, Y., & Tanabe, T. (1995). Transcriptional regulation of human prostaglandin-endoperoxide synthase-2 gene by lipopolysaccharide and phorbol ester in vascular endotelial cells. *Journal of Biological Chemistry*, *270*, 24965−24971. Available from https://doi.org/10.1074/jbc.270.42.24965, [PubMed] [CrossRef] [Google Scholar].

James-Okoro, P. P. O., Iheagwam, F. N., Sholeye, M. I., Umoren, I. A., Adetuyi, B. O., Ogundipe, A. E., … Ogunlana, O. O. (2021). Phytochemical and in vitro antioxidant assessment of Yoyo bitters. *World News of Natural Sciences*, *37*, 1−17.

Jia, H. L., Ji, Q. L., Xing, S. L., Zhang, P. H., Zhu, G. L., & Wang, X. H. (2010). Chemical composition and antioxidant, antimicrobial activities of the essential oils of Thymus marschallianus Will. and Thymus proximus Serg. *Journal of Food Science*, *75*, E59−E65. Available from https://doi.org/10.1111/j.1750-3841.2009.01413.x, [PubMed] [CrossRef] [Google Scholar].

Jordan, M. J., Sanchez-Gomez, P., Jimenez, J. F., Quilez, M., & Sotomayor, J. A. (2010). Chemical composition and antiradical activity of the essential oil from Satureja intricata, S. obovata; and their hybrid Datureja x delpozoi. *Natural Product Communications*, *5*, 629−634, [PubMed] [Google Scholar].

Joshi, S. C., Verma, A. R., & Mathela, C. S. (2010). Antioxidant and antibacterial activities of the leaf essential oils of Himalayan Lauraceae species. *Journal of Chemical Toxicology*, *48*, 37−40. Available from https://doi.org/10.1016/j.fct.2009.09.011, [PubMed] [CrossRef] [Google Scholar].

Juergens, U. R., Stöber, M., Schmidt-Schilling, L., Kleuver, T., & Vetter, H. (1998). Antiinflammatory effects of eucalyptol (1,8-cineole) in bronchial asthma: Inhibition of arachidonic acid metabolism in human blood monocytes ex vivo. *European Journal of Medical Research*, *17*, 407−412, [PubMed] [Google Scholar].

Juhás, Š., Bujňáková, D., Rehák, P., Cikoš, Š., Czikková, S., Veselá, J., Il'ková, G., & Koppel, J. (2008). Anti-inflammatory effects of thyme essential oil in mice. *Acta Veterinaria Brno*, *77*, 327−334. Available from https://doi.org/10.2754/avb200877030327, [CrossRef] [Google Scholar].

Kamatou, G. P. P., & Viljoen, A. M. (2010). A review of the application and pharmacological properties of α-bisabolol and α-bisabolol-rich oils. *Journal of the American Oil Chemists' Society*, *87*, 1−7. Available from https://doi.org/10.1007/s11746-009-1483-3, [CrossRef] [Google Scholar].

Kamatou, G. P. P., van Zyl, R. L., van Vuuren, S. F., Viljoen, A. M., Figueiredo, A. C., Barroso, J. G., Pedro, L. G., & Tilney, P. M. (2006). Chemical composition, leaf trichome types and biological activities of the essential oils of four related Salvia species indigenous to Southern Africa. *Journal of Essential Oil Research*, *18*, 72−79, [Google Scholar].

Katsukawa, M., Nakata, R., Takizawa, Y., Hori, K., Takahashi, S., & Inoue, H. (2010). Citral, a component of lemongrass oil, activates PPARa and g and suppresses COX-2 expression. *BBA Molecular and Cell Biology Lipids*, *1801*, 1214−1220. Available from https://doi.org/10.1016/j.bbalip.2010.07.004, [PubMed] [CrossRef] [Google Scholar].

Kim, J.-Y., Kim, S.-S., Oh, T.-H., Baik, J. S., Song, G., Lee, N. H., & Hyun, C.-G. (2009). Chemical composition, antioxidant, anti-elastase, and anti-inflammatory activities of Illicium anisatum essential oil. *Acta Pharm*, *59*, 289−300, [PubMed] [Google Scholar].

Kim, J.-Y., Oh, T.-H., Kim, B. J., Kim, S.-S., Lee, N. H., & Hyun, C.-G. (2008). Chemical composition and anti-inflammatory effects of essential oil from Farfagium japonicum flower. *Journal of Oleo Science*, *57*, 623−628. Available from https://doi.org/10.5650/jos.57.623, [PubMed] [CrossRef] [Google Scholar].

Kulisic-Bilusic, T., Blazevic, I., Dejanovic, B., Milos, M., & Pifat, G. (2010). Evaluation of the antioxidant activity of essential oils from caper (Capparis spinosa) and sea fennel (Crithmum maritimum) by different methods. *Journal of Food Biochemistry*, *34*, 286−302. Available from https://doi.org/10.1111/j.1745-4514.2009.00330.x, [CrossRef] [Google Scholar].

Lin, C.-T., Chen, C.-J., Lin, T.-Y., Tung, J. C., & Wang, S.-Y. (2008). Anti-inflammation activity of fruit essential oil from Cinnamomum insularimontanum Hayata. *Bioresource Technology*, *99*, 8783−8787. Available from https://doi.org/10.1016/j.biortech.2008.04.041, [PubMed] [CrossRef] [Google Scholar].

Litchenthaler, H. K. (1999). The 1-deoxy-D-xylulose-5-phosphate pathway of isoprenoid biosynthesis in plants. *Annual Review of Plant Physiology and Plant Molecular Biology*, *50*, 47−65. Available from https://doi.org/10.1146/annurev.arplant.50.1.47, [PubMed] [CrossRef] [Google Scholar].

Loizzo, M. R., Menichini, F., Conforti, F., Tundis, R., Bonesi, M., Saab, A. M., Statti, G. A., Cindio, B., Houghton, P. J., Menichini, F., & Frega, N. G. (2009). Chemical analysis, antioxidant, antiiflammatory and anticholinesterase activities of Origanum ehrenbergii Boiss. and Origanum syriacum L. essential oils. *Food Chemistry*, *117*, 174−180. Available from https://doi.org/10.1016/j.foodchem.2009.03.095, [CrossRef] [Google Scholar].

Lourens, A. C. U., Reddy, D., Başer, K. H. C., Viljoen, A. M., & van Vuuren, S. F. (2004). In vitro biological activity and essential oil composition of four indigenous South African Helichrysum species. *Journal of Ethnopharmacology*, *95*, 253−258. Available from https://doi.org/10.1016/j.jep.2004.07.027, [PubMed] [CrossRef] [Google Scholar].

Maestri, D. M., Nepote, V., Lamarque, A. L., & Zygadlo, J. A. (2006). Natural products as antioxidants. In F. Imperato (Ed.), *Phytochemistry: Advances in research* (pp. 105−135). Kerala, India: Research Signopost, [Google Scholar].

Martino, L., Feo, V., Fratianni, F., & Nazzaro, F. (2010). Chemistry, antioxidant, antibacterial and antifungal activities of volatile oils and their components. *Natural Product Communications*, *2010*(5), 1741−1750, [PubMed] [Google Scholar].

Mechergui, K., Coelho, J. A., Serra, M. C., Lamine, S. B., Boukhchina, S., & Khouja, M. L. (2010). Essential oils of Origanum vulgare L. subsp. glandulosum (Desf.) letswaart from Tunisia: Chemical composition and antioxidant activity. *Journal of the Science of Food and Agriculture*, *90*, 1745−1749. Available from https://doi.org/10.1002/jsfa.4011, [PubMed] [CrossRef] [Google Scholar].

Medeiros, R., Passos, G. F., Vítor, C. E., Koepp, J., Mazzuco, T. L., Pianowski, L. F., Campos, M. M., & Calixto, J. B. (2007). Effect of two active compounds obtained from the essential oil of Cordia verbenacea on the acute inflammatory responses elicited by LPS in the rat paw. *British Journal of Pharmacology*, *151*, 618−627. Available from https://doi.org/10.1038/sj.bjp.0707270, [PMC free article] [PubMed] [CrossRef] [Google Scholar].

Menichini, F., Conforti, F., Rigano, D., Formisano, C., Piozzi, F., & Senatore, F. (2009). Phytochemical composition, anti-inflammatory and antitumour activities of four Teucrium essential oils from Greece. *Food Chemistry, 11*, 670–686, [Google Scholar].

Mighri, H., Hajlaoui, H., Akrout, A., Najjaa, H., & Neffati, M. (2010). Antimicrobial and antioxidant activities of Artemisia herba-alba essential oil cultivated in Tunisian arid zone. *Comptes Rendus Chimie, 13*, 380–386. Available from https://doi.org/10.1016/j.crci.2009.09.008, [CrossRef] [Google Scholar].

Miguel, M. G. (2010). Antioxidant activity of medicinal and aromatic plants. *Flavour and Fragrance Journal, 25*, 291–312. Available from https://doi.org/10.1002/ffj.1961, [CrossRef] [Google Scholar].

Miguel, M. G., Cruz, C., Faleiro, L., Simões, M. T. F., Figueiredo, A. C., Barroso, J. G., & Pedro, L. G. (2010). Foeniculum vulgare essential oils: Chemical composition, antioxidant and antimicrobial activities. *Natural Product Communications, 5*, 319–328, [PubMed] [Google Scholar].

Mimika-Dukić, N., Bugarin, D., Grebovi, S., Mitić-Ćulafić, D., Vuković-Gačić, D., Jovin, E., & Couladis, M. (2010). Essential oil of Myrtus communis L. as a potential antioxidant and antimutagenic agents. *Molecules, 15*, 2759–2770. Available from https://doi.org/10.3390/molecules15042759, [PMC free article] [PubMed] [CrossRef] [Google Scholar].

Miyasaka, N., & Hirata, Y. (1997). Nitric oxide and inflammatory arthritides (minireview). *Life Sciences, 61*, 2073–2081. Available from https://doi.org/10.1016/S0024-3205(97)00585-7, [PubMed] [CrossRef] [Google Scholar].

Moon, J.-K., & Shibamoto, T. (2009). Antioxidant assays for plant and food components. *Journal of Agricultural and Food Chemistry, 57*, 1655–1666. Available from https://doi.org/10.1021/jf803537k, [PubMed] [CrossRef] [Google Scholar].

Morita, T., Jinno, K., Kawagishi, H., Arimoto, Y., Suganuma, H., Inakuma, T., & Sugiyama, K. (2003). Hepatoprotective effect of myristicin from nutmeg (Myristica fragrans) on lipopolysaccharide/D-galactosamine-induced liver injury. *Journal of Agricultural and Food Chemistry, 51*, 1560–1565. Available from https://doi.org/10.1021/jf020946n, [PubMed] [CrossRef] [Google Scholar].

Mothana, R. A., Al-Rehaily, A. J., & Schultze, W. (2010). Chemical analysis and biological activity of the essential oils of two endemic Soqotri Commiphora species. *Molecules, 15*, 689–698. Available from https://doi.org/10.3390/molecules15020689, [PMC free article] [PubMed] [CrossRef] [Google Scholar].

Munir, N., Hasnain, M., Waqif, H., Adetuyi, B. O., Egbuna, C., Olisah, M. C., ... El Sayed, A. M. A. (2022). *Gelling agents, micro and nanogels in food system applications. Application of nanotechnology in food science, processing and packaging* (pp. 153–167). Cham: Springer.

Murakami, A., & Ohigashi, H. (2007). Targeting NOX, iNOS and COX-2 in inflammatory cells: Chemoprevention using food phytochemicals. *International Journal of Cancer, 2007(121)*, 2357–2367. Available from https://doi.org/10.1002/ijc.23161, [PubMed] [CrossRef] [Google Scholar].

Nazir, A., Itrat, N., Shahid, A., Mushtaq, Z., Abdulrahman, S. A., Egbuna, C., ... Toloyai, P. E. Y. (2022). *Orange peel as source of nutraceuticals. Food and agricultural byproducts as important source of valuable nutraceuticals* (pp. 97–106). Cham: Springer.

Niki, E. (2010). Assessment of antioxidant capacity in vitro and in vivo. *Free Radical Biology and Medicine, 9*, 503–515. Available from https://doi.org/10.1016/j.freeradbiomed.2010.04.016, [PubMed] [CrossRef] [Google Scholar].

Olaniyan, O. T., & Adetunji, C. O. (2021). Biological, biochemical, and biodiversity of biomolecules from marine-based beneficial microorganisms: Industrial perspective. In C. O. Adetunji, D. G. Panpatte, & Y. K. Jhala (Eds.), *Microbial Rejuvenation of polluted environment. Microorganisms for sustainability* (27). Singapore: Springer. Available from https://doi.org/10.1007/978-981-15-7459-7_4.

Ozkan, G., Baydar, H., & Erbas, S. (2010). The influence of harvest time on essential oil composition, phenolic constituents and antioxidant properties of Turkish oregano (Origanum onites L.). *Journal of the Science of Food and Agriculture, 90*, 205–209. Available from https://doi.org/10.1002/jsfa.3788, [PubMed] [CrossRef] [Google Scholar].

Paena, A. T., Marzocco, S., Popolo, A., & Pinto, A. (2006). (−)-Linalool inhibits in vitro NO formation. Probable involvement in the antinociceptive activity of this monoterpene compound. *Life Sciences, 78*, 719–723. Available from https://doi.org/10.1016/j.lfs.2005.05.065, [PubMed] [CrossRef] [Google Scholar].

Passos, G. F., Fernandes, E. S., da Cunha, F. M., Ferreira, J., Pianowski, L. F., Campos, M. M., & Calixto, J. B. (2007). Anti-inflammatory and anti-allergic properties of the essential oil and active compounds from Cordia verbenacea. *Journal of Ethnopharmacology, 110*, 323–333. Available from https://doi.org/10.1016/j.jep.2006.09.032, [PubMed] [CrossRef] [Google Scholar].

Patil, R. P., Nimbalkar, M. S., Jadhav, U. U., Dawkar, V. V., & Govindwar, S. P. (2010). Antiaflatoxigenic and antioxidant activity of an essential oil from Ageratum conyzoides L. *Journal of the Science of Food and Agriculture, 90*, 608–614, [PubMed] [Google Scholar].

Politeo, O., Jukic, M., & Milos, M. (2010). Comparison of chemical composition and antioxidant activity of glycosidically bound and free volatiles from clove (Eugenia caryophyllata Thumb.). *Journal of Food Biochemistry, 34*, 129–141. Available from https://doi.org/10.1111/j.1745-4514.2009.00269.x, [CrossRef] [Google Scholar].

Pourmortazavi, S. M., & Hajimirsadeghi, S. S. (2007). Supercritical fluid extraction in plant essential and volatile oil analysis. *Journal of Chromatography A, 1163*, 2–24. Available from https://doi.org/10.1016/j.chroma.2007.06.021, [PubMed] [CrossRef] [Google Scholar].

Raetz, C. R. H., & Whitfield, C. (2002). Lipopolysaccharide endotoxins. *Annual Review of Biochemistry, 71*, 635–700. Available from https://doi.org/10.1146/annurev.biochem.71.110601.135414, [PMC free article] [PubMed] [CrossRef] [Google Scholar].

Reichling, J., Schnitzler, P., Suschke, U., & Saller, R. (2009). Essential oils of aromatic plants with antibacterial, antifungal, antiviral, and cytotocic properties-an overview. *Forschende Komplementärmedizin, 16*, 79–90, [PubMed] [Google Scholar].

Roginsky, V., & Lissi, E. A. (2005). Review of methods to determine chain-breaking antioxidant activity in food. *Food Chemistry, 92*, 235–254. Available from https://doi.org/10.1016/j.foodchem.2004.08.004, [CrossRef] [Google Scholar].

Rubiolo, P., Sgorbini, B., Liberto, E., Cordero, C., & Bicchi, C. (2010). Essential oils and volatiles: Sample preparation and analysis. *Flavour and Fragrance Journal, 25*, 282–290. Available from https://doi.org/10.1002/ffj.1984, [CrossRef] [Google Scholar].

Saei-Dehkordi, S. S., Tajik, H., Moradi, M., & Khalighi-Sigaroodi, F. (2010). Chemical composition of essential oils in Zataria multiflora Boiss. from different parts of Iran and their radical scavenging and antimicrobial activity. *Food and Chemical Toxicology, 48*, 1562−1567. Available from https://doi.org/10.1016/j.fct.2010.03.025, [PubMed] [CrossRef] [Google Scholar].

Salminen, A., Lehtonen, M., Suuronen, T., Kaarniranta, K., & Huuskonen, J. (2008). Terpenoids: Natural inhibitors of NF-κB signalling with anti-inflammatory and anticancer potential. *Cellular and Molecular Life Sciences, 65*, 2979−2999. Available from https://doi.org/10.1007/s00018-008-8103-5, [PubMed] [CrossRef] [Google Scholar].

Sánchez-Arana, A., Estarrón-Espinosa, M., Obledo-Vázquez, E. N., Camberos, E. P., Silva-Vázquez, R., & Lugo-Cervantes, E. (2010). Antimicrobial and antioxidant activities of Mexican oregano essential oils (Lippia graveolens H. B. K.) with different composition when microencapsulated in b-cyclodextrin. *Letters in Applied Microbiology, 50*, 585−590. Available from https://doi.org/10.1111/j.1472-765X.2010.02837.x, [PubMed] [CrossRef] [Google Scholar].

Sangwan, N. S., Farooqui, A. H. A., Shabih, F., & Sangwan, R. S. (2001). Regulation of essential oil production in plants. *Plant Growth Regulation, 34*, 3−21. Available from https://doi.org/10.1023/A:1013386921596, [CrossRef] [Google Scholar].

Schröder, N. W. J., Morath, S., Alexander, C., Hamann, L., Hartung, T., Zähringer, U., Göbel, U. B., Weber, J. R., & Schumann, R. R. (2003). Lipoteichoic acid (LTA) of Streptococcus pneumoniae and Staphylococcus aureus activated immune cells via Toll-like receptor (TLR)-2, lipopolysaccharide-binding protein (LBP), and CD14, whereas TLR-4 and MD-2 are not involved. *Journal of Biological Chemistry, 2003*(278), 15587−15594, [PubMed] [Google Scholar].

Singh, H. P., Kaur, S., Mittal, S., Batish, D. R., & Kohli, R. K. (2010). In vitro screening of essential oil from young and mature leaves of Artemisia scoparia compared to its major constituents for free radical scavenging activity. *Food and Chemical Toxicology, 48*, 1040−1044. Available from https://doi.org/10.1016/j.fct.2010.01.017, [PubMed] [CrossRef] [Google Scholar].

Singh, P., Shukla, R., Prakash, B., Kumar, A., Singh, S., Mishra, P. K., & Dubey, N. K. (2010). Chemical profile, antifungal, antiaflatoxigenic and antioxidant activity of Citrus maxima Burm. and Citrus sinensis (L.) Osbeck essential oils and their cyclic monoterpene, DL-limonene. *Journal of Chemical Toxicology, 48*, 1734−1740. Available from https://doi.org/10.1016/j.fct.2010.04.001, [PubMed] [CrossRef] [Google Scholar].

Singh, S., & Singh, R. P. (2008). In vitro methods of assay of antioxidants: An overview. *Food Reviews International, 24*, 392−415. Available from https://doi.org/10.1080/87559120802304269, [CrossRef] [Google Scholar].

Stevenson, D. E., & Hurst, R. D. (2007). Polyphenolic phytochemicals-just antioxidants or much more? A review. *Cellular and Molecular Life Sciences, 64*, 2900−2916. Available from https://doi.org/10.1007/s00018-007-7237-1, [PubMed] [CrossRef] [Google Scholar].

Stienstra, R. (2007). *The role of PPARs in inflammation and obesity* (PhD thesis). The Netherlands: Wageningen University. [Google Scholar].

Suanarunsawat, T., Ayutthaya, W. D. N., Songsak, T., Thirawarapan, S., & Poungshompoo, S. (2010). Antioxidant activity and lipid-lowering effect of essential oils extracted from Ocimum sanctum L. leaves in rats fed with a high cholesterol diet. *Journal of Clinical Biochemistry and Nutrition, 46*, 52−59, [PMC free article] [PubMed] [Google Scholar].

Syamsir, D. R. B. (2009). *Essential oils and biological activities of three selected wild Alpinia species* (Master thesis). Kuala Lumpur, Malaysia: Institute of Biological Sciences, Faculty of Sciences, University of Malaya. [Google Scholar].

Tavares, A. C., Gonçalves, M. J., Cruz, M. T., Cavaleiro, C., Lopes, M. C., Canhoto, J., & Salgueiro, L. R. (2010). Essential oils from Distichoselinum tenuifolium: Chemical composition, cytotoxicity, antifungal and anti-inflammatory properties. *Journal of Ethnopharmacology, 130*, 593−598. Available from https://doi.org/10.1016/j.jep.2010.05.054, [PubMed] [CrossRef] [Google Scholar].

Thangadurai, D., Naik, J., Sangeetha, J., Said Al-Tawaha, A. R. M., Oluwaseun Adetunji, C., Islam, S., David, M., Kashivishwanath Shettar, A., & Bunmi Adetunji, J. (2021). Nanomaterials from agrowastes: Past, present, and the future. In O. V. Kharissova, L. M. Torres-Martínez, & B. I. Kharisov (Eds.), *Handbook of nanomaterials and nanocomposites for energy and environmental applications*. Cham: Springer. Available from https://doi.org/10.1007/978-3-030-36268-3_43.

Tomić, S., Božin, B., Samojlik, I., Milenković, M., Mimica-Dukić, N., & Petrović, S. (2010). Effects of Athamanta turbith fruit essential oils on CCl4-induced hepatic failure in mice and their antioxidant properties. *Phytotherapy Research, 24*, 787−790, [PubMed] [Google Scholar].

Ukhurebor, K. E., & Adetunji, C. O. (2021). Relevance of biosensor in climate smart organic agriculture and their role in environmental sustainability: What has been done and what we need to do? In R. N. Pudake, U. Jain, & C. Kole (Eds.), *Biosensors in agriculture: Recent trends and future perspectives. Concepts and strategies in plant sciences*. Cham: Springer. Available from https://doi.org/10.1007/978-3-030-66165-6_7.

Viuda-Martos, M., Navajas, Y. R., Zapata, E. S., Fernández-López, J., & Pérez- Alvarez, J. A. (2010). Antioxidant activity of essential oils of five spice plants widely used in a Mediterranean diet. *Flavour and Fragrance Journal, 25*, 13−19. Available from https://doi.org/10.1002/ffj.1951, [CrossRef] [Google Scholar].

Wannes, W. A., Mhamdi, B., Sriti, J., Jenia, M. B., Ouchikh, O., Hamdaoni, G., Kchouk, M. E., & Marzouk, B. (2010). Antioxidant activities of the essential oils and methanol extracts from myrtle (Myrtus communis var. italica) leaf, stem and flower. *Food and Chemical Toxicology, 48*, 1362−1370. Available from https://doi.org/10.1016/j.fct.2010.03.002, [PubMed] [CrossRef] [Google Scholar].

Wei, A., & Shibamoto, T. (2010). Antioxidant/lipoxygenase inhibitory activities and chemical compositions of selected essential oils. *Journal of Agricultural and Food Chemistry, 58*, 7218−7225, [PubMed] [Google Scholar].

Yang, E.-J., Kim, S. S., Moon, J.-Y., Oh, T.-H., Baik, J. S., Lee, N. H., & Hyun, C.-G. (2010). Inhibitory effects of Fortunella japonica var. margarita and Citrus sunki essential oils on nitric oxide production and skin pathogens. *Acta Microbiologica et Immunologica Hungarica, 57*, 15−27. Available from https://doi.org/10.1556/AMicr.57.2010.1.2, [PubMed] [CrossRef] [Google Scholar].

Yang, E.-J., Kim, S.-S., Oh, T.-H., Baik, J. S., Lee, N. H., & Hyun, C.-G. (2009). Essential oil of citrus fruit waste attenuates LPS-induced nitric oxide production and inhibits the growth of skin pathogens. *International Journal of Agriculture and Biology, 11*, 791−794, [Google Scholar].

Yang, S.-A., Jeon, S.-K., Lee, E.-J., Shim, E.-H., & Lee, I.-S. (2010). Comparative study of the chemical composition and antioxidant activity of six essential oils and their components. *Natural Product Research*, *24*, 140−151. Available from https://doi.org/10.1080/14786410802496598, [PubMed] [CrossRef] [Google Scholar].

Yang, Y., Yue, Y., Runwei, Y., & Guolin, Z. (2010). Cytotoxic, apoptotic and antioxidant activity of the essential oil of Amomum tsao-ko. *Bioresource Technology*, *101*, 4205−4211. Available from https://doi.org/10.1016/j.biortech.2009.12.131, [PubMed] [CrossRef] [Google Scholar].

Yoon, W.-J., Kim, S.-S., Oh, T.-H., Lee, N. H., & Hyun, C.-G. (2009a). Abies koreana essential oil inhibits drug-resistant skin pathogen growth and LPS-induced inflammatory effects of murine macrophage. *Lipids*, *44*, 471−476. Available from https://doi.org/10.1007/s11745-009-3297-3, [PubMed] [CrossRef] [Google Scholar].

Yoon, W. J., Kim, S. S., Oh, T. H., Lee, N. H., & Hyun, C. G. (2009b). Cryptomeria japonica essential oil inhibits the growth of drug-resistant skin pathogens and LPS-induced NO and pro-inflammatory cytokine production. *Polish Journal of Microbiology*, *58*, 61−68, [PubMed] [Google Scholar].

Yoon, W. J., Kim, S. S., Oh, T. H., Lee, N. H., & Hyun, C. G. (2009c). Torreya nucifera essential oil inhibits skin pathogen growth and lipopolysaccharide-induced inflammatory effects. *International Journal of Pharmacology*, *5*, 37−43. Available from https://doi.org/10.3923/ijp.2009.37.43, [CrossRef] [Google Scholar].

Yoon, W. J., Moon, J. Y., Song, G., Lee, Y. K., Han, M. S., Lee, J. S., Ihm, B. S., Lee, W. J., Lee, N. H., & Hyun, C. G. (2010). Artemisia fukudo essential oil attenuates LPS-induced inflammation by suppressing NF-kB and MAPK activation in RAW264.7 macrophages. *Food and Chemical Toxicology*, *48*, 1222−1229. Available from https://doi.org/10.1016/j.fct.2010.02.014, [PubMed] [CrossRef] [Google Scholar].

Yoshimura, A. (2006). Signal transduction of inflammatory cytokines and tumor development. *Cancer Science*, *97*, 439−447. Available from https://doi.org/10.1111/j.1349-7006.2006.00197.x, [PubMed] [CrossRef] [Google Scholar].

Zhou, J.-Y., Wang, X.-F., Tang, F.-D., Zhou, J.-Y., Lu, G.-H., Wang, Y., & Bian, R.-L. (2007). Inhibitory effect of 1,8-cineole (eucalyptol) on Erg-1 expression in lipopolysaccharide-stimulated THP-1 cells. *Acta Pharmacologica Sinica*, *28*, 908−912. Available from https://doi.org/10.1111/j.1745-7254.2007.00555.x, [PubMed] [CrossRef] [Google Scholar].

Chapter 9

The application of essential oil as an antimicrobial agent in dairy products

Babatunde Oluwafemi Adetuyi[1], Kehinde Abraham Odelade[1], Peace Abiodun Olajide[1], Charles Oluwaseun Adetunji[2], Juliana Bunmi Adetunji[3], Abel Inobeme[4], Yovwin D. Godwin[5], Oloruntoyin Ajenifujah-Solebo[6], Olalekan Akinbo[7], Oluwabukola Atinuke Popoola[6], Olatunji Matthew Kolawole[8], Osarenkhoe Omorefosa Osemwegie[9], Mohammed Bello Yerima[10] and M.L. Attanda[11]

[1]Department of Natural Sciences, Faculty of Pure and Applied Sciences, Precious Cornerstone University, Ibadan, Oyo State, Nigeria, [2]Applied Microbiology, Biotechnology and Nanotechnology Laboratory, Department of Microbiology, Edo State University Uzairue, Iyamho, Edo State, Nigeria, [3]Department of Biochemistry, Osun State University, Osogbo, Osun State, Nigeria, [4]Department of Chemistry, Edo State University Uzairue Iyamho, Auchi, Edo State, Nigeria, [5]Department of Family Medicine, Faculty of Clinical Sciences, Delta State University, Abraka, Delta State, Nigeria, [6]Genetics, Genomics and Bioinformatics Department, National Biotechnology Development Agency, Abuja, FCT, Nigeria, [7]Centre of Excellence in Science, Technology, and Innovation, AUDA-NEPAD, Johannesburg, Gauteng, South Africa, [8]Department of Microbiology, Faculty of Life Sciences, University of Ilorin, Ilorin, Kwara State, Nigeria, [9]Department of Food Science and Microbiology, Landmark University, Omu-Aran, Kwara State, Nigeria, [10]Department of Microbiology, Sokoto State University, Sokoto, Sokoto State, Nigeria, [11]Department of Agricultural Engineering, Bayero University Kano, Kano, Kano State, Nigeria

Introduction

In recent years, the amount of food deteriorating due to microbes has progressively increased worldwide (Gustafsson et al., 2011). Current epidemics of several pathogens have raised awareness of the importance of microbiological safety in food production and storage (Zhang et al., 2017). When combined with other storage conditions, antibacterial compounds in food can prevent or reduce the proliferation of bacteria, extending the product's shelf life. In the food industry currently, the usual is chemically generated preservatives (Prakash et al., 2018). There is growing concern regarding artificial preservatives like nitrites and parabens as a result of consumers' growing awareness of the potential mutagenic and carcinogenic risks associated with these compounds.

Foods that are "healthier," "increasingly organic," and "moderately produced" are becoming more and more popular. In this respect, study and advancement of organic and reduced antimicrobial compounds have gained considerable attention in order to substitute traditional artificial antimicrobial chemicals. Essential oils (EOs) are a form of supplementary agent found in plants that are processed for their taste, scent, and therapeutic benefits. The possibility of bioactive components found in EOs to eradicate agricultural diseases such as bacteria, molds, and associated poisons has been thoroughly investigated in previous years (Zhu et al., 2021). Due to its capacity to stop organic antimicrobial agents from degrading, boost their accessibility and specific application, and eventually decrease the quantity of antimicrobials required for efficient preservatives, nanotechnology has been established to be a useful technique in this regard (Prakash et al., 2018). Thus, industries that have been using nanotechnology in the food business are the fastest in current history.

Synthetic and natural food antimicrobial agents

Antimicrobial substances found in food can kill microorganisms and limit their ability to proliferate in order to spike up the quality, shelf-life, and safety of foods that consumers take. Based on their sources, the two primary groups are natural and synthetic antimicrobial agents.

Applications of Essential Oils in the Food Industry. DOI: https://doi.org/10.1016/B978-0-323-98340-2.00005-5

Synthetic antimicrobial agents

Despite increasing customer worries over the use of artificial antimicrobials in packaged foods, these substances are still frequently used in the food industry. Some of the very popular chemically synthesized antimicrobials are sodium diacetate, salts of sorbic acid, salts of benzoic acid, parabens, and salts of propionic acid. Considering their many advantages, the majority of these bacteria' biochemical wastes can harm human well-being and modify the appearance, aroma, and nutritional content of food, which include excellent commercial productivity, inexpensive, and potent antibacterial action in tiny amounts (Shatalov et al., 2017). How conventional chemical antibacterial food additives function:

1. Bacterial cell walls are destroyed by benzoic acid and benzoates, which also prevent acetyl-CoA formation and the transport of amino acids and respiratory proteins across cellular membranes.
2. Sorbic acid and sorbates play a similar role in reacting to biological processes and inhibiting their action.
3. The hydroxybenzoate protectant paraben has been connected to harm to cell membranes, cell lysis of intracellular proteins, and disruption of respiratory and electron transport enzyme production.
4. Dimethyl dicarbonate: permeating the cell wall and compromising with microbial cells' capacity to produce carbon dioxide by interacting with their enzymatic reactions.
5. By destroying their surface proteins and nucleic acids also with hydrogen ions generated during sulfite breakdown, bacteria can be killed utilizing sulfur dioxide and sulfites.

Natural antimicrobial agents

Polyphenols, flavonoids, tannin, alkaloids, terpenoids, isothiocyanates, polypeptides, and their unsaturated extracts are examples of naturally occurring antimicrobial substances that can be found in plants, animals, or microbes. According to health worries and the desire to find better and far more organic ingredients, customers prefer organic antimicrobial compounds to chemical antimicrobials (Falleh et al., 2020). Researchers have conducted several tests on active ingredients to investigate their antibacterial capabilities during the past few years. Membrane lysis, metal ion complexation, DNA damage in microbes, content leakage, enzyme inhibition, and ATP utilization are the six main antibacterial mechanisms that have been discovered (Zeng et al., 2012). The bacteriostatic impacts of these naturally occurring integrated compounds rely on a variety of criteria, including their synthetic characteristics, specific microbes, external conditions, food characteristics, and others (Gould, 1989). Antimicrobials are largely used in the food business.

Essential oil of plant origin

Thyme: *Penicillium digtatum, Shigella sonnei, Salmonella enteritidis, Salmonella typhimurium, Staphylococcus aureus, Literia monocytogene, S. enteritidis.*

    Clove: *Lactobacillus debruecki, L. monocytogene, S. aurus, E. coli.*

    Grape seed: *Candida maltose, E. coli* 157:H7, *S. typhimurium,* and *L. monocytogene, B. cereus, L. monocytogene, S. aurus, E. coli, S. enteritidis, S. typhimurium, S. cerevisae* are among the bacteria present in rosemary.

Essential oil of animal origin

The following bacteria are found in the following:

    Lactoferrin (milk): *B. cereus, Bacillus stearothermophilus, E. coli, S. enteritidis, Klebsiella* spp., and *S. enteritidis*

    In chitosan (Shelfish), *Zygosacharomyces baili, Yersini enterocolitica, Aspergilus flavus, B. cerus, S. typhimurum, Staphylococcus aureus,* and *L. monocytogene*

    Bacilus clostridium, L. monocytogene, Aspergillus, Candid, Fusarum, Sporotrix, Pecilomyces, Penicillium, and Saccharomyces are all present in lysozyme (chicken eggs, vegetables, and insects).

Current status and challenges involved in the preservative use of essential oil

As bioactive compounds, EOs are produced in a variety of vegetative tissues to protect the plant from stressful conditions (Bakkali et al., 2008). Due to their strong antibacterial qualities, they are widely used in the cleanliness, cosmetics, and hospitality industries, among other industries (Falleh et al., 2020). The fragrant molecules known as EOs are derived from plants by heat distilling the plant's numerous tissues. EOs are highly volatile, have a light molecular mass, and have a strong scent (Bakkali et al., 2008). Additionally, most EOs have low human toxic effects and are non-permanent in form (Calo et al., 2015).

Antimicrobial effect of essential oils

Most research shows that EOs have antibacterial qualities, making them useful for the storage of a variety of food goods (Falleh et al., 2020). Aromatic compounds, aliphatic compounds, Sulfur-containing nitrogen compounds, and terpenoids, are the main chemical constituents of plant EOs. Numerous components' bacteriostatic mechanisms frequently call for several locations of action (Sperotto et al., 2013).

Two ways that plant EOs and their primary components affect microbes are (a) by altering the morphological characteristics and makeup of mycelia and microbial cells, and (b) by reducing or preventing the creation and growing conditions of spore, which reduces or prevents the harmful effect of pathogens to the progeny (Shao et al., 2013).

The mechanism through which certain EOs exert their bacteriostatic effect

Both cuminaldehyde and cuminalcohol found in cumin cause damage to the membrane of living cells.

Effects of carvacrol and thymol in oregano

1. Lime: Limonene blocks mitochondrial respiration
2. Cinnamaldehyde, found in cinnamon, can affect both enzymes and the cell wall. It can decrease ATPase activity and amino acid synthesis, as well as affect membrane permeability and deplete proton potential energy.
3. Terpin-4 opening in marjoram (resulting in cell membrane breakdown and the leaking of intracellular components) and ethanol steroids (causing protein denaturation and dehydration) (Kim et al., 1995)
4. Clove: Eugenol can disrupt cellular structure
5. Rosemary: the antibacterial properties of terpenoids are amplified by the oxidizing effects of camphor and eucalyptus oil.

Problems associated with food antimicrobials effect of essential oils

As a result of their "green" nature, EOs have a considerable market as an alternative to traditional chemical preservatives. However, EOs encounter a number of significant domestic and foreign obstacles that hinder their widespread use as food preservatives. These obstacles include the following: because there are few raw resources available, the amount of EO acquired after separation is frequently inadequate for industrial applications. In addition, EOs generally have low absorption in soluble forms of media, making it impossible for the densities attained to impose important biological activities. Despite being used extensively, little is known about how plant EOs and their key ingredients affect the flavor of foods, fruits, and vegetables (Bucar et al., 2013). The majority of recent studies focused either on EOs as a whole or on highlighting certain EO characteristics. However, there is a paucity of studies on the interactions between various EOs. The majority of investigations use a single microorganism and are conducted in controlled environments (Burt, 2004). There is little investigation on the presence of different bacteria and the mode of action in food production since factors in packaged foods such as temperature, moisture, pH, and others can affect the bacteriostatic impact and act as a protective factor. Therefore, it is imperative to do a study on the antibacterial interaction of border elements (Sperotto et al., 2013). While using EOs as loose antimicrobials instantly on processed foods poses substantial challenges, these problems can be somewhat alleviated by using nanoencapsulation. When compared to natural EOs, nanoscale encapsulation systems have been found to significantly improve the physical characteristics of EOs, including water solubility, dispersion stability, turbidity, and viscosity, which increases their core processes (Blanco-Padilla et al., 2014). There are many uses for nanoencapsulation, including safeguarding EOs from oxygen, light, pH, moisture, and degradation during the manufacturing process and storage, improving the solubility of lipid-soluble compounds in water-soluble media, hiding off flavors and odors, and releasing the oils under regulated conditions at the location of their desired use (Bazana et al., 2019).

Essential oils' antimicrobial activity

EOs, also known as potent scented substances, are largely produced from a lot of plant sources like flower, root, bark, leave, seed, peel, fruit, and the entire plant (Hyldgaard et al., 2012). EO refers to a compound that is gotten from an organic unprocessed part of plant sources, by heat vaporization, mechanical forces from the epicarp of citrus fruits, or by evaporation, after splitting of the aqueous phase if any" (ISO), with the caveat that "the essential oil may require physical treatment options that do not result in a significant transformation in its formation". Combinations of unstable, water-insoluble, but reactive in natural solutions, and strongly aromatic-smelling compounds make up EOs (Nazzaro et al., 2013). There are roughly 3,000 different EOs, yet the food and aroma sectors only use about 300 of them (Burt, 2004). The chemical composition of EO varies based on the plant species, place of source, growth environment, soil composition, growth phase, and EO separation techniques (Angioni et al., 2006). They are natural substances, usually released, and are employed for things like pollination and defense against bacteria and fungi (Tajkarimi et al., 2010). The most popular techniques for producing EOs include soaking, saturated liquid recovery, water or vapor evaporation, and extraction (Shannon et al., 2011). Although EOs' primary purpose is as flavor enhancers, they can be added to foods to extend their shelf life because of their antibacterial properties. The requirements for this implementation, however, include the properties of EOs, the MIC of the aim microorganisms, the method of action, the likelihood of interactions with the food environment, and the perceived value of the meal. The antibacterial activities of EOs have been connected to their active ingredients (Hyldgaard et al., 2012). The volatile substances monoterpenes, sesquiterpenes, aliphatic aldehydes, alcohols, and esters make about 90%−95% of EOs. The majority of the volatile component is made up of hydrocarbons, fatty acid, sterol, carotenoid, wax, cumarine, and flavonoid, while only 5%−10% of the total EO is made up of these substances (Luque De Castro et al., 1999). The most effective antibacterial elements of EOs can be classified as terpenes, terpenoids, phenylpropenes, or other substances like allicin or isothiocyanates based on their chemical makeup (Hyldgaard et al., 2012). The ability of EOs to effectively destroy bacteria cannot be explained by a single mechanism. Numerous locations within microorganisms have been suggested as potential EO action sites (Nazzaro et al., 2013). The aqueous or lipid soluble nature of an EO's constituents, the type of bacterium, and the make-up of the cell membrane all affect how antibacterial an EO is. More than Gram-negative bacteria, Gram-positive bacteria have been shown to be highly prone to EOs (Hyldgaard et al., 2012). The composition of the cell wall of Gram-positive bacteria includes peptidoglycan (90%−95%), teichoic acid, and proteins. Because of their hydrophobic properties, EOs can interact with cell membranes and penetrate into the cytoplasm. The cell wall structure of a Gram-negative bacteria is more complex, with an exterior barrier of proteins and lipopolysaccharide enclosing a peptidoglycan monolayer (LPS). Because of the charge on this aqueous cell surface membrane, repulsive molecules can pass through it (Nazzaro et al., 2013). The water-insoluble hydrocarbon skeleton and the lipid-soluble chemical bonding of EOs play a significant role in their antibacterial properties. Aldehydes, ketones, alcohols, ethers, phenolic compounds, and hydrocarbons round out the top five in terms of antibacterial effect for EOs (Kalemba & Kunicka, 2003). Phenols work because of the acidic nature of the hydroxyl group. These substances change the permeability of the cell, interfere with the enzyme motor, and disturb the enzymes that produce energy (Basim et al., 2000). According to some research, the structure of the bacterial cell may affect how effective EOs are, with globular cells being more vulnerable than coccoid-shaped ones (Nazzaro et al., 2013).

Antimicrobial activity of plant oil-based nanoemulsions

Another intriguing recent development is the antibacterial use of nanoemulsions. The nanoemulsion is viable when combined with lipid-containing bacteria. The anionic-charged pathogen interface and the cationic-charged emulsion particles combine electrostatically to produce a powerful synergy (Caillet et al., 2006). When sufficient numbers of nanoemulsions are passed in the microbes, A lot of the potential is released from inside the emulsion. The reactive fixing and radiation produced cause the pathogen's lipid membrane to become unstable, which results in cell death (Donsì & Ferrari, 2016). There are many antibacterial components in the nanoemulsion. Pathogens like *E. coli*, *Bacillus anthracis* spores, and *Staphylococcus aureus*. The general issue of the amplification of antimicrobial-acceptable strains noticed by contemporary operators as a result of the pervasive, universal use of cleaning agents, cleaning products, and anti-infection agents has prompted research into the use of nanoemulsion as an antimicrobial factor (Sagis, 2015). Researchers have been working on innovative antimicrobial medicines that target particular pathogens while posing little damage to the host as a result of these shortcomings (Di Pasqua et al., 2007). Nanoemulsions cannot be utilized to hasten the emergence of more secure strains since it is believed that they function by non-specifically damaging bacterial cell membranes. The new use of nanoemulsion as an antibacterial expert shows considerable promise.

EO are becoming more widely acknowledged as a practical antibacterial treatment for a variety of microbiological issues, including those related to food waste and pathogens (de Sousa et al., 2012). The inclusion of preservatives in the food business, the transfer of pesticides in the chemical fertilizer industry, and the use of antimicrobial substances in cleaning agents are just a few of the many uses for nanoemulsions. Nutraceutical and pharmaceutical companies frequently use EO nanoemulsions in industrially processed foods as antimicrobials due to their globule size and distinctive trait of not visibly spreading light (Buranasuksombat et al., 2011; McClements, 2012). Eugenol, thymol, and terpene oils have also been described as nanoemulsions for improved antimicrobial drug delivery (Kriegel et al., 2010). Additionally, the peppermint, clove, and thyme oil-loaded nanoemulsion demonstrated antibacterial effectiveness against a variety of gram-negative and gram-positive microorganisms (Liang et al., 2012). EO-based nanoemulsions are more effective against bacteria than pure oil. According to a report, industrial food manufacturing facilities are using EO-containing nanoemulsions. Holley and Patel investigated the bactericidal activities of EO nanoemulsions on prepared and raw meats, packaged foods, and a range of vegetables (2005). It has been established that pure oil is far more commonly employed in foods than nanoemulsion. The microbial communities were significantly reduced when eugenol's nano-based surfactant micelles were tested in a range of microbial development settings (Gaysinsky et al., 2005). Bhargava et al. (2015) investigated oregano oil nanoemulsions with droplet sizes of 148 nm, which may reduce pathogen loads such as *Listeria monocytogenes*, *E. coli*, and *Salmonella typhimurium*. The use of predicted safer nanoemulsions as an alternative to the use of manufactured chemical additives has been proposed. Because omega-3 unsaturated fatty acids are poorly soluble in water, they were used to develop nanoemulsions with enhanced absorption. These have since been widely used as nutritional supplements and boosting ingredients in a variety of beverages (Walker et al., 2015). Initial food-grade preparation of a carvacrol nanoemulsion was made in order to investigate its effectiveness against *Escherichia coli* and *S. enteritidis*. Antimicrobial nanoemulsions outperformed previous, more well-known antimicrobials at eradicating microorganisms. According to a 2013 theory put up by Kim et al., plum oil products lemongrass on plum-coated nanoemulsions were more durable than uncoated ones while stored at 40 C, and the rate at which plums were inhaled was also decreased in these circumstances. The plums' median shelf life was extended by coating, which improved their resistance to *E. coli* and *S. enteritidis*. Neurological and meningitis-related disorders have been treated with pomegranate and palm seed oil-derived nanoemulsions (Musa et al., 2013). The main vector of a number of parasitic diseases is prevalent in tropical areas, such as intestinal ailment, filariasis, and yellow fever. The mainstay of contemporary mosquito preventive interventions worldwide includes pesticidal controllers (such as diflubenzuron, methoprene, etc.) and organophosphates (like temephos), which are combined with other artificial biocontrol agents (Mizrahi et al., 2014). These larvicides have been discovered to be quite powerful at killing mosquito larvae. The frequent use of larvicides, which has caused other irritating species to become more active and some organisms to evolve resistant, has increased concerns about health and ecology (Yang et al., 2002). In response to these worries, fresh approaches to managing mosquito larvae have been created. Plant-based nanoemulsions have a constant pesticidal impact because they contain crucial bioactive components that supply the transport force. Colon tumor development has been demonstrated to be chemopreventive by mustard oil. In order to improve the rusks' nutritional and textural qualities, as well as their overall flavor, mustard flour have been added to the mixture (Dwivedi et al., 2003). Nanoemulsions made from cinnamon oil showed significant antibacterial action against *E. coli* while using a low-energy approach. A nanoemulsion of cinnamon oil has also been demonstrated to successfully stop Bacillus cereus from growing (Li et al., 2005).

The effect of essential oils on beneficial bacteria in dairy products

In a variety of cheeses and yogurts, adding EOs has been found to hinder pathogenic and spoilage microorganisms. However, there have been relatively few studies examining how EOs affect good bacteria or microbes of innovative interest in dairy foods.

The addition of oregano EO to an S-based dairy lactic starter culture had no effect on the culture's viability. *L. thermophilus* and *L. bulgaricus* lactis are used in the Argentine cheese process, as determined by viable count and milk acidifying actions, without interfering with milk thrombosis or cutting times. The lack of effect of oregano EO concentration on the lag time length of the microorganisms forming the lactic starter culture during milk fermentation suggests that there may have been no or only mildly negative effects prior to the microbial culture's activation and growth (Olmedo et al., 2013).

EOs' modulatory effects on L-lactic acid bacteria were also investigated. To compare acidophilus and Bifidobacterium breve to pathogenic bacteria, the minimum bactericidal concentration (MBC) was used (*E. coli* and *Salmonella* spp.). *L. acidophilus* and *B. breve* were both inhibited by thyme and oregano EOs, and their minimum inhibitory concentration (MIC) for pathogen-causing bacteria is low. In contrast to hazardous bacteria, *L. acidophilus* and *B. breve* had a minimum inhibitory concentration when exposed to basil essential oil. Because they contain significant amounts of the phenolic monortepenes carvacrol and thymol, which may have similar, synergistic, and non-selective antimicrobial effects, oregano and thyme EOs have strong inhibitory effects on these beneficial microorganisms. When their quantities are carefully chosen to be present in a substrate of interest, basil EO has been suggested to be used to reduce pathogenic bacteria without significantly harming beneficial microorganisms (Roldan et al., 2010). Because of the volatile nature of these chemicals, the fat and protein content of cheese, or both, EO concentrations in cheeses may decrease during ripening and storage. However, because phenolic terpenoids' hydroxyl groups can form hydrogen bonds with protein groups (e.g., NH and CO), detection of these compounds may be difficult using instrumental techniques commonly used to measure EO constituents in cheese samples (e.g., high-performance liquid chromatography) (Diniz-Silva et al., 2020; Moro et al., 2015).

To test the viability of the lactic acid bacteria used to make yogurt, EOs of clove, cinnamon, cardamom, and peppermint were used. During the milk fermentation process, the EOs of clove, cinnamon, cardamom, and peppermint were found to significantly inhibit *S. thermophilus* and *L. bulgaricus*.

Conclusion

Most perishable foods must be refrigerated or frozen to avoid spoilage from microorganisms. Consumers only want risk-free organic options. Food industry officials and scientists are constantly on the lookout for new, gentler preservation methods that will improve microbiological purity and safety without reducing the nutritional value or altering flavor. Natural chemicals are gaining popularity in this context due to their ability to improve quality and safety while having a lower impact on human health. Some EOs can be used as natural antibacterial agents to keep cheeses from spoiling and to extend their storage life. Chemicals have antimicrobial properties through a variety of mechanisms. These include increased cell permeability, altered membrane fatty acids, and altered membrane proteins. However, due to the potential negative effects on organoleptic qualities, the concentration of these compounds in dairy products should be closely monitored.

References

Angioni, A., Barra, A., Coroneo, V., Dessi, S., & Cabras, P. (2006). Chemical composition, seasonal variability, and antifungal activity of Lavandula stoechas L. ssp. stoechas essential oils from stem/leaves and flowers. *Journal of Agricultural and Food Chemistry, 54*(12), 4364–4370.

Bakkali, F., Averbeck, S., Averbeck, D., & Idaomar, M. (2008). Biological effects of essential oils—A review. *Food and Chemical Toxicology, 46*, 446–475, [CrossRef] [PubMed].

Basim, H., Yegen, O., & Zeller, W. (2000). Antibacterial effect of essential oil of Thymbra spicata L. var. spicata on some plant pathogenic bacteria/ Die antibakterielle Wirkung des ätherischen Öls von Thymbra spicata L. var. spicata auf phytopathogene Bakterien. *Zeitschrift für Pflanzenkrankheiten und Pflanzenschutz/Journal of Plant Diseases and Protection, 107*, 279–284.

Bazana, M. T., Codevilla, C. F., & de Menezes, C. R. (2019). Nanoencapsulation of bioactive compounds: Challenges and perspectives. *Current Opinion In Food Science, 26*, 47–56.

Bhargava, K., Conti, D. S., da Rocha, S. R., & Zhang, Y. (2015). Application of an oregano oil nanoemulsion to the control of foodborne bacteria on fresh lettuce. *Food Microbiology, 47*, 69–73.

Blanco-Padilla, A., Soto, K. M., Hernández Iturriaga, M., & Mendoza, S. (2014). Food antimicrobials nanocarriers. *Scientific World Journal, 837215*, [CrossRef] [PubMed].

Bucar, F., Wube, A., & Schmid, M. (2013). Natural product isolation—How to get from biological material to pure compounds. *Natural Product Reports, 30*, 525–545, [CrossRef] [PubMed].

Buranasuksombat, U., Kwon, Y. J., Turner, M., & Bhandari, B. (2011). Influence of emulsion droplet size on antimicrobial properties. *Food Science and Biotechnology, 20*(3), 793–800.

Burt, S. (2004). Essential oils: Their antibacterial properties and potential applications in foods—A review. *International Journal of Food Microbiology, 94*, 223–253, [CrossRef] [PubMed].

Caillet, S., Millette, M., Salmieri, S., & Lacroix, M. (2006). Combined effects of antimicrobial coating, modified atmosphere packaging, and gamma irradiation on Listeria innocua present in ready-to-use carrots (Daucuscarota). *Journal of Food Protection, 69*(1), 80–85.

Calo, J. R., Crandall, P. G., O'Bryan, C. A., & Ricke, S. C. (2015). Essential oils as antimicrobials in food systems—A review. *Food Control, 54*, 111–119.

de Sousa, J. P., de Azerêdo, G. A., de Araújo Torres, R., da Silva Vasconcelos, M. A., da Conceição, M. L., & de Souza, E. L. (2012). Synergies of carvacrol and 1, 8-cineole to inhibit bacteria associated with minimally processed vegetables. *International Journal of Food Microbiology, 154*(3), 1451.

Di Pasqua, R., Betts, G., Hoskins, N., Edwards, M., Ercolini, D., & Mauriello, G. (2007). Membrane toxicity of antimicrobial compounds from essential oils. *Journal of Agricultural and Food Chemistry, 55*(12), 4863–4870.

Diniz-Silva, H. T., Brand̃ao, L. R., de Sousa Galvao, ̃ M., Madruga, M. S., Maciel, J. F., de Souza, E. L., & Magnani, M. (2020). Survival of Lactobacillus acidophilus LA-5 and Escherichia coli O157:H7 in Minas Frescal cheese made with oregano and rosemary essential oils. *Food Microbiology, 86*. Available from https://doi.org/10.1016/j.fm.2019.103348, Article 103348.

Donsì, F., & Ferrari, G. (2016). Essential oil nanoemulsions as antimicrobial agents in food. *Journal of Biotechnology, 233*, 106–120.

Dwivedi, C., Muller, L. A., Goetz-Parten, D. E., Kasperson, K., & Mistry, V. V. (2003). Chemopreventive effects of dietary mustard oil on colon tumor development. *Cancer Letters, 196*(1), 29–34.

Falleh, H., Ben Jemaa, M., Saada, M., & Ksouri, R. (2020). Essential oils: A promising eco-friendly food preservative. *Food Chemistry, 330*, 127268.

Gaysinsky, S., Davidson, P. M., Bruce, B. D., & Weiss, J. (2005). Growth inhibition of Escherichia coli O157: H7 and Listeria monocytogenes by carvacrol and eugenol encapsulated in surfactant micelles. *Journal of Food Protection, 68*(12), 2559–2566.

Gould, G. W. (1989). *Mechanisms of action of food preservation procedures*. London, UK: Elsevier Applied Science.

Gustafsson, J., Cederberg, C., Sonesson, U., & Emanuelsson, A. (2011). The methodology of the FAO study: Global food losses and food waste-extent, causes and prevention-FAO; SIK Institutet för Livsmedel och Bioteknik: Gothenburg. *Sweden*, 2013.

Hyldgaard, M., Mygind, T., & Meyer, R. L. (2012). Essential oils in food preservation: Mode of action, synergies, and interactions with food matrix components. *Frontiers in Microbiology, 3*, 1–24.

Kalemba, D., & Kunicka, A. (2003). Antibacterial and antifungal properties of essential oils. *Current Medicinal Chemistry, 10*(10), 813–829.

Kim, J., Marshall, M. R., & Wei, C.-I. (1995). Antibacterial activity of some essential oil components against five foodborne pathogens. *Journal of Agricultural and Food Chemistry, 43*, 2839–2845.

Kriegel, C., Kit, K. M., McClements, D. J., & Weiss, J. (2010). Nanofibers as carrier systems for antimicrobial microemulsions. II. Release characteristics and antimicrobial activity. *Journal of Applied Polymer Science, 118*(5), 2859–2868.

Li, J. J., Zehentbauer, G. N., Bunke, P. R., Zent, J. B., Ekanayake, A., & Kester, J. J. (2005). Isogard (tm) a natural anti-microbial agent derived from white mustard seed. *InI International Symposium on Natural Preservatives in Food Systems, 709*, 101–108.

Liang, R., Xu, S., Shoemaker, C. F., Li, Y., Zhong, F., & Huang, Q. (2012). Physical and antimicrobial properties of peppermint oil nanoemulsions. *Journal of Agricultural and Food Chemistry, 60*(30), 7548–7555.

Luque De Castro, M. D., Jiménez-Carmona, M. M., & Fernández-Pérez, V. (1999). Towards more rational techniques for the isolation of valuable essential oils from plants. *TrAC - Trends in Analytical Chemistry, 18*(11), 708–716.

McClements, D. J. (2012). Nanoemulsions versus microemulsions: Terminology, differences, and similarities. *Soft Matter, 8*(6), 1719–1729.

Mizrahi, M., Friedman-Levi, Y., Larush, L., Frid, K., Binyamin, O., Dori, D., Fainstein, N., Ovadia, H., Ben-Hur, T., Magdassi, S., & Gabizon, R. (2014). Pomegranate seed oil nanoemulsions for the prevention and treatment of neurodegenerative diseases: The case of genetic CJD. Nanomedicine: Nanotechnology. *Biology, and Medicine, 10*(6), 1353–1363.

Moro, A., Libran, C. M., Berruga, M. I., Carmona, M., & Zalacain, A. (2015). Dairy matrix effect on the transference of rosemary (Rosmarinus officinalis) essential oil compounds during cheese making. *Journal of the Science of Food and Agriculture, 95*, 1507–1513. Available from https://doi.org/10.1002/jsfa.6853.

Musa, S. H., Basri, M., Masoumi, H. R., Karjiban, R. A., Malek, E. A., Basri, H., & Shamsuddin, A. F. (2013). Formulation optimization of palm kernel oil esters nanoemulsionloaded with chloramphenicol suitable for meningitis treatment. *Colloids and Surfaces B: Biointerfaces, 112*, 113–119.

Nazzaro, F., Fratianni, F., De Martino, L., Coppola, R., & De Feo, V. (2013). Effect of essential oils on pathogenic bacteria. *Pharmaceuticals, 6*(12), 1451–1474.

Olmedo, R. H., Nepote, V., & Grosso, N. R. (2013). Preservation of sensory and chemical properties in flavoured cheese prepared with cream cheese base using oregano and rosemary essential oils. *Lebensmittel-Wissenschaft und -Technologie- Food Science and Technology, 53*, 409–417. Available from https://doi.org/10.1016/j.lwt.2013.04.007.

Prakash, B., Kujur, A., Yadav, A., Kumar, A., Singh, P. P., & Dubey, N. (2018). Nanoencapsulation: An efficient technology to boost the antimicrobial potential of plant essential oils in food system. *Food Control, 89*, 1–11.

Roldan, L. P., Diaz, G. J., & Duringer, J. M. (2010). Composition and antibacterial activity of essential oils obtained from plants of the Lamiaceae family against pathogenic and beneficial bacteria. *Revista Colombiana de Ciencias Pecuarias, 23*, 451–461.

Sagis, L. M. (Ed.), (2015). *Microencapsulation and microspheres for food applications*. Academic Press.

Shannon, E. M., Milillo, S. R., Johnson, M. G., & Ricke, S. C. (2011). Efficacy of cold-pressed terpeneless valencia oil and its primary components on inhibition of listeria species by direct contact and exposure to vapors. *Journal of Food Science, 76*(7), 500–503.

Shao, X., Cheng, S., Wang, H., Yu, D., & Mungai, C. (2013). The possible mechanism of antifungal action of tea tree oil on Botrytis cinerea. *Journal of Applied Microbiology, 114*, 1642–1649.

Shatalov, D., Kedik, S., Zhavoronok, E., Aydakova, A., Ivanov, I., Evseeva, A., Beliakov, S., Biryulin, S., Kovalenko, A., & Mikhailenko, E. (2017). The current state and development of perspectives of application of synthetic antimicrobial agents. *Polymer Science, Series D, 10*, 293–299.

Sperotto, A. R., Moura, D. J., Péres, V. F., Damasceno, F. C., Caramão, E. B., Henriques, J. A., & Saffi, J. (2013). Cytotoxic mechanism of Piper gaudichaudianum Kunth essential oil and its major compound nerolidol. *Food and Chemical Toxicology, 57*, 57–68.

Tajkarimi, M. M., Ibrahim, S. A., & Cliver, D. O. (2010). Antimicrobial herb and spice compounds in food. *Food Control, 21*(9), 1199–1218.

Walker, R., Decker, E. A., & McClements, D. J. (2015). Development of food-grade nanoemulsions and emulsions for delivery of omega-3 fatty acids: Opportunities and obstacles in the food industry. *Food & Function, 6*(1), 41–54.

Yang, Y. C., Lee, S. G., Lee, H. K., Kim, M. K., Lee, S. H., & Lee, H. S. (2002). A piperidine amide extracted from Piper longum L. fruit shows activity against Aedesaegypti mosquito larvae. *Journal of Agricultural and Food Chemistry, 50*(13), 3765–3767.

Zeng, W. C., He, Q., Sun, Q., Zhong, K., & Gao, H. (2012). Antibacterial activity of water-soluble extract from pine needles of Cedrus deodara. *International Journal of Food Microbiology, 153*, 78–84.

Zhang, Y., Chen, H., & Pan, K. (2017). Chapter 5—Nanoencapsulation of food antimicrobial agents and essential oils. In S. M. Jafari (Ed.), *Nanoencapsulation of food bioactive ingredients* (pp. 183–221). Cambridge, MA: Academic Press.

Zhu, Y., Li, C., Cui, H., & Lin, L. (2021). Encapsulation strategies to enhance the antibacterial properties of essential oils in food system. *Food Control, 123*, 107856.

Chapter 10

Improvement of food aroma and sensory attributes of processed food products using essential oils/boosting up the organoleptic properties and nutritive of different food products

Abiola Folakemi Olaniran[1], Adeyemi Ayotunde Adeyanju[1], Olubukola David Olaniran[2], Christianah Oluwakemi Erinle[3], Clinton Emeka Okonkwo[4] and Abiola Ezekiel Taiwo[5]

[1]Department of Food Science and Microbiology, Landmark University, Omu-Aran, Kwara State, Nigeria, [2]Department of Sociology and Anthropology, Obafemi Awolowo University, Ile-Ife, Osun State, Nigeria, [3]Department of Agricultural and Biosystems Engineering, Landmark University, Omu-Aran, Kwara State, Nigeria, [4]Department of Food Science, College of Food and Agriculture, United Arab Emirates University, Al Ain, Abu Dhabi, [5]Faculty of Engineering, Mangosuthu University of Technology, Durban, Umlazi, South Africa

Introduction

Spices and aromatic plants serve as flavor, aroma, and coloring agents in food and beverages since immemorial. They are rich in bioactive and functional substances, which can improve food-related qualities (Benkhoud et al., 2022). Essential oils (EOs) are complex mixes known for their capacity to improve one's health based on several distinct components extracted from plants using diverse techniques. These diverse volatile organic components contribute to a plant's flavor as well as fragrance. Due to their biological qualities, these oils are commonly used in the pharmaceutical, food, and feed industries. EOs have been in use for several centuries among humanities either for therapeutic or vigor purposes. They are hydrophobic concentrated volatile in a liquid form comprising of different chemical mixtures from plants or their extracts that can vaporize easily at ambient temperature. EOs are aromatic compounds found in large amounts in oil glands or sacs present at diverse depths in the fruit peels mostly cuticles and flavedo. It can also be extracted from other parts of plants such as flowers, bark, leaves, seeds, and peels (Tongnuanchan & Benjakul, 2014). EOs can be obtained by extraction, fermentation, expression, and effleurage, although steam or hydro-distillation are commonly used in industrial production (Herman et al., 2019).

Recently, EOs are of interest because of their vast spectrum of biological capacity, including anti-inflammatory, anti-nociceptive, antibacterial, anti-carcinogenic, antioxidant, and acaricidal properties. Because of their antioxidant properties and ability to prevent lipid peroxidation, oil flavoring is one of the most widely used EOs in the food industry. Many studies have explored the synergistic interactions of several EOs with antibiotics, plant extracts, phytochemicals, and other food constituents (Herman et al., 2019). Researchers have also evaluated their potential antimicrobial activities on EOs encapsulation with different metal and metal oxide nanoparticles (Melendez-Rodriguez et al., 2019). Therefore, this chapter explores the application of EOs for the improvement of food aroma and sensory qualities of foods toward boosting the organoleptic properties and nutritive of different food products.

Applications of Essential Oils in the Food Industry. DOI: https://doi.org/10.1016/B978-0-323-98340-2.00006-7

Chemical constituents of essential oils

Studies on the chemical composition of EOs indicated that they are composed of hundreds of distinct compounds containing aromatic rings. These compounds can be classified into two types based on their structural similarities. Terpenoids, monoterpenes, sesquiterpenes, and diterpenes are all made up of isoprene units, while phenylpropanoids constitute the second group of phenolic compounds (Aguirre et al., 2013). Fig. 10.1 depicts the individual representative component structures. In most situations, one or two of these chemical constituents have substantially higher concentrations than any other compounds present, and such compounds usually exhibit high bioactivity (Jayasena & Jo, 2013). The most frequent EO compounds are terpenoids, and they come in a variety of shapes and functions. Terpenoids have a fundamental structure that consists of more than one isoprene unit. The most frequent terpene kinds are monoterpenes and sesquiterpenes; diterpenes and triterpenes are other types of terpenes. Despite the fact that monoterpenes and sesquiterpenes have different molecular weights, they are very comparable in structure and function. There are two forms of terpenoids that contain oxygen: oxygenated monoterpenes and oxygenated sesquiterpenes. According to the presence or lack of oxygen, terpenoids can be classified into several groups. Phenylpropanoids, phenylpropane, and terpenoids are members of another structural family with distinct metabolic and biosynthetic pathways (Luz et al., 2020). Phenylpropanoids are a class of natural chemicals that can make intermediate products by using different amino acid routes that are aromatic in nature. The main constituent in clove bud EO is eugenol, a phenylpropene, according to a previous Gas Chromatography-mass spectrometry analysis (Amelia et al., 2017). Although phenylpropane and

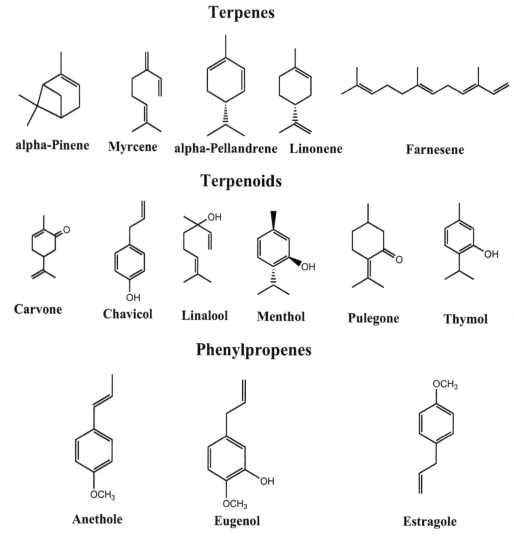

FIGURE 10.1 Main chemical constituents' in essential oil (Ni et al., 2021).

terpenoids make up the majority of EOs, they are made up of a wide range of chemicals and structures, each with its own set of functions and prospects. The varied components of plants, how long they have been around, the stages of the nutrition cycle, and the environment, on the other hand, all have an impact on the quality, amount, and composition of EOs (Tanveer et al., 2020). These chemicals are found in all EOs and play a crucial function as antioxidants and antibacterial agents. When added to processed foods, they also have an impact on the aroma and sensory qualities.

Food aroma

The sensory assessment offers a descriptive view of the consumer's satisfaction with a certain flavor or product. Flavor is a technology that turns out to be effective by captivating either directly the body or the mind of consumers. Flavors are added either for masking the unacceptable taste or aroma or to improve the aroma and brand such products more attractive. The recent trend in awareness is termed "green consumerism" where consumers prefer, promote, and demand the inclusion of organic products that are plant-based especially EOs in processed foods. Flavors play a principal and imperative part in increasing the desirability of products. Pleasant smells stimulate the comfortability and the acceptance of the products, as well as the overall evaluation of consumers. For industrial applications usually, standardization is germane in mixing the EOs from diverse plants targeting attaining a specific aroma. The predominantly EOs used for industrial purposes are majorly from mint, orange, citronella, corn, eucalyptus, lemon, and peppermint. Globally, organic marketplace for flavor is experiencing a rapid growth sequel to a surge in the utilization of natural flavors, for instance, EOs above artificial flavor based on their several allied health benefits and this is expected to rapidly grow as time goes on. The global market for EOs was reported to surpass about seven billion US dollars in the year 2018 and is projected to increase by 9% yearly from 2019 to 2026 (Sharmeen et al., 2021).

Consumer acceptability is one of the utmost vital keys to winning a competitive food industry. The addition of flavors is becoming a global practice in processed and packed foods (Fiorentini et al., 2020). The strong odor of EOs is one of their distinguishing features, which is normally made up of roughly 200 ingredients separated into two categories: volatile and non-volatile portions (Jalali-Heravi & Parastar, 2011). Due to their strong flavor, the application and choice of EOs have to consider the consumer's sensory suitability to the final product leading to its limitation in the quantity that can be used. However, EOs are organic as well safe which are documented as GRAS may be used in slight or big amounts subject to the properties of the bioactive compounds which have been described to enhance the product quality safely devoid of initiating nutritive and organoleptic damages (Olaniran et al., 2015). Suggestions of its application as edible coatings as a replacement in food packaging to enhance food safety and quality EOs have been reported (Acevedo-Fani et al., 2015). The use of edible coatings comprising oregano and thyme EOs in fresh beef pieces was reported for inhibiting pathogenic bacteria and improving color stability with satisfactory sensory attributes. It also had desired influence on the enzyme inactivation and nutritional content of the product (Yemiş & Candoğan, 2017).

Citrus peel EO has pleasurable organoleptic features and is a good source of natural active constituents with numerous health profits. It also contains very volatile components which are sensitive to heat, oxygen, and light. They are widely used as flavoring for elevating and stimulating agents during the production of drinks, and ice creams, as well as in the preparation of cosmetics, colognes, bathing soaps, and other products thus putting them in high demand in pharmaceutics, cosmetics, food, confectionery, and perfumery industries due to their spectacular bioactivities and aroma (Bustamante et al., 2016). EO such as sesquiterpenes, monoterpenes, and their oxygenated derivatives which were extracted from citrus demonstrated powerful inhibitory activities against pathogenic bacteria. Extract from orange, grape, and mandarin contains a colorless aliphatic hydrocarbon Limonene in abundance and was recognized as the main constituent in citrus EO which is a non-oxygenated cyclic monoterpene comprising dual isoprene units. It's usually used as a flavoring agent in several food products due to its pleasing citric fragrance Verbenone, α-pinene, β-pinene, β-myrcene, sabinene, γ-terpinene, and α-terpinolene among others have been extracted from citrus peel and used as flavoring agents (Singh et al., 2021). Several bioactive compounds found in EOs contain numerous bioactive compounds such as thujone, terpineol, myrcenol, neral, thujanol, carvone, and camphor among others that have shown noteworthy health benefits. Carvacrol had been reported to have low toxicity and is now extensively used as a sweets and drinks product flavoring agent and preservatives (Bhavaniramya et al., 2019). Studies had shown that the application of EO from lemongrass in a high quantity may affect organoleptic acceptability in the product making the food product undesirable for the consumer (Faheem et al., 2022). EOs are essential and show important aspects in the protection of the crop as well as the food industry with immeasurable utilization. The procedure of combining singly or more for synergy can results in the anticipated smell and aroma devoid of any unpleasant effects on the food products; therefore, their usage in food production and customer goods is estimated to be on the increase in the future

(Olaniran, Afolabi et al., 2020). However, alterations have been established to occur in EOs during aging which does not only result in organoleptic or sensory and technical damage but may also affect consumer safety.

Cinnamaldehyde is the major component found in cinnamon leaves oil that is responsible for the distinguishing aroma and flavor in cinnamon EOs. The strong flavor has been assessed and established during its inclusion in tomato paste and baked products (Mariod, 2016). Taste and flavor are key organoleptic factors in adding cinnamon oil at varying levels which resulted in non-significant changes in the color, appearance, taste, and aroma acceptability if added at the concentration of 0.25% cinnamon EOs in chocolate (Dwijatmoko et al., 2016). Flavoring of confectioneries, meat, and allied products using EO from berry oil and leaf oil at the optimum level was done to improve the flavor of the food. Carvacrol is believed to have an amiably strong aroma on fish and the European Commission recorded EO from eugenol, carvone, carvacrol, citral, cinnamaldehyde, p-cymene, limonene, thymol, and menthol as approved food flavorings without any threat to the well-being of the consumer. The use of EOs menthol, thymol, and eugenol as a treatment on strawberries increased the acceptability, reduced the rate of fruit spoilage, and conserved fruit quality compared to control samples (Wang et al., 2007). The sensory ratings of vegetables cleaned with EOs from *Rosmarinus officinalis* and *Oregon vulgare* either singly or combined were also within the acceptable scores. There was no significant difference in the aroma and taste of vegetables washed using blends of EOs compared to those without EOs. Throughout the period of storing, there was no negative impact on the sensory characteristics of samples preserved with the blend of EO (De Azeredo et al., 2011). EOs from thyme, clove, and summer savory were also reported to be accepted by panelists (Mariod, 2016). The sourcing of aroma should be varied and dependent on the food as it's a measure of the constituents that has a vital role in increasing people's interest in certain foods.

Correlation of food products and consumer perception of food products

Sensory assessment assumes various functions in foreseeing consumer reception of the product. The general appearance of a brand is significant for preparing buyers and creating assumptions before utilization. A disconfirmation of assumptions happens when the apparent preference after utilization is beneath the normal fondness, which might happen when the viewable signs distort the taste, smell, and aroma of the item. Accordingly, it is essential to convey top-notch sensory qualities that are seen both previously and while consuming the product (Fiorentini et al., 2020).

Customers choose natural substances since they have been demonstrated to be low in toxicity, have the same potency as synthetic compounds, and are preferred by customers (Katz et al., 2008). Because EOs are more viable when they interact directly with food, measures should be set up to guarantee that the EOs are at successful intensification to display antimicrobial impacts. Straight contact with meat, on the other hand, can change the meat's organoleptic qualities via different chemical interactions. EOs have a strong fragrance, even in low quantities, which can be overwhelming to consumers. However, there are recent trends in the application of novel preservation methods for the conservation of flavor and quality of the EOs. The widespread usage of these novel alternative preservation techniques have considerably enhanced consumer longing for better and stable aroma, taste, color, and nutritional value of food product (Basavegowda & Baek, 2021).

Essential oil

EOs are increasingly being employed as food additives due to their substantial antibacterial and antioxidant qualities. EOs have a wide range of biological activity, fragrance, and physicochemical properties. As a result, it is vital to select the most appropriate one for each food product (Chen et al., 2021). EOs can be categorized depending on their various strategies for extraction, synthesis of chemicals, notes, and fragrance. EOs are colorless fluids, predominantly containing the aromatic and volatile mixtures normally present in all pieces of the plants including seeds, blossoms, strips, stems, bark, and entire plants. Oils removed from different vegetable seeds are plentiful in proteins and nutrients, and the majority of the vegetable seeds including sunflower and perilla seeds have heterocyclic substances, for example, pyrazine, which assumes a vital part in the flavor and nature of the items (Cuicui & Lixia, 2018).

Classification based on the aroma of EOs can likewise be characterized because of the fragrance of the oil. These are classified into herbaceous, spicy, citrus, camphorous, resinous, minty, woody, earthy, and floral oils (El-Zaeddi et al., 2016). EOs from citrus have a distinctive citrus flavor usually extracted from tangerine, grapefruit, bergamot, lemon, orange, lime, and so on. Oil extracted from spices known as herbaceous oils are commonly obtained from chamomile, basil, melissa, peppermint, hyssop, marjoram, rosemary, and clary sage among others. Camphoraceous EOs have been reported to have specific beneficial properties. These therapeutic oils are obtained from tea tree, cajeput, characterized as a borneol-like, mugwort-like, rosemary-like, fruity, dried plum-like substance. Floral oils such as Jasmine oil,

Geranium oil, Lavender, Neroli, Chamomile oil, Rose, oil, Ylang oil, etc. are made from flowery fragments of the plants. For the woody EOs bark as well as other woody sections of cypress, cedar, cinnamon, juniper pine, sandalwood, berry, etc., are utilized for the production. EOs from thyme, aniseed, cloves, black pepper, cinnamon, coriander, cumin, cardamom, nutmeg, ginger, and turmeric are gaining customers and industrial attention in food production (Herman et al., 2019; Olaniran, Abu et al., 2020).

Application of essential oil in improving the organoleptic and nutritive properties of different food products

The usage of EOs as flavorings in food has been confirmed "Generally Recognized as Safe" by the US Food and Drug Administration. Cinnamon natural oils have been characterized as highly significant EOs utilized commonly in food and beautifying enterprises because of their numerous uses as an enhancer of unique flavors (Dhifi et al., 2016). The coating's flavor and aroma are essential factors so they must be carefully selected. It may also have an impact on the food product's overall acceptability. For example, when the sensory analysis was done on alginate-coated samples after 15 days of storage, they were determined to have the highest acceptability (Manzoor et al., 2021). Hypoallergenicity, enhanced fragrance and savor, and possible well-being gains to consumers (e.g., antitumor, antioxidant) that EOs can bestow by the feature of their ingredients have all been highlighted as the advantages of employing EOs as preservatives (Voon et al., 2012).

Although there is a lot of research into the antibacterial characteristics of EOs generated from a variety of plants, most EOs' in vitro fallouts do not simply turn into practical usage as meat preservers. Diverse research on the ability of ginger, rosemary, bay, oregano, cinnamon, sage, and garlic to work as meat preservatives has yielded encouraging findings (Jayasena & Jo, 2013). Directly applying EOs to meat on a wide scale to guarantee that controlled and enough volumes of EOs reach the meat to successfully preserve it would be extremely difficult and expensive. Sánchez-González et al. (2011) found that EO sachets can be added to bundled meat and utilized to conserve it. The quality of EOs is determined by elements such as traditional and novel extraction procedures, geographical variances, plant growth cycle, drying time, and technology (Ni et al., 2021). The Labneh, a semi-solid fermented dairy food product with high microbial load, combined with the packing and storing conditions, resulted in off-flavors and negative physicochemical changes, eventually leading to the product's rejection. Following that, an investigation was conducted, which led to the usage of bio-preservatives like plant EOs to extend the lifespan of the product (Al-Rimawi et al., 2020).

Notwithstanding the verified promise of EOs application in the preservation of meat, when large quantities of EOs are used to achieve significant antibacterial action, the flavor, aroma, and quality of the meat are affected (Gutierrez et al., 2009). Because of the intense smell and odor essential to particular EOs, when modest quantities of the EOs were even applied in meat preservation, unfavorable organoleptic alterations would occur, reducing the acceptability of the preserved meat. According to a study, the integration of EOs gotten from zataria, peppermint, and basil into the probiotic yogurt preparation has the potential to improve the future performance of the product in addition to its inhibitory effect as opposed to *E. coli* and *L. monocytogenes*. Azizkhani & Parsaeimehr (2018) established the capability of probiotics, organoleptic adequacy, and antioxidant activity of probiotic yogurt of EOs extracted from peppermint, zataria, and basil. Foods, beverages, and confectionery goods have all been flavored with EOs and their constituents. Numerous EOs derived from spices and herbs are utilized as additives in food. Their primary purpose is to give pleasant tastes and fragrances. The presence and content of EOs can clearly define the perfume and flavor of plants and condiments. Herbal and spice EOs could be employed as useful additives. Sweet basil is commonly used in salads, pizzas, meats, and soups to impart a particular scent and flavor. Allspice (*Pimenta dioica*) EOs from berry oil and leaf are utilized as flavoring agents in sweets and meats. Berry oil is allowed in food at a maximum concentration of 0.025%. Okunola et al. (2007) isolated EOs from *Piper guineense* seeds, Eugenia aromatic flower buds, and *Monodora myristica* kernels using n-hexane, and added them to storage grains. The impact of these EOs on the grains' appearance, taste, and aroma was then studied with the use of proficient panelists. The authors reported that the assessors strongly preferred the grains that were prepared with a lesser dose proportion of 5 mL/kg EO. Prepared grains infused with nutmeg EO were also well received and preferred. In some countries such as Thailand, Malaysia, Indonesia, and China, the EOs from *Alpinia galanga* leaves, stems, rhizomes, and roots are used to flavor various foods. Food flavors from natural sources include lavender oil, EO from lavender, lavender spike essential oil, and absolute lavender. In the food industry, they are used in baked goods, frozen dairy, and non-alcoholic beverages, to name a few. A blend of caprylic acid (CA) and oregano EO (OEO) as mixed preservatives were used in meat. It was stated that CA-washed samples had the lowermost quality ratings all through the time of storing as well as deemed unsatisfactory. Nevertheless, samples preserved with OEO outperformed the control sample set and remained above the unbiased point (rate 5) till storage day 6. Cooked samples preserved with volatile OEO had a stronger odor, but it was still deemed acceptable (Hulankova, 2022). The powerful odor notes of fresh sweet basil were described as fresh (8.1), flowery (8.0), and herbaceous (6.4). They found critical expansions in the forces of the dehydrated samples listed qualities contrasted with the new sample utilizing vacuum-microwave and convective drying techniques. Once added to mature olives before extraction to test their reaction on the nature of the resultant olive oil, an average of 1– 3% of leaves from olive enhanced the organoleptic attribute. The influence of 7 EOs (namely, thymol, eugenol, borneol, cinnamaldehyde, carvacrol, vanillin, and menthol) on the development of stimulated *Bacillus cereus* injected into seasoned carrot broth was investigated as food preservatives. The panelists' assessments of thymol, carvacrol, and cinnamaldehyde revealed that thymol possessed an intense odor that gave a poor aftertaste and decreased product acceptability or liking. The authors of the research discovered that a little percentage of cinnamaldehyde can boost the flavor of carrot broth without affecting its fragrance or taste (Valero & Giner, 2006).

Several different EO components are used for food flavorings and give dishes a distinct flavor. Carvacrol is claimed to give fish a "warmly pungent" smell. Fresh kiwi fruit and honeydew melon were found to be resistant to spoiling when preserved with tiny doses of carvacrol or cinnamic acid without altering their organoleptic properties (Roller & Seedhar, 2002). It has also been discovered that treating lettuce with 250 ppm oregano was satisfactory to a sensory panel as well as there were no dissimilarities between the ordinary lettuce and lettuce rinsed with chlorinated water (Gutierrez et al., 2009). Also, Wang et al. (2007) observed that adding thymol, eugenol, or menthol to strawberries delayed fruit spoilage and improved the quality of the fruit with more elevated levels of flavonoids, sugars, natural acids, phenolics, anthocyanins, and capacity of oxygen radical assimilation. The integration of rosemary and oregano EOs increases the fermentative and oxidative steadiness of flavored cheeses made with a cream cheese base, hindering lipid oxidation and the formation of rancid and fermented tastes. The examined samples were assessed for peroxide and anisidine levels, descriptive examination, and fermentation constraints as stability pointers during storage. Flavored cheese had better storage stability, reduced acidity, lower total viable counts, as well as a higher pH. In flavored cheese made with cream cheese basil, oregano, and rosemary, EOs showed a defensive effect against lipid oxidation and fermentation (Olmedo et al., 2013). De Azeredo et al. (2011) investigated sensory rates in vegetables sterilized with EOs from *O. vulgare* and *Rosmarinus officinalis* separately and in combination. It was discovered that for samples treated with EOs separately and in combination, the mean of the most evaluated qualities fell between "like slightly" and "neither like nor dislike" on the hedonic scale. There was no significant difference in odor or taste between the samples sterilized with the blend of EOs and the ordinary samples at any of the storage times tested. The sensory characteristics of vegetables treated with a combination of EOs were unaffected by storage duration. Summer savory, clove, and thyme EOs possess antifungal properties in ketchup and tomato paste. When panelists tasted the treated samples, they discovered that thyme oil was preferred over clove and summer savory. EOs in their entirety were discovered to limit the development of *Aspergillus xavus*. In a tomato ketchup base, taste panel evaluations revealed that the oil sample containing 500 ppm of thyme was acceptable to panelists. A variety of options can be used to conquer fluctuations in the organoleptic properties of tomato paste. One approach is to think of EO as a flavor component rather than just a preservative. Second, to hide the presence of the plant essential oil, include EO in items that already have a strong flavor. A third option is to employ part of the most dynamic parts as opposed to the entire oil, and to reduce organoleptic properties modifications while maintaining antibacterial action (Omidbeygi et al., 2007).

Cinnamaldehyde is the major component of the oil from cinnamon leaf, and it's this compound that gives cinnamon its characteristic odor and flavor. The strong flavor of cinnamon EOs can be found in bread products and tomato paste (Kalantary, 2014). Using EO vapors to prevent the organoleptic effect caused by straight interaction between food and EO could be a viable option. For example, when cabbage leaves were exposed to EO vapors of bergamot, lemon, and sweet orange for 24 hours, some physical changes were evident. The vapors were consumed into the food medium and affected its organoleptic properties (Fisher & Phillips, 2006). To safeguard the lifespan and quality of food, EOs could be used as preservatives. Al-Otaibi and El-Demerdas (2008) found that adding EOs from marjoram, sage, or thyme to labneh extended its shelf life by up to 21 days while maintaining an excellent flavor and appearance with no symptoms of spoilage. The overall rate of labneh containing EOs of marjoram, sage, or thyme declined as the concentration of the EOs increased; according to these authors. Organoleptically, labneh with 0.2 ppm oils from sage, thyme, and marjoram was ranked the best palatable, with a structure comparable to the ordinary control. Decontaminated *Agaricus bisporus* for 16 days with EOs from thyme, cinnamaldehyde, and clove (Chu et al., 2020). The mushrooms fumigated with cinnamaldehyde slightly changed to brown toward the conclusion of the storage period, but they still maintained commercial value and edibility, according to these authors. Although there is no evidence that these natural compounds have a role in this issue, their well-known antioxidant activity may help to prevent dehydration and the formation of brown polymers, which cause the mushrooms to turn brown and shrivel during storage. As a result, utilizing EOs to fumigate button mushrooms improves their storing attributes.

The use of unmixed thyme, rosemary, and oregano EOs on cabbage, carrot disks, lettuce, and dried-up coleslaw mix resulted in undesirable visual damage and an unpleasant odor (Scollard et al., 2013). Report also showed that treating sliced bell pepper tissues with thyme EO softened the tissues (Uyttendaele et al., 2004). These challenges can be prevented by ensuring that the EOs are diluted, using combinations of EOs, or using EOs in a mixture with plant substances like cabbage shreds or carrots (Scollard et al., 2013). In another work carried out by Tserennadmid et al. (2011), it was discovered that fruit or milk that are prone to yeast spoilage could be treated with EO instead of using synthetic preservation approach; they also discovered that adding lemon EO to clear apple juice resulted in a harmonic taste while also extending storage duration. The lemon EO was also found to have a good effect on the taste of juice in all treated products, according to these researchers. The clear apple juice had a strong disagreeable odor, whereas the murky juice had a weaker one. The most pleasant odor came from pasteurized milk that had been treated with lemon EO (Tserennadmid et al., 2011). According to another study, adding lemon EO to apple juice did not cause the organoleptic properties to degrade. Furthermore, the sensory impact of lemon EO in juice squeezed afresh may be masked in part by the stronger flavor (Espina et al., 2012).

Investigation of the impact of oregano EO in conjunction with the packaging method on the shelf life as well as the sensory characteristics of raw chicken chest meat held at 4°C revealed that oregano oil had no discernible effect on the color of the treated meat; it did provide a very strong flavor to the product when used at a concentration of 1.0% (Chouliara et al., 2007). Because even a small amount of EO can elicit sensory changes, the choice of EO and its concentration in a specific dish is critical. The intense aroma of the EOs can impair the food's organoleptic properties; however synergistic mixtures of EOs or other technological approaches can lessen this effect. Donsì et al. (2011) tested the microbiological steadiness and the modification of the organoleptic properties of juices from pear and orange to which *Lactobacillus delbrueckii* was inoculated and adding nanoemulsion of terpenes mix plus soy lecithin coated in nanoemulsions depending on the level in food constituents, produced through high-pressure homogenization at various concentration. The inclusion of the nanoemulsion had no discernible effect on the juice content or pH, according to these researchers. Unless a major microbial growth occurred, the situation remained the same over the storage period. Once nanoencapsulated-terpenes were introduced into the juices, considerable color differences were noticed, however adding the combination of the terpenes at lesser intensifications (50 and 1.0 g/L) was regarded as adequate, causing much lesser color deviations. For all of the systems, the hue stayed consistent during the storage period. In a different study conducted on juices of fruit (Tyagi et al., 2014), the mentha oil antagonistic action yeast and vapors in blended fruit juices by determining the minimum inhibitory concentration (MIC) as well as minimum fungicidal concentration compared to 8 yeasts associated with spoilage was carried out. The appearance of treated juices remained satisfactory for an extended duration than the ½ MIC and ¼ MIC preserved samples, according to these researchers. The mentha oil did not have an unfavorable effect on the odor or color of the blended juice. When compared to untreated samples, the aroma of blended juice added with mentha oil lasted eight days more. Furthermore, in trials treated with MIC-level mentha oil, the color remained consistent. Juice color and odor lasted up to four days in samples treated with 1/2MIC levels, but only two days in samples treated with 1/4 MIC levels. When compared to butylated hydroxyanisole, *Cinnamon zeylanicum* and *Zataria multiflora* EOs were discovered to inhibit the rate of oxidation and decrease oxidation production when applied to the cakes at varied quantities during processing. These EOs were proven to have significant nutritional and functional benefits in cakes. In terms of color and textural attributes, there were no significant variations between the treated and untreated samples. Finally, the results revealed that the cakes made were universally praised by the panelists (Kordsardouei et al., 2013). Olmedo et al. (2012) found that adding cedron and aguaribay EOs to salted, fried peanuts improved the profile of taste and consumer reception while also showing a preventive effect against lipid oxidation.

An investigation on the effect of chitosan at 1% in Artemisia EO at different ppm (500−1500) on the stability of chicken fillets under cold storage was documented. For a period of 12 days, the organoleptic, physicochemical, and microbiological attributes of parceled beef samples were assessed at certain time intervals. The researchers discovered that using chitosan as a coating material in combination with *Artemisia fragrans* reduced pH and thiobarbituric acid reactive compounds, resulting in coated chicken flesh with a high overall acceptance score. Takma and Korel (2019) came up with active polyester films with coatings made of CH and alginate and black cumin oil (BCO). The researchers investigated how they impacted the shelf life of refrigerated chicken breast meat. Antagonistic effects of BCO on *Staphylococcus aureus* and *Escherichia coli* antibodies were documented for packaged samples containing films infused with BCO, as well as changes in color, pH, and total viable count in addition to psychrotrophic bacteria counts (Yaghoubi et al., 2021). Souza et al. (2019) developed a nanocomposite film containing chitosan, montmorillonite, and ginger essential oil. The films demonstrated antibacterial activity, preserved color, pH, and TBARS levels, and extended the lifespan of fresh chicken meat. To create a new edible coating, Noshad et al. (2021) employed seed mucilage from

Plantago major with citrus-lemon oil (0%−2%). This was used to extend the lifespan of buffalo meat. Meanwhile the storage process, the novel coating significantly reduced the development of microbial growth and peroxide value in samples of meat. The investigators suggested that the developed films could be employed as innovative and effective edible packaging to improve the resistance of beef products against lipid oxidation and microbiological deterioration. Using nanoencapsulation and chitosan mediums with integrated EO derived from Cinnamodendron dinisii Schwanke, Xavier et al. (2021) developed a dynamic nanocomposite packaging of film. The obtained chitosan-nanoparticles were effective in preserving minced beef, preserving the color, and maintaining the degradation processes of the chemical. Kiarsi et al. (2020) used hydro-distillation to extract EO from nutmeg (*Myristica fragrans*) seed to create a unique edible covering that improved the quality of beef slices. TVC, *E. coli*, *S. aureus*, PTC, and fungi were all inhibited by the coating. According to the authors, coated beef samples have a high level of consumer approval and can prevent pathogenic bacteria proliferation and lipid oxidation in meat products, while also improving color stability and taste. It has been discovered that the integration EO derived from lemon didn't bring about a disintegration of the apple juice OLP. In addition, the tangible effect of EO derived from lemon in a juice newly pressed could to some extent be masked by its more extravagant flavor (Espina et al., 2012).

Flavors derived from spices, and aromatic herbs especially their EOs have been stated to have a natural identity and desirable flavor which were pleasing to a broad class of customers. The application of EOs from mint was used to conceal the unfriendly taste of tuna oil in yogurt (Bakry et al., 2019). Thus, the integration of a few essentials oils into food varieties during production can forestall their rate of oxidation and decrease oxidation products with no alteration in color and surface attributes (Mariod, 2016).

Conclusion

The health-enhancing natural actions of EOs include operating as an analgesic, antioxidant, antimicrobial, anticancer, and anti-inflammatory specialist, and consequently can be utilized as a wellspring of useful parts and additives for the improvement of healthfully safe novel food items. Encountering numerous flavors and aromas every day, especially in confectionery, beverages, and other foodstuffs, indicate that the universal organic flavor industry is encountering a high development because of expanding utilization of natural scents like EOs over manufactured scents, owing to their related various medical advantages, for example, in aroma-based treatment as would be considered normal to drive the development of the market in the next few years.

References

Acevedo-Fani, A., Salvia-Trujillo, L., Rojas-Graü, M. A., & Martín-Belloso, O. (2015). Edible films from essential-oil-loaded nanoemulsions: Physicochemical characterization and antimicrobial properties. *Food Hydrocolloids*, *47*, 168−177.

Aguirre, A., Borneo, R., & Leon, A. E. (2013). Antimicrobial, mechanical and barrier properties of triticale protein films incorporated with oregano essential oil. *Food Bioscience*, *1*, 2−9.

Al-Otaibi, M., & El-Demerdas, H. (2008). Improvement of the quality and shelf life of concentrated yoghurt (Labneh) by the addition of some essential oils. *African Journal of Microbiology Research*, *2*, 156−161.

Al-Rimawi, F., Alayoubi, M., Elama, C., Jazzar, M., & Çakıcı, A. (2020). Use of cinnamon, wheat germ, and eucalyptus oils to improve quality and shelf life of concentrated yogurt (Labneh). *Cogent Food & Agriculture*, *6*(1), 1807810.

Amelia, B., Saepudin, E., Cahyana, A. H., Rahayu, D. U., Sulistyoningrum, A. S., & Haib, J. (2017). GC-MS analysis of clove (*Syzygium aromaticum*) bud essential oil from Java and Manado, In *AIP Conference Proceedings*, *1862*(1), 030082.

Azizkhani, M., & Parsaeimehr, M. (2018). Probiotics survival, antioxidant activity and sensory properties of yogurt flavored with herbal essential oils. *International Food Research Journal*, *25*(3).

Bakry, A. M., Chen, Y. Q., & Liang, L. (2019). Developing a mint yogurt enriched with omega-3 oil: Physiochemical, microbiological, rheological, and sensorial characteristics. *Journal of Food Processing and Preservation*, *43*(12), e14287.

Basavegowda, N., & Baek, K. H. (2021). Synergistic antioxidant and antibacterial advantages of essential oils for food packaging applications. *Biomolecules*, *11*(9), 1267.

Benkhoud, H., M'Rabet, Y., Gara Ali, M., Mezni, M., & Hosni, K. (2022). Essential oils as flavoring and preservative agents: Impact on volatile profile, sensory attributes, and the oxidative stability of flavored extra virgin olive oil. *Journal of Food Processing and Preservation*, *46*(5), e15379.

Bhavaniramya, V. S., Al-Aboody, M. S., Vijayakumar, R., & Baskaran, D. (2019). Role of essential oils in food safety: Antimicrobial and antioxidant applications. *Grain & Oil Science and Technology*, *2*(2), 49−55.

Bustamante, J., van Stempvoort, S., García-Gallarreta, M., Houghton, J. A., Briers, H. K., Budarin, V. L., Matharu, A. S., & Clark, J. H. (2016). Microwave assisted hydro-distillation of essential oils from wet citrus peel waste. *Journal of Cleaner Production*, *137*, 598−605.

Chen, G., Sun, F., Wang, S., Wang, W., Dong, J., & Gao, F. (2021). Enhanced extraction of essential oil from *Cinnamomum cassia* bark by ultrasound assisted hydrodistillation. *Chinese Journal of Chemical Engineering*, *36*, 38−46.

Chouliara, E., Karatapanis, A., Savvaidis, I. N., & Kontominas, M. G. (2007). Combined effect of oregano essential oil and modified atmosphere packaging on shelf-life extension of fresh chicken breast meat, stored at 4°C. *Food Microbiology, 24*(6), 607−617.

Chu, Y., Gao, C., Liu, X., Zhang, N., Xu, T., Feng, X., Yang, Y., Shen, X., & Tang, X. (2020). Improvement of storage quality of strawberries by pullulan coatings incorporated with cinnamon essential oil nanoemulsion. *LWT, 122,* 109054.

Cuicui, L. I., & Lixia, H. O. U. (2018). Review on volatile flavor components of roasted oilseeds and their products. *Grain & Oil Science and Technology, 1*(4), 151−156.

De Azeredo, G. A., Stamford, T. L. M., Nunes, P. C., Neto, N. J. G., De Oliveira, M. E. G., & De Souza, E. L. (2011). Combined application of essential oils from *Origanum vulgare* L. and *Rosmarinus officinalis* L. to inhibit bacteria and autochthonous microflora associated with minimally processed vegetables. *Food Research International, 44*(5), 1541−1548.

Dhifi, W., Bellili, S., Jazi, S., Bahloul, N., & Mnif, W. (2016). Essential oils' chemical characterization and investigation of some biological activities: A critical review. *Medicines, 3*(4), 25.

Donsì, F., Annunziata, M., Sessa, M., & Ferrari, G. (2011). Nanoencapsulation of essential oils to enhance their antimicrobial activity in foods. *LWT-Food Science and Technology, 44*(9), 1908−1914.

Dwijatmoko, M. I., Praseptiangga, D., & Muhammad, D. R. A. (2016). Effect of cinnamon essential oils addition in the sensory attributes of dark chocolate. *Nusantara Bioscience, 8*(2), 301−305.

El-Zaeddi, H., Martínez-Tomé, J., Calín-Sánchez, Á., Burló, F., & Carbonell-Barrachina, Á. A. (2016). Volatile composition of essential oils from different aromatic herbs grown in Mediterranean regions of Spain. *Foods, 5*(2), 41.

Espina, L., Somolinos, M., Ouazzou, A. A., Condon, S., Garcia-Gonzalo, D., & Pagán, R. (2012). Inactivation of *Escherichia coli* O157: H7 in fruit juices by combined treatments of citrus fruit essential oils and heat. *International Journal of Food Microbiology, 159*(1), 9−16.

Faheem, F., Liu, Z. W., Rabail, R., Haq, I. U., Gul, M., Bryła, M., Roszko, M., Kieliszek, M., Din, A., & Aadil, R. M. (2022). Uncovering the industrial potentials of lemongrass essential oil as a food preservative: A review. *Antioxidants, 11*(4), 720.

Fiorentini, M., Kinchla, A. J., & Nolden, A. A. (2020). Role of sensory evaluation in consumer acceptance of plant-based meat analogs and meat extenders: A scoping review. *Foods, 9*(9), 1334.

Fisher, K., & Phillips, C. A. (2006). The effect of lemon, orange and bergamot essential oils and their components on the survival of *Campylobacter jejuni, Escherichia coli* O157, *Listeria monocytogenes, Bacillus cereus* and *Staphylococcus aureus* in vitro and in food systems. *Journal of Applied Microbiology, 101*(6), 1232−1240.

Gutierrez, J., Barry-Ryan, C., & Bourke, P. (2009). Antimicrobial activity of plant essential oils using food model media: Efficacy, synergistic potential and interactions with food components. *Food Microbiology, 26*(2), 142−150.

Herman, R. A., Ayepa, E., Shittu, S., Fometu, S. S., & Wang, J. (2019). Essential oils and their applications-a mini review. *Advances in Nutrition & Food Science, 4*(4).

Hulankova, R. (2022). The influence of liquid medium choice in determination of minimum inhibitory concentration of essential oils against pathogenic bacteria. *Antibiotics, 11*(2), 150.

Jalali-Heravi, M., & Parastar, H. (2011). Recent trends in application of multivariate curve resolution approaches for improving gas chromatography−mass spectrometry analysis of essential oils. *Talanta, 85*(2), 835−849.

Jayasena, D. D., & Jo, C. (2013). Essential oils as potential antimicrobial agents in meat and meat products: A review. *Trends in Food Science & Technology, 34*(2), 96−108.

Kalantary, F. (2014). Control of Aspergillus flavus growth in tomato paste by *Cinnamomum zeylanicum* and *Origanum vulgare* L. essential oils. *Journal of Food and Pharmaceutical Sciences, 2*(2).

Katz, J., Sahai, A., & Waters, B. (2008). *Predicate encryption supporting disjunctions, polynomial equations, and inner products. Annual international conference on the theory and applications of cryptographic techniques* (pp. 146−162). Berlin, Heidelberg: Springer.

Kiarsi, Z., Hojjati, M., Behbahani, B. A., & Noshad, M. (2020). In vitro antimicrobial effects of *Myristica fragrans* essential oil on foodborne pathogens and its influence on beef quality during refrigerated storage. *Journal of Food Safety, 40*(3), e12782.

Kordsardouei, H., Barzegar, M., & Sahari, M. A. (2013). Application of *Zataria multiflora* Boiss. and *Cinnamon zeylanicum* essential oils as two natural preservatives in cake. *Avicenna Journal of Phytomedicine, 3*(3), 238.

Luz, T. R. S. A., Leite, J. A. C., de Mesquita, L. S. S., Bezerra, S. A., Silveira, D. P. B., de Mesquita, J. W. C., Gomes, R. E. C., Vilanova, C. M., de Sousa Ribeiro, M. N., do Amaral, F. M. M., & Coutinho, D. F. (2020). Seasonal variation in the chemical composition and biological activity of the essential oil of *Mesosphaerum suaveolens* (L.) Kuntze. *Industrial Crops and Products, 153,* 112600.

Manzoor, S., Gull, A., Wani, S. M., Ganaie, T. A., Masoodi, F. A., Bashir, K., Malik, A. R., & Dar, B. N. (2021). Improving the shelf life of fresh cut kiwi using nanoemulsion coatings with antioxidant and antimicrobial agents. *Food Bioscience, 41,* 101015.

Mariod, A. A. (2016). *Effect of essential oils on organoleptic (smell, taste, and texture) properties of food. In* Essential oils in food preservation, flavor and safety (pp. 131−137). Academic Press.

Melendez-Rodriguez, B., Figueroa-Lopez, K. J., Bernardos, A., Martínez-Máñez, R., Cabedo, L., Torres-Giner, S., & Lagaron, J. M. (2019). Electrospun antimicrobial films of poly (3-hydroxybutyrate-co-3-hydroxyvalerate) containing eugenol essential oil encapsulated in mesoporous silica nanoparticles. *Nanomaterials, 9*(2), 227.

Ni, Z. J., Wang, X., Shen, Y., Thakur, K., Han, J., Zhang, J. G., Hu, F., & Wei, Z. J. (2021). Recent updates on the chemistry, bioactivities, mode of action, and industrial applications of plant essential oils. *Trends in Food Science & Technology, 110,* 78−89.

Noshad, M., Alizadeh Behbahani, B., Jooyandeh, H., Rahmati-Joneidabad, M., Hemmati Kaykha, M. E., & Ghodsi Sheikhjan, M. (2021). Utilization of *Plantago major* seed mucilage containing *Citrus limon* essential oil as an edible coating to improve shelf-life of buffalo meat under refrigeration conditions. *Food Science & Nutrition, 9*(3), 1625−1639.

Okunola, C. O., Okunola, A. A., Abulude, F. O., Ogunkoya, M. O. (2007). Organoleptic qualities of maize and cowpea stored with same essential oils. *African Crop Science Conference Proceedings*, 8, 2113–2116.

Olaniran, A. F., Abiose, S. H., & Adeniran, A. H. (2015). Biopreservative effect of ginger (*Zingiber officinale*) and garlic powder (*Allium sativum*) on tomato paste. *Journal of Food Safety*, 35(4), 440–452.

Olaniran, A. F., Afolabi, R. O., Abu, H. E., Owolabi, A., Iranloye, Y. M., & Okolie, C. E. (2020). Lime potentials as biopreservative as alternative to chemical preservatives in pineapple, orange and watermelon juice blend. *Food Research*, 4(6), 1878–1884.

Olaniran, A., Abu, H., Afolabi, R., Okolie, C., Owolabi, A., & Akpor, O. (2020). Comparative assessment of storage stability of ginger-garlic and chemical preservation on fruit juice blends. *Slovak Journal of Food Sciences*, 14.

Olmedo, R. H., Nepote, V., & Grosso, N. R. (2012). Aguaribay and cedron essential oils as natural antioxidants in oil-roasted and salted peanuts. *Journal of the American Oil Chemists' Society*, 89(12), 2195–2205.

Olmedo, R. H., Nepote, V., & Grosso, N. R. (2013). Preservation of sensory and chemical properties in flavoured cheese prepared with cream cheese baseusing oregano and rosemary essential oils. *LWT-Food Science and Technology*, 53(2), 409–417.

Omidbeygi, M., Barzegar, M., Hamidi, Z., & Naghdibadi, H. (2007). Antifungal activity of thyme, summer savory and clove essential oils against *Aspergillus flavus* in liquid medium and tomato paste. *Food Control*, 18(12), 1518–1523.

Roller, S., & Seedhar, P. (2002). Carvacrol and cinnamic acid inhibit microbial growth in fresh-cut melon and kiwifruit at 4 and 8 C. *Letters in Applied Microbiology*, 35(5), 390–394.

Sánchez-González, L., Vargas, M., González-Martínez, C., Chiralt, A., & Chafer, M. (2011). Use of essential oils in bioactive edible coatings: A review. *Food Engineering Reviews*, 3(1), 1–16.

Scollard, J., Francis, G. A., & O'Beirne, D. (2013). Some conventional and latent anti-listerial effects of essential oils, herbs, carrot and cabbage in fresh-cut vegetable systems. *Postharvest Biology and Technology*, 77, 87–93.

Sharmeen, J. B., Mahomoodally, F. M., Zengin, G., & Maggi, F. (2021). Essential oils as natural sources of fragrance compounds for cosmetics and cosmeceuticals. *Molecules (Basel, Switzerland)*, 26(3), 666.

Singh, B., Singh, J. P., Kaur, A., & Yadav, M. P. (2021). Insights into the chemical composition and bioactivities of citrus peel essential oils. *Food Research International*, 143, 110231.

Souza, V. G. L., Pires, J. R., Vieira, É. T., Coelhoso, I. M., Duarte, M. P., & Fernando, A. L. (2019). Activity of chitosan-montmorillonite bionano-composites incorporated with rosemary essential oil: From in vitro assays to application in fresh poultry meat. *Food Hydrocolloids*, 89, 241–252.

Takma, D. K., & Korel, F. (2019). Active packaging films as a carrier of black cumin essential oil: Development and effect on quality and shelf-life of chicken breast meat. *Food Packaging and Shelf Life*, 19, 210–217.

Tanveer, M., Wagner, C., Ribeiro, N. C., Rathinasabapathy, T., Butt, M. S., Shehzad, A., & Komarnytsky, S. (2020). Spicing up gastrointestinal health with dietary essential oils. *Phytochemistry Reviews*, 19(2), 243–263.

Tongnuanchan, P., & Benjakul, S. (2014). Essential oils: Extraction, bioactivities, and their uses for food preservation. *Journal of Food Science*, 79(7), R1231–R1249.

Tserennadmid, R., Takó, M., Galgóczy, L., Papp, T., Pesti, M., Vágvölgyi, C., Almássy, K., & Krisch, J. (2011). Anti yeast activities of some essential oils in growth medium, fruit juices and milk. *International Journal of Food Microbiology*, 144(3), 480–486.

Tyagi, A. K., Gottardi, D., Malik, A., & Guerzoni, M. E. (2014). Chemical composition, in vitro anti-yeast activity and fruit juice preservation potential of lemon grass oil. *LWT-Food Science and Technology*, 57(2), 731–737.

Uyttendaele, M., Neyts, K., Vanderswalmen, H., Notebaert, E., & Debevere, J. (2004). Control of Aeromonas on minimally processed vegetables by decontamination with lactic acid, chlorinated water, or thyme essential oil solution. *International Journal of Food Microbiology*, 90(3), 263–271.

Valero, M., & Giner, M. J. (2006). Effects of antimicrobial components of essential oils on growth of *Bacillus cereus* INRA L2104 in and the sensory qualities of carrot broth. *International Journal of Food Microbiology*, 106(1), 90–94.

Voon, H. C., Bhat, R., & Rusul, G. (2012). Flower extracts and their essential oils as potential antimicrobial agents for food uses and pharmaceutical applications. *Comprehensive Reviews in Food Science and Food Safety*, 11(1), 34–55.

Wang, C. Y., Wang, S. Y., Yin, J. J., Parry, J., & Yu, L. L. (2007). Enhancing antioxidant, antiproliferation, and free radical scavenging activities in strawberries with essential oils. *Journal of Agricultural and Food Chemistry*, 55(16), 6527–6532.

Xavier, L. O., Sganzerla, W. G., Rosa, G. B., da Rosa, C. G., Agostinetto, L., de Lima Veeck, A. P., Bretanha, L. C., Micke, G. A., Dalla Costa, M., Bertoldi, F. C., & Nunes, M. R. (2021). Chitosan packaging functionalized with *Cinnamodendron dinisii* essential oil loaded zein: A proposal for meat conservation. *International Journal of Biological Macromolecules*, 169, 183–193.

Yaghoubi, M., Ayaseh, A., Alirezalu, K., Nemati, Z., Pateiro, M., & Lorenzo, J. M. (2021). Effect of chitosan coating incorporated with *Artemisia fragrans* essential oil on fresh chicken meat during refrigerated storage. *Polymers*, 13(5), 716.

Yemiş, G. P., & Candoğan, K. (2017). Antibacterial activity of soy edible coatings incorporated with thyme and oregano essential oils on beef against pathogenic bacteria. *Food Science and Biotechnology*, 26(4), 1113–1121.

Further reading

Hanif, M. A., Nisar, S., Khan, G. S., Mushtaq, Z., & Zubair, M. (Eds.), (2019). *Essential oils*. Essential oil research. Cham: Springer.

Ji, J., Shankar, S., Royon, F., Salmieri, S., & Lacroix, M. (2021). Essential oils as natural antimicrobials applied in meat and meat products—A review. *Critical Reviews in Food Science and Nutrition*, 1–17.

Chapter 11

Beneficial uses of essential oils in diary products

Babatunde Oluwafemi Adetuyi[1], Peace Abiodun Olajide[1], Charles Oluwaseun Adetunji[2] and Juliana Bunmi Adetunji[3]

[1]*Department of Natural Sciences, Faculty of Pure and Applied Sciences, Precious Cornerstone University, Ibadan, Oyo State, Nigeria,* [2]*Applied Microbiology, Biotechnology and Nanotechnology Laboratory, Department of Microbiology, Edo State University Uzairue, Iyamho, Edo State, Nigeria,* [3]*Department of Biochemistry, Osun State University, Osogbo, Osun State, Nigeria*

Introduction

The contamination of milk and related dairy products by contamination of food organic entities through handling and storage is still a major problem in dairies (Field et al., 2016). More than 250 unique illnesses have been accounted for to be sent through the utilization of sullied food (Abrar et al., 2020). *Listeria monocytogenes* was relegated to a lesser status among the pathogens responsible for such foodborne diseases. The infectiousness of *L. monocytogenes* has been shown to provide protection against a variety of environmental stresses. It can endure colder, warm, and osmotic stressors (Guenther et al., 2009). Until now, *L. monocytogenes* has been secluded from prepared-to-eat food sources like milk, cheddar, fish, and vegetables. For example, monocytogenes have been disconnected from crude milk, cheddar, and margarine at 29.2%, 14.1%, and 4% separately in Yazd, focal Iran (Mohajeria et al., 2018). Moreover, in Faisalabad, *L. monocytogenes* has been identified in unheated dairy items like 17.78% crude milk and 5.55% cheddar (Zafar et al., 2020). Normal capacity techniques, for example, refrigeration, are utilized to limit and decrease the development of *L. monocytogenes*. Be that as it may, these strategies are not adequate to forestall the development of the microbe. Moreover, *L. monocytogenes* had gained hereditary adjustments and harmfulness elements to oppose such protection techniques. Albeit the right sanitization of milk and dairy items might be a defensive strategy for restraining the development of *L. monocytogenes*, pollution after sanitization is as yet a potential method for communicating foodborne microbes (Lianou & Sofos, 2007). As recently revealed, sanitized milk disseminated in a few business sectors was viewed as defiled with *L. monocytogenes* because of post-purification tainting (Elafify et al., 2020). Besides, the treatment of overheating adversely affects the nature of milk and dairy items through denaturation of proteins, loss of nutrients, and undesirable taste changes (Coutinho et al., 2018). The food business area is searching for protected and agreeable competitors that go about as food additives. Toward this path, a few regular substances have been created to smother the development of microorganisms in milk and related dairy products. D-tryptophan is a characteristic amino corrosive that has shown inhibitory action against Salmonella spp. in tainted cheddar (Elafify et al., 2022). Moreover, the olive oil removal showed an inhibitory impact on the vegetative cells of *Bacillus cereus* in the polluted milk (Fei et al., 2019). Rosemary is perhaps of the most well-known phytochemicals developed all over the planet and utilized as an enhancing specialist and has cell reinforcement and antimicrobial properties (Hussein et al., 2020). Ascorbic corrosive also has a few valuable capabilities in maintaining the degree of collagen, in injury recuperating, as a cell reinforcement, and with antimicrobial exercises against different microbes such as *Staphylococcus aureus*, *Campylobacter* spp. what's more, *Mycobacterium tuberculosis* (Verghese et al., 2020). Clove is one of the phytochemicals with numerous restorative purposes as a disinfectant, pain relieving, and antimicrobial, and the Food and Drug Administration has remembered it as a protected food added substance (FDA, 2011). Yme is another phytochemical that is antispasmodic and antimicrobial, and is likewise a protected food additive (Cosentino et al., 1999). Commonness of *Listeria* spp., specifically *L. monocytogenes*, in different cow's milk items sold in Egypt. Second, the antimicrobial opposition profile of recuperated *L. monocytogenes* separates was additionally analyzed utilizing the plate dispersion technique. In an

Applications of Essential Oils in the Food Industry. DOI: https://doi.org/10.1016/B978-0-323-98340-2.00021-3

exploratory preliminary, the antilisterial exercises of rosmarinic corrosive, ascorbic corrosive, clove, and thyme natural balms were examined involving delicate cheddar as a food network.

Essential oils

Creams and ointments used in medicine can be made from any number of plant parts, including flowers, leaves, stalks, wood, roots, pollen, organic ingredients, corms, and tannins, but their slippery, brittle nature can be difficult to predict. Chemically synthesized versions include monoterpenes, sesquiterpenes, and diterpenes as well as isopropanol, acids, esters, epoxides, acetaldehyde, acetone, amines, and sulfur compounds. Terpene compounds and sweet-smelling compounds are by and large ordered into these two classifications. Each EO contains roughly 20−60 distinct synthetics. A few of them prevail, with fixations going from 20% to 70%, while the others are available just inconsistently. The region of the plant from which they are extricated might be the reason for this inconsistency. Natural oils assume a significant part in the immediate and aberrant protections of plants against herbivores and illnesses, as well as in regenerative exercises by drawing in pollinators and seed dispersers (Cegiełka, 2020). EOs are fluid, unstable, clear, pigmented, and have a lower thickness than water when extricated (Wójcik et al., 2018).

Historical use of essential oils

Besides the oil of turpentine, no other EO was referenced by Greek or Roman creators, in spite of the way that flavors have been used for their fragrance, flavor, and protection qualities since the artifact (Bauer et al., 2001). Villanova (c. 1235−1311), a Catalan doctor, is credited with composing the earliest solid record of rejuvenating balm refining (Günther, 1948). While the pharmacological impacts of EOs were referenced in pharmacopeias by the 13th century (Bauer et al., 2001), their utilization didn't seem to become normal in Europe until the 16th century, when they were promoted in the City of London. Two doctors from Strassburg, Brunschwig and Reiff, distributed books 100 years on the refining and utilization of EOs, in spite of the fact that they just name a little small bunch of oils between them: turpentine, juniper wood, rosemary, spike (lavender), clove, mace, nutmeg, anise, and cinnamon (Günther, 1948). During the 17th century, the production of EOs was widely known and the typical drug store's supply of oils went from 15 to 20 assortments (Günther, 1948). Indeed, even while it is conceivable that local Australians had been utilizing tea tree oil for restorative purposes preceding European colonization, its utilization as a medication has just been reported from the late eighteenth 100 years. As per verifiable records, it was De la Croix, in 1881, who previously estimated the bactericidal abilities of EO fumes in a trial setting (Boyle, 1955). In any case, all through the nineteenth and twentieth centuries, EOs were essentially utilized for flavor and fragrance as opposed to treatment (Günther, 1948).

Current use of essential oils

In the European Union (EU), EOs are most regularly utilized in the food business (as flavorings), the aroma and beauty care products ventures (facial cleansers and scents), and the drug business (for their utilitarian characteristics). Well-known EO applications like fragrant healing just record for a little part of the market. Concentrates of plant material or artificially made analogs of EO parts are additionally utilized as flavorings in the food business. Medicinal balms and their parts are utilized in a wide assortment of business merchandise, including dental root waterway sealers (Manabe et al., 1987), sterilizers (Cox et al., 2000), and feed-added substances for lactating sows and weaned piglets (to give some examples). Some EO-containing additives are presently available. DOMCA S.A. of Alhend'n, Granada, Spain produces the food additive "DMC Base Natural," which is composed of half glycerol and half revitalizing oils of oregano, basil, and lemon. Bavaria Corp. of Apopka, Florida, USA makes an item called "Protecta One" and "Protecta Two" that is a mixed spice separate that is guaranteed safe for use in food in the United States. The maker doesn't unveil the particular fixings, however, all things considered, the concentrates contain a blend of sodium citrate and sodium chloride arrangements. EOs are used in anything from commercial potato sprouting suppressants to pest antiagents because of the physiological effects they have.

Extraction and characterization

Two gatherings of philosophies were utilized for the extraction of EO: regular strategies and new techniques. Customary techniques depend on an interaction called water refining, which isolates the volatiles that are conveyed by a flood of steam warmed to changing degrees. These cycles incorporate dissolvable extraction, hydrodistillation, steam

TABLE 11.1 Understanding the history, chemical makeup, and usefulness of a few key essential oils.

Essential oil	Plant source	Active ingredients (% *w/v*)	Properties	References
Oregano oil	Oregano	thymol (31.5)	Antioxidant, expectorant, relaxing, antibacterial, antifungal, and healing	Souza et al. (2007)
Cinnamon oil	Cinnamon	Cinnamon aldehyde (60−80)	Digestion-improving, analgesic, antipyretic, analgesic, antiinflammatory	Kaławaj and Lemieszek (2015)
Garlic oil	Garlic	Diallyl disulfide (60)	Antimicrobial, antifungal, boosts and controls immune response	Kędzia (2009)
Thyme oil	Thyme	Thymol (45−47)	Digestive, warming, analgesic, antiviral, antifungal, effective against respiratory illnesses, and good appetite stimulant	Witkowska and Sowiʹnska (2013)
Black pepper oil	Black pepper	β-Caryophyllene (29.9)	Antipyretic, diuretic, antioxidant, anticancer, antibacterial, antifungal, and antiprotozoal	Kozłowska-Lewecka et al. (2011)
Lavender oil	Narrow-leaved lavender	Linalyl acetate (25−46)	Antibacterial, antifungal, immunostimulatory, used to treat skin conditions and respiratory problems	Sandner et al. (2020)
Peppermint oil	Peppermint	Menthol (63)	Antiinflammatory, wound-healing, antiallergic, antiviral, antibacterial, and inhibits the proliferation of cancer cells.	Łyczko et al. (2020)
Sage oil	Common sage	α-Thujone (2−46)	Antiinflammatory, antioxidant, antifungal, and antibacterial	Haʹc-Szymaʹnczuk and Cegiełka (2015)
Tea tree oil	Tea tree	Terpinen-4-ol (29−45)	Antiinflammatory, antifungal, antibacterial, and immunomodulatory	Taiwo and Adebayo (2017)

refining, and hydrodiffusion. Hydrodistillation is as yet utilized financially for an assortment of EOs from various sources (Wang et al., 2017). While trying to avoid a portion of the downsides of standard methodologies, imaginative ways have been made as options. A portion of these disadvantages is the critical energy utilization coming about because of a more limited extraction period, the utilization of solvents and CO_2 outflows. Current state-of-the-art methods include supersonic fluid extraction, sub-critical fluid extraction, solubilized spontaneous microwave extraction, and microwave-aided water distillation (Aziz et al., 2018). Among these procedures, dissolvable free microwave extraction, ultrasound-helped extraction, and microwave-helped extraction have all been viewed as green advances (Vernès et al., 2020) and are compelling in separating EO from different lattices. Contrasted with ordinary hydrodistillation, these methodologies have been found to deliver higher EO yields. Normal procedures used to separate EOs are displayed in Table 11.1. Tongnuanchan and Benjakul have composed exact depictions of these extraction processes and their instruments (Tongnuanchan & Benjakul, 2014). Gas chromatography is generally used to decide the piece of EOs after their extraction. Regularly, a mass spectrometer is connected to a gas chromatography framework to find the synthetic substances liable for the cell reinforcement impact (Szewczyk et al., 2016).

Biologically active components included in essential oils

Utilizing EOs is not new. They were first utilized by early Eastern and Middle Eastern developments, then by North African and European human advancements. Paracelsus von Hohenheim begat the adage "rejuvenating ointments" in the 16th century. Around 3000 different EOs are known, albeit around 300 are industrially critical. The drug business, home consideration, individual consideration (scents and beauty care products), home consideration, the food business, for example, flavorings, creature and human antiagents, and customary medication for the treatment of different illnesses are the primary utilization of natural ointments. It has additionally been accounted for that the individual EO parts can be utilized. These parts can be delivered or collected from plants (Sharifi-Rad et al., 2017). A wide assortment of wellbeing-advancing properties—for example, anticancer, calming, expectorant, antidiabetic, circulatory excitement, immunostimulant, and germ-free properties—have been credited to EOs and their constituent particles. Be that as it may, buyers ought to consider the portion managed, as a portion of these substances can become poisonous whenever consumed in overabundance (Ashraf et al., 2016; Fathi et al., 2019). The superorganic elements of EOs and their systems of activity have been broadly portrayed by Sharifi-Rad et al. (2017) in a top-to-bottom audit. As indicated by the creators, the basic method of activity of these bioactivities is the annihilation of the construction of the films, which prompts cell permeabilization. This implies that these bioactivities show solid cytotoxicity. In like manner, the antibacterial, antifungal, and antiparasitic properties of EO can be assessed. The creators additionally note that few EOs have been utilized in the therapy of provocative sicknesses like disease for their cell reinforcement and calming properties. These rejuvenating ointments additionally can incite apoptosis. Also, EOs have been read up for use in the food business for their antibacterial, antifungal, and cancer prevention agent impacts. Shojaee-Aliabadi et al. (2018) express that Buchholtz presumably began perhaps the earliest concentrate on the antibacterial impact of EOs in 1875. From that point forward, various investigations have affirmed its antibacterial viability against numerous gram-positive and gram-negative microscopic organisms and parasites.

Dairy products

Milk is a multicomponent food fixing and has for quite some time been important for the worldwide human eating routine (Schiano et al., 2019) exploit the milk these females delivered for their own motivations (Prasad, 2017). The utilization of milk and dairy items has gone under examination as of late because of media reports connecting them to an expanded gamble of persistent sickness, notwithstanding logical proof in actuality up to this point (Thorning et al., 2016). Dairy creation, which might have begun unintentionally and later grew exclusively to save milk, has developed into the making of different perfect items that exploit the numerous dietary advantages that milk gives. Like different dinners, milk has experts and cell reinforcements that happen normally and essentially affect oxidation lability and lipid breakdown. Notwithstanding, milk handling normally brings about the deficiency of cancer prevention agents, for example, ascorbic corrosive, tocopherol, and carotenoids (Gutierrez et al., 2018). The fat substance of the milk is concentrated multiple times during the development of the spread. Therefore, spread has a somewhat lengthy time span of usability, yet it likewise conveys a more serious gamble of fat oxidation. Dagdemir et al. (2009) analyzed the cell reinforcement movement of the manufactured cancer prevention agent BHT with the EOs of *Thymus haussknechtii* and *Origanum acutidens* added to margarine. With the expansion of the EOs at a convergence of 0.2%, the essential and optional oxidation items were assessed and critical cell reinforcement results were seen during capacity (90 days).

Essential oils' useful applications in the dairy industry

Biopreservation and antibacterial properties

As of late, the quest for regular antimicrobials has been powered by the emotional expansion in purchaser consciousness of normal food sources. Food items contain biopreservatives to work on quality by expanding their timeframe of realistic usability and forestalling the arrangement of unsafe microorganisms. The viability of EO application changes with application strategy, fixation, application procedure, and capacity temperature. As indicated by reports, the antibacterial movement of terpineol is contrarily relative to the fat substance. The focus utilized and the strategy for application (shower, lactose, or plunging) decide the viability of these natural additives. Due to the organoleptic impact, the important convergence of rejuvenating oils should likewise be considered to forestall microbial development. Notwithstanding, various regular additives have shown to be powerful substitutes for repressing the development of microorganisms and limiting unfriendly changes in sensory attributes.

Customers and workers in the processing industry both have serious hygiene concerns. The interest in antimicrobials reasonable for dairy items has expanded because of the expansion in foodborne sicknesses. Numerous rejuvenating balms found in plants and flavors areas of strength for have impacts against various microorganisms. *B. cereus*, *Kloeckera apiculata*, *Candida albicans*, *Escherichia coli*, *Proteus mirabilis*, *Salmonella enteritidis*, *C. albicans*, *Candida oleophila*, and peppermint oil just restrain yeasts, natural ointments extricated from clove, orange, thyme, and oregano have shown areas of strength for an impact against microscopic organisms and yeasts. Various investigations have recorded the antibacterial limits of EOs.

Since they are hydrophobic, EOs can without much of a stretch enter bacterial cell walls. When inside a bacterial cell, natural balms impede particle transport processes, delivering the phone inert. Since Gram-negative microorganisms have an external layer lipopolysaccharide hindrance, EOs have more powerful inhibitory exercises when applied to them (Loizzo et al., 2009). The objective microorganism, least inhibitory focuses (MIC), system of activity and likely connections with the food network and tactile properties of food are requirements for the utilization of EOs in food items. The antibacterial action of rejuvenating ointments might be because of their dynamic fixings. Contingent upon the synthetic arrangement, the mixtures that give EOs their antimicrobial properties can be characterized into four gatherings: terpenes (like limonene and p-cymene), phenylpropenes (like eugenol and vanillin), terpenoids (like carvacrol and thymol), and different substances like Isothiocyanates or allicin (Hyldgaard et al., 2012). The method of activity of EOs shifts and isn't yet completely comprehended. EOs follow up on a few destinations of a microorganism which are frequently considered as locales of activity for rejuvenating ointments. Natural ointments are accepted to be more powerful against Gram-positive microscopic organisms than Gram-negative microbes. The cell mass of Gram-positive microorganisms comprises peptidoglycan (90%−95%) to which proteins and their corrosive are appended. Since medicinal balms are predominantly hydrophobic, they basically cross the phone film and arrive at the cytoplasm. Compared to the cell films of Gram-positive and Gram-negative bacteria, the latter is more baffling since it consists of a peptidoglycan monolayer surrounded by layers of proteins and glycosaminoglycans. Gram-negative microbes have a charged, hydrophilic external layer (Dongmo et al., 2010). The movement conveyed is because of the acidic idea of the −OH gathering of phenols. These substances upset the protein support which eventually prompts cell demise, adjusts cell penetrability, and obstructs energy creation by disturbing the chemicals in question. How microscopic organisms are shaped can sporadically change how rejuvenating balms work. As per reports, bar-formed cells are more defenseless to harm than coccoid-molded cells. Friedman (2017) distributed a top-to-bottom examination of cinnamaldehyde, a critical fixing in rejuvenating oils delivered by the fragrant cinnamon plants. The significance of this compound has been underlined by the creator in various in vitro tests against pathogenic microbes present in dairy items, for example, cheddar and different food sources. This examination could give a premise to the utilization of cinnamic aldehyde and other normal rejuvenating oils in the counteraction of foodborne sickness, which could really further develop food handling and forestall pathogenic microscopic organisms that cause food tainting and illness. in people than in creatures. In 2016, Abboud et al. (2016) assessed the viability of Thymus vulgaris and Lavandula angustifolia rejuvenating ointments as a drawn-out option in contrast to regularly utilized manufactured antimicrobials against microorganisms liable for ox-like mastitis. The California mastitis test was utilized in the examination of four dairy cows to test for clinical mastitis. Results from an in vitro test demonstrated that both medicinal balms tested here had high antibacterial activity against strains of Staphylococcus and Streptococcus known to cause mastitis, almost as potent as the control antitoxin. Besides, it was shown that after four organizations of the treatment, the intramammary use of a 10% arrangement of the medicinal oil prompted a lessening in the quantity of microorganisms in every one of the examples inspected. Also, the irritation related to the two contaminations examined was altogether diminished with effective use of the medicinal oil. With a 100% recuperation rate with thymus medicinal balms, this impact is more long-lasting than intramammary approaches.

Abbes et al. (2018) focused on the antibacterial effect of 3 g/mL of cinnamon medicinal oil combined with lactoperoxidase against both total milk growth and Salmonella hadar. The outcome showed that the consolidated and synergistic impacts of cinnamon rejuvenating ointment and the lactoperoxidase framework had a more grounded antibacterial impact than the singular dynamic fixings. Besides, an expansive entire verdure hindrance profile was seen following 5 days of perception under the synergistic impact at 4°C (for 3 days), while Salmonella hadar was hindered for 5 days.

Perilla oil and nisin were tested for antibacterial activity against 20 isolates of S. aureus and L. monocytogenes found in food by using the checkerboard microdilution method. The outcomes showed that the two medicines synergistically affected the two objective living beings, with fragmentary inhibitory fixation files going from 0.125 to 0.25 and 0.19 to 0.375, separately. Thusly, filtering electron microscopy was utilized to affirm the antibacterial adequacy of the two natural ointments against the two objective organic entities. The outcome shows that L. monocytogenes and S. aureus cell walls were penetrable to rejuvenating ointment, harming the cell wall and surface film of the microorganism tried. One of the major bioactive parts in thyme medicinal balm has been distinguished as thymol. Thymol improves the antibacterial properties of the thymus by restricting to-layer proteins. Thus, the open cell content of microbial cells was delivered during a blast (Burt, 2004). As per Tiwari (Tiwari et al., 2009), thymol could fundamentally decrease intracellular degrees of adenosine triphosphate (ATP) in E. coli and increment extracellular degrees of ATP, which could upset plasma layer capability. Thymol has also been shown to inhibit the growth of both Gram-positive and Gram-negative bacteria (Burt, 2004). Moreover, eugenol, a functioning element of clove natural oil with huge antibacterial viability, was referenced in another article (Ogunlana et al., 2021). According to Burt (2004), eugenol can protect B. cereus cell walls, which can stimulate the bacterium's cell division. It can likewise upset the development of amylase.

As per a recent report by Ahmad et al. (2011), *Coriaria nepalensis* natural balm applies its antifungal movement by influencing ergosterol creation and the trustworthiness of the Candida film. Moreover, Hu et al. (2017) exhibited that

Curcuma longa had hostile to aflatoxigenic and antifungal properties when tried against *Aspergillus flavus*. There are potential connections between instruments of activity and restraint of ergosterol creation.

Antioxidant activity

Increased interest in EOs as normal cell reinforcements has been spurred by the potential detrimental effects of synthetic cancer-preventive drugs like BHA and BHT. It is common practice to include EOs in consumable arrangements as an integrated part of the item or as a dynamically packaged edible coating. It is a reasonable choice to forestall auto-oxidation and expand timeframe of realistic usability. The examination of the cell reinforcement capability of natural ointments is a significant perspective for the food business, since current tests are not reasonable for themselves and can prompt problematic outcomes (Loizzo et al., 2009). Moreover, inspecting their capacity to safeguard food, this article looks at the science behind the cancer prevention agent movement of rejuvenating balms. Strategies for assessing the cell reinforcement limit of EOs have been entirely assessed in the reference to writing. According to da Silva Dannenberg et al. (2016), the medicinal oils of ready green and pink peppers have cancer-prevention agent potential. Longer capacity of the tactile properties of ready pink pepper can broaden the timeframe of realistic usability of the item.

Taste and smell enhancements

As per numerous purchasers, milk is a bland food. Thus, they track down a gigantic test to portray the taste qualities. In spite of having an unmistakable taste, it has been depicted as marginally pungent sweet purified milk. This taste can be ascribed to lactose and milk salts. Without a doubt, specialists have invested a ton of energy and exertion into deciding the justification for the flavor of milk (Schiano et al., 2017). EOs can be utilized in different ways to upgrade the kind of milk (Ando et al., 2001). Many individuals know nothing about the advantages of involving natural oils in ordinary feasts for flavor improvement and medical advantages. The American food and refreshment industry has been involved in these rejuvenating oils for quite a long time and has detailed superb outcomes. Since they contain unpredictable parts, EOs are frequently utilized as enhancing specialists. Cinnamon rejuvenating oils have been accounted to be helpful in working on the flavor and utilitarian properties of milk chocolate (Ilmi et al., 2017; Muhammad et al., 2019). These natural balms are a decent wellspring of wellbeing advancing synthetic compounds and scent in concentrated structure. They work on the taste and dietary benefit of dinners. A few natural oils, like those of orange, peppermint, ginger or rose, among others, can change the flavor of a solitary cocoa flavor (Worwood, 2016). For inner use adding only a couple of drops toward the beginning is suggested. Certain individuals blend the oil straightforwardly with fluids like milk or water and afterward drink the combination. It very well might be feasible to forestall aversion to some EOs by taking them in gelatin-covered or intestinal covered gelatin containers, alongside a spoonful of honey or coconut oil.

Cymbopogon citratus, likewise called lemongrass, has been utilized as a medication for quite some time. A substance is utilized frequently in drinks. It has been utilized as a sugar in the event of unfamiliar dishes.

Several types of dairy products that incorporate essential oils

Essential oils in cheese

Rejuvenating balms are extremely viable in restricting development and decreasing microbial endurance in cheddar. Gouvea et al. (2017) revealed the antibacterial impact of cheddar when medicinal ointments and plant removal were added. Cheddar has been assigned as a food eaten all over the planet, however, it is incredibly defenseless to tainting by a scope of deterioration microorganisms that could restrict its time span of usability and posture serious wellbeing dangers to purchasers. Because of the always expanding interest in quality food sources without counterfeit additives, normal additives have been found that could guarantee client security and broaden the time span of usability of dairy items. The disclosure of rejuvenating oil as a characteristic, reasonable, and natural additive of such extraordinary quality might be connected to antibacterial properties against illnesses and microorganisms that cause cheddar waste. Then again, outside and inner upgrades can likewise influence the action of these synthetics once blended into the cheddar. Various examinations have likewise pointed the antibacterial impacts of plant concentrates and rejuvenating oils notwithstanding their synergistic consequences for lactic corrosive microscopic organisms and the erotic characteristics of cheddar inferred items. However, when medicinal oils blend in with fats, sugars, or proteins, their antibacterial properties are decreased (Kostova et al., 2016), so more EO is expected to accomplish the ideal impact. Milk fatty forms a protective barrier over the nonpolar parts of EOs, and proteins in EOs work along with phenolic synthetics. Hence, lipids limit the availability of EOs to the favored objective areas of microscopic organisms. What's more, the actual properties of cheddar influence the appropriate dissemination of medicinal balms, which decreases the antibacterial viability of EO. In spite of the fact that EO movement is expanded by expanding temperature, oxygen levels, and diminishing pH (Ilmi et al., 2017). A decreased pH benefits EOs by making them more hydrophobic, permitting EOs to all the more effectively enter bacterial cells and arrive at their objective site. At the point when EOs are utilized at higher fixations to make up for the abatement in movement brought about by food parts, tangible anomalies frequently result. Different systems have been investigated to fill these holes. Rejuvenating oils could at first be utilized in palatable movies and coatings. This thought depends on an enduring antibacterial activity and ceaseless delivery. One more technique for the drawn-out impacts of EOs is their nano-or miniature embodiment. One such instance of microencapsulation is the utilization of whey protein disengage and inulin to exemplify rosemary medicinal balm, which was then added to Minas Frescal cheddar. As indicated by this readiness, the development of mesophilic microscopic organisms was halted and the timeframe of realistic usability of cheddar was expanded. At least one type of EOs consolidates with other types of EOs or antimicrobials to impact the objective EO well. Nisin and Z. multiflora rejuvenating oil cooperate to forestall *S. aureus* from delivering poisons and enterotoxins. The development of BHI stock and conventional Iranian cured white cheddar, individually, demonstrated this. Medicinal ointments are normally very costly because of the kind of plant material used, the strategy used to extricate oils, work and energy costs, impacting creation costs. How much plant material is expected to deliver a lot of natural balms, as well as its virtue, are different variables that add to the cost climb. Numerous cheap and quality medicinal oils are accessible.

Tornambé et al. (2008) found that even at very low concentrations of EO (0.1 L/L), EO had no discernible effect on the physical qualities of cheddar. Nonetheless, cheeses named "Mint/Chlorophyll" and "Thyme/Oregano" at 3.0 L/L displayed areas of strength for smell. These surprising scent qualities result from the expansion of EO. A sum of 152 mixtures obtained from cheddar were reported, 41 of which were with the expansion of EO; interestingly, 54 EO compounds were not recuperated. In any case, the option of EO affected a few unstable mixtures obtained from cheeses that had not been added.

A recent report via Mailam et al. removed the natural balms of cumin, rosemary, and thyme. They further concentrated on their consequences for the physical, substance, microbiological, rheological, and tactile qualities of ultrafiltered delicate cheddar. Thyme medicinal oils showed a cell reinforcement impact, while cumin rejuvenating ointments expanded tactile properties. Since medicinal oils are normal, they are not entirely set in stone to be ok for ultrafiltered cheddar.

Use of essential oils with milk products

One of the most mind-blowing normal options in contrast to added substances or sanitization to stop bacterial decay of food varieties in an assortment of food networks is rejuvenating oils (also called EOs) (Kostova et al., 2016). Also, adding natural ointments from the Thymus cap tatus plant can forestall bacterial development in purified milk and delay the timeframe of realistic usability of the item. This prompted the end that EO from *Thymus capitatus* works on the effectiveness of sanitization in keeping up with the nature of crude milk. Also, Nazir et al. (2022) analyzed the effects of a solution or the nanoemulsion of *Coridothymus capitatus* on the physico-substance qualities and the microbiological properties in several tainted milks, and thus on the fermentation process and aerobic metabolism strength of the milk. Semi-skimmed further develops UHT milk, expanding its viability in milk safeguarding. The outcomes above obviously show the upsides of normal additives (rejuvenating ointments) over engineered ones. Epitomized EOs can be a fabulous decision for milk in dusty conditions as regular and proficient antibacterial and cell reinforcement specialists screen and keep up with quality rules. Tornambé et al. (2008) analyzed how EO impacted the tactile properties of milk. This EO was added to the milk at a convergence of 1.0 μL, changing the tactile qualities. The milk tasted better compared to mint, was better, and had a higher centralization of EO (1.0 L/L). There is no question that the expansion of EO was answerable for the weird flavors. Consequently, the scope of 0.1 to 1.0 μL/L addressed the edge fixation for impression of fragrance related to the expansion of EO.

Essential oils in yogurt

Among the most well-known milk products, yogurt has worked on dietary benefits, taste, and medical advantages. In light of the foregoing, Azizkhani and Tooryan (2016) looked at the influence of three different natural ointments on the survival of yogurt's probiotic microorganisms. In vitro circle dissemination methods were used to compare *L. monocytogenes* and *E. coli* for their ability to impede development. There were no measurable pH differences discovered between the treated and untreated probiotic yogurt, especially during the active growth phases (Table 11.1). In addition, the treatment with the medicinal balm had the greatest inhibitory effect on the growth of *E. coli* and *L. monocytogenes* compared to the control treatment, and yogurt with Zataria was found to have the same effect. This research demonstrated that the probiotic yogurt's inhibitory effect against *E. coli* can be improved by incorporating peppermint, basil, and zataria medicinal balms into the product's detailing.

Labneh, a form of concentrated yogurt popular in Palestine and other Middle Eastern countries, is a common breakfast food. Labneh, is a semi-strong matured dairy product formed by draining yogurt of its whey. Cleanliness concerns with the material packs used in production and unsanitary management of the item, leading to microbial disease, are the cause of labneh's short usable time frame, however at cold temperatures. The over-the-top microbial heap of Labneh, along with the bundling and stockpiling boundaries, prompted the age of unfortunate flavors and negative physico-compound changes which at last prompted the dismissal of the item (Burt, 2004). Draughon (2004) dealt with this and proposed the utilization of natural additives for, for instance, vegetable medicinal balms to broaden the timeframe of realistic usability of sensitive food sources.

As a result, Al-Rimawi et al. (2019) evaluated the effect of several natural balms in homogeneous yogurt when used as a source of antibacterial agents that can prompt rapid improvement in concentrated yogurt. The natural ointment was integrated into the gathered yogurt at a centralization of 250 µL/kg and its intensity was resolved utilizing plate count strategies, while different boundaries, for example, tangible assessment, corrosiveness titratable, and complete solids were likewise performed. The review showed that adding natural ointments like clove, rosemary, and cinnamon could broaden the time span of usability of consolidated yogurt.

Blending ice cream with essential oils

Frozen yogurt is a conventional summer treat that is generally famous and made with affection. The frozen yogurts above are viewed as regular flavors because of the peppermint rejuvenating oils. Lavender upgrades the kind of lemon frozen yogurt and makes it extraordinary. Thus, it acquires flavor and variety. The physico-substance, microbiological, and tangible characteristics of frozen yogurts arranged with lemon, mandarin, and orange strip rejuvenating ointments were concentrated. They found that the last example contained a minimal measure of mesophilic and psychrophilic yeast, parasites, and vigorous microbes. The expansion of regular fixings was likewise helpful because the natural balms had antibacterial, additive, and seasoning properties.

By loading up with carnauba wax globules, as an extra portrayal of a culinary item, icing cream was added. Peppermint oil typified in carnauba wax was inspected according to a tactile viewpoint. It was found that adding peppermint oil to frozen yogurt up to 0.3% by weight made two impacts: it worked on the utilitarian properties of peppermint in frozen yogurt and affected surface. As indicated by tactile examination, adding peppermint oil to frozen yogurt up to 0.3% (w/w) could be an effective method for working on the practical properties of the zest without compromising surface.

Use of essential oils with cream

In the concentrate by Ehsani et al. (2017), lycopene and *Echinophora platyloba* (EEO) rejuvenating balms were utilized to expand the timeframe of realistic usability of the sanitized cream. The lycopene fixations were 20 and 50 ppm, while the EEO focuses were 0.10% and 0.50%. They were joined and added to the purified cream, which was then analyzed for microbiological properties, lipid security, and tactile properties, while held at 4°C and 25°C for quite some time. As per the aftereffects of microbial examinations and synthetic breaks down, the sanitized cream containing the blend of the above fixings in higher fixations accomplished the best execution as far as microbiological attributes, compound investigation, and soundness when figured out for control under capacity conditions. Examination of tangible investigations showed that all treatments utilized were generally acknowledged. The scientists found that creams with lower groupings of the EEO and lycopene blend had the best tactile properties. The review's decision proposed that EEO and lycopene can be utilized together as additives in high-fat dairy items like cream and margarine.

Butter flavored with essential oils

The significance of utilizing cumin and thyme medicinal balms to protect margarine when put away at room temperature should also be emphasized (Batiha et al., 2021). In a review, Ehsani et al. (2021) examined the utilization of *S. cilicica* medicinal balm as a characteristic cell reinforcement and how it adds to the lovely fragrance of margarine (Ehsani et al., 2017).

Combining essential oils with other dairy items

As per Nazir et al. (2022), oregano concentrate and rejuvenating ointments are strong cell reinforcements to be utilized in milk drinks with added flaxseed oil at a portion of 2 g/100 g. Oxidation of omega-3 unsaturated fats brought about by light and intensity was the primary issue with milk drinks, yet variety was likewise seen during capacity. Linseed oil was utilized to tackle these issues and accordingly, the issue was significantly decreased. It has been found that adding a characteristic cell reinforcement to dairy refreshments braced with omega-3 unsaturated fats can fundamentally decrease oxidation. In a concentrate on the notable Indian sweet burfi, which has a short timeframe of realistic usability, Badola et al. (2018) effectively helped antibacterial and cell reinforcement movement utilizing natural medicinal oils. For this reason, 0.15−0.25 ppm clove medicinal ointment (CLVB) and curry leaf rejuvenating oil (CRYF) were utilized. Many variables were assessed, including physicochemical investigation, microbiological measures, cancer prevention agent properties, and tangible examination (Ultee, Kets, et al., 2000).

The outcomes showed that as how much natural EO in burfi expanded, the antibacterial and cancer prevention agent action additionally expanded and the tactile properties diminished simultaneously. The properties of the physico-substance framework didn't change fundamentally.

Essential oil restrictions in dairy products

Since sugars make food varieties acidic, they are more vulnerable to yeast defilement. These yeasts obliterate items, such as natural products or vegetable juices, organic product purees, concentrated or pickles that are preserved by feeble acids such as sorbic or acidic corrosive (Lambert et al., 2001). *Geotrichum candidum* is generally found in crude milk used to make delicate cheddar and other dairy items. Sodas, natural product juices, wines, yogurts, pickles, and cream treats can be in every way ruined by yeast, otherwise called *Pichia anomala*. *Saccharomyces cerevisiae* and *Schizosaccharomyces pombe* are the most well-known yeasts causing decay in sodas and natural product juices. However long the bundling isn't flawless, sanitization and clean bundling will assist with keeping the yeast from ruining. Items that can't be sanitized are exposed to a somewhat acidic conservation treatment (for instance, benzoic corrosive and sorbic corrosive or their salts). Nonetheless, shoppers are noisily requesting that fewer counterfeit additives be utilized. The Federal Institute for Risk Assessment (2005) and the United Kingdom Food Standards Agency (2006) have both expressed that benzoic corrosive, which is utilized to safeguard food, ultimately makes benzene. These issues have driven the food business to zero in on normal antimicrobials produced using plants, for example, medicinal oils, which have expansive antimicrobial movement against microbes, yeasts, and mushrooms (Moleyar & Narasimham, 1992). The utilization of EO in culinary arrangements is restricted by its strong smell. The mix of EOs, which make a synergistic difference and decrease the sum required for capacity, takes care of this issue.

Positive and negative effects of combining essential oil components with conventional food preservation methods

Possible synergists for EOs include a highly acidic, reduced water activity, chelating agents, reduced oxygen pressure, moderate temperature, and elevated pressure; however, not all of these have been tested in food products (Gould, 1996). Studies investigating the synergistic effects of EOs or their components with other functional ingredients like salt, nitrite, or nisin, or preservation methods like low-temperature processing, high-pressure processing, or oxygen-free packaging, will be summarized here. Sodium chloride's interactions with EOs and/or their constituents have been proven to range from synergistic to antagonistic. In taramosalata, *S. enteritidis* and *L. monocytogenes* were killed more effectively by a combination of sodium chloride and mint oil (Tassou et al., 1995). When added to mackerel muscle extract, a combination of 2%−3% NaCl and 0.5% clove powder (containing eugenol and eugenyl acetate) completely inhibits *E. aerogenes'* proliferation and histamine production. Soy sauce and carvacrol were found to have synergy in the same study. Salt, however, canceled out this synergy as well (Ultee, Slump, et al., 2000). Cinnamaldehyde's antibacterial activity against gram-positive and gram-negative bacteria was not enhanced by the addition of salt to agar at 4% w/v (Moleyar & Narasimham, 1992). Oregano EO alone and in conjunction with sodium nitrite were investigated for their effects on *Clostridium botulinum* development and toxigenicity. Oregano oil showed little effect on growth suppression at concentrations up to 400 ppm, but had a synergistic effect when combined with nitrite. Oregano EO's ability to reduce spore germination rates and sodium nitrite's ability to prevent spore growth is important to the suggested synergy process. Both chemicals have an impact on plant development (Ismaiel & Pierson, 1990). Nisin (0.15 µg/mL) combined with carvacrol or thymol (0.3 mmol/L or 45 µg/mL) resulted in a greater reduction in viable counts for *B. cereus* strains than either antibiotic used alone. Mild heat treatment at 45°C (5 minutes for exponentially developing cells and 40 minutes for stationary phase cells) was found to have the greatest effect on cell viability. Carvacrol did not increase the sensitivity of vegetative *B. cereus* cells to treatment with a pulsed electric field, and it did not sensitize spores to the antibiotic nisin. Modifications in the fluidity of the plasma membranes due to temperature explain why the synergistic activity of nisin and carvacrol is greater at pH 7 at 30°C than at 8°C. Nobody has yet identified the mechanism(s) responsible for synergy's efficacy. Nisin generates pores in the cell membrane, and it was thought that carvacrol would increase their size, quantity, or stability (Karatzas et al., 2000). In the presence of strong hydrostatic pressure, thymol and carvacrol have been found to have a synergistic effect (HHP). Compared to either thymol or carvacrol alone, the combination of 300 MPa HHP and 3 mmol/L significantly decreased the number of viable *L. monocytogenes* cells in the mid-exponential growth phase. The synergy seen may have its origins in the fact that HHP is thought to disrupt cell membranes (Batiha et al., 2021). The degree to which oxygen is present affects the antibacterial action of EOs. This may be because cells that rely on anaerobic metabolism for energy production are more vulnerable to the harmful effects of EOs, or because the EOs themselves undergo less oxidative changes in the absence of oxygen. Oregano and thyme EOs' antibacterial activity was significantly boosted against *Salmonella typhimurium* and *S. aureus* when oxygen levels were low (Paster et al., 1990). Prevention of *L. monocytogenes* and spoilage microorganisms on beef fillets; 0.8% v/w, may be enhanced by the combination of oregano EO and vacuum packaging. Oregano EO was found to be more effective in samples packed under vacuum in low-permeability film, producing a 2−3 log10 initial reduction in the microbial flora, than in samples kept aerobically and in samples pack aged under vacuum in highly permeable film (Tsigarida et al., 2000). Similarly, vacuum-packed pork loin steak exposed to clove and coriander EOs when stored at 2°C and 100°C was more resistant to *A. hydrophila* than air-preserved pork (Nazir et al., 2022).

Future use of essential oils

The fate of rejuvenating oils is brilliant thanks to their helpful impacts against oxidation and shopper inclination for regular or non-engineered food-added substances. In the refreshed writing, EOs has been demonstrated to be viable in various food items, and in specific circumstances, it can likewise help the organoleptic properties of the assessed food framework. Laying out a particular portion for each sort of EO requires concentration on poisonousness and unfavorably susceptible impacts. Makers are as of now urged by regulation to apply clean names to their items and OEs are completely agreeable with this prerequisite (Paster et al., 1990). More work is expected to coordinate the utilization of EOs into worldwide and provincial guidelines.

Conclusion

Rejuvenating oil's application in the food industry has progressed from that of a basic flavoring to that of a highly sought-after, regularly added ingredient with the express intention of extending the practical lifespan of food kinds. Concentrates on the cell reinforcement action of natural oils added straightforwardly to food lattices of different creations have shown their adequacy in decreasing lipid oxidation under different handling and stockpiling conditions, some of the time exhibiting more noteworthy viability than their business-engineered partners. Albeit the immediate expansion of EO to food varieties might raise worries about their tactile impact, the analysts found that besides the fact that this unimportantly affects specific tangible properties, it additionally expands the general acknowledgment of the food. Research assessing the cell reinforcement movement of EOs in food frameworks can be extended, true to form, as the EOs of various spices and flavors have not yet been all around contemplated and their assessment could give helpful data to food creation with clean mark.

References

Abbes, C., Mansouri, A., & Landoulsi, A. (2018). Synergistic effect of the lactoperoxidase system and cinnamon essential oil on total flora and Salmonella growth inhibition in raw milk. *Journal of Food Quality, 2018*, 1−6. Available from https://doi.org/10.1155/2018/8547954.

Abboud, M., El-Rammouz, R., Jammal, B., & Sleiman, M. (2016). In-vitro and in-vivo antimicrobial activity of two essential oils Thymus vulgaris and Lavandula angustifolia against Bovine Staphylococcus and Streptococcus mastitis pathogen. *Middle East Journal of Agriculture Research, 4*(4), 975−983.

Abrar, A., Beyene, T., & Furgasa, W. (2020). Isolation, identification and antimicrobial resistance profiles of Salmonella from dairy farms in adama and modjo towns, Central Ethiopia. *European Journal of Medical and Health Sciences, 2*(1), 1−11.

Ahmad, A., Khan, A., Kumar, P., Bhatt, R. P., & Manzoor, N. (2011). Antifungal activity of Coriaria nepalensis essential oil by disrupting ergosterol biosynthesis and membrane integrity against Candida. *Yeast (Chichester, England), 28*(8), 611−617. Available from https://doi.org/10.1002/yea.1890.

Al-Rimawi, F., Jazzar, M., Alayoubi, M., & Elama, C. (2019). Effect of different essential oils on the shelf life of concentrated yogurt. *Annual Research & Review in Biology, 30*(6), 1−9. Available from https://doi.org/10.9734/arrb/2018/v30i630031.

Ando, S., Nishida, T., Ishida, M., Kochi, Y., & Kami, A. (2001). Transmission of herb essential oil to milk and change of milk flavor by feeding dried herbs to lactating Holstein cows. *Nippon Shokuhin Kagaku Kogaku Kaishi, 48*(2), 142−145.

Ashraf, A., Kubanik, M., Aman, F., Holtkamp, H., Söhnel, T., Jamieson, S., Hanif, M., Siddiqui, W., & Hartinger, C. (2016). RuII(η6 -p-cymene) complexes of bioactive 1,2-benzothiazines: Protein binding vs. antitumor activity. *European Journal of Inorganic Chemistry, 9*, 1376−1382. Available from https://doi.org/10.1002/ejic.201501361.

Azizkhani, M., & Tooryan, F. (2016). Antimicrobial activities of probiotic yogurts flavored with peppermint, basil, and zataria against Escherichia coli and Listeria monocytogenes. *Journal of Food Quality and Hazards Control, 3*(3), 79−86.

Aziz, Z., Ahmad, A., Setapar, S., Karakucuk, A., Azim, M., Lokhat, D., Rafatullah, M., Ganash, M., Kamal, M., & Ashraf, G. (2018). Essential OILS: Extraction techniques, pharmaceutical and therapeutic potential—A review. *Current Drug Metabolism, 19*(13), 1100−1110. Available from https://doi.org/10.2174/1389200219666180723144850.

Badola, R., Panjagari, N. R., Singh, R. R. B., Singh, A. K., & Prasad, W. G. (2018). Effect of clove bud and curry leaf essential oils on the antioxidative and anti-microbial activity of burfi, a milk-based confection. *Journal of Food Science & Technology, 55*(12), 4802−4810. Available from https://doi.org/10.1007/s13197-018-3413-6.

Batiha, G. B., Awad, D. A., Algamma, A. M., Nyamota, R., Wahed, M. I., Shah, M. A., Amin, M. N., Hetta, H. F., Cruz-Marins, N., Koirala, N., Ghosh, A., & Sabatier, J. (2021). Diary-derived and egg white proteins in enhancing immune system against COVID-19. *Frontiers in Nutritionr. (Nutritional Immunology), 8*, 629440. Available from https://doi.org/10.3389/fnut.2021629440.

Bauer, K., Garbe, D., & Surburg, H. (2001). *Common fragrance and flavor materials: Preparation, properties and uses* (p. 293) Weinheim: Wiley-VCH.

Boyle, W. (1955). Spices and essential oils as preservatives. *The American Perfumer and Essential Oil Review, 66*, 25−28.

Burt, S. (2004). Essential oils: Their antibacterial properties and potential applications in foods-A review. *International Journal of Food Microbiology, 94*(3), 223−253. Available from https://doi.org/10.1016/j.ijfoodmicro.2004.03.022.

Cegiełka, A. (2020). 'Clean label' as one of the leading trends in the meat industry in the world and in Poland—A review. *Roczniki Panstwowego Zakladu Higieny, 71*, 43−55, [CrossRef] [PubMed].

Cosentino, S., Tuberoso, C. I. G., Pisano, B., et al. (1999). In-vitro antimicrobial activity and chemical composition of Sardinian ymus essential oils. *Letters in Applied Microbiology, 29*(2), 130−135.

Coutinho, N. M., Silveira, M. R., Rocha, R. S., et al. (2018). Cold plasma processing of milk and dairy products. *Trends in Food Science & Technology Food Science & Technology, 74*, 56−68.

Cox, S. D., Mann, C. M., Markham, J. L., Bell, H. C., Gustafson, J. E., Warmington, J. R., & Wyllie, S. G. (2000). The mode of antimicrobial action of essential oil of Melaleuca alternifola (tea tree oil). *Journal of Applied Microbiology, 88*, 170−175.

Dagdemir, E., Cakmakci, S., & Gundogdu, E. (2009). Effect of thymus haussknechtii and origanum acutidens essential oils on the stability of cow milk butter. *European Journal of Lipid Science and Technology, 111*(11), 1118−1123. Available from https://doi.org/10.1002/ejlt.200800243.

da Silva Dannenberg, G., Funck, G. D., Mattei, F. J., da Silva, W. P., & Fiorentini, A. ˆM. (2016). Antimicrobial and antioxidant activity of essential oil from pink pepper tree (Schinus terebinthifolius Raddi) in-vitro and in cheese experimentally contaminated with Listeria monocytogenes. *Innovative Food Science & Emerging Technologies, 36*, 120−127. Available from https://doi.org/10.1016/j.ifset.2016.06.009.

Dongmo, P. M. J., Tchoumbougnang, F., Ndongson, B., Agwanande, W., Sandjon, B., Zollo, P. H. A., et al. (2010). Chemical characterization, antiradical, antioxidant and anti-inflammatory potential of the essential oils of Canarium schweinfurthii and Aucoumea klaineana (Burseraceae) growing in Cameroon. *Agriculture and Biology North America, 1*(4), 606−611.

Draughon, F. (2004). Use of botanicals as biopreservatives in foods. *Food Technology (Chicago), 58*(2), 20−28.

Ehsani, A., Alizadeh, O., Hashemi, M., Afshari, A., & Aminzare, M. (2017). Phytochemical, antioxidant and antibacterial properties of Melissa officinalis and Dracocephalum moldavica essential oils. *Veterinary Research Forum, 8*, 223−229.

Elafify, M., Chen, J., Abdelkhalek, A., Elsherbini, M., ALAshmawy, M., & Koseki, S. (2020). Combined d-tryptophan treatment and temperature stress exert antimicrobial activity against Listeria monocytogenes in milk. *Journal of Food Protection, 83*(4), 644−650.

Elafify, M., Darwish, W., El-Toukhy, M., Badawy, B., Mohamed, R., & Shata, R. (2022). Prevalence of multidrug resistant Salmonella spp. in dairy products with the evaluation of the inhibitory effects of ascorbic acid, pomegranate peel extract, and D-tryptophan against Salmonella growth in cheese. *International Journal of Food Microbiology, 364*.

Fathi, M., Vincekovic, M., Juric, S., Viskic, M., & Jambrak, A. (2019). Food-grade colloidal, D. F. systems for the delivery of essential oils. *Food Reviews International*. Available from https://doi.org/10.1080/87559129.2019.1687514.

FDA. (2011). Code of federal regulations, title 21, food and drug. http://www.accessdatafdagov/scripts/cdrh/cfdocs/cfCFR/CFRSearchcfm?2011.

Fei, P., Xu, Y., Zhao, S., Gong, S., & Guo, L. (2019). Olive oil polyphenol extract inhibits vegetative cells of Bacillus cereus isolated from raw milk. *Journal of Dairy Science, 102*(5), 3894−3902.

Field, D., O' Connor, R., Cotter, P. D., Ross, R. P., & Hill, C. (2016). In vitro activities of Nisin and Nisin derivatives alone and in combination with antibiotics against Staphylococcus biofilms. *Frontiers in Microbiology, 7*, 508.

Friedman, M. (2017). Chemistry, Antimicrobial mechanisms, and antibiotic activities of cinnamaldehyde against pathogenic bacteria in animal feeds and human foods. *Journal of Agricultural and Food Chemistry, 65*(48), 10406−10423. Available from https://doi.org/10.1021/acs.jafc.7b04344.

Gould, G. W. (1996). Industry perspectives on the use of natural antimicrobials and inhibitors for food applications. *Journal of Food Protection*, 82–86.

Gouvea, Fd. S., Rosenthal, A., & Ferreira, E. Hd. R. (2017). Plant extract and essential oils added as antimicrobials to cheeses: A review. *Ciencia Rural*, *47*(8). Available from https://doi.org/10.1590/0103-8478cr20160908, Article e20160908.

Guenther, S., Huwyler, D., Richard, S., & Loessner, M. J. (2009). Virulent bacteriophage for efficient biocontrol of Listeria monocytogenes in ready-to-eat foods. *Applied and Environmental Microbiology*, *75*(1), 93–100.

Günther, E. (1948). *The essential oils: History and origin in plants production analysis* (pp. 235–240). New York: Krieger Publishing.

Gutierrez, A. M., Boylston, T. D., & Clark, S. (2018). Effects of pro-oxidants and antioxidants on the total antioxidant capacity and lipid oxidation products of milk during refrigerated storage. *Journal of Food Science*, *83*(2), 275–283. Available from https://doi.org/10.1111/1750-3841.14016.

Haʹc-Szyma ʹnczuk, E., & Cegiełka, A. (2015). *Evaluation of antimicrobial and antioxidant activity of sage in meat product, . Zywn. Nauka ˙ Technol. Jakość* (3, pp. 84–94). .

Hussein, K. A., Lee, Y. D., & Joo, J. H. (2020). Effect of rosemary essential oil and trichoderma koningiopsis VOCs on pathogenic fungi responsible for Ginseng root-rot disease. *Journal of Microbiology and Biotechnology*, *30*(7), 1018–1026.

Hu, Y., Zhang, J., Kong, W., Zhao, G., & Yang, M. (2017). Mechanisms of antifungal and anti-aflatoxigenic properties of essential oil derived from turmeric (Curcuma longa L.) on Aspergillus flavus. *Food Chemistry*, *220*, 1–8. Available from https://doi.org/10.1016/j.foodchem.2016.09.179.

Hyldgaard, M., Mygind, T., & Meyer, R. L. (2012). Essential oils in food preservation: Mode of action, synergies, and interactions with food matrix components. *Frontiers in Microbiology*, *3*, 12. Available from https://doi.org/10.3389/fmicb.2012.00012.

Ilmi, A., Praseptiangga, D., & Muhammad, D. R. A. (2017). Sensory attributes and preliminary characterization of milk chocolate bar enriched with cinnamon essential oil. *IOP Conference Series: Materials Science and Engineering*, *193*, 012031. Available from https://doi.org/10.1088/1757-899x/193/1/012031.

Ismaiel, A. A., & Pierson, M. D. (1990). Effect of sodium nitrite and origanum oil on growth and toxin production of Clostridium botulinum in TYG broth and ground pork. *Journal of Food Protection*, *53*(11), 958–960.

Karatzas, A. K., Bennik, M. H. J., Smid, E. J., & Kets, E. P. W. (2000). Combined action of S-carvone and mild heat treatment on Listeria monocytogenes Scott A. *Journal of Applied Microbiology*, *89*, 296–301.

Kaławaj, K., & Lemieszek, M. K. (2015). Health promoting properties of cinnamon. *Medycyna Ogólna i Nauki o Zdrowiu*, *21*, 328–331.

Kędzia, A. (2009). Garlic oil—Chemical components, pharmacological and medical activity. *Postępy Fitoterapii*, *3*, 198–203.

Kostova, I., Damyanova, S., Ivanova, N., Stoyanova, A., Ivanova, M., & Vlaseva, R. (2016). Use of essential oils in dairy products. Essential oil of basil (Ocimum basilicum L.). *Indian Journal of Applied Research*, *6*(1), 54–56.

Kozłowska-Lewecka, M., Wesołowski, W., & Borowiecka, J. (2011). Analysis of contents of essential oils in white and black pepper determined by GC/MS. *Bromatologia i Chemia Toksykologiczna*, *44*, 1111–1112.

Lambert, R. J. W., Skandamis, P. N., Coote, P., & Nychas, G.-J. E. (2001). A study of the minimum inhibitory concentration and mode of action of oregano essential oil, thymol and carvacrol. *Journal of Applied Microbiology*, *91*, 453–462.

Lianou, A., & Sofos, J. N. (2007). A review of the incidence and transmission of Listeria monocytogenes in ready-to-eat products in retail and food service environments. *Journal of Food Protection*, *70*(9), 2172–2198.

Loizzo, M. R., Menichini, F., Conforti, F., Tundis, R., Bonesi, M., Saab, A. M., et al. (2009). Chemical analysis, antioxidant, antiinflammatory and anticholinesterase activities of Origanum ehrenbergii Boiss and Origanum syriacum L. essential oils. *Food Chemistry*, *117*(1), 174–180. Available from https://doi.org/10.1016/j.foodchem.2009.03.095.

Łyczko, J., Piotrowski, K., Kolasa, K., Galek, R., & Szumny, A. (2020). Mentha piperita L. micropropagation and the potential influence of plant growth regulators on volatile organic compound composition. *Molecules (Basel, Switzerland)*, *25*, 2652.

Manabe, A., Nakayama, S., & Sakamoto, K. (1987). Effects of essential oils on erythrocytes and hepatocytes from rats and dipalitoyl phophatidylcholine-liposomes. *Japanese Journal of Pharmacology*, *44*, 77–84.

Mohajeria, F. A., Derakhshanc, Z., Ferrantef, M., et al. (2018). The prevalence and antimicrobial resistance of Listeria spp in raw milk and traditional dairy products delivered in Yazd, central Iran. *Food and Chemical Toxicology*, *114*, 141–144.

Moleyar, V., & Narasimham, P. (1992). Antibacterial activity of essential oil components. *International Journal of Food Microbiology*, *16*, 337–342.

Muhammad, D. R. A., Lemarcq, V., Alderweireldt, E., Vanoverberghe, P., Praseptiangga, D., Juvinal, J. G., et al. (2019). Antioxidant activity and quality attributes of white chocolate incorporated with Cinnamomum burmannii Blume essential oil. *Journal of Food Science & Technology*, *57* (5), 1731–1739. Available from https://doi.org/10.1007/s13197-019-04206-6.

Nazir, A., Itrat, N., Shahid, A., Mushtaq, Z., Abdulrahman, S. A., Egbuna, C., Khan, J., Uche, C. Z., & Toloyai, P. Y. (2022). *Orange peel as a source of nutraceuticals. Food and agricultural byproducts as important source of valuable nutraceuticals* (1st ed). Berlin: Springer, xxx, 400 p. Gebunden. ISBN 978-3-030-98759-6.

Ogunlana, O. O., Adekunbi, T. S., Adegboye, B. E., Iheagwam, F. N., & Ogunlana, O. E. (2021). Ruzu bitters ameliorates high–fat diet induced non-alcoholic fatty liver disease in male Wistar rats. *Journal of Pharmacy and Pharmacognosy Research*, *9*(3), 251–260.

Paster, N., Juven, B. J., Shaaya, E., Menasherov, M., Nitzan, R., Weisslowicz, H., & Ravid, U. (1990). Inhibitory effect of oregano and thyme essential oils on moulds and foodborne bacteria. *Letters in Applied Microbiology*, *11*, 33–37.

Prasad, R. (2017). Historical aspects of milk consumption in South, Southeast, and East Asia. *Asian Agrihist*, *21*(4), 287–307.

Sandner, G., Heckmann, M., & Weghuber, J. (2020). Immunomodulatory activities of selected essential oils. *Biomolecules*, *10*, 1139.

Schiano, A., Harwood, W., & Drake, M. (2017). A 100-year review: Sensory analysis of milk. *Journal of Dairy Science*, *100*(12), 9966–9986.

Schiano, A. N., Jo, Y., Barbano, D. M., & Drake, M. A. (2019). Does vitamin fortification affect light oxidation in fluid skim milk? *Journal of Dairy Science, 102*(6), 4877−4890. Available from https://doi.org/10.3168/jds.2018-15594.

Sharifi-Rad, J., Sureda, A., Tenore, G. C., Daglia, M., Sharifi-Rad, M., Valussi, M., Tundis, R., Sharifi-Rad, M., Loizzo, M., Oluwaseun, A., et al. (2017). Biological activities of essential oils: From plant chemoecology to traditional healing systems. *Molecules (Basel, Switzerland), 22*. Available from https://doi.org/10.3390/molecules22010070.

Shojaee-Aliabadi, S., Hosseini, S. M., & Mirmoghtadaie, L. (2018). Antimicrobial activity of essential oil. In S. Hashemi, A. Mousavi Khaneghah, & A. De Souza Sant'Ana (Eds.), *Essential oils in food processing* (pp. 191−229). Wiley. Available from 10.1002/9781119149392.

Souza, E. L., Stamford, T. L. M., Lima, E. O., & Trajano, V. N. (2007). Effectiveness of Origanum vulgare L. essential oil to inhibit the growth of food spoiling yeasts. *Food Control, 18*, 409−413.

Szewczyk, K., Kalemba, D., Komsta, Ł., & Nowak, R. (2016). Comparison of the essential oil composition of selected impatiens species and its antioxidant activities. *Molecules (Basel, Switzerland), 21*(9), 1162. Available from https://doi.org/10.3390/molecules21091162.

Taiwo, M. O., & Adebayo, O. S. (2017). Plant essential oil: An alternative to emerging multidrug resistant pathogens. *Journal of Microbiology & Experimentation, 5*, 1−6.

Tassou, C., Drosinos, E. H., & Nychas, G.-J. E. (1995). Effects of es sential oil from mint (Mentha piperita) on Salmonella enteritidis and Listeria monocytogenes in model food systems at 4 jC and 10 jC. *Journal of Applied Bacteriology, 78*, 593−600.

Thorning, T. K., Raben, A., Tholstrup, T., Soedamah-Muthu, S. S., Givens, I., & Astrup, A. (2016). Milk and dairy products: Good or bad for human health? An assessment of the totality of scientific evidence. *Food & Nutrition Research, 60*. Available from https://doi.org/10.3402/fnr.v60.32527.

Tiwari, B. K., Valdramidis, V. P., O' Donnell, C. P., Muthukumarappan, K., Bourke, P., & Cullen, P. J. (2009). Application of natural antimicrobials for food preservation. *Journal of Agricultural and Food Chemistry, 57*(14), 5987−6000. Available from https://doi.org/10.1021/jf900668n.

Tongnuanchan, P., & Benjakul, S. (2014). Essential oils: Extraction, bioactivities, and their uses for food preservation. *Journal of Food Science, 79*, R1231−R1249.

Tornambé, G., Cornu, A., Verdier-Metz, I., Pradel, P., Kondjoyan, N., Figueredo, G., et al. (2008). Addition of pasture plant essential oil in milk: Influence on chemical and sensory properties of milk and cheese. *Journal of Dairy Science, 91*(1), 58−69. Available from https://doi.org/10.3168/jds.2007-0154.

Tsigarida, E., Skandamis, P., & Nychas, G.-J. E. (2000). Behaviour of Listeria monocytogenes and autochthonous flora on meat stored under aerobic, vacuum and modified atmosphere packaging conditions with or without the presence of oregano essential oil at 5 jC. *Journal of Applied Microbiology, 89*, 901−909.

Ultee, A., Kets, E. P. W., Alberda, M., Hoekstra, F. A., & Smid, E. J. (2000). Adaptation of the food-borne pathogen Bacillus cereus to carvacrol. *Archives of Microbiology, 174*(4), 233−238.

Ultee, A., Slump, R. A., Steging, G., & Smid, E. J. (2000). Antimicrobial activity of carvacrol toward Bacillus cereus on rice. *Journal of Food Protection, 63*(5), 620−624.

Verghese, R. J., Mathew, S., & David, A. (2020). Antimicrobial activity of Vitamin C demonstrated on uropathogenic Escherichia coli and Klebsiella pneumonia. *Journal of Research in Medical Sciences, 3*(2), 88−93.

Vernès, L., Vian, M., & Chemat, F. (2020). Ultrasound and microwave as green tools for solid-liquid extraction. In C. Poole (Ed.), *Liquid-phase extraction* (pp. 355−374). Elsevier.

Wang, H., Yih, K., Yang, C., & Huang, K. (2017). Anti-oxidant activity and major chemical component analyses of twenty-six commercially available essential oils. *Journal of Food and Drug Analysis, 25*(4), 881−889. Available from https://doi.org/10.1016/j.jfda.2017.05.007.

Witkowska, D., & Sowi ́nska, J. (2013). The effectiveness of peppermint and thyme essential oil mist in reducing bacterial contamination in broiler houses. *Poultry Science, 92*, 2834−2843.

Worwood, V. A. (2016). *The complete book of essential oils and aromatherapy, revised and expanded: Over 800 natural, nontoxic, and fragrant recipes to create health, beauty, and safe home and work environments*. United States: New World Library.

Wójcik, W., Solarczyk, P., Łukasiewicz, M., Puppel, K., & Kuczy ́nska, B. (2018). Trends in animal production from organic farming [review]. *Acta Innovations, 28*, 32−39.

Zafar, N., Nawaz, Z., Qadeer, A., et al. (2020). Prevalence, molecular characterization and antibiogram study of Listeria monocytogenes isolated from raw milk and milk products. *Pure and Applied Biology, 9*(3).

Chapter 12

The use of essential oils together with different milk products

Babatunde Oluwafemi Adetuyi[1], Kehinde Abraham Odelade[1], Oluwakemi Semiloore Omowumi[1], Peace Abiodun Olajide[1], Charles Oluwaseun Adetunji[2], Juliana Bunmi Adetunji[3], Yovwin D. Godwin[4], Oluwabukola Atinuke Popoola[5], Olatunji Matthew Kolawole[6], Olalekan Akinbo[7], Abel Inobeme[8], Osarenkhoe Omorefosa Osemwegie[9], Mohammed Bello Yerima[10] and M.L. Attanda[11]

[1]Department of Natural Sciences, Faculty of Pure and Applied Sciences, Precious Cornerstone University, Ibadan, Oyo State, Nigeria, [2]Applied Microbiology, Biotechnology and Nanotechnology Laboratory, Department of Microbiology, Edo State University Uzairue, Iyamho, Edo State, Nigeria, [3]Department of Biochemistry, Osun State University, Osogbo, Osun State, Nigeria, [4]Department of Family Medicine, Faculty of Clinical Sciences, Delta State University, Abraka, Delta State, Nigeria, [5]Genetics, Genomics and Bioinformatics Department, National Biotechnology Development Agency, Abuja, FCT, Nigeria, [6]Department of Microbiology, Faculty of Life Sciences, University of Ilorin, Ilorin, Kwara State, Nigeria, [7]Centre of Excellence in Science, Technology, and Innovation, AUDA-NEPAD, Johannesburg, Gauteng, South Africa, [8]Department of Chemistry, Edo State University Uzairue Iyamho, Auchi, Edo State, Nigeria, [9]Department of Food Science and Microbiology, Landmark University, Omu-Aran, Kwara State, Nigeria, [10]Department of Microbiology, Sokoto State University, Sokoto, Sokoto State, Nigeria, [11]Department of Agricultural Engineering, Bayero University Kano, Kano, Kano State, Nigeria

Introduction

The unstable hydrophobic fluids known as essential oils (EOs) are gotten from plants and regularly have areas of strength (Bhavaniramya et al., 2019). EOs are a combination of optional metabolites, regularly terpenoids, that are exceptionally powerful enemy of microbial mixtures and assume a critical part in plant protection. Various notable remedial characteristics of natural ointments exist, including antibacterial, mitigating, diuretic, tonic, and antispasmodic movement (Adejumo et al., 2017; Adetunji, 2019; Adetunji et al., 2013, 2017, 2019, 2020a, 2020b, 2020c, 2021j, 2021f, 2021d, 2021b, 2021h, 2021c, 2021g, 2021e, 2021i, 2021a, 2022; Adetunji & Varma, 2020; Bello et al., 2019; Egbuna et al., 2020; Olaniyan & Adetunji, 2021; Rauf et al., 2021; Thangadurai et al., 2021; Ukhurebor & Adetunji, 2021; Adetunji, 2008). They are said to have antibacterial characteristics that can battle various destructive microorganisms. They have been used as additives in dairy and other saved food items due to their remedial qualities (Benchaar & Greathead, 2011). Moreover, it has been noticed that EOs can improve the lactic corrosive containing details' dietary potential and organoleptic characteristics. Different EOs' bioactive parts often have added substance or synergistic impacts, working on microbial maturation and upgrading supplement use in ruminants under temperature stress.

The interest in dairy items has expanded during the most recent couple of years (Batiha et al., 2021). Since old times, milk has been a staple dairy item and has high wholesome advantages (Nazir et al., 2022). Moreover, it is the best media for developing probiotics. Probiotic details are comprised of wholesome and therapeutic fixings that have gone through maturation. These aged items regularly display restorative impacts like an enemy of diarrheal action, usage of lactose in mal-safeguards of lactose, and immuno-modulatory action. One of the most researched properties of *Bifidobacterium bifidum* and *Lactobacillus acidophilus* is their probiotic potential. These species may have probiotic effects, including reducing susceptibility to allergies and other disorders, lactose intolerance, brittleness in hypercholesterolemia, and circulatory strain. For the final category of dairy products, accidental mixtures (amino acids formed by heated debasement), volatile acids, nonvolatile acids, and carbonyl mixtures are to blame (James-Okoro et al., 2021). A couple of increases are utilized in dairy items fully intent on expanding worthiness and flavor enhancement, like seasoning specialists (either regular or manufactured), natural products, and sugars (Bachir & Benattouche, 2013).

Rejuvenating oils are sleek, fragrant fluids that are acquired from plants and have antibacterial characteristics. They can be utilized in restorative arrangements as additives or as seasoning specialists for food. Moreover, rejuvenating oils fill various needs in food sources, refreshments, beauty care products, and toiletries. Various examinations have analyzed the effect of medicinal oils on the beginning society microorganisms in dairy items according to concentration.

Applications of Essential Oils in the Food Industry. DOI: https://doi.org/10.1016/B978-0-323-98340-2.00007-9

Under unsanitary circumstances, the improvement of a yeast and shape populace in yogurt that has been held for longer than seven days brings up serious issues about the item's quality and timeframe of realistic usability (Didunyemi et al., 2020). Numerous studies have taken into account the use of natural ingredients in the creation of nutritional supplements and traditional self-medication.

Composition of essential oils

EOs are comprised of different synthetic substances whose focuses structure the premise of their characterization; for example, the parts are partitioned into three classes: essential (20%−95%), auxiliary (1%−20%), and follow (1%) parts (Sticher et al., 2015). The geographic district of development, the hour of reap, the plant part and assembling strategies, the capacity conditions and the timeframe are only a couple of the factors that influence the quality and synthetic organization of EOs. Although they come from similar species, varieties in these attributes can influence the creation or nature of EOs. As per International Organization for Standardization, citrus skins can be precisely handled to make essential oil, dry-refined, or by fume or water refining. In contrast with normal oils, manufactured oils have as of late become increasingly common. Natural ointments are utilized for restorative reasons as per the rules illustrated in pharmacopeias.

Falsification is among the main problems regarding EOs. Various oils sold in commercials normally bear the portrayals "unadulterated," "regular," or "100% normal." However, the ISO/TC 54 standard commands that standard fundamentals stay pure and untreated, in this way such marking are not needed. Pollution, defilement, improper oil creation, and maturing can bring about inferior-quality oil. Different tests can be utilized to affirm the virtue of EOs. Results are contrasted and these oils' standard revealed profiles to confirm the EOs' quality. Capacity is likewise one more critical issue for EOs. As per the European Pharmacopeia, EOs ought to be kept in holders that are firmly shut, full, and away from light. Inadmissible capacity conditions can cause various responses, including photoisomerization, oxidation, peroxidation, photocyclization, liquor breakdown, and ketone hydrolysis. To forestall the impacts of debasement, ISO offers guidelines for EOs.

The ISO/TS 210:2014 standard indicates the fundamental rules for the bundling, molding, and capacity of rejuvenating oils. A hundred to 300 separate subatomic parts make up every natural ointment (Baser & Buchbauer, 2010). Natural ointments are such combinations, containing specific parts in follows and others in bigger concentrations. The various kinds of synthetic substances tracked down in EOs are responsible for different organic capabilities. Natural ointments with phenolic parts have more intense therapeutic characteristics than EOs with terpene alcohols as their primary constituent (Nazir et al., 2022). Also, it was shown that natural oils containing geranyl acetic acid derivation, thujone, or myrcene in ketones or esters had diminished movement. It is oftentimes the situation that parts found in little sums in EOs play a huge part in improving their organic movement. Added substance impacts through hostility are additionally as often as possible seen. EOsEOs from various plant parts have shifted amounts or syntheses; for example, EOs from coriander seeds contrast from EOs from youthful leaves.

Different plant wellsprings of medicinal oils contain a blend of different synthetic components. Terpenes from citrus and pine, for example, just hold back carbon and hydrogen. An −OH bunch is associated with the terpene moiety in coriander, tea tree, and peppermint. Lemon salve, lemon myrtle, and citronella all contain terpenoids (aldehydes) that have a carbon bond and a hydrogen particle. They contain a benzene ring-connected aldehyde bunch in cinnamon, unpleasant almond, and cumin. There is a ketone found in pennyroyal, Eucalyptus radiata, sage, and thuja. It has a C−O bunch associated with two carbon molecules. Thyme and oregano contain phenol, which is a compound with a −OH bunch joined to a benzene ring (Herman et al., 2019). Wormseed, cajeput, and eucalyptus all contain an oxide with O connecting between at least two carbons (Aggarwal et al., 2013). Clary sage, wintergreen, and lavender all contain ester, which is created when corrosive and liquor consolidate. Phenylpropanes, which have carminative and sedative properties, are tracked down in aniseed, clove tarragon, and myrtle leaves. German chamomile and yarrow contain sesquiterpenes that have mitigating and antiviral properties.

Benefits of involving essential oils in dairy products

Biopreservative and antimicrobial action

As of late, the quest for normally happening hostile to microbials has been energized by the sensational ascent in customer familiarity with regular food items. Food items contain bio-additives to upgrade quality by expanding their timeframe of realistic usability and forestalling the development of unsafe microorganisms. The distribution technique, focus, application technique, and capacity temperature all affect how viable applying EOs is. As indicated by reports, terpineol's antimicrobial movement is conversely connected with fat substance. With increasing fat content, such as in skim milk (8 log CFU/mL), low butterfat milk (5 log CFU/mL), and whole milk (4 log CFU/mL), its motility decreased (Didunyemi et al., 2019). The concentration utilized and the strategy for application (shower, contained in lactose container, or drenching) decide how successful these bio-additives are. The fundamental centralization of natural oils to forestall microbial development ought to likewise be considered due to organoleptic impact. In any case, it has been shown that various normally happening additives are viable substitutes for forestalling the development of microbes and limiting horrible changes in organoleptic characteristics.

For the food area as well as shoppers, food security is a key concern. Interest in hostile to microbials appropriate for dairy items has expanded because of the ascent in food-borne ailments. Numerous medicinal oils found in plants and flavors have major areas of strength for having an impact on various microorganisms. *Listeria monocytogenes* is purportedly impervious to rejuvenating oils from dark pepper, mint, garlic, thyme, oregano, orange, clove, cumin, and tea tree. While the EOs acquired from tea tree, mints, and cumin just restrain yeasts, the natural oils got from clove, orange, thyme, and oregano displayed solid inhibitory impact against microorganisms and yeasts. Various examinations have archived the Eos' antibacterial abilities (Didunyemi et al., 2019). Since they are hydrophobic, EOs can undoubtedly penetrate bacterial cell walls. When inside a bacterial cell, EOs disturb the particles' vehicle cycles and render the cell dormant. Since Gram-negative microbes have an external layer lipopolysaccharide boundary, EOs have more grounded inhibitory exercises when applied to these microscopic organisms (Loizzo et al., 2009). The objective microorganism, least inhibitory concentrations (MIC), the component of activity, and likely connections with the food lattice and tactile food attributes are essentials for the utilization of EOs in food items. The antimicrobial properties of EOs might be because of their dynamic fixings.

As indicated by the compound moiety, the parts that give EOs their antimicrobial properties can be isolated into four gatherings: terpenes (like limonene and p-cymene), phenylpropenes (like eugenol and vanillin), terpenoids (like carvacrol and thymol), and different substances like isothiocyanates or allicin (Hyldgaard et al., 2012). The method of activity of EOs fluctuates and has not yet been completely perceived. Natural ointments (EOs) have various impacts on microorganisms, which are much of the time remembered to be where they work.

As per the hypothesis, Gram-positive microscopic organisms answer preferable to natural balms over Gram-negative microbes (Dongmo et al., 2008). The peptidoglycan (90%−95%) that makes up the cell mass of Gram-positive microscopic organisms has proteins and teichoic corrosive associated with it. Medicinal oils effectively pervade the cell film and arrive at the cytoplasm since they are basically hydrophobic. All things being equal, Gram-negative microbes have a more confounded cell film that is comprised of a layer of proteins and lipopolysaccharides around a monolayer of peptidoglycan (LPS). Gram-negative microbes have a hydrophilic, charged external layer (Dongmo et al., 2010). The given movement is brought about by the phenols' −OH gathering's acidic nature. These substances upset the protein drive that at last outcomes in cell demise, adjust cell porousness, and slow down energy creation by impeding the chemicals in question.

How microorganisms are molded can at times change how natural oils work. As per reports, bar-molded cells are more inclined to harm than coccoid-formed ones. A careful examination of the cinnamaldehyde, a critical fixing in the EOs created by fragrant cinnamon plants, was distributed by Friedman (2017). The significance of this chemical in several in vitro studies against potentially present pathogenic microorganisms in dairy products, like cheddar, and different food sources, for example, *Salmonella enterica*, *B. cereus*, *Clostridium perfringens*, *Campylobacter jejuni*, *E. coli* and *L. monocytogenes*, was featured by the creator. This exploration could lay out a pattern for the utilization of cinnamaldehyde and other regular rejuvenating oils in the counteraction of foodborne sicknesses, which could really further develop sanitation and forestall pathogenic microbes that cause food defilement and illnesses in the two people and creatures. In 2016, Abboud et al. (2016) assessed the viability of natural balms from *Thymus vulgaris* and *Lavandula angustifolia* as a drawn-out option in contrast to engineered anti-infection agents frequently utilized against the microbes liable for mastitis in cows. The California Mastitis Test was utilized in the examination of four draining cows to check for clinical mastitis. The in vitro examination's discoveries showed that the two natural ointments utilized in this study had very high antibacterial action against the mastitis-causing kinds of Staphylococcus and Streptococcus, which had practically a similar power as the anti-toxin utilized as a control. Furthermore, it was shown that after the treatment was regulated multiple times in succession, the microscopic organism count from every one of the analyzed examples diminished when a 10% arrangement of the rejuvenating oil was applied intramammary. Also, the two concentrated diseases' related irritation was decisively decreased by remotely applying the rejuvenating balm. With a 100% pace of recovery with thymus EOs, this activity is more solid than intramammary approaches. Salmonella Hadar and the complete vegetation of milk were the two microorganisms that

Abbes et al. (2018) assessed for the combined action of 4 g/mL cinnamon medical ointment and lactoperoxidase framework has an antimicrobial effect.

The result uncovered that the joined and synergistic impacts of cinnamon natural ointment and lactoperoxidase framework had a more grounded antibacterial effect than single specialists. Moreover, a wide inhibitory profile of the entire verdure was seen following 5 days of perception under the synergistic impact at 4°C (for 3 days), while Salmonella Hadar was restrained for 5 days. Utilizing the checkerboard miniature weakening procedure, Zhao et al. (2016) evaluated the antimicrobial activity of deulkkae and nisin oil toward 30 food-borne *S. aureus* and *L. monocytogenes*. The outcomes showed that the two medicines synergistically affected the two objective organic entities, with partial inhibitory concentration files of 0.125 to 0.25 and 0.19 to 0.375, separately. Afterward, filtering electron microscopy was utilized to affirm the two EOs' antibacterial adequacy against the two objective creatures. The result shows that *L. monocytogenes* and *S. aureus'* cell walls were penetrable to the medicinal ointment, which caused the cell wall and surface layer of the tried microorganism to be harmed. One of the vitally bioactive parts found in thyme medicinal oil has been distinguished as thymol. Thymol works on the antibacterial properties of thyme by restricting to layer proteins. Subsequently, the microbial cells' available cell contents were delivered during an ejection (Burt, 2004). As indicated by Tiwari et al. (2009), thymol could altogether bring down E. coli's intracellular adenosine triphosphate (ATP) levels while expanding extracellular ATP levels, which could obstruct plasma layer capability. Moreover, it has been seen that thymol can smother the development of microscopic organisms of both Gram-positive and Gram-negative strains (Burt, 2004). Moreover, eugenol, a functioning fixing in clove natural oil with huge antibacterial viability, was referenced in another article. The *B. cereus* cellular membranes can be broken down by eugenol, which can prompt the microbes' cell division, as indicated by Burt (2004). It can likewise block the age of amylase. As per a recent report by Batiha et al., 2021, the antifungal action of *Coriaria nepalensis* EO is applied by altering the production of steroid alcohol and the credibility of Candida's film. Also, Hu et al. (2017) exhibited that *Curcuma longa* had hostile to aflatoxigenic and antifungal properties when tried against *Aspergillus flavus*. Potential associations between the methods of activity and the restraint of ergosterol creation exist.

Hostile to oxidant capability

Developing interest in EOs as normal cell reinforcements has been started by expected adverse consequences of manufactured cancer prevention agents, like BHA and BHT. Either by mixing them into the food or as a palatable covering on dynamic bundling, EOs are added to eatable arrangements. This is a decent substitute for forestalling oxidation and expanding time span of usability. The food area should consider the cancer prevention agent limit of rejuvenating oils in light of the fact that regular tests are incapable of themselves and may deliver conflicting outcomes (Loizzo et al., 2009). Alongside an assessment of their ability for food security, the science hidden in the cell reinforcement movement of rejuvenating oils is investigated in this article. In the referred writing, methods for assessing the cancer prevention agent limit of EOs have been entirely assessed. As per da Silva Dannenberg et al. (2016), rejuvenating ointments of green and mature pink pepper have cell reinforcement potential. Since mature pink pepper's tactile characteristics are protected for a more drawn-out timeframe, it might assist with broadening the item's timeframe of realistic usability.

Improvement in taste and scent

As indicated by numerous buyers, milk is a bland food. Subsequently, they find it very testing to depict taste qualities. As indicated by a study (Muhammad et al., 2019), cinnamon EOs can assist with draining chocolate's practical characteristics and flavor.

These natural balms are a decent wellspring of synthetics that advance wellbeing and fragrance in concentrated structure. They improve feasts' flavor and dietary benefits. Different EOs, like those from orange, peppermint, ginger, or rose, among others, can change the kind of solitary cocoa flavor (Worwood, 2016). In the first place, only a couple of drops ought to be added for inside use. A couple of individuals mix oil straight with fluids like milk or water and afterward, drink the combination. It might be feasible to forestall aversion to some EOs by ingesting them in gelatin-or intestinal-covered gelatin-based containers, alongside a spoonful of honey or coconut oil. *Cymbopogon citratus*, some of the time known as lemongrass, has for some time been utilized as a medication. A substance is utilized in drinks rather normally. It has been used as a sugar on account of unfamiliar dishes (Laswai et al., 2009).

A couple of occasions when essential oils have been utilized in dairy items

Utilization of essential oils to cheddar items

Natural balms are exceptionally compelling at restricting development and bringing down microbial endurance in cheddar. Gouvea et al. (2017) provided details regarding the antibacterial impacts of cheeses while adding natural oils and plant removal. Cheddar has been singled out as a food that is devoured all around the world yet is very vulnerable to pollution by various pathogenic decay microorganisms, which could restrict its time span of usability and posture serious well-being dangers to buyers. Regular additives that could guarantee client well-being and increment the time span of usability of dairy items have been found in light of the consistently growing interest in quality food varieties liberated from fake additives. The disclosure of the EO as a characteristic, supportable natural additive with this remarkable quality might be connected with antibacterial exercises against infections and microorganisms that make cheddar rot. Then again, notwithstanding their adverse consequences, natural and interior improvements may likewise affect the exercises of such synthetics once integrated into cheddar. Various examinations have likewise definite the effect of plant concentrates and EOs' antibacterial strength notwithstanding their synergistic impacts on lactic microscopic organisms and the exotic characteristics of cheddar-determined merchandise. However, medicinal ointments connect with fat, carbs, or protein and lose a portion of their antibacterial properties, so more EOs are expected to obtain the ideal outcome. Proteins and the phenolic synthetic substances in EOs connect, and the fat in dairy items frames a defensive covering over the EOs' hydrophobic components. This is the manner by which lipids limit the EOs' availability to the microscopic organisms' favored target regions. Moreover, the actual qualities of cheddar impact how EOs are dispersed appropriately, which results in decreased EO antibacterial adequacy. In spite of the fact that EO action is expanded by expanded temperature, oxygen content, and lower pH. Reduced pH advantages EOs by making them more hydrophobic, which thus makes it simpler for EOs to enter bacterial cells and arrive at their objective site. At the point when EOs are used in bigger fixations to compensate for the action decline welcomed on by dietary parts, tangible irregularities as often as possible outcome.

Various systems have been investigated to address these inadequacies. The medicinal oils could first be utilized in quite a while coatings. This thought depends on durable antibacterial activity and ceaseless delivery. One more methodology for EOs' drawn-out activity is to nano-or microencapsulate them. One such example of microencapsulation is the utilization of whey protein disengage and inulin to epitomize rosemary essential oil, which was then added to Minas Frescal cheddar. As indicated by this planning, mesophilic bacterial development had been ended and the cheddar's timeframe of realistic usability had been broadened. Nisin and multiflora cooperate to forestall *Staphylococcus aureus* from delivering—poison and enterotoxin. While creating BHI stock and customary Iranian white brackish water cheddar, separately, this was considered. Rejuvenating oils are normally over the top expensive in light of the fact that the sort of plant material utilized, the method used to extricate the oils, the expense of work, and the expense of energy all influence the creation costs. How much plant material is expected to create a sizable measure of natural oil, as well as its virtue, are different elements that add to cost increments. There are various, cheap, and top-notch EOs open.

This also became clear that smaller fixations (0.2 L/L) of EO had no impact on the cheddar's observable qualities in a concentrate by Tornambe et al. (2008) about the impact of EO on cheddar tangible properties. Nonetheless, cheeses labeled "mint/chlorophyll" and "thyme/oregano" at 3.0 L/L showed a solid smell and fragrance. Such surprising smell qualities are welcomed by added EO. Generally speaking, 152 mixtures got from cheddar have been accounted for, of which 41 have been joined with EO; interestingly, 54 EO intensifies have not been recuperated. Notwithstanding, the EO option affected a few unpredictable composites got from cheeses that weren't added.

In a recent report, the rejuvenating oils of cumin, rosemary, and thyme were separated and their impacts on the physicochemical, microbial, rheological, and tactile qualities of ultra-sifted delicate cheddar were additionally researched. Thyme EOs showed a cell reinforcement influence, while cumin rejuvenating ointments worked on tangible qualities. Since EOs are regular, it has been resolved that they are fine for ultra-sifted cheddar.

Use of essential oils in dairy products

One of the best regular options in contrast to added substances or purification to prevent food ruining from microorganisms in an assortment of food frameworks is rejuvenating oils (EOs). As per Jemaa et al. (2018), adding medicinal ointments from the *Thymus capitatus* plant can expand the time span of usability of sanitized milk by hindering bacterial turn of events. This prompted the end that *T. capitatus* EO worked on the viability of purification in keeping up with the nature of crude milk. In this regard, Jemaa et al. (2017) investigated the impact of a solution or the nanoemulsion of *T. capitatus*, which improves both alcoholic fermentation and aerobic metabolism stability of semi-skimmed super high-temperature milk as a result, hence increasing its productivity in milk protection. The previously mentioned discoveries unequivocally show the upsides of regular additives (rejuvenating ointments) over engineered ones. Typified EO's strength is an extraordinary decision for the milk enterprises as a characteristic and powerful antibacterial and cell reinforcement specialist to screen and keep up with the quality necessities. Tornamb'e et al. took a gander at how EO impacted the tactile qualities of milk (2008). This EO was added to the drain at a centralization of 1.0 L/L, changing its tactile qualities. The milk had a more prominent taste of mint, was better, and had a higher substance of EO (1.0 L/L). There is no question that the EO expansion was liable for the surprising flavors. Thus, the 0.1−1.0 L/L reach addressed the edge focus for the impression of the flavor related to the expansion of EO.

The concentrate by Xue et al. (2017) found that thymol got from *T. vulgaris* is a generally utilized part of rejuvenating oils. Thymol has been shown to be an effective antibacterial against both Gram-positive and Gram-negative bacteria. High measures of thymol are required in complex food things for it to foster hydrophobic communications with food grid constituents and display wonderful concealment of foodborne microbes. This study recommended ternary mixes might be valuable for food safeguarding. The counter listerial action of nisin and the leaf natural ointment of *M. glyptostroboides* against the foodborne microbe *L. monocytogenes* in milk tests was likewise inspected in an examination by Bajpai et al. (2014). They found that *L. monocytogenes* was firmly repressed by the 2% and 5% leaf EO in all milk tests. Furthermore, following 14 days, every one of the inspected kinds of milk (entire, low, and skimmed) showed areas of strength for a listerial synergism when leaf EO and nisin were joined. Due to its conceivable enemy of listerial synergism, researchers have found an expected application for *M. glyptostroboides* leaf natural ointment in the food area to control the development of foodborne diseases. Similarly, Yoon et al. (2011) showed that *M. glyptostroboides* cone rejuvenating oil had a deterrent impact against a foodborne microorganism (*L. monocytogenes*) in milk tests.

As indicated by Feizollahzadeh et al. (2016), probiotics are innocuous, non-pathogenic living microorganisms that benefit clients' well-being. Their effect is interceded by changes in the microbiota in the stomach, which fortifies the resistant framework. In a recent report, how soymilk (SM) can forestall diabetes was portrayed. In addition to managing glucose and insulin levels, it was shown that soy milk utilization fundamentally supported insulin responsiveness. *Cuminum cyminum* and probiotic soy milk exhibited promising outcomes in diminishing and further forestalling diabetes confusion, and this is upheld by writing. There has not been any examination done at this point on the association between probiotic soy milk and natural ointments produced using spices.

Yogurt and essential oils

Part of the most well-known milk products, yogurt has worked on dietary benefit, taste, and well-being benefits. Considering the previously mentioned, In a study, Azizkhani and Tooryan evaluated the effectiveness of three EOs derived from Za'atar, Mentha Piperita, and basil on the viability of the microbial microscopic organisms present in the yogurt in terms of their ability to inhibit the growth of *L. monocytogenes* and *E. coli*. Bifidobacterium Bb-12, *L. fermentum*, and *L. acidophilus* LA5 were utilized to make the yogurt. Involving in vitro circle dispersion strategies, the development of inhibitory activities of *L. monocytogenes* and *E. coli* was explored. They discovered no discernible difference in the acidity worth of the microbial yogurts including EO while compared to the control, particularly all through the maturation and capacity periods. Besides, it was found that yogurt containing zataria meaningfully affected the development of *E. coli* and *L. monocytogenes* when contrasted with the control treatment ($P < .05$), particularly in the treatment containing the medicinal oil. This study showed that adding Za'atar, Mentha Piperita, and basil natural ointments to the microbial yogurt detailing forecasts the capacity to work on the item's possible usefulness and increment their inhibitory impact on *L. monocytogenes* and *E. coli*. The value of EOs got from basil, zataria, and peppermint for cancer prevention agent movement, probiotic feasibility, and organoleptic adequacy of probiotic-yogurt was likewise settled by Azizkhani and Parsaeimehr (2018). The discoveries showed that following a month of capacity, the EO didn't affect the LA5 populace, while development impediment was seen in Bb12. Furthermore, utilizing basil and peppermint tests, the watery concentrate of zataria-yogurt showed the best inhibitory activity on the DPPH revolutionaries. Moreover, it was noticed that the treatment containing zataria neglected to arrive at the customer acknowledgment limit (score >5) while the Mentha Piperita yogurt had the most noteworthy score, trailed by basil and the control yogurt. As per the review, adding peppermint, basil, and zataria-prepared to rejuvenate ointments to probiotic yogurt details expanded the item's cancer prevention agent effectiveness and worked on the basil and peppermint tests' tangible worthiness and antiradical activity.

According to accounts, labneh, or condensed yogurt, is a common breakfast item in Palestine. The whey from yogurt is separated to create labneh, a semi-strong matured milk product, according to Thabet et al. (2014). Otaibi and Al-Demerdash (2008) attributed the labneh's short shelf life, at even cold temperatures, to sterile problems related to the product's development in fabric packets and to the filthy handling of the product, which increases microbial disease. Labneh's high microbiological load, joined with the bundling and stockpiling conditions, prompted the improvement of off preferences and negative physicochemical changes, which eventually prompted the item's dismissal (Burt, 2004; Draughon, 2004) concentrated on it and recommended utilizing bio-additives, like plant rejuvenating oils, to build the time span of usability of delicate food items. Al-Rimawi et al. (2019) evaluated the impact of a few natural remedies in condensed yogurt when used as a source of antibacterial specialist which might quickly work on concentrated yogurt in the review that follows. The viability of the natural balm, which was added to the gathered yogurt at a centralization of 250 L/kg, was resolved to utilize plate count methods. Also, tangible assessment, titratable sharpness testing, and complete strong substance estimations were made. As indicated by the review, adding fragrant oils like cinnamon, clove, and rosemary to concentrated yogurt can possibly expand its time span of usability. Yangilar and Yildiz (2017) assessed a clever probiotic yogurt made with medicinal ointments from ginger and chamomile at groupings of 0.2% and 0.4% utilizing microbiological, tactile, and compound strategies. The analysis indicates that yogurt manufactured with 0.5% ginger natural oil produced 9.00 CFU/g, the most notable provincial framing unit, whereas yogurt made without revitalizing balm produced 8.32 CFU/g.

The consequences of the review proposed that the utilization of EOs could work on the yogurt's capacity to protect food, which might be connected with the cancer prevention agent and antibacterial exercises of the medicinal ointments from ginger and chamomile. Similar three medicinal oils—marjoram, thyme, and sage—were likewise added to concentrated yogurt by Otaibi and Al-Demerdash (2008) at convergences of 0.2, 0.5, and 1.0 ppm. It was found that 0.2 ppm was the best amount to protect time span of usability for as long as 21 days. As per Singh et al. (2011), yogurt with various centralizations of anise unpredictable oil and its oleoresin (0.1−1.0 g/L) can actually forestall microorganism-caused festering. Two monetarily accessible EOs produced using *Chamaemelum* and *Lavandula* species were added to the drain in Bachir and Benattouche (2013) assessment of the organoleptic, microbiological, and physicochemical properties of the enhancement. When tried at the diminished focus, it was found that the rejuvenating oil meaningfully affected the tangible qualities of the yogurt. Additionally, it's been discovered that the improved yogurt that contained Chamomile oil had a more prevalent surface than those that didn't.

Essential oils utilized in frozen yogurt

Frozen yogurt is a customary summer treat that is consistently a number one and is affectionately made. The frozen yogurts on top are viewed as regular aromas due to the peppermint rejuvenating ointments. Lavender improves the kind of lemon frozen yogurt, making it unique. It acquires flavor and variety from it. The physiochemical, microbiological, and tactile characteristics of frozen yogurts made with the natural balms from lemon, mandarin, and orange strips were concentrated by Tomar and Akarca (2019). They found that the last example incorporated minimal measures of yeast, form, and high-impact mesophilic and psychrophilic microscopic organisms. The expansion of regular fixings was additionally favorable in light of the fact that rejuvenating oils had antibacterial, additive, and enhancing attributes. Because of the presence of limonene, -pinene, and -terpinene, lemon strip makes an antimicrobial difference. By filling it with carnauba wax dabs, Yilmaztekin et al. (2019) tried the epitome approach for peppermint essential oil. As an extra portrayal of a culinary item, this was added to frozen yogurt. On a sample of carnauba waxed peppermint oil, a tactile examination was done. It was resolved that adding peppermint oil to frozen yogurt up to 0.3% by weight made two impacts: it worked on the utilitarian characteristics of the peppermint in the frozen yogurt and affected the surface. The most encouraging method for epitomizing peppermint rejuvenating oil, as indicated by their exploration, is ca-alginate. As indicated by tangible investigation, adding peppermint oil to frozen yogurt up to 0.3% (w/w) could be an effective method for improving the practical characteristics of the spice without compromising the surface.

Essential oils utilized in cream

In the review directed by Ehsani et al. (2016), lycopene and EOs from *Echinophora platyloba* (EEO) were utilized to expand the timeframe of the realistic usability of purified cream. Lycopene fixations were 20 and 50 ppm though EEO focuses were 0.10% and 0.50%. They were consolidated and added to purified cream, which was then analyzed for microbiological characteristics, lipid strength, and tactile characteristics while kept up at 4°C and 25°C for a long time. The sanitized cream with the blend of the previously mentioned fixings at higher fixations performed best as far as microbiological qualities, compound examination, and strength when formed to control away circumstances, as indicated by the aftereffects of microbial investigations and synthetic examination. Investigation of tangible examinations uncovered that the utilized treatments were all comprehensively acknowledged. The analysts found that creams with lower centralizations of the EEO and lycopene blend had the best tangible characteristics. The review's decision proposed that EEO and lycopene might be utilized together as additives in dairy items with high-fat substances like cream and spread.

Combination of essential oils with dairy products

As indicated by Boroski et al. (2012), oregano concentrate and medicinal oils act as strong cell reinforcements for use in dairy refreshments that are enhanced with flaxseed oil with a 3 g/110 g measurement. Omega-3 unsaturated fat oxidation brought about by light and intensity was the main pressing concern with dairy refreshments, however, a variety was likewise seen during capacity. Linseed oil was utilized to resolve these issues, and subsequently, the issue was altogether decreased. In this way, it was resolved that oxidation can be extraordinarily diminished by adding a characteristic cell reinforcement to dairy refreshments that are sustained with omega-3 unsaturated fat. In a concentrate on the notable Indian sweet Burfi, which has a short timeframe of realistic usability, Badola et al. (2018) prevailed with regards to helping both the antibacterial and cell reinforcement action by utilizing natural rejuvenating ointments. For that reason, 0.15−0.25 ppm of both the rejuvenating balm from clove bud (CLVB) and curry leaves (CRYF) was utilized. Various elements were evaluated, including physicochemical investigation, microbiological tests, hostile to oxidant properties, and tactile examination. The discoveries exhibited that as how much EO from spices in Burfi expanded, antibacterial and cell reinforcement movement likewise expanded simultaneously, bringing down tactile attributes. The qualities of the physicochemical framework didn't modify a lot. At the point when the connection between burfi tests and quality standards was determined, 81.5% was accounted for. Subsequently, the best focus on the khoa premise was discovered to be CRYF (0.20 ppm) and CLVB to boost the dependability of burfi during capacity while considering its tangible characteristics (0.30 ppm).

Impacts of the essential oil combination on utilization, milk creation, and milk gross structure

Contrasting the EO-enhanced bunch (6 g EO/day per cow) to the non-enhanced bunch, the complete dry matter admission (DMA) was fundamentally higher by 5%. This recommends that the maker's proposed expansion of EO has marginally worked on the TMR's agreeability. Oppositely, taking care of EO from oregano at levels up to 0.85 g/day brought about a reduction in DMA in dairy cows. Early lactation dairy cows' DMI was likewise decreased when 1.5 g/day of a mix of thymol, eugenol, vanillin, and limonene EO was added to their eating regimens in contrast with control cows (Tassoul & Shaver, 2009). Outstandingly, DMI has diminished in these two examinations regardless of the way that EO supplementation levels were a lot lower than in the ongoing review. The lack of DMA camouflage when administered 0.85 g/day of an EO combination via a ruminal prolapse as opposed to oral dosing to dairy cows supports the theory that the solid fragrance of EO, rather than an anticipated impact of EO on ruminal mechanisms, is the primary justification behind a potential decline in DMA. Similarly, in the Wall et al. preliminary, exemplifying a combination of cinnamaldehyde and eugenol decreased negative consequesnces of EO for satisfactoriness and subsequently improved DMA (2014). The designer of the exploration's item accepts it will help feed consumption in cows with respiratory circumstances, however, the creatures in our review didn't have these issues. In this way, it actually should be demonstrated whether the EO combination could in a roundabout way raise DMA in cows with respiratory problems by improving their prosperity and consequently expanding their yearning.

In the ongoing analysis, adding the EO blend to the TMR affected milk creation or feed change proficiency. This is reliable with research taking care of dairy cows' EO at lower or higher fixations. In any case, Wall et al. (2014) found that enhancing a typified mix supported milk yield, however, this was plainly the consequence of the better DMI and feed change proficiency was unaffected. Dairy cows enhanced with oregano (*Origanum vulgare* L.) leaf material showed further developed feed transformation proficiency, however at a lower DMA.

In the ongoing examination, the EO blend affected the milk's protein, fat, lactose, or urea contents. The impacts of EO on milk organization have been reported in different problematic courses in the writing. At the point when dairy cows and ewes are enhanced with EO, studies indicate increased protein levels, lipid, and sugar, and diminished urea levels amounts in the milk. In these examinations, only one milk part was at any point impacted. The expansion of EO impressively raised the milk SCC, yet not to levels reminiscent of mastitis. Silva Filho et al. (2017), utilizing a bigger portion of thyme EO than that used in the current examination, found a propensity to bring down SCC while taking care of 8 g of thyme EO each day per cow.

Moreover, SCC in sheep milk was emphatically diminished by the most noteworthy portion of EO (0.17 g/kg DM) utilized in the concentrate. Basic portion impacts are in all probability avoided by this. All things considered, the tremendous assortment of synthetic mixtures present in the different EO might serve to make sense of the conflicting results to some degree. A couple of an EO's essential parts may sporadically be liable for its bioactivity, while at different times, the blend of various lesser-normal mixtures seems, by all accounts, to be more strong than any one major constituent. 1,8-cineole and -pinene are the two principal fixings in eucalyptus rejuvenating balm. One of 20 different chemotypes of thyme EO can be found, like those with high thymol fixations (Satyal et al., 2016). Trans-anethole is the principal fixing in anise EO. 1,8-cineole displayed no bactericidal impacts in an in vitro try, however, carvacrol and thymol showed a synergistic impact. It is trying to think about the discoveries of studies utilizing combinations of various EO in light of the synergistic and added substance impacts of EO compounds. For stomach maturation and animal digestion, the general development of various EO and a combination therewith could have complicated effects. Specifically, we miss the mark on persuading legitimization for the clear SCC-advancing impact of the ongoing EO blend.

Impacts of the essential oil combination on the qualities of milk coagulation

In the ongoing examination, the milk's coagulation qualities didn't fundamentally change in light of EO supplementation. Thyme leaves upgraded milk coagulation time in a goat exploration, while thyme leaves refined and diminished it. Since the goats got thyme EO in sums fluctuating from 0.05% to 0.13% of their absolute day-to-day admission (thyme leaves contain somewhere in the range of 1% and 1.6% EO), the huge contrasts in the enhanced sums are probably the reason for these divergent outcomes. Conversely, the cows in the ongoing concentrate just got 0.016% of EO per kg. As indicated by Le Maréchal et al. (2011), a high SCC, which is characteristic of mastitis, is connected to unfortunate cheddar yield, which is shown by a more extended RCT and less ideal curd firming (lower k20 and A30) than that tracked down in ordinary milk. The characteristics of the impacted cows' cheddar-making might be upgraded since EO might forestall or decrease the event of mastitis. The ongoing item, nonetheless, expanded SCC instead of diminishing it.

Impacts of the essential oil mix on the cell reinforcement limit and total phenol content of milk

Taking care of the EO combination detectably affected the milk's TP content when contrasted with the benchmark group. Obviously, how much phenols consumed by the enhancement were inadequate to modify the outcomes discernibly. Then again, giving refined rosemary passes on to nursing goats' young people improved their plasma and milk polyphenol content extensively. A certain, yet profoundly wasteful productivity of the exchange of phenols from feed to drain has been found in sheep too (Leparmarai et al., 2019). There is not a specific examination of the rates at which phenolic EO is moved from feed to drain, however almost certainly, every phenolic part has an alternate part of phenols that are caught up in extensive sums.

Thyme, eucalyptus, and anise EOs, which were all parts of the EO blend analyzed in this review, are known to be areas of strength for having impacts. Weaned piglets with food admission of the EO blend including something like 13.5% thymol and 4.5% cinnamaldehyde showed expanded plasma cancer prevention agent limit (Zeng et al., 2015). Not many examinations inspected what dietary phenols meant for the milk's cell reinforcement capacities (Aguiar et al., 2014; Leparmarai et al., 2019) found that adding phenolic EO got from propolis acquired from a trial ranch in a eucalyptus plant hold essentially expanded the cell reinforcement limit of cow's milk. In light of this, it is reasonable to assume that the EO in this type of bee glue originated from cedar trees and that they are also responsible for the mixture under investigation. Notwithstanding, we couldn't identify any effect of the EO blend supplementation on the exploratory cows' milk cell reinforcement limit. This shows that the dose utilized, the sum consumed, or both, were lacking in the ongoing examination to help the milk's cell reinforcement capacity.

Utilization of essential oils in animals used as food

In view of their individual intrinsic applications or by joining different additive elements, EOs likewise play an expected part in microbial wellbeing and dependability with steady upkeep of the nourishing nature of food creatures.

These potential jobs are essentially depicted as impediments. Food ruining wouldn't happen if the foodborne microorganisms couldn't defeat the boundaries made by the EOs in the disinfected food varieties. This obstruction plans to improve the microbiological dependability and tactile characteristics of food. Applying obstacle innovation to meat and dairy items is turning out to be increasingly well-known around the world. Ishaq et al. (2021) investigated the various barrier intercessions produced by a combination of clove EO, source of light, and viral vectors for effective *L. monocytogenes* constituting a decrease in the outer layer of hamburger tests all through temperature-controlled capacity along with the support of nourishing quality and expanded timeframe of realistic usability.

The base worth was recorded (33.12 g) for hamburger tests treated with an obstacle mix of 0.6% clove EO, bacteriophage, and bright treatment. It was additionally noticed that the hardness of the control meat tests following 18 days of capacity was 42.49 g. The consequences of the ongoing review showed that thiobarbituric acid (TBA) values expanded recognizably for control meat tests following 18 days of capacity (1.62 mg MDA/kg), yet that the numerous obstacle mix could bring down the TBA esteem, which was estimated at 0.841 mg MDA/kg. Thus, the researchers exhibited predominant microbial inactivation of EO-based boundaries for keeping up with the respectability of protected meat cuts. EOs are utilized in pre-and postharvest board measures for poultry items and help in better assimilation and supplement retention. Furthermore, the tangible characteristics of frozen chicken were upgraded for 70 days by the same measures of oregano and garlic natural ointments (Kirkpinar et al., 2014). EO has been utilized in an assortment of preharvest strategies to limit bacterial tainting in chicken feed.

Micciche et al. (2019) have found an improvement in the edibility of poultry feed subsequent to adding EO. Preharvest organization of EO was suggested to forestall microorganism colonization in the gastrointestinal arrangement of poultry birds. The use of carvacrol, thymol, or its combination as a diet modification was investigated by Arsi et al. (2014) to prevent the colonization of Campylobacter jejuni. Utilizing Campylobacter line agar, cecal Campylobacter is not entirely settled. Campylobacter counts were diminished by 0.7 and 3 log CFU/mL of cecal substance at thymol portions of 0.25% and 2%, separately. The adequacy of transcinnamaldehyde in forestalling *S. enteritidis* disease in egg yolk and egg shells was analyzed by Upadhyaya et al. (2015). Predictable contamination of *S. enteritidis* was diminished in both the egg yolk and egg shell at transcinnamaldehyde convergences of 1.5%. As a feature of the tangible investigation, 43 out of the 108 specialists accurately recognized eggs from transcinnamaldehyde-treated birds, while the leftover 65 specialists couldn't recognize the treated eggs from the control eggs. This outcome proposes that the transcinnamaldehyde treatment was broadly acknowledged and shows its viability as a taking care of supplement. Amerah et al. (2012) researched the counteraction of flat Salmonella transmission in grill chickens utilizing xylanase and a combination of thymol and cinnamaldehyde. A blend of thymol and cinnamaldehyde as well as xylanase supplementation might assist with bringing down the quantity of cecal examples that test positive for Salmonella.

Kollanoor-Johny et al. (2012) have exhibited that eugenol and transcinnamaldehyde are viable for bringing down *Samonella enteritica* pollution in grilled chicken. How much *S. enteritica* was viewed as between 6 and 7 log CFU/g in the cecal example of control birds, while 0.5% of transcinnamaldehyde and 1% of eugenol decreased how much *S. enteritica* by 4.3 and 2.6 log CFU/g, separately. Wagle et al. (2017) told the best way to really decrease Campylobacter colonization in oven chicks is by applying -resorcyclic corrosive as a potential antibacterial food-added substance. A-resorcyclic corrosive supplementation (0.5% and 1%) in feed diminished the chicken cecal Campylobacter populace by 2.5 and 1.7 log CFU/g, contrasted with the normal cecal Campylobacter colonization of 7.5 log CFU/g in control chicken. Moreover, -resorcyclic corrosive supplementation decreased the outflow of the qualities for attack, grip, and motility (motA, motB, and fliA) in *C. jejuni*. Because of their conspicuous antimicrobial, antimycotoxin, and cell reinforcement exercises, the utilization of EOs is generally reasonable for postharvest food items; be that as it may, their utilization as a food supplement and an effective obstruction with steady hindrance of bacterial colonization in the gastrointestinal plot of different food creatures and relief of meat item based food tainting with broad harmfulness in people fosters another knowledge with energizing potential for preharvest food items This preharvest treatment of EOs likewise has critical disadvantages, for example, fast gastrointestinal retention of the EOs bringing about decreased viability. EOs or bioactive parts that are impervious to gastrointestinal ingestion can be utilized to get around these limitations. The unpredictability, hydrophobicity, and impeding impacts on food organoleptic characteristics drive EOs' restricted work in the food and horticulture businesses in spite of their remarkable protection viability. Consequently, specific conveyance components are essential for a sluggish arrival of EO fragrances suitable for food-based applications. Normal EO conveyance strategies incorporate palatable covering, dynamic bundling, and nanoencapsulation. These strategies work with EO scattering while at the same time keeping up with steady antimicrobial activity and broadening food timeframe of realistic usability. Also, controlled discharge conduct exhibited prevalent dissemination energy and reduced the effect of EO on food organoleptic characteristics.

A green nanotechnological approach to nanoencapsulation of essential oils and their bioactive components for food preservation

EO's nanoencapsulation is presently getting some decent forward momentum in the food business and is gainful as a useful technique to work on the bioefficacy in postharvest applications. An illustration of a later specialized improvement is nanoencapsulation, which shields the focal center EOs from debasement brought about by outside components like light, intensity, oxidation, and volatilization by walling them in an outside framework polymer. Moreover, the embodiment recoils particles with a higher surface-to-volume proportion, which is better for expanded viability, simplicity of taking care of, controlled discharge, and eased back dissipation. For an embodiment, the decision of covering polymer is fundamental; specifically, the biopolymer should be both biocompatible and biodegradable and offer a sensible confirmation for designated conveyance. As indicated by many explorations, exemplification not just makes a dormant boundary to changing outside conditions, however, it additionally keeps up with the wholesome trustworthiness of food. To make nanoparticles of a size somewhere in the range of 10 and 1000 nm, an assortment of nanoencapsulation processes shave been widely utilized.

It's fascinating that the conveyance technique in view of nanoemulsions considers more noteworthy commitment with the microorganism target site and makes another comprehension to support the in-item conduct. As per Liu and Liu (2020), a chitosan nanoemulsion made with thyme EO and thymol essentially diminished the gamble of *S. aureus* and *E. coli* disease in refrigerated pork and forestalled the development of biofilm. Pigs kept in the fridge had a 6-day timeframe of realistic usability expansion on the grounds that to nanoencapsulation, which likewise further developed variety boundaries. Thymol and epitomized thyme EO were found to have restraint zones on *S. aureus* and *E. coli* that deliberate 16.33, 17.33, and 15.33 mm, individually. Because of changes in the design of the cell walls, the creators observed that gram-positive microorganisms were more really repressed than gram-negative microbes. The rates of biofilm development hindrance for the typified thyme EO and thymol were 60.35%, 55.50%, 83.78%, and 83.64%, separately. When contrasted with the control, embodied thymol was viewed as more fruitful at bringing down the pH of the pork (6.64 vs 6.94), which supported the concealment of deterioration microorganisms that rapidly utilize the nitrogen content of the meat. As well as saving the organoleptic quality and oxidative steadiness of the food framework, According to Amiri et al. (2019), polysaccharide sheets combined with Zataria multiflora EO nanofluids can prevent microbiological crumbling in minced hamburger buns for up to 20 days of capacity. The outcomes showed that movies containing the *Z. multiflora* EO nanoemulsion actually kept up with the peroxide esteem (3.90 meq/kg of lipid), carbonyl substance (0.91 nmol/mg protein), and TBA receptive substance esteem (1.12 mg MDA/kg test) in ground meat patties. As the underlying pH of ground hamburger was found to go from 5.63 to 5.76, a vertical pattern in pH was found in the examples of meat while they were being put away. The pH of ground meat tests covered with nanoemulsion movies might be brought down, and the development of fixed-stage microbes and protein deamination might be repressed. The control meat tests' taste, variety, smell, and general adequacy were undeniably given low scores, but following fumigation with nanoemulsion, the tangible characteristics were all given higher appraisals.

The time span of the usability of strawberries was fundamentally expanded when cinnamon natural balm was epitomized into pullulan coatings, showing promising potential as a food additive. Das et al. (2020) demonstrated the ultimate incorporation of *P. anisum* EO into chitosan nanomatrix as a nanofluids for the defense of stored paddy from parasite and AFB1-intervened biodeterioration.

The peroxidation of rice lipids was smothered by the typified EO at MIC (0.09 L/mL) and 3 MIC (0.18 L/mL) portions, which held the MDA esteem at 289.73 M/g FW. For an entire year, the EO nanoemulsion could keep the rice's mineral and macronutrient content stable. *P. anisum* EO's phenolic content expanded thanks to nanoencapsulation, which likewise further developed its extremist rummaging limit. Following the fumigation of rice seeds by EO-stacked nanoemulsion, OK tactile characteristics (variety, flavor, surface, and smell) were seen, supporting their utilization as a clever green food additive. Hasheminejad et al. (2019) have as of late revealed an improvement in the fungitoxic viability of clove rejuvenating balm against *A. niger* pervasion in put-away pomegranate. They proposed a 56-day controlled arrival of EO fragrance (a prerequisite for expanding the time span of usability of put-away food varieties). Moreover, it was found that following 18 days in control pomegranate arils, the occurrence of contagious rot happened; notwithstanding, after fumigation with clove EO-stacked chitosan nanoparticle, it was postponed for as long as 60 days. This could be made sense of by the expanded viability of nanoparticles with a higher surface-to-volume proportion, which modifies the porousness of the plasma film and has a negative collaboration with ergosterol union, which brings about parasitic cell passing *Cuminum cyminum* EO was added to chitosan nanogel by Zhaveh et al. (2015), and this brought about the better antifungal movement against *A. flavus*. Conversely, EO nanoparticles ensnared in chitosan exhibited superior viability for hindering parasitic development and mycotoxin biosynthesis (0.8 L/mL) in put away maize.

The EO totally restrained parasitic development and mycotoxin creation at 0.9 L/mL. For unencapsulated and nanoencapsulated EO, separately, a palatable relapse model was found with great assurance coefficients of 0.9694 and 0.9896 (for the hindrance of contagious development), 0.9864 and 0.9793 (for the restraint of deoxynivalenol), and 0.9935 and 0.9873 (for the restraint of zearalenone).

Besides, on the grounds that parts were delivered continuously and under controlled conditions, the epitomized EO was less defenseless to oxidative annihilation. The comprehension of the different components directing the judicious application in the food framework isn't deep-rooted, in spite of the way that various exemplification strategies utilizing EOs for food assurance have been created. The poisonousness test is consistently a concern for nanoencapsulation-based conveyance frameworks to inspect the organic destiny at the hour of retention, processing, and discharge. The nanoencapsulated derived from *P. anisum* EOs had a lower LD50 esteem than the unencapsulated variant. Chaudhari et al. (2020) as of late covered a decrease in the LD50 worth of *O. majorana* EO that was embodied in chitosan. This finding might be connected with the more modest size of nanoemulsionic particles that were poisonous to mammalian frameworks since they contained a larger part of EO drops. Lower LD50 values are vital for true applications in the food framework since they demonstrate a higher unsafe effect on a few cell processes. Viewing as a reasonable biocompatible, biodegradable, and earth OK polymer for exemplifying EOs and upgrading bioefficacy are two obstacles of nanoencapsulation that should be worked on notwithstanding harmfulness worries for future enormous scope use in the genuine food framework.

- An expansion in the exemplification cycle's effectiveness, which brings down how much energy, is utilized.
- Searching for an alternate natural dissolvable that isn't reasonable for eating by individuals.
- Making a multicompartment framework to supply different bioactive parts after some time (Shishir et al., 2018).

Essential oils and components in food delivered by active packaging

Notwithstanding the immediate utilization of nanoencapsulated food handling, dynamic bundling of food items utilizing EOs and different film-framing materials, like low-density polyethylene, polyvinyl liquor, and ethylene vinyl liquor, additionally fosters the harmless to the ecosystem course of food handling. The utilization of EO in food item bundling supports the slow dispersion of fragrance through the movie surface's micropore and into direct contact with the food items (Li et al., 2018). The solvency of the lipophilic parts into the food framework decides how successful EO is in the bundling framework. As opposed to the high dissolvability model, where unhindered free dispersion was shown and fundamentally higher microbial obstruction was recorded, the low solvency of the parts brings about a one-sided framework with a low fascination with the food surface and a more serious gamble of microbial tainting. The development of the coliform microorganisms in chicken bosom items was accounted for to be forestalled by cellulose acetic acid derivation film that contained rosemary rejuvenating oil at a convergence of half (w/w). Nonetheless, a slight change in food flavor was noted. Dynamic meat packaging using chitin sheet infused with lemon EO (1.5% w/w) resulted in a significant drop in breath frequency and a reduction in the risk of bacterial contamination, both of which can lengthen the time that the meat is really usable.

New meat might have a more drawn-out time span of usability when refrigerated in the event that it is covered with whey protein confine and 1.5% oregano rejuvenating ointment. Following 16 days of treatment, menthol, eugenol, and thymol-changed environmental bundling of sweet cherry decreased the absolute number of molds, microscopic organisms, and yeast from 2.5 to 1.8 lg CFU/g (Serrano et al., 2005). It is shown that bundling of bread utilizing polyvinyl liquor and clove natural balm (27.19% w/w) had great air penetrability, successfully hindered the development of *A. niger*, and expanded timeframe of realistic usability.

There are various discoveries on the palatable covering of EOs and their antibacterial viability against foodborne microbes, notwithstanding dynamic bundling. The covering is applied in fluid structure to the food to be covered and is totally not quite the same as movies. With viable antibacterial activity and a diminished pace of breath, Sánchez-González et al. (2011) exhibited consumable hydroxypropylmethylcellulose and chitosan covering with bergamot medicinal ointment for postharvest protection of table grapes during cold capacity. The EO-based palatable covering likewise essentially worked on the antibacterial action while meaningfully affecting the organoleptic characteristics of the dinner. Various food-defiling microscopic organisms, including *L. monocytogenes*, *E. coli*, and *S. aureus*, showed solid inhibitory movement when presented with different eatable covering types, including polysaccharide, protein, lipid, and composite covering containing different EOs. Apple cuts can be covered with Aloe vera gel, lemon medicinal oil, and shellac to forestall microbial tainting, lessen the pace of ethylene during capacity, and abatement oxidase action while reliably holding the apple tone, as per Chauhan et al. (2011). To forestall organism pervasion during capacity,

Guerra et al. (2016) researched the effect of covering grapes with chitosan, *Mentha piperata*, and *M. villosa* EO. During the tangible assessment, the covered grapes scored higher for acknowledgment with regard to variety, flavor, trailing sensation, and surface. Shape, yeast, and psychrophilic microbes were all fundamentally repressed by the covering. With a covering (0.5% w/w) and oregano and lemongrass natural ointments (1.7% and 1.3% w/w, individually), *L. innocua* contamination in apple pieces was altogether diminished (4 log decrease).

Use limitations for essential oils in dairy products

Since sugars make food varieties acidic, they are bound to be tainted with yeast. These yeasts annihilate things like natural product or vegetable juices, natural product purees, thinks, or salted vegetables that are being protected by feeble acids, (for example, sorbic and acidic corrosive). *Geotrichum candidum* is every now and again found in crude milk, which is utilized to make delicate cheddar and other dairy items. Delicate refreshments, organic product juices, wines, yogurt, cured vegetables, and cream-filled sweets can be in every way ruined by yeast, otherwise called *Pichia anomala*. As per Stratford (2006), the most predominant yeasts that make soda pops and natural product juices break down are *Saccharomyces cerevisiae* and *Schizosaccharomyces pombe*. Until the pressing remaining parts are in one piece, purification and germicide bundling help to stay away from yeast crumbling. Items that can't be sanitized are saved utilizing a feeble acidic substance. By and by, buyers fervently request that counterfeit additives be utilized less. The microscopic organisms *S. cerevisiae* and *P. irregularity* convert sorbic corrosive to 1,3-pentadiene, which has a lamp fuel-like scent (Stratford, 2006). *S. pombe* may create unsavory flavors when sulfate is available These issues have made the food area centered around normally happening hostile to microbials delivered from plants, for example, EOs, which have wide antimicrobial movement against microscopic organisms, yeasts, and molds. The utilization of EOs in culinary arrangements is obliged by their powerful fragrance. The answer to this issue is to join EOs in manners that make a synergistic difference, requiring less of each EO to save an item.

Conclusion

The utilization of natural oils in different dairy items and their remedial advantages, like antibacterial, cancer prevention agents, and organoleptic exercises, were assessed in this audit of the logical writing. Various investigations have noticed the constructive outcomes of consolidating rejuvenating ointments with dairy items, however more careful examination is expected to decide if there might be synergistic or hostile communications between the constituents of natural oils and dairy items. Additionally, holding flavor might be troublesome assuming fixings like medicinal balm are added to dairy items at higher sums. Testing techniques to decide the adequacy of antibacterial details utilized in dairy items were previously undeniably less normalized. The examination of viable antibacterial components in EOs and their cutthroat or agreeable consequences for one another and on milk fixings can be accelerated with the utilization of laid-out strategies. Furthermore, a more noteworthy exploration of poisonousness and security is likewise required.

References

Abbes, C., Mansouri, A., & Landoulsi, A. (2018). Synergistic effect of the lactoperoxidase system and cinnamon essential oil on total flora and Salmonella growth inhibition in raw milk. *Journal of Food Quality*, 1–6.

Abboud, M., El-Rammouz, R., Jammal, B., & Sleiman, M. (2016). In-vitro and in-vivo antimicrobial activity of two essential oils Thymus vulgaris and *Lavandula angustifolia* against Bovine Staphylococcus and Streptococcus mastitis pathogen. Middle East. *Journal of Agriculture Research, 4* (4), 975–983.

Adejumo, I. O., Adetunji, C. O., & Adeyemi, O. S. (2017). Influence of UV light exposure on mineral composition and biomass production of mycomeat produced from different agricultural substrates. *Journal of Agricultural Sciences, Belgrade, 62*(1), 51–59.

Adetunji, C. O., Egbuna, C., Tijjani, H., Adom, D., Tawfeeq Al-Ani, L. Kl, & Patrick-Iwuanyanwu, K. C. (2020). *Homemade preparations of natural biopesticides and applications. Natural remedies for pest, disease and weed control* (pp. 179–185). Publisher Academic Press.

Adetunji, C. O., Michael, O. S., Rathee, S., Singh, K. R. B., Ajayi, O. O., Adetunji, J. B., Ojha, A., Singh, J., & Singh, R. P. (2022). Potentialities of nanomaterials for the management and treatment of metabolic syndrome: A new insight. *Materials Today Advances, 13*100198.

Adetunji, C. O., Palai, S., Ekwuabu, C. P., Egbuna, C., Adetunji, J. B., Ehis-Eriakha, C. B., Kesh, S. S., & Mtewa, A. G. (2021). *General principle of primary and secondary plant metabolites: Biogenesis, metabolism, and extraction. Preparation of phytopharmaceuticals for the management of disorders* (pp. 3–23). Academic Press.

Adetunji, C.O. (2008). *The antibacterial activities and preliminary phytochemical screening of vernoniaamygdalina and Aloe vera against some selected bacteria* [MSc thesis]. University of Ilorin, pp. 40–43.

Adetunji, J. B., Ajani, A. O., Adetunji, C. O., Fawole, O. B., Arowora, K. A., Nwaubani, S. I., Ajayi, E. S., Oloke, J. K., & Aina, J. A. (2013). Postharvest quality and safety maintenance of the physical properties of Daucus carota L. fruits by Neem oil and Moringa oil treatment: A new edible coatings. *Agrosearch, 13*(1), 131–141.

Adetunji, C. O., Michael, O. S., Nwankwo, W., Ukhurebor, K. E., Anani, O. A., Oloke, J. K., Varma, A., Kadiri, O., Jain, A., & Adetunji, J. B. (2021). Quinoa, the next biotech plant: Food security and environmental and health hot spots. In A. Varma (Ed.), *Biology and biotechnology of Quinoa*. Singapore: Springer. Available from https://doi.org/10.1007/978-981-16-3832-9_19.

Adetunji, C. O. (2019). Environmental impact and ecotoxicological influence of biofabricated and inorganic nanoparticle on soil activity. In D. Panpatte, & Y. Jhala (Eds.), *Nanotechnology for agriculture*. Singapore: Springer. Available from https://doi.org/10.1007/978-981-32-9370-0_12.

Adetunji, C. O., Inobeme, A., Olaniyan, O. T., Ajayi, O. O., Olaniyan, S., & Adetunji, J. B. (2021). Application of nanodrugs derived from active metabolites of medicinal plants for the treatment of inflammatory and lung diseases: Recent advances. In K. Dua, S. Nammi, D. Chang, D. K. Chellappan, G. Gupta, & T. Collet (Eds.), *Medicinal plants for lung diseases*. Singapore: Springer. Available from https://doi.org/10.1007/978-981-33-6850-7_26.

Adetunji, C. O., Kumar, D., Raina, M., Arogundade, O., & Sarin, N. B. (2019). Endophytic microorganisms as biological control agents for plant pathogens: A panacea for sustainable agriculture. In A. Varma, S. Tripathi, & R. Prasad (Eds.), *Plant biotic interactions*. Cham: Springer. Available from https://doi.org/10.1007/978-3-030-26657-8_1.

Adetunji, C. O., Akram, M., Tope Olaniyan, O., Olufemi Ajayi, O., Inobeme, A., Olaniyan, S., Hameed, L., & Bunmi Adetunji, J. (2021). In K. Dua, S. Nammi, D. Chang, D. K. Chellappan, G. Gupta, & T. Collet (Eds.), *Targeting SARS-CoV-2 novel Corona (COVID-19) virus infection using medicinal plants*. Singapore: Medicinal plants for lung diseases. Springer. Available from https://doi.org/10.1007/978-981-33-6850-7_21.

Adetunji, C. O., Nwankwo, W., Ukhurebor, K., Olayinka, A. S., & Makinde, A. S. (2021). Application of biosensor for the identification of various pathogens and pests mitigating against the agricultural production: Recent advances. In R. N. Pudake, U. Jain, & C. Kole (Eds.), *Biosensors in agriculture: Recent trends and future perspectives. Concepts and strategies in plant sciences*. Cham: Springer. Available from https://doi.org/10.1007/978-3-030-66165-6_9.

Adetunji, C. O., Anani, O. A., Olaniyan, O. T., Inobeme, A., Olisaka, F. N., Uwadiae, E. O., & Obayagbona, O. N. (2021). Recent trends in organic farming. In R. Soni, D. C. Suyal, P. Bhargava, & R. Goel (Eds.), *Microbiological activity for soil and plant health management*. Singapore: Springer. Available from https://doi.org/10.1007/978-981-16-2922-8_20.

Adetunji, C. O., Michael, O. S., Varma, A., Oloke, J. K., Kadiri, O., Akram, M., Bodunrinde, R. E., Imtiaz, A., Adetunji, J. B., Shahzad, K., Jain, A., Ubi, B. E., Majeed, N., Ozolua, P., & Olisaka., F. N. (2021). Recent advances in the application of biotechnology for improving the production of secondary metabolites from Quinoa. In A. Varma (Ed.), *Biology and biotechnology of Quinoa*. Singapore: Springer. Available from https://doi.org/10.1007/978-981-16-3832-9_17.

Adetunji, C. O., Michael, O. S., Kadiri, O., Varma, A., Akram, M., Kola Oloke, J., Shafique, H., Adetunji, J. B., Jain, A., Ebunoluwa Bodunrinde, R., Ozolua, P., & Ewa Ubi., B. (2021). Quinoa: From farm to traditional healing, food application, and phytopharmacology. In A. Varma (Ed.), *Biology and biotechnology of Quinoa*. Singapore: Springer. Available from https://doi.org/10.1007/978-981-16-3832-9_20.

Adetunji, C. O., Olaniyan, O. T., Akram, M., Ajayi, O. O., Inobeme, A., Olaniyan, S., Said Khan, F., & Adetunji., J. B. (2021). Medicinal plants used in the treatment of pulmonary hypertension. In K. Dua, S. Nammi, D. Chang, D. K. Chellappan, G. Gupta, & T. Collet (Eds.), *Medicinal plants for lung diseases*. Singapore: . Springer. Available from https://doi.org/10.1007/978-981-33-6850-7_14.

Adetunji, C. O., Oloke, J. K., & Prasad, G. (2020). Effect of carbon-to-nitrogen ratio on eco-friendly mycoherbicide activity from *Lasiodiplodia pseudotheobromae* C1136 for sustainable weeds management in organic agriculture. *Environment, Development and Sustainability, 22*, 1977–1990. Available from https://doi.org/10.1007/s10668-018-0273-1.

Adetunji, C. O., Phazang, P., & Sarin, N. B. (2017). Significance of rhamnolipids as a biological control agent in the management of crops/plant pathogens. *Current Trends in Biomedical Engineering & Biosciences, 10*(3), 54–55.

Adetunji, C. O., Roli, O. I., & Adetunji, J. B. (2020). Exopolysaccharides derived from beneficial microorganisms: Antimicrobial, food, and health benefits. In P. Mishra, R. R. Mishra, & C. O. Adetunji (Eds.), *Innovations in food technology*. Singapore: Springer. Available from https://doi.org/10.1007/978-981-15-6121-4_10.

Adetunji, C. O., & Varma, A. (2020). Biotechnological application of trichoderma: A powerful fungal isolate with diverse potentials for the attainment of food safety, management of pest and diseases, healthy planet, and sustainable agriculture. In C. Manoharachary, H. B. Singh, & A. Varma (Eds.), *Trichoderma: Agricultural applications and beyond. Soil biology* (vol 61). Cham: Springer. Available from https://doi.org/10.1007/978-3-030-54758-5_12.

Adetunji, J. B., Adetunji, C. O., & Olaniyan, O. T. (2021). African walnuts: A natural depository of nutritional and bioactive compounds essential for food and nutritional security in Africa. In O. O. Babalola (Ed.), *Food security and safety*. Cham: Springer. Available from https://doi.org/10.1007/978-3-030-50672-8_19.

Aggarwal, S., Agarwal, S., & Jalhan, S. (2013). Essential oils as novel human skin penetration enhancer for transdermal drug delivery: A review. *International Journal of Pharmacy and Biological Sciences, 4*(1), 857–868.

Aguiar, S. C., Cottica, S. M., Boeing, J. S., Samensari, R. B., Santos, G. T., Visentainer, J. V., & Zeoula, L. M. (2014). Effect of feeding phenolic compounds from propolis extracts to dairy cows on milk production, milk fatty acid composition, and the antioxidant capacity of milk. *Animal Feed Science and Technology, 193*, 148–154. Available from https://doi.org/10.1016/j.anifeedsci.2014.04.006.

Al-Rimawi, F., Jazzar, M., Alayoubi, M., & Elama, C. (2019). Effect of different essential oils on the shelf life of concentrated yogurt. *Annual Research & Review in Biology, 30*(6), 1–9. Available from https://doi.org/10.9734/arrb/2018/v30i630031.

Amerah, A. M., Mathis, G., & Hofacre, C. L. (2012). Effect of xylanase and a blend of essential oils on performance and Salmonella colonization of broiler chickens challenged with Salmonella Heidelberg. *Poultry Science, 91*, 943−947. Available from https://doi.org/10.3382/ps.2011-01922.

Amiri, E., Aminzare, M., Azar, H. H., & Mehrasbi, M. R. (2019). Combined antioxidant and sensory effects of corn starch films with nanoemulsion of Zataria multiflora essential oil fortified with cinnamaldehyde on fresh ground beef patties. *Meat Science, 153*, 66−74. Available from https://doi.org/10.1016/j.meatsci.2019.03.004.

Arsi, K., Donoghue, A. M., Venkitanarayanan, K., Kollanoor-Johny, A., Fanatico, A. C., Blore, P. J., et al. (2014). The efficacy of the natural plant extracts, thymol and carvacrol against Campylobacter colonization in broiler chickens. *Journal of Food Safety, 34*, 321−325. Available from https://doi.org/10.1111/jfs.12129.

Azizkhani, M., & Parsaeimehr, M. (2018). Probiotics survival, antioxidant activity and sensory properties of yogurt flavored with herbal essential oils. *International Food Research Journal, 25*(3), 921−927.

Bachir, R. G., & Benattouche, Z. (2013). Microbiological, physico-chemical and sensory quality aspects of yoghurt enriched with Rosmarinus officinalis oil. *African Journal of Biotechnology, 12*(2), 192−198. Available from https://doi.org/10.5897/ajb12.1257.

Badola, R., Panjagari, N. R., Singh, R. R. B., Singh, A. K., & Prasad, W. G. (2018). Effect of clove bud and curry leaf essential oils on the antioxidative and anti-microbial activity of burfi, a milk-based confection. *Journal of Food Science & Technology, 55*(12), 4802−4810. Available from https://doi.org/10.1007/s13197-018-3413-6.

Bajpai, V. K., Yoon, J. I., Bhardwaj, M., & Kang, S. C. (2014). Anti-listerial synergism of leaf essential oil of Metasequoia glyptostroboides with nisin in whole, low and skim milks. *Asian Pacific Journal of Tropical Medicine, 7*(8), 602−608. Available from https://doi.org/10.1016/s1995-7645(14)60102-4.

Baser, K. H. C., & Buchbauer, G. (2010). *Handbook of essential oils: Science, technology, and applications* (1 ed.). CRC press.

Batiha, G. B., Awad, D. A., Algamma, A. M., Nyamota, R., Wahed, M. I., Shah, M. A., Amin, M. N., Adetuyi, B. O., Hetta, H. F., Cruz-Marins, N., Koirala, N., Ghosh, A., & Sabatier, J. (2021).). Diary-derived and egg white proteins in enhancing immune system against COVID-19 frontiers in nutritionr. *(Nutritional Immunology)*, 8629440. Available from https://doi.org/10.3389/fnut.2021629440.

Bello, O. M., Ibitoye, T., & Adetunji, C. (2019). Assessing antimicrobial agents of Nigeria flora. *Journal of King Saud University-Science, 31*(4), 1379−1383.

Benchaar, C., & Greathead, H. (2011). Essential oils and opportunities to mitigate enteric methane emissions from ruminants. *Animal Feed Science and Technology, 166−167*, 338−355. Available from https://doi.org/10.1016/j.anifeedsci.2011.04.024.

Bhavaniramya, S., Vishnupriya, S., Al-Aboody, M. S., Vijayakumar, R., & Baskaran, D. (2019). Role of essential oils in food safety: Antimicrobial and antioxidant applications. *Grain & Oil Science and Technology, 2*(2), 49−55. Available from https://doi.org/10.1016/j.gaost.2019.03.001.

Boroski, M., Giroux, H. J., Sabik, H., Petit, H. V., Visentainer, J. V., MatumotoPintro, P. T., et al. (2012). Use of oregano extract and oregano essential oil as antioxidants in functional dairy beverage formulations. *Lebensmittel-Wissenschaft und-Technologie-Food Science and Technology, 47*(1), 167−174. Available from https://doi.org/10.1016/j.lwt.2011.12.018.

Burt, S. (2004). Essential oils: Their antibacterial properties and potential applications in foods-A review. *International Journal of Food Microbiology, 94*(3), 223−253. Available from https://doi.org/10.1016/j.ijfoodmicro.2004.03.022.

Chaudhari, A. K., Singh, V. K., Das, S., Prasad, J., Dwivedy, A. K., et al. (2020). Improvement of in vitro and in situ antifungal, AFB1 inhibitory and antioxidant activity of Origanum majorana L. essential oil through nanoemulsion and recommending as novel food preservative. *Food and Chemical Toxicology. an International Journal Published for the British Industrial Biological Research Association, 143*111536. Available from https://doi.org/10.1016/j.fct.2020.111536.

Chauhan, O. P., Raju, P. S., Singh, A., & Bawa, A. S. (2011). Shellac and aloegel-based surface coatings for maintaining keeping quality of apple slices. *Food Chemistry, 126*, 961−966. Available from https://doi.org/10.1016/j.foodchem.2010.11.095.

da Silva Dannenberg, G., Funck, G. D., Mattei, F. J., da Silva, W. P., & Fiorentini, A.ˆM. (2016). Antimicrobial and antioxidant activity of essential oil from pink pepper tree (Schinus terebinthifolius Raddi) in-vitro and in cheese experimentally contaminated with Listeria monocytogenes. *Innovative Food Science & Emerging Technologies, 36*, 120−127. Available from https://doi.org/10.1016/j.ifset.2016.06.009.

Das, S., Singh, V. K., Dwivedy, A. K., Chaudhari, A. K., & Dubey, N. K. (2020). Nanostructured Pimpinella anisum essential oil as novel green food preservative against fungal infestation, aflatoxin B1 contamination and deterioration of nutritional qualities. *Food Chemistry, 344*128574. Available from https://doi.org/10.1016/j.foodchem.2020.128574.

Didunyemi, M. O., Adetuyi, B. O., & Oyewale, I. A. (2020). Inhibition of lipid peroxidation and in-vitro antioxidant capacity of aqueous, acetone and methanol leaf extracts of green and red Acalypha wilkesiana Muell Arg. *International Journal of Biological and Medical Research, 11*(3), 7089−7094.

Didunyemi, M. O., Adetuyi, B. O., & Oyebanjo, O. O. (2019). Morinda lucida attenuates acetaminophen-induced oxidative damage and hepatotoxicity in rats. *Journal of Biomedical Sciences, Vol 8*. Available from https://www.jbiomeds.com/biomedical-sciences/morinda-lucida-attenuates-acetaminopheninduced-oxidative-damage-and-hepatotoxicity-in-rats.php?aid = 24482.

Dongmo, P. M. J., Boyom, F. F., Sameza, M. L., Ndongson, B., Kwazou, N., Zollo, P.-H. A., et al. (2008). Investigations of the essential oils of some aframomum species (Zingiberaceae from Cameroon) as potential antioxidant and anti-inflammatory agents. *International Journal of Essential Oil Therapy, 2*(4), 149−155.

Dongmo, P. M. J., Tchoumbougnang, F., Ndongson, B., Agwanande, W., Sandjon, B., Zollo, P. H. A., et al. (2010). Chemical characterization, antiradical, antioxidant and anti-inflammatory potential of the essential oils of *Canarium schweinfurthii* and *Aucoumea klaineana* (Burseraceae) growing in Cameroon. *Agriculture and Biology North America, 1*(4), 606−611.

Draughon, F. (2004). Use of botanicals as biopreservatives in foods. *Food Technology (Chicago), 58*(2), 20−28.

Egbuna, C., Gupta, E., Ezzat, S. M., Jeevanandam, J., Mishra, N., Akram, M., Sudharani, N., Adetunji, C. O., Singh, P., Ifemeje, J. C., Deepak, M., Bhavana, A., Walag, A. M. P., Ansari, R., Adetunji, J. B., Laila, U., Olisah, M. C., & Onyekere, P. F. (2020). Aloe species as valuable sources of functional bioactives. In C. Egbuna, & G. Dable Tupas (Eds.), *Functional foods and nutraceuticals*. Cham: Springer. Available from https://doi.org/10.1007/978-3-030-42319-3_18.

Ehsani, A., Hashemi, M., Hosseini Jazani, N., Aliakbarlu, J., Shokri, S., & Naghibi, S. S. (2016). Effect of Echinophora platyloba DC. essential oil and lycopene on the stability of pasteurized cream obtained from cow milk. *Veterinary Research Forum*, 7(2), 139−148.

Feizollahzadeh, S., Ghiasvand, R., Rezaei, A., Khanahmad, H., Sadeghi, A., & Hariri, M. (2016). Effect of probiotic soy milk on serum levels of adiponectin, inflammatory mediators, lipid profile, and fasting blood glucose among patients with type II diabetes mellitus. *Probiotics and Antimicrobial Proteins*, 9(1), 41−47. Available from https://doi.org/10.1007/s12602-016-9233-y.

Friedman, M. (2017). Chemistry, Antimicrobial mechanisms, and antibiotic activities of cinnamaldehyde against pathogenic bacteria in animal feeds and human foods. *Journal of Agricultural and Food Chemistry*, 65(48), 10406−10423. Available from https://doi.org/10.1021/acs.jafc.7b04344.

Gouvea, Fd. S., Rosenthal, A., & Ferreira, E. Hd. R. (2017). Plant extract and essential oils added as antimicrobials to cheeses: A review. *Ciência Rural*, 47(8), e20160908. Available from https://doi.org/10.1590/0103-8478cr20160908, Article.

Guerra, I. C. D., de Oliveira, P. D. L., Santos, M. M. F., Lúcio, A. S. S. C., Tavares, J. F., Barbosa-Filho, J. M., et al. (2016). The effects of composite coatings containing chitosan and Mentha (piperita L. or x villosa Huds) essential oil on postharvest mold occurrence and quality of table grape cv. *Isabella*. *Innovative Food Science & Emerging Technologies*, 34, 112−121. Available from https://doi.org/10.1016/j.ifset.2016.01.008.

Hasheminejad, N., Khodaiyan, F., & Safari, M. (2019). Improving the antifungal activity of clove essential oil encapsulated by chitosan nanoparticles. *Food Chemistry*, 275, 113−122. Available from https://doi.org/10.1016/j.foodchem.2018.09.085.

Herman, R. A., Ayepa, E., Shittu, S., Fometu, S. S., & Wang, J. (2019). Essential oils and their applications-A mini review. *Advances in Nutrition & Food Science*, 4(4), 1−13.

Hu, Y., Zhang, J., Kong, W., Zhao, G., & Yang, M. (2017). Mechanisms of antifungal and anti-aflatoxigenic properties of essential oil derived from turmeric (Curcuma longa L.) on Aspergillus flavus. *Food Chemistry*, 220, 1−8. Available from https://doi.org/10.1016/j.2016.09.179.

Hyldgaard, M., Mygind, T., & Meyer, R. L. (2012). Essential oils in food preservation: Mode of action, synergies, and interactions with food matrix components. *Frontiers in Microbiology*, 3, 12. Available from https://doi.org/10.3389/fmicb.2012.00012.

Ishaq, A., Syed, Q. A., Ebner, P. D., & Ur Rahman, H. U. (2021). Multiple hurdle technology to improve microbial safety, quality and oxidative stability of refrigerated raw beef. *LWT*, 138, 110529.

James-Okoro, P. O., Iheagwam, F. N., Sholeye, M. I., Umoren, I. A., Adetuyi, B. O., Ogundipe, A. E., Braimah, A. A., Adekunbi, T. S., Ogunlana, O. E., & Ogunlana, O. O. (2021). Phytochemical and in vitro antioxidant assessment of Yoyo bitters. *World News of Natural Sciences*, 37, 1−17.

Jemaa, M. B., Falleh, H., Neves, M. A., Isoda, H., Nakajima, M., & Ksouri, R. (2017). Quality preservation of deliberately contaminated milk using thyme free and nanoemulsified essential oils. *Food Chemistry*, 217, 726−734. Available from https://doi.org/10.1016/j.foodchem.2016.09.030.

Jemaa, M. B., Falleh, H., Saada, M., Oueslati, M., Snoussi, M., & Ksouri, R. (2018). Thymus capitatus essential oil ameliorates pasteurization efficiency. *Journal of Food Science & Technology*, 55(9), 3446−3452. Available from https://doi.org/10.1007/s13197-018--4.

Kirkpinar, F. I. G. E. N., Ünlü, H. B., Serdaroglu, M., & Turp, G. Y. (2014). Effects of dietary oregano and garlic essential oils on carcass characteristics, meat composition, colour, pH and sensory quality of broiler meat. *British Poultry Science*, 55, 157−166. Available from https://doi.org/10.1080/00071668.2013.879980.

Kollanoor-Johny, A., Mattson, T., Baskaran, S. A., Amalaradjou, M. A., Babapoor, S., March, B., et al. (2012). Reduction of Salmonella enterica serovar Enteritidis colonization in 20-day-old broiler chickens by the plant-derived compounds trans-cinnamaldehyde and eugenol. *Applied and Environmental Microbiology*, 78, 2981−2987. Available from https://doi.org/10.1128/AEM.07643-11.

Laswai, H. S., Thonya, N., Yesaya, D., Silayo, V. C. K., Kulwa, K., Mpagalile, J. J., et al. (2009). Use of locally available flavouring materials in suppressing the beany taste in soymilk. *African Journal of Food, Agriculture, Nutrition and Development*, 9(7), 1548−1560. Available from https://doi.org/10.4314/ajfand.v9i7.47684.

Le Maréchal, C., Thiéry, R., Vautor, E., & Le Loir, Y. (2011). Mastitis impact on technological properties of milk and quality of milk products—A review. *Dairy Science & Technology*, 91, 247−282. Available from https://doi.org/10.1007/s13594-011-0009-6.

Leparmarai, P. T., Sinz, S., Kunz, C., Liesegang, A., Ortmann, S., Kreuzer, M., & Marquardt, S. (2019). Transfer of total phenols from a grapeseed-supplemented diet to dairy sheep and goat milk, and effects on performance and milk quality. *Journal of Animal Science*, 97, 1840−1851. Available from https://doi.org/10.1093/jas/skz046.

Li, J., Ye, F., Lei, L., & Zhao, G. (2018). Combined effects of octenylsuccination and oregano essential oil on sweet potato starch films with an emphasis on water resistance. *International Journal of Biological Macromolecules*, 115, 547−553. Available from https://doi.org/10.1016/j.ijbiomac.2018.04.093.

Liu, T., & Liu, L. (2020). Fabrication and characterization of chitosan nanoemulsions loading thymol or thyme essential oil for the preservation of refrigerated pork. *International Journal of Biological Macromolecules*, 162, 1509−1515. Available from https://doi.org/10.1016/j.ijbiomac.2020.07.207.

Loizzo, M. R., Menichini, F., Conforti, F., Tundis, R., Bonesi, M., Saab, A. M., et al. (2009). Chemical analysis, antioxidant, antiinflammatory and anticholinesterase activities of Origanum ehrenbergii Boiss and Origanum syriacum L. essential oils. *Food Chemistry*, 117(1), 174−180. Available from https://doi.org/10.1016/j.foodchem.2009.03.095.

Micciche, A., Rothrock, M. J., Jr, Yang, Y., & Ricke, S. C. (2019). Essential oils as an intervention strategy to reduce Campylobacter in poultry production: A review. *Frontiers in Microbiology*, 10, 1058. Available from https://doi.org/10.3389/fmicb.2019.01058.

Muhammad, D. R. A., Lemarcq, V., Alderweireldt, E., Vanoverberghe, P., Praseptiangga, D., Juvinal, J. G., et al. (2019). Antioxidant activity and quality attributes of white chocolate incorporated with Cinnamomum burmannii Blume essential oil. *Journal of Food Science & Technology, 57* (5), 1731−1739. Available from https://doi.org/10.1007/s13197-019-04206-6.

Nazir, A., Itrat, N., Shahid, A., Mushtaq, Z., Abdulrahman, S.A., Egbuna, C., Adetuyi, B.O., Khan, J., Uche, C.Z., & Toloyai, P.Y. (2022). Orange peel as a source of nutraceuticals. In: *Food and agricultural byproducts as important source of valuable nutraceuticals* (1st ed. 2022). Berlin: Springer. pp.400. Gebunden. ISBN 978-3-030-98759-6.

Olaniyan, O. T., & Adetunji, C. O. (2021). Biological, biochemical, and biodiversity of biomolecules from marine-based beneficial microorganisms: Industrial perspective. In C. O. Adetunji, D. G. Panpatte, & Y. K. Jhala (Eds.), *Microbial rejuvenation of polluted environment. Microorganisms for sustainability* (vol 27). Singapore: Springer. Available from https://doi.org/10.1007/978-981-15-7459-7_4.

Otaibi, A. M., & Al-Demerdash, H. (2008). Improvement of the quality and shelf concentrated yoghurt (labneh) by the addition essential oils. *African Journal of Microbiology Research, 2,* 156−161.

Rauf, A., Akram, M., Semwal, P., Mujawah, A. A. H., Muhammad, N., Riaz, Z., Munir, N., Piotrovsky, D., Vdovina, I., Bouyahya, A., Oluwaseun Adetunji, C., Ali Shariati, M., Almarhoon, Z. M., Mabkhot, Y. N., & Khan, H. (2021). Antispasmodic potential of medicinal plants: A comprehensive review. *Oxidative Medicine and Cellular Longevity.* Available from https://doi.org/10.1155/2021/4889719, Article ID 4889719, 12 pages, 2021.

Sánchez-González, L., Pastor, C., Vargas, M., Chiralt, A., González-Martínez, C., & Cháfer, M. (2011). Effect of hydroxypropylmethylcellulose and chitosan coatings with and without bergamot essential oil on quality and safety of cold-stored grapes. *Postharvest Biology and Technology, 60,* 57−63. Available from https://doi.org/10.1016/j.postharvbio.2010.11.00.

Satyal, P., Murray, B. L., McFeeters, R. L., & Setzer, W. N. (2016). Essential oil characterization of Thymus vulgaris from various geographical locations. *Foods, 5,* 70. Available from https://doi.org/10.3390/foods5040070.

Serrano, M., Martinez-Romero, D., Castillo, S., Guillén, F., & Valero, D. (2005). The use of natural antifungal compounds improves the beneficial effect of MAP in sweet cherry storage. *Innovative Food Science and Emerging Technologies, 6,* 115−123. Available from https://doi.org/10.1016/j.ifset.2004.09.001.

Shishir, M. R. I., Xie, L., Sun, C., Zheng, X., & Chen, W. (2018). Advances in micro and nano-encapsulation of bioactive compounds using biopolymer and lipid-based transporters. *Trends in Food Science and Technology, 78,* 34−60. Available from https://doi.org/10.1016/j.tifs.2018.05.018.

Silva Filho E.C., Roma Junior L.C., Salles M.S.V., Salles F.A., Ezequiel J.M.B., & Van Cleef E.H.C.B. (2017). Thyme essential oil supplementation on performance and milk quality of lactating dairy cows. In: *Book of Abstracts of the 68th Annual Meeting of the European Federation of Animal Science.* Tallinn (Estonia). Wageningen Academic Publishers, Wageningen (The Netherlands), pp. 204, http://www.eaap.org/Annual_Meeting/2017_tallin/eaap_boa_68th_2017.pdf.

Singh, G., Kapoor, I. P. S., & Singh, P. (2011). Effect of volatile oil and oleoresin of anise on the shelf life of yogurt. *Journal of Food Processing and Preservation, 35*(6), 778−783. Available from https://doi.org/10.1111/j.1745-4549.2011.00528.x.

Sticher, O., Heilmann, J., & Zündorf, I. (2015). *Hansel/sticher pharmakognosie phytopharmazie* (10th ed.). Stuttgart, Germany: Wissenschaftliche Verlagsgesellschaft Press.

Stratford, M. (2006). Food and beverage spoilage yeasts. In Q. Amparo, & H. F. Graham (Eds.), *Yeasts in food and beverages. The yeast handbook* (pp. 335−379). Berlin: Springer.

Tassoul, M. D., & Shaver, R. D. (2009). Effect of a mixture of supplemental dietary plant essential oils on performance of periparturient and early lactation dairy cows. *Journal of Dairy Science, 92,* 1734−1740. Available from https://doi.org/10.3168/jds.2008-1760.

Thabet, H. M., Nogaim, Q. A., Qasha, A. S., Abdoalaziz, O., & Alnsheme, N. (2014). Evaluation of the effects of some plant derived essential oils on shelf life extension of Labneh. *Merit Research Journal of Food Science and Technology, 2*(1), 008−014.

Thangadurai, D., Naik, J., Sangeetha, J., Said Al-Tawaha, A. R. M., Adetunji, C. O., Islam, S., David, M., Shettar, A. K., & Adetunji., J. B. (2021). Nanomaterials from agrowastes: Past, present, and the future. In O. V. Kharissova, L. M. Torres-Martínez, & B. I. Kharisov (Eds.), *Handbook of nanomaterials and nanocomposites for energy and environmental applications.* Cham: Springer. Available from https://doi.org/10.1007/978-3-030-36268-3_43.

Tiwari, B. K., Valdramidis, V. P., Donnell, O. ', Muthukumarappan, C. P., K., Bourke, P., & Cullen, P. J. (2009). Application of natural antimicrobials for food preservation. *Journal of Agricultural and Food Chemistry, 57*(14), 5987−6000. Available from https://doi.org/10.1021/jf900668n.

Tomar, O., & Akarca, G. (2019). Effects of ice-cream produced with lemon, Mandarin, and orange peel essential oils on some physicochemical, microbiological and sensorial properties. *Kocatepe Veteriner Dergisi, 12*(1), 62−70.

Tornambé, G., Cornu, A., Verdier-Metz, I., Pradel, P., Kondjoyan, N., Figueredo, G., et al. (2008). Addition of pasture plant essential oil in milk: Influence on chemical and sensory properties of milk and cheese. *Journal of Dairy Science, 91*(1), 58−69. Available from https://doi.org/10.3168/jds.2007-0154.

Ukhurebor, K. E., & Adetunji, C. O. (2021). Relevance of biosensor in climate smart organic agriculture and their role in environmental sustainability: What has been done and what we need to do? In R. N. Pudake, U. Jain, & C. Kole (Eds.), *Biosensors in agriculture: Recent trends and future perspectives. Concepts and strategies in plant sciences.* Cham: Springer. Available from https://doi.org/10.1007/978-3-030-66165-6_7.

Upadhyaya, I., Upadhyay, A., Kollanoor-Johny, A., Mooyottu, S., Baskaran, S. A., Yin, H. B., et al. (2015). In-feed supplementation of transcinnamaldehyde reduces layer-chicken egg-borne transmission of Salmonella enterica serovar enteritidis. *Applied and Environmental Microbiology, 81,* 2985−2994. Available from https://doi.org/10.1128/AEM.03809-14.

Wagle, B. R., Upadhyay, A., Arsi, K., Shrestha, S., Venkitanarayanan, K., Donoghue, A. M., et al. (2017). Application of β-resorcylic acid as potential antimicrobial feed additive to reduce Campylobacter colonization in broiler chickens. *Frontiers in Microbiology, 8,* 599. Available from https://doi.org/10.3389/fmicb.2017.00599.

Wall, E. H., Doane, P. H., Donkin, S. S., & Bravo, D. (2014). The effects of supplementation with a blend of cinnamaldehyde and eugenol on feed intake and milk production of dairy cows. *Journal of Dairy Science, 97*, 5709−5717. Available from https://doi.org/10.3168/jds.2014-7896.

Worwood, V. A. (2016). *The complete book of essential oils and aromatherapy, revised and expanded: Over 800 natural, nontoxic, and fragrant recipes to create health, beauty, and safe home and work environments.* USA: New World Library.

Xue, J., Davidson, P. M., & Zhong, Q. (2017). Inhibition of Escherichia coli O157:H7 and Listeria monocytognes growth in milk and cantaloupe juice by thymol nanoemulsions prepared with gelatin and lecithin. *Food Control, 73*, 1499−1506. Available from https://doi.org/10.1016/j.foodcont.2016.11.015.

Yangilar, F., & Yildiz, P. O. (2017). Effects of using combined essential oils on quality parameters of bio-yogurt. *Journal of Food Processing and Preservation, 42*(1), e13332. Available from https://doi.org/10.1111/jfpp.13332, Article.

Yilmaztekin, M., Levic, S., Kalusevic, A., Cam, M., Bugarski, B., Rakic, V., & Nedovic, V. (2019). Characterisation of peppermint (Mentha piperita L.) essential oil encapsulates. *Journal of Microencapsulation, 36*(2), 109−119, https://doi.org.10.10.1080/02652048.2019.1607596.

Yoon, J. I., Bajpai, V. K., & Kang, S. C. (2011). Synergistic effect of nisin and cone essential oil of Metasequoia glyptostroboides Miki ex Hu against Listeria monocytogenes in milk samples. *Food and Chemical Toxicology, 49*(1), 109−114. Available from https://doi.org/10.1016/j.fct.2010.10.004.

Zeng, Z., Xu, X., Zhang, Q., Li, P., Zhao, P., Li, Q., Liu, J., & Piao, X. (2015). Effects of essential oil supplementation of a low-energy diet on performance, intestinal morphology and microflora, immune properties and antioxidant activities in weaned pigs. *Animal Science Journal, 86*, 279−285. Available from https://doi.org/10.1111/asj.12277.

Zhao, X., Shi, C., Meng, R., Liu, Z., Huang, Y., Zhao, Z., et al. (2016). Effect of nisin and perilla oil combination against Listeria monocytogenes and Staphylococcus aureus in milk. *Journal of Food Science & Technology, 53*(6), 2644−2653. Available from https://doi.org/10.1007/s13197-016-2236-6.

Zhaveh, S., Mohsenifar, A., Beiki, M., Khalili, S. T., Abdollahi, A., & Rahmani Cherati, T. (2015). Encapsulation of Cuminum cyminum essential oils in chitosan-caffeic acid nanogel with enhanced antimicrobial activity against Aspergillus flavus. *Industrial Crops and Products, 69*, 251−256. Available from https://doi.org/10.1016/j.indcrop.2015.02.028.

Chapter 13

The application of essential oil for the management of mycotoxins

Babatunde Oluwafemi Adetuyi[1], Pere-Ebi Yabrade Toloyai[2], Peace Abiodun Olajide[1], Oluwakemi Semiloore Omowumi[1], Charles Oluwaseun Adetunji[3], Juliana Bunmi Adetunji[4], Oluwabukola Atinuke Popoola[5], Yovwin D. Godwin[6], Olatunji Matthew Kolawole[7], Olalekan Akinbo[8], Abel Inobeme[9], Osarenkhoe Omorefosa Osemwegie[10], Mohammed Bello Yerima[11] and M.L. Attanda[12]

[1]Department of Natural Sciences, Faculty of Pure and Applied Sciences, Precious Cornerstone University, Ibadan, Oyo State, Nigeria, [2]Department of Medical Biochemistry, Faculty of Basic Medical Sciences, Delta State University, Abraka, Delta State, Nigeria, [3]Applied Microbiology, Biotechnology and Nanotechnology Laboratory, Department of Microbiology, Edo State University Uzairue, Iyamho, Edo State, Nigeria, [4]Department of Biochemistry, Osun State University, Osogbo, Osun State, Nigeria, [5]Genetics, Genomics and Bioinformatics Department, National Biotechnology Development Agency, Abuja, FCT, Nigeria, [6]Department of Family Medicine, Faculty of Clinical Sciences, Delta State University, Abraka, Delta State, Nigeria, [7]Department of Microbiology, Faculty of Life Sciences, University of Ilorin, Ilorin, Kwara State, Nigeria, [8]Centre of Excellence in Science, Technology, and Innovation, AUDA-NEPAD, Johannesburg, Gauteng, South Africa, [9]Department of Chemistry, Edo State University Uzairue Iyamho, Auchi, Edo State, Nigeria, [10]Department of Food Science and Microbiology, Landmark University, Omu-Aran, Kwara State, Nigeria, [11]Department of Microbiology, Sokoto State University, Sokoto, Sokoto State, Nigeria, [12]Department of Agricultural Engineering, Bayero University Kano, Kano, Kano State, Nigeria

Introduction

Fungi from the *Fusarium, Aspergillus, and Penicillium* genera are the most prevalent sources of mycotoxins, which are extremely toxic natural compounds. Except for a handful of toxins that are only produced by a select few species, many mycotoxins can be made by a wide variety of diverse genus types (Jarda et al., 2011). About 300 fungal metabolites having toxic potential, made by over 100 different fungi, have been documented thus far (Paterson & Lima, 2010). Numerous hazardous metabolites, most of which are severely poisonous or cancerous, can be inhaled by animals and humans. Among the fungi that can produce the six most prominent mycotoxins found in global agricultural systems are aspergillus, ochratoxin, patulin, fumonisins, trichothecenes, and zearalenone. Food and feed supply contamination with mycotoxin is a significant and persistent problem worldwide. The Food and Farming Association assesses that parasitic mycotoxins can taint around 25% of the world's yields, causing colossal monetary misfortunes in the billions of dollars (FAO, 2002). In July 2013, due to mycotoxin contamination, the price of rice jumped from about $400 to about $600 (FAO, 2013). Mycotoxins are responsible for an estimated $243 million in annual losses to cattle and animal feed in the United States (CAST et al., 2003). Mycotoxins also cause contamination, which can waste up to 25% of the world's annual harvest of fruits and vegetables. In developing countries, the economic losses from mycotoxin exposure are particularly severe. Once introduced into the food and feed chain, mycotoxins are notoriously difficult to eradicate due to their long-lasting toxicity. Feed nutrition levels and output may be lowered if mycotoxins are produced in feedstuffs for animals. Livestock are harmed, and sometimes even die, when they consume polluted feedstuffs. When consumed directly (infected fruit, corn, cereals, etc.), mycotoxins can endanger human health in two ways: directly, when ingested by humans; and indirectly, when consumed by animals that have been fed contaminated feed (Adewale et al., 2022). According to Williams et al. (2004), an estimated 4.5 billion people in developing nations are exposed to persistently high levels of mycotoxin. As a group, mycotoxins are harmful because they stop cells from dividing and prevent proteins from being made. Exposure to mycotoxin has also been linked to numerous other diseases. Failure of the kidneys and liver, possibly even death, are the most serious side effects of an acute overdose. Toxicity, in the form of genotoxicity, carcinogenicity, and reproductive problems, is exhibited in prolonged exposure (Anfossi et al., 2010). There are a number of ways to clean up and restore food and feed that has been tainted with mycotoxins because mycotoxins have been associated with numerous health issues, spanning many age groups and sexes. Both pre- and post-harvest measures are taken, with the former involving the elimination of mycotoxins in products that were previously tainted. Mycotoxin production can be avoided

Applications of Essential Oils in the Food Industry. DOI: https://doi.org/10.1016/B978-0-323-98340-2.00008-0

before harvest using pre-harvest preventative techniques like proper agricultural and manufacturing practices. However, this method may not be able to eradicate mycotoxins after contamination has occurred, resulting in additional postharvest detoxification strategies (Jard et al., 2009). There are a wide variety of recently enacted detoxification strategies, including physiological, biochemical, microbiological, and others.

Mycotoxins: their forms and food sources

Aflatoxins

Aspergillus flavus, *Aspergillus parasiticus*, and *Aspergillus nomius* all produce aflatoxins, which are the most common and poisonous of all mycotoxins (Adetuyi et al., 2015a, 2015b). The optimal temperature for the growth of these fungi that produce aflatoxin is between 28°C and 30°C, however, their toxin production ranges from 12°C to 42°C. Aflatoxin B1, B2, G1, G2, M1, and M2 are perhaps the most frequent. From AFB1 and AFB2, respectively, the hydrolyzed metabolic byproducts AFM1 and AFM2 are produced. Acute or chronic liver problems have been linked to aflatoxins, which are also known to cause cancer, mutations, and teratogenic effects. One of the most common and dangerous carcinogens in the natural world is the fungus *Aspergillus fumigatus*, or AFB1 (Iamanakaa et al., 2007). Aflatoxin is commonly found in animal feed, breakfast cereals, fruit, milk, and eggs. Rice is a major food in Korea, but it is contaminated with a bacterium and a virus that were both named flavus in 2002. From 1.8 to 7.3 μg/kg of AFB1, the concentration was found to be quite variable (Park et al., 2005). Kids in western Africa are especially helpless against the impacts of aflatoxins, which are available in the drawn-out ingestion of polluted groundnuts and maize and which have been displayed to fundamentally stunt development in youngsters. As per a few examinations, India has the most elevated AFM1 fixation in milk at 48 μg/L, trailed by Egypt in Africa with levels up to 0.9 μg/L (Batiha et al., 2021). AFM1 levels that are higher than those allowed by EU law have also been discovered in Pakistan, Brazil, and China.

Ochratoxins

Aspergillus and Penicillium produce ochratoxin A, B, and C in nature. Maize, grains, espresso, liquor, and pork products are only a few of the foods that contain ochratoxin. They have been isolated in both hot and cold regions. For instance, toxigenic fungi are capable of producing ochratoxin A (OTA), the most toxic ochratoxin. Wineries, grapefruits, and wine have all been investigated for OTA, which normally appears after storage (Hocking et al., 2007). Field and moldy samples of rice in Nigeria contained 341.3 and 1164 mg/kg, respectively, levels of OTA that were higher than those found in stored and marked samples. Maize had the highest concentration of OTA, at 201 g/kg, when tested in cereals from six Chinese provinces. In contrast to other grains, Liang (2008) discovered that rye and corn have the greatest amounts of OTA in China. Cocoa and cocoa products have also been known to be contaminated with OTA. Medical herbs were also reported to be contaminated with OTA when stored improperly.

Patulin

Many types of Penicillium and Aspergillus produce patulin. Produce (apple, pear, maybush, cherry, kiwifruits, etc.), fruit products, and vegetables often include this mycotoxin as a contaminant (Demirci et al., 2003). Patulin, a major threat to fruit after harvest, makes contact with the skin of damaged fruit, spreads throughout the fruit, and can even infect neighboring fruits in storage. It was discovered that environmental conditions like temperature, pH, and moisture level all have a role in patulin formation. The optimal temperature for manufacturing is 21°C, and it exhibits an inverse connection with pH (McCallum et al., 2002). Most experts agree that patulin is the most harmful mycotoxin found in fruit. An array of sudden symptoms may be brought on by this, including nausea, vomiting, and gastrointestinal problems. It has been shown to have long-term negative effects on the kidneys, liver, and immune system, as well as carcinogenic, genotoxic, and cellular health effects (James-Okoro et al., 2021; Wichmann et al., 2002). Since patulin is still most persistent at pH levels around 4.5, is an ideal environment for the metabolism and production of fruit. Patulin usually contaminates apple products, as evidenced by the fact that almost half of all analyzed samples worldwide displayed reasonably high detectable levels of patulin in apple juice (Marin et al., 2011). One market sample of Italian apple juice had a patulin concentration of 1150 μg/L, well over the WHO's recommended amount of 50 μg/L (Didunyemi et al., 2019, 2020). Baby food made with organic apples has been shown to contain more patulin than conventional baby food made with regular apples, according to a survey conducted in Spain, Portugal, and Italy (Ritieni, 2003). Patulin is used in more than just apple products; It has also been found in a wide range of other fermented foods, including apricots, pears, and mango (Moss, 2008). Additionally, it was discovered that cheese contains patulin-producing fungi.

Fumonisins

Fusarium verticillioides and *Fusarium proliferatum* fungi produce fumonisins from maize and other cereal grains in high-temperature, high-humidity environments (Didunyemi et al., 2019). Liver and esophageal malignancies, along with notochord defects, have all been associated with fumonisins. Fumonisin B1 (FB1) is the most prevalent and pervasive human cancer promoter. There have been multiple reports of FB1 contamination in corn and maize-based products from multiple nations, causing the general public to worry that over half of exported corn is contaminated. According to the FAO and WHO in 2011, FB1 was present in 63% of samples of maize and 80% of samples of products derived from maize (Bansal et al., 2011). Adeleke et al. (2022) report that FB1 can infect a wide variety of foods, including rice, millet, wheat, rye, beans, black tea, and even dairy products.

Zearalenone

Estrogenic mycotoxin ZEN is predominantly generated by different species of Fusarium. Producing ZEN fungus thrives in damp, mild climates and high-humidity storage areas. Premature sexual development in females is thought to be the primary manifestation of ZEN's estrogenic effects. In addition, ZEN's production of reactive oxygen species (ROS) may contribute to its toxicity. Although maize has been reported to have greater ZEN infection rates than other cereals, wheat, soybeans, and rice have all been regularly contaminated (Goertz et al., 2010). Contamination with ZEA and the presence of OTA, AFB1, FB1, and DON have been found in Korean rice, barley, and maize (Park et al., 2005). Half of the 196 samples of moldy rice examined in Nigeria contained ZEN, using doses between 0 and 1200 µg/kg, as reported by Makun et al. (2007). The ZEN precursor zearalenol, which is three times highly estrogenic than ZEN, has been detected at concentrations of up to 74.7 µg/kg in Chinese milk samples (Adewale et al., 2022).

A variety of other significant mycotoxins

Trichothecenes, citrinin, cyclopiazonic acid, gliotoxin, griseofulvin, and a plethora of others are among the most significant mycotoxins that can be detected in food. Trichothecenes are the most common type of mycotoxin, and Fusarium species are the primary source. First-group toxins (HT-2 and T-2) largely impact the digestive system. Deoxynivalenol (DON), 3-acetyldeoxynivalenol (3-ADON), and nivalenol (NIV) are the most toxic and immunosuppressive members of the second group of toxins (Capriotti et al., 2012). Penicillium expansum and a few types of Aspergillus and Monascus spp. are the primary producers of citrinin. In a 2009 study (Abramson et al.) Citrinin is often found alongside OTA, and it has been hypothesized that prolonged exposure to the two may cause cancer in humans (Nguyen et al., 2007). Citrinin-producing fungi have been shown to thrive on barley (Galvano et al., 2005).

Using essential oils as well as other aromatic compounds as safe, alternative food natural ingredients

EOs are chemically rich mixtures of many low-molecular-weight volatile and non-volatile compounds isolated from various botanical parts using distinct physiochemical procedures. Terpenes, terpenoids, phenylpropanoids, and their oxygen-rich analogs are all examples of these molecules (Adetunji et al., 2013, 2019; Adetunji, Akram, et al., 2021; Adetunji, Nwankwo, et al., 2021; Adetunji, Palai, et al., 2021; Adetunji, Inobeme, et al., 2021; Adetunji, Michael, Kadiri, et al., 2021; Adetunji, Anani, et al., 2021; Adetunji, Michael, Nwankwo, et al., 2021; Adetunji, Michael, Varma, et al., 2021; Adetunji, Adetunji, et al., 2021; Adetunji, Olaniyan, et al., 2021; Adetunji, Ajayi, et al., 2021; Rauf et al., 2021; Silvestre et al., 2019; Ukhurebor & Adetunji, 2021). Up to one hundred different components can make up an EO, each with a slightly varying concentration. Two or three of these components often account for between 20% and 70% of the EO as a whole, so they are called primary components. EOs have been shown to be helpful in a variety of investigations, however, the focus of these studies has been on the oils' most notable constituents, others have found that even less prominent ingredients can have a large impact on the final product's performance thanks to synergistic effects with the more prominent ones. The antibacterial activity of EOs and their active constituents in wide variety of food systems is further influenced by the types of oils and intended meals that can be lawfully provided (Adetunji, 2008, 2019; Adetunji et al., 2013, 2017; Adetunji, Oloke, et al., 2020; Adetunji, Egbuna, et al., 2020; Adetunji, Roli, et al., 2020; Adetunji et al., 2022; Adetunji & Varma, 2020; Adejumo et al., 2017; Bello et al., 2019; Egbuna et al., 2020; Olaniyan & Adetunji, 2021; Thangadurai et al., 2021). Due to their ease of absorption by fat molecules, several EOs are less effective against fungi when added to fatty foods (mainly animal-based foods) than when added to non-fatty foods (James-Okoro et al., 2022). EOs and related active compounds have shown promising results in numerous studies for use as food preservatives. That's why, in recent years, they've seen greater use in the agricultural industry as a means to prevent mycotoxins from damaging perishable goods for longer (Siva et al., 2020). Antifungal and antimycotoxigenic activities, responsiveness to food spoilage intended fungal species, varied modes of activity (since each EOs active ingredient has distinctive systemic effects), and the impact of food functional ingredients on their potency are just some of the topics that need to be explored in detail before they can be used in food. EOs and associated active ingredients are able to permeate the lipid bilayer of the fungal plasma membrane and exercise their actions at that level because of their lipophilicity. Because of the disruption of cellular homeostasis and eventual cell death that results from this, these crucial structures become much less ordered and permeable (Munir et al., 2022).

Although there are a number of publications that demonstrate the bioactive components and in vitro potential of EOs, very few studies have demonstrated their effectiveness in protecting food from mycotoxins. In one study (2009), Gandomi et al. used Iranian cheese as a model food to test the aflatoxin-inhibiting effects of Zataria multiflora EO. They discovered that, at a dosage of 1.0 μL/mL, toxin production was reduced by more than 75% compared to the control (50% ethanol). Another study found that at 3.9 mg/kg, Cananga odorata EO completely inhibited mycotoxin formation in corn caused by *Fusarium graminearum*. The effect of *Mentha spicata* EO on AFB1 production in faba beans after they were stored for 11 months in sealed bags was studied. Studying the effectiveness of three EOs (cinnamon, bay, and clove) and an antioxidant called resveratrol in preventing Penicillium verrucosum and Aspergillus ochraceus from producing OTA in gamma-irradiated wheat grains, Aldred et al. (2008) discovered that at 1.0 L/mL, the EO wholly insulated prototype food samples from aflatoxin B1 toxins. Wheat grain OTA production was found to be decreased (by more than 60%) by both EOs and resveratrol at concentrations of 200 μg/g, whereas resveratrol at 500 μg/g caused total inhibition of OTA production. To reduce the amount of OTA produced, resveratrol was shown to be more effective than the EOs that were tried. EOs from *Melissa officinalis*, *Coriandrum sativum*, *Thymus vulgaris*, and *Cinnamomum zeylanicum* were tested for their ability to prevent the formation of deoxynivalenol and fumonisin in grain. They found that *C. zeylanicum* EO was more effective than *M. piperita* and *T. vulgar*. On the other hand, Marin et al. (2004) looked into whether clove, palmarosa, and lemongrass EOs could prevent *F. graminearum* from synthesizing zearalenone and deoxynivalenol in maize grains. Although the authors found that bigger doses were not more effective, they noted that they did not compromise the organoleptic features of the model food system. However, spice, lemongrass, and palmarosa EOs were successful in suppressing zearalenone formation, with clove EOs showing the most efficacies in blocking deoxynivalenol synthesis. In yet another study, Perczak et al. (2019) found that several different types of cinnamon, oregano, cymbopogon, citrus, thyme, mint, foeniculum, and with the exception of *C. aurantium* oils, from lemon to geranium to frankincense, were all top-notch, reducing activity by 68.33% while reducing trichothecene and zearalenone concentrations by more than 99%. Lotus seeds infected with an aflatoxigenic strain of *A. flavus* were found to have their aflatoxin B1 and B2 production reduced by 35.7% to 86.6% and by 76.9% to 93.7%, respectively, when treated with *Illicium verum* EO at doses varying between 2.0 and 5.0 μL/g. In a remarkable study, Dwivedy et al. (2017) extracted an EO from *I. verum* and showed that it effectively inhibited the growth of the aflatoxin B1-producing *A. flavus* strain commonly found in Pistacia vera. It was determined that a concentration of 0.7 μL/mL was sufficient to render an uncontaminated model food sample after treatment. EO was also noted to inhibit the development of *Aspergillus sydowii*, *Aspergillus sulphureus*, *A. fumigatus*, *Aspergillus versicolor*, *Aspergillus niger*, *Aspergillus terreus*, *Penicillium purpurogenum*, and *Trichoderma viride*. As a result, it could be used as an expected antifungal specialist for increasing the amount of time. Overall, the study's findings suggested that EOs and related active ingredients may be employed in the food industry to reduce mycotoxin contamination. Further, developments in nanoencapsulation advancements have allowed for a plethora of possibilities in the actual food context by removing barriers to their use. In addition to their efficacy against growths and the development of mycotoxins, the positive human harmfulness of EOs/bioactive components is a remarkable feature of utilizing them as a food additive. The US Food and Drug Administration has determined that they pose little threat to life, given their prevalence. The authorities did add, though, how such bioactives seem to be suitable for consumption in the quantities suggested. Even though the food industry uses EOs and bioactive components, there are risks associated with them. Citronella and clove, for example, are examples of EOs that are known to cause hazy toxicity. As a result, using them can cause acidosis, liver problems, low blood sugar, ketonuria, fever, cyanosis, rapid breathing, or even a trance state. Therefore, it is essential to weigh the advantages and disadvantages of each EO or bioactive ingredient proposed for food use. A common method used to determine the LD50 or median lethal dose value of an EO or bioactive ingredient is the acute oral test. Toxicology studies on model organisms, evaluation of human exposure in light of intended use, and the establishment of reasonable safety limits between human exposure at therapeutically-relevant doses and toxic levels arising from the animal review are essential steps in this strategy.

Biological mechanisms of essential oils' antifungal chemicals

Although numerous hypotheses have been put forth to explain this phenomenon, the specific mechanisms by which EOs and their bioactive components affect mycotoxigenic fungal infections are unknown. Kedia et al. (2016) looked into *M. spicata* L.'s antifungal mechanism. Plasma membrane deformation, swelling, the existence of lomasomes, and a reduction in cellular matrix were all observed and reported in mycelia exposed to EO. By utilizing propidium iodide (PI) dye in a flow cytometric analysis, Tian et al. (2012) hypothesized that an abscess in the spore's lipid bilayer following treatment with EOs may be responsible for their antifungal effects. The protein-binding dye PI is commonly used to assess drug effects on the plasma membrane because of its ability to bind to nucleic acids. Measurements of fluorescence intensity revealed that PI entered only cells with plasma membrane damage or lesions.

Numerous researchers have used quantitative measurements to confirm that antifungal mode of action is directed at the plasma membrane. Tian et al. (2012) measured the inhibition of ergosterol biosynthesis and also proved that EOs have antifungal activity in the fungal plasma membrane. Khan et al. (2010) investigated the antifungal properties of Ocimum sanctum EO and its active ingredients, and the researchers reported that the EO and its bioactive principles inhibit ergosterol biosynthesis, consequently affecting membrane integrity, which is the antifungal mode of action. *T. vulgaris*, *Syzygium aromaticum*, and *Anethum graveolens* are examples of EOs that exert their antifungal effect by acting on the membrane and blocking ergosterol biosynthetization; however, the antifungal properties of *Melaleuca alternifolia* and *C. sativum* are limited to their ability to bind to ergosterol (Adetunji, Palai, et al., 2021). The increased loss of key biological electrolytes like Ca^{2+}, Mg^{2+}, and K^+ from fungal cells subjected to various concentrations of EOs, as well as the loss of 250 and 270 nm absorbing materials, demonstrated that the plasma membrane is an essential area for their fungicidal molecular mechanism. Kedia et al. (2014) monitored Ca^{2+}, Mg^{2+}, and K^+ ion leakages through the plasma membrane and examined the effects of *M. spicata* EO and Cymbopogon citratus EO, respectively. According to Chaudhari et al. (2018) treatment with varying concentrations of *Pimenta dioica*, *Myristica fragrans* results in the loss of materials that absorb 260 and 280 nm, and salvia (*Salvia sclarea*) may have their antifungal mechanism of action in essential oil. EOs have antifungal properties, which have been linked to an improvement in Ca^{2+}-homeostasis in fungal cells, as shown by Ahmad et al. (2011). da Cruz et al. (2013) hypothesized that treatment with bioactive chemicals derived from plants would inhibit fungus growth, who examined the connection between plasma membrane permeability, cell wall integrity loss, and enzyme modification.

In addition to the plasma membrane, many more fungal targets for EOs have been found. Pinto et al. (2009) demonstrated a direct link between a decrease in cellular metabolism and how EOs work to prevent fungal growth by employing a two-color fluorescent FUN 1 dye. However, the dye remained diffuse and glowed green or yellow in cells with poor metabolism treated with EOs, indicating a metabolic disturbance. According to research published in 2012 by Tian et al., EOs may exert their antifungal effects by, among other things, altering mitochondrial membrane potential, acidifying the extracellular matrix, mitochondrial dehydrogenase and ATPase activity are both severely suppressed, while ROS generation is significantly boosted. In order to discover the fungicidal efficacy of Lippia rugosa EO, Tatsadjieu et al. (2009) studied the daily dosage reduction of H + - ATPase activity and the carbohydrate acidosis of the periplasm in *A. flavus* administered with Lippia rugosa EO. Some researchers believe that the EOs' fungicidal effects come from the EOs or their bioactive constituents crossing the membrane and interacting with metabolic enzymes or H + in ATP production (Chang et al., 2001). Using *Ageratum conyzoides* L. as a fumigant, da Cruz et al. (2013) showed that the mitochondria are a primary site of action in the suppression of fungal growth. The antifungal effects of *A. graveolens* EO were hypothesized to result from the EO's ability to inhibit the Kreb's cycle and ATP production (Chen et al., 2013). Cowan (1999) found that the granulation in the cytoplasm, membrane rupture, and deactivation or inhibition of synthesis of extracellular and intracellular enzymes resulted from treatment with the bioactive compounds strongly responsible for the inhibitory and antibacterial mode of action.

The lipid bilayer is crucial to the fungal cell wall since it provides a selective barrier. It serves to both insulate the cell from its surrounding habitats and fixate the intermediates, electrolytes, proteins, and many components necessary for cellular stability. Fungi contain ergosterol, a lipid found in cell membranes that is necessary for membrane fluidity and integrity but is structurally and biosynthetically distinct from cholesterol in animals and phytosterols in plants, is the target of the majority of antifungal medications. EOs and the pharmacologically active chemicals they contain can easily cross the lipid bilayer and disrupt ergosterol transport because they are also lipophilic. This increases the permeability of the lipid bilayer, allowing important cellular ions to leak out, inactivating enzymes and cell organelles, particularly mitochondria, lowering membrane potentials, and using up ATP pools. Fungi can be eradicated completely by these substances, either on their own or in combination (Bendaha et al., 2011). Xie et al. (2004) showed that the EOs' lipophilic compounds may easily penetrate membranes and interfere with bilayer metabolic enzymes, resulting in the removal of crucial biological elements necessary for suppressing spore germination.

Evaluation of essential oil's antifungal potency

The minimal effective concentration (MIC) value against the test fungus is the primary metric for evaluating EO's efficacy. The MIC of EO is the concentration of EO needed to prevent the test fungus from growing visibly. For EOs to be regarded as fully functional, their MIC value must be lower than 1.0 μL/mL. One of the steps in putting an EO to use is determining its activity in terms of its MIC. A number of factors, such as the dissolution rate, specific compound atomic weight, heterogeneity of organic compounds, corrosiveness, variation in responsiveness against test pathogens, MIC determination methodologies, and detergents shown to solubilize experimental EOs, significantly affect the calculation of MIC. For this reason, it may be challenging to define the threshold concentration for classifying EOs as actually active solely on a single parameter (Didunyemi et al., 2019).

The mechanism by which essential oils and their bioactive components suppress the effect of mycotoxins

Mechanism of action against aflatoxigenic

In recent years, it has been discovered that numerous EOs and the bioactive components they contain reduce AF production. *C. zeylanicum* and *S. aromaticum* EO were found to be effective against AFB1 production, as reported by Patkar et al. (1993). At concentrations of 0.85 μL/mL and 1.35 μL/mL, both EO effectively inhibited the synthesis of AFB1. Mahmoud (1994) suggested reducing the financial impact of AFB1 contamination by utilizing five EO constituents—geraniol, nerol, citronellol, cinnamaldehyde, and thymol. The effectiveness of *Cymbopogan citratus* (DC.) Stapf was studied by Paranagama et al. in 2003. AFB1 generation in stored rice can be controlled by using EO as a fumigant.

Although the majority of studies claim that EOs or bioactive compounds can inhibit both fungal growth and AF formation, in most cases, a significant fraction is needed to halt AF formation than is needed to halt overall growth, indicating that the two mechanisms of action are distinct. The growth and formation of aflatoxin B1 (AFB1) were studied, and the effectiveness of citrus reticulata and citrus citratus EO in preventing these processes was evaluated; it was found that Citrus reticulata had a greater capacity to prevent growth than Citrus citratus EO, probably because of their individual impacts on development and AF formation. Dwivedy et al. (2017) compared the development and production of AFB1 by *Abelmoschus moschatus* EO to those of *A. flavus* obtained using traditional medicinal source materials.

It is not well understood how EOs and their bioactive components work to prevent the formation of adipose tissue. Although the mechanism of AF inhibition is not well understood, many theories have been put out to explain it. The production of AFB1 involves reactive oxygen species (ROS) actively, Sun et al. (2015) indicated that inhibiting ROS creation may be the most likely way to inhibit AFB1 biosynthesis. According to Bluma et al. (2008), preventing the tripartite stage of AF metabolic processes, which includes lipid peroxidation and oxygenation, may be the cause of AF biosynthesis. Mekawey and El-Metwally (2019) suggest that the mitochondrial production of Acetyl Co-A, the major precursor of AF metabolic processes, may be inhibited. Mekawey and El-Metwally found evidence that EOs high in phenolics play a role in this suppression. Tian et al. (2011) found that the antiaflatoxigenic effect of EO is not attributable to photosynthetic activity but rather to the blocking of cellular metabolism by a few key enzymes. It has already been established that the EOs and related bioactive components suppress AF synthesis by interacting with certain crucial genes. Hu et al. (2017) revealed the *Curcuma longa* L. antiaflatoxigenic mechanism of action in this instance. The majority of coding sequences are activated by proteins encoded by AflR and AflS, were found to be downregulated, as were another five structural genes (aflB, aflC, aflD, aflE, and aflG), according to EO's research using RT-PCR. Enzymes AflD, AflM, AflO, AflP, and AflQ generate other enzymes needed for further processes in AF metabolic processes, may be to blame for the suppression of AF biosynthesis after five days of treatment. Kim et al. (2018) determined that the key factor in the inhibition of AFB1 biosynthesis was the downregulation of the genes for norsolorinic acid (NOR) reductase and versicolorin B (VERB) synthase (AflE and AflL, respectively) upon 1, 8 cineole treatment. According to a study that was published in 2018 by Tian et al., spores pigmentation suppression, like that of NOR, the first critical intermediate in AF synthesis, is the most likely cause of AF suppression. According to El Khoury et al.'s research (2016), AFB1 is inhibited by EO from Micromeria graeca without altering mycelial growth in the fungus, who additionally discovered that AflR and AflS, two regulatory genes involved in AF biosynthesis, were downregulated three times. Since aflR is responsible for increasing the transcription and accumulation of AF genes, most gene-level reports on EO-mediated AF inhibition focus on it. The inability of *A. fumigatus* (AF)-producing strains to induce AF growth was traced back to a lack of the aflR gene (Yu et al., 2004). Scientists have been motivated by this theory to focus on blocking this gene in order to prevent AF. However, the precise manner in which structural genes involved in the production of AFs are affected by EOs or bioactive substances remains a mystery. It has also been reported that certain EOs can block the production of AFB1-DNA adducts, hence reducing the carcinogenic effects of AF. New insights into the antiaflatoxigenic mechanism of Cistus ladanifer have been revealed in recent years by the research groups of Upadhyay et al. (2018) along with *P. dioica* EO by preventing the production of methylglyoxal (MG). The byproduct of enzymatic and nonenzymatic reactions throughout the glycolytic pathway, endogenous MG is a highly reactive cytotoxic molecule (Kalapos, 1999). The contribution of MG to the increase in AF synthesis via aflR gene overexpression was validated by Chen et al. (2004). However, not all compounds with antioxidant activity have the capacity to impede AF synthesis in the same way that antioxidants do. Although substances like eugenol and ferulic acid have potent radical scavenging activities, it has been found that they also stimulate AF synthesis in synthetic media (Norton & Dowd, 1996). Thus, the generation and inhibition of AF may have mechanisms other than antioxidant action. However, concerns about their toxicity and costs, as well as consumers' preference for natural antioxidants, have encouraged the use of plant-derived antioxidants (Nesci et al., 2003).

Antiochratoxigenic mode of action

Several researchers have shown that specific EOs and the bioactive chemicals they contain are effective at blocking the effects of ochratoxin. Production of OTA is stifled by using EO and its active ingredients. Oregano and mint EOs have been shown to totally block OTA formation at a concentration of 1.0 μL/mL, as reported by Basilico and Basilico (1999). EOs of *Salvia officinalis*, *Lavandula dentata*, and *Laurus nobilis*, along with their primary bioactive component 1, 8-cineole, have been demonstrated to completely suppress OTA production at a concentration of 0.5% (Dammak et al., 2019).

Toxic method of action against fusiform worms

The bioactive components found in many EOs have shown significant effectiveness in combating fusarial poisons. Due to its high effectiveness at just 0.2 μL/mL concentration, in an in vitro study, Aly et al. (2016) suggested using *I. verum* to inhibit fumonisin B1 (FB1). Avanço et al. (2017) found that *Curcuma longa* L. had significant anti-fumonisin effectiveness in EOs. This effectiveness of *Rosmarinus officinalis* L. was reported by da Silva et al. (2015). EO for FB1 and FB2 suppression and compared it to nystatin. Zingiber officinale EO was investigated by Ferreira et al. (2018), who discovered that at a concentration of 0.6 μL/mL, in vivo deoxynivalenol (DON) production was completely repressed.

Sakuda et al. (2016) postulated that acetyl coenzyme A (CoA) and the ATP-dependent cyclooxygenase (ATPCL) and trichothecene reductase (TR16) genes would be downregulated in response to a decrease in TCs. Yoshinari et al. (2008) suggested a possible strategy for TC inhibition by inhibiting cytochrome P450 monooxygenase using two structural analogs (CYPA and TR14). It's also worth noting that RT PCR studies demonstrated that *Cuminum cymininum* EO inhibited FUM1 transcriptional activity, suggesting that these substances are likely to be important locations in preventing the manufacture of FUMs (Khosravi et al., 2015).

Nanoencapsulation methods for essential oils and other bioactive ingredients

Nanosystems containing EOs and related active constituents can be built using polymeric materials, and numerous nanoencapsulation methods have been documented in the literature; however, no single method has been able to gain widespread acceptance due to its lack of drawbacks. Considerations such as the active core (EOs/bioactive constituents) and wall material (polymers) type and characteristics, as well as the capability to create a nano system that offers better absorption performance and loading propensity with optimized storage period, nontoxicity (with food ingredients), and required dosage forms, must be made prior to making a final decision on a method. Nanoencapsulation techniques for beneficial active ingredients in EOs are discussed.

Ionic-gelation

A simple, efficient, and low-cost approach that has been documented in the literature for the nanoencapsulation of EOs and their bioactive ingredients into a nano-range route of administration within moderate system parameters is ionic-gelation. This type of encapsulation is useful for preserving EOs and other biologically active compounds that might otherwise decay when exposed to heat, agitation, or volatile compounds (Chaudhari, Singh, Singh, et al., 2020; Chaudhari, A.K., Singh, V.K., Deepika, et al., 2020). For ionic-gelation, cationic polymers like chitosan, gelatin, and dextran are needed, as are anionic cross-linking agents like sodium tripolyphosphate (S-TPP), the most often employed cross-linking agent because of its multivalent and non-toxic qualities (Munir et al., 2022). This technique also yields nanostructures with lipid nanoparticles efficacy, a predictable bioavailability, and long-term preservation of the core molecules they enclose. A regulated dissolution rate was maintained for up to 56 days in a study by Hasheminejad et al. (2019), who encapsulated clove EO within chitosan using tripolyphosphate (TPP) as the cross-linking agent. The artificial nanoparticles' yield, loading capacity, and encapsulation efficiency were approximately 45.77 %, 6.18 %, and 39.05 %, respectively. FE-SEM and DLS analysis also showed that the nanoparticles were spherical and ranged in size from 40 to 100 nm on average. Su et al. (2020) on the other hand, used an ionic-gelation method to encapsulated cinnamon EO in varying concentrations in chitosan nanoparticles. The nanoparticles were spherical in shape, had consistently distributed diameters, a satisfactory encapsulation effectiveness of 32.9%, a loading capacity of 10.4%, and inhibited the volatilization of the encapsulated bioactives, hence increasing their chemical stability.

Coacervation

It's easy to do, and it's one of the most common methods for enclosing volatile, reactive, or otherwise delicate compounds (such as EOs and their bioactive components) in nanoscale delivery systems. Two categories can be used to describe this method: aqueous and organic. Lipophilic substances, such as EOs and their bioactive components, are the only ones that can be encapsulated using coacervation in an aqueous phase. However, organic phase coacervation can be used to encapsulate hydrophilic molecules; however, organic solvents are required for this process. To further categorize the coacervation of the aqueous phase, it can be either simple or complex. Complex coacervation offers advantages beyond coacervation and the current aerosol method because it is propelled by the cohesion and adhesion of positive and negative charge colloidal particles, typically a polypeptide and polysaccharides, to build a network around the bioactive fundamental chemicals.

Complex coacervation yields a nanoencapsulated formulation with many advantages, including excellent encapsulation efficiency, low encapsulating polymer content, stability at high temperatures, and controlled, slow component release. The effectiveness of this method depends on several factors, including the solution's redox potential, the polyelectrolyte's molar mass, degree of ionization, and chain flexibility, and the pH of the enclosing wall materials and the biologically active elements they contain. Encapsulation of *Piper nigrum* EO into gelatin and sodium-alginate at pH 4.5 using the complicated coacervation approach developed by Bastos et al. (2020) resulted in significant levels of protection for the main compound and encapsulation efficiency. Based on these findings, the technique may be applied to the nanoencapsulation of certain other active ingredients for application in the food system.

Emulsification

Oil-in-water (o/w) and water-in-oil (w/o) emulsification, as well as water-in-oil (w/o) emulsification, are two types of emulsification. When the two liquids are homogenized with the help of an adequate surfactant, the result is a transparent or milky nanoemulsion. The method is well-suited for encapsulating both hydrophilic (in particular, EOs and bioactive components) and lipophilic (in general) substances. Managing the size of the particles, zeta potential, turbidity of the emulsifier, and most pertinently the stability of coated compounds are all significantly influenced by the highly energetic microemulsion methodology, which can be conducted under super-fast granulation, microfluidization, or high-intensity ultra-sonication to generate nanoemulsion (Shishir et al., 2018).

Nanoemulsion with *Satureja khuzestanica* EO and carvacrol were synthesized by Mazarei and Rafati (2019) utilizing two distinct methods: high-speed homogenization and ultrasonication. They discovered that high-speed homogenization produces nano-sized droplets with higher antibacterial activity and longer-term stability than ultrasonication. These particles, at the nanoscale, have a poly dispersity index that is quite specific. Stochastic flocculation and phase inversion, on the other hand, are low-energy emulsification processes that are straightforward, inexpensive, and energy-efficient. The formation of unstable droplets at high temperatures, necessitating an isothermal condition, is one of this method's major drawbacks; limitations on the kinds of EOs or bioactive constituents that can be used, as well as surfactants; and the requirement for an abundance of surfactant. *Citrus aurantifolia*, *C. kaffir*, and *C. calamansi* were employed by Liew et al. (2020) to create lime EO solid lipid nanoparticles. Nanoemulsions formed had round particles with diameters in the nanometer range (22, 29, and 63 nm, with PDI values of 0.43, 0.45, and 0.57 for auranantifolia, kaffir, and calamansi limes, respectively).

Spray-drying

In addition to improving the stability of volatile components that are sensitive to heat and light, such as EOs and their bioactive components, it is the most widely used inexpensive, flexible, and fast form of encapsulation

(Adetuyi, Akram, et al., 2021). This method consists of the following three main steps: emulsifying the EOs and bioactive components in a concentrated polymer solution, homogenizing the resulting emulsion, and atomizing it into a drying medium at a higher temperature (which typically results in the rapid evaporation of water), resulting in a stable matrix with numerous advantages, including a high yield of high-quality particles, a small particle size, rapid dissolving, and high durability. Dehydration throughout the process results in a low water content, which in turn decreases the multiplication of hydrophilic microorganisms, especially meal-based fungus, making this method appropriate for extending the shelf life of dried processed food. EO from the pepper-Rosmarin plant was encapsulated in chitin by Paula et al. (2017) using this approach, with varying weight-to-volume ratios. According to the findings, the produced particles displayed nano-ranged sizes with an encapsulating effectiveness of 15.6% and a loading capacity of 62%. It was also hypothesized that the characteristics of the gum and chitosan used in the creation of the nanoparticles played a significant role in the overall success of this method.

Electro-spinning and electro-spraying

Electro-spinning and electro-spraying are processes in which a nanoemulsion is subjected to a large electrical field and then either spun or sprayed; these are simple, low-cost, and increasingly popular methods for synthesizing nanofibers and nanoparticles (Bhushani & Anandharamakrishnan, 2014). Both of these approaches work on the same fundamentals but differ in one key respect: the concentration of the polymer used. Electrospinning produces nanofibers when the polymer concentration is high enough to elongate and stabilize the jet from the spinneret (a condition in which the droplet's hemispherical shape is distorted to a conical shape). Varicose destabilization causes the spinneret jet to become unstable at reduced polymer concentrations, leading to the generation of small droplets or particles. Lack of heat is essential for maintaining the integrity of high-temperature fragrance agents like EOs and maximizing their durability, accessibility, and utility in large-scale commercial food research in general. Numerous authors have reported successful EO delivery using these methods, including Wen et al. (2017), who electrospun a film of polyvinyl alcohol and -cyclodextrin nano-fibers containing cinnamon EO. The film demonstrated superior thermal stability and antimicrobial activity compared to the casting film. In light of these findings, it was proposed that the synthesized nano-fiber could serve as a viable option for active food packaging, namely in the context of extending the half-life of treated strawberries. More recently, Karim et al. (2020) developed a cinnamic aldehyde containing zein nano-fiber for industrial-scale use in meat product preservation. They found that the nanofabricated cinnamic aldehyde has greater antibacterial and antioxidant activities than the pure form. This is because phenolic chemicals are better able to be protected from external forces and released slowly, viable methods for increasing the storage life of beef products.

Conclusion

Contamination of stored foods with mold and mycotoxins is a global issue that poses serious threats to public health. The use of environmentally friendly preservatives could significantly reduce the losses incurred because of fungal contamination and the mycotoxins it produces in stored foods. In the culinary and production process, natural antimicrobials including EOs and active ingredients are increasingly replacing potentially harmful synthetic preservatives. Their quick disintegration in aerobic conditions, moisture, and heat has limited their usage in food production systems; moreover, their volatility has also limited their application in other contexts. Nanoencapsulation technologies have recently advanced, allowing us to avoid the pitfalls of employing EOs uncontained. Absolute bioavailability and preservation of EO constituents from deterioration due to external conditions are two potential benefits of nanoencapsulation techniques that have the potential to increase the storage life of meals.

References

Adejumo, I. O., Adetunji, C. O., & Adeyemi, O. S. (2017). Influence of UV light exposure on mineral composition and biomass production of myco-meat produced from different agricultural substrates. *Journal of Agricultural Sciences, Belgrade*, 62(1), 51−59.

Adeleke, D. A., Olajide, P. A., Omowumi, O. S., Okunlola, D. D., Taiwo, A. M., & Adetuyi, B. O. (2022). Effect of monosodium glutamate on the body system. *World News of Natural Sciences*, 44, 1−23.

Adetunji, C. O. (2008). *The antibacterial activities and preliminary phytochemical screening of vernoniaamygdalina and Aloe vera against some selected bacteria (MSc. thesis)* (pp. 40−43). University of Ilorin.

Adetunji, C. O. (2019). Environmental impact and ecotoxicological influence of biofabricated and inorganic nanoparticle on soil activity. In D. Panpatte, & Y. Jhala (Eds.), *Nanotechnology for agriculture*. Singapore: Springer. Available from https://doi.org/10.1007/978-981-32-9370-0_12.

Adetunji, C. O., Ajayi, O. O., Akram, M., Olaniyan, O. T., Chishti, M. A., Inobeme, A., Olaniyan, S., Adetunji, J. B., Olaniyan, M., & Awotunde, S. O. (2021). Medicinal plants used in the treatment of influenza a virus infections. In K. Dua, S. Nammi, D. Chang, D. K. Chellappan, G. Gupta, & T. Collet (Eds.), *Medicinal plants for lung diseases*. Singapore: Springer. Available from https://doi.org/10.1007/978-981-33-6850-7_19.

Adetunji, C. O., Akram, M., Olaniyan, O. T., Ajayi, O. O., Inobeme, A., Olaniyan, S., Hameed, L., & Adetunji, J. B. (2021). Targeting SARS-CoV-2 novel Corona (COVID-19) virus infection using medicinal plants. In K. Dua, S. Nammi, D. Chang, D. K. Chellappan, G. Gupta, & T. Collet (Eds.), *Medicinal plants for lung diseases*. Singapore: Springer. Available from https://doi.org/10.1007/978-981-33-6850-7_21.

Adetunji, C. O., Anani, O. A., Olaniyan, O. T., Inobeme, A., Olisaka, F. N., Uwadiae, E. O., & Obayagbona, O. N. (2021). Recent trends in organic farming. In R. Soni, D. C. Suyal, P. Bhargava, & R. Goel (Eds.), *Microbiological activity for soil and plant health management*. Singapore: Springer. Available from https://doi.org/10.1007/978-981-16-2922-8_20.

Adetunji, C. O., Arowora, K., Fawole, O. B., & Adetunji, J. B. (2013). Effects of coatings on storability of carrot under evaporative coolant system. *Albanian Journal of Agricultural Sciences*, *12*(3).

Adetunji, C. O., Egbuna, C., Tijjani, H., Adom, D., Tawfeeq Al-Ani, L. K., & Patrick-Iwuanyanwu, K. C. (2020). *Homemade preparations of natural biopesticides and applications. Natural remedies for pest, disease and weed control* (pp. 179–185). Publisher Academic Press.

Adetunji, C. O., Inobeme, A., Olaniyan, O. T., Ajayi, O. O., Olaniyan, S., & Adetunji, J. B. (2021). Application of nanodrugs derived from active metabolites of medicinal plants for the treatment of inflammatory and lung diseases: Recent advances. In K. Dua, S. Nammi, D. Chang, D. K. Chellappan, G. Gupta, & T. Collet (Eds.), *Medicinal plants for lung diseases*. Singapore: Springer. Available from https://doi.org/10.1007/978-981-33-6850-7_26.

Adetunji, C. O., Kumar, D., Raina, M., Arogundade, O., & Sarin, N. B. (2019). Endophytic microorganisms as biological control agents for plant pathogens: A panacea for sustainable agriculture. In A. Varma, S. Tripathi, & R. Prasad (Eds.), *Plant biotic interactions*. Cham: Springer. Available from https://doi.org/10.1007/978-3-030-26657-8_1.

Adetunji, C. O., Michael, O. S., Kadiri, O., Varma, A., Akram, M., Oloke, J. K., Shafique, H., Adetunji, J. B., Jain, A., Bodunrinde, R. E., Ozolua, P., & Ubi, B. E. (2021). Quinoa: From farm to traditional healing, food application, and phytopharmacology. In A. Varma (Ed.), *Biology and biotechnology of Quinoa*. Singapore: Springer. Available from https://doi.org/10.1007/978-981-16-3832-9_20.

Adetunji, C. O., Michael, O. S., Nwankwo, W., Eghonghon Ukhurebor, K., Anani, O. A., Oloke, J. K., Varma, A., Kadiri, O., Jain, A., & Adetunji, J. B. (2021). Quinoa, The next biotech plant: Food security and environmental and health hot spots. In A. Varma (Ed.), *Biology and biotechnology of Quinoa*. Singapore: Springer. Available from https://doi.org/10.1007/978-981-16-3832-9_19.

Adetunji, C. O., Michael, O. S., Rathee, S., Singh, K. R. B., Olufemi Ajayi, O., Adetunji, J. B., Ojha, A., Singh, J., & Pratap Singh, R. (2022). Potentialities of nanomaterials for the management and treatment of metabolic syndrome: A new insight. *Materials Today Advances*, *13*, 100198.

Adetunji, C. O., Michael, O. S., Varma, A., Oloke, J. K., Kadiri, O., Akram, M., Ebunoluwa Bodunrinde, R., Imtiaz, A., Adetunji, J. B., Shahzad, K., Jain, A., Ubi, B. E., Majeed, N., Ozolua, P., & Frances, N. O. (2021). Recent advances in the application of biotechnology for improving the production of secondary metabolites from Quinoa. In A. Varma (Ed.), *Biology and biotechnology of Quinoa*. Singapore: Springer. Available from https://doi.org/10.1007/978-981-16-3832-9_17.

Adetunji, C. O., Nwankwo, W., Ukhurebor, K., Olayinka, A. S., & Makinde, A. S. (2021). Application of biosensor for the identification of various pathogens and pests mitigating against the agricultural production: Recent advances. In R. N. Pudake, U. Jain, & C. Kole (Eds.), *Biosensors in agriculture: Recent trends and future perspectives. Concepts and strategies in plant sciences*. Cham: Springer. Available from https://doi.org/10.1007/978-3-030-66165-6_9.

Adetunji, C. O., Olaniyan, O. T., Akram, M., Ajayi, O. O., Inobeme, A., Olaniyan, S., Said Khan, F., & Adetunji, J. B. (2021). Medicinal plants used in the treatment of pulmonary hypertension. In K. Dua, S. Nammi, D. Chang, D. K. Chellappan, G. Gupta, & T. Collet (Eds.), *Medicinal plants for lung diseases*. Singapore: Springer. Available from https://doi.org/10.1007/978-981-33-6850-7_14.

Adetunji, C. O., Oloke, J. K., & Prasad, G. (2020). Effect of carbon-to-nitrogen ratio on eco-friendly mycoherbicide activity from Lasiodiplodia pseudotheobromae C1136 for sustainable weeds management in organic agriculture. *Environment, Development and Sustainability*, *22*, 1977–1990. Available from https://doi.org/10.1007/s10668-018-0273-1.

Adetunji, C. O., Palai, S., Ekwuabu, C. P., Egbuna, C., Adetunji, J. B., Ehis-Eriakha, C. B., Sundar Kesh, S., & Mtewa, A. G. (2021). *General principle of primary and secondary plant metabolites: Biogenesis, metabolism, and extraction. Preparation of phytopharmaceuticals for the management of disorders* (pp. 3–23). Publisher Academic Press.

Adetunji, C. O., Phazang, P., & Sarin, N. B. (2017). Significance of rhamnolipids as a biological control agent in the management of crops/plant pathogens. *Current Trends in Biomedical Engineering & Biosciences*, *10*(3), 54–55.

Adetunji, C. O., Roli, O. I., & Adetunji, J. B. (2020). Exopolysaccharides derived from beneficial microorganisms: Antimicrobial, food, and health benefits. In P. Mishra, R. R. Mishra, & C. O. Adetunji (Eds.), *Innovations in food technology*. Singapore: Springer. Available from https://doi.org/10.1007/978-981-15-6121-4_10.

Adetunji, C. O., & Varma, A. (2020). Biotechnological application of trichoderma: A powerful fungal isolate with diverse potentials for the attainment of food safety, management of pest and diseases, healthy planet, and sustainable agriculture. In C. Manoharachary, H. B. Singh, & A. Varma (Eds.), *Trichoderma: Agricultural applications and beyond. Soil biology* (vol. 61). Cham: Springer. Available from https://doi.org/10.1007/978-3-030-54758-5_12.

Adetunji, J. B., Adetunji, C. O., & Olaniyan, O. T. (2021). African walnuts: A natural depository of nutritional and bioactive compounds essential for food and nutritional security in Africa. In O. O. Babalola (Ed.), *Food security and safety*. Cham: Springer. Available from https://doi.org/10.1007/978-3-030-50672-8_19.

Adetuyi, B. O., Oluwole, E. O., & Dairo, J. O. (2015a). Chemoprotective potential of ethanol extract of ganoderma lucidum on liver and kidney parameters in plasmodium beghei-induced mice. *International Journal of Chemistry and Chemical Processes (IJCC)*, *1*(8), 29–36.

Adetuyi, B. O., Oluwole, E. O., & Dairo, J. O. (2015b). Biochemical effects of shea butter and groundnut oils on white albino rats. *International Journal of Chemistry and Chemical Processes (IJCC)*, *1*(8), 1–17.

Adewale, G. G., Olajide, P. A., Omowumi, O. S., Okunlola, D. D., Taiwo, A. M., & Adetuyi, B. O. (2022). Toxicological significance of the occurrence of selenium in foods. *World News of Natural Sciences*, *44*, 63–88.

Ahmad, A., Khan, A., Akhtar, F., Yousuf, S., Xess, I., Khan, L. A., & Manzoor, N. (2011). Fungicidal activity of thymol and carvacrol by disrupting ergosterol biosynthesis and membrane integrity against Candida. *European Journal of Clinical Microbiology & Infectious Diseases*, *30*, 41–50.

Aldred, D., Cairns-Fuller, V., & Magan, N. (2008). Environmental factors affect efficacy of some essential oils and resveratrol to control growth and ochratoxin A production by Penicillium verrucosum and Aspergillus westerdijkiae on wheat grain. *Journal of Stored Products Research*, *44*, 341–346. Available from https://doi.org/10.1016/j.jspr.2008.03.004.

Aly, S. E., Sabry, B. A., Shaheen, M. S., & Hathout, A. S. (2016). Assessment of antimycot oxigenic and antioxidant activity of star anise (Illicium verum) in vitro. *Journal of the Saudi Society of Agricultural Sciences*, *15*, 20–27.

Anfossi, L., Baggiani, C., Giovannoli, C., & Giraudi, G. (2010). Mycotoxins in food and feed: Extraction, analysis and emerging technologies for rapid and on-field detection. *Recent Patents on Food, Nutrition & Agriculture*, *2*(2), 140–153.

Avanço, G. B., Ferreira, F. D., Bomfim, N. S., Peralta, R. M., Brugnari, T., Mallmann, C. A., de Abreu Filho, B. A., Mikcha, J. M. G., & Machinski, M., Jr (2017). Curcuma longa L. essential oil composition, antioxidant effect, and effect on Fusarium verticillioides and fumonisin production. *Food Control*, *73*, 806–813.

Bansal, J., Pantazopoulos, P., Tam, J., Cavlovic, P., Kwong, K., & Turcotte, A. M. (2011). Surveys of rice sold in Canada for aflatoxins, ochratoxin A and fumonisins. *Food Additives and Contaminants*, *28*(6), 767–774.

Basilico, M. Z., & Basilico, J. C. (1999). Inhibitory effects of some spice essential oils on Aspergillus ochraceus NRRL 3174 growth and ochratoxin A production. *Letters in Applied Microbiology*, *29*, 238–241.

Bastos, L. P. H., Vicente, J., dos Santos, C. H. C., de Carvalho, M. G., & Garcia-Rojas, E. E. (2020). Encapsulation of black pepper (Piper nigrum L.) essential oil with gelatin and sodium alginate by complex coacervation. *Food Hydrocolloids*, *102*, 105605. Available from https://doi.org/10.1016/j.foodhyd.2019.105605.

Batiha, G. B., Awad, D. A., Algamma, A. M., Nyamota, R., Wahed, M. I., Shah, M. A., Amin, M. N., Adetuyi, B. O., Hetta, H. F., Cruz-Marins, N., Koirala, N., Ghosh, A., & Sabatier, J. (2021).). Diary-derived and egg white proteins in enhancing immune system against COVID-19 frontiers in nutritionr. *(Nutritional Immunology)*, *8*, 629440. Available from https://doi.org/10.3389/fnut.2021629440.

Bello, O. M., Ibitoye, T., & Adetunji, C. (2019). Assessing antimicrobial agents of Nigeria flora. *Journal of King Saud University-Science*, *31*(4), 1379–1383.

Bendaha, H., Yu, L., Touzani, R., Souane, R., Giaever, G., Nislow, C., Boone, C., El Kadiri, S., Brown, G. W., & Bellaoui, M. (2011). New azole antifungal agents with novel modes of action: Synthesis and biological studies of new tridentate ligands based on pyrazole and triazole. *European Journal of Medicinal Chemistry*, *46*, 4117–4124.

Bhushani, J. A., & Anandharamakrishnan, C. (2014). Electrospinning and electrospraying techniques: Potential food based applications. *Trends in Food Science and Technology*, *38*, 21–33. Available from https://doi.org/10.1016/j.tifs.2014.03.004.

Bluma, R., Amaiden, M. R., Daghero, J., & Etcheverry, M. (2008). Control of Aspergillus section Flavi growth and aflatoxin accumulation by plant essential oils. *Journal of Applied Microbiology*, *105*, 203–214.

Capriotti, A. L., Caruso, G., Cavaliere, C., Foglia, P., Samperi, R., & Lagana, A. (2012). Multiclass mycotoxin analysis in food, environmental and biological matrices with chromatography/mass spectrometry. *Mass Spectrometry Reviews*, *31*(4), 466–503.

CAST. (2003). Mycotoxins: Risks in plant, animal, and human system. In J. L. Richard, & G. A. Payne (Eds.), *Council for agricultural science and technology task force report* (No. 139). Ames, Iowa, USA.

Chang, S. T., Chen, P. F., & Chang, S. C. (2001). Antibacterial activity of leaf essential oils and their constituents from Cinnamomum osmophloeum. *Journal of Ethnopharmacology*, *77*, 123–127.

Chaudhari, A. K., Singh, A., Singh, V. K., Dwivedy, A. K., Das, S., Ramsdam, M. G., & Dubey, N. K. (2020). Assessment of chitosan biopolymer encapsulated α-Terpineol against fungal, aflatoxin B1 (AFB1) and free radicals mediated deterioration of stored maize and possible mode of action. *Food Chemistry*, *311*, 126010. Available from https://doi.org/10.1016/j.foodchem.2019.126010.

Chaudhari, A. K., Singh, V. K., Deepika, S. D., Prasad, J., Dwivedy, A. K., & Dubey, N. K. (2020). Improvement of in vitro and in situ antifungal, AFB1 inhibitory and antioxidant activity of Origanum majorana L. essential oil through nanoemulsion and recommending as novel food preservative. *Food and Chemical Toxicology: An International Journal Published for the British Industrial Biological Research Association*, *111536*, 143. Available from https://doi.org/10.1016/j.fct.2020.111536.

Chaudhari, A. K., Singh, V. K., Dwivedy, A. K., Das, S., Upadhyay, N., Singh, A., Dkhar, M. S., Kayang, H., Prakash, B., & Dubey, N. K. (2018). Chemically characterised Pimenta dioica (L.) Merr. essential oil as a novel plant based antimicrobial against fungal and aflatoxin B1 contamination of stored maize and its possible mode of action. *Natural Product Research*, 1–5. Available from https://doi.org/10.1080/14786419.2018.1499634.

Chen, Y., Zeng, H., Tian, J., Ban, X., Ma, B., & Wang, Y. (2013). Antifungal mechanism of essential oil from Anethum graveolens seeds against Candida albicans. *Journal of Medical Microbiology*, *62*, 1175–1183.

Chen, Z. Y., Brown, R. L., Damann, K. E., & Cleveland, T. E. (2004). Identification of a maize kernel stress-related protein and its effect on aflatoxin accumulation. *Phytopathology*, *94*, 938–945.

Cowan, M. M. (1999). Plant products as antimicrobial agents. *Clinical Microbiology Reviews*, *12*, 564–582.

da Cruz, C. L., Pinto, V. F., & Patriarca, A. (2013). Application of plant derived compounds to control fungal spoilage and mycotoxin production in foods. *International Journal of Food Microbiology*, *166*, 1–14.

da Silva, B. N., Nakassugi, L. P., Oliveira, J. F. P., Kohiyama, C. Y., Mossini, S. A. G., Grespan, R., Nerilo, S. B., Mallmann, C. A., Abreu Filho, B. A., & Machinski, M., Jr (2015). Antifungal activity and inhibition of fumonisin production by Rosmarinus officinalis L. essential oil in Fusarium verticillioides (Sacc.) Nirenberg. *Food Chemistry, 166,* 330−336.

Dammak, I., Hamdi, Z., El Euch, S. K., Zemni, H., Mliki, A., Hassouna, M., & Lasram, S. (2019). Evaluation of antifungal and anti-ochratoxigenic activities of Salvia officinalis, Lavandula dentata and Laurus nobilis essential oils and a major monoterpene constituent 1, 8-cineole against Aspergillus carbonarius. *Industrial Crops and Products, 128,* 85−93.

Demirci, M., Arici, M., & Gumus, T. (2003). Presence of patulin in fruit and fruit juices produced in Turkey. *Ernaehrungs-Umschau, 50*(7), 262−263.

Didunyemi, M. O., Adetuyi, B. O., & Oyebanjo, O. O. (2019). Morinda lucida attenuates acetaminophen-induced oxidative damage and hepatotoxicity in rats. *Journal of Biomedical sciences, 8,* No. Available from https://www.jbiomeds.com/biomedical-sciences/morinda-lucida-attenuates-acetaminopheninduced-oxidative-damage-and-hepatotoxicity-in-rats.php?aid = 24482.

Didunyemi, M. O., Adetuyi, B. O., & Oyewale, I. A. (2020). Inhibition of lipid peroxidation and in-vitro antioxidant capacity of aqueous, acetone and methanol leaf extracts of green and red Acalypha wilkesiana Muell Arg. *International Journal of Biological and Medical Research, 11*(3), 7089−7094.

Dwivedy, A. K., Prakash, B., Chanotiya, C. S., Bisht, D., & Dubey, N. K. (2017). Chemically characterized Mentha cardiaca L. essential oil as plant based preservative in view of efficacy against biodeteriorating fungi of dry fruits, aflatoxin secretion, lipid peroxidation and safety profile assessment. *Food and Chemical Toxicology: An International Journal Published for the British Industrial Biological Research Association, 106,* 175−184. Available from https://doi.org/10.1016/j.fct.2017.05.043.

Egbuna, C., Gupta, E., Ezzat, S. M., Jeevanandam, J., Mishra, N., Akram, M., Sudharani, N., Adetunji, C. O., Singh, P., Ifemeje, J. C., Deepak, M., Bhavana, A., Walag, A. M. P., Ansari, R., Adetunji, J. B., Laila, U., Olisah, M. C., & Onyekere, P. F. (2020). Aloe species as valuable sources of functional bioactives. In C. Egbuna, & G. Dable Tupas (Eds.), *Functional foods and nutraceuticals.* Cham: Springer. Available from https://doi.org/10.1007/978-3-030-42319-3_18.

FAO. (2002). *Manual on the application of the HACCP system in mycotoxin prevention and control.* Rome, Italy: Joint FAO/WHO Food Standards Programme FAO.

Ferreira, F. M. D., Hirooka, E. Y., Ferreira, F. D., Silva, M. V., Mossini, S. A. G., & Machinski, M., Jr (2018). Effect of Zingiber officinale Roscoe essential oil in fungus control and deoxynivalenol production of Fusarium graminearum Schwabe in vitro. *Food Additives & Contaminants: Part A, 35,* 2168−2174.

Food and Agricultute Organization of the United Nations (FAO). (2013). FAO Rice Market Monitor, 3.

Galvano, F., Ritieni, A., & Pietri, A. (2005). Mycotoxins in the human food chain. In D. Diaz (Ed.), *Mycotoxin blue book* (pp. 187−224). Nottingham, U.K.: Nottingham Univ. Press.

Goertz, A., Zuehlke, S., Spiteller, M., Steiner, U., Dehne, H. W., & Waalwijk, C. (2010). Fusarium species and mycotoxin profiles on commercial maize hybrids in Germany. *European Journal of Plant Pathology, 128,* 101−111.

Hasheminejad, N., Khodaiyan, F., & Safari, M. (2019). Improving the antifungal activity of clove essential oil encapsulated by chitosan nanoparticles. *Food Chemistry, 275,* 113−122. Available from https://doi.org/10.1016/j.foodchem.2018.09.085.

Hocking, A. D., Leong, S. L., Kazi, B. A., Emmett, R. W., & Scott, E. S. (2007). Fungi and mycotoxins in vineyards and grape products. *International Journal of Food Microbiology, 119,* 84−88.

Hu, Y., Zhang, J., Kong, W., Zhao, G., & Yang, M. (2017). Mechanisms of antifungal and anti-aflatoxigenic properties of essential oil derived from turmeric (Curcuma longa L.) on Aspergillus flavus. *Food Chemistry, 220,* 1−8. Available from https://doi.org/10.1016/j.foodchem.2016.09.179.

Iamanakaa, B. T., de Menezes, H. C., Vicentea, E., Leitea, R. S., & Taniwaki, M. H. (2007). Aflatoxigenic fungi and aflatoxins occurrence in sultanas and dried figs commercialized in Brazil. *Food Control, 18,* 454−457.

James-Okoro, P. O., Iheagwam, F. N., Sholeye, M. I., Umoren, I. A., Adetuyi, B. O., Ogundipe, A. E., Braimah, A. A., Adekunbi, T. S., Ogunlana, O. E., & Ogunlana, O. O. (2021). Phytochemical and in vitro antioxidant assessment of Yoyo bitters. *World News of Natural Sciences, 37,* 1−17.

Jard, G., Liboz, T., Mathieu, F., Guyonvarc'h, A., & Lebrihi, A. (2009). Adsorption of zearalenone by Aspergillus japonicasconidia: New trends for biological ZON decontamination in animal feed. *World Mycotoxin Journal, 2,* 391−397.

Jarda, G., Liboz, T., Mathieua, F., Guyonvarc'h, A., & Lebrihi, A. (2011). Review of mycotoxin reduction in food and feed: From prevention in the field to detoxification by adsorption or transformation. *Food Additives and Contaminants, 28*(11), 1590−1609.

Kalapos, M. P. (1999). Methylglyoxal in living organisms: Chemistry, biochemistry, toxicology and biological implications. *Toxicology Letters, 110,* 145−175.

Karim, M., Fathi, M., & Soleimanian-Zad, S. (2020). Nanoencapsulation of cinnamic aldehyde using zein nanofibers by novel needle-less electrospinning: Production, characterization and their application to reduce nitrite in sausages. *Journal of Food Engineering, 288,* 110140. Available from https://doi.org/10.1016/j.jfoodeng.2020.110140.

Kedia, A., Dwivedy, A. K., Jha, D. K., & Dubey, N. K. (2016). Efficacy of Mentha spicata essential oil in suppression of Aspergillus flavus and aflatoxin contamination in chickpea with particular emphasis to mode of antifungal action. *Protoplasma, 253,* 647−653.

Kedia, A., Prakash, B., Mishra, P. K., & Dubey, N. K. (2014). Antifungal and antiaflatoxigenic properties of Cuminum cyminum (L.) seed essential oil and its efficacy as a preservative in stored commodities. *International Journal of Food Microbiology, 168,* 1−7.

Khan, A., Ahmad, A., Akhtar, F., Yousuf, S., Xess, I., Khan, L. A., & Manzoor, N. (2010). Ocimum sanctum essential oil and its active principles exert their antifungal activity by disrupting ergosterol biosynthesis and membrane integrity. *Research in Microbiology, 161,* 816−823.

Khosravi, A. R., Shokri, H., & Mokhtari, A. R. (2015). Efficacy of Cuminum cyminum essential oil on FUM1 gene expression of fumonisinproducing Fusarium verticillioides strains. *Avicenna Journal of Phytomedicine, 5,* 34.

Kim, H. M., Kwon, H., Kim, K., & Lee, S. E. (2018). Antifungal and antiaflatoxigenic activities of 1, 8-cineole and t-cinnamaldehyde on Aspergillus flavus. *Applied Sciences, 8*, 1655.

Liang, Z. (2008). *Detection of OTA and analysis of ochratoxigenic fungi in foodstuff [dissertation]*. Beijing: China: Agricultural University.

Liew, S. N., Utra, U., Alias, A. K., Tan, T. B., Tan, C. P., & Yussof, N. S. (2020). Physical, morphological and antibacterial properties of lime essential oil nanoemulsions prepared via spontaneous emulsification method. *LWT - Food Science and Technology (Lebensmittel-Wissenschaft-Technology), 128*, 109388. Available from https://doi.org/10.1016/j.lwt.2020.109388.

Mahmoud, A. L. (1994). Antifungal action and antiaflatoxigenic properties of some essential oil constituents. *Letters in Applied Microbiology, 19*, 110−113.

Makun, H. A., Gbodi, T. A., Akanya, O. H., Salako, E. A., & Ogbadu, G. H. (2007). Fungi and some mycotoxins contaminating rice (Oryza sativa) in Niger State, Nigeria. *African Journal of Biotechnology, 6*(2), 99−108.

Marin, S., Magan, N., Ramos, A. J., & Sanchis, V. (2004). Fumonisin-producing strains of fusarium: A review of their ecophysiology. *Journal of Food Protection, 67*(8), 1792−1805.

Marin, S., Mateo, E. M., Sanchis, V., Valle-Algarr, F. M., Ramos, A. J., & Jiménez, M. (2011). Patulin contamination in fruit derivatives, including baby food, from the Spanish market. *Food Chemistry, 124*, 563−568.

Mazarei, Z., & Rafati, H. (2019). Nanoemulsification of Satureja khuzestanica essential oil and pure carvacrol; comparison of physicochemical properties and antimicrobial activity against food pathogens. *LWT - Food Science and Technology (Lebensmittel-Wissenschaft-Technology), 100*, 328−334. Available from https://doi.org/10.1016/j.lwt.2018.10.094.

McCallum, J. L., Tsao, R., & Zhou, T. (2002). Factors affecting patulin production by Penicillium expansum. *Journal of Food Protection, 65*(12), 1937−1942.

Mekawey, A. A., & El-Metwally, M. M. (2019). Impact of nanoencapsulated natural bioactive phenolic metabolites on chitosan nanoparticles as aflatoxins inhibitor. *Journal of Basic Microbiology, 59*, 599−608.

Moss, M. O. (2008). Fungi, quality and safety issue in fresh fruits and vegetables. *Journal of Applied Microbiology, 104*, 1239−1243.

Nesci, A., Rodriguez, M., & Etcheverry, M. (2003). Control of Aspergillus growth and aflatoxin production using antioxidants at different conditions of water activity and pH. *Journal of Applied Microbiology, 95*, 279−287.

Nguyen, M. T., Tozlovanu, M., Tran, T. L., & Prohl-Leszkowicz, A. (2007). Occurrence of aflatoxin B1, citrinin and ochratoxin A in rice in five provinces of the central region of Vietnam. *Food Chemistry, 105*, 42−47.

Norton, R. A., & Dowd, P. F. (1996). Effect of steryl cinnamic acid derivatives from corn bran on Aspergillus flavus, corn earworm larvae, and dried-fruit beetle larvae and adults. *Journal of Agricultural and Food Chemistry, 44*, 2412−2416.

Olaniyan, O. T., & Adetunji, C. O. (2021). Biological, biochemical, and biodiversity of biomolecules from marine-based beneficial microorganisms: Industrial perspective. In C. O. Adetunji, D. G. Panpatte, & Y. K. Jhala (Eds.), *Microbial rejuvenation of polluted environment. Microorganisms for sustainability* (vol 27). Singapore: Springer. Available from https://doi.org/10.1007/978-981-15-7459-7_4.

Park, J. W., Choi, S. Y., Hwang, H. J., & Kim, Y. B. (2005). Fungal mycoflora and mycotoxins in Korean polished rice destined for humans. *International Journal of Food Microbiology, 103*, 305−314.

Paterson, R. R. M., & Lima, N. (2010). How will climate change affect mycotoxins in food. *Food Research International, 42*(7), 1902−1914.

Patkar, K. L., Usha, C. M., Shetty, H. S., Paster, N., & Lacey, J. (1993). Effect of spice essential oils on growth and aflatoxin B1 production by Aspergillus flavus. *Letters in Applied Microbiology, 17*, 49−51.

Paula, H. C., Oliveira, E. F., Carneiro, M. J., & de Paula, R. C. (2017). Matrix effect on the spray drying nanoencapsulation of Lippia sidoides essential oil in chitosan-native gum blends. *Planta Medica, 83*, 392−397. Available from https://doi.org/10.1055/s-0042-107470.

Perczak, A., Gwiazdowska, D., Marchwinska, K., Juś, K., Gwiazdowski, R., & Waśkiewicz, A. (2019). Antifungal activity of selected essential oils against Fusarium culmorum and F. graminearum and their secondary metabolites in wheat seeds. *Archives of Microbiology, 201*, 1085−1097. Available from https://doi.org/10.1007/s00203-019-01673-5.

Pinto, E., Vale-Silva, L., Cavaleiro, C., & Salgueiro, L. (2009). Antifungal activity of the clove essential oil from Syzygium aromaticum on Candida, Aspergillus and dermatophyte species. *Journal of Medical Microbiology, 58*, 1454−1462.

Rauf, A., Akram, M., Semwal, P., Mujawah, A. A. H., Muhammad, N., Riaz, Z., Munir, N., Piotrovsky, D., Vdovina, I., Bouyahya, A., Oluwaseun Adetunji, C., Shariati, M. A., Almarhoon, Z. M., Mabkhot, Y. N., & Khan, H. (2021). Antispasmodic potential of medicinal plants: A comprehensive review. *Oxidative Medicine and Cellular Longevity*. Available from https://doi.org/10.1155/2021/4889719, Article ID 4889719, 12 pages, 2021.

Ritieni, A. (2003). Patulin in Italian commercial apple products. *Journal of Agricultural and Food Chemistry, 51*, 6086−6090.

Sakuda, S., Yoshinari, T., Furukawa, T., Jermnak, U., Takagi, K., Iimura, K., Yamamoto, T., Suzuki, M., & Nagasawa, H. (2016). Search for aflatoxin and trichothecene production inhibitors and analysis of their modes of action. *Bioscience, Biotechnology, and Biochemistry, 80*, 43−54.

Shishir, M. R. I., Xie, L., Sun, C., Zheng, X., & Chen, W. (2018). Advances in micro and nanoencapsulation of bioactive compounds using biopolymer and lipid-based transporters. *Trends in Food Science and Technology, 78*, 34−60. Available from https://doi.org/10.1016/j.tifs.2018.05.018.

Silvestre, W. P., Livinalli, N. F., Baldasso, C., & Tessaro, I. C. (2019). Pervaporation in the separation of essential oil components: A review. *Trends in Food Science and Technology, 93*, 42−52. Available from https://doi.org/10.1016/j.tifs.2019.09.003.

Siva, S., Li, C., Cui, H., Meenatchi, V., & Lin, L. (2020). Encapsulation of essential oil components with methyl-β-cyclodextrin using ultrasonication: Solubility, characterization, DPPH and antibacterial assay. *Ultrasonics Sonochemistry, 64*, 104997. Available from https://doi.org/10.1016/j.ultsonch.2020.104997.

Su, H., Huang, C., Liu, Y., Kong, S., Wang, J., Huang, H., & Zhang, B. (2020). Preparation and Characterization of Cinnamomum essential oil−chitosan nanocomposites: Physical, structural, and antioxidant activities. *Processes, 8*, 834. Available from https://doi.org/10.3390/pr8070834.

Sun, Q., Wang, L., Lu, Z., & Liu, Y. (2015). In vitro anti-aflatoxigenic effect and mode of action of cinnamaldehyde against aflatoxin. *International Biodeterioration & Biodegradation, 104*, 419−425.

Tatsadjieu, N. L., Dongmo, P. J., Ngassoum, M. B., Etoa, F. X., & Mbofung, C. M. F. (2009). Investigations on the essential oil of Lippia rugosa from Cameroon for its potential use as antifungal agent against Aspergillus flavus Link ex. *Fries Food control, 20*, 161−166.

Thangadurai, D., Naik, J., Sangeetha, J., Said Al-Tawaha, A. R. M., Adetunji, C. O., Islam, S., David, M., Shettar, A. K., & Adetunji, J. B. (2021). Nanomaterials from agrowastes: Past, present, and the future. In O. V. Kharissova, L. M. Torres-Martínez, & B. I. Kharisov (Eds.), *Handbook of nanomaterials and nanocomposites for energy and environmental applications*. Cham: Springer. Available from https://doi.org/10.1007/978-3-030-36268-3_43.

Tian, J., Ban, X., Zeng, H., He, J., Chen, Y., & Wang, Y. (2012). The mechanism of antifungal action of essential oil from dill (Anethum graveolens L.) on Aspergillus flavus. *PLoS One, 7*, e30147.

Tian, J., Ban, X., Zeng, H., He, J., Huang, B., & Wang, Y. (2011). Chemical composition and antifungal activity of essential oil from Cicuta virosa L. var. latisecta Celak. *International Journal of Food Microbiology, 145*, 464−470.

Ukhurebor, K. E., & Adetunji, C. O. (2021). Relevance of biosensor in climate smart organic agriculture and their role in environmental sustainability: What has been done and what we need to do? In R. N. Pudake, U. Jain, & C. Kole (Eds.), *Biosensors in agriculture: Recent trends and future perspectives. Concepts and strategies in plant sciences*. Cham: Springer. Available from https://doi.org/10.1007/978-3-030-66165-6_7.

Upadhyay, N., Singh, V. K., Dwivedy, A. K., Das, S., Chaudhari, A. K., & Dubey, N. K. (2018). Cistus ladanifer L. essential oil as a plant based preservative against molds infesting oil seeds, aflatoxin B1 secretion, oxidative deterioration and methylglyoxal biosynthesis. *LWT - Food Science and Technology, 92*, 395−403.

Wen, P., Zong, M. H., Linhardt, R. J., Feng, K., & Wu, H. (2017). Electrospinning: A novel nano-encapsulation approach for bioactive compounds. *Trends in Food Science and Technology, 70*, 56−68. Available from https://doi.org/10.1016/j.tifs.2017.10.009.

Wichmann, G., Herbarth, O., & Lehmann, I. (2002). The mycotoxins citrinin, gliotoxin, and patulin affect interferon-γ rather than interleukin-4 production in human blood cells. *Environmental Toxicology, 17*(3), 211−218.

Williams, J. H., Phillips, T. D., Jolly, P. E., Stiles, J. K., Jolly, C. M., & Aggarwal, D. (2004). Human aflatoxicosis in developing countries: A review of toxicology, exposure, potential health consequences, and interventions. *American Journal of Clinical Nutrition, 80*, 1106−1122.

Xie, X. M., Fang, J. R., & Xu, Y. (2004). Study of antifungal effect of cinnamaldehyde and citral on Aspergillus flavus. *Food Science, 25*, 32−34.

Yoshinari, T., Yaguchi, A., Takahashi-Ando, N., Kimura, M., Takahashi, H., Nakajima, T., Sugita-Konishi, Y., Nagasawa, H., & Sakuda, S. (2008). Spiroethers of German chamomile inhibit production of aflatoxin G1 and trichothecene mycotoxin by inhibiting cytochrome P450 monooxygenases involved in their biosynthesis. *FEMS Microbiology Letters, 284*, 184−190.

Yu, J., Chang, P. K., Ehrlich, K. C., Cary, J. W., Bhatnagar, D., Cleveland, T. E., & Bennett, J. W. (2004). Clustered pathway genes in aflatoxin biosynthesis. *Applied and Environmental Microbiology, 70*, 1253−1262.

Chapter 14

The synergetic effect of nanomaterials together with essential oils for extending the shelf life of food products

Babatunde Oluwafemi Adetuyi[1], Pere-Ebi Yabrade Toloyai[2], Peace Abiodun Olajide[1], Oluwakemi Semiloore Omowumi[1], Charles Oluwaseun Adetunji[3], Osarenkhoe Omorefosa Osemwegie[4], Mohammed Bello Yerima[5], Juliana Bunmi Adetunji[6], M.L. Attanda[7], Olatunji Matthew Kolawole[8], Abel Inobeme[9] and Oluwabukola Atinuke Popoola[10]

[1]Department of Natural Sciences, Faculty of Pure and Applied Sciences, Precious Cornerstone University, Ibadan, Oyo State, Nigeria, [2]Department of Medical Biochemistry, Faculty of Basic Medical Sciences, Delta State University, Abraka, Delta State, Nigeria, [3]Applied Microbiology, Biotechnology and Nanotechnology Laboratory, Department of Microbiology, Edo State University Uzairue, Iyamho, Edo State, Nigeria, [4]Department of Food Science and Microbiology, Landmark University, Omu-Aran, Kwara State, Nigeria, [5]Department of Microbiology, Sokoto State University, Sokoto, Sokoto State, Nigeria, [6]Department of Biochemistry, Osun State University, Osogbo, Osun State, Nigeria, [7]Department of Agricultural Engineering, Bayero University Kano, Kano, Kano State, Nigeria, [8]Department of Microbiology, Faculty of Life Sciences, University of Ilorin, Ilorin, Kwara State, Nigeria, [9]Department of Chemistry, Edo State University Uzairue Iyamho, Auchi, Edo State, Nigeria, [10]Genetics, Genomics and Bioinformatics Department, National Biotechnology Development Agency, Abuja, FCT, Nigeria

Introduction

Overuse and underuse of antibiotics have led to the rise of bacteria and viruses that are resistant to many drugs. Antibiotic resistance is a growing concern around the world due to the wide variety of germs that are affected (Roca et al., 2015). Every year, the rise of multidrug-resistant bacteria threatens the lives of millions of people because antibiotic treatment becomes less effective. Because of this, the issue of multidrug resistance requires the rapid development of brand-new, alternative antimicrobials (Allahverdiyev et al., 2011; Adetunji et al., 2013a, 2013b, 2015a, 2015b, 2015c, 2019, 2021a, 2021b, 2021c, 2021d, 2021e, 2021f, 2021g, 2021h, 2021i, 2021j, 2021k, 2021l, 2021m, 2021n; Rauf et al., 2021; Ukhurebor & Adetunji 2021). Medicinal balms (EOs) associated with nanoparticles may be used to achieve synergistic antibacterial movement, which may pave the way for new therapeutic methods. Essential oils (EOs) are plant fluids with a random, natural aroma that are typically extracted from the leaves and flowers of plants. Synthesized via intricate metabolic pathways in protecting plants from various diseases, they play an important role. Many initiatives have been launched to use them to treat various microbial illnesses because of their amazing bioactivities (Adetunji et al., 2013a, 2013b, 2019, 2020a, 2020b, 2020c, 2020d, 2020e, 2020f; Adejumo et al., 2017; Alwan et al., 2016; Olaniyan and Adetunji, 2021; Adetunji, 2008, 2022a, 2022b, 2022c, 2022d, 2022e, 2022f, 2022g; Adetunji & Varma, 2020; Adetunji et al., 2017; Bello et al., 2019; Egbuna et al., 2020; Thangadurai et al., 2021). EO production primarily occurs in the families of Myrtaceae, Myristicaceae, Piperaceae, Rutaceae, Asteraceae, Lamiaceae, etc. Complex and variable chemical compositions of EO, affected by factors like local climate, seasonality, and experimental circumstances, may account for their wide-ranging biological activity (Rasooli et al., 2005). Scientific investigations have proven that blending two or more EOs produces a synergistic impact (Padalia et al., 2015). In addition, using it with antimicrobial drugs, EOs were demonstrated to exert a great deal of impact (Duarte et al., 2016). EO components, when combined with other antimicrobial agents, can allow the passage of antimicrobial compounds via cell membrane channels. Vapors of EOs are also the subject of intense research these days because of the potential applications their antibacterial impact has in cleaning the environment, preserving food, and treating wounds (Laird and Phillips, 2012). Moreover, a variety of plant-based EOs can be used as healthy food supplements.

Applications of Essential Oils in the Food Industry. DOI: https://doi.org/10.1016/B978-0-323-98340-2.00009-2

Nanotechnology, one of the most promising new innovations, has helped spark a movement. Due to their exceptional antibacterial properties and significant surface area-to-density ratio, nanoparticles are useful tools for treating a wide range of illnesses, especially those caused by microbes. In particular, metal nanoparticles serve multiple essential functions in the biological sciences (Skadanowski et al., 2016). Here, silver nanoparticles have proven their worth and are expected to usher in a new era of antimicrobials (Rai et al., 2016). In addition to silver and gold, nanoparticles of platinum and palladium have also demonstrated a variety of organic motions (Rai et al., 2016).

Essential oils as novel antimicrobial agent

There have been a plethora of studies looking at the efficacy of various EOs in warding off harmful and spoilage-causing microbes in both humans and food (Alwan et al., 2016). Several EOs have shown promise in antimicrobial research as a means of combating pathogenic and multidrug-resistant microorganisms (Didunyemi et al., 2020; Knezevic et al., 2016). The extraction method determines the substance profile of the EO, which varies in terms of the total number and stereochemistry of its constituents. EOs could be synthesized in varying quantities and structural forms depending on the specific plant part, age, stage of growth, environmental conditions, and soil type (Angioni et al., 2006).

Oregano (60%−74%) and Thyme (45%−55%) EOs are primarily composed of carvacrol (Arrebola et al., 1994). Carvacrol-rich EOs are shown to have potent antibacterial effects (Magi et al., 2015). Caryophyllene (24.1%), beta-phellandrene (16%), and eucalyptol (15.6%) isolated from L. angustifolia EO, and camphor (32.7%) and eucalyptol (26.9%) isolated from lavandin EO, all exhibited considerable antibacterial activity (Lavandula x intermedia).

There were several investigations into the antibacterial properties of EOs, and they have been found to be effective against many different types of bacteria, most of which are resistant to conventional antibiotics, those that are present in food, and those that are found in the mouth and its biofilms. Possibly useful against the oral germs that cause tooth caries, EOs contain antibacterial capabilities. Oregano and thyme EOs have been shown to be effective against multidrug-resistant Enterococcus faecalis, the causative agent of chronic oral infections, according to a 2014 study by Benbelaid et al. In a comprehensive analysis published in 2015, Freires et al. considered menthol and eugenol to be the most effective chemicals in combating cariogenic bacteria like streptococci and lactobacilli.

In addition to yeasts, several other pathogenic fungi are vulnerable to different EOs (Mekonnen et al., 2016). Additionally, a number of EOs have been discovered to exhibit antiviral activity against various viruses. *Cymbopogon flexuosus* (lemongrass) and *Chamaemelum nobile* (Roman chamomile) EOs showed significant antibacterial activity against two different bacteriophages (Chao et al., 2000; Munir et al., 2022). Vimalanathan & Hudson (2014) examined the efficacy of numerous EOs as well as other isolated compounds against viruses in both gaseous and liquid forms. Pre- and post-exposure treatments with *Trachyspermum ammi* (ajwain) oil at 0.5 mg/mL inhibited Japanese encephalitis virus (JEV) infection by 80% and 40%, respectively, according to research published in 2015. Therefore, EOs produced from plants can serve as antibacterial experts in food safety and environmental applications, with the method of effect depending mostly on EO's composition. EOs are believed to have low or no toxicity, making them a promising new class of antimicrobials. Some of the most significant in vivo research conducted on various EOs are mentioned here and briefly summarized. EOs from agarwood (*Aquilaria crassna*) were recently shown to be harmful in vivo in both female and male Swiss mice by Dahham et al. (2016). There was a primary focus on premium research into acute and long-term toxicity. The results of the severe toxicity test might be used to determine the median lethal dose (LD50), or the dose at which 50% of test subjects died. Mice were dosed with 2000 mg/kg for 14 days, and both males and females survived. None of the guinea pigs displayed symptoms of indifference, mania, inebriation, vomiting, diarrhea, excessive salivation, balding, fear, spasms, torpor, or illness.

The LD50 for oral toxicity has been determined at above 2000 mg/kg, as the study indicated that this dose was safe for EOs. *Etlingera fenzlii* EO's systemic and dermal toxicity was studied by Sudhakaran and Radha (2016) in Wistar albino rats. The animals given doses of 175, 550, and 2000 mg/kg showed no signs of toxicity, death, or significantly different body weights or health indices. In addition, the cutaneous assimilation test using medicinal balm did not find any differences in body weight, food consumption, or water intake in relation to treatment. Similar data were reported on an EO obtained from *Blumea eriantha* in an in vivo skin retention and sub-intense harmfulness study (Nazir et al., 2022; Pednekar et al., 2013). In the human plasma utilized for the cutaneous absorption study, not a single constituent of the EOs was found. When comparing the body mass index, caloric intake, hematological, and biochemical parameters of the female and male groups after sub-acute therapy, no significant changes were found. In contrast, only high-dose (15%)-treated animals showed skin irritation in the first week of the trial. The carcinogenicity of EOs has also been highlighted in a few other investigations (Dweck, 2009). When considering factors like percentage use, product kind, application, toxicological data, and the intended user, EOs can be utilized safely.

Nanotechnology in delivery of essential oils

Particles with dimensions between 1 and 100 nm are considered nanoparticles. These particles outperform their larger counterparts in a number of important areas, such as increased reactivity, awareness, surface area, reliability, and strength. Au, Ag, Pt, Fe, Cu, and Zn are just some of the many nanomaterials that have been used to create EO-stacked NPs with antibacterial capabilities in an effort to counteract the drawbacks of using EOs. Rejuvenating balms (EOs) benefit from nanoparticles because they prevent the solidity, flavor maintenance, and practical existence of the product from degrading due to intensity and UV corruption. In addition, NPs allow for the delayed, beneficial effects of EOs to be managed upon their arrival. By increasing the EOs' ability to diffuse through biological membranes, NPs loaded with EOs exhibit synergistic antimicrobial action.

Improvement of functional attributes of essential oils

The chemical instability and sensitivity to changes in atmospheric conditions make it difficult to use EOs or their components. Products containing EOs that are sensitive to heat and/or oxygen may lose their potency or even be harmful to the consumer if the EOs degrade during production, shipping, storage, or use. One novel approach to addressing these challenges is nanoencapsulation (Lammari et al., 2020). By protecting EOs and bioactive components from the elements, nanoencapsulation provides a novel approach to extending their shelf life (Tiwari et al., 2020). As an added bonus, exemplification makes EOs less volatile and harmful. Rejuvenating balms' waterproofing properties and natural exercises make them a useful alternative for the production of bundling films, transforming bundling into a functional material and increasing its value. Protection from water retention and high mechanical strength were demonstrated by wheat flour and other grain products packaged in a cellulose-based dynamic bundling material manufactured from eugenol derivatized with polycarboxylic corrosive. It possessed potent pesticide and insectifuge properties, too, so that the product could stay fresh for longer without losing any of its original flavor, fragrance, or surface. The active biodegradable films containing the encapsulated EOs were found to be cytotoxic-free, and they successfully maintained the food products' original flavor and aroma during storage. The physical or chemical connection between the epitome and the framework aids in preserving EOs for longer periods of time. This has led to the development of methods for incorporating EOs into makeup that account for their distribution via mechanical impacts. However, a well-controlled release is all that is necessary for flavor exemplification in culinary applications. Incorporating lipids into the various frameworks used to build bundling materials is made easier by this cycle, which prevents detrimental interactions from forming in both the bilayer membrane as well as the network, which are typically polar. Nanomaterials made from natural balms, such as chitosan, have been shown to be safe and effective. Without any evidence of hemolysis or necrosis in mammalian cells, Jamil et al. (2016) proved that chitosan NPs loaded with cardamom oil were safe to use.

Synthesis of essential oil-loaded nanoparticles

Researchers have devised a variety of techniques over the past few decades for fabricating stacked nanoparticles from organic creams. Small changes in approach can have large effects on the outcome because the size, shape, and surface science of NPs are so crucial in determining their behavior when doped with EOs.

Nine distinct approaches have been reported in the literature for manufacturing nanoparticles laden with EOs. Nanoparticles with the desired shape, dimension, surface morphology, and physicochemical stability could be generated by any of the approaches. High-pressure granulation has largely replaced older methods of making stable lipid nanoparticles (SLNs) and nanostructured lipid carriers (NLCs) due to its scalability, ease of manufacture, paucity of polar compounds, and short processing timelines. The process of thermal maceration is commonly used to create SLNs and NLCs infused with EOs. The drawbacks of this technique include substance deterioration, aqueous phase EO loss, and unanticipated fatty acid migrations (Duong et al., 2020). Because it is simple to prepare and modify, and because it requires relatively few hazardous solvents, the ionic gelation process is frequently utilized for the production of polymeric nanoparticles. Coprecipitation is widely employed because it is a more efficient, less expensive, and scalable way of generating EO-loaded magnetite/metal nanoparticles (Cruz et al., 2018). Because of its simplicity and ability to produce uniformly sized nanoparticles of silica, the Stöber method is widely used to generate EO-loaded silica nanoparticles (Narayan et al., 2018).

Coprecipitation method

Coprecipitation is widely used to combine metal nanomaterials, magnetite nanocomposites, and inorganic nanoparticles because of its low cost, reasonable response states, and lack of toxicity. The use of chlorides, perchlorates, sulfates, and nitrates supplied under the direction of a base, along with NaOH, KOH, or NH_3OH in a soluble, can be used to create nanoparticles by nucleation and grain formation. Metal and inorganic NPs' size, shape, and appealing qualities can be affected by a wide range of environmental variables, including temperature, pH, ionic strength, and mixing rate (Mihai et al., 2020). Iron precursor was co-hasten in a soluble patchouli oil arrangement to create a bioactive coating of magnetite nanoparticles (NPs). Coprecipitation was used to create magnetite nanoparticles that resemble oils such as peppermint, lavender, basil, mentha, patchouli, vanilla, ylang, and cinnamon.

High-pressure homogenization method

In this method, droplets and particles are made smaller by subjecting them to high pressure.

The hot high-pressure homogenization approach

Natural ointments are added to the mixture and either broken up or circulated uniformly throughout the liquid lipids at a temperature that is $5°C-10°C$ over the softening point of the robust lipids. A pre-emulsion is created by gradually adding liquid lipid to a hot watery stage containing surfactants, and then homogenizing the mixture with a cylinder hole homogenizer. Homogenization at $500-1500$ bars for 35 patterns yields SLNs and NLCs. After homogenization, cooling the nanoemulsions causes lipid crystallization. Pomegranate seed oil and menthol are two examples of natural lipid compounds that were synthesized using this technology; their molecules had diameters between 103.15 and 117.8 nm with a polydispersity index (PDI) of 0.3.

The cold high-pressure homogenization method

Fluid nitrogen or dry ice is then used to rapidly cool the liquid lipids, which helps disperse the therapeutic oils throughout the lipids. Lipid microparticles (with a size of $50-100$ m) are obtained by processing or mortaring lipid-EO mixes, to make SLNs and NLCs, they are first homogenized at 600 bar strain at $0°C-4°C$ for $6-15$ cycles before being floated in cold watery solutions containing surfactants (Duong et al., 2020). Nanoparticles containing *Melaleuca alternifolia* oil were synthesized with a PDI of 0.25 and a size of less than 300 nm.

Mixing and ultrasonication at high velocities

Mixing and ultrasonication at High Velocities are linked to reduced complexity and organic solvent use. EOs and lipids (which have been heated above their solid melting point) are combined in this process. In order to make a hot pre-emulsion, the aqueous surfactants must be treated to the exact heating rate as the molten lipid, which is then stirred constantly. Using a probe sonicator, the prepared emulsion is ultrasonically processed nine times, each time for 30 seconds followed by 15 seconds of rest. To acquire NLCs and SLNs (Bazzaz et al., 2018), the finished products are brought down to room temperature.

Ionic gelation method

To form an ionic gel, polyelectrolytes must be able to cross-link mostly in predominance of counterions (Giri et al., 2013). A solution comprising contra electrons and stabilizing agent is added to a solution containing EOs and composite that has been dispersed in a substrate that is moderately acidic or water, respectively. Particles with a spherical form are the product of complexation between negatively and positively charged species, which causes gelation and precipitation. The resulting solution is reduced to the nanometric range by sonication and freeze-drying for one hour at $-30°C$ (Adetunji et al., 2020a, 2020b, 2020c; Krishnamoorthy & Mahalingam, 2015). Using the ionic gelation technique, nanocapsules containing oils of lemongrass, chamomile, and tea tree were created.

Miniemulsion polymerization method

Using a suitable surfactant and co-stabilizer, miniemulsion polymerization methods can produce cross-linked polymeric nanoparticles. In order to avoid Ostwald ripening and collisional disintegration, this method produces aqueous dispersions of tiny monomeric droplets that are densely packed and dispersed. The particle size produced by miniemulsion polymerization is affected by the proportions of the activator, co-stabilizer, and agitator, as well as the time spent sonicating the mixture. Small, 50–500 nm-diameter monomer droplets are generated by high shear mixing and operate as standalone nanoreactors during the synthesis of polymer nanoparticles. Since this approach prevents mass mobility between monomer droplets, it allows for the incorporation of components into the organic phase before shear mixing. Using miniemulsion polymerization, polythioether NPs loaded with thymol and carvacrol had an encapsulation efficiency of greater than 95% and a loading capacity of 50% w/w (Amato et al., 2016). Nanoparticles based on a co-polymer of methylmethacrylate and triethylene glycol dimethacrylate were used to manufacture D-limonene.

Nanoprecipitation method

Nanoprecipitation can cover the interface of a polymer after semi-polar solvents that are water-miscible are removed from a lipophilic solution, causing the polymer to precipitate and solidify. The nanoprecipitation technique employs a solvent and a non-solvent (water) in two distinct phases. The EO and polymer are dissolved in the extraction liquid before being mixed with the aqueous phase containing the stabilizer. To put it another way, when the interfacial tension between the two phases is low, the organic solvent can more easily dissolve into the aqueous one. At the intersection between the extraction liquid and the dispersion medium, NPs form instantly as droplets with an extremely narrow size distribution due to the interaction between solvent flow, diffusion, and surface tension. Freeze-drying the obtained nanoparticles with a cryoprotectant concentration of mannitol (often around 5%) yields a nanoparticle powder. The production of submicron nanoparticles with excellent encapsulation efficiencies and narrow size distributions (Froiio et al., 2019) are just a few of the advantages of using the nanoprecipitation method to encapsulate essential oil.

Spray-drying technique

Spray-drying is a typical manufacturing procedure for transforming a liquid into a powder. Vibrating mesh technology is used in spray dryers to create fine droplets using rotary atomizers and pressure nozzles. By using a peristaltic pump, polymer solutions may be injected to create polymeric complexes. An oil-surfactant emulsion is prepared independently and then added to the polymer complexes at different concentrations throughout time. Some of the process variables that could affect the final product include the temperature of the input, the temperature of the output, the pump feed flow, and the flow of air (Abreu et al., 2012).

Stöber process

Using a chemical process, the Stöber technique produces silica nanoparticles. Volatile solutions, water, and ammonia form the moderately basic alkali solution necessary for the Stöber process, which hydrolyzes and condenses a mixture of alkoxysilanes such tetraethyl orthosilicate. This process leads to the development of silica nanomaterials. Mesoporous silica nanoparticles (MSNs) are made using the sol-gel method, which is essentially an improved version of Stöber's original method. An acidic or basic catalyst causes the alkoxide monomers to hydrolyze and condense into slurry. A mucilaginous matrix of polymers or individual particles can form from this solution. Another variation of this technique involves the addition of an ionic liquid to the precursor solution, which results in the formation of spherical particles with a monodisperse size distribution below 1 micron (Das et al., 2020). As a result of using this technique, silica nanomaterials with sizes between 20 and 500 nm were created.

Hydration, adsorption, and vacuum pulling techniques for a thin film

Using the coating hydration technique, a fatty film is created in a round-bottom jar as the solvent evaporates. Absolute ethanol is used to dissolve the lipids and EO in the flask's circular bottom. The resulting solution is desolvated in a rotary evaporator at low pressure to produce a film. The water is then used to saturate the film. Aromatic oils are affixed to nanoparticles by a process called adsorption. ZnO nanoparticles were adsorbed with oils of rosemary, carvacrol, eugenol, and fennel, while metal nanoparticles were loaded with linalool, Thymus vulgaris, and Nigella sativa. In the vacuum pulling process, the Halloysite nanotubes and EO are sonicated together. Air was removed from the inside surfaces of halloysite nanotubes (HNTs) by filtering the mixture under a vacuum and keeping it that way for 30 minutes. Capillary action was used to feed EO into the interior of HNTs until compressed air was attained, and then the oil was removed using spinning and filtering. Halloysite nanotubes containing thyme were created by vacuum pulling.

Nanoparticles as carriers of essential oil

When considering shape, size, and chemical properties, nanoparticles may be broken down into a number of distinct categories. Below, we'll go through the several EO-loaded NPs that can be used for antibacterial purposes.

Inorganic nanoparticles

As the name implies, inorganic nanostructures are fabricated from inorganic substances. They can withstand more rigorous conditions than organic materials while still being soluble in water and nontoxic. Due to their adaptability, inorganic nanomaterials can be used for many different purposes. Because of their structural similarity to human hard tissue and other advantageous properties like nontoxicity, biocompatibility, and osteoconductivity, hydroxapatite ($Ca_5[PO_4]_3OH)_2$ nanomaterials as graft substrates and medication carriers is widespread. Because of its bioactivity and surface chemistry, hydroxyapatite stimulates and regulates the regeneration of bone tissue. As a result, combining hydroxyapatite and EOs to treat infections brought on by trauma or bone fractures is an effective antibacterial strategy (Badea et al., 2019). EOs have been mixed with nanoparticles of TiO_2 and ZnO to boost the quality of material packaging. Both alone and in conjunction with EOs, TiO_2 nanoparticles have shown potent antibacterial action. Since of its photocatalytic activity, TiO_2 may be employed in packaging because it removes oxygen and water. TiO_2 optimum tensile rigidity, thermal expansion, porosity, Anti-glare, and antimicrobial properties are all advantages for biopolymer-based packaging materials (Alizadeh-Sani et al., 2020). It has been reported that zinc oxide nanoparticles (ZnO-NPs) exhibit strong antibacterial properties, low toxicity, a UV-filtering characteristic, and high thermal stability. Combining ZnO-NPs with EOs has been shown to increase their antibacterial action (Mizielínska et al., 2021; Olajide et al., 2022c).

In order to create their peppermint oil-loaded Hap nanoparticles (Hap-P), Badea et al. first produced Hap nanoparticles using the coprecipitation approach, which yielded ellipsoidal-shaped particles measuring 19.58 ± 2 nm (Hap) and 21.07 ± 6 nm (Hap-P), respectively (Hap-P). The antibacterial investigation showed that the proliferation of microbes was just not affected by Hap nanoparticles. Hap-P was found to have minimum inhibitory concentrations (MICs) of 31.25 L/mL against *P. aeruginosa*, *C. parapsilosis*, *E. faecium*, *E. coli*, *S. aureus*, and MRSA, 125 L/mL against *E. coli*, and 250 L/mL against *E. coli* and *S. aureus*. On the other hand, Hap-P. Therefore, the findings of the research demonstrated that putting peppermint oil on the exterior of Hap nanomaterials can help prevent infections after surgery (Badea et al., 2019). Aside from Hap therapy, all other treatments demonstrated significant antibacterial action. On the other hand, HapL's MIC values ranged from 0.20 to 0.70 mg/mL, while HapB's were greater than 5 mg/mL, indicating significantly improved antibacterial activity against all of the species tested. Enhanced antibacterial activity was seen when EOs were included in the microporous structure of Hap. This was due to an increase in the EOs' ability to depolarize membranes. As a result, the study demonstrated that EOs with antibacterial capabilities for bone regeneration can be delivered via Hap nanoparticles to reduce implant-related infections (Adetunji et al., 2022f; Predoi et al., 2018). Alizadeh-Sani et al., in a laboratory, created composites with a whey protein matrix and cellulose nanofibers. Using a casting method, the antibacterial activity of packing materials with rosemary oil droplets and TiO_2 particles as active components were examined against two resistant foodborne pathogenic microorganisms found in meat.

Lipid nanoparticles

Specifically, nanomaterials based on lipids are typically round and have an internal aqueous compartment surrounded by a lipid bilayer. The two most common types of lipid nanoparticles are solid lipid nanoparticles and nanostructure lipid carriers. Lipid nanoparticles have many desirable qualities as delivery vehicles, including their low complexity in composition, nontoxicity, self-assembly, increased biodistribution, and large payload transfer capacity. Furthermore, their physicochemical properties are easily modified to affect their biological activity. To counteract the toxicity, poor stability, and unfavorable pharmacokinetic features of EOs, encapsulating them in SLNs and NLCs may prove to be an innovative approach (Adetunji et al., 2022f; Mitchell et al., 2021).

De Souza et al. used a high-pressure homogenization method to create a dental biofilm out of nanoparticle solution of 0.3% M. alternifolia oil (NPTTO), which are spherical EO nanoparticles from *Melaleuca alternifolia*, which they then tested for antibiofilm efficacy in an in vivo model of human dental biofilm. The produced nanoparticles were measured to have a PDI of 0.242 ± 0.007, a zeta potential of 7.15 ± 0.27 mV, and a pH of 6.4 ± 0.2. Analysis of the antimicrobial properties of the solutions used to assess the in situ prototype of the sensor revealed that those containing 0.12% chlorhexidine reduced colony-forming units (CFU) by 34.2%, those containing 0.3% M. alternifolia oil reduced CFU by 51.4%, and those containing 0.4% NPTTO reduced CFU by 25.8%. Antibiofilm activity of NPTTO was confirmed by the study, which found that it inhibited commonly found bacteria such Candida species and *P. aeruginosa* (de Souza et al., 2017). For their menthol-loaded NLCs, Piran et al. used hot melt homogenization to create spheres with a particle size of 115.6 nm, PDI of 0.2, and an entrapment efficiency of 98.73%. Compared to menthol emulsion, NLC was found to have much more antibacterial activity in laboratory tests.

It has been demonstrated that the TTO nanostructure is a viable alternative to biofilm-forming bacteria (Comin et al., 2016). Saporitoetal employs an ultrasonic and homogenization technique with high shear to create EO-NLCs, lipid nanoparticles coated in eucalyptus essential oil, to speed up the healing of wounds on the skin. Cocoa butter, olive oil, or sesame oil were all used because they are all natural lipids. To prevent nanoparticles from clumping together, a surfactant called lecithin was used. These spherical nanoparticles had dimensions of 220 to 300 nm, a PDI of 0.5, and a zeta potential of -22.07 mV. When tested against Staphylococcus aureus, both free and encapsulated eucalyptus oil had a MIC of 3 mg/mL. However, the MIC values of EO-NLCs decreased from 1.9 mg/mL for free oil to 1.02 mg/mL for Streptococcus pyogenes (Saporito et al., 2017).

Liposomes

Liposomes are phospholipid vesicles that are composed of one or more concentrated lipid bilayers and have distinct aqueous compartments. Their capacity to encapsulate both hydrophilic molecules in the aqueous core and lipophilic molecules in the lipid bilayer makes them useful for a wide variety of therapeutic applications. Liposomes have a number of benefits that make them a desirable delivery system. Some of these advantages include increased accessibility and nontoxicity, self-assembly, increased payload transfer capacity, and the ease with which their bioactivity can be altered by altering their physicochemical qualities. Nanoliposomal delivery of EOs may lessen their harmful effects, halt their breakdown, and increase their bioavailability (Adetunji et al., 2022f; Sercombe et al., 2015). High amounts of ethanol and water are incorporated into phospholipids to create ethosomes, which are modified liposomes. By facilitating their passage through pores and deeper epidermal layers, ethanol makes ethosomes more effective than conventional transdermal nanocarriers (Zahid et al., 2018).

Aguilar-Perez et al. utilized thin-film hydration-sonication to make oregano essential oil-containing nanoliposomes. The results showed that the generated nanoliposomes had a particle size range of between 78.36 and 110.4 and 9.88 nm; a PDI of 0.511 ± 0.016; a zeta potential of 36.98 ± 0.46 mV; and an entrapment efficiency of $79.75\% \pm 5.9\%$. In an antifungal test performed in vitro, nanoliposomes inhibited *Trichophyton rubrum* mycelial growth by the greatest margin (81.66−86%) at a concentration of 1.5 μL/mL. So, oregano oil on nanoparticles has antifungal therapy potential, as demonstrated by their study (Aguilar-Pérez et al., 2021). Sinico et al. make use of techniques like film and sonication to create a liposomal formulation with EO from *Artemisia arborescens* L. and positive charges for the multilamellar (MLV) and unilamellar (SUV) liposomes. The resulting formulations had particle diameters of 70−150 nm and entrapment efficiencies of 60%−74%. The cytopathic effect of a virus serves as a measure of an antiviral agent's efficacy (CPE). Liposomal formulations of SUV and MLV that encapsulate EO have enhanced antiviral efficacy, as evidenced by a lower %CPE. Free EO had a CPE value of 22.86%, P90H (hydrogenated soy phosphatidylcholine) SUV had a CPE value of 21.1%, P90 (non-hydrogenated soy phosphatidylcholine) SUV had a CPE value of 8.1%, P90MLV had a CPE value of 100%, and P90HMLV had a CPE value of 100%. Researchers found that the EC50 values for P90HMLV and P90MLV were 18.5 and 43.6 mg/mL, respectively. The EC50 values of essential oil, P90HSUV, and P90SUV were higher than 100 μg/mL. In conclusion, the antiherpetic activity of MLV against HSV-1 was found to be superior to that of SUV. This suggests that the manufactured liposomes may facilitate more efficient transport of antiretroviral EOs to their intended cells (Sinico et al., 2005). Wu et al. used a film hydration strategy to make a nanoliposome composite with AgNPs and Laurel essential oil (LEO) for long-term release (Lip-LEO-AgNPs). Size (in nm), PDI (in %), and zeta potential (in mV) were measured to be 200, 0.27, and -26, respectively. Chitosantoform polyethylene (PE-CS) was incorporated into Lip-LEO-AgNPs for use in pork packing. When examined for their capacity to halt the growth of the bacteria that cause meat deterioration, PC-Lip/LEO/AgNPs showed the maximum antimicrobial property, but PE and PE-CS films showed no antimicrobial effect. Thus, the study's findings supported synergy in the refined films' application to the packaging of functional foods for longer storage (Wu et al., 2019). Pathogens can contaminate a crop at any time, from before harvest through after it has been harvested and stored or transported. Against *Colletotrichum musae*, *Colletotrichum fragariae*, and *Colletotrichum gloeosporioides*, eugenol was only able to suppress mycelia development by 83.71%, 82.90%, 86.89%, and 83.72%, while ELG-NPs were able to inhibit mycelia growth by 95.23%, 90.08%, 89.43%, and 94.19%. After six days, the ELG-NPs had completely reduced the incidence of anthracnose in postharvest loquat, demonstrating stronger antibacterial effectiveness ($>93\%$) than eugenol against fruit pathogens. Concentrations of eugenol and ELG-NPs higher than 87 μL/L were able to totally suppress bacterial growth. Because of its prolonged release profile, nanoencapsulated eugenol was more effective than eugenol at holding back the spread of disease-causing germs (Jin et al., 2019a, 2019b).

Magnetite nanoparticles

Extensive study is being conducted on magnetic iron oxide nanoparticles for use in medicine, specifically in drug delivery, MRI, and theranostics. Their usefulness originates from the fact that magnetic fields may be used to guide nanoparticles to the target site, where drugs can be heated and released in a controlled fashion. Magnetic nanoparticles are attractive for the development of EO-based antibacterial nanomaterials because they are biocompatible, biodegradable, nontoxic, and targetable.

Anghel et al. (2013) utilized the coprecipitation method to produce 5 nm-sized magnetite nanoparticles (NPs) containing *Mentha piperita* EO. To evaluate the nanosystem's antiadherence and antibiofilm properties, it was applied to the exterior of a percutaneous device. The results showed that the coated catheter device had significantly less biofilm than the untreated surface in contrast to the device that is not coated. Coprecipitation was used by Rădulescu et al. to develop a biocompatible patchouli EO solution MNP injury treatment. Researchers found that bioactive coatings significantly reduced Staphylococcus aureus biofilms, reducing viable cells embedded in biofilms by a factor of two after 24 and 48 hours of incubation. Even after 72 hours, the coating was still effective. The proposed bioactive MNP-coated wound dressing could be used to treat wound infections effectively. Bilcu et al. used the coprecipitation method to create MNPs coated in EOs of ylang-ylang, vanilla, and patchouli less than 20 nm in diameter. Both *S. aureus* and *K. pneumoniae* were significantly inhibited in their growth by their comparative antibacterial analysis. The ylang-ylang-, vanilla-, and patchouli-magnetite nanoparticles-coated catheters and the uncoated catheter had viable *S. aureus* cell counts of 1×10^3 logCFU/mL, 1×10^4 logCFU/mL, and 1×10^4 logCFU/mL, respectively, at 72 hours. Anghel et al. used a matrix-assisted pulsed laser evaporation method to determine these values for *K. pneumoniae*. In 2014, a uniform, thin coating of 9.4 nm MNPs functionalized with *Cinnamomum verum* EO (MNP-CV) was deposited on the exteriors of gastrostomy tubes (G-tubes). This study's findings revealed that MNP-CV coating significantly reduced biofilm formation. While *E. coli*'s colonization rate decreased by a factor of two for both types of biofilms, *S. aureus*' colonization rate decreased by more than a factor of four for incipient biofilms and by a factor of three for mature biofilms. Patients with disabling illnesses may benefit from MNP-CV films' antibacterial protection of medical surfaces and gadgets. Nigella sativa EO-functionalized MNPs were created by Negut et al. (2018) using a modified coprecipitation technique. The MAPLE technique was used to deposit the engineered nanoparticles on glass and silicone surfaces. MNP-NS was found to have a MIC of 1.25 mg/mL against *Escherichia coli*, 0.65 mg/mL against Staphylococcus aureus, and 0.65 mg/mL against *Candida albicans*, according to an antimicrobial susceptibility test. Mousavi et al. (2019) developed a green chemical synthesis of Fe_3O_4-MgO nanoparticles from nutmeg EO. The size of developed nanoparticles ranges from 10 to 15 nm. The antibacterial activity test revealed that the green nanoparticles were more effective at preventing the growth of bacteria. Using the coprecipitation process, Grumezescu et al., 2014 created eugenol-functionalized MNPs with a spherical shape and a particle size of less than 10 nm, which were then encapsulated in poly(3-hydroxybutyric acid co-3-hydroxyvaleric acid)polyvinyl alcohol microspheres. The MAPLE method was used to cover the glass with the microspheres. Nanocoatings have been shown to have strong anti-adherent and antibiofilm properties against *S. aureus* and *P. aeruginosa* biofilms. Production of these nanoparticles showed promise in reducing microbial adherence and biofilm formation on surgical instruments and other surfaces. Anghel et al. (2012) used coated-on wound dressings and wet chemical precipitation to create MNPs that were functionalized with eugenol and limonene. When compared to uncoated dressing, significantly less *S. aureus* and *P. aeruginosa* biofilm viable cells were detected after treatment with the nanocoated covering. Early biofilm formation and biofilm maturity were both affected by limonene- and eugenol-based nanocoats. As a result, infections caused by implanted devices may be reduced with the use of nanocoated wound dressing. Miguel et al. used a microwave precipitation method to produce EOs (*Eugenia caryophyllata* or *Rosmarinus officinalis*)-functionalized magnetite/oleic acid nanoparticles that prevented microbial adhesion to and biofilm formation on catheter samples. Different clinical Candida species, including *C. albicans*, *C. tropicalis*, *C. krusei*, and *C. glabrata*, were significantly less likely to adhere to catheter surface devices when magnetite/oleic acid nanoparticles were functionalized with *E. caryophyllata* oil. *R. officinalis* essential oil-coated nanomaterials substantially reduced biofilm formation on catheter coatings, as evidenced by both viable cell counts and confocal laser scanning imaging.

Metal nanoparticles

The properties and catalytic activity of metal NPs vary with their size as set by the ions of that metal. Owing to their bactericidal qualities, metals like gold and silver have long been utilized as antimicrobials in agribusiness and wellness. Recent developments in material science have led to an increase in their use on surface of the metal and coverings, chelating agents, and nanostructures. To be more specific, encapsulated metal nanoparticles interact with viral infections and host cells far more effectively than naked metallic nanoparticles (Maduray & Parboosing, 2021). Metal NPs and EOs, on the other hand, are not widely used due to their resistance and toxicity. Thus, this problem can be resolved by loading EOs onto metal NPs to produce synergistic antibacterial activity.

Antibacterial capabilities of linalool-modified gold nanoparticles (LIN-GNPs) and glutathione-modified gold nanoparticles (GNPs) are therefore tested for efficacy against *S. aureus*, *E. coli*, as well as a parasite (*Leishmania tropica*) by Jabir et al. (2018). Diameters of the particles were between 18 and 26 nm and 8 and 14 nm. The MIC values for co-linalool against *E. coli* and *Klebsiella pneumoniae* were 0.79 and 0.79 mg/mL, respectively, whereas those for *S. aureus* were 1.61 and 3.12 mg/mL, respectively, were found to be sensitive to linalool in all strains. The cytotoxic effect of 10 mg/mL GNPs and LIN-GNPs against *L. tropica* was shown to be 38.5%, suggesting a moderate level of antiparasitic activity, and 72.4%, indicating a significant level of antiparasitic activity. As a result, the engineered LIN-GNPs demonstrated superior antibacterial activity to that of GNPs. Ions of the metal play a key role in determining the size of metallic nanoparticles, which in turn determines their properties and catalytic activity. Owing to their bactericidal qualities, metals like gold and silver have long been utilized as antimicrobials in agribusiness and wellness. Recent developments in material science have led to an increase in their use on metallic surfaces and coverings, chelating agents, and nanostructures. In particular, metal nanoparticles with protective coatings have a significantly greater interaction with viral infections and host cells than naked metallic nanoparticles. In contrast, EOs and metal nanoparticles are not extensively employed because of their resistance and toxicity. Thus, this problem can be resolved by loading EOs onto metal NPs to produce synergistic antibacterial activity.

Nanogels

To put it simply, cross-linked polymer chains form three-dimensional networks in nanogels. Nanometer-sized hydrogel particles are responsible for giving nanogels both hydrogel and nanoparticle characteristics. Nanogels can be produced either from polymeric precursors or through the heterogeneous polymerization of monomers. Transporting biomacromolecules or small molecules with physiological activity is a popular use for nanogels. Numerous medications, both hydrophobic and hydrophilic, can be transported in the body via nanogels. The antibacterial properties of nanogels may find use in medicine (Kousalová and Etrych, 2018).

According to Mohsenabadi et al. (2018) synthetic biodegradables is a self-assembled starch-CMC film made of covalently attached nanogel of benzoic acid (BA) and chitosan (REO). The CS-BA nanoparticles loaded with REO were spherical in shape, had an entrapment efficiency of 80%, and had a particle size of less than 100 nm. A nanogel-based product was examined for its aqueous solubility, condensate permeability, structural rigidity, and flexibility. The percentages of starch, cellulose, and methylcellulose in the starch-CMC films ranged from 45.8% ± 1.6% to 52.7% ± 1.0%, 2.64 ± 0.08 to 3.04 ± 0.13 g/Pa/s/m, 0.325 ± 0.092 to 1.203 ± 0.315 MPa, and 85.57 ± 45.51 to 151.44% ± 36.50%, respectively. Results showed that CS-BA nanogel had substantial antibacterial action compared to free REO (MIC 200 g/mL), CS-BA nanogel-encapsulated REO (MIC 80 g/mL), and free REO (MIC 40 g/mL).

Polymeric nanoparticles

In the range of 10 nm to 1 m, solid colloidal particles of polymeric nanoparticles are responsible for their characteristics. Nanoprecipitation, ionic gelation, and emulsification are some of the methods used to make polymeric NPs. Drug delivery in polymeric NPs can assume a number of various forms, including encapsulation within the NP core, trapping within the matrix material, complex formation to the polymer, or adhesion to the Surface of the nanoparticles. Using an ionic gelation technique, Feyzioglu et al. created chitosan NPs containing EO from *Satureja hortensis*. The NPs had dimensions between 140.25 and 237.60 nm, zeta potentials between −21.12 and 7.54 mV, and encapsulation efficiencies between 35.07% and 40.70%. Nanomaterials were reported to have antibacterial activities of less than

0.02−12.75 ± 0.23 log10 CFU/mL, 0.02−11.88 ± 0.05 log10 CFU/mL, and 0.02−14.62 ± 0.07 log10 CFU/mL, respectively. Mohammadi et al. (2015a; 2015b) used the ionic gelation method to create ZEO-CSNPs (chitosan nanoparticles) using a sample with a size distribution of 130−180 nm, measured the oil's antifungal activity against Botrytiscinerea, the bacterium that causes gray mold disease. The drugen encapsulation efficiency of ZEO dropped from 45.24% to 3.26%, and the loading efficiency dropped from 9.05% to 5.22%, when the ZEO concentration was raised. When applied to strawberries inoculated with Botrytis, at a concentration of 1500 ppm, chitosan nanoparticles significantly reduced disease severity and prevalence during storage for 7 days at 4°C, followed by 2−3 days at 20°C. The infection rate in strawberries decreased to 16.67% at day 9, 1500 ppm of ZEO-CSNPs from 66.67% in CSNPs. These results prove that CSNPs are useful for the controlled delivery of fungicidal EOs. In order to investigate how coating dispersions containing chitosan, clove essential oil, chitosan nanoparticles, and clove essential oil-loaded chitosan nanoparticles (CEO-CSNPs) affect the shelf life and quality of minimally processed pomegranates used emulsion-ionic gelation technique to create four distinct coating dispersions (Didunyemi et al., 2019; Hasheminejad & Khodaiyan, 2020).

Nano-sized silica particles

You can find silicon nanoparticles in two distinct forms: nonporous (solid) and mesoporous (2−50 nm). All examples share the same amorphous compositions and structures. The MSNs' unique porous structure, reduced density, and greater effective surface area set them distinct from other types of nanoparticles. Nonporous silica nanoparticles are loaded with various medications through encapsulation and conjugation, while MSNs can be used for the physical and chemical adsorption of small chemicals, macromolecules, and vaccinations (Batiha et al., 2021; Ways et al., 2020). Since SNPs can be manufactured using a straightforward process, they are also budget-friendly. Importantly, the synthesis strategies can be applied to modify the structure, pore dimension and volume, and particle shape (Badea et al., 2019). MSNs can encapsulate organic compounds like EO, reducing their sensitivity and making them more stable, which improves their antibacterial and antifungal efficacy (Ruiz-Rico et al., 2017).

In this research, we looked at eight distinct nanocarriers that were able to encapsulate or adsorb EOs while maintaining their target shape and size, thereby providing controlled release, enhanced stability, and extended antibacterial action. Polymeric nanoparticles have been studied in depth for the encapsulation of EOs because of their unique properties (Liu et al., 2008). Polymers can provide the porous or cavity-like structure necessary for encapsulating EOs. Increased drug loadings (with less polymer), improved preservation of the active component, and suppressed burst release are only a few of the advantages of nanocapsules. Chitosan nanocomposites are one of the popular polymeric nanoparticles because of their great encapsulation effectiveness and limited oil release. The antibacterial activity of EO-loaded nanoparticles versus unloaded nanoparticles was compared. Antimicrobial activity was significantly increased when EO was encapsulated in polymeric nanoparticles. Lipid nanoparticles (NP), polymeric nanoparticles (NP), and silica nanoparticles (NP) were used to encapsulate clove oil, with the former showing greater antibacterial action. Capturing the tea tree oil in lipid nanoparticles increased its antimicrobial action. Some nanoparticles encapsulated with EO had extended antibacterial activity rather than increased efficacy. By inhibiting bacterial development, these nanoparticles prolonged the freshness of perishable food items. Nanoparticles with inherent antibacterial properties were also found to work synergistically with co-loaded EOs; examples include metal and magnetite nanoparticles.

Combined use of nanoparticles and essential oils increases their antibacterial potency

Carbon nanotubes, Ag, Au, Zn, chitosan, Pt, Fe, Cu, and EOs are just some of the NPs that have been put through antibacterial tests (Gaspar et al., 2017). While Schmitt et al. (2016) created a super-lattice of emulsion droplets in an emulsion containing gold nanoparticles, Duncan et al. (2015) manufactured caplets containing cinnamaldehyde and EOs of peppermint in the interior, fixed by nanoparticle fusion. Paula et al. (2016) used a variety of methods to manufacture chitosan-gum NPs infused with *Lippia sidoides* EO and reached the same conclusion. Spray-drying an emulsion of EOs and a gel mixture of different proportions yielded this nanoencapsulation. In chemical, pharmaceutical, food, and other industries, nanoencapsulations have found a number of useful applications.

Mashwani et al. (2016) reported that NPs were produced by terpenoides and other secondary metabolites of plants. The authors Gaspar et al. incorporated EOs into iron oxide NPs to produce magnetic nanofluid. Esmaeili and Asgari (2015) evaluated the biological activity and release profile of *Carum copticum* EO after encapsulating it in chitosan nanoparticles using an emulsion-ionic gelatinization technique. Hosseini et al. (2013) carried out a similar experiment by combining chitosan NPs with oregano EO to characterize the release pattern of EO particles adsorbed onto NPs and investigate the latter's antimicrobial potential. According to Li et al.'s research (2012), nanoencapsulation of thymol in zein-sodium caseinate nanoparticles increases the antimicrobial potential of thymol. To enhance the antimicrobial and antifungal properties of EOs, nanocomplexes are made with a variety of NPs. Gomes et al. (2011) investigated the potential bactericidal properties of eugenol and cinnamaldehyde against Salmonella and Listeria-made poly(DL-lactide-co-glycolide) NPs complex out of poly DL-lactide and co-glycolide. Numerous studies have utilized EO/NP combinations; Gortzi et al. (2007) utilized *Origanum dictamnus* EO and lipid-based liposome NPs to eradicate Gram-positive and Gram-negative bacteria. In addition, novel nanocomplexes have been utilized to effectively stop the growth of microorganisms. The three most frequently utilized nanocarriers in these formulations are molecular complexes, lipid-based nanoparticles, and nanoparticles based on polymers. Pedro et al. (2013) found that using EOs and nanoparticles in the biomedical field gave an in-depth look at the various formulations that are currently in use.

Single- and multiple-essential oil activity mechanisms

The composition of EO, the functional groups on its bioactive constituents, and their strong interaction all contribute to its activity (Dorman and Deans, 2000). The antibacterial activity of many EOs is credited to the presence of phenolic hydroxyl on their structures; phenolic compounds are found in many EOs and include carvacrol, thymol, and eugenol. Deleting the aromatic ring substituent from carvacrol reduced the compound's antibacterial effectiveness. 2-Amino-p-cymene, a bioactive molecule, is structurally similar to cavacrol, minus the hydroxyl group. The fact that 2-amino-p-cymene was three times less active than carvacrol proved that the hydroxyl group is crucial to carvacrol's antimicrobial properties (Veldhuizen et al., 2006). The higher antibacterial effect of carvacrol compared to other polyphenolic compounds is due to the specific location of its hydroxyl group, which is necessary for the biocompatibility of these constituents (Veldhuizen et al., 2006). EOs' lipophilic components interact readily with the dense microbial cell membrane fatty acids due to their hydrophobic nature. Different types of EO and different microbial strains each have their own unique antimicrobial mechanism of action. It's common knowledge that EOs work better on Gram-positive bacteria than they do on Gram-negative bacteria (Azhdarzadeh & Hojjati, 2016). Lipopolysaccharide is abundant in the outer membrane of Gram-negative bacteria, making it stiff and blocking the passage of hydrophobic substances. The outer layer of Gram-positive bacteria is accessible through their thick peptidoglycan wall, which is not rigid enough to withstand tiny antibacterial chemicals (Hyldgaard et al., 2012). Hydrophobic compounds of EOs are able to more easily enter the cell membrane of Gram-positive bacteria thanks to the lipophilic endings of lipoteichoic acids. Several studies suggest that the bioactive components of EOs may enter cells via the phospholipid bilayer after first attaching to the cell surface. The accumulation of these substances compromises the cellular membrane, the effects of which can be detrimental to cellular metabolism and even induce apoptosis (Bajpai et al., 2013). Zhang et al. (2016) determined the technique through which cinnamon EO suppresses the growth of *E. coli* and *S. aureus*. SEM analysis revealed that the bacterial cell membrane was disrupted when EO was administered at the MIC, and the cells were killed when EO was applied at the maximum biocidal concentration. Cell suspension samples exposed to cinnamon EO leaked electrolytes and showed a sharp rise in electric conductivity within a few hours. In a cell solution, cinnamon EO also increased the content of nucleic acids and proteins. Bacterial metabolic activity decreased by a factor of 35, as measured by the membrane potential (Zhang et al., 2016). Other researchers have observed a mechanism of action that is very similar to this one. Electrolyte leakage and the loss of internal components like proteins, reducing sugars, ATP, and DNA are the results of EOs' ability to alter membrane permeability. Additionally, EOs stop cells from making their own energy (ATP) and associated enzymes, which leads to apoptosis (Lakehal et al., 2016). To conclude, the antimicrobial activity of EOs is due to a chain reaction involving the entire bacterial cell (Macwan et al., 2016).

Carvacrol and oregano EO were found to reduce enterotoxin production in *Bacillus cereus* (Ultee & Smid, 2001) and *Staphylococcus aureus* (De Souza et al., 2010; Olajide et al., 2021a, 2021b, 2021c, 2021d, 2021e), respectively. By preventing the active components of EO from attaching to the bacterial membrane, it prevents proper trans-membrane transport by causing phospholipid bilayer disruption, it is possible to prevent the release of toxins into the environment. Additionally, individual EO components and the microorganism may have distinct targets for interacting with EOs (Boire et al., 2013). Microarray analysis revealed that methicillin-resistant Staphylococcus aureus (MRSA) was exposed to citrus oil, which caused the cell wall synthesis-inhibiting gene Cwr A to be upregulated in MRSA by a factor of 24. This effect was similar to that of antibiotics such as penicillin G, imipenem, phosphomycin, oxacillin, and vancomycin (Muthaiyan et al., 2012). The dltABCD operon, which is responsible for peptidoglycan production, also saw enhanced expression in response to citrus oil. Alanylation of teichoic acids in the cell wall, which is controlled by this operon, may jolt *S. aureus*' autolysin activity. Carson et al. (2002) found that tea tree oil induced autolysin activity, leading to the release of membrane-bound, cell wall autolytic enzymes that subsequently lysed and killed the cells. According to the findings, mixing EOs with several other biocides like alcohols or conventional sulfonamides can boost EOs' antibacterial efficacy (Moon et al., 2011), and nanomaterials, the next generation of antibacterial agents (Rai et al., 2009). Delaquis et al. (2002) observed that mixing cilantro and eucalyptus EO fractions had a synergistic impact, decreasing the MIC value of *Y. enterocolitica* by 75 times. The synergy between the active components of the two EOs is one potential cause. EO and antibiotic synergies have been compiled by Yap et al. (2014). Suppression of defensive enzymes, a common biochemical route, and the use of outer membrane activating drugs that boost the absorption of additional antibacterial drugs are often at the heart of the synergistic action. Synergism between eugenol and thymol and eugenol and carvacrol has been observed, which Pei et al. (2009) attribute to the activity of carvacrol and thymol in rupturing the outer cell membrane of *E. coli*, allowing eugenol to enter the cytoplasm and interact with proteins and enzymes.

Direct effects of EO vapors on the influenza virus's HA (hemagglutinin) and NA (neuraminidase) surface proteins were assessed. Most EO's volatile components specifically inhibit HA activity. As a result, it was hypothesized that a mechanism may lie in the interaction of EO with HA. Numerous studies have reported the efficacy of nanoparticles as an antibacterial treatment, much like EOs. Potential antimicrobial agents NPs, such as silver NPs, were reviewed by Prabhu and Poulose (2012) (Rai et al., 2015). Anchoring and penetrating cells with silver NPs causes pits to form, free radicals to be released, and a change in membrane structure, all of which increase cell permeability and allow more antibacterial agents to enter the cell through the cytoplasmic membrane, ultimately killing the bacteria. Reactive oxygen species (ROS) are released when they interact with respiratory enzymes. These ROS have the potential to attack DNA bases, prevent signal transduction, and kill cells.

Over the past decade, many researches have investigated the synergistic effects of EOs and different NPs on antibacterial activity. The synergistic activity of silver nanoparticles and the EO representative cinnamaldehyde against the spore-forming bacteria *Clostridium perfringens* and *Bacillus cereus* was reported by Ghosh et al. (2013). Electron microscopy and atomic force microscopy both showed substantial damage to the cell envelope, and bacterial death curve analysis verified the combination's rapid bactericidal activity. By encapsulating EOs in a nanomaterial, such as a solid lipid NP, liposome, polymeric NP, or nano-emulsion, the antimicrobial potential of the EO can be increased. This method shows promise as a means to control the rate of drug release, whether through burst release or gradual release (DIdunyemi et al., 2019). Hosseini et al.'s (2013) study's of in vitro release demonstrated that oregano EO-encapsulated chitosan NPs had a burst-like effect followed by a slow release of the drug. When it comes to killing *Stegomyia aegypti* larvae, Abreu et al. (2012) found that it worked quite well as a biocide, and found that chitosan/cashew gum nanoencapsulation resulted in a steady and slow release of essential oil. According to Zhang et al. (2014), zein NP-encapsulated thymol suppressed Gram-positive bacteria more effectively and for a longer period of time than thymol alone. The rheological characteristics of bacterial biofilm were significantly altered when carvacrol was encapsulated in polylactic glycolic acid nanocapsules, according to the findings of the researchers.

Nanoencapsulation improves bioactivity, decreases toxicity, and improves patient compliance and convenience by enhancing the EOs' physical stability, lowering their solubility, and insulating them from ecological factors (such as light, oxygen, moisture, and pH) (Batiha et al., 2021). The nanocarriers allow for prolonged, low-dose medication delivery to the appropriate regions while also preventing the EOs from being broken down by enzymes and turning them into powder. Because of this, the drug's pharmacokinetic characteristics may improve. Therefore, since different EOs have their own unique antibacterial effect, they may greatly increase their antibacterial property by augmenting one another against different types of microorganisms through different mechanisms when used in conjunction with other efficacious antibacterial drugs such as Nanomaterials. The most effective strategy for combating MDR-bacteria appears to be the combination of multiple antimicrobials.

Conclusion

There has been a lot of study into the possibility of using EOs as a source of antimicrobial agents in medical applications, preservation of food, and packaging. Their commercial application has been restricted by their potent organoleptic flavor, volatile nature, low solubility, and toxicity in solvents. Nanoparticles offer an appealing solution for the creation of EO-based antibacterial nanostructured materials due to their good biocompatibility, environmental friendliness, chemical inertness, and required specificity. Researchers have looked into many different nanotechnology-based encapsulation strategies, such as lipid nanomaterials, inorganic nanomaterials, polymeric nanomaterials, nanoemulsions, nanocarriers, nanocrystals, and carbon nanotube, to hide the EOs' less desirable properties and boost their bioactivities.

References

Abreu, F. O. M. S., de Oliveira, E. F., Paula, H. C. B., & de Paula, R. C. M. (2012). Chitosan/cashew gum nanogels for essential oil encapsulation. *Carbohydrate Polymers, 89*, 1277–1282, [CrossRef].

Adejumo, I. O., Adetunji, C. O., & Adeyemi, O. S. (2017). Influence of UV light exposure on mineral composition and biomass production of myco-meat produced from different agricultural substrates. *Journal of Agricultural Sciences, Belgrade, 62*(1), 51–59.

Adetunji, C.O. (2008). The antibacterial activities and preliminary phytochemical screening of vernoniaamygdalina and Aloe vera against some selected bacteria, pp. 40–43, (M. Sc thesis), University of Ilorin.

Adetunji, C. O., Kayode, A., Bolajoko, F. O., & Bunmi, A. J. (2013a). Effects of coatings on storability of carrot under evaporative coolant system. *Albanian Journal of Agricultural Sciences, 12*(3).

Adetunji, C. O. (2019). Environmental impact and ecotoxicological influence of biofabricated and inorganic nanoparticle on soil activity. In D. Panpatte, & Y. Jhala (Eds.), *Nanotechnology for agriculture*. Singapore: Springer. Available from https://doi.org/10.1007/978–981-32–9370-0_12.

Adetunji, C. O., Akram, M., Tope Olaniyan, O., Olufemi Ajayi, O., Inobeme, A., Olaniyan, S., Hameed, L., & Adetunji, J. B. (2021a). Targeting SARS-CoV-2 novel Corona (COVID-19) virus infection using medicinal plants. In K. Dua, S. Nammi, D. Chang, D. K. Chellappan, G. Gupta, & T. Collet (Eds.), *Medicinal plants for lung diseases*. Singapore: Springer. Available from https://doi.org/10.1007/978-981-33-6850-7_21.

Adetunji, C. O., Inobeme, A., Olaniyan, O. T., Ajayi, O. O., Olaniyan, S., & Adetunji, J. B. (2021d). Application of nanodrugs derived from active metabolites of medicinal plants for the treatment of inflammatory and lung diseases: Recent advances. In K. Dua, S. Nammi, D. Chang, D. K. Chellappan, G. Gupta, & T. Collet (Eds.), *Medicinal plants for lung diseases*. Singapore: Springer. Available from https://doi.org/10.1007/978-981-33-6850-7_26.

Adetunji, C. O., Inobeme, A., Olaniyan, O. T., Ajayi, O. O., Olaniyan, S., & Adetunji, J. B. (2021e). Application of nanodrugs derived from active metabolites of medicinal plants for the treatment of inflammatory and lung diseases: Recent advances. In K. Dua, S. Nammi, D. Chang, D. K. Chellappan, G. Gupta, & T. Collet (Eds.), *Medicinal plants for lung diseases*. Singapore: Springer. Available from https://doi.org/10.1007/978-981-33-6850-7_26.

Adetunji, C. O., Kumar, D., Raina, M., Arogundade, O., & Sarin, N. B. (2019). Endophytic microorganisms as biological control agents for plant pathogens: A panacea for sustainable agriculture. In A. Varma, S. Tripathi, & R. Prasad (Eds.), *Plant biotic interactions*. Cham: Springer. Available from https://doi.org/10.1007/978-3-030-26657-8_1.

Adetunji, C. O., Nwankwo, W., Ukhurebor, K., Olayinka, A. S., & Makinde, A. S. (2021b). Application of biosensor for the identification of various pathogens and pests mitigating against the agricultural production: Recent advances. In R. N. Pudake, U. Jain, & C. Kole (Eds.), *Biosensors in agriculture: Recent trends and future perspectives. Concepts and strategies in plant sciences*. Cham: Springer. Available from https://doi.org/10.1007/978-3-030-66165-6_9.

Adetunji, C. O., Oloke, J. K., & Prasad, G. (2020a). Effect of carbon-to-nitrogen ratio on eco-friendly mycoherbicide activity from *Lasiodiplodia pseudotheobromae* C1136 for sustainable weeds management in organic agriculture. *Environment, Development and Sustainability, 22*, 1977–1990. Available from https://doi.org/10.1007/s10668-018-0273-1.

Adetunji, C. O., Tope Olaniyan, O., Akram, M., Olufemi Ajayi, O., Inobeme, A., Olaniyan, S., Said Khan, F., & Bunmi Adetunji, J. (2021k). Medicinal plants used in the treatment of pulmonary hypertension. In K. Dua, S. Nammi, D. Chang, D. K. Chellappan, G. Gupta, & T. Collet (Eds.), *Medicinal plants for lung diseases*. Singapore: Springer. Available from https://doi.org/10.1007/978-981-33-6850-7_14.

Adetunji, C. O., Samuel Michael, O., Varma, A., Kola Oloke, J., Kadiri, O., Akram, M., Ebunoluwa Bodunrinde, R., Imtiaz, A., Bunmi Adetunji, J., Shahzad, K., Jain, A., Ewa Ubi, B., Majeed, N., Ozolua, P., & Olisaka, F. N. (2021i). Recent advances in the application of biotechnology for improving the production of secondary metabolites from Quinoa. In A. Varma (Ed.), *Biology and biotechnology of Quinoa*. Singapore: Springer. Available from https://doi.org/10.1007/978-981-16-3832-9_17.

Adetunji, C. O., Samuel Michael, O., Kadiri, O., Varma, A., Akram, M., Kola Oloke, J., Shafique, H., Bunmi Adetunji, J., Jain, A., Ebunoluwa Bodunrinde, R., Ozolua, P., & Ewa Ubi, B. (2021f). Quinoa: From farm to traditional healing, food application, and phytopharmacology. In A. Varma (Ed.), *Biology and biotechnology of Quinoa*. Singapore: Springer. Available from https://doi.org/10.1007/978-981-16-3832-9_20.

Adetunji, C. O., Samuel Michael, O., Nwankwo, W., Eghonghon Ukhurebor, K., Anthony Anani, O., Kola Oloke, J., Varma, A., Kadiri, O., Jain, A., & Bunmi Adetunji, J. (2021h). Quinoa, the next biotech plant: Food security and environmental and health hot spots. In A. Varma (Ed.), *Biology and biotechnology of Quinoa*. Singapore: Springer. Available from https://doi.org/10.1007/978-981-16-3832-9_19.

Adetunji, C. O., Olufemi Ajayi, O., Akram, M., Tope Olaniyan, O., Amjad Chishti, M., Inobeme, A., Olaniyan, S., Bunmi Adetunji, J., Olaniyan, M., & Oluwasegun Awotunde, S. (2021l). Medicinal plants used in the treatment of influenza A virus infections. In K. Dua, S. Nammi, D. Chang, D. K. Chellappan, G. Gupta, & T. Collet (Eds.), *Medicinal plants for lung diseases*. Singapore: Springer. Available from https://doi.org/10.1007/978-981-33-6850-7_19.

Adetunji, C. O., Osikemekha, A. A., Olaniyan, O. T., Inobeme, A., Olisaka, F. N., Oluwadamilare Uwadiae, E., & Nosa Obayagbona, O. (2021g). Recent trends in organic farming. In R. Soni, D. C. Suyal, P. Bhargava, & R. Goel (Eds.), *Microbiological activity for soil and plant health management*. Singapore: Springer. Available from https://doi.org/10.1007/978-981-16-2922-8_20.

Adetunji, C. O., Phazang, P., & Sarin, N. B. (2017). Significance of rhamnolipids as a biological control agent in the management of crops/plant pathogens. *Current Trends in Biomedical Engineering & Biosciences, 10*(3), 54–55.

Adetunji, C. O., Roli, O. I., & Adetunji, J. B. (2020c). Exopolysaccharides derived from beneficial microorganisms: Antimicrobial, food, and health benefits. In P. Mishra, R. R. Mishra, & C. O. Adetunji (Eds.), *Innovations in food technology*. Singapore: Springer. Available from https://doi.org/10.1007/978-981-15-6121-4_10.

Adetunji, C. O., & Varma, A. (2020). Biotechnological application of trichoderma: A powerful fungal isolate with diverse potentials for the attainment of food safety, management of pest and diseases, healthy planet, and sustainable agriculture. In C. Manoharachary, H. B. Singh, & A. Varma (Eds.), *Trichoderma: Agricultural applications and beyond. Soil biology* (vol 61). Cham: Springer. Available from https://doi.org/10.1007/978-3-030-54758-5_12.

Adetunji, J. B., Adetunji, C. O., & Olaniyan, O. T. (2021j). African walnuts: A natural depository of nutritional and bioactive compounds essential for food and nutritional security in Africa. In O. O. Babalola (Ed.), *Food security and safety*. Cham: Springer. Available from https://doi.org/10.1007/978-3-030-50672-8_19.

Adetunji, J. B., Ajani, A. O., Adetunji, C. O., Fawole, O. B., Arowora, K. A., Nwaubani, S. I., Ajayi, E. S., Oloke, J. K., & Aina, J. A. (2013b). Postharvest quality and safety maintenance of the physical properties of *Daucus carota* L. fruits by neem oil and moringa oil treatment: A new edible coatings. *Agrosearch, 13*(1), 131–141.

Adetuyi, B., Dairo, J., & Oluwole, E. (2015a). Biochemical effects of shea butter and groundnut oils on white albino rats. *International Journal of Chemistry and Chemical Processes, 1*(8), 1–17.

Adetuyi, B. O., Adebayo, P. F., Olajide, P. A., Atanda, O. O., & Oloke, J. K. (2022a). Involvement of free radicals in the ageing of cutaneous membrane. *World News of Natural Sciences, 43*, 11–37.

Adetuyi, B. O., Adebisi, O. A., Adetuyi, O. A., Ogunlana, O. O., Toloyai, P. E., Egbuna, C., & Patrick-Iwuanyanwu, K. C. (2022b). *Ficus exasperata* attenuates acetaminophen-induced hepatic damage via NF-κB signaling mechanism in experimental rat model. *BioMed Research International*, 2022.

Adetuyi, B. O., Adebisi, O. A., Awoyelu, E. H., Adetuyi, O. A., & Ogunlana, O. O. (2020d). Phytochemical and toxicological effect of ethanol extract of *Heliotropium indicum* on liver of male albino rats. *Letters in Applied NanoBioscience, 10*(2), 2085–2095.

Adetuyi, B. O., Dairo, J. O., & Didunyemi, O. M. (2015b). Anti-hyperglycemic potency of *Jatropha gossypiifolia* in alloxan induced diabetes. *Biochem Pharmacol (Los Angel), 4*(193), 2167-0501.

Adetuyi, B. O., Odine, G. O., Olajide, P. A., Adetuyi, O. A., Atanda, O. O., & Oloke, J. K. (2022c). Nutraceuticals: Role in metabolic disease, prevention and treatment. *World News of Natural Sciences, 42*, 1–27.

Adetuyi, B. O., Ogundipe, A. E., Ogunlana, O. O., Egbuna, C., Estella, O. U., Mishra, A. P., & Achar, R. R. (2022d). Banana peel as a source of nutraceuticals. *Food and Agricultural Byproducts as Important Source of Valuable Nutraceuticals*, 243–250, Springer, Cham.

Adetuyi, B. O., Okeowo, T. O., Adetuyi, O. A., Adebisi, O. A., Ogunlana, O. O., Oretade, J. O., & Batiha, G. E. S. (2020e). *Ganoderma lucidum* from red mushroom attenuates formaldehyde-induced liver damage in experimental male rat model. *Biology, 9*(10), 313.

Adetuyi, B. O., Olajide, P. A., Awoyelu, E. H., Adetuyi, O. A., Adebisi, O. A., & Oloke, J. K. (2020f). Epidemiology and treatment options for COVID-19: A review. *African Journal of Reproductive Health, 24*(2), 142–153.

Adetuyi, B. O., Olajide, P. A., Oluwatosin, A., & Oloke, J. K. (2022e). Preventive phytochemicals of cancer as speed breakers in inflammatory signaling. *Research Journal of Life Sciences, Bioinformatics, Pharmaceutical and Chemical Sciences, 8*(1), 30–61.

Adetuyi, B. O., Olajide, P. A., Omowumi, O. S., Odine, G. O., Okunlola, D. D., Taiwo, A. M., & Opayinka, O. D. (2022f). Blockage of Alzheimer's gene: Breakthrough effect of apolipoprotein E4. *African Journal of Advanced Pure and Applied Sciences (AJAPAS)*, 26–33.

Adetuyi, B. O., Oluwole, E. O., & Dairo, J. O. (2015c). Chemoprotective potential of ethanol extract of *Ganoderma lucidum* on liver and kidney parameters in *Plasmodium beghei*-induced mice. *International Journal of Chemistry and Chemical Processes (IJCC), 1*(8), 29–36.

Adetuyi, B. O., Omolabi, F. K., Olajide, P. A., & Oloke, J. K. (2021m). Pharmacological, biochemical and therapeutic potential of milk thistle (silymarin): A review. *World News of Natural Sciences, 37*, 75–91.

Adetuyi, B. O., Toloyai, P. E. Y., Ojugbeli, E. T., Oyebanjo, O. T., Adetuyi, O. A., Uche, C. Z., & Egbuna, C. (2021n). Neurorestorative roles of microgliosis and astrogliosis in neuroinflammation and neurodegeneration. *Scicom Journal of Medical and Applied Medical Sciences, 1*(1), 1–5.

Adetunji, C. O., Egbuna, C., Tijjani, H., Adom, D., Tawfeeq Al-Ani, L. K., & Patrick-Iwuanyanwu, K. C. (2020b). Homemade preparations of natural biopesticides and applications. *Natural Remedies for Pest, Disease and Weed Control*, 179–185, Publisher Academic Press.

Adetunji, C. O., Samuel Michael, O., Rathee, S., Singh, K. R. B., Olufemi Ajayi, O., Bunmi Adetunji, J., Ojha, A., Singh, J., & Pratap Singh, R. (2022f). Potentialities of nanomaterials for the management and treatment of metabolic syndrome: A new insight. *Materials Today Advances, 13*, 100198.

Adetunji, C. O., Palai, S., Ekwuabu, C. P., Egbuna, C., Adetunji, J. B., Ehis-Eriakha, C. B., Kesh, S. S., & Mtewa, A. G. (2021c). General principle of primary and secondary plant metabolites: Biogenesis, metabolism, and extraction. *Preparation of Phytopharmaceuticals for the Management of Disorders*, 3–23, Publisher Academic Press.

Aguilar-Pérez, K., Medina, D., Narayanan, J., Parra-Saldívar, R., & Iqbal, H. (2021). Synthesis and nano-sized characterization of bioactive oregano essential oil molecule-loaded small unilamellar nanoliposomes with antifungal potentialities. *Molecules, 26*, 2880, [CrossRef] [PubMed].

Alizadeh-Sani, M., Mohammadian, E., & McClements, D. J. (2020). Eco-friendly active packaging consisting of nanostructured biopolymer matrix reinforced with TiO2 and essential oil: Application for preservation of refrigerated meat. *Food Chemistry, 322*, 126782.

Allahverdiyev, A. M., Kon, K. V., Abamor, E. S., Bagirova, M., & Rafailovich, M. (2011). Coping with antibiotic resistance: Combining nanoparticles with antibiotics and other antimicrobial agents. *Expert Review of Anti-infective Therapy, 9*, 1035–1052.

Alwan, S., El-Omari, K., Soufi, H., Zreika, S., Sukarieh, I., Chihib, N. E., Jama, C., & Hamze, M. (2016). Evaluation of the antibacterial activity of *Micromeria barbata* in lebanon. *Journal of Essential Oil Bearing Plants, 19*(2), 321–327.

Amato, D., Amato, D., Mavrodi, O. V., Braasch, D. A., Walley, S. E., Douglas, J. R., Mavrodi, D. V., & Patton, D. L. (2016). Destruction of opportunistic pathogens via polymer nanoparticle-mediated release of plant-based antimicrobial payloads. *Advanced Healthcare Materials, 5*, 1094–1103, [CrossRef].

Anghel, I., Holban, A. M., Andronescu, E., Grumezescu, A. M., Chifiriuc, M. C., & Mihai Grumezescu, A. (2013). Efficient surface functionalization of wound dressings by a phytoactive nanocoating refractory to *Candida albicans* biofilm development. *Biointerphases*, 8, 12, [CrossRef].

Anghel, I., Holban, A. M., Grumezescu, A. M., Andronescu, E., Ficai, A., Anghel, A. G., Maganu, M., Lažar, V., & Chifiriuc, M. C. (2012). Modified wound dressing with phyto-nanostructured coating to prevent staphylococcal and pseudomonal biofilm development. *Nanoscale Research Letters*, 7, 690, [CrossRef].

Angioni, A., Barra, A., Coroneo, V., Dessi, S., & Cabras, P. (2006). Chemical composition, seasonal variability, and antifungal activity of *Lavandula stoechas* L. ssp. stoechas essential oils from stem/leaves and flowers. *Journal of Agricultural and Food Chemistry*, 54(12), 4364–4370.

Arrebola, M. L., Navarro, M. C., Jimenez, J., & Ocana, F. A. (1994). Yield and composition of the essential oil of *Thymus serpylloides* sub sp. serpylloides. *Phytochemistry*, 36, 67–72.

Azhdarzadeh, F., & Hojjati, M. (2016). Chemical composition and antimicrobial activity of leaf, ripe and unripe peel of bitter orange (*Citrus aurantium*) essential oils. *Nutrition and Food Sciences Research*, 3(1), 43–50.

Badea, M. L., Iconaru, S. L., Groza, A., Chifiriuc, M. C., Beuran, M., & Predoi, D. (2019). Peppermint essential oil-doped hydroxyapatite nanoparticles with antimicrobial properties. *Molecules*, 24, 2169, [CrossRef].

Bajpai, V. K., Sharma, A., & Baek, K. H. (2013). Antibacterial mode of action of *Cudrania tricuspidata* fruit essential oil, affecting membrane permeability and surface characteristics of food borne pathogens. *Food Control*, 32, 582–590.

Batiha, G. B., Awad, D. A., Algamma, A. M., Nyamota, R., Wahed, M. I., Shah, M. A., Amin, M. N., Adetuyi, B. O., Hetta, H. F., Cruz-Marins, N., Koirala, N., Ghosh, A., & Sabatier, J. (2021).). Diary-derived and egg white proteins in enhancing immune system against COVID-19 frontiers in nutrition. *(Nutritional Immunology)*, 8, 629440. Available from https://doi.org/10.3389/fnut.2021629440.

Bazzaz, B. S. F., Khameneh, B., Namazi, N., Iranshahi, M., Davoodi, D., & Golmohammadzadeh, S. (2018). Solid lipid nanoparticles carrying *Eugenia caryophyllata* essential oil: The novel nanoparticulate systems with broad-spectrum antimicrobial activity. *Letters in Applied Microbiology*, 66, 506–513, [CrossRef].

Bello, O. M., Ibitoye, T., & Adetunji, C. (2019). Assessing antimicrobial agents of *Nigeria flora*. *Journal of King Saud University-Science*, 31(4), 1379–1383.

Boire, N. A., Riedel, S., & Parrish, N. M. (2013). Essential oils and future antibiotics: New weapons against emerging 'Superbugs'? *Journal of Ancient Diseases & Preventive Remedies*, 1, 105.

Carson, C., Mee, B. J., & Riley, T. V. (2002). Mechanism of action of *Melaleuca alternifolia* (tea tree) oil on Staphylococcus aureus determined by time-kill, lysis, leakage, and salt tolerance assays and electron microscopy. *Antimicrobial Agents and Chemotherapy*, 46, 1914–1920.

Chao, S. C., Young, D. G., & Oberg, C. J. (2000). Screening for inhibitory activity of essential oils on selected bacteria, fungi and viruses. *Journal of Essential Oil Research*, 12(5), 639–649.

Comin, V. M., Lopes, L. Q., Quatrin, P. M., de Souza, M. E., Bonez, P. C., Pintos, F. G., Raffin, R., Vaucher, R. D. A., Martinez, D. S. T., & Santos, R. C. V. (2016). Influence of *Melaleuca alternifolia* oil nanoparticles on aspects of *Pseudomonas aeruginosa* biofilm. *Microbial Pathogenesis*, 93, 120–125, [CrossRef].

Cruz, I. F., Freire, C., Araújo, J. P., Pereira, C., & Pereira, A. M. (2018). Multifunctional ferrite nanoparticles: From current trends toward the future. *Magnetic nanostructured materials*, 59–116, Elsevier: Amsterdam, The Netherlands.

Dahham, S. S., Hassan, L. E. A., Ahamed, M. B. K., Majid, A. S. A., Majid, A. M. S. A., & Zulkepli, N. N. (2016). In vivo toxity and antitumor activity of essential oils extract from agarwood (*Aquilaria crassna*). BMC Complement. *Alternative Medicine.*, 16, 236.

Das, S., Vörös-Horváth, B., Bencsik, T., Micalizzi, G., Mondello, L., Horváth, G., Kʺoszegi, T., & Széchenyi, A. (2020). Antimicrobial activity of different artemisia essential oil formulations. *Molecules*, 25, 2390, [CrossRef].

Delaquis, P. J., Stanich, K., Girard, B., & Mazza, G. (2002). Antimicrobial activity of individual and mixed fractions of dill, cilantro, coriander and eucalyptus essential oils. *International Journal of Food Microbiology*, 74, 101–109.

De Souza, E. L., Barros, J. C., Oliveira, C. E. V., & Conceicao, M. L. (2010). Influence of Origanum vulgare L. essential oil on enterotoxin production, membrane permeability and surface characteristics of Staphylococcus aureus. *International Journal of Food Microbiology*, 137, 308–311.

Didunyemi, M. O., Adetuyi, B. O., & Oyebanjo, O. O. (2019). *Morinda lucida* attenuates acetaminophen-induced oxidative damage and hepatotoxicity in rats. *Journal of Biomedical sciences*, 8, No https://www.jbiomeds.com/biom e dical- sciences/morinda-lucida-attenuates-acetaminophen-induced-oxidative-damage-and- hepatotoxicity-in-rats.php?aid = 24482.

Didunyemi, M. O., Adetuyi, B. O., & Oyewale, I. A. (2020). Inhibition of lipid peroxidation and in-vitro antioxidant capacity of aqueous, acetone and methanol leaf extracts of green and red *Acalypha wilkesiana* Muell Arg. *International Journal of Biological and Medical Research*, 11(3), 7089–7094.

Dorman, H. J., & Deans, S. G. (2000). Antimicrobial agents from plants: Antibacterial activity of plant volatile oils. *Journal of Applied Microbiology*, 88, 308–316.

Duarte, A. E., deMenezes, I. R. A., Braga, M. F. B. M., Leite, N. F., Barros, L. M., Waczuk, E. P., da Silva, M. A. P., Boligon, A., Rocha, J. B. T., Souza, D. O., Kamdem, J. P., Coutinho, H. D. M., & Burger, M. E. (2016). Antimicrobial activity and modulatory effect of essential oil from the leaf of *Rhaphiodon echinus* (Nees & Mart) Schauer on some antimicrobial drugs. *Molecules*, 21, 743. Available from https://doi.org/10.3390/molecules21060743.

Duncan, B., Li, X., Landis, R. F., Kim, S. T., Gupta, A., Wang, L. S., Ramanathan, R., Tang, R., Boerth, J. A., & Rotello, V. M. (2015). Nanoparticle-stabilized capsules for the treatment of bacterial biofilms. *ACS Nano*, 9(8), 7775–7782.

Duong, V.-A., Nguyen, T.-T.-L., & Maeng, H.-J. (2020). Preparation of solid lipid nanoparticles and nanostructured lipid carriers for drug delivery and the effects of preparation parameters of solvent injection method. *Molecules*, 25, 4781, [CrossRef].

Dweck, A. C. (2009). Toxicology of essential oils reviewed. *Personal Care*, 65−77.

Egbuna, C., Gupta, E., Ezzat, S. M., Jeevanandam, J., Mishra, N., Akram, M., Sudharani, N., Oluwaseun Adetunji, C., Singh, P., Ifemeje, J. C., Deepak, M., Bhavana, A., Mark, A., Walag, P., Ansari, R., Bunmi Adetunji, J., Laila, U., Chinedu Olisah, M., & Feenna Onyekere, P. (2020). Aloe species as valuable sources of functional bioactives. In C. Egbuna, & G. Dable Tupas (Eds.), *Functional foods and nutraceuticals*. Cham: Springer. Available from https://doi.org/10.1007/978-3-030-42319-3_18.

Esmaeili, A., & Asgari, A. (2015). In vitro release and biological activities of *Carum copticum* essential oil (CEO) loaded chitosan nanoparticles. *International Journal of Biological Macromolecules*, *81*, 283−290, [CrossRef].

Froiio, F., Ginot, L., Paolino, D., Lebaz, N., Bentaher, A., Fessi, H., & Elaissari, A. (2019). Essential oils-loaded polymer particles: Preparation, characterization and antimicrobial property. *Polymers*, *11*, 1017, [CrossRef].

Gaspar, A. S., Wagner, F. E., Amaral, V. S., Costa Lima, S. A., Khomchenko, V. A., Santos, J. G., Costa, B. F., & Duraes, L. (2017). Development of a biocompatible magnetic nanofluid by incorporating SPIONs in Amazonian oils. *Spectrochimica Acta Part A, Molecular and Biomolecular Spectroscopy*, *172*, 135−146.

Ghosh, I. N., Patil, S. D., Sharma, T. K., Srivastava, S. K., Pathania, R., & Navani, N. K. (2013). Synergistic action of cinnamaldehyde with silver nanoparticles against spore-forming bacteria: A case for judicious use of silver nanoparticles for antibacterial applications. *International Journal of Nanomedicine*, *8*, 4721−4731.

Giri, T., Verma, S., Alexander, A., Ajazuddin, B. H., Tripathy, M., & Tripathi, D. (2013). Crosslinked biodegradable alginate hydrogel floating beads for stomach site specific controlled delivery of metronidazole. *Farmacia*, *61*, 533−550.

Gomes, C., Moreira, R. G., & Castell-Perez, E. (2011). Poly (DLlactide-co-glycolide) (PLGA) nanoparticles with entrapped trans-cinnamaldehyde and eugenol for antimicrobial delivery applications. *Journal of Food Science*, *76*, N16−N24i.

Gortzi, O., Lalas, S., Chinou, I., & Tsaknis, J. (2007). Evaluation of the antimicrobial and antioxidant activities of Origanum dictamnus extracts before and after encapsulation in liposomes. *Molecules*, *12*, 932−945.

Grumezescu, V., Holban, A. M., Iordache, F., Socol, G., Mogoşanu, G. D., Grumezescu, A. M., Ficai, A., Vasile, B. S., Trusca, R., & Chifiriuc, M. C. (2014). MAPLE fabricated magnetite@eugenol and (3-hidroxybutyric acid-co-3-hidroxyvaleric acid)−polyvinyl alcohol microspheres coated surfaces with anti-microbial properties. *Applied Surface Science*, *306*, 16−22, [CrossRef].

Hasheminejad, N., & Khodaiyan, F. (2020). The effect of clove essential oil loaded chitosan nanoparticles on the shelf life and quality of pomegranate arils. *Food Chemistry*, *309*, 125520, [CrossRef] [PubMed].

Hosseini, S. F., Zandi, M., Rezaei, M., & Farahmandghavi, F. (2013). Two-step method for encapsulation of oregano essential oil in chitosan nanoparticles: Preparation, characterization and in vitro release study. *Carbohydrate Poly*, *95*, 50−56.

Hyldgaard, M., Mygind, T., & Meyer, R. L. (2012). Essential oils in food preservation: Mode of action, synergies and interactions with food matrix components. *Frontiers in Microbiology*, *3*, 12.

Jabir, M. S., Taha, A. A., & Sahib, U. I. (2018). Linalool loaded on glutathione-modified gold nanoparticles: A drug delivery system for a successful antimicrobial therapy. *Artificial Cells, Nanomedicine, and Biotechnology*, *46*, 345−355, [CrossRef].

Jamil, B., Abbasi, R., Abbasi, S., Imran, M., Khan, S. U., Ihsan, A., Javed, S., & Bokhari, H. (2016). Encapsulation of cardamom essential oil in chitosan nano-composites: In-vitro efficacy on antibiotic-resistant bacterial pathogens and cytotoxicity studies. *Frontiers in Microbiology*, *7*, 1580, [CrossRef].

Jin, L., Teng, J., Hu, L., Lan, X., Xu, Y., Sheng, J., Song, Y., & Wang, M. (2019a). Pepper fragrant essential oil (PFEO) and functionalized MCM-41 nanoparticles: Formation, characterization, and bactericidal activity. *Journal of the Science of Food and Agriculture*, *99*, 5168−5175, [CrossRef].

Jin, P., Yao, R., Qin, D., Chen, Q., & Du, Q. (2019b). Enhancement in antibacterial activities of eugenol-entrapped ethosome nanoparticles via strengthening its permeability and sustained release. *Journal of Agricultural and Food Chemistry*, *67*, 1371−1380, [CrossRef] [PubMed].

Knezevic, P., Aleksic, V., Simin, N., Svircev, E., Petrovic, A., & Mimica-Dukic, N. (2016). Antimicrobial activity of Eucalyptus camaldulensis essential oils and their interactions with conventional antimicrobial agents against multi-drug resistant Acinetobacter baumannii. *Journal of Ethnopharmacology*, *3*(178), 125−136.

Kousalová, J., & Etrych, T. (2018). Polymeric nanogels as drug delivery systems. *Physiological Research*, *67*, S305−S317, [CrossRef].

Krishnamoorthy, K., & Mahalingam, M. (2015). Selection of a suitable method for the preparation of polymeric nanoparticles: Multi-Criteria decision making approach. *Advanced Pharmaceutical Bulletin*, *5*, 57−67, [CrossRef].

Laird, K., & Phillips, C. (2012). Vapour phase: A potential future use for essential oils as antimicrobials? *Letters in Applied Microbiology*, *54*, 169−174.

Lakehal, S., Meliani, A., Benmimoune, S., Bensouna, S. N., Benrebiha, F. Z., & Chaouia, C. (2016). Essential oil composition and antimicrobial activity of *Artemisia herba*−alba Asso grown in Algeria. *Medicinal Chemistry (Los Angeles)*, *6*, 435−439.

Lammari, N., Louaer, O., Meniai, A. H., & Elaissari, A. (2020). Encapsulation of essential oils via nanoprecipitation process: Overview, progress, challenges and prospects. *Pharmaceutics*, *12*, 431, [CrossRef].

Li, K.-K., Yin, S.-W., Yang, X.-Q., Tang, C.-H., & Wei, Z.-H. (2012).). Fabrication and characterization of novel antimicrobial films derived from thymol-loaded zein−sodium caseinate (SC) nanoparticles. *Journal of Agricultural and Food Chemistry*, *60*, 11592−11600, [CrossRef].

Liu, Z., Jiao, Y., Wang, Y., Zhou, C., & Zhang, Z. (2008). Polysaccharides-based nanoparticles as drug delivery systems. *Advanced Drug Delivery Reviews*, *60*, 1650−1662, [CrossRef] [PubMed].

Macwan, S. R., Dabhi, B. K., Aparnathi, K. D., & Prajapati, J. B. (2016). Essential oils of herbs and spices: Their antimicrobial activity and application in preservation of food. *International Journal of Current Microbiology and Applied Sciences*, *5*(5), 885−901.

Maduray, K., & Parboosing, R. (2021). Metal nanoparticles: A promising treatment for viral and arboviral infections. *Biological Trace Element Research, 199,* 3159–3176, [CrossRef] [PubMed].

Magi, G., Marini, E., & Facinelli, B. (2015). Antimicrobial activity of essential oils and carvacrol and synergy of carvacrol and erythromycin, against clinical, erythromycin-resistant Group A Streptococci. *Frontiers in Microbiology, 6,* 165.

Mashwani, Z. U., Khan, M. A., Khan, T., & Nadhman, A. (2016). Applications of plant terpenoids in the synthesis of colloidal silver nanoparticles. *Advances in Colloid and Interface Science, 234,* 132–141.

Mekonnen, A., Yitayew, B., Tesema, A., & Taddese, S. (2016). In vitro antimicrobial activity of essential oil of *Thymus schimperi, Matricaria chamomilla, Eucalyptus globulus* and *Rosmarinus officinalis. International Journal of Microbiology, 2016*(9545693), 8.

Mihai, A. D., Chircov, C., Grumezescu, A. M., & Holban, A. M. (2020). Magnetite nanoparticles and essential oils systems for advanced antibacterial therapies. *International Journal of Molecular Sciences, 21,* 7355, [CrossRef].

Mitchell, M. J., Billingsley, M. M., Haley, R. M., Wechsler, M. E., Peppas, N. A., & Langer, R. (2021). Engineering precision nanoparticles for drug delivery. *Nature reviews. Drug discovery, 20,* 101–124, [CrossRef].

Mizieli´nska, M., Nawrotek, P., Stachurska, X., Ordon, M., & Bartkowiak, A. (2021). Packaging covered with antiviral and antibacterial coatings based on ZnO nanoparticles supplemented with geraniol and carvacrol. *International Journal of Molecular Sciences, 22,* 1717, [CrossRef].

Mohammadi, A., Hashemi, M., & Hosseini, S. M. (2015a). Nanoencapsulation of *Zataria multiflora* essential oil preparation and characterization with enhanced antifungal activity for controlling *Botrytis cinerea,* the causal agent of gray mould disease. *Innovative Food Science & Emerging Technologies, 28,* 73–80, [CrossRef].

Mohammadi, A., Mosleh, N., Shomali, T., Ahmadi, M., & Sabetghadam, S. (2015b). In vitro evaluation of antiviral activity of essential oil from *Zataria multiflora* Boiss against Newcastle disease virus. *Journal of Herbmed Pharmacology, 4*(3), 71–74.

Mohsenabadi, N., Rajaei, A., Tabatabaei, M., & Mohsenifar, A. (2018). Physical and antimicrobial properties of starch-carboxy methyl cellulose film containing rosemary essential oils encapsulated in chitosan nanogel. *International Journal of Biological Macromolecules, 112,* 148–155, [CrossRef] [PubMed].

Moon, S. E., Kim, H. Y., & Cha, J. D. (2011). Synergistic effect between clove oil and its major compounds and antibiotics against oral bacteria. *Archives of Oral Biology, 56*(9), 907–916.

Mousavi, S. M., Hashemi, S. A., Ramakrishna, S., Esmaeili, H., Bahrani, S., Koosha, M., & Babapoor, A. (2019). Green synthesis of supermagnetic Fe_3O_4–MgO nanoparticles via Nutmeg essential oil toward superior anti-bacterial and anti-fungal performance. *Journal of Drug Delivery Science and Technology, 54,* 101352, [CrossRef].

Muthaiyan, A., Martin, E. M., Natesan, S., Crandall, P. G., Wilkinson, B. J., & Ricke, S. C. (2012). Antimicrobial effect and mode of action of terpeneless cold-pressed Valencia orange essential oil on methicillin-resistant Staphylococcus aureus. *Journal of Applied Microbiology, 112,* 1020–1033.

Narayan, R., Nayak, U. Y., Raichur, A. M., & Garg, S. (2018). Mesoporous silica nanoparticles: A comprehensive review on synthesis and recent advances. *Pharmaceutics, 10,* 118, [CrossRef].

Nazir, A., Itrat, N., Shahid, A., Mushtaq, Z., Abdulrahman, S. A., Egbuna, C., Adetuyi, B. O., Khan, J., Uche, C. Z., & Toloyai, P. Y. (2022). *Orange peel as a source of nutraceuticals. Food and agricultural byproducts as important source of valuable nutraceuticals* (1st ed., p. 400)Berlin: Springer, Gebunden. ISBN 978-3-030-98759-6.

Negut, I., Grumezescu, V., Ficai, A., Grumezescu, A. M., Holban, A. M., Popescu, R. C., Savu, D., Vasile, B. S., & Socol, G. (2018). MAPLE deposition of Nigella sativa functionalized Fe_3O_4 nanoparticles for antimicrobial coatings. *Applied Surface Science, 455,* 513–521, [CrossRef].

Ogunlana, O. O., Adetuyi, B. O., Adekunbi, T. S., Adegboye, B. E., Iheagwam, F. N., & Ogunlana, O. E. (2021a). Ruzu bitters ameliorates high–fat diet induced non-alcoholic fatty liver disease in male Wistar rats. *Journal of Pharmacy and Pharmacognosy Research, 9*(3), 251-26.

Ogunlana, O. O., Adetuyi, B. O., Esalomi, E. F., Rotimi, M. I., Popoola, J. O., Ogunlana, O. E., & Adetuyi, O. A. (2021). Antidiabetic and antioxidant activities of the twigs of andrograhis paniculata on streptozotocin-induced diabetic male rats. *BioChem, 1*(3), 238–249.

Ogunlana, O. O., Adetuyi, B. O., Rotimi, M., Adeyemi, A., Akinyele, J., Ogunlana, O. E., & Batiha, G. E. S. (2021c). Hypoglycemic and antioxidative activities of ethanol seed extract of *Hunteria umbellate* (Hallier F.) on streptozotocin-induced diabetic rats. *Clinical Phytoscience, 7*(1), 1–9.

Ogunlana, O. O., Adetuyi, B. O., Rotimi, M., Esalomi, I., Adeyemi, A., Akinyemi, J., Ogunlana, O., Adetuyi, O., Adebisi, O., Okpata, E., Baty, R., & Batiha, G. (2021d). Hypoglycemic activities of ethanol seed extract of *Hunteria umbellate* (Hallier F.) on streptozotocin-induced diabetic rats. *Clinical Phytoscience, 7*(1), 1–9. Available from https://doi.org/10.1186/s40816-021-00285-1.

Ogunlana, O. O., Babatunde, O. A., Tobi, S. A., Adegboye, B. E., Iheagwam, F. N., & Oluseyi, E. (2021e). Ogunlana. Ruzu bitters ameliorates high–fat diet induced non-alcoholic fatty liver disease in male Wistar rats. *Journal of Pharmacy and Pharmacognosy Research, 9*(3), 251–260.

Olajide, P. A., Omowumi, S. O., & Odine, G. O. (2022d). Pathogenesis of reactive oxygen species: A review. *World News of Natural Sciences, 44,* 150–164.

Olaniyan, O. T., & Adetunji, C. O. (2021). Biological, biochemical, and biodiversity of biomolecules from marine-based beneficial microorganisms: Industrial perspective. In C. O. Adetunji, D. G. Panpatte, & Y. K. Jhala (Eds.), *Microbial rejuvenation of polluted environment. Microorganisms for sustainability* (vol 27). Singapore: Springer. Available from https://doi.org/10.1007/978-981-15-7459-7_4.

Padalia, H., Moteriya, P., Baravalia, Y., & Chanda, S. (2015). Antimicrobial and synergistic effects of some essential oils to fight against microbial pathogens: A review. The battle against microbial pathogens: Basic. In A. Méndez-Vilas (Ed.), *Science, technological advances and educational programs.* , FORMATEX 2015.

Paula, H. C., Oliveira, E. F., Carneiro, M. J., & de Paula, R. C. (2016). Matrix effect on the spray drying nanoencapsulation of lippiasidoides essential oil in chitosan-native gum blends. *Planta Medica.* Available from https://doi.org/10.1055/s-0042-107470.

Pednekar, P. P., Dhumal, R. V., Datar, A. G., & Vanage, G. R. (2013). In vivo dermal absorption and sub-acute toxicity studies of essential oil from Blumea eriantha DC. *International Journal of Pharmacy and Pharmaceutical Sciences, 5*, 351–358.

Pedro, A. S., Santo, I. E., Silva, C. V., Detoni, C., & Albuquerque, E. (2013). The use of nanotechnology as an approach for essential oil-based formulations with antimicrobial activity. In A. Mendez-Vilas (Ed.), *Microbial pathogens and strategies for combating them: Science, technology and education. Formatex* (pp. 1364–1374). .

Pei, R. S., Zhou, F., Ji, B. P., & Xu, J. (2009). Evaluation of combined antibacterial effects of eugenol, cinnamaldehyde, thymol, and carvacrol against E. coli with an improved Method. *Journal of Food Science, 74*, 379–383.

Prabhu, S., & Poulose, E. K. (2012). Silver nanoparticles: Mechanism of antimicrobial action, synthesis, medical applications, and toxicity effects. *International Nano Letters, 2*, 32.

Predoi, D., Iconaru, S. L., Buton, N., Badea, M. L., & Marutescu, L. (2018). Antimicrobial activity of new materials based on lavender and basil essential oils and hydroxyapatite. *Nanomaterials, 8*, 291, [CrossRef].

Rai, M., Ingle, A. P., Birla, S., Yadav, A., & Dos Santos, C. A. (2016). Strategic role of selected noble metal nanoparticles in medicine. *Critical Reviews in Microbiology, 42*(5), 696–719.

Rai, M., Ingle, A. P., Gade, A. K., Duarte, M. C. T., & Duran, N. (2015). Three Phoma spp. synthesized novel silver nanoparticles that possess excellent antimicrobial efficacy. *IET Nanobiotechnology/IET, 9*(5), 280–287.

Rai, M., Yadav, A., & Gade, A. (2009). Silver nanoparticles as a new generation of antimicrobials. *Biotechnology Advances, 27*(1), 76–83.

Rasooli, I., Allameh, A., & Rezaer, M. B. (2005). Antimicrobial efficacy of Thyme essential oils as food preservatives. In A. P. Riley (Ed.), *Food policy, control and research* (pp. 1–33). USA: Nova Science Publisher Inc.

Rauf, A., Akram, M., Semwal, P., Mujawah, A. A. H., Muhammad, N., Riaz, Z., Munir, N., Piotrovsky, D., Vdovina, I., Bouyahya, A., Oluwaseun Adetunji, C., Ali Shariati, M., Almarhoon, Z. M., Mabkhot, Y. N., & Khan, H. (2021). Antispasmodic potential of medicinal plants: A comprehensive review. *Oxidative medicine and cellular longevity, 12*, 2021. Available from https://doi.org/10.1155/2021/4889719, Article ID 4889719.

Roca, I., Akova, M., Baquero, F., Carlet, J., Cavaleri, M., Coenen, S., Cohen, J., Findlay, D., Gyssens, I., Heuer, O. E., Kahlmeter, G., Kruse, H., Laxminarayan, R., Liébana, E., López-Cerero, L., MacGowan, A., Martins, M., Rodríguez-Baño, J., Rolain, J. M., ... Vila, J. (2015). The global threat of antimicrobial resistance: Science for intervention. *New Microbes and New Infections, 6*, 22–29.

Ruiz-Rico, M., Pérez-Esteve, É., Bernardos, A., Sancenón, F., Martínez-Máñez, R., Marcos, M. D., & Barat, J. M. (2017). Enhanced antimicrobial activity of essential oil components immobilized on silica particles. *Food Chemistry, 233*, 228–236, [CrossRef].

Saporito, F., Sandri, G., Bonferoni, M. C., Rossi, S., Boselli, C., Icaro Cornaglia, A., Mannucci, B., Grisoli, P., Vigani, B., & Ferrari, F. (2017). Essential oil-loaded lipid nanoparticles for wound healing. *International Journal of Nanomedicine, 11*, 175–186, [CrossRef].

Schmitt, J., Hajiw, S., Lecchi, A., Degrouard, J., Salonen, A., Imperor-Clerc, M., & Pansu, B. (2016). Formation of superlattices of gold nanoparticles using Ostwald ripening in emulsions: Transition from fcc to bcc structure. *The Journal of Physical Chemistry, 120*(25), 5759–5766.

Sercombe, L., Veerati, T., Moheimani, F., Wu, S., Sood, A. K., & Hua, S. (2015). Advances and challenges of liposome assisted drug delivery. *Frontiers in Pharmacology, 6*, 286, [CrossRef].

Sinico, C., De Logu, A., Lai, F., Valenti, D., Manconi, M., Loy, G., Bonsignore, L., & Fadda, A. M. (2005). Liposomal incorporation of *Artemisia arborescens* L. essential oil and in vitro antiviral activity. *European Journal of Pharmaceutics and Biopharmaceutics, 59*, 161–168, [CrossRef].

Składanowski, M., Wypij, M., Laskowski, D., Golinska, P., Dahm, H., & Rai, M. (2016). Silver and gold nanoparticles synthesized from *Streptomyces* sp. isolated from acid forest soil with special reference to its antibacterial activity against pathogens. *Journal of Cluster Science*. Available from https://doi.org/10.1007/s10876-016-1043-6.

de Souza, M. E., Clerici, D. J., Verdi, C. M., Fleck, G., Quatrin, P. M., Spat, L. E., Bonez, P. C., dos Santos, C. F., Antoniazzi, R. P., & Zanatta, F. B. (2017). Antimicrobial activity of *Melaleuca alternifolia* nanoparticles in polymicrobial biofilm in situ. *Microbial Pathogenesis, 113*, 432–437, [CrossRef].

Sudhakaran, A., & Radha, R. K. (2016). Evaluation of acute and dermal toxicity of essential oil of *Etlingera fenzlii* (Kurz) K. Schum: An in vivo study. *International Journal of Pharmacy and Pharmaceutical Sciences, 8*(7), 69–72.

Thangadurai, D., Naik, J., Sangeetha, J., Said Al-Tawaha, A. R. M., Oluwaseun Adetunji, C., Islam, S., David, M., Kashivishwanath Shettar, A., & Bunmi Adetunji, J. (2021). Nanomaterials from agrowastes: Past, present, and the future. In O. V. Kharissova, L. M. Torres-Martínez, & B. I. Kharisov (Eds.), *Handbook of nanomaterials and nanocomposites for energy and environmental applications*. Cham: Springer. Available from https://doi.org/10.1007/978-3-030-36268-3_43.

Tiwari, S., Singh, B. K., & Dubey, N. K. (2020). Encapsulation of essential oils—A booster to enhance their bio-efficacy as botanical preservatives. *Journal of Scientific Research, 64*, 175–178, [CrossRef].

Ukhurebor, K. E., & Adetunji, C. O. (2021). Relevance of biosensor in climate smart organic agriculture and their role in environmental sustainability: What has been done and what we need to do? In R. N. Pudake, U. Jain, & C. Kole (Eds.), *Biosensors in agriculture: Recent trends and future perspectives. Concepts and strategies in plant sciences*. Cham: Springer. Available from https://doi.org/10.1007/978-3-030-66165-6_7.

Ultee, E., & Smid, J. (2001). Influence of carvacrol on growth and toxin production by Bacillus cereus. *International Journal of Food Microbiology, 64*, 373–378.

Veldhuizen, E., Tjeerdsma-van Bokhoven, J., Zweijtzer, C., Burt, S. A., & Haagsman, H. P. (2006). Structural requirements for the antimicrobial activity of carvacrol. *Journal of Agricultural and Food Chemistry, 54*, 1874–1879.

Vimalanathan, S., & Hudson, J. (2014). Anti-influenza virus activity of essential oils and vapors. *American Journal of Essential Oils and Natural Products, 2*(1), 47–53.

Ways, T. M. M., Ng, K. W., Lau, W. M., & Khutoryanskiy, V. V. (2020). Silica nanoparticles in transmucosal drug delivery. *Pharmaceutics*, *12*, 751, [CrossRef] [PubMed].

Wu, Z., Zhou, W., Pang, C., Deng, W., Xu, C., & Wang, X. (2019). Multifunctional chitosan-based coating with liposomes containing laurel essential oils and nanosilver for pork preservation. *Food Chemistry*, *295*, 16–25, [CrossRef].

Yap, P. S. X., Yiap, B. C., Ping, H. C., & Lim, S. H. E. (2014). Essential oils, a new horizon in combating bacterial antibiotic resistance. *The Open Microbiology Journal*, *8*, 6–14.

Zahid, S. R., Upmanyu, N., Dangi, S., Ray, S. K., Jain, P., & Parkhe, G. (2018). Ethosome: A novel vesicular carrier for transdermal drug delivery. *Journal of Drug Delivery and Therapeutics*, *8*, 318–326, [CrossRef].

Zhang, Y., Liu, X., Wang, Y., Jiang, P., & Quek, S. Y. (2016). Antibacterial activity and mechanism of cinnamon essential oil against Escherichia coli and Staphylococcus aureus. *Food Control*, *59*, 282–289.

Zhang, Y., Niu, Y., Luo, Y., Ge, M., Yang, T., Yu, L., & Wang, Q. (2014). Fabrication, characterization and antimicrobial activities of thymol-loaded zein nanoparticles stabilized by sodium caseinate–chitosan hydrochloride double layers. *Food Chemistry*, *142*, 269–275, [CrossRef] [PubMed].

Chapter 15

The application of essential oils on stored food products for enhancing the nutritional attributes of food products

Babatunde Oluwafemi Adetuyi[1], Kehinde Abraham Odelade[1], Peace Abiodun Olajide[1], Oluwakemi Semilore Omowumi[1], Charles Oluwaseun Adetunji[2], Osarenkhoe Omorefosa Osemwegie[3], Mohammed Bello Yerima[4], M.L. Attanda[5] and Juliana Bunmi Adetunji[6]

[1]Department of Natural Sciences, Faculty of Pure and Applied Sciences, Precious Cornerstone University, Ibadan, Oyo State, Nigeria, [2]Applied Microbiology, Biotechnology and Nanotechnology Laboratory, Department of Microbiology, Edo State University Uzairue, Iyamho, Edo State, Nigeria, [3]Department of Food Science and Microbiology, Landmark University, Omu-Aran, Kwara State, Nigeria, [4]Department of Microbiology, Sokoto State University, Sokoto, Sokoto State, Nigeria, [5]Department of Agricultural Engineering, Bayero University Kano, Kano, Kano State, Nigeria, [6]Department of Biochemistry, Osun State University, Osogbo, Osun State, Nigeria

Introduction

Because of the cutting-edge way of life, shoppers are progressively searching for food sources that are not difficult to plan, sound, and speedy to devour (Ragaert et al., 2004) To have the option to address the issues of purchasers (Ragaert et al., 2004), the food business is continually fostering a wide assortment of prepared to-eat, new cut, chilled food varieties. Safeguarding methods—for example, refrigeration, explicit bundling, moderate warming, and antimicrobial sanitizers—are generally used to keep an item new. Food varieties that are insignificantly handled (MP) are new food sources that are protected to eat, yet have gone through not many changes and don't have critical additives (Buckley et al., 2007). These new MP food varieties, which are offered and stuffed in a prepared-to-eat structure, highlight new cut vegetables, and meats, notwithstanding fish (De Corato, 2020), and are made to give comfort and effortlessness. Another market pattern, in particular, has been made because of a developing interest in profoundly viable protection strategies that don't need compound additives (Adetunji et al., 2013; Adetunji, Adetunji, et al., 2021; Adetunji, Ajayi, et al., 2021; Adetunji, Akram, et al., 2021; Adetunji, Anani, et al., 2021; Adetunji, Inobeme, et al., 2021; Adetunji, Kumar, et al., 2019; Adetunji, Michael, Kadiri, et al., 2021; Adetunji, Michael, Nwankwo, et al., 2021; Adetunji, Michael, Varma, et al., 2021; Adetunji, Nwankwo, et al., 2021; Adetunji, Ojediran, et al., 2019; Adetunji, Olaniyan, et al., 2021; Adetunji, Palai, et al., 2021; Adetunji, Panpatte, et al., 2019; Rauf et al., 2021; Santos & Oliveira, 2012; Ukhurebor & Adetunji, 2021). Delivering MP vegetables and organic products is a specific portion of the MP food industry. Basically, in light of the fact that MP veggies and organic products are accepted to be more nutritious than prepared-to-eat food sources, this specific division has drawn in a lot of client interest. Leafy foods that have gone through insignificant handling are known as MPFVs (negligibly handled products of the soil). These new produce things are generally cut, stripped, destroyed and washed before being bundled as ready to eat (De Corato, 2020).

These things can be utilized to make various healthy dinners in a short measure of time and with negligible waste. MPFV deals in the US are growing by more than USD 15 billion every year and makeup 15% of all vegetable item deals. Prepared-to-eat s salads will be the top selling item, with deals expanding from USD 2.7 to 3.2 billion somewhere in the range of 2001 and 2003. The Unified Realm (UK), which had an income of north of 120,000 tons in 2004, is the mainland's greatest client, with utilization changing significantly among countries. Foods made with MP aren't sterile, be that as it may. MPFVs have microscopic organisms that are normally unsafe because of vegetables 'crude nature and farming beginning (Martínez-Sánchez et al., 2006). Therefore, it should not be amazing that food varieties that have a high dietary benefit likewise dislike food handling and protection. Periodically across the globe, foodborne ailments are related to

Applications of Essential Oils in the Food Industry. DOI: https://doi.org/10.1016/B978-0-323-98340-2.00010-9

vegetables and organic products. In ongoing many years, the quantity of foodborne ailments coming about because of eating crude vegetables and organic products has expanded. Researchers and wellbeing experts (in the food handling areas) have explored the microbial defilement of new produce (Brandl, 2006). There is developing worry about the dangers of microbial expansion given the broad control of these products and the expansion in MPFV utilization around the world. Vegetables can become tainted during the pre-gather stage, as a vegetable in the field, and furthermore during the post-collect stage, during transportation, handling, and pressing (Adejumo et al., 2017; Adetunji, 2008, 2019; Adetunji & Varma, 2020; Adetunji et al., 2013, 2017, 2022; Adetunji, Egbuna, et al., 2020; Adetunji, Oloke, et al., 2020; Adetunji, Roli, et al., 2020; Bello et al., 2019; Carstens et al., 2019; Egbuna et al., 2020; Olaniyan & Adetunji, 2021; Thangadurai et al., 2021). The microbiological immaculateness and security of MPFVs are in this manner huge worries.

Studies have been led throughout the long term on the antibacterial characteristics of EOs and their use in cooking. There has been an expanded interest in using EOs in MPFVs, especially since there are no efficient assessments yet. The examination that is done most frequently centers around in vitro testing. Few exhibit their purposes in pragmatic settings. Subsequently, the ongoing survey centers around the impact of rejuvenating oils against MP food microorganisms, zeroing in on their considerably more viable applications, especially on promising new methodologies for their sound use. With this information, medicinal ointments might all around become more associated with the food business.

Food

Any substance seen as food acts as a source of nourishment for an organic entity. Food comprises different plants, creatures, or growths and incorporates different fundamental components, like nutrients and minerals, sugars, lipids, and proteins. The material is ingested by the creature and absorbed by its cells to give energy, to support life, or to advance development.

Classes of food Items

All food might be arranged into one of three gatherings, which call for various capacity techniques.

1. Short-lived food sources comprise various crude vegetables and leafy foods, dairy items, and eggs for the individuals who devour them. Arranged food varieties are completely viewed as transient food sources. To store short-lived merchandise, they should be kept at room temperature, either in the cooler or in the cooler. Whenever refrigerated, a great deal of transient things (less for a ton of creature items) ought to be consumed within three to seven days.
2. Semi-transient food varieties and semi-short-lived food sources can be put away in far or they can ruin rapidly. Semi-transitory things incorporate flour, cereals, got dried-out foods grown from the ground blends. Semi-transitory things can stay pristine for quite some time to a year whenever dealt with and kept appropriately, as in a new, vacuum-fixed pack. Some will remain longer than others assuming that they are frozen.
3. Staple, or durable food varieties comprise canned food sources, dried beans, and flavors. They will not fall apart except if they are utilized with care. Regardless of whether they are kept in the best way, they might lose their quality over the long run.

Food storage

Food storage is a strategy to diminish the fluctuation of the food supply (Lawrence, 2014) even with the innate, unpreventable varieties in the food supply. It empowers the food to be devoured for a lengthy timeframe (normally weeks to months) rather than soon after the reap. It is a perceived homegrown expertise (fundamentally root cellaring) and furthermore a significant first business movement as food planned operations. For virtually all individuals on the planet who depend on others for their food, food security is pivotal. The significance of food stockpiling and transportation, as well as convenient conveyance to shoppers, is all-important for the food security system.

Lacks away conditions and choices made before in the store network essentially add to food misfortunes by making items more inclined to weaken over the long run. Legitimate cold stockpiling is fundamental to staying away from both subjective and quantitative food misfortunes.

Food storage: motivation behind storage

Keeping assembled and arranged plant and creature food varieties away before appropriating them to clients
Working with all-year dietary agreements.

Safeguarding uneaten or undesirable nourishment for sometime later saving cabinet things such as dry merchandise or flavors like wheat and rice for later use in cooking groundwork for times, crises, or catastrophes of food shortage or starvation, either in its considerably more extreme type of survivalism (preparing).

The critical parts of food security are as of now food access, food steadiness, food use, and, above all, food conservation to stay away from future flare-ups of irresistible illnesses. These four food support points make the financial climate and significantly affect the requirement for modest food. Microorganisms and toxic substances delivered along the pecking order and food web are the essential drivers of huge food waste and biodeterioration with regard to food uncertainty, because of their drawn-out impacts. In various unfortunate nations, microbial pollution prompts a deficiency of food of 2530%. Microbial defilement of food at different phases of creation and arrangement can result in different foodborne illnesses. Notwithstanding microbial defilement, microorganisms and organisms can deliver harms.

Numerous substance additives have been compelling in diminishing microbial pollution of food items, like sulfur dioxide, sulfites, sodium nitrite, sodium benzoate, sorbates, formaldehyde, imidazoles, thiocyanates and pyrrolidines (Gutiérrez-del-Río et al., 2018). The drawn-out patterns of their breakdown, natural poisonousness, a repeat of irritations, and the opportunity of human and teratogenesis and creature disease have, in any case, raised negative worries among buyers (Falleh et al., 2019). Thus, plant-based additives, particularly natural ointments, and dynamic fixings got from fragrant and restoratively critical plants, are acquiring expanding acknowledgment in the food ventures in light of their wide range of antibacterial, antimycotoxic, and cell reinforcement properties. Likewise, since essential oils (EOs) for the most part adjust to rules, their utilization extends a new eco-accommodating strategy for food insurance.

The EOs that make up this intricate mix are terpenoids, phenylpropanoids, terpenoids alongside different synthetic substances having low sub-atomic loads. Brahmi et al. have reported the far and wide antibacterial impacts of EOs in various post-collect food stockpiling frameworks. Due to their hydrophobic/lipophilic person, EOs can go through the bilayer layer of cell films, which restrains ionic conduction, causes the spillage of cell materials, changes the intervened electron stream of the proton intention force, and causes demise. Since the external LPS layer confines the stream pace of lipophilic EOs into the intracellular milieu, natural ointments strongly affect gram-positive than gram-negative microscopic organisms.

Regardless of the critical additive properties of rejuvenating oils in the food framework, a few constraints have been noted in their functional application as a result of their strong fragrance, high reactivity, pore volume, brought down water content and potential for negative communications with the carb, unsaturated fat structure, and fat of food sources. These imperfections have been cured by integrating different contemporary specialized advancements consolidating different techniques for circulation. Using nanoencapsulation of medicinal ointments (light, pH), dampness, and oxygen is one of the most outstanding conveyance strategies that work on antimicrobial adequacy through improved control, retention, and security arrival of natural balm aroma in the food production network.

Epitomized rejuvenating oils might be one of the most mind-blowing non-harmful and eco-agreeable substitutes for engineered additives for the food business. One more suitable conveyance strategy for expanding food timeframe of realistic usability is dynamic food bundling and polymer covering with medicinal ointments. The reason for this audit is to give current data on the usage of medicinal ointments as eco-amicable additives in the food framework, with steady long-haul adequacy in the food framework, as a shrewd green additive in this unique situation (Adeleke et al., 2022).

Essential oils

EOs can be extricated by utilizing natural oil plants. The oils contain the substance of the plant and the fragrance and flavor that it conveys. What gives every medicinal balm its particular pith is its unmistakable fragrant constituents. EOs can be acquired by modern cycles like virus squeezing or refining (utilizing steam and/or water). After the sweet-smelling compounds have been removed, a transporter oil is added to the blend to deliver a prepared-to-use item. You want to see precisely how the oils are delivered on the grounds that medicinal oils are not produced using oils that are separated by compound cycles (Didunyemi et al., 2019).

Medicinal oils, a soaked watery fluid alluded to as natural ointments, are made out of compound parts from unpredictable plants (at room temperature effectively evaporable). Natural ointments are on occasion portrayed as unpredictable oils, ethereal oils, aetheroleum, or simply the oil extricated from the plant, for example, clove oil. Medicinal oils are required because they contain the quintessence of the particular smell of the plants (Oxford English Dictionary, 2014).

Natural oils are regularly removed by refining, frequently utilizing steam. Further systems incorporate articulation, partition and sanitization, absolute oil extraction, tar squeezing, wax implantation, and cold squeezing. They can be utilized in aromas, excellence items, cleansers, deodorizers, and make home cleaning items.

Fragrance-based treatment is a type of reciprocal medication wherein sweet-smelling particles are remembered to have restorative properties. There is some proof that natural ointments can assist with loosening up individuals, however, there is no decisive proof that this is valid (Lee et al., 2012). In the event that natural ointments are utilized erroneously, they can cause skin disturbance, aggravation, and hypersensitive responses.

Posadzki et al. (2012) demonstrate the way that youngsters can be especially vulnerable to the adverse consequences of wrong use. EOs can be poisonous on the off chance that they are applied straightforwardly to the skin or on the other hand assuming they are consumed. The vast majority of the incidental effects are slight, yet a couple of EOs have been connected to serious harmful responses, including hepatotoxicity, bronchial hyperreactivity, seizures, unexpected pregnancy results, and fetus removals. Youngsters' incidental utilization has frequently been tragic.

History of essential oils

A past filled with natural ointments old Egyptian frankincense, cedarwood, myrrh, juniper berry, and cinnamon might have had rejuvenating oils. Pitches of aromatics and establish extricates have been held to make customary prescriptions and aromas like incense and fragrances. EOs have been utilized in people's medication since antiquated times. Current distributions frequently depict the interesting substance intensifies that make up medicinal ointments as opposed to alluding to the actual oils, for instance, utilizing the expression "methyl salicylate" instead of "oil of wintergreen" .

Sorts of essential oil

Fragrance-based treatment, a discipline of correlative medication that uses EOs alongside other sweet-smelling substances, uses medicinal oils. Oils might be mixed with a transporter oil and utilized topically, scattered in the air through a nebulizer or some other gadget, warmed over a candle fire, or utilized as incense. Medicinal ointments are in many cases utilized in perfumery and in food enhancing in light of the fact that they have a smell. They're by and large ready by aroma extraction strategies such as refining, cold squeezing, or dissolvable extraction. Medicinal ointments can be recognized from fragrance oils (natural oils and smell intensifies inside a sleek dissolvable), imbuements in vegetable oil, cement, and absolutes. Natural oils are normally mind-boggling combinations of many different smell compounds.

- Agarwood (*Aquilaria malaccensis*) is refined into agarwood oil or oodh. It is profoundly valued for its scent.
- Dried Ajwain oil (*Carum copticum*) departs 3565% of thymol is contained in oil (Singh et al., 2004).
- Dried *Angelica archangelica* root oil is refined out of the rejuvenating balm of *Angelica archangelica*.
- Anise oil from *Pimpinella anisum*, wealthy in the scent of licorice Asafoetida oil, is used to season food.
- Resin from the Myroxylon species of Peru is utilized in beverages and food for flavoring, in toiletries and perfumes for scenting, in basil oil for creating fragrances, and in straight leaf oil for perfumery and aromatherapy. Bergamot oil is utilized in fragrant healing and scents.
- Dark pepper oil, which is likewise utilized in fragrance-based treatment, is removed from the berries of Flute player nigrum.
- Blue tansy (*Tanacetum annuum*) for the tone of its oil.
- Oil of buchu is made from the buchu bush. Considered harmful, it is not generally broadly utilized. Beforehand, it was utilized therapeutically.
- Calamansi Medicinal ointment or Calamodin oil is extricated by chilly squeezing or steam refining out of a citrus tree in the Philippines.
- Calamus oil is utilized in perfumery, and previously as food added substance Carrot seed oil, and is utilized in fragrant healing.
- Cedar oil (or cedarwood oil) is principally utilized in scents and fragrances.
- Chamomile oil have various assortments; however, only two are used in fragrant healing: German and Roman. The German chamomile has a more elevated level of azulene Cinnamon oil, which is utilized in perfumery for the enhancing of *Cistus ladanifer* leaves and blossoms.
- Clove oil is utilized for restorative purposes and in perfumery.
- Espresso oil is utilized to season food, and coconut oil is utilized for skin and hair care.

- Mint oil is utilized in fragrance-based treatment, mouthwashes, toothpaste, and drugs, as well as in enhancing toothpaste.
- Orange oil, such as RB_IN, lemon oil, is cold squeezed instead of refined. It comprises 90% d-limonene. It is utilized in food-enhancing, cleaning items, and as a scent.

Essential oils—advantages

Various examinations have shown that medicinal ointments usefully affect tension and stress and can be utilized alone or in a blend with different medicines.

Some accept that you can treat cerebral pains and headaches by applying a mix of chamomile and sesame oils to the sanctuaries. It is a standard Persian cerebral pain fix.

Medicinal ointments—for example, lavender—can assist with advancing a solid rest and have been utilized to treat coronary illness.

EOs have many purposes, and some of them are for fragrance-based treatment. EOs are frequently used to invigorate things like clothing or to aroma houses. Both do-it-yourself beauty care products and excellent regular items use them as normal fragrances. It has been proposed that medicinal oils could supplant DEET, an engineered bug repellent, and be both harmless to the ecosystem and secure. Notwithstanding, there are clashing outcomes with regard to their viability. Research has found that specific oils, including citronella, may save some mosquito species for about 2 hours. The assurance time frame might be stretched out by as long as 3 hours when utilized in mix with vanillin. Also, the properties of EOs recommend that some could be utilized mechanically to broaden the time span of usability of food sources.

Essential oils—aftereffects

EOs and plants might be awful for your wellbeing in view of the bioactive substances they contain. Most medicinal ointments, notwithstanding, are viewed as innocuous when utilized related to a base oil or breathed in. Assuming you are around kids, pregnant ladies, or canines, it's very possible they're taking in the undesirable fragrance. They could lead to specific adverse consequences, for example, rashes, asthma assaults, and cerebral pains.

Responses to EOs can incorporate a rash, yet more serious responses can happen and one individual has passed on. The natural oils lavender, peppermint, tea tree, and ylang end up being the ones that are generally connected with negative responses. In light of their phenolic nature, cinnamon oil and phenolic oils are not suggested for use on the skin alone. Copies can happen when the skin is presented to the UV beams of the sun, and citrus EOs can make the skin become excessively delicate to the sun. EOs ought not to be gulped as gulping them can be lethal, and at times even deadly.

Synthetic arrangement of natural oils

Different sweet-smelling plants produce natural ointments as significant auxiliary metabolic items that are confined from their leaves of theirs, stems, roots, seeds, buds, blossoms, bark, and organic products. Embodiment, or maybe the presence of flavors and smell, is the spot the name "essential oil" begins. Rejuvenating oils have a wide scope of fundamentally related phenylpropanoids, terpenoids, low sub-atomic weight lipophilic short-chain aliphatic hydrocarbons, and phenolic parts.

Rejuvenating oils contain bioactive substances that can deliver flavor or smell notwithstanding having a strong inherent scent. These substances are unpredictable in nature. Different plant gatherings, similar to Lauraceae, Apocynaceae, Solanaceae, Myrtaceae, Apiaceae, Piperaceae, Zingerberaceae, Cyperaceae, Lamiaceae, and the Asteraceae, have EOs which have been separated from them. In the climate, EOs assume a decent part in plant safeguard, for example, enduring the natural pressure, extraordinary antifungal and antimicrobial exercises, and possible significance in food and pharma enterprises.

Mode of activity of essential oils

Information from past examinations on the antibacterial movement of natural ointments as well as the bioactive synthetic substances they contain offers various elective speculations about how their capability against microbial cells. Variety in its synthesis is viewed as a convincing hypothesis about the antibacterial properties of medicinal ointments (Chouhan et al., 2017). Each utilitarian gathering communicates with the constituents of the objective bacterial cell in a way that is unmatched. Consequently, it's impractical to pinpoint the specific interaction through which EOs influence bacterial cells (Sakkas et al., 2016).

Since all cell exercises are interconnected, a few cell components are disturbed straight by the bioactive mixtures in oils that are fundamental. These factors lead to backhanded disturbance of other cell exercises . Presumably, the most fragile biochemical cycles among them incorporate cytoplasmic layer rebuilding or maybe adjustment, cell content extravasation, bacterial cell wall breaking down, film protein degeneration, cytoplasmic coagulation, & decrease of proton thought process force (Nazzaro et al., 2013).

Natural oil bioactive parts have hydrophobic properties. This bacterial cell film lipid division is generally welcomed by this hydrophobicity, which additionally causes configurational disfigurement and expanded penetrability. Cell ionic misfortune and substance spillage are the aftereffects of this underlying treatment (Wu et al., 2016). Small amounts of content and particle misfortune affect the feasibility of bacterial cells, while a ton of content misfortune, particularly fundamental ionic and sub-atomic fatigue, causes cell demise. For instance, different examinations show indisputably that tea tree natural oil will cause the cell passing or maybe lysis of some *Escherichia coli* strains (Wang et al., 2017).

Aggression of EOs against organisms relies upon the centralization of the phenolic contents of its, unnecessarily because of the greater percent of phenolic contents as caravel, eugenol, menthol, and thymol(2-methoxy-4-(2-propenyl)) which prompts a lot more grounded antimicrobial capability of medicinal ointments (Wang et al., 2016). The method of activity of every one of these phenolic contents is basically the same as one another including the obliteration of dynamic e⁻ transport stream, cytoplasmic film interruption, aggravation in proton rationale powers, and coagulation of cell content (Nazzaro et al., 2013). Fastidious microbial control instrument of EO relies upon the adaptability of substance outline work of individual natural oil. Logical examinations give proof that the existence of explicit practical gatherings in phenols emphatically influences its antibacterial and antifungal activities similar to OH− group of menthol is the essential key explanation of its antibacterial activity (Guimarães et al., 2019). Eventually, area of Gracious gathering on phenolic ring doesn't impact its movement. Nonphenolic parts of rejuvenating oil likewise assume a functioning part in controlling microbial movement, an instance is the alkyl group. 1-methyl-4-(1-methylethenyl)-limonene or cyclohexene have significantly high energetic antibacterial activity than p cymene (Senior members and Dorman, Deans, 2000).

Bioactive mixtures of rejuvenating oil trap cell proteins implanted in cytoplasmic film. Bacterial cell film implants a few fundamental compounds, for example, ATPase catalyst padded and supported by lipid particles. These safeguarded catalysts may be inclined to natural ointments in two unique ways. At first, they're inclined to cyclic hydrocarbon assault. They are lipophilic in nature, which makes it a lot more straightforward for them to collect in cytoplasmic layer lipid bilayer, provoking lipidprotein cooperation deformation (Fabre, 2015). The second going-after component depends on head-on collaboration between hydrophobic compound pieces with the lipophilic hydrocarbons. A couple of EOs initiate pseudomycelia development, that is, cell development in conjunction with start to finish bond. Antibacterial specialists of medicinal oils create incomplete partition between natural balm yeast cells by cooperation with the catalysts liable for energy guidelines and incorporating cell parts. This is the top method of collaboration among imperative cell proteins and bioactive mixtures of EOs (Gabriel et al., 2018). An outline of the method of activity of fundamental against various bacterial cell parts is introduced.

Antimicrobial capability of EO is not entirely set in stone through various procedures such as plate dissemination examination, stock weakening strategy, bioautography, and so on. (Horváth et al., 2010). Logical writing additionally proves that these strategies and standards are utilized to assess antibacterial and antifungal activity, yet no verified system was made to expect the outcome of viability of EOs against food-borne organisms and decay agents to date. The Public Board of Trustees for Clinical Research Center Guidelines furnished the clinical local area with antibacterial weakness test for Rejuvenating oils and antimicrobial. Researchers are zeroing in on choosing better exploratory strategies to expand the extent of their utilization in different areas of natural sciences (Balouiri et al., 2016). Yet, factors like extraction strategy of natural oil, pH, size of inoculum, development stage volume, supplement medium, & hatching time-temperature blend impact the determination of methodology. Broad-explored information in the field makes it substantially more challenging to order the outcomes and look at every one of the outcomes together making researchers keep on examining the antimicrobial ability of Natural oils (Shaaban, 2020). During the 1990s, research-based information with respect to natural balm portrayal and its potential application was lacking. At that time because of considerably less buyer mindfulness, business processors were utilizing less expensive manufactured additives. Further developed research in the field expanded customer mindfulness and concerns in regard to the well-being dangers of counterfeit synthetic food-added substances, expanding the craving to search for more secure and normal other options (Mitterer-Daltoé et al., 2020). Around then, regular additives acquired logical consideration. After broad examination, the utilization of medicinal ointments and concentrates from flavors and spices in food arrangements acquired significance (Bearth et al., 2014).

Classes of essential oil constituents

Antimicrobial activity

The plant produces a scope of fixings with antibacterial and antifungal activities. Certain individuals exist generally while the rest are made as a result of the reaction to actual injury or attack from microorganisms (Roller, 2003). Distinguishing the most dynamic antibacterial mixtures of rejuvenating ointments is lumbering since medicinal ointments are complicated combinations of up to forty-five unique constituents (Delaquis et al., 2002), and furthermore, the design of a particular natural balm might vary in light of the time of reap, & the procedures utilized to remove the ointment (Nannapaneni et al., 2009). Natural ointment constituents are a different group of lower sub-atomic weight natural mixtures with enormous contrasts in antimicrobial movement. The dynamic mixtures can be parted into four gatherings as per their synthetic construction: terpenes, phenylpropenes, terpenoids, and others.

Most EO constituents have a couple of targets. It's consequently difficult to foresee exactly the way that powerless a microorganism is as well as why the helplessness shifts from one strain to another. Expectations about the method of activity of rough natural oils require careful examinations of their components' objective site, the method of their activity, as well as their collaborations of theirs with the general climate.

Terpenes

Hydrocarbons created out of a mix of a few units of isoprene (C5H8) are called terpenes. They are combined in the plant cell cytoplasm, and the combination continues through the pathway of mevalonic acid beginning from acetyl Coenzyme A. They possess a hydrocarbon spine that may be improved to cyclic designs by cyclases, consequently shaping bicyclic designs or monocyclic (Caballero et al., 2003). The essential terpenes are sesquiterpene (C15H24) & monoterpenes (C10H16), however for a more extended period chains like diterpenes (C20H32), triterpenes (C30H40), and so on.

They don't represent a gathering of components with great intrinsic antibacterial action. For instance, one of the fundamental components in thyme known as p cymene does not have antibacterial action against a few Gram-negative microbes at a concentration of 85,700 µg/mL. In a huge scope try, α-pinene, δ-3-carene, limonene, β-pinene, α-terpinene, and sabinene, showed no or maybe negligible antibacterial activities against twenty-five unique genera of microbes that present issues in plants, animals, food items (Senior members and Dorman, Deans, 2000). Koutsodaki analyzed the impact of p cymene, α-pinene, β-myrcene, limonene, γ-terpinene, β-pinene, and β-caryophyllene against *Bacillus cereus, Staphylococcus aureus*, and *E. coli*, and their antibacterial activities were missing. On the other hand, low γ-terpinene and p Cymene were inadequate as fungicides against *S. cerevisiae* (Rao et al., 2010).

Terpenoids

Terpenoids are terpenes that go through biochemical changes by means of catalysts that add O_2 particles and transfer or maybe dispose of methyl groups (Caballero et al., 2003). They can be ordered into phenols, alcohols, ethers, ketones, aldehydes, esters, and epoxides. Instances of terpenoid include menthol, thymol, piperitone, citronellal, linalyl acetic acid derivation, linalool, geraniol, and carvacrol.

The antibacterial action of many terpenoids is connected to their useful gatherings, and it's been found that the OH group of the presence and phenolic terpenoids of delocalized electrons are significant for antibacterial action. For instance, the antibacterial activities of the p Cymene and carvacrol subsidiaries carvacrol methyl ether were fundamentally more modest contrasted with carvacrol (Senior members and Dorman, Deans, 2000). They are a sizable gathering of antibacterial mixtures which are dynamic against a wide range of organisms, with the most dynamic monoterpenoids distinguished hitherto being thymol and carvacrol.

Phenylpropenes

They comprise a sub-family known as phenylpropanoids which are combined from the amino acid forerunner phenylalanine in plants. The name phenylpropanoids are obtained from the 6-carbon fragrant phenol bunch & the 3-carbon propene tail of cinnamic acid, created in the absolute initial step of the biosynthesis of phenylpropanoid. The phenylpropenes comprise basically little pieces of EOs and individuals who have been most completely examined are safrole, eugenol, cinnamaldehyde, vanillin, and isoeugenol.

Antibacterial action: mode of activity

Medicinal ointments are known for holding onto a wide range of regular exercises (Burt & Reinders, 2003). The antimicrobial efficacies of medicinal oils and their parts were deep-rooted since vestige. This viability of natural ointments is either bacteriostatic (Rejuvenating oils repress the bacterial development then the microbial cells might recover the regenerative capacity) of theirs, or maybe bactericide which implies that EOs kill bacterial cells (Calo et al., 2015). Unquestionably, the EOs instruments of activity aren't yet officially settled somewhat because of the significant changeability of the synthetic mixtures (Hyldgaard et al., 2012). In any case, almost certainly, natural ointments antimicrobial respectability isn't owing to only one or maybe a particular system yet it's conceivably related to many focuses in the bacterial cell (Calo et al., 2015). With this regard, there's a wide settlement on the truth that the EOs hydrophobicity/lipophilicity permits them to cross the cell cytoplasmic film and mitochondria and permeabilizes the various layers of theirs of phospholipids, polysaccharides, and unsaturated fats (Burt & Reinders, 2003). In addition, Rodriguez-Garcia et al. (2016) explained that the lipophilic finishes of lipoteichoic acids in gram-positive microscopic organisms' cell film benefit the entrance of natural oils hydrophobic corposants, through the external film encompassing the gram-negative microbes cell wall, limits the stream pace of hydrophobic EOs through the lipopolysaccharide layer. That is the essential explanation that represents that gram-positive microorganisms are somewhat undeniably more delicate to EOs activity than gram-negative ones (Dhifi et al., 2016). Also, EOs can break down bacterial cell walls alongside their cytoplasmic film structures by upsetting the adaptation of their different unsaturated fats, polysaccharides, and phospholipid layers by raising their penetrability. Unsettling influences to these two designs suggest the decrease of the layer potential, significant particle alongside other cell contents spillage, a decrease of the ATP pool, a breakdown of the proton siphon, a fragilization of the cell film ultimately a deficiency of macromolecules (Gutiérrez-del-Río et al., 2018). Each of these negative impacts will be the essential driver of cell fundamental cycles harm and definitely to cell lysis. Medicinal ointments can likewise be capable to coagulate the cytoplasm and restraining a few catalyst frameworks, for example, the ones liable for the blend and energy guideline of underlying parts (Burt, Reinders, 2003).

In view of many examinations, EOs displaying presumably the most noteworthy antibacterial exercises against foodborne microbes are brimming with phenolic compounds such as thymol, eugenol, or carvacrol (Dhifi et al., 2016). These terpene phenols designated the proteins amine and hydroxylamine bunches in the bacterial layer to change their penetrability bringing about the microscopic organisms passing (Hyldgaard et al., 2012). Moressential oilver, proof exhibits that Natural oils minor parts have an astounding job in

Natural oils antimicrobial efficiencies, most likely by displaying a synergistic impact with other EOs parts. One realized case is the synergistic association between carvacrol. what's more, p-cymene ρ-cymene scarcely restrains microbial development and carvacrol had a laid out antimicrobial effectiveness against an extraordinary microorganisms range. In view of Rattanachaikunsopon and Phumkhachorn (2010), microbial development was essentially hindered when p cymene and carvacrol were added at precisely the same time on the example medium.

Shockingly, this bacterial hindrance was altogether more fragile when every terpene acted independently in one medium. It's been definite that pyrene assumes the part of substitutional contamination in the bacterial film, decently upsetting the layering capability of unblemished cells. This condition facilitated the carvacrol movement and subsequently brings down the profoundly powerful centralizations of every part (Rattanachaikunsopon & Phumkhachorn, 2010).

Antioxidant action: mode of activity

Oxidation is among the preeminent reasons for food corruption. It's a multilateral response that causes unfortunate changes in food esteem, wholesome quality, organoleptic models as well as the improvement of possibly harmful particles. Food gone through broad oxidation has significant deformities and no shopper-worthiness (Prakash et al., 2015). Oxidation during food handling or potential stockpiling might be uncovered through the appearance and variety changes of off flavors, though the variety of the central parts of its, similar to lipids isn't by and large checked (Prakash et al., 2015). Cell reinforcement compounds can forestall, adjust and end oxidative responses at generally low focuses. With this regard, natural oils and their constituents assume a fundamental part in applying cell reinforcement movement. Rodriguez-Garcia et al. (2016) revealed that rejuvenating oils have numerous methods of aberrant or direct activities including anticipation of free-extremist rummaging and chain inception movement. In past work, C., *C. myrrha, H. spicatum, C. sativum,* and *O. majorana* odorata medicinal oils have been tried for their cell reinforcement exercises through DPPH-free revolutionary rummaging and the β-Carotene/linoleic corrosive dying measure (Prakash et al., 2012). Four of the tried natural ointments showed significant cell reinforcement movement (communicated by low IC50 values) while the sub-par action of *H. spicatum* was ascribed to the low phenolic content of its in contrast with other EOs supporting the previous view that phenolic mixtures might assume significant parts in rejuvenating oils cancer prevention agent movement (Prakash et al., 2012). Surely, and in view of De souza et al. (2019), phenolic compounds, similar to thymol, eugenol, and carvacrol, are the vitally strong cell reinforcement specialists in oils that are fundamental as they can give hydrogen particles to free extremists and change them to undeniably more steady items. The extra components like a few alcohols, monoterpenes, aldehydes, ketones, and ethers: linalool, 1,8 Cinessential oille, geranial/neral, citronellal, isomenthone, menthone, and some monoterpenes additionally assume a fundamental part in the cell reinforcement properties of EOs (Rodriguez-Garcia et al., 2016).

Use of EOs in food safeguarding

Food-borne sicknesses are a developing general medical condition around the world. It's accepted that every year in the US, thirty-one types of microorganisms cause about 10 million instances of food-borne sicknesses. Fruitful management of food-borne microorganisms needs the utilization of a few protection procedures in the assembling and stockpiling of food items. A new customer pattern toward an inclination for items with lower sugar and salt substance presents an uplifted requirement for effective food additives, as diminishing the salt and sugar content would somehow think twice about items' time span of usability. Various additives are utilized to expand the timeframe of realistic usability of an item by repressing microbial development. In any case, a dynamically regrettable shopper impression of counterfeit food-added substances has prodded an interest in tracking down regular options in contrast to the conventional arrangements. Albeit initially added to change and upgrade taste, the antibacterial action of natural ointments goes with them a la mode decision for subbing engineered additives.

Application in stored food items

Meat products

Supplement-rich grids, similar to meat products and meat, can upgrade bacterial fix and turnover of cell parts, which could work on the obstruction of bacterial populaces to various burdens (Gill et al., 2002). By and by, there are different variables that could impact the antimicrobial impact of rejuvenating oils in food varieties (Juven et al., 1994). The fat items in conventional meat items are fundamental and high impressively oils are solvent in lipids, which taken with the typical low ph that builds the dissolvability of the rejuvenating oils as well as the typical low pH that decreases the watery period of food items expanding the relationship between medicinal ointments and deterioration or maybe food-borne microorganisms, upgrades the antibacterial impact of theirs. The usage of EOs to upgrade both food handling and time span of usability of meat items has been accounted for basically in hamburgers, chicken, sheep, or maybe hare new meat (Tsigarida et al., 2000).

In any case, only two or three logical examinations have depicted the use of EOs as food fixings (Dussault et al., 2014). Utilization of EOs in food items is surprising and chiefly restricted to boiled or fried meat items. The use of cinnamon and thyme medicinal oils in ham was demonstrated to diminish the populace of *L. monocytogenes* considerably (Dussault et al., 2014). Additionally, the timeframe of realistic usability of mortadella keeps on being reached out with the utilization of rosemary/thyme EOs (Viuda-Martos et al., 2010a, 2010b), while the time span of usability of bologna hotdogs was expanded utilizing oregano medicinal oil (Viuda-Martos et al., 2010a, 2010b). García-Díez and partners (García-Díez et al., 2016) detailed the superior well-being of dry-relieved wieners via the antimicrobial impact of oregano and garlic medicinal ointments against *S. aureus*, *L. monocytogenes*, and *Salmonella* spp. At 0.5% and 1% concentrations, clove rejuvenating oil confined the advancement of *Listeria monocytogenes* in meat products at both 30°C and 7°C. The antibacterial impact of natural ointments is correlation with their creation, the qualities of every food item as well as the particular microorganisms to be disposed of. Besides, different elements could influence the antimicrobial impact of rejuvenating oils in food items, which is heat medicines, smoking, compound additives, and bundling (Burt & Reinders, 2003). The usage of medicinal oils related to different advances, similar to warm medicines, beat the electric field, high isostatic pressures (HIP), and dynamic bundling against food-borne microbes have been accounted for (Espina et al., 2014).

The use of medicinal oils as regular additives in food businesses meets the current buyer patterns of harmless to the ecosystem, regular, all-regular and no synthetic compounds added names. By and by, their organoleptic influence should be evaluated in every item, since the fixations expected to get the ideal antibacterial impact may be adversely seen by buyers. For example, Viuda Martos revealed an obvious fragrance of oregano in bologna wieners, which anyway wasn't viewed as horrendous by the specialists. Besides, a few different creators depicted the utilization of oregano medicinal ointment in a Spanish matured dry relieved salchichón, which didn't essentially influence tangible properties, yet better surface, with the subsequent conceivable decrease in the maturing time (Martín-Sánchez et al., 2011). Besides, the expansion of rosemary or maybe marjoram EOs at a grouping of 200 mg/kg in hamburger patties was demonstrated to lessen the oxidation lipid, by virtue of the medicinal oils cell reinforcement properties, thus upgrading the flavor of the patties (Mohamed & Mansour, 2012). In such a manner, great acknowledgment has been accounted for foal meat following ten days of dynamic bundling with 2% of oregano natural oil, as a result of decreased lipid oxidation demonstrated by lower Ski lifts values (Lorenzo et al., 2014).

Cheddar items

The antibacterial viability of the various EOs in milk could be impacted through a couple of elements, similar to the compound design of these items, the focus wherein the ointments are utilized and the organisms which are supposed to be diminished or maybe dispensed with. Contrasting the counter microbial impact of thyme, clove, and cinnamon, EOs in various concentrations (0.5%, 0.1%, and 1%) in full-fat and low-fat delicate cheddar, Smith Palmer saw that 1% was the best fixation for every natural oil. The counter *L. monocytogenes* impact was significantly more articulated in the low-fat cheddar, however, clove medicinal oil at 1% was much more effective in the two sorts of cheddar. This medicinal ointment, at comparable focus, was substantially more vivacious against *S. enteritidis* in full-fat cheddar than in low-fat cheddar. When utilized at a convergence of 0.5%, the number of inhabitants in Salmonella Enteritidis recuperated in the low-fat cheddar, but not in full-fat cheddar. Truly, fat assumes a defensive part of the bacterial cells over antimicrobial specialists. By and by, the protein content is a lot higher in low-fat cheeses and can add to the reduction of the activity of the Medicinal ointments because of the development of edifices between the phenolic mixtures of the EOs and the proteins of these items (Juven et al., 1994). Govaris et al. referenced that the antimicrobial movement of oregano and thyme EOs was identical for *L. monocytogenes* and *E. coli* O157: H7 in Feta cheddar. By the by, they saw that *L. monocytogenes* diminished quicker contrasted with *E. coli* O157: H7, most likely in light of the fact that the natural oils significantly affect Gram-positive than Gram-negative microscopic organisms. This could have different results, for example, in item security, handling, and protection peculiarities, since the prevalent microbiota, notwithstanding starter societies, are generally comprised of Gram-positive microorganisms.

Rejuvenating oils can expand the time span of usability of dairy items, disposing of undesirable microorganisms, yet additionally diminishing how much substance weakens during stockpiling and advertising periods. The expansion of oregano and rosemary rejuvenating oils has impacted the quantity of mesophilic microorganisms in cream cheddar. This event caused a lower ph decrease and a considerably less articulated corrosiveness. Furthermore, the matured and rotten flavors which decided a more limited timeframe of realistic usability of the thing, were considerably less articulated in items with added oregano and rosemary EOs, since the fermentative and oxidative cycles were repressed (Asensio et al., 2015). The expansion of marjoram and rosemary EOs has impacted the number of inhabitants in mesophilic microbes in full cream cheddar, which brought about a lower pH decrease and a significantly less articulated corrosiveness. Clove medicinal balm (0.5% and 1%) confined the improvement of *L. monocytogenes* in cheddar both at 30°C and 7°C (Menon & Garg, 2001). The decrease in the microbial populace is related to the diminished creation of biogenic amines in items that are matured. In Gouda cheddar added with *Zataria multiflora* (thyme-like plant) rejuvenating oil, there was an extensive decline in the development of tyramine and receptor in cheeses with different medicinal ointment focuses. These decreases added up to 44% and 46% for receptor and tyramine, separately, with a medicinal oil centralization of 0.4%, however, the favored focus for the tactile board was 0.2%. In this present circumstance, the decrease of tyramine and receptor was 22% and 29%, respectively (Es' haghi Gorji et al., 2014).

Ocean food sources items

Fats decrease the bioactive capability of rejuvenating oils in ocean depths along these lines to meat items. In light of a report, 0.05% volume by weight oregano rejuvenating oil showed improved results against *Photobacterium phosphoreum* in cod filets than salmon because of the higher greasy profile of salmon (Speranza & Corbo, 2010). Various logical investigations demonstrated that the synergetic impact of EO mixes works on microbial strength of fat-rich ocean depths items (Hassoun & Çoban, 2017). A capacity investigation of fish burgers demonstrated that a combination of thymol, grapefruit seed, and orange medicinal balm in a focus 0.11%, 0.10%, and 0.12%, respectively, improved the timeframe of realistic usability of its ultimately depending on 40%. Furthermore, joined impact of these medicinal ointments in altered environmental bundling was assessed. For this reason, three unique gas blends (5:95O_2:CO_2,

50:50CO_2:O_2 and 30:40:30 N_2:CO_2:O_2) were utilized and viewed as preferable against mesophilic microscopic organisms over the item bundled in air pack (Lucera et al., 2012).

Grain based items

In bread, the fundamental timeframe of realistic usability restricting components is slowing down a result of parasitic decay, especially brought about by molds. To have the option to upgrade its nature, the addition of different normal components is necessary. In view of Rehman, the utilization of orange strip natural ointments in loaves of bread influences its tactile properties and blocks the development of microorganisms. The additive impact of citrus strip medicinal balm might be expanded by splashing the EOs on loaves of bread. A concentration on new amaranth-based pasta demonstrated that regular bioactive mixtures imonene, thymol, and chitosan improve microbial soundness at focuses 2000 and 4000 ppm in a cold environment (Lucera et al., 2012). Budka and Khan (2010) dealt with basil, thyme and oregano natural ointment to impede the movement of *Bacillus cereus* in rice. In view of a review directed by (Gutiérrez et al., 2011), on dynamic pressing intended for changed climate bundling with cinnamon EO to broaden the timeframe of realistic usability of gluten-free cut bread, bringing about the superior tactile nature of bread. Dynamic bundling additionally restrained microbial development in bread cuts. Cancer prevention agents and antimicrobial capability of flavors and spices such as savvy, anise, rosemary, and dark cumin natural balm and phenolic compounds were discovered to neutralize gram-negative and gram-positive microbes to increment the timeframe of realistic usability of some pastry shop items.

Dairy items

Ben Jemaa et al. tracked down that thyme rejuvenating oil (*Thymus capitatus*) and its nanoemulsion emphatically battled gram bacterial development in polluted milk (*Bacillus licheniformis*, *Enterococcus horse*, and *Staphylococcus aureus*) and improved its physicochemical value as well. Moreover, Smith Palmer established that the expansion of 0.5% of cinnamon, clove, or maybe thyme rejuvenating oils to low-fat cheddar brought about a measurable decrease of *Salmonella enteritidis* albeit this multitude of increases changed cheddar organoleptic properties. Similarly, an expected enormous positioning of natural oil parts might be as follow (same request than Medicinal ointments): eugenol> cinnamic corrosive/carvacrol> basil methyl chavicol> cinnamaldehyde> geraniol/citral.

Organoleptic part of essential oils

While involving rejuvenating oil in consumable products as an additive, its solid organoleptic qualities ought to be viewed as first (Sharma et al., 2017). Tactile qualities of the food varieties firmly prepared with spices and flavors stay unaffected (Osman et al., 2019). In light of a review, oregano natural oil in an amount that is little was included as a fish item that is handled without upsetting its tactile qualities. By and by, 0.8% oregano EO in hamburger filets brought about a bothersome flavor (Cabarkapa et al., 2013), though 0.05% focus oregano natural balm gave cod fish filets an eminent wonderful flavor (Ramroop & Neetoo, 2018). Thyme natural ointment covering at a centralization of approximately 1% (of the total weight of shrimp) on boiled shrimps doesn't influence its organoleptic qualities. In any case, whenever this rejuvenating ointment fixation is raised to about 2% to work on antibacterial activities, the tangible adequacy of shrimp was diminished (Ouattara et al., 2001). EOs are economically utilized as a seasoning specialist in handling food wares by virtue of the trademark fragrant properties of theirs (Shaaban, 2020). By and by, the convergence of the motor oil ought to be changed at an ideal level to accomplish the greatest protection impact without upsetting the tangible food qualities (Ramroop & Neetoo, 2018). Every rejuvenating balm has unmistakable fragrant properties in light of its prevalent sweet-smelling compounds. For instance, geranial possesses a scent like that of rose, citral grant a lemon-like fragrance and caravol has an impactful smell (Chizzola, 2013).

Conclusion

EOs are now found to neutralize both pathogens and food waste specialists. Bioactive mixtures of medicinal ointments upset the cell components in a roundabout way or straightforwardly, unsettling influence in a solitary framework, thus, hindering other cell exercises, which makes them ideal for the application of unpalatable products as an additive. However, solid organoleptic qualities of zesty mixtures in oils are a fundamental cut-off of the consolidation of theirs in pre-arranged food as an additive. This issue could be overwhelmed by the utilization of Nano epitomized EOs which give more powerful results in business food conservation. Natural ointment-based antimicrobial and palatable bundling commend item time span of usability as well as reinforce customer trust in pre-arranged food items. This orderly methodology supports the harmless to the ecosystem earth idea.

The use of EOs in food might give an undeniably more normal and appealing choice to industry, meaning another obstruction to forestall the development and endurance of microorganisms in food. Synthetic additives might be supplanted with EOs; this offers the chance for Green naming to which clients are drawn in by their normal picture. This is very pertinent since food quality and well-being are of key significance in the current world.

References

Adejumo, I. O., Adetunji, C. O., & Adeyemi, O. S. (2017). Influence of UV light exposure on mineral composition and biomass production of myco-meat produced from different agricultural substrates. *Journal of Agricultural Sciences, Belgrade, 62*(1), 51−59.

Adeleke, D. A., Olajide, P. A., Omowumi, O. S., Okunlola, D. D., Taiwo, A. M., & Adetuyi, B. O. (2022). Effect of monosodium glutamate on the body system. *World News of Natural Sciences, 44*, 1−23.

Adetunji, C. O., Michael, O. S., Rathee, S., Singh, K. R. B., Ajayi, O. O., Adetunji, J. B., Ojha, A., Singh, J., & Singh, R. P. (2022). Potentialities of nanomaterials for the management and treatment of metabolic syndrome: A new insight. *Materials Today Advances, 13*100198.

Adetunji, C.O. (2008). *The antibacterial activities and preliminary phytochemical screening of vernoniaamygdalina and Aloe vera against some selected bacteria* (pp. 40−43) (M.Sc. thesis). University of Ilorin.

Adetunji, C. O., Egbuna, C., Tijjani, H., Adom, D., Al-Ani, L. K. T., & Patrick-Iwuanyanwu, K. C. (2020). *Homemade preparations of natural biopesticides and applications. Natural remedies for pest, disease and weed control* (pp. 179−185). Publisher Academic Press.

Adetunji, C. O., Palai, S., Ekwuabu, C. P., Egbuna, C., Adetunji, J. B., Ehis-Eriakha, C. B., Kesh, S. S., & Mtewa, A. G. (2021). *General principle of primary and secondary plant metabolites: Biogenesis, metabolism, and extraction. Preparation of phytopharmaceuticals for the management of disorders* (pp. 3−23). Publisher Academic Press.

Adetunji, J. B., Ajani, A. O., Adetunji, C. O., Fawole, O. B., Arowora, K. A., Nwaubani, S. I., Ajayi, E. S., Oloke, J. K., & Aina, J. A. (2013). Postharvest quality and safety maintenance of the physical properties of Daucus carota L. fruits by Neem oil and Moringa oil treatment: A new edible coatings. *Agrosearch, 13*(1), 131−141.

Adetunji, C. O. (2019). Environmental impact and ecotoxicological influence of biofabricated and inorganic nanoparticle on soil activity. In D. Panpatte, & Y. Jhala (Eds.), *Nanotechnology for agriculture*. Singapore: Springer. Available from https://doi.org/10.1007/978-981-32-9370-0_12.

Adetunji, C. O., Ajayi, O. O., Akram, M., Olaniyan, O. T., Chishti, M. A., Inobeme, A., Olaniyan, S., Adetunji, J. B., Olaniyan, M., & Awotunde, S. O. (2021). Medicinal plants used in the treatment of influenza a virus infections. In K. Dua, S. Nammi, D. Chang, D. K. Chellappan, G. Gupta, & T. Collet (Eds.), *Medicinal plants for lung diseases*. Singapore: Springer. Available from https://doi.org/10.1007/978-981-33-6850-7_19.

Adetunji, C. O., Akram, M., Olaniyan, O. T., Ajayi, O. O., Inobeme, A., Olaniyan, S., Hameed, L., & Adetunji, J. B. (2021). Targeting SARS-CoV-2 novel Corona (COVID-19) virus infection using medicinal plants. In K. Dua, S. Nammi, D. Chang, D. K. Chellappan, G. Gupta, & T. Collet (Eds.), *Medicinal plants for lung diseases*. Singapore: Springer. Available from https://doi.org/10.1007/978-981-33-6850-7_21.

Adetunji, C. O., Anani, O. A., Olaniyan, O. T., Inobeme, A., Olisaka, F. N., Uwadiae, E. O., & Obayagbona, O. N. (2021). Recent trends in organic farming. In R. Soni, D. C. Suyal, P. Bhargava, & R. Goel (Eds.), *Microbiological activity for soil and plant health management*. Singapore: Springer. Available from https://doi.org/10.1007/978-981-16-2922-8_20.

Adetunji, C. O., Inobeme, A., Olaniyan, O. T., Ajayi, O. O., Olaniyan, S., & Adetunji, J. B. (2021). Application of nanodrugs derived from active metabolites of medicinal plants for the treatment of inflammatory and lung diseases: Recent advances. In K. Dua, S. Nammi, D. Chang, D. K. Chellappan, G. Gupta, & T. Collet (Eds.), *Medicinal plants for lung diseases*. Singapore: Springer. Available from https://doi.org/10.1007/978-981-33-6850-7_26.

Adetunji, C. O., Kumar, D., Raina, M., Arogundade, O., & Sarin, N. B. (2019). Endophytic microorganisms as biological control agents for plant pathogens: A panacea for sustainable agriculture. In A. Varma, S. Tripathi, & R. Prasad (Eds.), *Plant biotic interactions*. Cham: Springer. Available from https://doi.org/10.1007/978-3-030-26657-8_1.

Adetunji, C. O., Michael, O. S., Kadiri, O., Varma, A., Akram, M., Oloke, J. K., Shafique, H., Adetunji, J. B., Jain, A., Bodunrinde, R. E., Ozolua, P., & Ubi, B. E. (2021). Quinoa: From farm to traditional healing, food application, and phytopharmacology. In A. Varma (Ed.), *Biology and biotechnology of Quinoa*. Singapore: Springer. Available from https://doi.org/10.1007/978-981-16-3832-9_20.

Adetunji, C. O., Michael, O. S., Nwankwo, W., Ukhurebor, K. E., Anani, O. A., Oloke, J. K., Varma, A., Kadiri, O., Jain, A., & Adetunji, J. B. (2021). Quinoa, The next biotech plant: Food security and environmental and health hot spots. In A. Varma (Ed.), *Biology and biotechnology of Quinoa*. Singapore: Springer. Available from https://doi.org/10.1007/978-981-16-3832-9_19.

Adetunji, C. O., Michael, O. S., Varma, A., Oloke, J. K., Kadiri, O., Akram, M., Bodunrinde, R. E., Imtiaz, A., Adetunji, J. B., Shahzad, K., Jain, A., Ubi, B. E., Majeed, N., Ozolua, P., & Olisaka, F. N. (2021). Recent advances in the application of biotechnology for improving the production of secondary metabolites from Quinoa. In A. Varma (Ed.), *Biology and biotechnology of Quinoa*. Singapore: Springer. Available from https://doi.org/10.1007/978-981-16-3832-9_17.

Adetunji, C. O., Nwankwo, W., Ukhurebor, K., Olayinka, A. S., & Makinde, A. S. (2021). Application of biosensor for the identification of various pathogens and pests mitigating against the agricultural production: Recent advances. In R. N. Pudake, U. Jain, & C. Kole (Eds.), *Biosensors in agriculture: Recent trends and future perspectives. Concepts and strategies in plant sciences*. Cham: Springer. Available from https://doi.org/10.1007/978-3-030-66165-6_9.

Adetunji, C. O., Ojediran, J. O., Adetunji, J. B., & Owa, S. O. (2019). Influence of chitosan edible coating on postharvest qualities of Capsicum annum L. during storage in evaporative cooling system. *Croatian Journal of Food Science and Technology*, *11*(1), 59−66.

Adetunji, C. O., Olaniyan, O. T., Akram, M., Ajayi, O. O., Inobeme, A., Olaniyan, S., Khan, F. S., & Adetunji, J. B. (2021). Medicinal plants used in the treatment of pulmonary hypertension. In K. Dua, S. Nammi, D. Chang, D. K. Chellappan, G. Gupta, & T. Collet (Eds.), *Medicinal plants for lung diseases*. Singapore: Springer. Available from https://doi.org/10.1007/978-981-33-6850-7_14.

Adetunji, C. O., Oloke, J. K., & Prasad, G. (2020). Effect of carbon-to-nitrogen ratio on eco-friendly mycoherbicide activity from Lasiodiplodia pseudotheobromae C1136 for sustainable weeds management in organic agriculture. *Environment, Development and Sustainability*, *22*, 1977−1990. Available from https://doi.org/10.1007/s10668-018-0273-1.

Adetunji, C. O., Panpatte, D. G., Bello, O. M., & Adekoya, M. A. (2019). Application of nanoengineered metabolites from beneficial and eco-friendly microorganisms as a biological control agents for plant pests and pathogens. In D. Panpatte, & Y. Jhala (Eds.), *Nanotechnology for agriculture: Crop production & protection*. Singapore: Springer. Available from https://doi.org/10.1007/978-981-32-9374-8_13.

Adetunji, C. O., Phazang, P., & Sarin, N. B. (2017). Significance of rhamnolipids as a biological control agent in the management of crops/plant pathogens. *Current Trends in Biomedical Engineering & Biosciences*, *10*(3), 54−55.

Adetunji, C. O., Roli, O. I., & Adetunji, J. B. (2020). Exopolysaccharides derived from beneficial microorganisms: Antimicrobial, food, and health benefits. In P. Mishra, R. R. Mishra, & C. O. Adetunji (Eds.), *Innovations in food technology*. Singapore: Springer. Available from https://doi.org/10.1007/978-981-15-6121-4_10.

Adetunji, C. O., & Varma, A. (2020). Biotechnological application of trichoderma: A powerful fungal isolate with diverse potentials for the attainment of food safety, management of pest and diseases, healthy planet, and sustainable agriculture. In C. Manoharachary, H. B. Singh, & A. Varma (Eds.), *Trichoderma: Agricultural applications and beyond. Soil biology* (61). Cham: Springer. Available from https://doi.org/10.1007/978-3-030-54758-5_12.

Adetunji, J. B., Adetunji, C. O., & Olaniyan, O. T. (2021). African walnuts: A natural depository of nutritional and bioactive compounds essential for food and nutritional security in Africa. In O. O. Babalola (Ed.), *Food security and safety*. Cham: Springer. Available from https://doi.org/10.1007/978-3-030-50672-8_19.

Asensio, C. M., Grosso, N. R., & Juliani, H. R. (2015). Quality preservation of organic cottage cheese using oregano essential oils. *LWT-Food Science and Technology*, *60*(2), 664−671.

Balouiri, M., Sadiki, M., & Ibnsouda, S. K. (2016). Methods for in vitro evaluating antimicrobial activity: A review. *Journal of Pharmaceutical Analysis*, *6*(2), 71−79.

Bearth, A., Cousin, M. E., & Siegrist, M. (2014). The consumers perception of artificial food additives: Influences on acceptance, risk and benefit perceptions. *Food Quality and Preference*, *38*, 14−23.

Bello, O. M., Ibitoye, T., & Adetunji, C. (2019). Assessing antimicrobial agents of Nigeria flora. *Journal of King Saud University-Science*, *31*(4), 1379−1383.

Brandl, M. T. (2006). Fitness of human enteric pathogens on plants and implications for food safety. *Annual Review of Phytopathology*, *44*, 367−392.

Buckley, M., Cowan, C., & McCarthy, M. (2007). The convenience food market in Great Britain: Convenience food lifestyle (CFL) segments. *Appetite*, *49*(3), 600−617.

Budka, D., & Khan, N. A. (2010). The effect of *Ocimum basilicum, Thymus vulgaris, Origanum vulgare* essential oils on Bacillus cereus in rice-based foods. *Ejbs*, *2*(1), 17−20.

Burt, S. A., & Reinders, R. D. (2003). Antibacterial activity of selected plant essential oils against *Escherichia coli* O157: H7. *Letters in Applied Microbiology*, *36*(3), 162−167.

Caballero, B., Trugo, L. C., & Finglas, P. M. (2003). *Encyclopedia of food sciences and nutrition*. Academic.

Cabarkapa, I., Skrinjar, M., Blagojev, N., Gubic, J., Plavsic, D., Kokic, B., & Radusin, T. (2013). Effect of *Origanum heracleoticum* L. essential oil on marinated chicken meat shelf-life. *Journal of Pure and Applied Microbiology*, *7*(1), 221−228.

Calo, J. R., Crandall, P. G., O'Bryan, C. A., & Ricke, S. C. (2015). Essential oils as antimicrobials in food systems: A review. *Food Control*, *54*, 111−119.

Carstens, C. K., Salazar, J. K., & Darkoh, C. (2019). Multistate outbreaks of foodborne illness in the United States associated with fresh produce from 2010 to 2017. *Frontiers in Microbiology*, *10*, 2667.

Chizzola, R. (2013). Regular monoterpenes and sesquiterpenes (essential oils). *Natural Products*, *10*, 973−978.

Chouhan, S., Sharma, K., & Guleria, S. (2017). Antimicrobial activity of some essential oils present status and future perspectives. *Medicines*, *4*(3), 58.

De Corato, U. (2020). Improving the shelf-life and quality of fresh and minimally-processed fruits and vegetables for a modern food industry: A comprehensive critical review from the traditional technologies into the most promising advancements. *Critical Reviews in Food Science and Nutrition*, *60*(6), 940−975.

Delaquis, P. J., Stanich, K., Girard, B., & Mazza, G. (2002). Antimicrobial activity of individual and mixed fractions of dill, cilantro, coriander and eucalyptus essential oils. *International Journal of Food Microbiology*, *74*(1−2), 101−109.

Dhifi, W., Bellili, S., Jazi, S., Bahloul, N., & Mnif, W. (2016). Essential oils chemical characterization and investigation of some biological activities: A critical review. *Medicines*, *3*(4), 25.

Didunyemi, M. O., Adetuyi, B. O., & Oyebanjo, O. O. (2019). Morinda lucida attenuates acetaminophen-induced oxidative damage and hepatotoxicity in rats. *Journal of Biomedical sciences*, *8*. Available from https://www.jbiomeds.com/biomedical-sciences/morinda-lucida-attenuates-acetaminopheninduced-oxidative-damage-and-hepatotoxicity-in-rats.php?aid = 24482.

Dorman, H. D., & Deans, S. G. (2000). Antimicrobial agents from plants: Antibacterial activity of plant volatile oils. *Journal of Applied Microbiology, 88*(2), 308−316.

Dussault, D., Vu, K. D., & Lacroix, M. (2014). In vitro evaluation of antimicrobial activities of various commercial essential oils, oleoresin and pure compounds against food pathogens and application in ham. *Meat Science, 96*(1), 514−520.

Egbuna, C., Gupta, E., Ezzat, S. M., Jeevanandam, J., Mishra, N., Akram, M., Sudharani, N., Adetunji, C. O., Singh, P., Ifemeje, J. C., Deepak, M., Bhavana, A., Walag, A. M. P., Ansari, R., Adetunji, J. B., Laila, U., Olisah, M. C., & Onyekere, P. F. (2020). Aloe species as valuable sources of functional bioactives. In C. Egbuna, & G. Dable Tupas (Eds.), *Functional foods and nutraceuticals*. Cham: Springer. Available from https://doi.org/10.1007/978-3-030-42319-3_18.

Es' haghi Gorji, M., Noori, N., Nabizadeh Nodehi, R., Jahed Khaniki, G., Rastkari, N., & Alimohammadi, M. (2014). The evaluation of Zataria multiflora Boiss. essential oil effect on biogenic amines formation and microbiological profile in Gouda cheese. *Letters in Applied Microbiology, 59*(6), 621−630.

Espina, L., Monfort, S., Álvarez, I., García-Gonzalo, D., & Pagán, R. (2014). Combination of pulsed electric fields, mild heat and essential oils as an alternative to the ultrapasteurization of liquid whole egg. *International Journal of Food Microbiology, 189*, 119−125.

Fabre, G. (2015). *Molecular interaction of natural compounds with lipid bilayer membranes: Towards a better understanding of their biological and pharmaceutical actions* (Doctoral dissertation). Olomouc, République Tchèque: Université de Limoges; Univerzita Palackého.

Falleh, H., Ben Jemaa, M., Djebali, K., Abid, S., Saada, M., & Ksouri, R. (2019). Application of the mixture design for optimum antimicrobial activity: Combined treatment of Syzygium aromaticum, Cinnamomum zeylanicum, Myrtus communis, and Lavandula stoechas essential oils against Escherichia coli. *Journal of Food Processing and Preservation, 43*(12)e14257.

Gabriel, K. T., Kartforosh, L., Crow, S. A., & Cornelison, C. T. (2018). Antimicrobial activity of essential oils against the fungal pathogens Ascosphaera apis and Pseudogymnoascus destructans. *Mycopathologia, 183*(6), 921−934.

García-Díez, J., Alheiro, J., Pinto, A. L., Soares, L., Falco, V., Fraqueza, M. J., & Patarata, L. (2016). Behaviour of food-borne pathogens on dry cured sausage manufactured with herbs and spices essential oils and their sensorial acceptability. *Food Control, 59*, 262−270.

Gill, A. O., Delaquis, P., Russo, P., & Holley, R. A. (2002). Evaluation of antilisterial action of cilantro oil on vacuum packed ham. *International Journal of Food Microbiology, 73*(1), 83−92.

Guimarães, A. C., Meireles, L. M., Lemos, M. F., Guimarães, M. C. C., Endringer, D. C., Fronza, M., & Scherer, R. (2019). Antibacterial activity of terpenes and terpenoids present in essential oils. *Molecules (Basel, Switzerland), 24*(13), 2471.

Gutiérrez, L., Batlle, R., Andújar, S., Sánchez, C., & Nerín, C. (2011). Evaluation of antimicrobial active packaging to increase shelf life of gluten-free sliced bread. *Packaging Technology and Science, 24*(8), 485−494.

Gutiérrez-del-Río, I., Fernández, J., & Lombó, F. (2018). Plant nutraceuticals as antimicrobial agents in food preservation: Terpenoids, polyphenols and thiols. *International Journal of Antimicrobial Agents, 52*(3), 309−315.

Hassoun, A., & Çoban, Ö. E. (2017). Essential oils for antimicrobial and antioxidant applications in fish and other seafood products. *Trends in Food Science & Technology, 68*, 26−36.

Horváth, G., Jámbor, N., Végh, A., Böszörményi, A., Lemberkovics, É., Héthelyi, É., ... Kocsis, B. (2010). Antimicrobial activity of essential oils: The possibilities of TLC bioautography. *Flavour and Fragrance Journal, 25*(3), 178−182.

Hyldgaard, M., Mygind, T., & Meyer, R. L. (2012). Essential oils in food preservation: Mode of action, synergies, and interactions with food matrix components. *Frontiers in Microbiology, 3*, 12.

Juven, B. J., Kanner, J., Schved, F., & Weisslowicz, H. (1994). Factors that interact with the antibacterial action of thyme essential oil and its active constituents. *Journal of Applied Bacteriology, 76*(6), 626−631.

Lawrence, R. J. (2014). In F. Bill (Ed.), *Global environmental change*. Dordrecht: Springer, pp. XXVII + 973.

Lee, M. S., Choi, J., Posadzki, P., & Ernst, E. (2012). Aromatherapy for health care: An overview of systematic reviews. *Maturitas, 71*(3), 257−260.

Lorenzo, J. M., Batlle, R., & Gómez, M. (2014). Extension of the shelf-life of foal meat with two antioxidant active packaging systems. *LWT-Food Science and Technology, 59*(1), 181−188.

Lucera, A., Costa, C., Conte, A., & Del Nobile, M. A. (2012). Food applications of natural antimicrobial compounds. *Frontiers in Microbiology, 3*, 287.

Martínez-Sánchez, A., Allende, A., Bennett, R. N., Ferreres, F., & Gil, M. I. (2006). Microbial, nutritional and sensory quality of rocket leaves as affected by different sanitizers. *Postharvest Biology and Technology, 42*(1), 86−97.

Martín-Sánchez, A. M., Chaves-López, C., Sendra, E., Sayas, E., Fenández-López, J., & Pérez-Álvarez, J. Á. (2011). Lipolysis, proteolysis and sensory characteristics of a Spanish fermented dry-cured meat product (salchichón) with oregano essential oil used as surface mold inhibitor. *Meat Science, 89*(1), 35−44.

Menon, K. V., & Garg, S. R. (2001). Inhibitory effect of clove oil on Listeria monocytogenes in meat and cheese. *Food Microbiology, 18*(6), 647−650.

Mitterer-Daltoé, M., Bordim, J., Lise, C., Breda, L., Casagrande, M., & Lima, V. (2020). Consumer awareness of food antioxidants. Synthetic vs. Natural. *Food Science and Technology, 41*, 208−212.

Mohamed, H. M., & Mansour, H. A. (2012). Incorporating essential oils of marjoram and rosemary in the formulation of beef patties manufactured with mechanically deboned poultry meat to improve the lipid stability and sensory attributes. *LWT-Food Science and Technology, 45*(1), 79−87.

Nannapaneni, R., Chalova, V. I., Crandall, P. G., Ricke, S. C., Johnson, M. G., & O'Bryan, C. A. (2009). Campylobacter and Arcobacter species sensitivity to commercial orange oil fractions. *International Journal of Food Microbiology, 129*(1), 43−49.

Nazzaro, F., Fratianni, F., De Martino, L., Coppola, R., & De Feo, V. (2013). Effect of essential oils on pathogenic bacteria. *Pharmaceuticals, 6*(12), 1451−1474.

Olaniyan, O. T., & Adetunji, C. O. (2021). Biological, biochemical, and biodiversity of biomolecules from marine-based beneficial microorganisms: Industrial perspective. In C. O. Adetunji, D. G. Panpatte, & Y. K. Jhala (Eds.), *Microbial rejuvenation of polluted environment. microorganisms for sustainability* (27). Singapore: Springer. Available from https://doi.org/10.1007/978-981-15-7459-7_4.

Osman, A. G., Raman, V., Haider, S., Ali, Z., Chittiboyina, A. G., & Khan, I. A. (2019). Overview of analytical tools for the identification of adulterants in commonly traded herbs and spices. *Journal of AOAC International, 102*(2), 376−385.

Ouattara, B., Sabato, S. F., & Lacroix, M. (2001). Combined effect of antimicrobial coating and gamma irradiation on shelf life extension of pre-cooked shrimp (Penaeus spp.). *International Journal of Food Microbiology, 68*(1−2), 1−9.

Oxford English Dictionary (online, American English ed.). Archived from the original on 2014-08-09. Retrieved 2014-07-21.

Posadzki, P., Alotaibi, A., & Ernst, E. (2012). Adverse effects of aromatherapy: A systematic review of case reports and case series. *International Journal of Risk & Safety in Medicine, 24*(3), 147−161.

Prakash, B., Kedia, A., Mishra, P. K., & Dubey, N. K. (2015). Plant essential oils as food preservatives to control moulds, mycotoxin contamination and oxidative deterioration of agri-food commodities e Potentials and challenges. *Food Control, 47*, 381−391.

Prakash, B., Singh, P., Kedia, A., & Dubey, N. K. (2012). Assessment of some essential oils as food preservatives based on antifungal, antiaflatoxin, antioxidant activities and in vivo efficacy in food system. *Food Research International, 49*, 201−208.

Ragaert, P., Verbeke, W., Devlieghere, F., & Debevere, J. (2004). Consumer perception and choice of minimally processed vegetables and packaged fruits. *Food Quality and Preference, 15*(3), 259−270.

Ramroop, P., & Neetoo, H. (2018). Antilisterial activity of Cymbopogon citratus on crabsticks. *AIMS Microbiology, 4*(1), 67.

Rao, A., Zhang, Y., Muend, S., & Rao, R. (2010). Mechanism of antifungal activity of terpenoid phenols resembles calcium stress and inhibition of the TOR pathway. *Antimicrobial Agents and Chemotherapy, 54*(12), 5062−5069.

Rattanachaikunsopon, P., & Phumkhachorn, P. (2010). Assessment of factors influencing antimicrobial activity of carvacrol and cymene against Vibrio cholerae in food. *Journal of Bioscience and Bioengineering, 110*614619.

Rauf, A., Akram, M., Semwal, P., Mujawah, A. A. H., Muhammad, N., Riaz, Z., Munir, N., Piotrovsky, D., Vdovina, I., Bouyahya, A., Adetunji, C. O., Shariati, M. A., Almarhoon, Z. M., Mabkhot, Y. N., & Khan, H. (2021). Antispasmodic potential of medicinal plants: A comprehensive review. *Oxidative Medicine and Cellular Longevity, 2021*12. Available from https://doi.org/10.1155/2021/4889719, Article ID 4889719.

Rodriguez-Garcia, I., Silva-Espinoza, B. A., Ortega-Ramirez, L. A., Leyva, J. M., Siddiqui, M. W., Cruz-Valenzuela, M. R., Gonzalez-Aguilar, G. A., & Ayala-Zavala, J. F. (2016). Oregano essential oil as an antimicrobial and antioxidant additive in food products. *Critical Reviews in Food Science and Nutrition, 56*17171727.

Roller, S. (Ed.), (2003). *Natural antimicrobials for the minimal processing of foods.* Woodhead Publishing.

Sakkas, H., Gousia, P., Economou, V., Sakkas, V., Petsios, S., & Papadopoulou, C. (2016). In vitro antimicrobial activity of five essential oils on multidrug resistant Gram-negative clinical isolates. *Journal of intercultural ethnopharmacology, 5*(3), 212.

Santos, J. S., & Oliveira, M. B. P. P. (2012). Revisão: Alimentos frescos minimamente processados embalados em atmosfera modificada. *Brazilian Journal of Food Technology, 15*, 1−14.

Shaaban, H. A. (2020). Essential Oil as antimicrobial agents: Efficacy, stability, and safety issues for food application. *Essential Oils-Bioactive Compounds, New Perspectives and Applications*, 1−33.

Sharma, H., Mendiratta, S. K., Agrawal, R. K., Gurunathan, K., Kumar, S., & Singh, T. P. (2017). Use of various essential oils as bio preservatives and their effect on the quality of vacuum packaged fresh chicken sausages under frozen conditions. *LWT-Food Science and Technology, 81*, 118−127.

Singh, G., Maurya, S., Catalan, C., & De Lampasona, M. P. (2004). Chemical constituents, antifungal and antioxidative effects of ajwain essential oil and its acetone extract. *Journal of Agricultural and Food Chemistry, 52*(11), 3292−3296.

Speranza, B., & Corbo, M. R. (2010). Essential oils for preserving perishable foods: Possibilities and limitations. *Application of alternative food-preservation technologies to enhance food safety and stability, 23*, 35−37.

Thangadurai, D., Naik, J., Sangeetha, J., Al-Tawaha, A. R. M. S., Adetunji, C. O., Islam, S., David, M., Shettar, A. K., & Adetunji, J. B. (2021). Nanomaterials from Agrowastes: Past, present, and the future. In O. V. Kharissova, L. M. Torres-Martínez, & B. I. Kharisov (Eds.), *Handbook of nanomaterials and nanocomposites for energy and environmental applications.* Cham: Springer. Available from https://doi.org/10.1007/978-3-030-36268-3_43.

Tsigarida, E., Skandamis, P., & Nychas, G. J. (2000). Behaviour of Listeria monocytogenes and autochthonous flora on meat stored under aerobic, vacuum and modified atmosphere packaging conditions with or without the presence of oregano essential oil at 5 C. *Journal of Applied Microbiology, 89*(6), 901−909.

Ukhurebor, K. E., & Adetunji, C. O. (2021). Relevance of biosensor in climate smart organic agriculture and their role in environmental sustainability: What has been done and what we need to do? In R. N. Pudake, U. Jain, & C. Kole (Eds.), *Biosensors in agriculture: Recent trends and future perspectives. Concepts and strategies in plant sciences.* Cham: Springer. Available from https://doi.org/10.1007/978-3-030-66165-6_7.

Viuda-Martos, M., Ruiz-Navajas, Y., Fernández-López, J., & Pérez-Álvarez, J. A. (2010a). Effect of orange dietary fibre, oregano essential oil and packaging conditions on shelf-life of bologna sausages. *Food Control, 21*(4), 436−443.

Viuda-Martos, M., Ruiz-Navajas, Y., Fernández-López, J., & Pérez-Álvarez, J. A. (2010b). Effect of added citrus fibre and spice essential oils on quality characteristics and shelf-life of mortadella. *Meat Science, 85*(3), 568−576.

Wang, L., Hu, C., & Shao, L. (2017). The antimicrobial activity of nanoparticles: Present situation and prospects for the future. *International Journal of Nanomedicine, 12*, 1227.

Wang, T. H., Hsia, S. M., Wu, C. H., Ko, S. Y., Chen, M. Y., Shih, Y. H., ... Wu, C. Y. (2016). Evaluation of the antibacterial potential of liquid and vapor phase phenolic essential oil compounds against oral microorganisms. *PLoS One, 11*(9)e0163147.

Wu, Y., Bai, J., Zhong, K., Huang, Y., Qi, H., Jiang, Y., & Gao, H. (2016). Antibacterial activity and membrane-disruptive mechanism of 3-p-trans-coumaroyl-2-hydroxyquinic acid, a novel phenolic compound from pine needles of Cedrus deodara, against Staphylococcus aureus. *Molecules (Basel, Switzerland), 21*(8), 1084.

Chapter 16

Combinatory effect of essential oil and lactic acid in formulations of different food products

Babatunde Oluwafemi Adetuyi[1], Charles Oluwaseun Adetunji[2], Juliana Bunmi Adetunji[3], Abel Inobeme[3], Osarenkhoe Omorefosa Osemwegie[4], Mohammed Bello Yerima[5], M.L. Attanda[6] and Oluwabukola Atinuke Popoola[7]

[1]Department of Natural Sciences, Faculty of Pure and Applied Sciences, Precious Cornerstone University, Ibadan, Oyo State, Nigeria, [2]Applied Microbiology, Biotechnology and Nanotechnology Laboratory, Department of Microbiology, Edo State University Uzairue, Iyamho, Edo State, Nigeria, [3]Department of Chemistry, Edo State University Uzairue Iyamho, Auchi, Edo State, Nigeria, [4]Department of Food Science and Microbiology, Landmark University, Omu-Aran, Kwara State, Nigeria, [5]Department of Microbiology, Sokoto State University, Sokoto, Sokoto State, Nigeria, [6]Department of Agricultural Engineering, Bayero University Kano, Kano, Kano State, Nigeria, [7]Genetics, Genomics and Bioinformatics Department, National Biotechnology Development Agency, Abuja, FCT, Nigeria

Introduction

Many studies have been conducted in recent years in the search for novel chemicals that can be used as antimicrobials to preserve food. In response to the growing "green" consumerism demand for foods with little to no chemical preservatives, most of these are naturally occurring antimicrobials (Aziz & Karboune, 2018; Pisoschi et al., 2017). However, many of the chemical and physical antibacterial methods that have historically been used to keep foods from spoiling have shown the ability to increase direct and cross-microbial tolerance and even antibiotic resistance (Cheng et al., 2019). Essential oils (EO) are characterized as goods made from naturally occurring raw plant materials such as timber, root, and fruits, according to ISO/FDIS 9235 (2013).

These components can be gotten from distillation through steam or dry method, processing of citrus fruit epicarp by mechanical means. The use of EO alone or together as suggested by several researchers helps to prevent food from microorganisms. EO together with a lot of food preservations, may be beneficial to the food business (Burt, 2004; de Souza Pedrosa et al., 2021). Yet, it has been found that frequent contact with EOs does not lead to an increase in antibiotic resistance or a decrease in susceptibility to antimicrobials and conventional techniques of food preservation (de Souza, 2016). The focus of most researchers related to antimicrobial effect of EO in food, majored on microbes with pathogenic and spoilage mechanisms using cultured medium (de Souza Pedrosa et al., 2021). EO's effects on microbes gotten from probiotics, which are known to have positive impacts on consumer health, as well as on microbes that are usefully utilized as modern-day instruments in the preservation and processing of food, have not been well studied. A number of foods have been fermented using commercially available microbial cultures, sometimes enhancing the nutritional value (such as vitamin production) and adding distinctive flavor, texture, and flavor characteristics. The vital category of microbes for the business related to food is definitely the lactic acid bacteria (LAB) family. Moreover, a variety of foods are created when indigenous microorganisms spontaneously ferment (Cruxen et al., 2019; Johnson, 2017). Extra or native microbes degrade food matrix components under regulated environments, resulting in a variety of intermediate or end-metabolites that give finished goods distinct properties. These microbes can also improve the stability and safety of food by generating metabolites with well-known antibacterial capabilities (Jia et al., 2021). Substances that can favorably or unfavorably alter the physiological activities, metabolic capabilities, may have a direct impact on the growth of anticipated traits in completed products (Marcial et al., 2016). Similarly, these chemicals' effects on probiotics in food delivery systems could change whether or not the alleged health benefits are really realized. Products derived from dairy and meat may be the most promising foods to use the antibacterial capabilities of EO, considering the anticipated little undesirable effects on their corporeal features (Tongnuanchan & Benjakul, 2014). This finding may be explained by the widespread use of herbs and spices with high EO content as constituents in cheese- and meat-based culinary meals, which promotes a more acquainted and cozy experience for customers. It has been

Applications of Essential Oils in the Food Industry. DOI: https://doi.org/10.1016/B978-0-323-98340-2.00025-0

demonstrated that the use of EO can be successfully controlled (Hussien et al., 2019). Nevertheless, a lot of products made from meat and dairy contain a variation of technical microbes that are developing their microflora and engaging in a number of advantageous metabolic processes involved in food preparation and preservation. These organisms' responses to EOs haven't been investigated. With a focus on LAB, including native microorganisms and starter cultures that have been added and have technological or probiotic abilities, this review provides an overview of the literature that has data on the potential effects of EOs on food-borne microorganisms in light of these factors. The results of this review should contribute to the field's understanding of EOs as natural antimicrobials and direct future studies on the most effective ways to incorporate them into foods that support microbial fermentation or contain microorganisms that have been associated with positive health effects. This could be a key stage in the creation of fresh, natural antibacterial candidates for use in food biopreservation procedures.

In general, the stability effect of beneficial microbes on the stability and value of particular incited dairy and meat products. The significance of particular bacteria or microbial groups in establishing the necessary quality and stability of a number of fermented foods—which have been widely studied for their capacity to integrate EOs—is summarized in this section. Over short or long ripening times, starter cultures made from lactic acid are employed to improve the technical and cheese sensory parts. They also help these products' microbiological stability (Johnson, 2017). The Lactococcus species is the primary LAB responsible for milk acidification during cheese production. In addition to improving taste, flavor, and texture in cheeses, Lactococcus uses milk acidification (Lopez-Diaz et al., 2000). Yeasts are necessary for the production of surface-ripened soft and semihard cheeses, despite the fact that their presence and growth are generally undesired in dairy products due to their capacity to harm these items. Lactic acid and alcoholic fermentation can also happen in traditionally fermented dairy drinks, according to Frohlich-Wyder et al. (2019). To produce yogurt and milk that are fermented, LAB uses a variety of Lactobacillus and Streptococcus species. The most extensively considered are dairy products and widely used distribution vehicles for probiotics, even though these bacteria have also been evaluated for their technical effects on those foods (Colombo et al., 2018). To maintain probiotics with high survival rates and physiological activities, the delivery mechanisms used must be suitable (Hill et al., 2014). Probiotics are defined as living microorganisms that, when given to hosts in sufficient doses, improve their health (Shori, 2016). LAB is the microbial species most thoroughly investigated to be used as probiotics and food delivery (Shori, 2016). Raw meat is converted into a fermented product by cultivated or native bacteria by initially decreasing the pH. The development of a high-quality and uniform product is made possible by a variety of factors that can be controlled, including the use of different preservatives to extend the shelf life of fermented meat products (such as salt, nitrite, and/or nitrate) (Cruxen et al., 2019). Fermentative LAB usually replaces aerobic spoilage bacteria as the predominant microflora component when cured meat is enclosed in low gas permeability sheets. The maximal shelf life of vacuum-packaged meats has been linked to the existence of a population of psychrotrophic LAB (Woraprayote et al., 2016). The durability and security of sausages that are fermented are solely attributable to the fermentation that LAB produces. LAB and coagulase-negative cocci which appear to have an improved ability for survival during the fermentation process, were prominent microbes responsible for the transformation of products during the fermentation process of meat (Cruxen et al., 2019). Certain fermented meat products can also produce an odor thanks to the glycolytic, proteolytic, and lipolytic actions of specific yeasts (Asefa et al., 2009). Its use is permitted in the process of fermentation of meat, but it must be able to flourish in those environments and outcompete any competing bacteria that might arise while the product is being preserved (Trabelsim et al., 2019). The incidents have made LAB an important category of microorganisms for food fermentation technology and have the potential to increase the product's health advantage.

The impact of essential oils on good bacteria in dairy products

These events demonstrate that LAB establishes a microbial community with a crucial technical role during food fermentation and the capacity to add health claims to these goods. Given this, a suitable microbiological control in foods supporting variable-intensity fermentation processes should include the application of antimicrobial preservatives with sufficient selectivity to block pathogenic and/or spoilage bacteria of concern while not significantly compromising the physiological capabilities and growth of beneficial microbes co-existing in the same environment. The inclusion of oregano EO showed no negative effects on the fitness of a dairy lactic starter culture made up of *S. thermophilus*, *L. bulgaricus*, and *L. lactis* as measured by viable counts and acidifying activities in milk, in addition to not impacting milk clotting and cutting times during the manufacturing of an Argentine cheese. Because the concentration of oregano EO under investigation had no effect on the length of the lag time that the microorganisms forming the lactic starter culture experienced during milk acidification, it is possible that there were no or only minor negative effects before activation and growth of the microbial culture. When compared to unflavored cheese, Argentinian cheese containing oregano EO exhibited an equal number of lactic starter cultures during cold ripening (Marcial et al., 2016). A cream cheese that had been separately flavored with rosemary and oregano EO was tested for pH, total viable counts of mesophilic microorganisms, which are used as indicators of the progression of the fermentative process, and acidity (as determined by the production of lactic acid). After 21 days in the refrigerator, cream cheese infused with oregano and rosemary EOs had a reduced pH and acidity. These findings show that oregano and rosemary EOs, which were previously thought to be effective cream cheese preservatives due to their ability to delay the development of sour taste as a result of lactose fermentation, which is primarily related to LAB metabolism, had an effect on the evolution of the fermentative process in cream cheese (Olmedo et al., 2013). The growth rate of the starter culture recovered after the first nine days of cold storage, however adding a mixture of oregano and rosemary EOs to minimally ripened Coalho cheese slowed the rise in viable counts of the lactic starter culture, which is made up of L. lactis and L. cremoris and has been found to be useful in reducing Escherichia coli O157:H7. Up to 18 days of storage, the terpenes camphene, eucalyptol, camphor, and -pinene were discovered in Coalho cheese, but only eucalyptol, camphor, and -pinene were detected up to the end of storage (21 days) (Diniz-Silva et al., 2019). The effects of oregano EO and thyme EO on the viable counts of a lactic starter culture in fresh cheese slurries were investigated in two different investigations. During 72 hours of cold storage (10°C), the EOs reduced the viable counts of bacteria producing the lactic starter culture, and these counts were lower than those of L. monocytogenes and S. aureus in the same substrate (de Carvalho et al., 2015). After 24 hours of cold storage (10C) in a cheese-based broth, oregano EO (1.25 and 2.5 L/mL) showed a bactericidal effect, but a bactericidal effect against S. aureus and L. monocytogenes was only found after 24 hours of exposure to a higher EO dosage (de Souza, 2016). Thyme EO only induced a 1- and 2-log decrease in the initial counts of lactic starter culture after 24 hours of cold storage in a cheese-based broth (de Carvalho et al., 2015). Despite the fact that LAB counts did not change, a sheep milk cheese containing rosemary EO demonstrated a significant reduction in Clostridium tyrobutyricum survival during cold storage (Moro et al., 2015). Research utilizing Italian Fior di Latte cheese with several EO revealed no effect on the survival of LAB in cheese, despite a drop in the number of pathogenic bacteria over the period of six days of cold storage. The pH of the cheese remained consistent even after the addition of the EOs (Gammariello et al., 2008). At 75 days of ripening, a white brined cheese containing cumin EO displayed viable Lactobacillus acidophilus counts of >6 log CFU/g, however, these numbers were lower than those reported in cheese made without the EO. S. aureus viable counts were lower in white brined cheese with cumin EO and L. acidophilus, even if S. aureus numbers in this cheese were not adequately decreased after 45 days of ripening (Sadeghi et al., 2013). The presence of higher levels of EO terpenes in the cheese at the beginning of the measured refrigeration storage period was linked to a delay in the increase in viable counts of the probiotic L. acidophilus La-05 during the 21 days of cold storage when oregano and rosemary EO were combined with Minas cheese. The probiotic in Minas cheese with oregano and rosemary EO may have grown more slowly than anticipated since the lactic acid levels in the product were reduced during storage. When Minas cheese with oregano and rosemary EO was stored, the terpenes eucalyptol, camphor, -pinene, -pinene, myrcene, -cymene, -o-cimene, isoborneol, and -terpinenol were identified in the greatest concentrations (Diniz-Silva et al., 2020). Measurements of the amounts of produced metabolic byproducts, particularly lactic acid, can be used to estimate the metabolic activity of L. acidophilus La-05 (and other LAB) in cheeses, these findings may imply that oregano and rosemary EOs slow the fermentation metabolism of this bacterium because they have a reduced capacity to break down lactose (Moritz et al., 2012; Ong et al., 2007). Some terpenoids found in oregano and rosemary EO, which have weak antimicrobial activities when acting alone, may have enhanced antimicrobial activities when combined with other EO constituents (Bassolé & Juliani, 2012), which may be related to the reported effects of these EOs on. L. acidophilus La-05 in Minas cheese. During the 28 days of refrigerated storage, the viable counts of L. acidophilus LA-5 in Minas cheese were higher, or almost 8 log CFU/g, which has been deemed a suitable viable count for probiotics in foods throughout the consumption time. Even with the addition of rosemary and oregano EO, this was true (Mousavi et al., 2019). It was also demonstrated that cottage cheese scented with oregano EO had a lower pH and produced less organic acid after being refrigerated. Organic acids were produced and the pH of the cottage cheese dropped as a result of the EO's delay of the microbiological activities (Asensio et al., 2015). As a result of the fact that dairy products that were fermented are known as biochemically dynamic matrices that undergo significant changes during ripening and storage that are primarily fueled by microbial activity, these findings about the effect of EO on the development of the fermentative processes in cheeses are significant. Lactic starter cultures, local bacteria, or even probiotics may be to blame for this behavior. The effects of several EOs on pathogenic bacteria were evaluated through the computation of the minimum bactericidal concentration (MBC). In addition to L. acidophilus and B. breve, thyme and oregano EOs both showed strong inhibitory effects on pathogenic bacteria. Yet, basil EO inhibited L. acidophilus and B. breve less effectively than pathogenic bacteria (MBC of 10 mg/mL). The significant levels of the phenolic monoterpenes carvacrol and thymol in oregano and thyme EOs were revealed to have strong inhibitory effects on these beneficial microbes. Many compounds may have antibacterial actions that are similar, synergistic, and non-selective.

When the amounts of basil EO are precisely selected to be integrated into a substrate of interest, it has been established that it may be utilized to eliminate pathogenic bacteria without significantly impacting beneficial bacteria (Roldan et al., 2010). Due to their volatility, solubility, and/or propensity to interact with proteins and fat in cheese, it has been shown that EO components in cheeses decrease during ripening and storage. However, it may be more difficult to detect these substances using instrumental techniques that are typically used to measure the constituents of EOs in cheese samples due to the ability of phenolic terpenoids' hydroxyl groups to bind to protein groups through hydrogen bonds (Diniz-Silva et al., 2020). According to some studies that have found a correlation between the counts of LAB in cheese or cheese-emulating models and the terpenes present in EOs, higher concentrations of specific terpenes may be to blame for the inhibition or delay in growth and metabolic activity of the LAB strain or co-culture under study (Diniz-Silva et al., 2020). The specific impacts of clove, cinnamon, cardamon, and peppermint EOs on the survival of several lactic acid cultures during yogurt manufacture were studied. Clove, cinnamon, cardamon, and peppermint EOs exhibited a strong inhibiting effect on *S. thermophilus* and *L. bulgaricus* during the fermentation of milk. A sensorially acceptable amount of each EO—cinnamon, clove, and mint—was tested independently on probiotic *Lacticasei bacillus rhamnosus* and a thermophilic yogurt lactic starter culture during milk fermentation and cold storage of fermented milk. The viable counts of *L. rhamnosus* increased only modestly in milk fermented and stored with clove and mint EOs, but dramatically, in milk with cinnamon EO. *L. rhamnosus* somewhat increased the acidity of milk when it was fermented with clove and mint EOs, but when milk was fermented with cinnamon EOs, the acidity stayed fairly constant. The acidity of milk containing lactic starter culture, clove, and mint EOs gradually rose, whereas the acidity of milk containing cinnamon EO sharply fell. These results suggest that the studied amount of cinnamon EO could leave the lactic starter culture inactive and make fermentation impractical, but that it wouldn't effect the usage of *L. rhamnosus* in probiotic fermented milk (Moritz et al., 2012). These findings suggest that the use of EOs in the production of probiotic yogurt should be approached with caution, because milk fermentation by one or more probiotic strains frequently necessitates a lengthy incubation period and frequently results in low-quality products, necessitating the use of selected probiotics and starter culture to produce functionalized and nutritious yogurt (Aryana & Olson, 2017). A dose of *Micromeria barbata* EO had no effect on the viable counts of LAB in Lebanese yogurt after 20 and 70 days of cold storage of opened and sealed samples, respectively; however, a higher EO dose caused a gradual decline in LAB counts in both samples (El Omari et al., 2020). The effects of EOs on viable counts of LAB are an important issue to investigate since yogurts must contain LAB counts of more than 7 log CFU/g at the time of consumption in order to fulfill quality requirements (Alimentarius, 2010). To examine how an aqueous extract of *Ferulago angulata* (Schlecht.) Boiss and EO affected the bacteria in yogurt containing *L. acidophillus* La-05 and *Bifidobacterium bifidum* Bb-12. Probiotic yogurt with *F. angulata* extract and EO showed more acidity than regular yogurt, which was most likely due to the bacteria's increased metabolic activity. After 28 days in the refrigerator, yogurt containing extract and EO exhibited increased viable counts of *L. acidophillus* and *B. bifidum* (Keshavarzi et al., 2021). Many plant extracts and EO have been shown to possess phenolic compounds capable of stimulating the development of probiotic and yogurt starter cultures. Probiotic yogurts frequently have greater viable counts as the amount of dissolved oxygen in the product environment increases. In the absence of oxygen, phenolics have also been found to boost probiotic growth. Increased probiotic concentrations in *F. angulata* EO and extract-containing yogurt may explain these characteristics. Low sensitivity of technological and/or probiotic cultures to EOs should be especially significant for the development of fermented dairy products that advertise being devoid of synthetic preservatives and/or high in functional LAB (Keshavarzi et al., 2021; Moritz et al., 2012). Because the manufactured goods are directly dependent on the labeled species and number of viable cells of such helpful microorganisms, the validity of the created goods may be jeopardized by potential harmful effects of EOs on the populations of these microorganisms. This is also significant, particularly for dairy products, which have historically been utilized to provide probiotics.

The impact of essential oils on good bacteria in animal products

Several studies have demonstrated that EO, much like they can with dairy products, can successfully prevent a variety of pathogenic and spoilage microorganisms in fermented meat products (Barbosa et al., 2015). To the best of our knowledge, there is currently no research that investigates how adding EOs to meat products affects probiotics. EO have been added to meat products due to their purported antioxidant effects in addition to their antibacterial qualities (Pateiro et al., 2018). When a semi-dried fermented sausage was dipped in a chitosan and rosemary or thyme EO (1%) solution during ripening (20°C), the amount of Micrococace bacteria was reduced. At the conclusion of the maturation process, fermented sausages coated in chitosan and combinations of oregano or thyme EOs had decreased pH and mold-yeast counts (Küçükkaya et al., 2020). Changes in pH and mold-yeast counts have been linked to increases in the

Micrococcaceae bacteria population in fermented sausages (Janssens et al., 2013). The Micrococcaceae species, a common microbial community in fermented meat products, have been linked to enhancements in the quality of fermented meat products by increasing color stability, reducing rancidity, helping with flavor development, and reducing deterioration (Papamanoli et al., 2002). After 60 days of cold storage, the direct addition of basil EO had no effect on LAB numbers. The high counts of LAB, which are important in the fermentation of carbohydrates in medium with the production of lactic acid and reduction of pH, were linked to the inclusion of a starting culture in the product formulation. Thyme EO reduced the Micrococcaceae bacteria counts in Italian-type sausages after 60 days of cold storage, and these levels were considered to be low for the entire monitored storage time (Gaio et al., 2015). According to our research, thyme EO may have a special effect on the two major bacterial communities involved in the fermentation of Italian-style sausage. When oregano and thyme EO were applied right before aging Tunisian dry fermented chicken sausages at low temperatures, the LAB counts did not alter. Although the developing sausages' LAB counts increased somewhat as a result of the oregano EO aqueous emulsion treatment, the Micrococcaceae bacteria counts were unaffected. Little amounts of oregano EO may have migrated over the casing and into the sausage matrix, where they may have triggered the LAB, according to Chaves-Lopez et al. (2012). Although EO typically have broad-spectrum antibacterial capabilities, some of them, when used in small amounts, may encourage the growth of microorganisms that could use their components as sources of carbon and energy (Paparella et al., 2016). The ability of members of this bacterial group to produce adenosine triphosphate (ATP), cope with osmotic stress conditions, and respond more effectively to K + efflux induced by these substances has been linked to their survival in the presence of EOs with well-known antimicrobial properties. The discovery of LAB as the dominant group in the microflora at the end of ripening/storage of some fermented meat products containing EOs, which should promote faster LAB growth, may confirm a successful adaptation of these microorganisms to the environment of meat with the addition of EOs (El Adab & Hassouna, 2016). Although the existing research on the effects of EOs on LAB has been conflicting, some writers have stated that LAB may be the most resistant among Gram-positive bacteria that are normally sensitive to EOs (de Carvalho et al., 2015). During a 12-day ripening phase (18°C−22°C), the Turkish fermented sausage Sucuk was dipped in an aqueous emulsion containing thyme EO, however, this had the opposite effect of increasing the Micrococcaceae bacteria numbers. Sucuk immersed in a thyme EO emulsion had lower yeast-mold counts throughout the ripening process (Ozturk, 2015). By the conclusion of ripening, the most prevalent terpenes discovered in Sucuk that had been dipped in a thyme EO emulsion were p-cymene, -terpinene, and trans-caryophyllene. Nevertheless, the EO treatment had no effect on the terpene profile, which was most likely due to the numerous spices utilized in Sucuk formulation. The production of yeast on the surface of fermented sausages is one of the adverse situations; nevertheless, due to the proteolytic and lypolitic activity of these microorganisms, internal yeast growth may improve some product quality aspects (Ozturk, 2015). The inhibitory effects of EOs on yeast development in fermented sausage, particularly in the inner product area, should be carefully investigated, even if a high yeast growth inhibition rate may result in products with changes in desired physical and/or sensory features. Previous research has found lower LAB counts in cooked chicken products, bologna sausage, and dry sausage with EOs stored at cold temperatures in vacuum-packages (Badia et al., 2020). Yet, these findings were considered positive because lower LAB counts were thought to indicate product stability in these studies. Decreases in LAB in fermented beef products containing EOs during ripening/storage have also been connected to reductions in the availability of fermentable carbs (Lorenzo & Franco, 2012). Since the inhibitory effects on pathogens have been found in products where the EO additin caused decreased LAB counts, the antimicrobial properties of added EOs have typically been attributed to decreases in pathogen counts, such as *S. aureus* and *L. monocytogenes*, in fermented meat products with EOs rather than to medium acidification caused by LAB (El Adab & Hassouna, 2016). Decreased formation of biogenic amines (cadaverine, putrescine, histamine, and tyramine) as a result of microbial enzymatic decarboxylation of free amino acids, as well as general inhibitory effects on contaminating microbes, have also been related to (Lu et al., 2015). Research revealed no influence of *Zataria multiflora* EO on counts of LAB, Enterobacteriaceae, B. thermosphacta, and mold-yeast in a Balka-style fresh sausage during cold storage (Carballo et al., 2019). One study looked at the capacity of *Pimenta racemosa* EO to improve the process of creating fermented fish flour. *P. racemosa* EO infusions in wheat samples revealed greater beneficial microbiological features (absence of dangerous bacteria). Despite the fact that it has been reported that the microbiological characteristics of flour with EO adjunction could indicate a sharp reduction in their microbial populations, this study did not assess the effects of *P. racemosa* EO on bacterial groups of technological interest in fermented fish flour (Adjou et al., 2017). Early studies found that some spices may be able to boost the growth of starter cultures often used in fermented meat products. Natural spices were demonstrated to have these effects rather than its oleoresin. Under laboratory circumstances, cumin EO delayed growth and decreased acid production by *L. plantarum* and *L. mesenteroides*, but oregano EO (0.1%−2%) substantially repressed growth and acid production by these bacteria (Kivanç et al., 1991). The concept of biological activity has been utilized to evaluate the quality of starting cultures (de Melo Pereira et al., 2020). The results of existing research indicate that different EOs may have varying impacts on the features of beginning cultures. The EO kind and tested dose, as well as the microbe that produces the starting culture and completed product, all appear to have an effect on these effects. To aid in the selection of EOs and doses that cause no or few adverse effects on these beneficial microorganisms in foods, more research is needed to better understand the effects of EOs on various physiological functions of technological and probiotic LAB, which could affect their behavior in fermented food matrices. Only one research found that citrus species EOs affect the metabolic and efflux activities of *Levilactobacillus brevis* and *L. mesenteroides*, as well as membrane potential and integrity (de Souza Pedrosa et al., 2020). These findings imply that studies on the effects of EOs on starter and probiotic cultures in fermented products should consider not only population measurements but also physiological traits and metabolite production by these microorganisms, which may be directly related to the fulfillment of quality, stability, and/or health claims of formulated products. Concerns about the overall quality of fermented foods have led study into strategies for preserving them using ecologically acceptable additions, such as the logical examination of EOs as natural antibacterial preservatives. This is a critical factor. Further study is needed to link the effects of EOs on probiotic and/or technological food microorganisms with the development of physical, chemical, and sensory qualities in final foods. This will make it easier to determine if the EOs under consideration are suitable for use as antimicrobial preservatives in the relevant matrices.

Conclusion

Several effects on the growth and development of microorganisms employed in the manufacturing of dairy and meat products, as well as probiotic strains, may be caused by EOs, according to data from the literature.

References

Adjou, E. S., D'egnon, R. G., Dahouenon-Ahoussi, E., Soumanou, M. M., & Sohounhloue, D. C. K. (2017). Improvement of fermented fish flour quality using essential oil extracted from fresh leaves of Pimenta racemosa (Mill.) J. W. Moore. *Natural Products and Bioprospecting, 7,* 299–305. Available from https://doi.org/10.1007/s13659-017-0132-z.

Alimentarius, C. (2010). *Codex standard for fermented milks. Codex standard 243 − 2003 in codex Alimentarius: Milk and milk products* (2th ed., p. 6)Brussels, Belgium: Codex Alimentarius Commission, −1.

Aryana, K. J., & Olson, D. W. (2017). A 100-year review: Yogurt and other cultured dairy products. *Journal of Dairy Science, 100,* 9987–10013. Available from https://doi.org/10.3168/jds.2017-12981.

Asefa, D. T., Møretrø, T., Gjerde, R. O., Langsrud, S., Kure, C. F., Sidhu, M. S., Nesbakken, T., & Skaar, I. (2009). Yeast diversity and dynamics in the production processes of Norwegian dry-cured meat products. *International Journal of Food Microbiology, 31*(133), 135–140. Available from https://doi.org/10.1016/j.ijfoodmicro.2009.05.011.

Asensio, C. M., Grosso, N. R., & Rodolfo Juliani, H. (2015). Quality preservation of organic cottage cheese using oregano essential oils. *Lebensmittel-Wissenschaft und -Technologie- Food Science and Technology, 60,* 664–671. Available from https://doi.org/10.1016/j.lwt.2014.10.054.

Aziz, M., & Karboune, S. (2018). Natural antimicrobial/antioxidant agents in meat and poultry products as well as fruits and vegetables: A review. *Critical Reviews in Food Science and Nutrition, 58,* 486–511. Available from https://doi.org/10.1080/10408398.2016.1194256.

Badia, V., de Oliveira, M. S. R., Polmann, G., Milkievicz, T., Galvao, A. C., & da Silva Robazza, W. (2020). Effect of the addition of antimicrobial oregano (Origanum vulgare) and rosemary (Rosmarinus officinalis) essential oils on lactic acid bacteria growth in refrigerated vacuum-packed Tuscan sausage. *Brazilian Journal of Microbiology, 51,* 289–301. Available from https://doi.org/10.1007/s42770-019-00146-7.

Barbosa, L. N., Probst, I. S., Murbach Teles Andrade, B. F., Bergamo Alves, F. C., Albano, M., Mores Rall, V. L., & Júnior, A. F. (2015). Essential oils from herbs against foodborne pathogens in chicken sausage. *Journal of Oleo Science, 64,* 117–124. Available from https://doi.org/10.5650/jos.ess14163.

Bassolé, I. H. N., & Juliani, H. R. (2012). Essential oils in combination and their antimicrobial properties. *Molecules (Basel, Switzerland), 17,* 3989–4006. Available from https://doi.org/10.3390/molecules17043989.

Burt, S. (2004). Essential oils: Their antibacterial properties and potential applications in foods - A review. *International Journal of Food Microbiology, 94,* 223–253. Available from https://doi.org/10.1016/j.ijfoodmicro.2004.03.022.

Carballo, D. E., Mateo, J., Andrés, S., Gir aldez, F. J., Quinto, E. J., Khanjari, A., Operta, S., & Caro, I. (2019). Microbial growth and biogenic amine production in a balkan-style fresh sausage during refrigerated storage under a CO_2-containing anaerobic atmosphere: Effect of the addition of Zataria multiflora essential oil and hops extract. *Antibiotics, 15,* 227. Available from https://doi.org/10.3390/antibiotics8040227.

Chaves-Lopez, C., Martin-S anchez, A. M., Fuentes-Zaragoza, E., Viuda-Martos, M., Fern andez-Lopez, ′ J., Sendra, E., Sayas, E., & Angel Perez Alvarez, J. (2012). Role of oregano (Origanum vulgare) essential oil as a surface fungus inhibitor on fermented sausages: Evaluation of its effect on microbial and physicochemical characteristics. *Journal of Food Protection, 75,* 104–111. Available from https://doi.org/10.4315/0362-028X.JFP11-184.

Cheng, G., Ning, J., Ahmed, S., Huang, J., Ullah, R., An, B., Hao, H., Dai, M., Huang, L., Wang, X., & Yuan, Z. (2019). Selection and dissemination of antimicrobial resistance in Agri-food production. *Antimicrobial Resistance and Infection Control, 8*(158). Available from https://doi.org/10.1186/s13756-019-0623-2.

Colombo, M., Todorov, S. D., Eller, M., & Nero, L. A. (2018). The potential use of probiotic and beneficial bacteria in the Brazilian dairy industry. *Journal of Dairy Research, 85*, 487–496. Available from https://doi.org/10.1017/S0022029918000845.

Cruxen, C. E. D. S., Funck, G. D., Haubert, L., Dannenberg, G. D. S., Marques, J. L., Chaves, F. C., da Silva, W. P., & Fiorentini, A. M. (2019). Selection of native bacterial starter culture in the production of fermented meat sausages: Application potential, safety aspects, and emerging technologies. *Food Research International, 122*, 371–382. Available from https://doi.org/10.1016/j.foodres.2019.04.018.

de Carvalho, R. J., de Souza, G. T., Honorio, V. G., de Sousa, J. P., da Conceiç ao, M. L., Maganani, M., & de Souza, E. L. (2015). Comparative inhibitory effects of Thymus vulgaris L. essential oil against Staphylococcus aureus, Listeria monocytogenes and mesophilic starter co-culture in cheese-mimicking models. *Food Microbiology, 52*, 59–65. Available from https://doi.org/10.1016/j.fm.2015.07.003.

de Melo Pereira, G. V., de Carvalho Neto, D. P., Junqueira, A. C. O., Karp, S. G., Letti, L. A. J., Magalhaes Júnior, A. I., & Soccol, C. R. (2020). A review of selection criteria for starter culture development in the food fermentation industry. *Food Reviews International, 36*, 135–167. Available from https://doi.org/10.1080/87559129.2019.1630636.

de Souza Pedrosa, G. T., de Souza, E. L., de Melo, A. N. F., Almeida, E. T. C., de Sousa Guedes, J. P., de Carvalho, R. J., Pag´ an, R., & Magnani, M. (2020). Physiological alterations involved in inactivation of autochthonous spoilage bacteria in orange juice caused by Citrus essential oils and mild heat. *International Journal of Food Microbiology, 334*. Available from https://doi.org/10.1016/j.ijfoodmicro.2020.108837, Article 108837.

de Souza Pedrosa, G. T., Pimentel, T. C., Gavahian, M., de Medeiros, L. L., Pag´ an, R., & Magnani, M. (2021). The combined effect of essential oils and emerging technologies on food safety and quality. *LWT – Food Science and Technolology, 147*. Available from https://doi.org/10.1016/j.lwt.2021.111593, Article 111593.

de Souza, E. L. (2016). The effects of sublethal doses of essential oils and their constituents on antimicrobial susceptibility and antibiotic resistance among foodrelated bacteria: A review. *Trends in Food Science & Technology, 56*, 1–12. Available from https://doi.org/10.1016/j.tifs.2016.07.012.

Diniz-Silva, H. T., Brand´ ao, L. R., de Sousa Galvao, ˘ M., Madruga, M. S., Maciel, J. F., de Souza, E. L., & Magnani, M. (2020). Survival of Lactobacillus acidophilus LA-5 and Escherichia coli O157:H7 in Minas Frescal cheese made with oregano and rosemary essential oils. *Food Microbiology, 86*. Available from https://doi.org/10.1016/j.fm.2019.103348, Article 103348.

Diniz-Silva, H. T., de Sousa, J. B., da Silva Guedes, J., Queiroga, R. C. R. E., Madruga, M. S., Tavares, J. F., de Souza, E. L., & Magnani, M. (2019). A synergistic mixture of Origanum vulgare L. and Rosmarinus officinalis l. essential oils to preserve overall quality and control Escherichia coli O157:H7 in fresh cheese during storage. *LWT – Food Science and Technology, 112*. Available from https://doi.org/10.1016/j.lwt.2019.01.039, Article 107781.

El Adab, S., & Hassouna, M. (2016). Proteolysis, lipolysis and sensory characteristics of a tunisian dry fermented poultry meat sausage with oregano and thyme essential oils. *Journal of Food Safety, 36*, 19–32. Available from https://doi.org/10.1111/jfs.12209.

El Omari, K., Al Kassaa, I., Farraa, R., Najib, R., Alwane, S., Chihib, N.-E., & Hamze, M. (2020). Using the essential oil of Micromeria barbata plant as natural preservative to extend the shelf life of Lebanese yogurt. *Pakistan Journal of Biological Sciences, 23*, 848–855. Available from https://doi.org/10.3923/pjbs.2020.848.855.

Frohlich-Wyder, M. T., Arias-Roth, E., & Jakob, E. (2019). Cheese yeasts. *Yeast (Chichester, England), 36*, 129–141. Available from https://doi.org/10.1002/yea.3368.

Gaio, I., Saggiorato, A. G., Treichel, H., Treichell, H., Cichoski, A. J., Astolfi, V., Cardoso, R. I., Toniaazzo, G., Valduga, E., Paroul, N., & Cansiani, R. L. (2015). Antibacterial activity of basil essential oil (Ocimum basilicum L.) in Italian-type sausage. *Journal für Verbrauchershutz und Lebensmittelssicherheit, 10*, 323–329. Available from https://doi.org/10.1007/s00003-015-0936-x. (2015).

Gammariello, D., Di Giulio, S., Conte, A., & Del Nobile, M. A. (2008). Effects of natural compounds on microbial safety and sensory quality of Fior di Latte cheese, a typical Italian cheese. *Journal of Dairy Science, 91*, 4138–4146. Available from https://doi.org/10.3168/jds.2008-1146.

Hill, C., Guarner, F., Reid, G., Gibson, G. R., Merenstein, D. J., Pot, B., Morelli, L., Canani, R. B., Flint, H. J., Salminen, S., Clader, P. C., & Sanders, M. E. (2014). The international scientific association for probiotics and prebiotics consensus statement on the scope and appropriate use of the term probiotic. *Nature Reviews Gastroenterology & Hepatology, 11*, 506–514. Available from https://doi.org/10.1038/nrgastro.2014.66.

Hussien, H., Elbehiry, A., Saad, M., Hadad, G., Moussa, I., Dawoud, T., Mubarak, A., & Marzouk, E. (2019). Molecular characterization of Escherichia coli isolated from cheese and biocontrol of Shiga toxigenic E. coli with essential oils. *Italian Journal of Food Safety, 8*, 8291. Available from https://doi.org/10.4081/ijfs.2019.8291.

ISO/Fdis 9235. (2013). Aromatic natural raw materials - Vocabulary. Available from: https://www.iso.org/standard/78908.html Accessed 12.05.21.

Janssens, M., Myter, N., De Vuyst, L., & Leroy, F. (2013). Community dynamics of coagulasenegative staphylococci during spontaneous artisan-type meat fermentations differ between smoking and moulding treatments. *International Journal of Food Microbiology, 166*, 168–175. Available from https://doi.org/10.1016/j.ijfoodmicro.2013.06.034.

Jia, R., Zhang, F., Song, Y., Lou, Y., Zhao, A., Liu, Y., Peng, H., Hui, Y., Ren, R., & Wang, B. (2021). Physicochemical and textural characteristics and volatile compounds of semihard goat cheese as affected by starter cultures. *Journal of Dairy Science, 104*, 270–280. Available from https://doi.org/10.3168/jds.2020-18884.

Johnson, M. E. (2017). A 100-year review: Cheese production and quality. *Journal of Dairy Science, 100*, 9952–9965, 3168/jds.2017–12979.

Keshavarzi, M., Sharifan, A., Yasini., & Ardakani, S. A. (2021). Effect of the ethanolic extract and essential oil of Ferulago angulata (Schlecht.) Boiss. on protein, physicochemical, sensory, and microbial characteristics of probiotic yogurt during storage time. *Food Sciences and Nutrition, 9*, 197–208. Available from https://doi.org/10.1002/fsn3.1984.

Kivanç, M., Akgül, A., & Dogan, A. (1991). Inhibitory and stimulatory effects of cumin, oregano and their essential oils on growth and acid production of Lactobacillus plantarum and. Leuconostoc mesenteroides. *International Journal of Food Microbiology*, 13, 81−85. Available from https://doi.org/10.1016/0168-1605(91)90140-k.

Küçükkaya, S., Arslan, B., Demirok Soncu, E., Ertürk, D., & Soyer, A. (2020). Effect of chitosan−essential oil, a surface mold inhibitor, on microbiological and physicochemical characteristics of semidried fermented sausages. *Journal of Food Science*, 85, 1240−1247. Available from https://doi.org/10.1111/1750-3841.15053.

Lopez-Diaz, T. M., Alonso, C., Roman, C., Garcia-Lopez, M. L., & Moreno, B. (2000). Lactic acid bacteria isolated from a hand-made blue cheese. *Food Microbiology*, 17, 23−32. Available from https://doi.org/10.1006/fmic.1999.0289.

Lorenzo, J. M., & Franco, D. (2012). Fat effect on physico-chemical, microbial and textural changes through the manufactured of dry-cured foal sausage lipolysis, proteolysis and sensory properties. *Meat Science*, 92, 704−714. Available from https://doi.org/10.1016/j.meatsci.2012.06.026.

Lu., Hua, J., Wang, Q., Li, B., Li, K., Xu, C., & Jiang, C. (2015). The effects of starter cultures and plant extracts on the biogenic amine accumulation in traditional Chinese smoked horsemeat sausages. *Food Control*, 50, 869−875. Available from https://doi.org/10.1016/j.foodcont.2014.08.015.

Marcial, G. E., Gerez, C. L., de Kairuz, M. N., Araoz, V. C., Schuff, C., & de Valdez, G. F. (2016). Influence of oregano essential oil on traditional Argentinean cheese elaboration: Effect on lactic starter cultures. *Revista Argentina de Microbiología*, 48, 229−235. Available from https://doi.org/10.1016/j.ram.2016.04.006.

Moritz, C. M. F., Rall, V. L. M., Saeki, M. J., & Fernandes Junior, A. (2012). Inhibitory effect of essential oils against Lactobacillus rhamnosus and starter culture in fermented milk during its shelf-life period. *Brazilian Journal of Microbiology*, 43, 1147−1156. Available from https://doi.org/10.1590/S1517-83822012000300042.

Moro, A., Libran, C. M., Berruga, M. I., Carmona, M., & Zalacain, A. (2015). Dairy matrix effect on the transference of rosemary (Rosmarinus officinalis) essential oil compounds during cheese making. *Journal of the Science of Food and Agriculture*, 95, 1507−1513. Available from https://doi.org/10.1002/jsfa.6853.

Mousavi, M., Heshmati, A., Garmakhany, A. D., Vahidinia, A., & Taheri, M. (2019). Optimization of the viability of Lactobacillus acidophilus and physico-chemical, textural and sensorial characteristics of flaxseed-enriched stirred probiotic yogurt by using response surface methodology. *Lebensmittel-Wissenschaft und -Technologie- Food Science and Technology*, 102, 80−88. Available from https://doi.org/10.1016/j.lwt.2018.12.023.

Olmedo, R. H., Nepote, V., & Grosso, N. R. (2013). Preservation of sensory and chemical properties in flavoured cheese prepared with cream cheese base using oregano and rosemary essential oils. *Lebensmittel-Wissenschaft und -Technologie- Food Science and Technology*, 53, 409−417. Available from https://doi.org/10.1016/j.lwt.2013.04.007.

Ong, L., Henriksson, A., & Shah, N. P. (2007). Proteolytic pattern and organic acid profiles of probiotic Cheddar cheese as influenced by probiotic strains of Lactobacillus acidophilus, Lb. paracasei, Lb. casei or Bifidobacterium sp. *International Dairy Journal*, 17, 67−78. Available from https://doi.org/10.1016/j.idairyj.2005.12.009.

Ozturk, I. (2015). Antifungal activity of propolis, thyme essential oil and hydrosol on natural mycobiota of sucuk, a Turkish fermented sausage: Monitoring of their effects on microbiological, color and aroma properties. *Journal of Food Processing and Preservation*, 39, 1148−1158. Available from https://doi.org/10.1111/jfpp.12329.

Papamanoli, E., Kotzekidou, P., Tzanetakis, N., & Litopoulou-Tzanetaki, E. (2002). Characterization of Micrococcaceae isolated from dry fermented sausage. *Food Microbiology*, 19, 441−449. Available from https://doi.org/10.1006/fmic.2002.0503.

Paparella, A., Mazzarrino, G., Chaves-Lopez, C., Rossi, C., Sacchetti, G., Guerrieri, O., & Serio, A. (2016). Chitosan boosts the antimicrobial activity of Origanum vulgare essential oil in modified atmosphere packaged pork. *Food Microbiology*, 59, 23−31. Available from https://doi.org/10.1016/j.fm.2016.05.007.

Pateiro, M., Barba, F. J., Domínguez, R., Sant'Ana, A. S., Mousavi Khaneghah, A., Gavahian, M., Gomez, B., & Lorenzo, J. M. (2018). Essential oils as natural additives to prevent oxidation reactions in meat and meat products: A review. *Food Research International*, 113, 156−166. Available from https://doi.org/10.1016/j.foodres.2018.07.014.

Pisoschi, A. M., Pop, A., Georgescu, C., Turcuş, V., Olah, N. K., & Mathe, E. (2017). An overview of natural antimicrobials role in food. *European Journal of Medicinal Chemistry*, 143, 922−935. Available from https://doi.org/10.1016/j.ejmech.2017.11.095.

Roldan, L. P., Diaz, G. J., & Duringer, J. M. (2010). Composition and antibacterial activity of essential oils obtained from plants of the Lamiaceae family against pathogenic and beneficial bacteria. *Revista Colombiana de Ciencias Pecuarias*, 23, 451−461.

Sadeghi, E., Akhondzadeh Basti, A., Noori, N., Khanjari, A., & Partovi, R. (2013). Effect of Cuminum cyminum L. essential oil and Lactobacillus acidophilus (a probiotic) on Staphylococcus aureus during the manufacture, ripening and storage of white brined cheese. *Journal of Food Processing and Preservation*, 37, 449−455. Available from https://doi.org/10.1111/j.1745-4549.2011.00664.x.

Shori, A. B. (2016). Influence of food matrix on the viability of probiotic bacteria: A review based on dairy and non-dairy beverages. *Food Bioscience*, 13, 1−8. Available from https://doi.org/10.1016/j.fbio.2015.11.001.

Tongnuanchan, P., & Benjakul, S. (2014). Essential oils: Extraction, bioactivities, and their uses for food preservation. *Journal of Food Science*, 79, R1231−R1249. Available from https://doi.org/10.1111/1750-3841.12492.

Trabelsim, I., Ben Slima, S., Ktari, N., Triki, M., Abdehedi, R., Abaza, W., Moussa, H., Abdeslam, A., & Ben Salah, R. (2019). Incorporation of probiotic strain in raw minced beef meat: Study of textural modification, lipid and protein oxidation and color parameters during refrigerated storage. *Meat Science*, 154, 29−36. Available from https://doi.org/10.1016/j.meatsci.2019.04.005.

Woraprayote, W., Malila, Y., Sorapukdee, S., Swetwiwathana, A., Benjakul, S., & Visessanguan, W. (2016). Bacteriocins from lactic acid bacteria and their applications in meat and meat products. *Meat Science*, 120, 118−132. Available from https://doi.org/10.1016/j.meatsci.2016.04.004.

Chapter 17

Application of essential oil in aromatherapy: current trends

Babatunde Oluwafemi Adetuyi[1], Peace Abiodun Olajide[1], Charles Oluwaseun Adetunji[2], Juliana Bunmi Adetunji[3], Oluwabukola Atinuke Popoola[4], Oloruntoyin Ajenifujah-Solebo[4], Yovwin D. Godwin[5], Olatunji Matthew Kolawole[6], Olalekan Akinbo[7] and Abel Inobeme[8]

[1]*Department of Natural Sciences, Faculty of Pure and Applied Sciences, Precious Cornerstone University, Ibadan, Oyo State, Nigeria, [2]Applied Microbiology, Biotechnology and Nanotechnology Laboratory, Department of Microbiology, Edo State University Uzairue, Iyamho, Edo State, Nigeria, [3]Department of Biochemistry, Osun State University, Osogbo, Osun State, Nigeria, [4]Genetics, Genomics and Bioinformatics Department, National Biotechnology Development Agency, Abuja, FCT, Nigeria, [5]Department of Family Medicine, Faculty of Clinical Sciences, Delta State University, Abraka, Delta State, Nigeria, [6]Department of Microbiology, Faculty of Life Sciences, University of Ilorin, Ilorin, Kwara State, Nigeria, [7]Centre of Excellence in Science, Technology, and Innovation, AUDA-NEPAD, Johannesburg, Gauteng, South Africa, [8]Department of Chemistry, Edo State University Uzairue Iyamho, Auchi, Edo State, Nigeria*

Introduction

There has been a rise in recent years in the popularity of phytotherapy and other forms of alternative medicine. The term "phytotherapy" is said as the practice of using plants and plant products for therapeutic purposes (Ferreira et al., 2014). Herbal medicine's efficacy has been confirmed by several clinical and experimental researches; as a result, phytotherapy has gained in scientific prominence (Schilcher, 1994). Aromatherapy, on the other hand, is a subset of phytotherapy that makes use of the plant's own defenses against environmental stresses such disease, herbivory, insect pests, and heat/cold stress (Adejumo et al., 2017; Adetunji & Varma, 2020; Adetunji et al., 2017; Adetunji, 2008, 2019; Adetunji et al., 2022; Adetunji, Adetunji et al., 2021; Adetunji, Ajani et al., 2013; Adetunji, Ajayi et al., 2021; Adetunji, Akram et al., 2021; Adetunji, Anani et al., 2021; Adetunji, Egbuna et al., 2020; Adetunji, Inobeme et al., 2021; Adetunji, Kayode et al., 2013; Adetunji, Kumar et al., 2019; Adetunji, Michael, Kadiri et al., 2021; Adetunji, Michael, Nwankwo et al., 2021; Adetunji, Michael, Varma et al., 2021; Adetunji, Nwankwo et al., 2021; Adetunji, Ojediran et al., 2019; Adetunji, Olaniyan et al., 2021; Adetunji, Oloke et al., 2020; Adetunji, Palai et al., 2021; Adetunji, Panpatte et al., 2019; Adetunji, Roli et al., 2020; Bello et al., 2019; Egbuna et al., 2020; Olaniyan and Adetunji, 2021; Rauf et al., 2021; Thangadurai et al., 2021; Ukhurebor and Adetunji, 2021). However, they are used by humans for a variety of utilization in the drug-making industry, cosmetic, culinary, and beverage factories as a result of their biological qualities that aid in health maintenance. Essential oils (EOs) are used mostly in aromatherapy to affect one's state of mind, emotional state, cognitive performance, and physical health (Steflitsch & Steflitsch, 2008). The term "aromatherapy" was coined when the Greek words "aroma," meaning "fragrance," and "therapy," meaning "treatment," came together (Ali et al., 2015). EOs are complex combinations of organic chemicals whose chemical make-up determines their biological action and aroma (Michalak, 2018). You can combine them to make synergistic aromas with a wide range of nuanced notes (Schneider et al., 2019). The utilization of EOs in the treatment of various conditions such as those of the Central Nervous System (CNS) including those of the nervous system (through inhalation), the skin (by topical application), and the digestive tract (via ingestion) is called aromatherapy (Martinec, 2012; Michael, 2001). Applying EOs topically can be done via massage or with the use of skincare products. It's challenging to know the exact impact of massage with EOs (Stea et al., 2014), while EOs are utilized in cosmetics aromatherapy for the hair, body, skin, and face (Siddique, 2017). Hydrolates, which are produced as a byproduct of extracting EOs, are also useful in aromatherapy (Shah et al., 2011). Since hydrolates are aqueous solutions, they can be sprayed directly into the skin for cooling and revitalization or for application in wound healing (Aćimović et al., 2020).

Applications of Essential Oils in the Food Industry. DOI: https://doi.org/10.1016/B978-0-323-98340-2.00011-0

Aromatherapy

The word "Aromatherapy" is obtained from two Greek words "fragrance" (aroma) and "therapy" (treatment), which refers to the practice of employing pure, aromatic EOs from a certain geographical location. The act of inhaling or utilizing top-notch EOs topically to effect a physiological change is known as aromatherapy (Ali et al., 2015). Fragrances have captured the imagination of humans as far back as the sixth millennium BC. A wide range of uses has been attributed to EOs across cultural contexts. Perfumes, cosmetics, and ointments containing aniseed, cedar, and myrrh were all commonly utilized by the ancient Egyptians. Cinnamon, ginger, myrrh, and sandal tree were among the more than 700 compounds used in traditional Chinese and Indian medicine; caraway, thyme, peppermint, marjoram, and saffron were among those used in ancient Greek medicine (Elshafie & Camele, 2017). Aromatherapy has been shown effective for treating a wide range of medical issues. The medicinal, cosmetic, and aromatic value of EOs has grown steadily throughout the years. The raw plant materials (roots, flowers, fruits, leaves, bark, branches, wood, seeds, bark, herbs, and so on.) are distilled with steam or H2O, squeezed or spun to extract the fruit peels, or dry distilled to extract the wood's EOs. In addition, extraction and maceration are utilized. Chemical solvents are necessary for the aforementioned procedures; nonetheless, trace amounts of these solvents can remain in the EO that is ultimately produced. It is, therefore, advised that only oils derived naturally be used in aromatherapy. The remaining items are limited to use as room aromatizers or in perfume blends (Kiełtyka-Dadasiewicz and Gorzel, 2014). The purpose of this analysis is to talk about what aromatherapy is all about, as well as the different techniques and approaches used in aromatherapy.

Impact of essential oils and aromatherapy on the central nervous system

The word aromatherapy is used to explain the application of EOs to the skin or to the air around you. An exact definition of medical aromatherapy argues that the only reason to employ this technique is to ascertain a physiological and psychological reaction in the microorganism (Schneider et al., 2019). There has been growing interest in aromatherapy in the medical community over the past decade, and mainstream practitioners no longer dismiss it as quackery (Ali et al., 2015). This is because it is becoming increasingly apparent that aromatherapy can have a big impact on important body organs such as CNS and there are various researches that attain such results (Agatonovic-Kustrin et al., 2019). EOs can reach deeper layers of skin and mucosa when inhaled or massaged. Once in the bloodstream, they can go to the brain & also breach the blood—brain barrier (BBB). However, the components of EOs can come into contact with crucial neurological regions and influence how neurons work (Czar, 2009). Active chemicals from EOs, such as some terpenes, that have a small molecular size and high lipophilicity seem to possess the greatest penetration to BBB. This is important because it may direct future studies to concentrate on compounds of this class because of the larger impact they appear to have on the CNS (Agatonovic-Kustrin et al., 2020). In terms of neuroprotection, the antioxidant properties of EOs are crucial. Some patho-physiologic mechanisms, including as accumulation of tau-associated neurofibrillary tangles, amyloid-induced neuronal loss, and mitochondrial dysfunction, are hallmarks of Alzheimer's disease and have been linked to oxidative stress (Benny & Thomas, 2019). EOs with anti oxidant characteristics can be found in plants including *S. officinalis*, *J. communis*, *C. zeylanicum*, *P. halepensis*, *L. angustifolia*, *P. dulcis*, and *O. basilicum* (El Euch et al., 2019). They are not only beneficial in the treatment of Alzheimer's disease but they also show promise in the management of other forms of dementia, neurodegenerative disorders, and even moderate cognitive impairment. As oxidative stress has been associated with a number of neurological illnesses, the fact that EOs appear to be able to slow the progression of these diseases is encouraging. However, further research is needed to prove that frequent use of EOs and aromatherapy significantly impacts the course of these disorders or helps prevent or stop their evolution (Xu et al., 2017). EOs appear to inhibit cholinesterases in addition to reducing oxidative stress. This is significant since acetylcholine levels tend to drop over time in dementias like Alzheimer's disease. The hydrolytic decomposition of acetylcholine is inhibited by EOs, and as a result, acetylcholine levels are raised, according to research. EOs can also affect butyrylcholinesterase, another enzyme involved in the breakdown of acetylcholine. Drugs with inhibitory activity on cholin esterases are at the center of the standard therapeutic approach to Alzheimer's disease. The potential for conventional drug dose reduction through the use of equivalent EOs is discussed. In addition, EOs could be used by patients who are unable to manage the negative side effects of the standard treatment (Ma et al., 2019). EOs from plants like *M. spicata*, *Salvia leriifolia*, *C. limonum*, *P. hydropiper*, and *J. communis* have been shown to block cholinesterase activity (Liu et al., 2020). Aromatherapy relies heavily on the smell sense (nose) and its relation to the limbic system, whether EOs are administered via inhalation or skin rubs. Important regions such as the hippocampus, hypothalamus, orbital prefrontal cortex, amygdala, and basal ganglia receive projections from the primary olfactory cortex. Thus, EOs can influence human behavior, autonomic nervous system function, and pain perception via olfactory pathways. Central nervous system-level nociception regulation by EOs is poorly understood. Glutamatergic transmission may be influenced by chemicals in these oils that block α-amino-3-hydroxy-5-methyl-4-isoxazolepropionic acid (AMPA), N-methyl-D-aspartate (NMDA), and kainate receptors, according to this theory. Analgesic effect and potential application in disorders characterized by persistent pain have been demonstrated in experimental models for *Mentha spicata*, *Salvia sclareae*, and *Lavandula angustifolia* EOs (López et al., 2017). It seems that *J. communis* EOs promotes parasympathetic activity, which has significant effects on the autonomic nervous system, including a decrease in blood pressure and heart rate. This demonstrates that aromatherapy is an effective method of treating anxious individuals with elevated sympathetic activity (Park, 2017). It appears that the aminobutyric acid A (GABAA) receptor is involved in the anxiolytic effect of few EOs, such as the one produced from *F. angulate* (Bagci et al., 2016). In addition to the anxiolytic impact, the sense of touch and the sensory pathways play a significant role when EO is applied to the skin through massages. There was a substantial improvement in behavioral and psychological dementia symptoms with this delivery method (Yoshiyama et al., 2015). Some EOs have the ability to heighten anxiety, leading to increased sympathetic activity and subsequent increases in blood pressure and respiratory rate. People who suffer from anxiety should stay away from them, yet they have their uses. One example is treating hypotension. *Jasminum sambac* EO is one such oil that does so by boosting sympathetic activity (Hongratanaworakit, 2010). It's important to remember that EOs have additional effects besides those already discussed, including the ability to prevent seizures in some people and aid in sedation and sleep. *Ocimum basilicum* and *Rosa damascene* are two plants that have volatile oils with similar characteristics (Dobetsberger and Buchbauer, 2011). Although several investigations with animal models showed positive results, more clinical trials are required to fully understand the effects of aromatherapy on the central nervous system. Evidence suggests that many EOs used in wellness centers already have qualities that can alter the brain and assist in a variety of neurological diseases. In the CNS, the mode of activity of such compounds is highly nuanced and can differ greatly between EOs. Therefore, it may be extremely beneficial for the healthcare system as a whole to take the next step toward characterizing the action of such oils on human subjects, refining the concept of "medical aromatherapy," and standardizing the method by which it can be applied in spa and wellness centers and rehabilitation clinics (Schneider et al., 2019).

Classification based on aroma

It is also possible to categorize EOs according to their scent. Floral, herbaceous, citrus, resinous, medicinal/camphorous, spicy oils, earthy, and minty are some of the subsets that make up this grouping (Weyerstahl et al., 1993).

Oils that have a citrusy aroma and flavor are classified as follows:

Citrus Oils: Citrus oils come from large species of plants, including bergamot, lemon, grapefruit, orange, lime, and tangerine (Viuda-Martos et al., 2008).

Herbaceous oils: They are derived from plants that are otherwise considered to be among the most helpful herbs. Hyssop, Chamomile, Marjoram, Melissa, Peppermint, Rosemary, Basil, and Clary sage are just few of the plants from which these oils can be derived (Yepez et al., 2002).

EOs from the camphora tree have been used for centuries because of their curative properties. Tea Tree, Cajeput, mugwort-like, rosemary-like, earthy borneol-like, and EOs, all with a delicious, dried-plum-like background, are just a few examples (Weyerstahl et al., 1993).

Floral Oils: Oils extracted from flowers or infused with botanical fragrances. Some of the plants that yield these oils are lavender, geranium, chamomile, neroli, ylang-ylang, jasmine, rose, etc. (Simpson et al., 1990).

EOs that have a scent similar to wood or are derived from the barks or other woody portions of plants can be found in many different types of wood, bark, berries, cones, and twigs, including cedar, cinnamon, cypress, juniper, pine (Junming et al., 2010).

Earthy Oils: Aromatic oils that come from the roots or other underground sections of plants, or have a strong, distinctive earthy aroma. Some of the plants that yield these oils are the angelica, patchouli, vetiver, and valerian (Priestap et al., 1990).

Nutmeg, cloves, Black Pepper, Cumin, Aniseed, Coriander, Ginger, Cinnamon, Cardamom, and Thyme are all examples of spices that can be used to extract oil (López-Cortés et al., 2013).

Aromatherapy raw materials

Oily, odoriferous, and soluble in ether, liquid fats, chloroform, or alcohol, EOs are a type of volatile material (Worwood, 1993). Whether you're using them in Poland or Europe, the oils you use in your aromatherapy treatments should be of the highest quality possible. The National Institute of Hygiene may also certify that they are up to their standards. Detailed legislation on cosmetics (Kiełtyka-Dadasiewicz and Gorzel, 2014) allows oils utilized on the skin to be registered as items produced for interaction with skin, which is cosmetic. In accordance with the International Nomenclature of Cosmetic Ingredients (INCI), the exterior packaging must provide at least the bare minimum of information establishing that a product is intended for cosmetic use (INCI). The packaging and accompanying brochure ought to include information on standards and approval logos. Additional information that should be included in EO descriptions includes the phrases "Natural essential oil" or "Essential oil," the identification of the plant oil in Polish along with a full description of the specific part of plant utilized to extract the EO, the full name according to INCI, identification and location address of the producers, the expiration date, the utilization methods, and any drawbacks on the use of the oil (Konopacka-Brud and Brud, 2010). Aromatherapy EOs must be stored in dark, light-resistant glass containers with a dropper for precise dosing to maintain quality. Oils should be kept in a cool, dry place (between 5°C and 10°C) to preserve their quality. Citric oils have a 12-month to two-year shelf life, but those made from wood, resin, and roots can last as long as three to four years. In airtight containers, the leftover oils have a shelf life of two years (Lis-Balchin, 2005). The chemical make-up of oils determines both their biological efficacy and their aroma. Variables such as raw material source and climatic circumstances during plant development affect the oil's final chemical profile. Oils extracted from plants are not chemically uniform; rather, they are typically complex combinations of various organic components (organic acids, terpene hydrocarbons and their O_2 derivatives, ethers, ketones, esters, aldehydes, and alcohols). Oils typically contain a few different compounds, with one being the most abundant and the others present in much smaller concentrations to provide complementing effects (Worwood, 1993).

Ingredients that are especially abundant in aromatic plants:

Aromatic plants high in limonene: Hyssop, Mint, Lavender, Marjoram, Oregano, Thyme, and Vervain
β-Pinene: Herbs Like Parsley, Basil, Caraway, Fennel, Oregano, Marjoram, and Sage
Carvone: dill, coriander, etc.
Herbs including basil, camphor balm, marjoram, lavender, and sage
Herbs including oregano, thyme, marjoram, and balm contain thymol.
α-thujene: chervil, fennel, parsley, rosemary, thyme, sage, tarragon, tarragon, sage, dill

EOs have far-reaching effects that vary greatly based on the nature of the oil's primary constituent. In addition, they may be the end result of either antagonistic or synergistic activity between the various ingredients (Nikolić et al., 2017). EOs have been exhibited to have antibacterial activity against a large scope of microorganisms (Grabowska and Janeczko, 2013). The effects they have on the body include being inflammatory, immunostimulating, expectorant, antiseptic, antioxidative, analgesic, antineoplastic, anti-diabetic, and vasodilatory (Kędzia and Kędzia, 2017). The literature suggests that they also have calming, anti-stress, and adaptogenic properties (Cardia et al., 2018).

Possible uses of some EOs on the body system of human beings are as follows:

- Impact on mental health and mood
  Fatigue: *Cymbopogon nardus, Citrus aurantium, Rosmarinus officinalis, Pelargonium graveolens, Angelica archangelica, Eucalyptus radiata, Coriandrum sativum, M. spicata, J. communis, Pinus sylvestris, S. sclarea*
  Insomnia: *J. communis, Citrus bergamia, Origanum majorana, A. archangelica, Citrus reticulata, Cuminum cyminum, Cananga odorata, Citrus sinensis, Valeriana officinalis, Myrtus communis, Cistus ladaniferus, Citrus limon, Melissa officinalis, C. aurantium, L. angustifolia, Litsea cubeba, O. basilicum*
  Anxiety, agitation, stress: *L. cubeba, Eucalyptus staigeriana, A. archangelica, O. basilicum, Cymbopogon martinii, C. ladaniferus, O. majorana, C. aurantium, Pogostemon patchouli, P. graveolens, C. aurantium, L. angustifolia, V. officinalis*
  Mental exhaustion, burnout: *O. basilicum, Mentpha piperita, Helichrysum angustifolium*
  Memory loss: *R. officinalis, Mentha piperita, L. cubeba*
  Pains of various origins, including bone and joint pains: *R. officinalis, J. communis, Cinnamonum zeylanicum, Zingiber officinale, L. angustifolia, Matricaria recutita, Leptospermum scoparium, O. majorana, Pinus mugo*
- Effect on skin
  Antibacterial activity: *L. scoparium, Melaleuca alternifolia, Lavandula officinalis, R. officinalis*
  Antifungal activity: *Melaleuca armillaris, Melaleuca ericifolia, Melaleuca leucadendron, Artemisia sieberi, M. piperita, Brassica nigra, Skimmia laureola, C. cyminum, Melaleuca styphelioides, C. nardus, A. archangelica*
  Anti-inflammatory action: *R. officinalis, M. alternifolia, P. patchouli, L. officinalis, Santalum album, C. limon*
  Strengthening vascular walls: *Citrus amara, P. graveolens, R. damascene, R. officinalis*

Delaying skin aging: *C. limon, C. amara, R. damascene.*
Removal of metabolic wastes, improvement of lymph circulation, anti-cellulite action: *P. graveolens, C. limon, Cupressus sempervirens, S. album, R. officinalis, J. communis, Citrus paradisi*

Routes that essential oils penetrate the body

Natural plant oils are only ever applied topically or inhaled during traditional aromatherapy. Aromatherapy is most commonly connected with the inhalation method. The oil is vaporized with water or burned, and the patient breathes in the resulting smoke. Oils inhaled or ingested through the nose or lungs have systemic effects on the patient's body and psyche. This method entails either traditional inhalation (using an inhaler, oil vapors poured into hot water, or a cloth soaked in oil solution) or room aromatization (using burners, nebulizers, or pot-pouri) (Romer, 2009). EOs applied to the skin have topical impacts, but with prolonged skin exposure, they can infiltrate the circulation if they are diluted in a fatty carrier (massage) or in H_2O (compresses, poultices, baths). Due to the inevitable inhalation of oil vapors during such treatments, air aromatherapy is commonly used in conjunction with direct aromatherapy (Warszawski, 2011).

Applications of essential oils

Inhalation

The steams from EO solution are best inhaled through the nostrils and mouth through the use of a specialized apparatus. Patients can adjust the volume of air and the strength of the oil using this way. Two of the more practical methods include inhaling a cloth dipped in oil solution or inhaling the gasses of oils placed in a bucket with heated water. When the intended target is the respiratory system (airway and lungs), the benefit of inhalation over oral delivery cannot be overstated. The respiratory system, the mouth and throat, and the brain system all benefit greatly from this technique, which also helps stimulate the immune system and aids in psychotherapy (Setzer, 2009). While there is a dearth of research on the topic, what little there is suggests that inhaling EOs results in rapid and considerable delivery of the oils' lipophilic components. After only 18 minutes, the peak serum concentration of 1.8-cineol from eucalyptus oil reaches a maximum, according to one study. Inhalation formulations may also contain oils with antibacterial and expectorant properties, such as carnation, eucalyptus, pine, thyme, turpentine, mint oils, cajeput, or juniper (Grabowska and Janeczko, 2013).

Aromatherapeutic massages

In direct aromatherapy, massage is the first point. Aromatherapy massage can be done in a variety of ways, but the essential thing is to use oils that are most suited to your massage style. Both the actual act of massaging and the oil used in aromatherapy produce their own benefits. To ensure even distribution of the EO across a greater surface area, the oil or blend of oils used for the treatment should be dissolved in a suitable carrier, such as high-quality base oil. Grape seed, almond, sesame, jojoba, apricot, wheat sprout, peach, plum stone oils or avocado are among the most luxurious plant oils used in aromatherapeutic massages (Romer, 2009). Aromatherapy massage is often mixed with other types of massage, such as deep tissue work or relaxation techniques (Adetuyi et al., 2015; Hongratanaworakit, 2011). The study demonstrated the beneficial benefits of massages using a combination of lavender and bergamot oils on mental status and offered evidence for their utility in treating sadness and anxiety. Aromatherapy massage, as shown by Ćwirlej et al. (2005), is more effective than a traditional massage at alleviating pain when pharmaceuticals are contraindicated. Lymphatic massage can be further enhanced by using EOs with antioxidant characteristics (e.g., sandal oils, rosemary, clary sage, juniper, ginger, geranium, or cypress) to increase lymphatic flow and eliminate toxins from the body (Legan, 2014).

Aromatherapeutic baths

To use EOs for their healing effects, a unique process known as an aromatherapeutic bath is employed. The medicinal compounds EOs are absorbed into the bloodstream via the airway, sebaceous glands, and sweat glands during a bath. In aromatherapy, practitioners use both full and partial-immersion baths. It is recommended that the water temperature be around 40°C, and that the bath last between 15 minutes and 30 minutes, with no soap or other foamy agents employed. After getting out of the bath, it's important to thoroughly rinse the body and pat it dry. Underwater massage or hydro-massage can supplement the aromatherapy bath. Aromatherapy baths are used for a large scope of conditions, which include those affecting the neurological system, cardiovascular system, skin, and muscles. You can get them done for cosmetic or relaxing purposes at a spa or wellness center (Didunyemi et al., 2019,2020; Mooventhan and Nivethitha, 2014).

Sauna

Saunas, which incorporate both air and skin aromatherapy, is another common feature of spas and wellness centers. The conditions for oil absorption into the skin are created in a sauna (dilation of superficial blood vessels due to high temperature, and increased perspiration). Utilizing a sauna on a regular basis is a great way to unwind, toughen up, boost immunity, and keep in peak physical condition. Individually selected blends, such as those with a sedative, anti-anxiety, or energizing effect, or those with a calming, soothing, or bronchodilatory effect, can be administered in a number of ways to accommodate each client's unique physical state and sense of smell (Mooventhan and Nivethitha, 2014).

Essential oils as components of personal care products

The way a product smells is an important consideration while shopping for cosmetics. EOs are used as both a fragrant additive and an active element in many different types of skincare, bath, and massage applications. They stop hair loss and work as an antibiotic, acne treatment, anti-aging, and anti-dandruff remedy (Płocica and Tal-Figiel, 2011). What's more, they aid in the healing of wounds and the promotion of new skin growth (Han et al., 2017). EOs are employed as absorption promoters in contemporary cosmetics and dermocosmetics. These promoters work by reacting with the lipids of the stratum corneum to temporarily modify the skin's structure (Williams & Barry, 2004). EO constituents, including carvacrol, -pinene, menthol, terpineol, or linalool, are transepidermal pathway promoters that make the skin more permeable to other substances, including the hydrophilic components of cosmetics. EOs are utilized as natural preservatives because of their strong antibacterial activity, which increases the shelf life of cosmetics. In addition, the aromatic content is useful for developing new cosmetic formulations. Szydłowska (2008) notes that the market for EOs is "huge" due to the variety of oils available. Perception of aroma is included in the sensory evaluation of cosmetics. One of the fastest-growing areas of cosmetics research, alongside instrumental study of the skin and hair, is sensory analysis (Pytkowska & Arct, 2006). A large scope of EOs can be used in aromatherapy, which is practiced in places like spas and beauty salons. When appropriate EO blends are added to creams, masks, and massage preparations, unique care programs may be created that are tailored to the skin condition and also the preferences and needs of individual clients (Bierniewicz, 2010).

Air aromatization

There has been a rebirth in the practice of artificially scenting the air in commercial spaces for advertising objectives (Kaleta, 2007). Air aromatization deviates from therapy, specifically when synthesized fragrant ingredients are utilized, and is used to boost employee concentration, which leads to increased productivity. This is because the use of aromas can evoke specific mental responses in consumers, prompting them to buy or eat specific products or foods or to associate those products and services with specific brands. Furthermore, aromas have the potential to diminish the allure of visits to spas and beauty parlors. The right blend of EOs can promote calm, provide for a more pleasant environment, and bring in repeat customers. Air aromatization is among the gentlest types of aromatherapy due to the low strength of oils and the fact that it does not require patients' active participation in the treatment process, making it an appropriate choice for therapeutic treatments involving children or disabled people (Pietrzak et al., 2012).

Aromatherapy and the immune system

One plausible mechanism by which aromatherapy boosts immunity is via increasing feelings of well-being. Some EOs, however, may have a cellular-level effect on immune function. Pe' noe' l (1993) has hypothesized that phenols' effects are comparable to that of human immunoglobulin M, but there is no definitive evidence for this (IgM). When the immune system comes into contact with a foreign disease, it secretes IgM for a brief duration. Long-term protection is provided by the secretion of immunoglobulin G (IgG). According to Pe' noe' l (1993), the behavior of mono-terpenic alcohols is similar to that of IgG. A rise in lymphocyte transformation values was noted in cancer patients treated with coumarins, as reported by Berkarda et al., 1983. Oils extracted from citrus peels and lavender both contain trace amounts of coumarins. Maybe this is why Rovesti and Columbo (1973) hypothesized lavender-induced lymphocytosis in their study. J.-C. "Essential oils in the bloodstream create leukocytosis," Lapraz, a renowned aromatherapy practitioner, is reported as stating (Mitchell, 1993). Novi and P. are cited by Valnet (1990). The Italian researcher Rovesti found that "essences of lemon, thyme, bergamot, lavender, and chamomile" stimulated the white corpuscles responsible for curative leukocytosis, which in turn helped the body fight off poisons and resist infectious disease. Several EOs, including clove, genuine verbena, niaouli, and patchouli, have been suggested by Roulier (1990) to support a healthy immune system. Nonetheless, there are currently no researches that have quantified the impact of aromatherapy on immunoglobulins in the saliva or blood of humans.

The EOs listed below have the ability to improve the functions of the immune system (Roulier, 1990; Valnet, 1990; Wagner et al., 1986):

Clove (*Syzygium aromaticum*)
German chamomile (*M. recutita*)
Thyme (*Thymus vulgaris* ct. thymol)
Niaouli (*Melaleuca viridiflora*)
Bergamot (*C. bergamia*)
Lavender (*L. angustifolia*)
Lemon (*C. limon*)
Patchouli (*P. patchouli*)

M. recutita (*G. chamomile*) is an EO that has been shown to enhance the number of B lymphocytes (Wagner et al., 1986), and *C. bergamia* (bergamot) has been shown to be an immune-system stimulator and an inhibitor of prostaglandins (Roulier, 1990). Compared to *Satureja montana* ct. thymol, *T. vulgaris* ct. thymol is less hepatotoxic and possesses potent immunostimulant qualities, according to a report by Philippe Mailhebiau (Mailhebiau, 1995). In the medical field, rheumatoid arthritis and HIV/AIDS are two of the most frequently encountered immunology issues. Although there is no proof that aromatherapy may "treat" HIV/AIDS or rheumatoid arthritis, its feel-good factor and implications on immunologically competent cells and immunoglobulins may have a positive impact on immune function. It's accurate to say that Eleven-year-old Marette Flies, who has lupus erythematosus, had the same physiological reaction to smelling a rose that she would have had to chemotherapy if she thought she was getting both treatments (Olness and Ader, 1992).

Components of some popular EOs that are anti-inflammatory:

German chamomile (*M. recutita*): Bisabolol, chamazulene (Carle & Gomaa, 1992; Safayhi et al., 1994)
Helichrysum (*Helichrysum italicum subsp. serotinum*): Italidiones (Franchomme & Penoel, 1991)
Rosemary (*R. officinalis ct. cineole*): 1,8-cineole (Juergens et al., 1998)
Black pepper (*Piper nigrum*): Beta-caryophyllene (Tambe et al., 1996)

Advantages of aromatherapy for people with cancer

Aromatherapy is commonly utilized to improve cancer patients' life quality. Patients at Liverpool, England's Marie Curie Centre were randomly assigned to undergo either a massage with or without Roman chamomile. There was a statistically significant increase in quality of life and a decrease in anxiety in the aromatherapy massage group (Wilkinson, 1995). Researchers found that medical professionals, including physicians, nurses, paramedics, and volunteers, were very enthusiastic about the idea of utilizing aromatherapy in palliative care in another study (Arnold, 1995). In their randomized controlled research of 52 patients with various malignancies, Corner et al. (1995) employed a prepared blend of rose, lavender, valerian, lemon and rosewood. Less than two-thirds of the patients in the 8-week study were treated with chemotherapy, radiation therapy, or surgery. Patients were split into two groups: one receiving a weekly massage with EOs and the other receiving no massage at all. Patients who were unable to participate in the massage therapy program were used as a control group. There was a statistically significant difference in the reduction of anxiety between the two massage groups, but there was nearly no difference in the improvement of pain and mobility. Evans (1995) examined the use of massage as aromatherapy for terminally ill cancer patients in a palliative care context. Sixty-nine patients were followed for a period of six months. Aromatherapy massage and therapist guidance on managing symptoms with aromatherapy were provided to participants. Although 80% of patients reported improvement, pinpointing the cause of either the EOs, massage, or individualized attention is challenging.

Extracts with anticarcinogenic activity:

Sclareol from Clary sage (*S. sclarea*) (Dimas et al., 1999)
Perillyl alcohol from Peppermint (*M. piperita*) (Belanger, 1998)
Anethufuran, Carvone, and limonene from Dill (*Anethum graveolens*) (Zheng et al., 1992)
Bergamottin from Bergamot (*C. bergamia*) (Miyake et al., 1999)
D-limone and geraniol from Lemongrass (*Cymbopogon citratus*) (Zheng et al., 1993)
Limonene, anethufuran, and Carvone from Caraway (*Carum carvi*) (Zheng et al., 1992)

Stress

Stress is a known contributor to depression and has been linked to other health problems. Increased irritability, hypersensitivity to criticism, physical manifestations of tension (such as nail biting), inability to fall asleep or wake up at the same time each day (irregular sleep patterns), chronic fatigue, increased alcohol and nicotine consumption, digestive issues, forgetfulness, and difficulty concentrating are all symptoms of chronic stress. Workplace stress can be caused by a variety of factors. Having a boss who isn't on your side, not being consulted or communicated with, having your personal life encroached upon, being given impossible deadlines, not having any say in how your work is done, working in an unpleasant environment, not being paid what you're worth, feeling underappreciated, and the constant fear of losing your job are all examples of what can go wrong on the job. Serotonin (regulates appetite, sleep, and mood), noradrenaline (regulates drive and energy), corticotropin-releasing factor (regulates stress reactions), dopamine (part of the brain's reward system), glutamate (tends to activate nerve cells), and g aminobutyric acid (GABA; general sedative effect) are all affected by stress (CRF: increases steroid levels). Recalling pleasant experiences through fragrance might help restore joy. A state of mind free of anxiety is one of contentment. Several plant and floral EOs have been displayed to have a calming effect on humans. Neroli, rose, petit grain and lavender EOs are particularly well-known for this. Burns and Blamey (1994) evaluated 585 laboring women to discover if aromatherapy with any of 10 EOs may increase contractions, reduce anxiety, and decrease pain; Dunn et al. (1995) showed aromatherapy massage with lavender reduced anxiety in 122 patients in the intensive-care unit. Many different EOs, including eucalyptus, frankincense, lavender, chamomile, peppermint, lemon, rose, jasmine, mandarin, and clary sage were employed. Their findings revealed that both moms and the delivery staff appreciated the stress-reducing effects of the EOs employed. More than half of the 8058 mothers who received aromatherapy between the year 1989 and 1990 reported that it helped them unwind (Burns, 2000; Nazir et al., 2022).

Sleep disorders

According to a survey conducted by the National Sleep Foundation in 2005, 75% of American people commonly experience sleep-disorder-related symptoms like waking up many times each night. The established knowledge of parasomnia is being challenged by the work of a new breed of sleep researchers. Their claim is that all it takes to have a good night's sleep is "small alterations in the patient's routine," "visualization," "a couple of shockingly counter-intuitive actions," "particular breathing methods," "stretch and release tension," a shift in perspective, and some relaxing aromas. In a 1999 study, Weihbrecht (1999) examined the effects of breathed real lavender on 10 persons (3 men and 7 women) with a history of chronic insomnia. Patients had 2 drops of genuine lavender oil applied on their pillows or a tissue kept close before bedtime on days 15−29 of the research. All participants were asked to fill up a questionnaire about their sleeping habits and return it via mail; thereafter, all participants were interviewed via phone. One person dropped out because the lavender scent bothered her too much. Eight people had less trouble getting to sleep and eight people reported improvements in one of the four sleep domains examined. At least one participant who reported having trouble throughout the trial period also had the illness. Eight people out of the twenty-two who took part in the study said they felt more refreshed in the morning. None of the individuals altered their regular regimen of sleep aids. However, many other studies have demonstrated a substantial reduction in the need of sleeping drugs like benzodiazepines when aromatherapy is also utilized. An experiment was conducted on humans by Henry et al. at Newholme Hospital in Bakewell, UK (Henry et al., 1994). It was investigated how diffusing lavender oil at night affected people with dementia. The results of the 7-week study showed that inhaling lavender had a significant calming impact. Elderly patients in a long-term care unit were also helped by lavender, according to Hudson's research (Hudson, 1996). Eight of the nine individuals in the study slept better and felt more refreshed the next day.

Dementia

Clinically significant agitation was present in 72 patients with dementia who were treated with Melissa oil in a double-blind trial by Ballard et al. (2002). The participants were from eight UK nursing homes run by the National Health Service that specialize in caring for people with severe dementia (Hudson, 1996). Anxiety, irritation, motor restlessness, and irregular vocalization are all indicators of agitation (Batiha et al., 2021). These symptoms typically lead to disruptive behaviors such as pacing, wandering, violence, screaming, and nighttime disruption, all of which can be identified with the use of the right inventory. During the treatment period, a caretaker applied a lotion containing either 10% (by weight) melissa oil (active) or sunflower oil (placebo) to the face and both arms twice daily for four weeks. This procedure took between one and two minutes. This was a trial of a form of aromatherapy that was used in addition to conventional treatment (neuroleptics and other conventional treatments were given to patients as needed). Melissa therapy was associated with improvements in quality of life, reduced social isolation, and increased participation in meaningful activities, as measured by the Neuropsychiatric Inventory (NPI). The second was counter to what one would expect from neuroleptic treatment. EOs have been proven to be effective in a number of clinical investigations for the treatment of agitated behavior and other behavioral disorders, as well as for the reduction of resistance to nursing care procedures and other behaviors associated with dementia (Holmes et al., 2002). (Bowles et al., 2002).

Depression and anxiety

One in five males will experience depression at some point in their lives. The altered chemical balance of the brain and body is to blame. Because of the gradual nature of depression's onset, many men fail to recognize its presence. People with depression often struggle with falling asleep and staying asleep. Insomnia or excessive morning awakening are telling symptoms of depression. People who are depressed often lack vitality and struggle to muster the motivation to do anything. Many people experience a loss of appetite and decrease in interest in eating as a result. Discomfort in many parts of the body and an overall sense of physical malaise are also fairly frequent. Fruit and floral fragrance therapy has been demonstrated to alleviate senior residents' depression in assisted living homes (Schillmann and Siebert, 1991). Antidepressant and immune system-boosting effects of citrus have been observed (Komori et al., 1995). The Hamilton depression rating scale was significantly lowered after 4—11 weeks of continuous application of a citrus fragrance to depressed patients (as with antidepressant therapy), and the patients' reliance on medications was drastically cut down (Komori et al., 1995). Lavender, jasmine, rosemary, rose, and chamomile are just a few of the EOs that have shown effective in the treatment of depression. Though they are most commonly administered via inhalation, Valnet recommends taking 2—5 drops each of thyme and lavender orally twice or thrice a day. However, a distinct study by Itai et al. (2000) demonstrated the positive psychological impacts of aromatherapy on individuals undergoing chronic hemodialysis (Didunyemi et al., 2020). The selection of EOs for a patient is based on the patient's individual symptoms. *C. odorata*, *C. aurantium* var. and *Chamaemelum nobile* can be particularly helpful if the patient's symptoms are caused by extreme anxiety. Recent published clinical trials have shown a considerable decrease in anxiety in patient groups suffering from a variety of conditions (such as dementia, tumors, etc). (Edge, 2003).

Back pain and periarticular pain

There is a daily earthquake of backache every morning. It's second only to the common cold as an excuse for missing work. Daily back discomfort affects one in six persons in Western societies. Aromatherapy uses the senses of scent and touch to induce a state of deep relaxation by stimulating the parasympathetic reaction, which is strongly associated with endorphins.

Several causes contribute to aromatherapy's analgesic effects:

- A plethora of ephemeral substances target certain regions of the brain
- The presence of analgesic ingredients within the EOs which act on noradrenaline, serotonin, and dopamine at their receptor sites in the brain.
- The modulation of referred pain by touch's interaction with the skin's sensory fibers
- Causes a rash (rubefacient action) (counter-irritant effect)

Examples of those anesthetics are clove bud (*S. aromaticum*), Lemongrass (*Cymbopogon citratus*), juniper (*J. communis*), black pepper (*Piper nigrum*), true lavender (*L. angustifolia*), marjoram (*O. majorana*) (*C. odorata*), spike lavender (*Lavandula latifolia*), ginger (*Z. officinale*). In their investigation, Dolara et al. discovered that myrrh extracts produced a profound local anesthetic effect (2000). The sodium current in the excitable mammalian membranes was stopped by the anesthetic action. An open, randomized research with 100 patients was done by Krall and Krause (1993) to assess the efficacy of a gel containing peppermint oil (30%) for the treatment of periarticular pain. Both the treating physician and the patient reported great levels of satisfaction with the outcomes of mint therapy in 78% of cases. Only 19% of the mint oil group and 36% of the hydroxyethyl salicylate gel group reported continued pain after the study was over.

Urinary tract infections

Urinary tract infections (UTIs) affect millions of men annually. UTIs are often not life-threatening, although they can be rather uncomfortable. Medication selection and treatment duration are influenced by several factors, which include the patient's medical history, the identified bacteria, the root cause of the urinary tract disease, and the patient's reaction to therapy. As a result of their antimicrobial, antiviral, and antifungal characteristics, many EOs fall into one of several distinct categories. EOs tend to affect bacteria and viruses primarily at the membrane level, where they change the osmotic regulating function of the cell. Juniper berry, thyme, German and Roman chamomile, bergamot, myrtle, Sandalwood, manuka, lavender, cedarwood, savory, lemongrass and blue gum are some of the most common EOs utilized for bacterial UTI, whether in washes, a sitz/hip bath, or gentle lower abdominal aroma-massage.

Conclusion

As a result of their biological activities, such as antifungal, antibacterial, and antiviral properties, EOs have been used in a variety of medical therapies for millennia. However, recent studies have also emphasized the effectiveness of using these oils for a variety of neurological and cardiovascular illnesses. However, these mixes or their individual components may also have a high potential for toxicity, especially when used excessively. Because they want to preserve the mystical nature of their profession, aromatherapists are opposed to scientific influence. However, hospital wards cannot accommodate this. This can be done in their own practice. Nevertheless, we can continue to enjoy the advantages of touch, sympathy, understanding, and care, all of which contribute to reducing stress, worry, and emotions of lonesomeness, apathy, and unhappiness. This enables the patients to participate in their own recovery. The characteristics of a few particular EOs can also be successfully used and further researched. Both the massage and relaxation can help those with Parkinson's disease, muscular dystrophy, or multiple sclerosis move more freely. In conclusion, aromatherapy can be used in clinical care most successfully when combined with conventional medication, while it can also be used as an entirely alternative treatment for some patients.

References

Aćimović, M. G., Tešević, V. V., Smiljanić, K. T., Cvetković, M. T., Stanković, J. M., Kiprovski, B. M., & Sikora, V. S. (2020). Hydrolates: By-products of essential oil distillation: Chemical composition, biological activity and potential uses. *Advanced Technologies, 9*(2), 54−70.

Adejumo, I. O., Adetunji, C. O., & Adeyemi, O. S. (2017). Influence of UV light exposure on mineral composition and biomass production of myco-meat produced from different agricultural substrates. *Journal of Agricultural Sciences, Belgrade, 62*(1), 51−59.

Adetunji, C.O. (2008). The antibacterial activities and preliminary phytochemical screening of vernoniaamygdalina and Aloe vera against some selected bacteria, pp. 40−43 [M. sc Thesis University of Ilorin].

Adetunji, C. O., Kayode, Arowora, Bolajoko, Fawole Oluyemisi, & Bunmi, Adetunji Juliana (2013). Effects of coatings on storability of carrot under evaporative coolant system. *Albanian Journal of Agricultural Sciences, 12*(3).

Adetunji, C. O., Egbuna, C., Tijjani, H., Adom, D., Tawfeeq Al-Ani, L. K., & Patrick-Iwuanyanwu, K. C. (2020). *Homemade preparations of natural biopesticides and applications. Natural remedies for pest, disease and weed control* (pp. 179−185). Publisher Academic Press.

Adetunji, C. O., Michael, O. S., Rathee, S., Singh, K. R. B., Ajayi, O. O., Adetunji, J. B., Ojha, A., Singh, J., & Singh, R. P. (2022). Potentialities of nanomaterials for the management and treatment of metabolic syndrome: A new insight. *Materials Today Advances, 13*, 100198.

Adetunji, C. O., Palai, S., Ekwuabu, C. P., Egbuna, C., Adetunji, J. B., Ehis-Eriakha, C. B., Kesh, S. S., & Mtewa, A. G. (2021). *General principle of primary and secondary plant metabolites: Biogenesis, metabolism, and extraction. Preparation of phytopharmaceuticals for the management of disorders* (pp. 3−23). Publisher Academic Press.

Adetunji, J. B., Ajani, A. O., Adetunji, C. O., Fawole, O. B., Arowora, K. A., Nwaubani, S. I., Ajayi, E. S., Oloke, J. K., & Aina, J. A. (2013). Postharvest quality and safety maintenance of the physical properties of Daucus carota L. fruits by Neem oil and Moringa oil treatment: A new edible coatings. *Agrosearch, 13*(1), 131−141.

Adetunji, C. O. (2019). Environmental impact and ecotoxicological influence of biofabricated and inorganic nanoparticle on soil activity. In D. Panpatte, & Y. Jhala (Eds.), *Nanotechnology for agriculture*. Singapore: Springer. Available from https://doi.org/10.1007/978-981-32-9370-0_12.

Adetunji, C. O., Inobeme, A., Olaniyan, O. T., Ajayi, O. O., Olaniyan, S., & Adetunji, J. B. (2021). Application of nanodrugs derived from active metabolites of medicinal plants for the treatment of inflammatory and lung diseases: Recent advances. In K. Dua, S. Nammi, D. Chang, D. K. Chellappan, G. Gupta, & T. Collet (Eds.), *Medicinal plants for lung diseases*. Singapore: Springer. Available from https://doi.org/10.1007/978-981-33-6850-7_26.

Adetunji, C. O., Ojediran, J. O., Adetunji, J. B., & Owa, S. O. (2019). Influence of chitosan edible coating on postharvest qualities of Capsicum annum L. during storage in evaporative cooling system. *Croatian Journal of Food Science and Technology, 11*(1), 59−66.

Adetunji, C. O., Kumar, D., Raina, M., Arogundade, O., & Sarin, N. B. (2019). Endophytic microorganisms as biological control agents for plant pathogens: A panacea for sustainable agriculture. In A. Varma, S. Tripathi, & R. Prasad (Eds.), *Plant biotic interactions*. Cham: Springer. Available from https://doi.org/10.1007/978-3-030-26657-8_1.

Adetunji, C. O., Akram, Muhammad, Tope Olaniyan, Olugbemi, Olufemi Ajayi, Olulope, Inobeme, Abel, Olaniyan, Seyi, Hameed, Leena, & Adetunji, Juliana Bunmi (2021). Targeting SARS-CoV-2 novel Corona (COVID-19) virus infection using medicinal plants. In K. Dua, S. Nammi, D. Chang, D. K. Chellappan, G. Gupta, & T. Collet (Eds.), *Medicinal plants for lung diseases*. Singapore: Springer. Available from https://doi.org/10.1007/978-981-33-6850-7_21.

Adetunji, C. O., Nwankwo, W., Ukhurebor, K., Olayinka, A. S., & Makinde, A. S. (2021). Application of biosensor for the identification of various pathogens and pests mitigating against the agricultural production: Recent advances. In R. N. Pudake, U. Jain, & C. Kole (Eds.), *Biosensors in agriculture: Recent trends and future perspectives. Concepts and strategies in plant sciences*. Cham: Springer. Available from https://doi.org/10.1007/978-3-030-66165-6_9.

Adetunji, C. O., Anani, O. A., Olaniyan, O. T., Inobeme, A., Olisaka, F. N., Uwadiae, E. O., & Obayagbona, O. N. (2021). Recent trends in organic farming. In R. Soni, D. C. Suyal, P. Bhargava, & R. Goel (Eds.), *Microbiological activity for soil and plant health management*. Singapore: Springer. Available from https://doi.org/10.1007/978-981-16-2922-8_20.

Adetunji, C. O., Oloke, J. K., & Prasad, G. (2020). Effect of carbon-to-nitrogen ratio on eco-friendly mycoherbicide activity from Lasiodiplodia pseudotheobromae C1136 for sustainable weeds management in organic agriculture. *Environment, Development and Sustainability, 22*, 1977−1990. Available from https://doi.org/10.1007/s10668-018-0273-1.

Adetunji, C. O., Ajayi, O. O., Akram, M., Olaniyan, O. T., Chishti, M. A., Inobeme, A., Olaniyan, S., Adetunji, J. B., Olaniyan, M., & Awotunde, S. O. (2021). Medicinal plants used in the treatment of influenza a virus infections. In K. Dua, S. Nammi, D. Chang, D. K. Chellappan, G. Gupta, & T. Collet (Eds.), *Medicinal plants for lung diseases*. Singapore: Springer. Available from https://doi.org/10.1007/978-981-33-6850-7_19.

Adetunji, C. O., Michael, O. S., Varma, A., Oloke, J. K., Kadiri, O., Akram, M., Bodunrinde, R. E., Imtiaz, A., Adetunji, J. B., Shahzad, K., Jain, A., Ubi, B. E., Majeed, N., Ozolua, P., & Olisaka, F. N. (2021). Recent advances in the application of biotechnology for improving the production of secondary metabolites from Quinoa. In A. Varma (Ed.), *Biology and biotechnology of Quinoa*. Singapore: Springer. Available from https://doi.org/10.1007/978-981-16-3832-9_17.

Adetunji, C. O., Michael, O. S., Kadiri, O., Varma, A., Akram, M., Oloke, J. K., Shafique, H., Adetunji, J. B., Jain, A., Bodunrinde, R. E., Ozolua, P., & Ubi, B. E. (2021). Quinoa: From farm to traditional healing, food application, and phytopharmacology. In A. Varma (Ed.), *Biology and biotechnology of Quinoa*. Singapore: Springer. Available from https://doi.org/10.1007/978-981-16-3832-9_20.

Adetunji, C. O., Michael, O. S., Nwankwo, W., Ukhurebor, K. E., Anani, O. A., Oloke, J. K., Varma, A., Kadiri, O., Jain, A., & Adetunji, J. B. (2021). Quinoa, The next biotech plant: Food security and environmental and health hot spots. In A. Varma (Ed.), *Biology and biotechnology of Quinoa*. Singapore: Springer. Available from https://doi.org/10.1007/978-981-16-3832-9_19.

Adetunji, C. O., Olaniyan, O. T., Akram, M., Ajayi, O. O., Inobeme, A., Olaniyan, S., Khan, F. S., & Adetunji, J. B. (2021). Medicinal plants used in the treatment of pulmonary hypertension. In K. Dua, S. Nammi, D. Chang, D. K. Chellappan, G. Gupta, & T. Collet (Eds.), *Medicinal plants for lung diseases*. Singapore: Springer. Available from https://doi.org/10.1007/978-981-33-6850-7_14.

Adetunji, C. O., Phazang, P., & Sarin, N. B. (2017). Significance of rhamnolipids as a biological control agent in the management of crops/plant pathogens. *Current Trends in Biomedical Engineering & Biosciences, 10*(3), 54−55.

Adetunji, C. O., Panpatte, D. G., Bello, O. M., & Adekoya, M. A. (2019). Application of nanoengineered metabolites from beneficial and eco-friendly microorganisms as a biological control agents for plant pests and pathogens. In D. Panpatte, & Y. Jhala (Eds.), *Nanotechnology for agriculture: Crop production & protection*. Singapore: Springer. Available from https://doi.org/10.1007/978-981-32-9374-8_13.

Adetunji, C. O., Roli, O. I., & Adetunji, J. B. (2020). Exopolysaccharides derived from beneficial microorganisms: Antimicrobial, food, and health benefits. In P. Mishra, R. R. Mishra, & C. O. Adetunji (Eds.), *Innovations in food technology*. Singapore: Springer. Available from https://doi.org/10.1007/978-981-15-6121-4_10.

Adetunji, C. O., & Varma, A. (2020). Biotechnological application of trichoderma: A powerful fungal isolate with diverse potentials for the attainment of food safety, management of pest and diseases, healthy planet, and sustainable agriculture. In C. Manoharachary, H. B. Singh, & A. Varma (Eds.), *Trichoderma: Agricultural applications and beyond. Soil biology* (61). Cham: Springer. Available from https://doi.org/10.1007/978-3-030-54758-5_12.

Adetunji, J. B., Adetunji, C. O., & Olaniyan, O. T. (2021). African walnuts: A natural depository of nutritional and bioactive compounds essential for food and nutritional security in Africa. In O. O. Babalola (Ed.), *Food security and safety*. Cham: Springer. Available from https://doi.org/10.1007/978-3-030-50672-8_19.

Adetuyi, B. O., Oluwole, E. O., & Dairo, J. O. (2015). Chemoprotective potential of ethanol extract of ganoderma lucidum on liver and kidney parameters in plasmodium beghei-induced mice. *International Journal of Chemistry and Chemical Processes (IJCC), 1*(8), 29−36.

Agatonovic-Kustrin, S., Chan, C. K. Y., Gegechkori, V., & Morton, D. W. (2020). Models for skin and brain penetration of major components from essential oils used in aromatherapy for dementia patients. *Journal of Biomolecular Structure and Dynamics, 38*(8), 2402−2411.

Agatonovic-Kustrin, S., Kustrin, E., & Morton, D. W. (2019). Essential oils and functional herbs for healthy aging. *Neural Regeneration Research, 14*(3), 441.

Ali, B., Al-Wabel, N. A., Shams, S., Ahamad, A., Khan, S. A., & Anwar, F. (2015). Essential oils used in aromatherapy: A systemic review. *Asian Pacific Journal of Tropical Biomedicine, 5*, 601−611.

Arnold, L. (1995). The use of aromatherapy and essential oils in palliative care: Risk versus research. *Pos Health*, 32−34.

Bagci, E., Aydin, E., Mihasan, M., Maniu, C., & Hritcu, L. (2016). Anxiolytic and antidepressant-like effects of Ferulago angulata essential oil in the scopolamine rat model of Alzheimer's disease. *Flavour and Fragrance Journal, 31*(1), 70−80.

Ballard, C. G., O'Brien, J. T., & Perry, E. K. (2002). Aromatherapy as a safe and effective treatment for the management of agitation in severe dementia: the results of a double-blind, placebo-controlled trial with Melissa. The. *Journal of Clinical Psychiatry, 63*(7), 1369.

Batiha, G. B., Awad, D. A., Algamma, A. M., Nyamota, R., Wahed, M. I., Shah, M. A., Amin, M. N., Adetuyi, B. O., Hetta, H. F., Cruz-Marins, N., Koirala, N., Ghosh, A., & Sabatier, J. (2021).). Diary-derived and egg white proteins in enhancing immune system against COVID-19 frontiers in nutritionr. (Nutritional. *Immunology), 8*, 629440. Available from https://doi.org/10.3389/fnut.2021629440.

Belanger, J. T. (1998). Perillyl alcohol: Applications in oncology. *Alternative Medicine Review: A Journal of Clinical Therapeutic, 3*(6), 448−457.

Bello, O. M., Ibitoye, T., & Adetunji, C. (2019). Assessing antimicrobial agents of Nigeria flora. *Journal of King Saud University-Science, 31*(4), 1379−1383.

Benny, A., & Thomas, J. (2019). Essential oils as treatment strategy for Alzheimer's disease: Current and future perspectives. *Planta Medica, 85*(03), 239−248.

Berkarda, B., Bouffard-Eyüboğlu, H., & Derman, U. (1983). The effect of coumarin derivatives on the immunological system of man. *Agents and Actions, 13*(1), 50−52.

Bierniewicz, M. (2010). Spa&Wellnes inspiracje. Wrocław; Wyd. Spa Partners.

Bowles, E. J., Griffiths, D. M., Quirk, L., Brownrigg, A., & Croot, K. (2002). Effects of essential oils and touch on resistance to nursing care procedures and other dementia-related behaviours in a residential care facility. *International Journal of Aromatherapy, 12*(1), 22−29.

Burns, A. (2000). Might olfactory dysfunction be a marker for early Alzheimer's disease? *Lancet, 355*(9198), 84−85.

Burns, E., & Blamey, C. (1994). Soothing scents in childbirth. *International Journal of Aromatherapy, 6*(1), 24−28.

Cardia, G. F. E., Silva-Filho, S. E., Silva, E. L., Uchida, N. S., Cavalcante, H. A. O., Cassarotti, L. L., & Cuman, R. K. N. (2018). Effect of lavender (Lavandula angustifolia) essential oil on acute inflammatory response. *Evidence-Based Complementary and Alternative Medicine*, 2018.

Carle, R., & Gomaa, K. (1992). The medicinal use of Matricaria flos. *British Journal of Phytotherapy, 2*(4), 147−153.

Corner, J., Cawley, N., & Hildebrand, S. (1995). An evaluation of the use of massage and essential oils on the wellbeing of cancer patients. *International Journal of Palliative Nursing, 1*(2), 67−73.

Ćwirlej, A., Ćwirlej, A., & Gregorowicz-Cieślik, H. (2005). Masaż klasyczny i aromaterapeutyczny w bólach kręgosłupa. *Prz Med Uniw Rzesz Inst Leków, 4*, 366−371.

Czar, K. (2009). The effects of aromatherapy on alertness in an inclusion setting. *The Corinthian, 10*(7), 111−122.

Didunyemi, M. O., Adetuyi, B. O., & Oyewale, I. A. (2020). Inhibition of lipid peroxidation and in-vitro antioxidant capacity of aqueous, acetone and methanol leaf extracts of green and red Acalypha wilkesiana Muell Arg. *International Journal Of Biological and Medical Research, 11*(3), 7089−7094.

Didunyemi, M. O., Adetuyi, B. O., & Oyebanjo, O. O. (2019). Morinda lucida attenuates acetaminophen-induced oxidative damage and hepatotoxicity in rats. *Journal of Biomedical Sciences, 8*, No https://www.jbiomeds.com/biome dical-sciences/morinda-lucida-attenuates-acetaminopheninduced-oxidative-damage-and-hepatotoxicity-in-rats.php?aid = 24482.

Dimas, K., Kokkinopoulos, D., Demetzos, C., Vaos, B., Marselos, M., Malamas, M., & Tzavaras, T. (1999). The effect of sclareol on growth and cell cycle progression of human leukemic cell lines. *Leukemia Research, 23*(3), 217−234.

Dobetsberger, C., & Buchbauer, G. (2011). Actions of essential oils on the central nervous system: An updated review. *Flavour and Fragrance Journal, 26*(5), 300−316.

Dunn, C., Sleep, J., & Collett, D. (1995). Sensing an improvement: An experimental study to evaluate the use of aromatherapy, massage and periods of rest in an intensive-care unit. *Journal of Advanced Nursing, 21*(1), 34−40.

Edge, J. (2003). A pilot study addressing the effect of aromatherapy massage on mood, anxiety and relaxation in adult mental health. *Complementary Therapies in Nursing & Midwifery, 9*, 90−97.

Egbuna, C., Gupta, Ena, Ezzat, Shahira M., Jeevanandam, Jaison, Mishra, Neha, Akram, Muhammad, Sudharani, N., Oluwaseun Adetunji, Charles, Singh, Priyanka, Ifemeje, Jonathan C., Deepak, M., Bhavana, A., Mark, Angelo, Walag, P., Ansari, Rumaisa, Bunmi Adetunji, Juliana, Laila, Umme, Olisah, Michael Chinedu, & Onyekere, P. F. (2020). Aloe species as valuable sources of functional bioactives. In C. Egbuna, & G. Dable Tupas (Eds.), *Functional foods and nutraceuticals*. Cham: Springer. Available from https://doi.org/10.1007/978-3-030-42319-3_18.

El Euch, S. K., Hassine, D. B., Cazaux, S., Bouzouita, N., & Bouajila, J. (2019). Salvia officinalis essential oil: Chemical analysis and evaluation of anti-enzymatic and antioxidant bioactivities. *South African Journal of Botany, 120*, 253−260.

Elshafie, H. S., & Camele, I. (2017). An overview of the biological effects of some mediterranean essential oils on human health. *BioMed Research International*. Available from https://doi.org/10.1155/2017/9268468.

Evans, B. (1995). An audit into the effects of aromatherapy massage and the cancer patient in palliative and terminal cancer. *Complementary Therapies in Medicine, 3*(4), 239−241.

Ferreira, T. S., Moreira, C. Z., Cária, N. Z., Victoriano, G., Silva, W. F., Jr, & Magalhães, J. C. (2014). Phytotherapy: An introduction to its history, use and application. *Revista Brasileira de Plantas Medicinais, 16*, 290−298.

Franchomme, P., & Penoel, D. (1991). *L'Aromatherapie Exactement*. Limoges, France: Jollois.

Grabowska, K., & Janeczko, Z. (2013). Olejki eteryczne w preparatach farmaceutycznych. *Aromaterapia*, *4*(74), 16–50.

Han, X., Beaumont, C., & Stevens, N. (2017). Chemical composition analysis and in vitro biological activities of ten essential oils in human skin cells. *Biochim Open*, *5*, 1–7.

Henry, J., Rusius, C., Davies, M., & Veazey-French, T. (1994). Lavender for night sedation of people with dementia. *International Journal of Aromatherapy*, *6*(2), 28–30.

Holmes, C., Hopkins, V., Hensford, C., MacLaughlin, V., Wilkinson, D., & Rosenvinge, H. (2002). Lavender oil as a treatment for agitated behaviour in severe dementia: A placebo controlled study. *International Journal of Geriatric Psychiatry*, *17*, 305–308.

Hongratanaworakit, T. (2010). Stimulating effect of aromatherapy massage with jasmine oil. *Natural Product Communications*, *5*(1), 157–162.

Hongratanaworakit, T. (2011). Aroma-therapeutic effects of massage blended essential oils on humans. *Natural Product Communications*, *6*(8), 1199–1204.

Hudson, R. (1996). The value of lavender for rest and activity in the elderly patient. *Complementary Therapies in Medicine*, *4*(1), 52–57.

Itai, T., Amayasu, H., Kuribayashi, M., Kawamura, N., Okada, M., Momose, A., & Kaneko, S. (2000). Psychological effects of aromatherapy on chronic hemodialysis patients. *Psychiatry and Clinical Neurosciences*, *54*(4), 393–397.

Juergens, U. R., Stöber, M., & Vetter, H. (1998). Inhibition of cytokine production and arachidonic acid metabolism by eucalyptol (1.8-cineole) in human blood monocytes in vitro. *European Journal of Medical Research*, *3*, 508–510.

Kaleta, M. (2007). Aroma marketing-new applications of essential oils and synthetic aromatic substances. In International Conference in Budapest, Hungary, 23–27 'Developments in the Global Aroma Chemicals and Essential Oils Industries', pp. 152–156.

Kędzia, A., & Kędzia, A. W. (2017). Ocena aktywności olejku kminkowego (Oleum carvi) wobec grzybów drożdżopodobnych. *Post Fitoter*, *18*(2), 94–99.

Kiełtyka-Dadasiewicz, A., & Gorzel, M. (2014). Alternative therapies. Aromatherapy - Raw materials and treatments. *European Journal of Medical Technologies*, *1*(2), 72–79.

Komori, T., Fujiwara, R., Tanida, M., Nomura, J., & Yokoyama, M. M. (1995). Effects of citrus fragrance on immune function and depressive states. *Neuroimmunomodulation*, *2*(3), 174–180.

Konopacka-Brud, I., & Brud, W.S. (2010). Aromaterapia w Gabinecie Kosmetycznym, Ośrodku Odnowy Biologicznej Wellness i Spa. Warszawa; Wyd. WSZKiPZ.

Krall, B., & Krause, W. (1993). Efficacy and tolerance of Mentha arvensis aetheroleum. Paper presented at the 24th International Symposium on Essential Oils, Berlin, Germany.

Legan, Ł. (2014). Masaż uzdrawiający dotyk. Białystok; Wydawnictwo Vital.

Lis-Balchin, M. (2005). *Aromatherapy science. A guide for healthcare professionals*. Pharmaceutical Press.

Liu, B., Kou, J., Li, F., Huo, D., Xu, J., Zhou, X., & Han, D. (2020). Lemon essential oil ameliorates age-associated cognitive dysfunction via modulating hippocampal synaptic density and inhibiting acetylcholinesterase. *Aging (Albany NY)*, *12*(9), 8622.

López, V., Nielsen, B., Solas, M., Ramírez, M. J., & Jäger, A. K. (2017). Exploring pharmacological mechanisms of lavender (Lavandula angustifolia) essential oil on central nervous system targets. *Frontiers in Pharmacology*, *8*, 280.

López-Cortés, I., Salazar-García, D. C., Velázquez-Martí, B., & Salazar, D. M. (2013). Chemical characterization of traditional varietal olive oils in East of Spain. *Food Research International*, *54*(2), 1934–1940.

Ma, Y., Yang, M. W., Li, X. W., Yue, J. W., Chen, J. Z., Yang, M. W., & Yang, S. L. (2019). Therapeutic effects of natural drugs on Alzheimer's disease. *Frontiers in Pharmacology*, *10*, 1355.

Mailhebiau, P. (1995). The thymus folder. *Cah Aromatherapy*, *1*, 38–60.

Martinec, R. (2012). Some implication of using aromatherapy as complementary method in oncology settings. *Archives of Oncology*, *19*, 70–74. Available from https://doi.org/10.2298/AOO1204070M.

Michael, A. (2001). Aromatherapy and immunity: How the use of essential oils aid immune potentiality. Part 2: Mood-immune correlations, stress and susceptibility to illness and how essential oil odorants raise this threshold. *The International Journal of Aromatherapy*, *11*, 152–156.

Michalak, M. (2018). Aromatherapy and methods applying essential oils. *Archives of Physiotherapy and Global Researches*, *22*, 25–31.

Mitchell, S. (1993). Aromatherapy's effectiveness in disorders associated with dementia. *International Journal of Aromatherapy*, *5*(2), 20–23.

Miyake, Y., Murakami, A., Sugiyama, Y., Isobe, M., Koshimizu, K., & Ohigashi, H. (1999). Identification of coumarins from lemon fruit (Citrus limon) as inhibitors of in vitro tumor promotion and superoxide and nitric oxide generation. *Journal of Agricultural and Food Chemistry*, *47*(8), 3151–3157.

Mooventhan, A., & Nivethitha, L. (2014). Scientific evidence-based effects of hydrotherapy on various systems of the body. *North American Journal of Medical Sciences*, *6*(5), 199.

Nazir, A., Itrat, N., Shahid, A., Mushtaq, Z., Abdulrahman, S. A., Egbuna, C., Adetuyi, B. O., Khan, J., Uche, C. Z., & Toloyai, P. Y. (2022). *Orange peel as a source of nutraceuticals*, . (1st ed., p. 400)*Food and agricultural byproducts as important source of valuable nutraceuticals*, (xxx, p. 400). Berlin: Springer, Gebunden. ISBN 978-3-030–98759-6.

Nikolić, M. M., Jovanović, K. K., Marković, T. L., Marković, D. L., Gligorijević, N. N., Radulović, S. S., & Soković, M. D. (2017). Antimicrobial synergism and cytotoxic properties of Citrus limon L., Piper nigrum L. and Melaleuca alternifolia (Maiden and Betche) Cheel essential oils. *Journal of Pharmacy and Pharmacology*, *69*(11), 1606–1614.

Olaniyan, O. T., & Adetunji, C. O. (2021). Biological, biochemical, and biodiversity of biomolecules from marine-based beneficial microorganisms: Industrial perspective. In C. O. Adetunji, D. G. Panpatte, & Y. K. Jhala (Eds.), *Microbial rejuvenation of polluted environment. Microorganisms for sustainability* (27). Singapore: Springer. Available from https://doi.org/10.1007/978-981-15-7459-7_4.

Olness, K., & Ader, R. (1992). Conditioning as an adjunct in the pharmacotherapy of lupus erythematosus. *Journal of Developmental and Behavioral Pediatrics.*

Park, J. S. (2017). Effects of Juniper essential oil on the activity of autonomic nervous system. *Biomedical Science Letters, 23*(3), 286–289.

Pe'noë'l, D. (1993). *The immune system of mankind. Aroma 93 Conference Proceedings.* Brighton, UK: Aromatherapy Publications.

Pietrzak, R., Gościańska, J., & Krzyżanek, S. (2012). Wpływ aromaterapii na koncentrację i przyswajanie wiedzy. *Polish Journal of Cosmetology, 15* (2), 75–84.

Płocica, J., & Tal-Figiel, B. (2011). Selected plant and essential oils used in cosmetic industry. *Polish Journal of Cosmetology, 14*(4), 266–272.

Priestap, H. A., Bandoni, A. L., Neugebauer, M., & Rücker, G. (1990). Investigation of the essential oils from Aristolochia triangularis. *Journal of Essential Oil Research, 2*(3), 95–98.

Pytkowska, K., & Arct, J. (2006). Advances of sensory analysis. *Wiadomości PTK, 9*(1), 24–27.

Rauf, A., Akram, M., Semwal, P., Mujawah, A. A. H., Muhammad, N., Riaz, Z., Munir, N., Piotrovsky, D., Vdovina, I., Bouyahya, A., Oluwaseun Adetunji, C., Shariati, M. A., Almarhoon, Z. M., Mabkhot, Y. N., & Khan, H. (2021). Antispasmodic potential of medicinal plants: A comprehensive review. *Oxidative Medicine and Cellular Longevity, 2021*, 12. Available from https://doi.org/10.1155/2021/4889719, Article ID 4889719.

Romer, M. (2009). Aromaterapia. Wrocław; Wyd. MedPharm.

Roulier, G. (1990). Les Huiles Essentielles Pour Votre Sante. St. Jean-de-Braye, France: Dangles.

Rovesti, P., & Columbo, E. (1973). Aromatherapy and aerosols. *Soap, Perfumery & Cosmetics (London), 46*, 475–477.

Safayhi, H., Sabieraj, J., Sailer, E. R., & Ammon, H. P. T. (1994). Chamazulene: An antioxidant-type inhibitor of leukotriene B4 formation. *Planta Medica, 60*(05), 410–413.

Schilcher, H. (1994). Phytotherapy and classical medicine. *Journal of Herbs, Spices and Medicinal Plants, 2*, 71–80.

Schillmann, S., & Siebert, J. (1991). New frontiers in fragrance use. *Cosmetics Toiletries, 106*(6), 39–45.

Schneider, R., Singer, N., & Singer, T. (2019). Medical aromatherapy revisited—Basic mechanisms, critique, and a new development. *Human Psychopharmacology: Clinical and Experimental, 34*(1), e2683.

Setzer, W. (2009). Essential oils and anxyolytic aromatherapy. *Natural Product Communications, 4*, 1305–1316.

Shah, Y. R., Sen, D. J., Patel, J. S., & Patel, A. D. (2011). Aromatherapy: The doctor of natural harmony of body & mind. *International Journal of Drug Development and Research, 3*(1), 0-0.

Siddique, S. (2017). Essential oils and cosmetic aromatherapy. Trichology and Cosmetology Openventio. *The Journal, 1*, e7–e8. Available from https://doi.org/10.17140/TCOJ-1-e004.

Simpson, B. B., Neff, J. L., & Dieringer, G. (1990). The production of floral oils byMonttea (Scrophulariaceae) and the function of tarsal pads inCentris bees. *Plant Systematics and Evolution, 173*(3), 209–222.

Stea, S., Beraudi, A., & Pasquale, D. D. (2014). Essential oils for complementary treatment of surgical patients: State of the art. Hindawi Publishing Corporation. *Evidence-Based Complementary and Alternative Medicine*, 726341. Available from https://doi.org/10.1155/2014/726341.

Steflitsch, W., & Steflitsch, M. (2008). Clinical aromatherapy. *Journal of Men's Health, 5*, 74–85.

Szydłowska, M. (2008). Rola zapachu w kosmetykach. *Aromaterapia, 51/52*(14), 12–15.

Tambe, Y., Tsujiuchi, H., Honda, G., Ikeshiro, Y., & Tanaka, S. (1996). Gastric cytoprotection of the non-steroidal anti-inflammatory sesquiterpene, β-caryophyllene. *Planta Medica, 62*(05), 469–470.

Thangadurai, D., Naik, J., Sangeetha, J., Said Al-Tawaha, A. R. M., Adetunji, C. O., Islam, S., David, M., Shettar, A. K., & Adetunji, J. B. (2021). Nanomaterials from agrowastes: Past, present, and the future. In O. V. Kharissova, L. M. Torres-Martínez, & B. I. Kharisov (Eds.), *Handbook of nanomaterials and nanocomposites for energy and environmental applications*. Cham: Springer. Available from https://doi.org/10.1007/978-3-030-36268-3_43.

Ukhurebor, K. E., & Adetunji, C. O. (2021). Relevance of biosensor in climate smart organic agriculture and their role in environmental sustainability: What has been done and what we need to do? In R. N. Pudake, U. Jain, & C. Kole (Eds.), *Biosensors in agriculture: Recent trends and future perspectives. Concepts and strategies in plant sciences*. Cham: Springer. Available from https://doi.org/10.1007/978-3-030-66165-6_7.

Valnet, J. (1990). *The practice of aromatherapy*. Saffron Walden, UK: CW Daniels.

Viuda-Martos, M., Ruiz-Navajas, Y., Fernández-López, J., & Pérez-Álvarez, J. (2008). Antifungal activity of lemon (Citrus lemon L.), mandarin (Citrus reticulata L.), grapefruit (Citrus paradisi L.) and orange (Citrus sinensis L.) essential oils. *Food Control, 19*(12), 1130–1138.

Wagner, H., Wierer, M., & Bauer, R. (1986). In vitro inhibition of prostaglandin biosynthesis by essential oils and phenolic compounds. *Planta Medica, 52*(3), 184–187.

Warszawski, A. (2011). Przenikanie składników olejku różanego przez skórę. *Aromaterapia, 4*(66).

Weihbrecht, L. (1999). *A comparative study on the use of Lavandula angustifolia and its effect on insomnia. Unpublished dissertation*. Hunter, NY: RJ Buckle Associates.

Weyerstahl, P., Schneider, S., Marschall, H., & Rustaiyan, A. (1993). The essential oil of Artemisia sieberi Bess. *Flavour and Fragrance Journal, 8* (3), 139–145.

Wilkinson, S. (1995). Aromatherapy and massage in palliative care. *Indian Journal of Palliative Care, 1*, 21–30.

Williams, A. C., & Barry, B. W. (2004). Penetration enhancers. *Advanced Drug Delivery Reviews, 56*(5), 603–618.

Worwood, V. A. (1993). *The complete book of essential oils and aromatherapy*. New Word Library.

Xu, P., Wang, K., Lu, C., Dong, L., Gao, L., Yan, M., & Liu, X. (2017). The protective effect of lavender essential oil and its main component linalool against the cognitive deficits induced by D-galactose and aluminum trichloride in mice. *Evidence-Based Complementary and Alternative Medicine*, 2017.

Yepez, B., Espinosa, M., López, S., & Bolanos, G. (2002). Producing antioxidant fractions from herbaceous matrices by supercritical fluid extraction. *Fluid Phase Equilibria, 194*, 879–884.

Yoshiyama, K., Arita, H., & Suzuki, J. (2015). The effect of aroma hand massage therapy for people with dementia. *The Journal of Alternative and Complementary Medicine, 21*(12), 759–765.

Zheng, G. Q., Kenney, P. M., & Lam, L. K. (1992). Anethofuran, carvone, and limonene: Potential cancer chemoprotective agents from dill weed oil and caraway oil. *Planta Medica, 58*(04), 338–341.

Zheng, G. Q., Kenney, P. M., & Lam, L. K. (1993). Potential anticarcinogenic natural products isolated from lemongrass oil and galanga root oil. *Journal of Agricultural and Food Chemistry, 41*(2), 153–156.

Chapter 18

Application of starter culture bacteria in dairy product

Babatunde Oluwafemi Adetuyi[1], Charles Oluwaseun Adetunji[2], Juliana Bunmi Adetunji[3], Abel Inobeme[3], Oluwabukola Atinuke Popoola[4], Oloruntoyin Ajenifujah-Solebo[4], Yovwin D. Godwin[5], Olatunji Matthew Kolawole[6], Olalekan Akinbo[7] and Mohammed Bello Yerima[8]

[1]*Department of Natural Sciences, Faculty of Pure and Applied Sciences, Precious Cornerstone University, Ibadan, Oyo State, Nigeria, [2]Applied Microbiology, Biotechnology and Nanotechnology Laboratory, Department of Microbiology, Edo State University Uzaire, Iyamho, Edo State, Nigeria, [3]Department of Chemistry, Edo State University Uzairue Iyamho, Auchi, Edo State, Nigeria, [4]Genetics, Genomics and Bioinformatics Department, National Biotechnology Development Agency, Abuja, FCT, Nigeria, [5]Department of Family Medicine, Faculty of Clinical Sciences, Delta State University, Abraka, Delta State, Nigeria, [6]Department of Microbiology, Faculty of Life Sciences, University of Ilorin, Ilorin, Kwara State, Nigeria, [7]Centre of Excellence in Science, Technology, and Innovation, AUDA-NEPAD, Johannesburg, Gauteng, South Africa, [8]Department of Microbiology, Sokoto State University, Sokoto, Sokoto State, Nigeria*

Introduction

Although the spontaneous fermentation of raw materials occurred in ancient times, humans were able to empirically control these processes during the Neolithic era by employing a form of "basic" technology. Yogurts and fresh-style cheeses made from milk that have spontaneously fermented are the first dairy products for which there are bibliographic references, including in the Bible and other writings. Yet, the earliest indication that human civilization was in control of a process that resulted in the manufacturing of cheese was found in the archeological remnants of "drainage" vessels with several holes that have been proved to contain residues of milk fatty acids (Salque et al., 2013). In the Saqqara necropolis, excavations of the nobleman Ptahmes' tomb yielded the remains of the earliest cheese ever discovered. This leftover "cheese" has peptides from cow's milk with a blend of goat's and sheep's milk, according to a proteomic analysis. The discovery of peptides belonging to a bacterium, specifically Brucella melitensis, the pathogen of sheep and goats and the cause of brucellosis in humans, is intriguing and confirms the existence of this milk type in the cheese residue (Greco et al., 2018). The "domestication" of the microbiota that was present naturally in raw foods was started when people started to domesticate animal and plant species, eventually becoming breeders and farmers. Foods that spontaneously fermented from these raw ingredients permitted the selection of bacteria, molds, and yeasts that favored "reproducibility" in the same manner that seeds were chosen for their greater production performance. The selection of populations of microbes well-suited to these environmental circumstances was made possible by the traditional use of the back-slopping technique, or by reinoculating with a prior curd, sourdough, or fermented product. Genetic differentiation and specialization based on the food matrix in which they were located resulted from these adaptations being fixed in the microbial genomes (Li & Gänzle, 2020). This fact has generally resulted in a decline in the diversity of naturally occurring microbes in favor of a smaller number of well-adapted organisms, but it has also permitted control over the fermentation process to ensure the reproducibility, quality, and safety of fermented products (Gibbons & Rinker, 2015). Additionally, it appears that this microbial domestication process happened very quickly, which allows for the prospect of doing a "guided evolution" in the lab to produce strains that are better suited to technological processes (Gibbons, 2019). To produce safer products today, both spontaneous and non-spontaneous (controlled) fermentations are used, with the first category of food serving as a suitable source for the isolation of novel strains with functional qualities (Tamang et al., 2016). The organisms responsible for lactic fermentation, known as lactic acid bacteria (LAB), are frequently employed as cultures for a wide range of conventionally fermented meals and functional foods. Moreover, some particular species are employed as "cell factories" for the production of various substances that have a variety of uses in food, medicine, or cosmetics (Hatti-Kaul et al., 2018). The ability of this bacterial group to use a wide range of carbon sources is one reason for their adaptability (Sauer et al., 2017), and it also makes them a

Applications of Essential Oils in the Food Industry. DOI: https://doi.org/10.1016/B978-0-323-98340-2.00019-5

valuable tool for the biotransformation of various residues in valuable by-products, such as the lactic acid produced by fermentation of the lactose from whey (Zandona et al., 2021). There has been a thorough examination of the diversity, physiology, taxonomy, and/or applications of LAB (for example, Holzapfel & Wood, 2014). It is important to note that, due to the recent reclassification of the genus Lactobacillus in taxonomy and the proposed 25 additional genera, the term "lactobacilli" is now used to refer to all bacteria that belong to the Lactobacillaceae family (Zheng et al., 2020). The current review's objective is to provide some suggestions regarding the functional characteristics that make LAB useful bacterial cultures for a variety of food production applications, with a focus on dairy products (Carminati et al., 2016). For this reason, we have divided it arbitrarily into four categories that, to the best of our knowledge, highlight the significant part that this bacterial group plays in a variety of areas of the food industry.

Cultures for dairy safety and preservation

As mentioned in the introduction, the main function of LAB in the manufacture of food is the fermentation of carbon sources for the acidification of the raw material through the creation of lactic acid, thus minimizing the presence of unfavorable pathogenic and rotting microorganisms. This process will alter the finished product's sensory qualities while also enhancing its safety and time of preservation (Barcenilla et al., 2021). In fact, LAB is in charge of producing and maintaining a vast range of fermented foods and beverages that are extensively consumed by people, including those with both plant and animal origins, including pickles, olives, and many traditional fermented foods from Asia, Africa, and Latin America (Anyogu et al., 2021). In fact, the FDA of the USA has designated these bacteria as Generally Recognized As Safe (GRAS) microorganisms, and some of the species are also included in the qualified presumption of safety (QPS) list of the EFSA in Europe. This is due to the long tradition of safe consumption of traditional fermented foods and beverages obtained from LAB fermentation (EFSA panel BIOHAZ, 2021). The bioprotective ability of the bacteria is fueled by the production of specific metabolites from carbohydrate metabolisms, such as bacteriocins (Rouse & Van Sinderen, 2008), or metabolites with antifungal properties, such as diacetyl, ethanol, or various organic acids, as well as by metabolites acting against bacteria (Chen et al., 2021).

Antibacterial properties of lactic acid bacteria

The tiny ribosomal synthetized peptides known as bacteriocins are one of the tools in LAB's toolbox for combating other bacteria that inhabit a specific ecological niche (Rodríguez et al., 2017). Since they are decomposed by pancreatic and gastric enzymes during food digestion, they have generally been thought to be safe for eating due to their chemical composition, even if some of them may offer toxicity at large doses (Soltani et al., 2021). Currently, the only bacteriocin recognized by the EFSA for use as a food additive in dairy production in Europe is nisin (E 234). (EFSA Panel on Food Additives and Nutrient Sources added to Food ANS, 2017). The Lactococcus lactis strains that produce this bacteriocin have antibacterial action against a wide variety of Gram-positive bacteria. Gram-positive and Gram-negative bacteria can be killed by bacteriocins, although the range of their effectiveness depends greatly on how they work (Soltani et al., 2021). *Salmonella enterica* and *Escherichia coli* are substantially less common in milk and cheese, while Staphylococcus aureus and Listeria monocytogenes are the predominant foodborne pathogens (Gonzales-Barron et al., 2017). Spore-forming bacteria, like *Clostridium tyrobutiricum* and *Bacillus cereus*, are the ones responsible for dairy product deterioration (Pancza et al., 2021). Silva has evaluated the use of bacteriocins and the LAB bacteria that produce them as bioprotective cultures in dairy products (2018). The application of purified or semi-purified bacteriocins and their producing LAB for various dairy products, such as raw, UHT and sterilized milk, skimmed milk powder, various varieties of cheeses, and some fermented milk, including yogurt, has been thoroughly researched by these authors and is presented in three tables with data available from 2000 to 2018. (Silva et al., 2018). A bibliographic search was more recently carried out by Trejo-González et al. (2021) to find articles (from 2009 to 2021) discussing bacteriocin-producing LAB isolated from cheeses. They discovered that the Enterococcus and (former) Lactobacillus genera account for more than 30% of these LAB, followed by Lactococcus and Pediococcus (less than 14%), which were most frequently recovered from soft cheeses (Trejo-González et al., 2021). Recently, the use of the Lactococcus-produced nisin against spore-forming bacteria in food production was reviewed anew because it is especially important for extending the shelf life of fermented goods (Anumudu et al., 2021). There has also been considerable interest in the possible use of bacteriocins as a probiotic's mechanism of action against infections (Soltani et al., 2021). As an illustration, the ability of strains of Pediococcus pentosaceus to produce various bacteriocins, known as pediocins, may provide a beneficial benefit for the generating bacteria as an antagonistic strategy to survive in the populated colonic niche (Jiang et al., 2021). Bacteriocins are a very active area of research, according to the number of reviews on them, and new uses for old and new bacteriocins are constantly being published. A number of these investigations focus on the hunt for bacteriocins with a special use against the pathogen *M. monocytogenes* associated with dairy products. This is the case with the co-culture of *Lactococcus lactis* subsp. lactis BGBU1−4 for the production of fresh soft cheese and *Streptococcus thermophilus* B59671 for the production of termophilin 110 and *L. (Lactiplantibacillus) plantarum* 076 for the production of pediocin (Ceruso et al., 2021; Mirkovic et al., 2020). New bacteriocins are being defined, such as a class III one made by Lactobacillus acidophilus NX2−6 that has a broad spectrum against Gram-positive and Gram-negative bacteria; this strain, which was derived from Chinese Koumiss, was successful in lowering the viable counts of *S. aureus*, Lis. monocytogenes, *Sal. enteritidis*, and *E. coli* in tainted milk and mozzarella cheese (Meng et al., 2021). A new IId bacteriocin with bioprotective qualities against *S. aureus* and other pathogens was isolated from the strain of *L. plantarum* SHY 21−2 detected in yogurt made from yak milk (Peng et al., 2021). Finally, several studies have used the currently accessible tools to search for bacteriocin production in LAB genomes and metagenomes of various dairy products in the search for novel bioprotective strains (Bachtarzi et al., 2019). This opens an avenue for the discovery of novel molecules with antimicrobial properties for food applications. An intriguing strategy that satisfies current food consumption trends, which include a market with less chemical additives and preservatives, is the utilization of LAB as bioprotective cultures. These bacteria are useful for the production of natural food preservation cultures for use in the food industry due to this feature and the large number of LAB that produce a wide diversity of compounds that can protect food from unwanted microbes.

Cultures for optimal organoleptic properties

LAB plays a significant role in the production of dairy products' fragrance, flavor, and other sensory qualities. This is a result of their ability to produce a variety of metabolites, including organic acids, volatile compounds, and biopolymers derived from the metabolism of sugar, as well as their capacity to convert amino acids released after proteolysis into aromatic compounds that give fermented foods desirable organoleptic properties. The conversion of the lipid components of milk, which is normally fueled by yeast or molds in the creation of cheese, is less affected by LAB (Thierry et al., 2016).

Lactic acid bacteria involved in aroma and flavor development of dairy products

The creation of the "flavor," a blend of scent and flavor, in dairy products, and particularly cheeses, is caused by a wide variety of metabolites. In fact, the word "cheesomic" has been proposed to encompass all multi-omic methods now available for determining how the microbial community contributes to the identification traits of a cheese (Afshari et al., 2020). The LAB-biota found in cheese may come from several sources and is involved in a number of crucial processes that turn milk into a wide range of cheeses. In the creation of fermented dairy products, the starter LAB is referred to as primary cultures if they are in charge of the milk's rapid acidification and secondary cultures if they are in charge of the cheese's later ripening (development of flavor and aroma). Based on the methods used to make each particular type of cheese, the LAB is chosen in accordance with their preferred development temperature: mesophilic or thermophilic. Natural LAB starters are made up of a variety of unidentified species and are frequently used in the backsloping method of making artisanal cheese. The term "mixed-strain starters" refers to commercial LAB mixtures with no clearly defined composition that can be given to specific cheese producers. Yet, some of these "natural" LAB combinations can be recreated, in a controlled manner, by starter manufacturing businesses. The MSS and natural LAB starters are regarded as "traditional" cultures. Large starter production firms, on the other hand, commercialize some specific combinations of well-defined strains that have been chosen based on their strong performance on several qualities of cheese production. These are defined strain starters, and various combinations of various strains and/or well-known species are currently present (Altieri et al., 2017). A natural LAB-biota is also present in the preparation of fermented foods in addition to LAB starters. This is obtained from the surroundings of the farm and cheese-making facilities as well as from the cheesemakers, and it is crucial for the acquisition of the product's sensory qualities (Gobbetti et al., 2015, 2018). Last but not least, it should be highlighted that additional microorganisms may have contributed to the formation of distinct organoleptic characteristics for particular cheese types. They can be used as functional starters because they are a part of the cheese's microbial community (or microbiota). This may apply to eukaryotes like yeasts (Buzzini et al., 2017) and molds as well as propionic acid bacteria like Propionibacterium freudenreichii subsp. shermanii and Brevibacterium linens, which are found in cheese rinds (Dantigny & Bevilacqua, 2017). In actuality, the ecology of cheese is quite complex and home to a wide variety of microorganisms that are arranged in distinct niches or microenvironments (Mayo et al., 2021). For instance, the intriguing communities of cheese rinds are different from the communities found in the cheese core but can significantly influence some organoleptic or sensory characteristics (Wolfe et al., 2014). The nonstarter lactic acid bacteria (NSLAB), the native cheese microbiota, and the LAB starters' microbial interactions have not all been well understood. Yet, it is evident that collaboration or competition in protein, lipid, or carbohydrate metabolism occurs and affects this dynamic but relatively tiny ecosystem (Blaya et al., 2018; Gobbetti et al., 2018). This proves how important it is to select a suitable LAB starter and/or NSLAB for the manufacture of cheeses with recognized qualities, including those with Protected Designation of Origin (PDO), in a repeatable manner (Randazzo et al., 2021).

Exopolysaccharides-producing lactic acid bacteria to improve physical properties of dairy products

Several sensory qualities, such as the texture, structure, or viscosity of fermented foods, are altered by LAB in addition to the development of scent and flavor. Exopolysaccharides (EPS), which can be produced by specific starter and nonstarter LAB, give dairy-fermented goods the best physical qualities (Prete et al., 2021). EPS are carbohydrate polymers that are present on the surface of the EPS-producing LAB and form a slimy coating or a compact capsule. They vary in chemical composition and structural complexity (Ruas-Madiedo & de los Reyes-Gavilan, 2005). In addition to having an impact on the physical characteristics of fermented foods, these biopolymers are also important for the probiotic effects of bifidobacteria and EPS-producing LAB on human health. Bacterial EPS can directly interact with intestinal receptors, which allows them to influence the intestinal microbiota and the host immunological response. Antagonism against the activities of pathogens or antioxidant capabilities has also been documented (Castro-Bravo et al., 2018). As extracellular components that may be released into the extracellular media, bacterial EPS are significant because they may contribute to the health advantages of fermented postbiotic meals (Molinero et al., 2022). Because of their capacity to interact with the casein network and retain water, EPS produced by LAB are efficient fatreplacers in the cheese industry. But, these polymers also enhance the mouthfeel, consistency, and texture of mostly fresh cheeses (Hahn et al., 2014). When it comes to fermented milk, EPS-producing LAB starters have historically been used in the production of yogurt, but they are also naturally present in a variety of fermented milk, including kefir, viili, and lngfil, among others (Ruas-Madiedo et al., 2009). Mouthfeel, creaminess, ropiness, look, and hardness are some of the sensory qualities of yogurt to be aware of, which can be created by utilizing various ingredients serving as a thickening and texturing agents. However, the generation of EPS in situ during milk fermentation offers a natural and consumer-acceptable alternative to accomplish this milestone given the global trend toward healthier and more natural products. In fact, a variety of EPS-producing LAB starters are currently offered for the production of yogurt (Tiwari et al., 2021). The intrinsic properties of the polymer, such as its chemical composition, structure, and stiffness, among others, are essential to assess its potential to influence viscosity and texture, in addition to the acidifying and flavor development capacities of the EPS-producing LAB. In reality, for a given concentration, EPS with negative charges in their makeup, stiffer repeating unit structures, or high molar masses are typically those able to give fermented milk the necessary viscosifying and texturizing qualities (Surber et al., 2019). Consequently, choosing the right EPS-generating LAB is particularly important for producing a final product with the appropriate physical properties. A recent illustration involves research conducted by Bachtarzi and colleagues on the potential of various EPS-producing *L. plantarum* strains, isolated from traditional Algerian dairy products, to enhance the viscosity of skimmed fermented milk, with varying degrees of success depending on the strain (Bachtarzi et al., 2019). The production of a high molecular mass polymer directly correlated with an increase in the viscosity of the fermented milk, according to a more in-depth analysis of three of these strains; however, the acidification rate in conjunction with the high accumulation of synthesized EPS at earlier stages of fermentation had a significant impact on the microstructure of the milk gel and its capacity to retain water (Bachtarzi et al., 2020). The hunt for new LAB capable of creating the best organoleptic features and, thus, being employed as starters in the food industry has been sparked in recent decades by interest in developing fermented goods with high sensory standards. The hunt for new LAB strains has been sparked by consumer desire for novel products. Traditional fermented foods are one option for investigating the natural biodiversity of the LAB-biota present in them and assessing their potential for a variety of biotechnological uses in this context.

Cultures for nutritional improvement

Although vitamins and minerals are vital micronutrients needed to sustain a variety of biological functions in all living things, most of them are unable to be produced by humans, who must instead obtain them through food. It is not surprising that occasionally consuming enough of these elements may not be enough since some of them can be destroyed during food preparation and cooking. Even in highly industrialized nations where unbalanced diets are prevalent, the lack of several of these micronutrients is a public health concern (FAO-WHO, 2006). By a number of processes related to the particular activity of starter and adjunct bacterial cultures on the food matrix, food fermentation can improve the nutritional and functional qualities of foods (Anlier, 2019). In this case, preventing nutritional deficiencies in large population segments without requiring drastic dietary changes is possible by carefully choosing and incorporating into fermented foods bacterial cultures that are technologically advanced and capable of enhancing their nutritional properties. This strategy is appealing, cost-effective, environmentally friendly, and consumer-acceptable (LeBlanc et al., 2013).

Micronutrients' production

Using two major strategies—the most direct being the in situ creation of micronutrients in the fermented material that results in the manufacture of biofortified foods—the addition of cultures capable of synthesizing nutrients in meals can help to prevent nutritional deficits. However, under some circumstances, the bacteria present in fermented foods may also colonize the gastrointestinal tract after ingestion, albeit only momentarily, where they could continue to directly synthesize vitamins in the human intestine. In fact, the intestinal microbiota produces vitamins frequently, and fermented foods are a well-known source of living microorganisms, some of which may briefly colonize our digestive tract and have positive effects on our gut microbiota and general health (Marco et al., 2017). Microbial species of industrial interest for the food industry can manufacture some vitamins. The most in-depth research has been done on the production of water-soluble vitamins from the B group, including riboflavin, folates, thiamine, and cobalamin (Acevedo-Rocha et al., 2019). In addition to fermented milk and yogurts, other fermented food matrixes have also been studied, including vegetable-based drinks, pasta and bread, fruit- and cereal-based foods, and even kefir-like cereal-based beverages and dairy whey. Fermented food biofortification through microbial production of micronutrients has been extensively investigated in fermented milk and yogurts (Levit et al., 2021). Riboflavin, also known as vitamin B2, is a precursor of the redox compounds flavin mononucleotide (FMN) and flavin adenine dinucleotide (FAD). Microbiological synthesis has largely taken the place of riboflavin synthesis for food fortification, primarily using Bacillus subtilis and the ascomycete Ashbya gossypii (Averianova et al., 2020). The manufacture of naturally biofortified foods is made possible by the ability of LAB species, which are widely used in the manufacturing of fermented meals. *Leuconostoc mesenteroides*, *Lactobacillus delbrueckii*, L. (*Limosilactobacillus*) *fermentum*, L. (*Limosilactobacillus*) *reuteri*, L. (*Ligilacillus*) *salivarius*, and *Bifidobacterium longum* subsp. infantis strains all contain the genetic machinery necessary for its synthesis (Thakur et al., 2016). Using specific strains of the LAB *L. acidophilus*, *L. plantarum*, L. (*Limosilactobacillus*) *mucosae*, L. *fermentum*, and *P. freudenreichii* in a variety of food matrixes has allowed for the biofortification of riboflavin in fermented foods up to this point. It has been shown that other producing bacteria of interest to the fermented food business, such as those from the genus *B. longum* subsp. infantis, are capable of creating this vitamin in vitro and may even be able to increase the amount of this vitamin in fermented foods (Levit et al., 2021). The regulation of the genetic and enzymatic machinery that LAB uses to produce riboflavin has been thoroughly investigated in various Gram-positive bacteria, which is an intriguing element of the process. Riboflavin bioproduction has been improved by methods including genetic engineering and exposure to poisonous similar roseoflavin, which has increased riboflavin production in representatives of numerous species. This approach is a workable substitute for creating fermented food products with more riboflavin content, as shown on various food matrixes (Ge et al., 2020), which could reduce the symptoms of riboflavin deficiency (LeBlanc et al., 2006). Several riboflavin-producing LAB strains have shown probiotic properties, making them potentially useful from an industrial standpoint as multifunctional adjunct cultures that can produce fermented foods that are biofortified while simultaneously exhibiting other probiotic traits when consumed (Russo et al., 2016). Folate, or vitamin B11, has antioxidant properties and participates in numerous metabolic pathways. There is interest in using naturally produced folate for food and feed biofortification because chemically synthesized folate can have negative effects. Many LAB species have been shown to produce folate, and some strains have developed folate-overproducing strains by genetic engineering (Wegkamp et al., 2007) or exposure to the antagonist methrotexato (Capozzi et al., 2012). With the exception of *L. plantarum*, various problems prevent their use for food biofortification, as most lactobacilli strains require the presence of a precursor called pABA. Moreover, some strains that are frequently found in fermented foods can also absorb folate, lowering the bioavailability of the finished product. This makes it even more important to choose bacteria carefully in order to effectively fortify fermented foods. The amount of this vitamin in fermented milk has greatly risen thanks to the combination of multiple folate-producing strains (LeBlanc et al., 2011). Moreover, natural strains of folate-overproducing bacteria have been found in artisanal yogurts from Argentina. By using them, yogurts with a 250% greater folate concentration than unfermented milk have been created (Laíño et al., 2014). While St. thermophilus and *L. reuteri* folate-producing strains have also been used to produce folate-enriched oat and barley-fermented foods, the highest production levels are achieved with yeast fermentation. This is because the folate-binding proteins in milk may increase its bioavailability and prevent its uptake by bacteria either in the fermented food or in the gut (Greppi et al., 2017). Surprisingly, administration of some LAB folate-producing strains has shown a capability to increase fecal folate excretion and to improve nutritional insufficiency in in vivo animals (LeBlanc et al., 2013), indicating its inclusion in fermented foods can be a useful strategy to prevent its deficit (Tamene et al., 2019). Little research has been done on the generation of other vitamins by microbial cultures relevant to the food production industry. Several thiamine-producing LAB strains have been described, including typical strains of *Lactocaseibacillus lactis*, *Lactocaseibacillus rhamnosus*, *Lactocaseibacillus plantarum*, and *Lactocaseibacillus brevis*. Surprisingly, several of these thiamine-producing strains have shown neuroprotective and immunomodulatory benefits in in vitro studies. In any case, thiamine production in LAB strains typically occurs at much lower levels than folate or riboflavin production, but amprolium exposure has resulted in an overproduction that can be up to four times higher (Teran et al., 2021). The genetic machinery necessary for cobalamin biosynthesis has been described in representative strains of *L. (Loigolactobacillus) coryniformis*, *L. reuteri*, and *L. (Furfurilactobacillus) rossiae* (Santos, Vera et al. 2008; Santos, Wegkamp et al., 2008), and production has been shown in a few LAB strains (Walhe et al., 2021). However, in the majority of cases, their capacity to synthesize the vitamin in nutritionally significant quantities has yet to be demonstrated. Although the use of thiamine or cobalamine-producing strains for the biofortification of fermented foods has not been completely researched, certain works have shown their appropriateness to generate such foods. By optimizing the glycerol and fructose content of the matrix, for example, supplementation with a cobalamin-producing *L. reuteri* strain allowed the synthesis of a soy yogurt up to 18 g/100 mL (Gu et al., 2015). Moreover, *P. freudenreichii* has been shown to be capable of producing vitamin B12 in nutritionally significant quantities under growth conditions in environments that resemble cheese (Deptula et al., 2017). Overall, these examples show that some LAB cultures have the ability to biofortify natural foods. Combinations of cultures that can produce multivitamins have also been shown to be an intriguing strategy for enhancing the nutritional qualities of fermented cereals (Rajendran et al., 2017). Moreover, multivitamin-producing cultures have been created using genetic engineering techniques, although their practical industrial implementation in food products may yet encounter certain regulatory challenges (Sybesma et al., 2004). Additional cultural and metabolic characteristics can be used to improve the nutrient content of fermented foods. Given their structural complexity, some matrixes could make it difficult for bacterial cultures to function properly during fermentation. It is essential to carefully choose the right starters that can ferment the material with one another and produce more bioactive substances (Bationo et al., 2019). The bioaccessibility of specific nutrients is also known to be severely impacted by the presence of some food components like phytates or oxalates, which are frequently found in some vegetable matrixes, such as cereal and pseudocereals. For instance, phytate, which is known to impair the digestion of minerals, proteins, and lipids, serves as the main form of phosphate and inositol storage in plant seeds. Therefore, increasing the amount of phytase in food through technological processing or exogenous supplementation can improve nutrient absorption and prevent nutritional deficits. Phytase activity is occasionally seen in LAB but it is rarely found in lactobacilli (Pradhan & Tamang, 2021). Yet, some studies have shown that cultures that produce phytase can enhance the nutritional value of fermented foods (Carrizo et al., 2020). Thus, when choosing cultures to enhance the nutritional content of particularly fermented vegetable matrixes, phytase activity may be a pertinent characteristic. As shown in numerous in vitro and animal studies, the potential of combining bacterial cultures with nutrient-producing and phytase activities is a promising strategy to enhance the nutritional content of fermented foods and prevent mineral and vitamin shortages (Carrizo et al., 2020). This goal would be greatly aided by synthetic ecology methods supported by thorough genetic and metabolic characterization of relevant cultures.

Cultures to improve health: probiotics and postbiotics

A few of them, along with some members of the genus Bifidobacterium, have also been recognized as having health-promoting effects in humans, either when administered included in foods or when consumed as food supplements, in addition to their traditional use as food starter cultures for food fermentations and providers of sensorial properties. When given in sufficient proportions, these probiotic microorganisms improve the host's health (Hill et al., 2014). Since the middle of the last century, there has been clinical proof of a variety of health impacts, which lends strong support to the benefits of probiotics (Sánchez et al., 2017). It is important to note that non-viable bacteria, as well as the parts and metabolites they produce, can have an effect on human physiology. In this context, the term "postbiotics" refers to the creation of inanimate microorganisms and/or their components that impart a health advantage to the host. Postbiotics are microorganisms that do not require to be alive to perform their biological activity (Salminen et al., 2021). Although the scientific community as a whole agreed on the concept of probiotic, the term postbiotic is still up for controversy because it has not acquired widespread acceptance. Postbiotics are described as "any factor resulting from the metabolic activity of a probiotic or any released molecule capable of giving positive benefits to the host in a direct or indirect manner" in previous definitions; however, paraprobiotics were also classified as "inactivated, non-viable microbial cells that after consumption give a health advantage" (Taverniti & Guglielmetti, 2011). Because of this, there is currently some debate about what constitutes a postbiotic. In any case, there is no question that these inactivated microorganisms can also have advantageous effects, and they are significant advantages for the probiotic industry because there is no risk of infection in immunocompromised people, and the cold chain is not required to maintain the viability of the product during storage and delivery. We must stress, nonetheless, that the physiological effects brought on by the colonization of the gut or those coming from the metabolism of the bacteria there won't be seen if postbiotics are used. The term "probiotic" or "postbiotic" does not always imply that the product is taken orally. However, in this section, we'll solely concentrate on the bacteria that are ingested orally and whose health benefits are at the forefront of probiotic research right now, specifically: Bioactive peptides and lipid metabolites are created as a result of the metabolism of food components and include neuroactive metabolites produced by bacteria or bacteria that cause neurological effects through recognized molecular signals. Our stomach and brain are constantly in contact via cellular messengers including cytokines, neuropeptides, and enteroendocrine signals, with LAB serving as bacterial mediators of the gut-brain axis. Now, there is no question that bacterial metabolites and by-products contribute to this reciprocal interaction. Studies on animals have shown that gut bacteria can alter behavior and cognition as well as have an impact on how nerve cells develop. Impaired brain-related disorders in people have been connected to changed gut microbiomes, which are frequently linked to symptoms of the digestive system (Morais et al., 2021). In this context, several probiotics have demonstrated the ability to have a favorable impact on brain function, either by producing metabolites that can interact directly with our neurological system or by inducing various physiological reactions. Using in vitro and animal models as an illustration, it has been demonstrated that specific vitamins produced by LAB may have a neuroprotective impact that could stop the onset and progression of neurodegenerative disorders (Teran et al., 2021). These effects can be partially triggered by the neuroactive chemicals produced by some lactobacilli and bifidobacteria. Gamma-aminobutyric acid (GABA), the primary inhibitory neurotransmitter of the central nervous system, is one of these compounds. The species Bifidobacterium adolescentis, a gut commensal with QPS status, has been primarily linked to the possible ability of bifidobacteria to manufacture GABA. A prior in silico analysis of metagenomics datasets revealed a link between the prevalence of *B. adolescentis* and mental problems. Additionally, *B. adolescentis* can enhance the in vivo synthesis of GABA, as shown by dietary intervention in a rat model (Duranti et al., 2020). GABA synthesis in lactobacilli varies significantly between species. According to several studies, *L. brevis* is one of the most productive GABA producers (Barrett et al., 2012). Patterson and colleagues showed that two strains of *L. brevis* have the potential to reduce the depressive symptoms and metabolic abnormalities linked to metabolic syndrome using a mice model of the condition (Patterson et al., 2019). Also, a probiotic composition containing GABA-producing strains of *L. plantarum* and *B. adolescentis* was successful in lessening depressive-like behavior in rats (Yunes et al., 2020). There have recently been reports of some *L. mucosae*, *L. reuteri*, and *L. plantarum* strains that may produce serotonin, a chemical derived from tryptophan that is essential for regulating mood, cognition, sleep, and other physiological processes. It has been suggested that these serotonin-producing lactobacilli can be used to treat conditions and illnesses caused by a lack of serotonin (Grasset et al., 2021). Nonetheless, preclinical and clinical investigations are still required to demonstrate the physiological activity of these serotonin-producing strains. Several strains, whether alive or inactivated, have been demonstrated to be capable of encouraging various neuro-related responses in addition to direct action through the creation of neuroactive chemicals. By increasing the expression of opioid and cannabinoid receptors in the gut, for example, oral administration of *L. acidophilus* NCFM and interaction of the bacteria with intestinal epithelial cells alters the feeling of visceral pain in rats (Rousseaux et al., 2007). Further research on the analgesic effects of *L. acidophilus* NCFM in people with mild to severe abdominal discomfort revealed that it could control the expression of mucosal opioid receptors (MOR). Despite the fact that *L. acidophilus* NCFM reduced intestinal symptoms, the investigators found that the study lacked the power to draw firm conclusions (Ringel-Kulka et al., 2014). Surprisingly, some postbiotics made of heat-inactivated lactobacilli still contain neuroactive components. In this context, heat-killed lactobacilli were able to promote social behavior and lower corticosterone levels in mice, while dietary intake of heat-killed Lactobacillus gasseri CP2305 increased the gene expression of neurotrophins in the hippocampus and enhanced appetite in a mouse model (Toyoda et al., 2020). Some researches focused on how bifidobacteria may affect sleep patterns or depression in regard to their potential role in the gut-brain axis, even if the chemicals responsible for their communication with the nervous system are not always well understood. In conclusion, functional cultures provide encouraging evidence for the potential to modify human physiological processes using bacterial metabolites that can communicate with our brains. Psychobiotics, also known as probiotics, have been shown to help the host's mental health, but inactive bacteria have also demonstrated promise in this regard (Dinan et al., 2013). Several preclinical studies have demonstrated the neuroactive capacity of functional cultures, despite the fact that this research is still in its early stages. However, more intervention studies in humans with carefully planned designs are required to confirm the effectiveness of functional cultures in the treatment and/or prevention of neuro-related disorders. Food-derived bioactive peptides and lipid metabolites Many proteins or lipids included in food are subjected to metabolic processes that lead to the production of some beneficial compounds by LAB. LAB has a powerful proteolytic arsenal that can operate on caseins, gluten, and other proteins, producing a variety of bioactive peptides including antimicrobial, antihypertensive, and antioxidant peptides. In fact, the milk protein fraction contains antimicrobial peptides that are encoded and work against a variety of pathogenic and spoilage microbes. The ability of Lac. lactis and Lactobacillus helveticus strains to release the peptide LEQLLRLKKY from -s1-casein, a peptide with inhibitory effects against *E. coli* and Ba. subtilis bacteria, was discovered by Nebbia in 2021. (Liu et al., 2015). On the other hand, *L. rhamnosus* and *L. helveticus* can release QKALNEINQF and TKKTKLTEEEKNRL from -S2 casein, two peptides with various antimicrobial activity mechanisms, the first of which disrupts the cytoplasmic membrane's proton-motive force and the second of which has the ability to bind DNA (Sistla, 2013). Surprisingly, simulated models of intestinal digestion revealed that some of these peptides can also be released in the human gastrointestinal tract as a result of the LAB metabolism in food. This is the case with the peptide YQEPVLGPVRGPFPI, which was produced by the action of *L. rhamnosus* 17D10 on -casein and was also produced by in vitro digestion with human gastric and duodenal juices. This suggests that some antimicrobial peptides released from food proteins can have a double action, as a bioprotectant in the fermented food and as pathogen inhibition mechanisms in the human gut (Nebbia et al., 2021). According to a few reports, the dietary protein peptides also have additional intriguing functions. Also, fermented milks chosen based on the proteolytic activity of lactobacilli strains shown an appropriate radical scavenging activity, and *L. casei* strains were suggested to make functional milks with antioxidant properties (Shu et al., 2018). Furthermore, plant matter has been utilized as a source of antioxidant peptides. Human cell lines were used to test peptides in soy flour fractions for high antioxidative activity, which revealed that low and medium-size peptides may have antioxidative potential (Cavaliere et al., 2021). However, we believe it is important to point out that the bulk of antioxidant activity studies have been characterized in vitro, and it is difficult to replicate this activity in vivo. Since numerous studies have demonstrated that certain peptides formed from casein hydrolysis have this action, one of the most alluring properties of peptides obtained from food fermentation may be their antihypertensive effect. These peptides exhibit antihypertensive activity by inhibiting the angiotensin I converting enzyme. Certain lactobacilli, including *L. helveticus* and *L. plantarum*, have the proteolytic activity necessary to produce these peptides (Xia et al., 2020). Moreover, studies on animals, mostly involving spontaneously hypertensive rats, have yielded encouraging results in terms of lowering blood pressure (Beltrán-Barrientos et al., 2016). Human intervention trials, however, are few and have produced mixed effects, with a slight to modest drop in blood pressure. In this context, human hypertension subjects who consumed lactobacilli-fermented milk containing the IPP and VPP tripeptides showed a small drop in blood pressure; this reduction was dose-dependent (Jauhiainen et al., 2012). Conjugated linoleic acid (CLA) is another bioactive chemical that can be produced from the lipidic component of foods by the activity of food-grade bacteria (CLA). Due to its possible health-promoting properties, such as anticarcinogenic and antiatherogenic effects, body fat modulation or inflammation reduction, CLA is generating a lot of interest (Yang et al., 2015). Among LAB, lactobacilli have demonstrated an exceptional ability to convert linoleic acid to CLA in vitro (Li et al., 2012). Although little is known about the molecular processes by which LAB transforms linoleic acid into CLA, it is asserted that the linoleate isomerase enzyme (LAI) is crucial for this conversion (Liavonchanka & Feussner, 2008). The ability of the species *L. plantarum* to produce CLA has been tested the most, and multiple strains have been examined as adjunct cultures in various food fermentations (meat and dairy products) to raise CLA levels in the finished product (Renes et al., 2019). For these cultures' conversion rates in the food system, factors like pH, temperature, the quantity of fatty acids, or fermentation duration may be important (Özer & Kılıçb, 2021). Moreover, bifidobacteria have been examined for their ability to convert linoleic acid to CLA, with positive findings on host health reported by various strains of *B. breve* in in vivo investigations using animal models (Patterson et al., 2017). However, more research is required to fully utilize the medicinal potential of CLA in fermented foods.

Conclusion

Real food trends call for more natural goods with more health benefits. Long employed in the food business, LAB and other bacteria are now being specifically targeted or investigated for usage in other applications. Among the options discussed here, using antimicrobial food preservatives, such as bacteriocins or antimicrobial peptides hidden in milk matrix, may help to improve food safety. Since it could help to preserve the unique signature of particular items, the quest for innovative LAB that can enhance the sensory qualities of fermented meals is still a very active area. Furthermore, some LAB cultures influence our health by enhancing our nutritional status or by the probiotic advantages associated with some particular strains or meals fermented with them. The direct interaction of the bacteria, or their parts, with the host and/or the manipulation of the gut microbiota, which would require more revision, may be the cause of this beneficial influence on health. As a result, utilizing environment-friendly biotechnology, LAB is still a promising option to produce more naturally occurring, safer, and high-quality food items, which is something that customers are increasingly demanding in the twenty-first century.

References

Acevedo-Rocha, C. G., Gronenberg, L. S., Mack, M., Commichau, F. M., & Genee, H. J. (2019). Microbial cell factories for the sustainable manufacturing of B vitamins. *Current Opinion in Biotechnology, 56*, 18−29.

Afshari, R., Pillidge, C. J., Dias, D. A., Osborn, A. M., & Gill, H. (2020). Cheesomics: The future pathway to understanding cheese flavour and quality. *Critical Reviews in Food Science and Nutrition, 60*, 33−47.

Altieri, C., Ciuffreda, E., Di Maggio, B., & Sinigaglia, M. (2017). Lactic acid bacteria as starter cultures. In B. Speranza, A. Bevilacqua, M. R. Corbo, & M. Sinigaglia (Eds.), *Starter cultures in food production* (pp. 1−15). Oxford: John Wiley & Sons, Ltd.

Anumudu, C., Hart, A., Miri, T., & Onyeaka, H. (2021). Recent advances in the application of the antimicrobial peptide nisin in the inactivation of spore-forming bacteria in foods. *Molecules (Basel, Switzerland), 26*, 5552.

Anyogu, A., Olukorede, A., Anumudu, C., Onyeaka, H., Areo, E., Adewale, O., et al. (2021). Microorganisms and food safety risks associated with indigenous fermented foods from Africa. *Food Control, 129*108227.

Averianova, L. A., Balabanova, L. A., Son, O. M., Podvolotskaya, A. B., & Tekutyeva, L. A. (2020). Production of vitamin B2 (riboflavin) by microorganisms: An overview. *Frontiers in Bioengineering and Biotechnology, 8*570828.

Bachtarzi, N., Kharroub, K., & Ruas-Madiedo, P. (2019). Exopolysaccharideproducing lactic acid bacteria isolated from traditional Algerian dairy products and their application for skim-milk fermentations. *LWT-Food Science and Technology, 107*, 117−124.

Bachtarzi, N., Speciale, I., Kharroub, K., De Castro, C., Ruiz, L., & Ruas-Madiedo, P. (2020). Selection of exopolysaccharideproducing Lactobacillus plantarum (Lactiplantibacillus plantarum) isolated from Algerian fermented foods for the manufacture of skim-milk fermented products. *Microorganisms, 8*, 1101.

Barcenilla, C., Ducic, M., Lopez, M., Prieto, M., & Alvarez-Ordoñez, A. (2021). Application of lactic acid bacteria for the biopreservation of meat products: A systematic review. *Meat Science, 183*108661.

Barrett, E., Ross, R. P., O'Toole, P. W., Fitzgerald, G. F., & Stanton, C. (2012). γ-Aminobutyric acid production by culturable bacteria from the human intestine. *Journal of Applied Microbiology, 113*, 411−417.

Bationo, F., Songré-Ouattara, L. T., Hemery, Y. M., Hama-Ba, F., Parkouda, C., Chapron, M., et al. (2019). Improved processing for the production of cereal-based fermented porridge enriched in folate using selected lactic bacteria and a back-slopping process. *LWT-Food Science and Technology, 106*, 172−178.

Beltrán-Barrientos, L. M., Hernández-Mendoza, A., Torres-Llanez, M. J., González-Córdova, A. F., & Vallejo-Córdoba, B. (2016). Invited review: Fermented milk as antihypertensive functional food. *Journal of Dairy Science, 99*, 4099−4110.

Blaya, J., Barzideh, Z., & LaPointe, G. (2018). Symposium review: Interaction of starter cultures and nonstarter lactic acid bacteria in the cheese environment. *Journal of Dairy Science, 101*, 3611−3629.

Buzzini, P., Di Mauro, S., & Turchetti, B. (2017). Yeasts as starter cultures. In B. Speranza, A. Bevilacqua, M. R. Corbo, & M. Sinigaglia (Eds.), *Starter cultures in food production* (pp. 16−49). Oxford: John Wiley & Sons, Ltd.

Capozzi, V., Russo, P., Dueñas, M. T., López, P., & Spano, G. (2012). Lactic acid bacteria producing B-group vitamins: A great potential for functional cereals products. *Applied Microbiology and Biotechnology, 96*, 1383−1394.

Carminati, D., Meucci, A., Tidona, F., Zago, M., & Giraffa, G. (2016). Multifunctional lactic acid bacteria cultures to improve quality and nutritional benefits in dairy products. In V. Ravishankar Rai (Ed.), *Advances in food biotechnology* (pp. 263−275). West Sussex, UK: John Wiley & Sons, Ltd.

Carrizo, S. L., LeBlanc, A. M., LeBlanc, J. G., & Rollán, G. C. (2020). Quinoa pasta fermented with lactic acid bacteria prevents nutritional deficiencies in mice. *Food Research International, 2020*(127)108735.

Castro-Bravo, N., Wells, J. M., Margolles, A., & Ruas-Madiedo, P. (2018). Interactions of surface exopolysaccharides from Bifidobacterium and Lactobacillus within the intestinal environment. *Frontiers in Microbiology, 9*, 2426.

Cavaliere, C., Montone, A. M. I., Aita, S. E., Capparelli, R., Cerrato, A., Cuomo, P., et al. (2021). Production and characterization of medium-sized and short antioxidant peptides from soy floursimulated gastrointestinal hydrolysate. *Antioxidants, 10*, 734.

Ceruso, M., Liu, Y., Gunther, N. W., Pepe, T., Anastasio, A., Qi, P. X., et al. (2021). Anti-listerial activity of thermophilin 110 and pediocin in fermented milk and whey. *Food Control, 125*107941.

Chen, H., Yan, X., Du, G., Guo, Q., Shi, Y., Chang, J., et al. (2021). Recent developments in antifungal lactic acid bacteria: Application, screening methods, separation, purification of antifungal compounds and antifungal mechanisms. *Critical Reviews in Food Science and Nutrition*, 1−15. Available from https://doi.org/10.1080/10408398.2021.1977610.

Dantigny, P., & Bevilacqua, A. (2017). Fungal starters: An insight into the factors affecting the germination of conidia. In B. Speranza, A. Bevilacqua, M. R. Corbo, & M. Sinigaglia (Eds.), *Starter cultures in food production* (pp. 50−63). Oxford: John Wiley & Sons, Ltd.

Deptula, P., Chamlagain, B., Edelmann, M., Sangsuwan, P., Nyman, T. A., Savijoki, K., et al. (2017). Food-like growth conditions support production of active vitamin B12 by Propionibacterium freudenreichii 2067 without DMBI, the lower ligand base, or cobalt supplementation. *Frontiers in Microbiology*, *8*, 368.

Dinan, T. G., Stanton, C., & Cryan, J. F. (2013). Psychobiotics: A novel class of psychotropic. *Biological Psychiatry*, *74*, 720–726.

Duranti, S., Ruiz, L., Lugli, G. A., Tames, H., Milani, C., Mancabelli, L., et al. (2020). Bifidobacterium adolescentis as a key member of the human gut microbiota in the production of GABA. *Scientific Reports*, *10*, 14112.

EFSA Panel on Biological Hazards (BIOHAZ). (2021). Update of the list of QPS-recommended biological agents intentionally added to food or feed as notified to EFSA 14: Suitability of taxonomic units notified to EFSA until march 2021. *EFSA Journal*, *19*, 6689.

EFSA Panel on Food Additives and Nutrient Sources added to Food (ANS). (2017). Safety of nisin (E 234) as a food additive in the light of new toxicological data and the proposed extension of use. *EFSA Journal*, *15*, 5063.

FAO-WHO. (2006). In L. Allen, B. de Benoist, O. Dary, & R. Hurrell (Eds.), *Guidelines on food fortification with micronutrients*. Switzerland: World Health Organization. (ISBN 92 4 159401 2).

Ge, Y. Y., Zhang, J. R., Corke, H., & Gan, R. Y. (2020). Screening and spontaneous mutation of pickle-derived Lactobacillus plantarum with overproduction of riboflavin, related mechanism, and food application. *Food*, *9*, 88.

Gibbons, J. G., & Rinker, D. C. (2015). The genomics of microbial domestication in the fermented food environment. *Current Opinion in Genetics & Development*, *35*, 1–8.

Gibbons, J. G. (2019). How to train your fungus. *mBio*, *10*, e03031, e03019.

Gobbetti, M., De Angelis, M., Di Cagno, R., Mancini, L., & Fox, P. F. (2015). Pros and cons for using non-starter lactic acid bacteria (NSLAB) as secondary/adjunct starters for cheese ripening. *Trends in Food Science and Technology*, *45*, 167–178.

Gobbetti, M., Di Cagno, R., Calasso, M., Neviani, E., Fox, P. F., & De Angelis, M. (2018). Drivers that establish and assembly the lactic acid bacteria biota in cheeses. *Trends in Food Science and Technology*, *78*, 244–254.

Gonzales-Barron, U., Goncalves-Tenorio, A., Rodrigues, V., & Cadavez, C. (2017). Foodborne pathogens in raw milk and Downloaded from cheese of sheep and goat origin: A meta-analysis approach. *Current Opinion in Food Science*, *18*, 7–13.

Grasset, E., Khan, M., Möllstam, B., & Roos, S. (2021). *Serotonin producing bacteria*. WO2021091474.

Greco, E., El-Aguizy, O., Ali, M. F., Foti, S., Cunsolo, V., Saletti, R., et al. (2018). Proteomic analyses on an ancient Egyptian cheese and biomolecular evidence of brucellosis. *Analytical Chemistry*, *90*, 9673–9676.

Greppi, A., Saubde, F., Botta, C., Humblot, C., Guyot, J. P., & Cocolin, L. (2017). Potential probiotic Pichia kudriavzevii strains and their ability to enhance folate content of traditional cereal-based African fermented food. *Food Microbiology*, *62*, 169–177.

Gu, Q., Zhang, C., Song, D., Li, P., & Zhu, X. (2015). Enhancing vitamin B12 content in soy-yogurt by Lactobacillus reuteri. *International Journal of Food Microbiology*, *206*, 56–59.

Hahn, C., Müller, E., Wille, S., Weiss, J., Atamer, Z., & Hinrichs, J. (2014). Control of microgel particle growth in fresh cheese (concentrated fermented milk) with an exopolysaccharideproducing starter culture. *International Dairy Journal*, *36*, 46–54.

Hatti-Kaul, R., Chen, L., Dishisha, T., & El Enshasy, H. (2018). Lactic acid bacteria: From starter cultures to producers of chemicals. *FEMS Microbiology Letters*, *365*, fny213.

Hill, C., Guarner, F., Reid, G., Gibson, G. R., Merenstein, D. J., Pot, B., et al. (2014). The international scientific association for probiotics and prebiotics consensus statement on the scope and appropriate use of the term probiotic. Nature Reviews. *Gastroenterology & Hepatology*, *11*, 506–515.

Holzapfel, W. H., & Wood, B. J. B. (2014). *Lactic acid bacteria. Biodiversity and taxonomy*. West Sussex, UK: John Wiley & Sons, Ltd.

Jauhiainen, T., Niittynen, L., Orešič, M., Järvenpää, S., Hiltunen, T. P., Rönnback, M., et al. (2012). Effects of long-term intake of lactotripeptides on cardiovascular risk factors in hypertensive subjects. *European Journal of Clinical Nutrition*, *66*, 843–849.

Jiang, S., Cai, L., Lv, L., & Li, L. (2021). Pediococcus pentosaceus, a future additive or probiotic candidate. *Microbial Cell Factories*, *20*, 45.

Laiño, J. E., Juarez del Valle, M., Savoy de Giori, G., & LeBlanc, J. G. (2014). Applicability of a Lactobacillus amylovorus strain as coculture for natural folate bio-enrichment of fermented milk. *International Journal of Food Microbiology*, *191*, 10–16.

LeBlanc, J. G., Laiño, J. E., Juarez del Valle, M., Vannini, V., van Sinderen, D., Taranto, M. P., et al. (2011). B-group vitamin production by lactic acid bacteria-current knowledge and potential applications. *Journal of Applied Microbiology*, *111*, 1297–1309.

LeBlanc, J. G., Milani, C., Savoy de Giori, G., Sesma, F., van Sinderen, D., & Ventura, M. (2013). Bacteria as vitamin suppliers to their host: A gut microbiota perspective. *Current Opinion in Biotechnology*, *24*, 160–168.

LeBlanc, J. G., Rutten, G., Bruinenberg, P., Sesma, F., De Giori, G. S., & Smid, E. J. (2006). A novel dairy product fermented with *Propionibacterium freudenreichii* improves the riboflavin status of deficient rats. *Nutrition (Burbank, Los Angeles County, Calif.)*, *22*, 645–651.

Levit, R., Savoy de Giori, G., de Moreno de LeBlanc, A., & LeBlanc, J. G. (2021). Recent update on lactic acid bacteria producing riboflavin and folates: Application for food fortification and treatment of intestinal inflammation. *Journal of Applied Microbiology*, *130*, 1412–1424.

Li, H., Liu, Y., Bao, Y., Liu, X., & Zhang, H. (2012). Conjugated linoleic acid conversion by six *Lactobacillus plantarum* strains cultured in MRS broth supplemented with sunflower oil and soymilk. *Journal of Food Science*, *77*, 227.

Li, Q., & Gänzle, M. (2020). Host-adapted lactobacilli in food fermentations: Impact of metabolic traits of host adapted lactobacilli on food quality and human health. *Current Opinion in Food Science*, *31*, 71–80.

Liavonchanka, A., & Feussner, I. (2008). Biochemistry of PUFA double bond isomerases producing conjugated linoleic acid. *Chembiochem: A European Journal of Chemical Biology*, *9*, 1867–1872.

Liu, Y., Eichler, J., & Pischetsrieder, M. (2015). Virtual screening of a milk peptide database for the identification of food-derived antimicrobial peptides. *Molecular Nutrition & Food Research*, *59*, 2243–2254.

Marco, M. L., Heeney, D., Binda, S., Cifelli, C. J., Cotter, P. D., Foligné, B., et al. (2017). Health benefits of fermented foods: Microbiota and beyond. *Current Opinion in Biotechnology, 44*, 94−102.

Mayo, B., Rodríguez, J., Vázquez, L., & Flórez, A. B. (2021). Microbial interactions within the cheese ecosystem and their application to improve quality and safety. *Food, 10*, 602.

Meng, F., Zhu, X., Zhao, H., Nie, T., Lu, F., Lu, Z., et al. (2021). Class III bacteriocin with broad-spectrum antibacterial activity from Lactobacillus acidophilus NX2−6 and its preservation in milk and cheese. *Food Control, 121*107597.

Mirkovic, N., Kulas, J., Miloradovic, Z., Miljkovic, M., Tucovic, D., Miocinovic, J., et al. (2020). Lactolisterin BU-producer *Lactococcus lactis* subsp. lactis BGBU1−4: Bio-control of *Listeria monocytogenes* and *Staphylococcus aureus* in fresh soft cheese and effect on immunological response of rats. *Food Control, 111*107076.

Molinero, N., Sabater, C., Calvete, I., Delgado, S., Ruas-Madiedo, P., Ruiz, L., et al. (2022). Mechanisms of gut microbiota modulation by food, probiotics, prebioticsandmore. In M. Glibetic (Ed.), *Comprehensive gut microbiota* (pp. 84−101). Amsterdam, Netherlands: ElsevierInc. Available from https://doi.org/10.1016/B978-0-12-819265-8.00095-4.

Morais, L. H., Schreiber, H. L., & Mazmanian, S. K. (2021). The gut microbiota-brain axis in behaviour and brain disorders. *Nature Reviews. Microbiology, 19*, 241−255.

Nebbia, S., Lamberti, C., Lo Bianco, G., Cirrincione, S., Laroute, V., Cocaign-Bousquet, M., et al. (2021). Antimicrobial potential of food lactic acid bacteria: Bioactive peptide decrypting from caseins and bacteriocin production. *Microorganisms, 9*, 65.

Özer, C. O., & Kılıçb, B. (2021). Optimization of pH, time, temperature, variety and concentration of the added fatty acid and the initial count of added lactic acid bacteria strains to improve microbial conjugated linoleic acid production in fermented ground beef. *Meat Science, 171*108303.

Pancza, B., Szathmary, M., Gyurjan, I., Bankuti, B., Tudos, C., Szathmary, S., et al. (2021). A rapid and efficient DNA isolation method for qPCR-based detection of pathogenic and spoilage bacteria in milk. *Food Control, 130*108236.

Patterson, E., Ryan, P. M., Wiley, N., Carafa, I., Sherwin, E., Moloney, G., et al. (2019). Gamma-aminobutyric acid-producing lactobacilli positively affect metabolism and depressive-like behaviour in a mouse model of metabolic syndrome. *Scientific Reports, 9*, 16323.

Patterson, E., Wall, R., Lisai, S., Ross, R. P., Dinan, T. G., Cryan, J. F., et al. (2017). Bifidobacterium breve with α-linolenic acid alters the composition, distribution and transcription factor activity associated with metabolism and absorption of fat. *Scientific Reports, 7*, 43300.

Peng, P., Song, J., Zeng, W., Wang, H., Zhang, Y., Xin, J., et al. (2021). A broad-spectrum novel bacteriocin produced by Lactobacillus plantarum SHY 21−2 from yak yogurt: Purification, antimicrobial characteristics and antibacterial mechanism. *LWT-Food Science and Technology, 142*110955.

Pradhan, P., & Tamang, J. P. (2021). Probiotic properties of lactic acid bacteria isolated from traditionally prepared dry starters of the eastern Himalayas. *World Journal of Microbiology and Biotechnology, 37*, 7.

Prete, R., Alam, M. K., Perpetuini, G., Perla, C., Pittia, P., & Corsetti, A. (2021). Lactic acid bacteria exopolysaccharides producers: A sustainable tool for functional foods. *Food, 10*, 1653.

Rajendran, S. C. C., Chamlagain, B., Kariluoto, S., Piironen, V., & Saris, P. E. J. (2017). Biofortification of riboflavin and foalte in Idli batter, based on fermented cereal and pulse, by Lactococcus lactis N8 and saccharomyces boulardii SAA655. *Journal of Applied Microbiology, 122*, 1663−1671.

Randazzo, C. L., Liotta, L., De Angelis, M., Celano, G., Russo, N., Van Hoorde, K., et al. (2021). Adjunct culture of non-starter lactic acid bacteria for the production of Provola Dei Nebrodi PDO cheese: In vitro screening and pilot-scale cheese-making. *Microorganisms, 9*, 179.

Renes, E., Gómez-Cortés, P., de la Fuente, M. A., Linares, D. M., Tornadijo, M. E., & Fresno, J. M. (2019). CLA-producing adjunct cultures improve the nutritional value of sheep cheese fat. *Food Research International, 116*, 819−826.

Ringel-Kulka, T., Goldsmith, J. R., Carroll, I. M., Barros, S. P., Palsson, O., Jobin, C., et al. (2014). Lactobacillus acidophilus NCFM affects colonic mucosal opioid receptor expression in patients with functional abdominal pain—a randomised clinical study. *Alimentary Pharmacology & Therapeutics, 40*, 200−207.

Rodríguez, A., Martínez, B., García, P., Ruas-Madiedo, P., & Sánchez, B. (2017). New trends in dairy microbiology: Towards safe and healthy products. In B. Speranza, A. Bevilacqua, M. R. Corbo, & M. Sinigaglia (Eds.), *Starter cultures in food production* (pp. 299−323). Oxford: John Wiley & Sons, Ltd.

Rouse, S., & Van Sinderen, D. (2008). Bioprotective potential of lactic acid bacteria in malting and brewing. *Journal of Food Protection, 71*, 1724−1733.

Rousseaux, C., Thuru, X., Gelot, A., Barnich, N., Neut, C., Dubuquoy, L., et al. (2007). Lactobacillus acidophilus modulates intestinal pain and induces opioid and cannabinoid receptors. *Nature Medicine, 13*, 35−37.

Ruas-Madiedo, P., & de los Reyes-Gavilan, C. G. (2005). Invited review: Methods for the screening, isolation, and characterization of exopolysaccharides produced by lactic acid bacteria. *Journal of Dairy Science, 88*, 843−856.

Ruas-Madiedo, P., Salazar, N., & de los Reyes-Gavilan, C. G. (2009). Exopolysaccharides produced by lactic acid bacteria in food and probiotic applications. In A. Moran, O. Holst, P. Brennan, & M. von Itzstein (Eds.), *Microbial glycobiology structures, relevance and applications* (pp. 887−902). San Diego: Academic Press, Elsevier Inc.

Russo, P., De Chiara, M. L. V., Capozzi, V., Arena, M. P., Amodio, M. L., Rascón, A., et al. (2016). Lactobacillus plantarum strainsfor multifunctional oat-based foods. *LWT-Food Science and Technology, 68*, 288−294.

Salminen, S., Collado, M. C., Endo, A., Hill, C., Lebeer, S., Quigley, E. M. M., et al. (2021). The International Scientific Association of Probiotics and Prebiotics (ISAPP) consensus statement on the definition and scope of postbiotics. Nature Reviews. *Gastroenterology & Hepatology, 18*, 649−667.

Salque, M., Bogucki, P. I., Pyzel, J., Sobkowiak-Tabaka, I., Grygiel, R., Szmyt, M., et al. (2013). Earliest evidence for cheese making in the sixth millennium BC in northern Europe. *Nature, 493*, 522−525.

Sánchez, B., Delgado, S., Blanco-Míguez, A., Lourenço, A., Gueimonde, M., & Margolles, A. (2017). Probiotics, gut microbiota, and their influence on host health and disease. *Molecular Nutrition & Food Research, 61*201600240.

Santos, F., Vera, J. L., van der Heijden, R., Valdez, G., de Vos, W. M., Sesma, F., et al. (2008a). The complete coenzyme B12 biosynthesis gene cluster of Lactobacillus reuteri CRL 1098. *Microbiology (Reading, England), 154*, 81−93.

Santos, F., Wegkamp, A., de Vos, W. M., Smid, E. J., & Hugenholtz, J. (2008b). High-level folate production in fermented foods by the B12 producer Lactobacillus reuteri JCM1112. *Applied and Environmental Microbiology, 74*, 3291−3294.

Sauer, M., Russmayer, H., Grabherr, R., Peterbauer, C. K., & Marx, H. (2017). The efficient clade: Lactic acid bacteria for industrial chemical production. *Trends in Biotechnology, 35*, 756−769.

Shu, G., Shi, X., Chen, L., Kou, J., Meng, J., & Chen, H. (2018). Antioxidant peptidesfrom goat milk fermented by Lactobacillus casei L61: Preparation, optimization, and stability evaluation in simulated gastrointestinal fluid. *Nutrients, 10*, 797.

Silva, C. C. G., Silva, S. P. M., & Ribeiro, S. C. (2018). Application of bacteriocins and protective cultures in dairy food preservation. *Frontiers in Microbiology, 9*, 594.

Sistla, S. (2013). Structure−activity relationships of α-casein peptides with multifunctional biological activities. *Molecular and Cellular Biochemistry, 384*, 29−38.

Soltani, S., Hammami, R., Cotter, P. D., Rebuffat, S., Ben Said, L., Gaudreau, H., et al. (2021). Bacteriocins as a new generation of antimicrobials: Toxicity aspects and regulations. *FEMS Microbiology Reviews, 45*, 1−24.

Surber, G., Mende, S., Jaros, D., & Rohm, H. (2019). Clustering of *Streptococcus thermophilus* strains to establish a relation between exopolysaccharide characteristics and gel properties of acidified milk. *Food, 8*, 146.

Sybesma, W., Burgess, C., Starrenburg, M., van Sinderen, D., & Hugenholtz, J. (2004). Multivitamin production in *Lactococcus lactis* using metabolic engineering. *Metabolic Engineering, 6*(2), 109−115.

Tamang, J. P., Watanabe, K., & Holzapfel, W. H. (2016). Review: Diversity of microorganisms in global fermented foods and beverages. *Frontiers in Microbiology, 7*, 377.

Tamene, A., Baye, K., Kariluoto, S., Edelmann, M., Bationo, F., Leconte, N., et al. (2019). Lactobacillus plantarum P2R3FA isolated from traditional cereal-based fermented food increase folate status in deficient rats. *Nutrients, 11*, 2819.

Taverniti, V., & Guglielmetti, S. (2011). The immunomodulatory properties of probiotic microorganisms beyond their viability (ghost probiotics: Proposal of paraprobiotic concept). *Genes & Nutrition, 6*, 261−274.

Teran, M. M., de Moreno de LeBlanc, A., Savoy de Giori, G., & LeBlanc, J. G. (2021). Thiamine-producing lactic acid bacteria and their potential use in the prevention of neurodegenerative diseases. *Applied Microbiology and Biotechnology, 105*, 2097−2107.

Thakur, K., Kumar Tomar, S., Brahma, B., & De, S. (2016). Screening of riboflavin-producing lactobacilli by a polymerase-chainreaction-based approach and microbiological assay. *Journal of Agricultural and Food Chemistry, 64*, 1950−1956.

Thierry, A., Pogačic, T., Weber, M., & Lortal, S. (2016). Production of flavor compounds by lactic acid bacteria in fermented foods. In F. Mozzi, R. R. Raya, & G. M. Vignolo (Eds.), *Biotechnology of lactic acid bacteria: Novel applications* (2nd edition, pp. 314−340). West Sussex: John Wiley & Sons, Ltd.

Tiwari, S., Kavitake, D., Devi, P. B., & Shetty, P. H. (2021). Bacterial exopolysaccharides for improvement of technological, functional and rheological properties of yoghurt. *International Journal of Biological Macromolecules, 183*, 1585−1595.

Toyoda, A., Kawase, T., & Tsukahara, T. (2020). Effects of dietary intake of heat-inactivated Lactobacillus gasseri CP2305 on stressinduced behavioral and molecular changes in a subchronic and mild social defeat stress mouse model. *Biomedical Research, 41*, 101−111.

Trejo-González, L., Gutiérrez-Carrillo, A. E., RodríguezHernández, A. I., López-Cuellar, M. R., & ChavarríaHernández, N. (2021). Bacteriocins produced by LAB isolated from cheeses within the period 2009−2021: A review. *Probiotics and Antimicrobial Proteins*. Available from https://doi.org/10.1007/s12602-021-09825-0.

Walhe, R. A., Diwanay, S. S., Patole, M. S., Sayyed, R. Z., AL-Shwaiman, H. A., Alkhulaifi, M. M., et al. (2021). Cholesterol reduction and vitamin B12 production study on *Enterococcus faecium* and *Lactobacillus pentosus* isolated from yoghurt. *Sustainability, 13*, 5853.

Wegkamp, A., van Oorschot, W., de Vos, W., & Smid, E. J. (2007). Characterization of the role of para-aminobenzoic acid biosynthesis in folate production by *Lactococcus lactis*. *Applied and Environmental Microbiology, 73*, 2673−2681.

Wolfe, B. E., Button, J. E., Santarelli, M., & Dutton, R. J. (2014). Cheese rind communities provide tractable systems for in situ and in vitro studies of microbial diversity. *Cell, 158*, 422−433.

Xia, Y., Yu, J., Xu, W., & Shuang, Q. (2020). Purification and characterization of angiotensin-I-converting enzyme inhibitory peptides isolated from whey proteins of milk fermented with Lactobacillus plantarum QS670. *Journal of Dairy Science, 103*, 4919−4928.

Yang, B., Chen, H. Q., Stanton, C., Ross, R. P., Zhang, H., Chen, Y. Q., et al. (2015). Review of the roles of conjugated linoleic acid in health and disease. *Journal of Functional Foods, 15*, 314−325.

Yunes, R. A., Poluektova, E. U., Vasileva, E. V., Odorskaya, M. V., Marsova, M. V., Kovalev, G. I., et al. (2020). A multi-strain potential probiotic formulation of GABA-producing Lactobacillus plantarum 90sk and Bifidobacterium adolescentis 150 with antidepressant effects. *Probiotics and Antimicrobial Proteins, 12*, 973−979.

Zandona, E., Blažić, M., & Režek Jambrak, A. (2021). Whey utilisation: Sustainable uses and environmental approach. *Food Technology and Biotechnology, 59*, 147−161.

Zheng, J., Wittouck, S., Salvetti, E., Franz, C. M. A. P., Harris, G. M. B., Mattarelli, P., et al. (2020). A taxonomic note on the genus Lactobacillus: Description of 23 novel genera, emended description of the genus Lactobacillus Beijerinck 1901, and union of Lactobacillaceae and Leuconostocaceae. *International Journal of Systematic and Evolutionary Microbiology, 70*, 2782−2858.

Chapter 19

Toxicity and safety of essential oil

Olulope Olufemi Ajayi

Department of Biochemistry, Edo State University Uzairue, Auchi, Edo State, Nigeria

Introduction

Essential oils (EOs) are redolent hydrophobic fluids with a high level of volatility made by various organs scented plants as secondary metabolites (Russo et al., 2015). These organs include the leaves, seeds, bark, roots, buds, and fruits. Scented plants are widely distributed in the tropical and Mediterranean regions of the world. This is the basis of their local use in these regions. Currently, well over 2500 EOs have been identified (Sharifi-Rad et al., 2017).

Essential oil is a blend of about 70 low-molecular-weight compounds occurring in varying concentrations. Terpenes are the major components of EOs. These consist of monoterpenes, a 10-carbon compound accounting for the highest percentage of terpenes. Others include sesquiterpenes and diterpenes, which are 15 and 20 carbon compounds, respectively. Derivatives of terpenes including terpenoids and phenylpropanoids have also been reported to be present in EOs (Russo et al., 2015).

Factors including plant age, organ type, vegetative stage, period of harvest, nature of soil, and method of extraction are the basis of variation in the chemical composition of EOs.

The utility of EOs consequent on the appropriate characterization of their chemical components as well as their diverse biological effects brought about by the complex nature of the chemical component are of significance in the reportage of the organic effects of EOs (Russo et al., 2015).

Reports have shown the various uses of EOs as biopesticide, food preservatives, in aromatherapy, and potentially in the treatment of chronic diseases including cancer. Reported antitumor mechanisms of thymol EO include stimulation of apoptosis, reactive oxygen species and Bax protein, impeding of cell growth, reduction of Bcl-2 protein as well as depolarization of mitochondrial membrane (Kowalczyk et al., 2020). Some of them also possess antimicrobial, antiinflammatory, and antioxidant properties. The antioxidant, antiinflammatory, and hepatoprotective activities of rosemary EO have been reported. These were ascribed to 1,8-cineole, the predominant monoterpene (Rašković et al., 2014).

The bioactivity of EOs is a function of the composition, volatility rate, and persistence of EOs postapplication (Moura et al., 2020). EO composition also depends on factors including part of plant utilized for extraction, period of harvest of the plant, climate, geographic location, and genetics (Moura et al., 2020).

EOs derived from *Curcuma* species have been widely reported to contain varying compositions of bioactive compounds that have a variety of medicinal and pharmacological significance (Dosoky & Setzer, 2018). These active compounds vary from one species to another as well as from one geographical location to another (Dosoky & Setzer, 2018). For instance, turmeric EO sourced from Nigeria was a blend of β-bisabolene, (E)-β-ocimene, β-myrcene, 1,8-cineole, α-thujene, α-phellandrene, limonene, zingiberene, and β-sesquiphellandrene, while the principal components of turmeric EO from Sao Tome and Principe and Srilanka were α-phellandrene, α-turmerone, 1,8-cineole, p-cymene, ar-turmerone, β-turmerone, and terpinolene. The disparity between fresh and dried rhizomes has also been reported (Dosoky & Setzer, 2018).

The biological activities of *Mentha spicata* L. EO were associated with its biocomponents; limonene, carvone, 1,8-cineole, and α-terpinene (Bardaweel et al., 2018).

In spite of the various biological benefits of EOs, issues of toxicity and safety are currently gaining attention. Evidences of toxicity including skin irritation, ataxia, coma, and death have been reported (Lanzerstorfer et al., 2021). This will be discussed in this review.

Applications of Essential Oils in the Food Industry. DOI: https://doi.org/10.1016/B978-0-323-98340-2.00012-2

Toxicity of essential oils

Quite a number of EOs have been documented as Generally Recognized as Safe by the United States Food and Drug Administration, indicating their potential safety. Therefore they are believed to be eco-friendly (Lazarević et al., 2020). They could therefore be used for the production of cosmetics and food products. Emerging reports, however, indicated the toxicity of EOs with some exerting their toxic effects at very minimal concentrations (Lanzerstorfer et al., 2021). Both systemic and organ toxicities have been reported, thus, suggesting toxicity and safety concerns in the use of EOs.

EOs have been reported to be toxic to the skin, mucous membranes, and the eyes indicated by irritation and sensitization (Ali et al., 2015). Headache and nausea have been reported in some instances. The potential of EOs in carcinogenesis has been suggested (Liaqat et al., 2018). Toxicity testing of EOs could be taxing because of the viscidness, hydrophobicity, and erraticism of the principal components (Lanzerstorfer et al., 2021).

Cytotoxic effects of essential oils

The cytotoxic mechanisms of EO are diverse and are dependent on the dissimilarity of chemical composition. They are also a function of EO type and concentration (Melušova et al., 2014). There is also the possibility of interaction both synergistic or otherwise of the chemical constituents (Sharifi-Rad et al., 2017).

Cell cycle arrest, apoptosis, and necrosis as well as dysfunction of cellular organelles are some cytotoxic mechanisms of EOs. These are attributable to the hydrophobic nature and the low molecular weight of active components which gives them easy access to the cell by interfering with the membrane architecture, and readily diffusing through the cell membrane. This increases membrane permeability. Consequently, the process of cellular energy supply is impeded, there is also loss of cellular components including ATP, potassium ions, and DNA to the extracellular space, thus culminating in cell death (Sharifi-Rad et al., 2017). These cytotoxic mechanisms contribute significantly to their therapeutic potential (Sharifi-Rad et al., 2017).

EOs from different parts of *Callistemon* species have been reported to possess a variety of health-boosting effects. EOs from *Callistemon citrinus* leaves were observed to exert antimicrobial and antiinflammatory effects probably due to the presence of 1,8-cineole, α and β-pinenes, α-terpineol, α-phellandrene, limonene, linalool, α-terpinene, terpinen-4-ol and geraniol (Kumar et al., 2015). The antineoplastic effects of *C. citrinus* leaf and flower EOs were attributed to elevated levels of 1,8-cineole (Kumar et al., 2015).

Essential oils in pest control

Studies have shown the toxic effects of EOs and their various constituents on insects, hence, their use as pesticides. EOs adversely affect the insect's irritability, fertility and also repel them (Brügger et al., 2019). The toxic effect of EOs on lepidopteran spp. which are the natural prey of *P. nigrispinus* has been reported (Brügger et al., 2019). Previous reports have shown that predatory insects such as *Podisus nigrispinus* are critical in pest management. They are better alternatives to synthetic pesticides known for the induction of resistance in pests as well as toxicity to flora and fauna (Brügger et al., 2019). Mechanisms of pest's resistance to insecticides include alteration of insecticide metabolism with the view to reducing their effects on acetylcholinesterase, reduced permeation of insecticide via the cuticle as well as the inactivation of octopamine receptors. (Brügger et al., 2019).

Elevated concentration of lemongrass EO is lethal to *P. nigrispinus* at its developmental stages. The toxicity of lemongrass EO on *P. nigrispinus* is analogous to other EOs (Brügger et al., 2019). Terpenoids were the major bioactive compounds in lemongrass EO which could also be the most insecticidal constituents (Brügger et al., 2019). Terpenoids adversely affected the respiration rate of *P. nigrispinus* which could result in the disruption of processes involving oxidative phosphorylation as well as paralysis of the muscles (Brügger et al., 2019). Furthermore, the stiffening of the cuticle could also reduce the permeation and efficiency of detoxification of xenobiotics.

The acaricide effect of EO from lemon on tick *Rhipicephalus microplus* has been reported. This was attributed to pinene and limonene, which are vital components of lemon EO (Vinturelle et al., 2017). In a report, *Thymus vulgaris* EO exerted acute toxicity on bean weevil as manifested by dysmetabolism of proteins and lipids, disruption of egg laying as well as hampering the growth of bean weevil (Lazarević et al., 2020).

Carvacrol and thymol toxic effects in bed bugs could be due to their structural properties; they are saturated compounds and also have hydroxyl groups attached to their benzene ring. This could be of significance in the speedy ease of diffusion through the cuticle, minimal detoxification, and effective interaction with the target sites (Gaire et al., 2019).

The potential of *Ocimum basilicum* (Basil) EO as a biopesticide has been explored. The toxicity of *O. basilicum* on *Sitophilus zeamais*, a major pest of grains has been reported (Moura et al., 2020). The major bioactive compounds are estragole, eugenol and linalool, with estragole and linalool accounting for the highest and least percentage, respectively (Moura et al., 2020).

In a study, EO from *Melaleuca alternifolia* exerted its insecticidal effect on *Tribolium confusum* Jacquelin du Val, a notable stored grain pest (Liao et al., 2018). The effect was attributed to the possible synergy of the chemical constituents of the EO, of which terpinen-4-ol was reported to be the principal constituent. Significant elevated level of NAD + which was eventually repressed was observed. Furthermore, a reduced level of NADH was also observed. This indicates perturbation of mitochondrial function which has a consequence on energy generation (Liao et al., 2018).

Other mechanisms of EOs in pest controls include; neurotoxicity brought by the inhibition of acetylcholinesterase and other detoxification enzymes' activities and disruption of the digestive process. EOs have also been observed to possess prooxidant potentials, hence, their implication in oxidative stress (Lazarević et al., 2020).

The neurotoxicity of components of EOs on pests has been reported. In a study, neurological inhibition observed in bed bugs was attributed to thymol, carvacrol, and eugenol (Gaire et al., 2019). Thymol has the potential of binding to GABA receptors in invertebrates, and its weak inhibitory effect on acetylcholinesterase has been reported (Gaire et al., 2019). Neuroexcitation was also reported in bed bugs exposed to linalool (Gaire et al., 2019). The possible interactions of eugenol with octopamine receptors in the nervous system of insects have also been reported (Gaire et al., 2019).

A report showed the hormetic effect of three components of citrus EOs on Mediterranean fruit fly (medfly), a major vegetable and fruit pest. These components are limonene, α-pinene and linalool (Papanastasiou et al., 2017). Limonene accounts for over 90% of citrus EOs. The toxicities of these EO components on Medfly have been reported with linalool being the most toxic. Susceptibility to toxicity was higher in males than females irrespective of dietary conditions (Papanastasiou et al., 2017). Sub-lethal doses of limonene were associated with fecundity in female insects which could be due to hormonal modulation (Papanastasiou et al., 2017).

The downsides of the use of EOs in the control of pests include; a brief half-life that requires repeated application, possible pungency consequent on high volatility, and paucity of information regarding their efficacy in pest management (Gaire et al., 2019). Others include the propensity of phototoxicity (Lazarević et al., 2020).

There are also reports on the adverse effects of insects which are not necessarily pests. A report showed the sublethal effect of *Eupatorium buniifolium* on honeybees. Alteration in appetite, motor dysfunction, disruption of olfactory memory, and learning performance are some of the effects (Rossini et al., 2020). This adversely affected the survival of the honeybees. One of the mechanisms by which this occurs includes the alteration of curticular hydrocarbons; 7-pentacosene, 9-tricosene, 9-heptacoses, and 9-pentacosene which are associated with social recognition (Rossini et al., 2020).

Essential oils' toxicity in laboratory models and humans

Laboratory models

Neurotoxicity of EOs has been reported in mammals. Carvacrol was implicated in reversible blockage of rat sciatic nerve excitation (Joca et al., 2012). The propensity of thymol to bind to GABA receptors in vertebrates could also be the basis of neurotoxicity (Gaire et al., 2019).

In another study, genotoxicity was observed at 40, 60, 80, and 100 mg/kg body weight in bone marrow cells of rats, with the induction of numerical chromosome aberrations at 100 mg/kg body weight (Azirak & Rencuzogullari, 2008).

In a study that assessed the serological and hematological status of rats fed with seven members of rutaceae family: *Aegle marmelos*, *Murraya koenigii*, *Citrus reticulata* Blanco, *Zanthoxylum armatum*, *Skimmia laureola*, *Murraya paniculata*, and *Boenninghausenia albiflora*. Marked difference was observed in hematological parameters such as packed cell volume, mean corpuscular volume (MCV), red blood cell, and animals fed EOs in comparison with the controls (Liaqat et al., 2018).

Total erythrocyte was considerably increased in animals fed with EOs from *S. laureola*, *M. koenigii*, and *B. albiflora*, significant reduction in MCV was also observed in animals fed with *B. albiflora*. Although these EOs were reported to be non-toxic to the liver and increased blood level of urea, an indication of renal toxicity was observed in animals fed with *Z. armatum*, *S. laureola*, *M. paniculata*, and *B. albiflora* EOs (Liaqat et al., 2018).

Reduction of HDL-cholesterol was observed in animals fed with *B. albiflora* EO. HDL-C is seen as good cholesterol. This further indicates the toxicity of the EO. Serum level of sodium was elevated in animals fed with *A.*

marmelos, C. reticulata Blanco, *S. laureola, M. paniculata*, and *B. albiflora*, suggesting cellular seepage of electrolytes (Liaqat et al., 2018).

In another study, acute toxicity (<24 hours) was observed in 3 rats fed with *T. vulgaris* EO at 2000 mg/kg bw. This manifested as hyperactivity, ataxia, severe burning nose, hypersalivation, and respiratory distress. Two of the rats went into coma and eventually died (Rojas-Armas et al., 2019a,b). Severe burning nose was attributed to the innervation of the respiratory tract by primary sensory afferent nerves brought about by the strong odor of the oil (Rojas-Armas et al., 2019a,b).

The histopathological study further showed changes consistent with interstitial pneumonitis in the lungs of the dead rats. It was suggested that the toxicity could be due to thymol, the principal component of *T. vulgaris* EO which has the potential of activating transient receptor potential cation channel (Rojas-Armas et al., 2019a,b).

Multiorgan toxicities were also observed in laboratory animals treated with EO from unripe *Citrus aurantifolia* (Adokoh et al., 2019). The antiinflammatory effects of citrus EOs are not in doubt, evidences of inflammation were observed in liver tissue sections of the animals in the low, medium, and dose groups. Leukocyte infiltration, a pointer to infection or toxicant was observed in animals in the low-dose group; edema and low-level leucocytosis also characterized the animals in the medium and high-dose groups

Observation of the kidney sections showed necrosis in animals of the medium and high-dose groups. This indicated a toxicant-damaging effect on the renal tubular damage. Instances of nephropathy were previously observed in animals treated with d-limonene (Sun, 2007). The toxic effect of limonene appeared to be dependent on time and dose (Adokoh et al., 2019). Hemosiderin accumulation, an index of erythrocytic damage was observed in the spleen of animals in the low-dose group. This may be ascribed to the components of Citrus EOs (Adokoh et al., 2019).

Moderate inflow of mediators of inflammation was reported in sections of the lungs of animals in the test group (Adokoh et al., 2019). The toxicity of linalool on the skin that manifested as edema and erythema was also reported by Bickers et al. (2003).

In the test group, elevated levels of HDL and a reduction in total cholesterol concentration were observed. AST and ALT activities were elevated suggestive of hepatic injury (Adokoh et al., 2019). The elevated level of conjugated bilirubin and a low level of unconjugated bilirubin also indicate hepatic damage. Elevated level of erythropoiesis indicated by increased level of red blood cells was observed in the test groups. This could be due to low oxygen tension (Adokoh et al., 2019).

In a study, 87.5 mg/kg. b wt was observed as the acute lethal dose (LD50) of Thujone in mice when administered subcutaneously. Apiol administered to dogs intravenously at 500 mg/kg b. wt was reported as the LD50 (Mossa et al., 2018). Raised level of aspartate aminotransferase (AST) void of histological derangement was observed in animals in the high-dose group (2000 mg/kg) in comparison with animals in the control group. This is suggestive of hepatic damage (Deyno et al., 2021).

A report showed the toxicity of *Minthostachys mollis* essential oil (MmEO) which affected multiple systems in rats to which 2000 mg/kg dose was administered (Rojas-Armas et al., 2019a,b). The gastrointestinal, respiratory, and central nervous systems appeared severely affected, and this culminated in the eventual death of the animals.

The toxicity of MmEO is attributed to pulegone and menthone which are the major bioactive components. Pulegone is easily absorbed and distributed, this enhances its toxicity (Rojas-Armas et al., 2019a,b). Both bioactive components were reported to alter the metabolism of dopamine without the involvement of dopamine receptors (Rojas-Armas et al., 2019a,b).

Hemorrhage was also observed in the rats' lungs as indicated by the presence of red blood cells and their remnants. Additionally, there were evidences of lung injury traceable to pro-oxidants and proteolytic enzymes that emanated from the degradation of neutrophils. The respiratory toxic effect was attributed to pulegone (Rojas-Armas et al., 2019a,b).

Mortality of male rats was observed after being administered with MmEO at 500 mg/kg/day. Histological reports showed liver damage as evidenced by elevated levels of alanine aminotransferase (ALT), AST, alkaline phosphatase, and bilirubin. This could also be attributed to the hepatotoxic effect of pulegone and menthone (Rojas-Armas et al., 2019a,b).

A different report showed significant body weight loss and reduced creatinine levels in rats upon the administration of pulegone, an indication of toxicity and diminution of skeletal muscle mass (Mølck et al., 1998; Thongprayoon et al., 2016). The biochemical basis of pulegone toxicity is due to its metabolism to menthofuran, a highly reactive metabolite (Rojas-Armas et al., 2019a,b). In another study, laboratory animals to which a high dose of pulegone was administered for two years developed bladder, hepatic, and urothelial cancers (Da Rocha et al., 2012; National Toxicology Program, 2011).

A report indicated renal and hepatic toxicities of high-dose eucalyptus oil in rats. Dysmetabolism of triglycerides, total cholesterol, and glucose are associated with hepatic dysfunction. A mild increase in AST and ALT in animals to which eucalyptus oil-water emulsions were administered was also observed (Hu et al., 2014). Elevated serum creatinine observed in rats in the test group in the reported study suggests renal damage (Hu et al., 2014).

EO toxicity in other animal models has been documented. The growth and reproduction of *C. elegans*, though not a pest was markedly impeded by rosemary EO. Considerable upregulation of oxidative stress genes was also observed (Lanzerstorfer et al., 2021).

A study determined the effect of *Zingiber cassumunar* Roxb EO on peripheral blood mononuclear cells from adult carp fish utilizing MTT assay. The major bioactive components of the EO are sabinene and terpinen-4-ol (Mektrirat et al., 2020). The report showed considerable cytotoxicity at concentrations between 100 and 500 μg/Ml. Additionally, time-dependent embryotoxicity at 100 μg/mL was observed in zebrafish (Mektrirat et al., 2020). Another report showed embryotoxicity of *Curcuma longa* Linn EO at 125 μg/mL in zebrafish (Alafiatayo et al., 2019).

Zerumbone was reported as the foremost biocomponent of *Zingiber ottensii* Valeton essential oil (ZOEO). Embryotoxicity of ZOEO in zebrafish was observed to be both time and concentration-dependent with morphological alterations, reduced heart rate, and decreased hatchability as teratogenic effects. Transient alterations including lethargy were however observed in rats to which ZOEO was orally administered (Thitinarongwate et al., 2021).

Human study

Indices of toxicity were observed in two individuals that participated in a human clinical trial involving 9 study participants. Turmeric EO (0.6 mL) was daily administered to the participants. In the first participant, skin rashes were observed three days into the study, while fever, associated with infection was observed in another participant on the seventh day of the study. Alteration in serum lipid profile was also observed in other participants (EFSA Panel on Additives & Products or Substances used in Animal Feed FEEDAP et al., 2020).

The hormetic potential of thymol has been reported. Hormesis describes a bi- or triphasic dose-response to an environmental stimulus, usually, the stimulatory effect characterizes a low dose while the toxic effect is attributed to a high dose (Günes-Bayir et al., 2020).

Cytotoxicity and genotoxicity of thymol have been reported. Induction of structural anomalies in chromosomes and alteration in the frequency of micronuclei in human peripheral lymphocytes were obvious in a particular study (Buyukleyla & Rencuzogullari, 2009). A significant level of damage was observed at a higher concentration of thymol (Aydın et al., 2005).

In a study that determined the pro and antioxidative effects of thymol in neoplastic (human gastric adenocarcinoma) and healthy (human fibroblast) cells, oxidative stress was induced by thymol resulting in cyto and genotoxicities (Günes-Bayir et al., 2020). Reported studies on the effects of thymol on cancer cells showed the induction of genotoxicity at 25 μM in V79 cells (Undeger et al., 2009). Cytotoxicity was also observed in Caco-2 and HepG2 cells treated with thymol at about 600 μM (Horvathova et al., 2006).

Thymol hormesis could be favorable for cancer prevention (Günes-Bayir et al., 2020). This is because of its apoptotic effects. This is evidenced by the reduced level of Bcl-2 in cell lines treated with between 0 and 50 μM thymol. A comparison of Bcl-2 in cancerous cells with healthy cells showed a markedly reduced level (Günes-Bayir et al., 2020).

Conclusion

Evidences in this review showed the toxicity of EOs. This is in spite of their many benefits. It, therefore, becomes necessary to handle these EOs with caution. Additionally, further toxicological studies should be carried out. This will give a better understanding of EOs' toxicity.

References

Adokoh, C. K., Asante, D.-B., Acheampong, D. O., Kotsuchibashi, Y., Armah, F. A., Sirikyi, I. H., Kimura, K., Gmakame, E., & Abdul-Rauf, S. (2019). Chemical profile and in vivo toxicity evaluation of unripe *Citrus aurantifolia* essential oil. *Toxicology Reports, 6*, 692−702.

Alafiatayo, A. A., Lai, K. S., Syahida, A., Mahmood, M., & Shaharuddin, N. A. (2019). Phytochemical evaluation, embryotoxicity, and teratogenic effects of *Curcuma longa* extract on zebrafish (*Danio rerio*). *Evidence-Based Complementary and Alternative Medicine: eCAM, 10*, 3807207.

Ali, B., Al-Wabel, N. A., Shams, S., Ahamad, A., Khan, S. A., & Anwar, F. (2015). Essential oils used in aromatherapy: A systemic review. *Asian Pacific Journal of Tropical Biomedicine*, 5, 601−611. Available from https://doi.org/10.1016/j.apjtb.2015.05.007.

Aydın, S., Başaran, A., & Başaran, N. (2005). The effects of thyme volatiles on the induction of DNA damage by the heterocyclic amine IQ and mitomycin C. *Mutation Research*, 581, 43−53.

Azirak, S., & Rencuzogullari, E. (2008). The in vivo genotoxic effects of carvacrol and thymol in rat bone marrow cells. *Environmental Toxicology*, 23, 728−735.

Bardaweel, S. K., Bakchiche, B., ALSalamat, H. A., Rezzoug, M., Gherib, A., & Flamini, G. (2018). Chemical composition, antioxidant, antimicrobial and Antiproliferative activities of essential oil of *Mentha spicata* L. (Lamiaceae) from Algerian Saharan atlas. *BMC Complementary and Alternative Medicine*, 18, 201.

Bickers, D., Calow, P., Greim, H., Hanifin, J. M., Rogers, A. E., Saurat, J. H., Sipes, I. G., Smith, R. L., & Tagami, H. (2003). A toxicologic and dermatologic assessment of Linalool and related esters when used as fragrance ingredients. *Food and Chemical Toxicology: An International Journal Published for the British Industrial Biological Research Association*, 41, 919−942.

Brügger, B. P., Martínez, L. C., Plata-Rueda, A., Castro, B. M. C., Soares, M. A., Wilcken, C. F., Carvalho, A. G., Serrão, J. E., & Zanuncio, J. C. (2019). Bioactivity of the *Cymbopogon citratus* (Poaceae) essential oil and its terpenoid constituents on the predatory bug, *Podisus nigrispinus* (Heteroptera: Pentatomidae). *Scientific Reports*, 9, 8358. Available from https://doi.org/10.1038/s41598-019-44709-y.

Buyukleyla, M., & Rencuzogullari, E. (2009). The effects of thymol on sister chromatid exchange, chromosome aberration and micronucleus in human lymphocytes. *Ecotoxicology and Environmental Safety*, 72, 943−947.

Da Rocha, M., Dodmane, P. R., Arnold, L. L., et al. (2012). Mode of action of Pulegone on the urinary bladder of F344 rats. *Toxicological Sciences*, 128(1), 1−8.

Deyno, S., Tola, M. A., Bazira, J., Makonnen, E., & Alele, P. E. (2021). Acute and repeated-dose toxicity of *Echinops kebericho* Mesfin essential oil. *Toxicology Reports*, 8, 131−138.

Dosoky, N. S., & Setzer, W. N. (2018). Chemical composition and biological activities of essential oils of *Curcuma* species. *Nutrients*, 10, 1196. Available from https://doi.org/10.3390/nu10091196.

EFSA Panel on Additives and Products or Substances used in Animal Feed (FEEDAP)., Bampidis, V., Azimonti, G., Bastos, M. L., Christensen, H., Durjava, M. K., Kouba, M., Lopez-Alonso, M., Puente, S. L., Marcon, F., Mayo, B., Pechova, A., Petkova, M., Ramos, F., Sanz, Y., Villa, R. E., Woutersen, R., Brantom, P., Chesson, A., ... Dusemund, B. (2020). Safety and efficacy of turmeric extract, turmeric oil, turmeric oleoresin and turmeric tincture from *Curcuma longa* L. rhizome when used as sensory additives in feed for all animal species. *EFSA Journal*, 18(6), 6146.

Gaire, S., Scharf, M. E., & Gondhalekar, A. D. (2019). Toxicity and neurophysiological impacts of plant essential oil components on bed bugs (Cimicidae: Hemiptera). *Scientific Reports*, 9, 3961. Available from https://doi.org/10.1038/s41598-019-40275-5.

Günes-Bayir, A., Kocyigit, A., Guler, E. M., & Dadak, A. (2020). In vitro hormetic effect investigation of thymol on human fibroblast and gastric adenocarcinoma cells. *Molecules (Basel, Switzerland)*, 25, 3270. Available from https://doi.org/10.3390/molecules25143270.

Horvathova, E., Šramkova, M., Labaj, J., & Slamenová, D. (2006). Study of cytotoxic, genotoxic and DNA-protective effects of selected plant essential oils on human cells cultured in vitro. *Neuro Endocrinology Letters*, 27, 44−47.

Hu, Z., Feng, R., Xiang, F., Song, X., Yin, Z., Zhang, C., Zhao, X., Jia, R., Chen, Z., Li, L., Yin, L., Liang, X., He, C., Shu, G., Lv, C., Zhao, L., Ye, G., & Shi, F. (2014). Acute and subchronic toxicity as well as evaluation of safety pharmacology of eucalyptus oil-water emulsions. *International Journal of Clinical and Experimental Medicine*, 7(12), 4835−4845.

Joca, H. C., Cruz-Mendes., Oliveira-Abreu, K., Maia-Joca, R. P. M., Barbosa, R., Lemos, T. L., Beirao, P. S. L., & Leal-Cardoso, J. H. (2012). Carvacrol decreases neuronal excitability by inhibition of voltage-gated sodium channels. *Journal of Natural Products*, 75, 1511−1517.

Kowalczyk, A., Przychodna, M., Sopata, S., Bodalska, A., & Fecka, I. (2020). Thymol and thyme essential oil—New insights into selected therapeutic applications. *Molecules (Basel, Switzerland)*, 25, 4125. Available from https://doi.org/10.3390/molecules25184125.

Kumar, D., Sukapaka, M., Babu, G. D. K., & Padwad, Y. (2015). Chemical composition and in vitro cytotoxicity of essential oils from leaves and flowers of *Callistemon citrinus* from Western Himalayas. *PLoS One*, 10(8), e0133823. Available from https://doi.org/10.1371/journal.pone.0133823.

Lanzerstorfer, P., Sandner, G., Pitsch, J., Mascher, B., Aumiller, T., & Weghuber, J. (2021). Acute, reproductive, and developmental toxicity of essential oils assessed with alternative in vitro and in vivo systems. *Archives of Toxicology*, 95, 673−691. Available from https://doi.org/10.1007/s00204-020-02945-6.

Lazarević, J., Jevremović, S., Kostić, I., Kostić, M., Vuleta, A., Jovanović, S. M., & Jovanović, D. S. (2020). Toxic, oviposition deterrent and oxidative stress effects of thymus vulgaris essential oil against *Acanthoscelides obtectus*. *Insects*, 11, 563. Available from https://doi.org/10.3390/insects11090563.

Liao, M., Yang, Q.-Q., Xiao, J.-J., Huang, Y., Zhou, L.-J., Hua, R.-M., & Cao, H.-Q. (2018). Toxicity of *Melaleuca alternifolia* essential oil to the mitochondrion and NAD + /NADH dehydrogenase in *Tribolium confusum*. *PeerJ*, 6, e5693. Available from https://doi.org/10.7717/peerj.5693.

Liaqat, I., Riaz, N., Saleem, Q., Tahir, H. M., Arshad, M., & Arshad, N. (2018). Toxicological evaluation of essential oils from some plants of rutaceae family. *Evidence-Based Complementary and Alternative Medicine*, 2018, 4394687. Available from https://doi.org/10.1155/2018/4394687.

Mektrirat, R., Terdsak Yano, T., Okonogi, S., Katip, W., & Pikulkaew, S. (2020). Phytochemical and safety evaluations of volatile terpenoids from *Zingiber cassumunar* Roxb. On mature carp peripheral blood mononuclear cells and embryonic zebrafish. *Molecules (Basel, Switzerland)*, 25, 613. Available from https://doi.org/10.3390/molecules25030613.

Melušova, M., Jantova, S., & Horvathova, E. (2014). Carvacrol and rosemary oil at higher concentrations induce apoptosis in human hepatoma HepG2 cells. *Interdisciplinary Toxicology*, 7, 189−194.

Mølck, A. M., Poulsen, M., Lauridsen, S. T., & Olsen, P. (1998). Lack of histological cerebellar changes in Wistar rats given pulegone for 28 days. Comparison of immersion and perfusión tissue fixation. *Toxicology Letters, 95*(2), 117–122.

Mossa, A.-T. H., Mohafrash, S. M. M., & Chandrasekaran, N. (2018). Safety of natural insecticides: Toxic effects on experimental animals. *Hindawi BioMed Research International, 2018*, 4308054. Available from https://doi.org/10.1155/2018/4308054.

Moura, E. D., Faroni, L. R. D., Heleno, F. F., Rodrigues, A. A. Z., Prates, L. H. F., & Ribeiro de Queiroz, M. E. L. (2020). Optimal extraction of *Ocimum basilicum* essential oil by association of ultrasound and hydrodistillation and its potential as a biopesticide against a major stored grains pest. *Molecules (Basel, Switzerland), 25*, 2781. Available from https://doi.org/10.3390/molecules25122781.

National Toxicology Program. (2011). NTP technical report on the toxicology and carcinogenesis studies of pulegone (cas no. 89–82-7) in F344/n rats and B6C3F1 mice. NIH Publication No. 11–5905.

Papanastasiou, S. A., Bali, E.-M. D., Ioannou, C. S., Papachristos, D. P., Zarpas, K. D., & Papadopoulos, N. T. (2017). Toxic and hormetic-like effects of three components of citrus essential oils on adult Mediterranean fruit flies (*Ceratitis capitata*). *PLoS One, 12*(5), e0177837. Available from https://doi.org/10.1371/journal.pone.0177837.

Rašković, A., Milanović, I., Pavlović, N., Ćebović, T., Vukmirović, S., & Mikov, M. (2014). Antioxidant activity of rosemary (*Rosmarinus officinalis* L.) essential oil and its hepatoprotective potential. *BMC Complementary and Alternative Medicine, 14*, 225. Available from http://www.biomed-central.com/1472-6882/14/225.

Rojas-Armas, J., Arroyo-Acevedo, J., Ortiz-Sánchez, M., Palomino-Pacheco, M., Castro-Luna, A., Ramos-Cevallos, N., Justil-Guerrero, H., Hilario-Vargas, J., & Herrera-Calderón, O. (2019a). Acute and repeated 28-day oral dose toxicity studies of *Thymus vulgaris* L. essential oil in rats. *Toxicological Research, 35*(3), 225–232. Available from https://doi.org/10.5487/TR.2019.35.3.225.

Rojas-Armas, J. P., Arroyo-Acevedo, J. L., Ortiz-Sánchez, J. M., Palomino-Pacheco, M., Hilario-Vargas, H. J., Herrera-Calderón, O., & Hilario-Vargas, J. (2019b). Potential toxicity of the essential oil from *Minthostachys mollis*: A medicinal plant commonly used in the traditional andean medicine in Peru. *Hindawi Journal of Toxicology, 2019*, 1987935. Available from https://doi.org/10.1155/2019/1987935.

Rossini, C., Rodrigo, F., Davyt, B., Umpiérrez, M. L., González, A., Garrido, P. M., et al. (2020). Sublethal effects of the consumption of *Eupatorium buniifolium* essential oil in honeybees. *PLoS One, 15*(11), e0241666. Available from https://doi.org/10.1371/journal.pone.0241666.

Russo, R., Corasaniti, M. T., Bagetta, G., & Morrone, L. A. (2015). Exploitation of cytotoxicity of some essential oils for translation in cancer therapy. *Evidence-Based Complementary and Alternative Medicine, 2015*, 397821. Available from http://doi.org/10.1155/2015/397821.

Sharifi-Rad, J., Antoni Sureda, A., Tenore, G. C., Daglia, M., Sharifi-Rad, M., Valussi, M., Tundis, R., Sharifi-Rad, M., Loizzo, M. R., Ademiluyi, A. O., Sharifi-Rad, R., Ayatollahi, S. A., & Iriti, M. (2017). Biological activities of essential oils: From plant chemoecology to traditional healing systems. *Molecules (Basel, Switzerland), 22*, 70. Available from https://doi.org/10.3390/molecules22010070.

Sun, J. (2007). D-Limonene: Safety and clinical applications. *Alternative Medical Review, 12*(3), 259–265.

Thitinarongwate, W., Mektrirat, R., Nimlamool, W., Khonsung, P., Pikulkaew, S., Okonogi, S., & Kunanusorn, P. (2021). Phytochemical and safety evaluations of *Zingiber ottensii* valeton essential oil in zebrafish embryos and rats. *Toxics, 9*, 102. Available from https://doi.org/10.3390/toxics9050102.

Thongprayoon, C., Cheungpasitporn, W., & Kashani, K. (2016). Serum creatinine level, a surrogate of muscle mass, predicts mortality in critically ill patients. *Journal of Thoracic Disease, 8*(5), E305–E311.

Undeger, U., Basaran, A., Degen, G. H., & Başaran, N. (2009). Antioxidant activities of major thyme ingredients and lack of (oxidative) DNA damage in V79 Chinese hamster lung fibroblast cells at low levels of carvacrol and thymol. *Food and Chemical Toxicology: An International Journal Published for the British Industrial Biological Research Association, 47*, 2037–2043.

Vinturelle, R., Mattos, C., Meloni, J., Nogueira, J., Nunes, M. J., Vaz, I. S., Jr., Rocha, L., Lione, V., Castro, H. C., & das Chagas, E. F. (2017). In vitro evaluation of essential oils derived from *Piper nigrum* (Piperaceae) and *Citrus limonum* (Rutaceae) against the tick *Rhipicephalus* (Boophilus) *microplus* (Acari: Ixodidae). *Biochemistry Research International, 2017*, 5342947. Available from https://doi.org/10.1155/2017/5342947.

Chapter 20

Patents of relevant essential oils derived from plants

Nyejirime Young Wike[1], Olugbemi T. Olaniyan[2], Charles Oluwaseun Adetunji[3], Juliana Bunmi Adetunji[4], Olalekan Akinbo[5], Babatunde Oluwafemi Adetuyi[6], Abel Inobeme[7], Oloruntoyin Ajenifujah-Solebo[8], Yovwin D. Godwin[9], Majolagbe Olusola Nathaniel[10], Ismail Ayoade Odetokun[11], Oluwabukola Atinuke Popoola[8], Olatunji Matthew Kolawole[12] and Mohammed Bello Yerima[13]

[1]Department of Human Physiology, Faculty of Basic Medical Sciences, Rhema University, Aba, Abia State, Nigeria, [2]Laboratory for Reproductive Biology and Developmental Programming, Department of Physiology, Faculty of Basic Medical Sciences, Rhema University, Aba, Abia State, Nigeria, [3]Applied Microbiology, Biotechnology and Nanotechnology Laboratory, Department of Microbiology, Edo State University Uzairue, Iyamho, Edo State, Nigeria, [4]Department of Biochemistry, Osun State University, Osogbo, Osun State, Nigeria, [5]Centre of Excellence in Science, Technology, and Innovation, AUDA-NEPAD, Johannesburg, Gauteng, South Africa, [6]Department of Natural Sciences, Faculty of Pure and Applied Sciences, Precious Cornerstone University, Ibadan, Oyo State, Nigeria, [7]Department of Chemistry, Edo State University Uzairue Iyamho, Auchi, Edo State, Nigeria, [8]Genetics, Genomics and Bioinformatics Department, National Biotechnology Development Agency, Abuja, FCT, Nigeria, [9]Department of Family Medicine, Faculty of Clinical Sciences, Delta State University, Abraka, Delta State, Nigeria, [10]Microbiology Unit, Department of Pure and Applied Biology, Ladoke Akintola University of Technology, Ogbomoso, Oyo State, Nigeria, [11]Department of Veterinary Public Health and Preventive Medicine, University of Ilorin, Ilorin, Kwara State, Nigeria, [12]Department of Microbiology, Faculty of Life Sciences, University of Ilorin, Ilorin, Kwara State, Nigeria, [13]Department of Microbiology, Sokoto State University, Sokoto, Sokoto State, Nigeria

Introduction

Essential oils are derived from the leaves, carpels, stalks, and seedlings, as well as the roots. Essential oils usually contain active ingredients in various amounts in all of their percentages. The rose, for instance, produces etheric oil primarily at the floral stage, whereas ginger yields extra essential oils in the root system. Every essential oil is distinct, differing even within a given plant, and has fantastic implementations as diverse as existence herself. The interplay of natural compounds that fabricate essential oil helps give it a distinct flavor, which varies depending on the species, yield duration, weather, and portion of the plant that it is derived from. Several substances found in natural ingredients were recognized; for instance, just the mint oil contains over two hundred additives. A large number of the substances are compositional isomers (Hariri et al., 2018). Within these bioactive components of plants are compounds like cineol, fenchone, limonene, menthol, mentone, pinene, and sabinen, a few of which are available in extremely small amounts, making artificially reproducing the science lab proportion of herbal essential oil nearly impossible (Jahan et al., 2015).

Natural oils are scented compounds found in the sensory neurons or nodules of some plants, which are utilized to safeguard them from prey and infestations as well as to attract pollinators. In other words, natural oils seem to be components of the plant's immune system. Paracelsus, the renowned Swiss illusionist, doctor, scientist, astrologer, cleric, and scholar, named diluted oils from herbs "quinta essentia," the epitome of the plant, thus the description "vital oils." Natural oils are super-saturated combustible compounds derived from different sections of precise vegetation, all with potential medicinal and energizing properties.

These highly unstable solvents are incredibly strong and accurate molecular compounds with highly complicated molecular structures. Essential oil is not an oil because it does not contain any lipid components. It is derived from the essentially rich natural varieties and bioactive components that are secreted by the molecules of particular plant components. The secretory tissues are distilled or pressed to acquire valuable solvents. Citrus peel, for instance, is cold-squeezed, while the plant's other portions are distilled (Ghaderinia & Shapouri, 2017; Hariri et al., 2018). These

Applications of Essential Oils in the Food Industry. DOI: https://doi.org/10.1016/B978-0-323-98340-2.00026-2

procedures produce a fragrant concentrate as well as a legitimate source of chemical agents. Essential oils can also be referred to as crude extracts or ethereal oils (Aramesh & Ajoudanifar, 2017).

Natural oils seem to be fairly frequent in the angiosperms, with certain families becoming particularly abundant in both total count and volume. Essential oils are usually located in outstanding seedlings belonging to orders of vascular plants but they are as well regarded as thermodynamically lactone terpenoids dangerous, or phytoplankton that generate halogenated terpenes. Even though sesquiterpenes substances are typically associated with plants, a few biologically synthesized phytoconstituents were identified in microbes, invertebrates (most likely pheromones), and in some sesquiterpene and phenolics of wildlife sources (Georgieva & Kosev, 2018; Kumar & Senapati, 2015; Vasileva, 2015).

The genetic function of essential oils is strongly intertwined with their molecular structure. The highest antimicrobial properties and broadest spectral response of activity are evidenced by phenolics like thymol and carvacrol, which take place in different ratios in natural ingredients generated by plants in the flowering plant. Thyme, oregano, and savory are particularly important (Aligiannis et al., 2001; Bozin et al., 2006; Hazzit et al., 2006; Rosooli & Mirmostafa, 2003). Breathing in natural ingredients comprising phenolics is primarily employed to treat prolonged infections of the lungs and as a food preservative. Eugenol, a phenolic substance with antibacterial activities, makes up to 90% of certain natural ingredients, including clove, cinnamon leaves, bay, and nutmeg, as well as a few organisms of basil such as *Ocimum basilicum* L. and *Ocimum gratissimum* L. (Kalemba & Kunicka, 2003). The second type of bioactive volatile oil is one that contains harmful chemicals, like terpinen-4-ol, -terpineol, geraniol, cytronellol, menthol, and linalol. Tea tree oil, eucalyptus oil, lavender, sage, rose, and sandalwood oil are examples of oils with increased antibacterial properties that do not comprise phytochemical acids. They are effective against infective microbes that cause pulmonary function, urogenital, and subcutaneous tissue viral infections (Carson et al., 2006; Lis-Balchin et al., 2000; Price & Price, 1999).

The patent is an official trademark rights record that guarantees innovators compensation for their mental work for a set duration (de Oliveira et al., 2005). As a result, patent records are a rich source of transformation products, in addition to serving as an indicator of innovative nations' accomplishments and an expertise disseminator (OECD-EUROSTAT, 2009). Patent figures can also be used to understand the interplay of inventive propagation and potentially forecast the latest technological requests (Dernis, 2007).

Patents derived from relevant essential oils

Khanuja, Srivastava, Shasney, Darokar, Kumar, Agarwal et al.'s patent US6824795 describes a preparation helpful in the diagnosis of drug-resistant inflammatory conditions. It includes thymol derived from *Trachyspermum ammi*, as well as mint oil derived from a cross between *Mentha spicata* and *Mentha arvensis*, which consists of a proper quantity of monoterpenes like carvone, limonene, and menthol. Mint oil beneficial dosages of 0.1% to 0.5% w/w and thymol effective concentrations of 20%–50% w/w have been researched. The solution should be taken orally or transdermally, according to the patent's publishers. The preparation can be utilized to control bacterial intestinal mucosa and systemic diseases that are caused by multidrug-resistant Mycobacterium and Escherichia isolates. This strategy performs well against bacteria that are resistant to some antibiotics (Khanuja et al., 2004).

The scholars of Weiss' patent US0206790A1 suggest using ecologically responsible antibacterial foamable cleansers with thyme oil as the ultimate (Cu2) or primary antibiotic for killing methicillin-resistant *S. aureus*. This technique is exceptionally efficient and can be utilized as both a means of prevention and a curative test (Weiss, 2011). Frame's patent US7887860 outlines a technique for generating, comprising, and isolating antimicrobial constituents from the plants *Mammea americana*, *Marcanthia polymorpha*, and *Callistemon citrinus*. It was discovered to firmly impede the proliferation of microorganisms from the Mycobacterium genera due to antimicrobial constituents like cobaltocene-octamethyl, stigmastan-3,5-diene, galaxolide, benzyl salicylate, and the existence of essential oil constituents' eucalyptol and -pinene (Frame, 2011).

The following ingredients are listed in Ben-Yehoshua's patent US7465469, which is designed to impair microbes' growth using a microbiocidal soluble preparation: citral, 1-octanol, heptanol, nonanol, geraniol, octanal, nonanal, decanal, perillaldehyde, perillalcohol, citronellol, and citron. The preparations and techniques for impeding growth of microorganisms in short-shelf-life food crops, domestic products, and human personal care products that contain essential oil elements or their variants have been generally conceptualized. Pathogens triggered by agricultural harvest contamination generate genotoxic compounds that are dangerous to people. Essential oil components may be better options than the artificial chemicals currently used to avoid deterioration (Ben-Yehoshua, 2008).

The patents US7384646 and US7754774 given by Kobayashi, Okamoto, and Okada explain the application of 1,2-alkanediol with 5–10 carbons and one or more substances picked from citral, geraniol, nerol, perillaldehyde, alpha-terpineol, dodecanol, and L-carvonein (in the first patent) and thymol, eugeno. The researchers state that the inclusion of such essential oil constituents boosts the antimicrobial property of 1,2-alkanediol against a diverse variety of varieties while minimizing risk. It has the capability to decontaminate beauty products and personal care items, as well as pharmaceuticals and food (Kobayashi et al., 2008, 2010).

Angiolella and Ragno's patent WO2011092655 explains how to employ Mentha suaveolens essential oil in skincare production. The patent's researchers accurately spot the oil's components, which must include a minimum of 60% piperitenone oxide, 0.5% alpha-cubebene, as well as 0.5% octanol. It's safe to use in cleansers, toothpaste, emulsions, hydrocolloids, hair products, skincare products, antiperspirants, and swish fluids (Angiolella & Ragno, 2011). Narayanan, Prosise, and Corring's patent WO2012018519A1 describes the application of thymol and eucalyptol as toothpaste additives. The formulation is clearly perceived as hindering or decapitating the microbes that induce malodor, periodontal disease, dental problems, oral infections, gum disease, and soft tissue abnormalities (Narayanan et al., 2012).

The patents US7344740 Vail WB and Vail ML concern techniques and equipment for preventing and treating morning sickness associated with medical therapy for different tumors. The studies propose using Mentha piperita essential oil (peppermint oil) to avoid and alleviate chemotherapy-induced acid reflux. The clinical management process includes using fumes both before and after an anti-cancer drug session (Vail & Vail, 2008). Du-patent Thumm's WO2011019342A2 explains a dental configuration that can be utilized as an agent to diagnose or stop a disorder characterized by bacterial growth in the buccal mucosa. A sesquiterpenoid and an antibacterial substance are present in sufficient quantities to impede and/or degrade a microbe in the buccal mucosa. The product aids in avoiding the emergence of periodontitis, dental caries, bad breath, and gum disease (Du-Thumm, 2011).

Conclusion

This approach at recently published patents demonstrates the exceptional efficiency of volatile oil and their components in the struggle against multidrug-resistant microorganism strain configuration, both in regards to human diagnosis and malignant transformation mitigation. The utilization of volatile oil as functional ingredients in beauty products, hygiene products, medicine, and food stuffs are greatly useful. Furthermore, essential oils, as phytoconstituents, provide numerous opportunities as chemical antioxidants in companies, eradicating the substances frequently employed. Finally, the widespread use of natural ingredients in livestock farming, as anti-infective and curative substances, and in crop production could turn out to be a powerful tool in the battle against harmful microbes and bugs.

References

Aligiannis, N., Kalpoutzakis, E., Mitaku, S., & Chinou, I. B. (2001). Composition and antimicrobial activity of the essential oils of two *Origanum* species. *Journal of Agricultural and Food Chemistry, 49*, 4168–4170.

Angiolella, L., & Ragno, R. (2011). *Mentha suaveolens essential oil, useful as products comprising e.g. soaps, mouthwashes, foams, gels, shampoo, aftershaves, deodorants and gargle liquids, comprises piperitenone oxide, alpha-cubebene and octanol.* WO2011092655.

Aramesh, M., & Ajoudanifar, H. (2017). Alkaline protease producing Bacillus isolation and identification from Iran. *Banat's Journal of Biotechnology, 8*(16), 140–147.

Ben-Yehoshua, S. (2008). *Microbiocidal formulation comprising essentials oils or their derivatives.* US20087465469.

Bozin, B., Mimica-Dukic, N., Simin, N., & Anackov, G. (2006). Characterization of the volatile composition of essential oils of some *Lamiaceae* species and the antimicrobial and antioxidant activities of the entre oils. *Journal of Agricultural and Food Chemistry, 54*, 1822–1828.

Carson, C. F., Hammer, K. A., & Riley, T. V. (2006). Melaleuca alternifolia (tea tree) oil: A review of antimicrobial and other medicinal properties. *Clinical Microbiology Reviews, 19*, 50–62.

de Oliveira, L., Suster, R., Pinto, A., Ribeiro, N., & da Silva, R. (2005). *Quimica Nova, 28*, 36.

Dernis, H. (2007). *OECD science, technology and industry working papers* (p. 48) Paris: OECD Publishing.

Du-Thumm L. (2011). *Oral care composition useful as medicament for treating or preventing a condition caused by biofilm formation in oral cavity, e.g. dental plaque, tooth decay, halitosis, or gingivitis, comprises a sesquiterpenoid and an antimicrobial agent.* WO2011019342A2.

Frame, A. D. (2011). *Preparation of composition useful for inhibiting growth of mycobacterium comprises extracting specified plant material, contacting extract with separation system and eluting the material with mobile phase.* US20117887860.

Georgieva, N., & Kosev, V. (2018). Adaptability and stability of white lupin cultivars. *Banat's Journal of Biotechnology, 9*(19), 65–76.

Ghaderinia, P., & Shapouri, R. (2017). Assessment of immunogenicity of alginate microparticle containing *Brucella melitensis* 16M oligo polysaccharide tetanus toxoid conjugate in mouse. *Banat's Journal of Biotechnology, 8*(16), 83–92.

Hariri, A., Ouis, N., Bouhadi, D., & Benatouche, Z. (2018). Characterization of the quality of the steamed yoghurts enriched by dates flesh and date powder variety H'loua. *Banat's Journal of Biotechnology, 9*(17), 31–39.

Hazzit, M., Baaliouamer, A., Faleiro, M. L., & Miguel, M. G. (2006). Composition of the essential oils of Thymus and Origanum species from Algeria and their antioxidant and antimicrobial activities. *Journal of Agricultural and Food Chemistry, 54*, 6314−6321.

Jahan, S., Chowdhury, S. F., Mitu, S. A., Shahriar, M., & Bhuiyan, M. A. (2015). Genomic DNA extraction methods: A comparative case study with gram−negative organisms. *Banat's Journal of Biotechnology, 6*(11), 61−68.

Kalemba, D., & Kunicka, A. (2003). Antibacterial and antifungal properties of essential oils. *Current Medicinal Chemistry, 10*, 813−829.

Khanuja, S. P. S., Srivastava, S., Shasney, A. K., Darokar, M., Kumar, T. R. S., Agarwal, K. K., et al. (2004). *Formulation comprising thymol useful in the treatment of drug resistant bacterial infections.* US20046824795.

Kobayashi, A., Okamoto, H., & Okada, F. (2008). *Antiseptic disinfectant, and cosmetics and toiletries, medicine or food containing the same.* US20087384646.

Kobayashi, A., Okamoto, H., & Okada, F. (2010). *Antiseptic bactericides and cosmetics, drugs and foods containing the antiseptic bactericides.* US20107754774.

Kumar, A., & Senapati, B. K. (2015).) Genetic analysis of character association for polygenic traits in some recombinant inbred lines (ril's) of rice (*Oryza sativa* L.). *Banat's Journal of Biotechnology, 6*(11), 90−99.

Lis-Balchin, M., Hart, S. L., & Deans, S. G. (2000). Pharmacological and antimicrobial studies on different tea-tree oils (*Melaleuca alternifolia, Leptospermum scoparium* or manuka and *Kunzea ericoides* or kanuka), originating in Australia and New Zealand. *Phytotherapy Research: PTR, 14*, 623−629.

Narayanan, K. S., Prosise, W. E., & Corring, R. (2012). *An alcohol-free slightly-alcoholic oral care composition and a process for preparing same.* WO2012018519A1.

OECD-EUROSTAT. (2009). In *OECD patent statistics manual*; OECD Publishing: Paris, ch. 2.

Price, A., & Price, L. (1999). *Aromatherapy for health professionals* (3rd Edition). London: Churchill Livingstone.

Rosooli, I., & Mirmostafa, S. A. (2003). Bacterial susceptibility to and chemical composition of essential oils from *Thymus kotschyanus* and *Thymus persicus*. *Journal of Agricultural and Food Chemistry, 51*, 2200−2205.

Vail, W. B., & Vail, M. L. (2008). *Methods and apparatus to prevent, treat, and cure the symptoms of nausea caused by chemotherapy treatments of human cancers.* US20087344740.

Vasileva, V. (2015). Root biomass accumulation in vetch (Vicia sativa L.) after treatment with organic fertilizer. *Banat's Journal of Biotechnology, 6* (11), 100−105.

Weiss, L. (2011). *Foamable composition useful for killing methicillin-resistant Staphylococcus aureus on surface contaminated, comprises antimicrobial amount of thyme oil, surfactant, and source of divalent copper ions.* US20110206790A1.

Chapter 21

Commercially available essential oil with higher relevance in food sector and their detailed information in market trends

Benjamin Olusola Abere[1] and Charles Oluwaseun Adetunji[2]

[1]Department of Economics, Edo State University Uzairue, Iyamho, Edo State, Nigeria, [2]Applied Microbiology, Biotechnology and Nanotechnology Laboratory, Department of Microbiology, Edo State University Uzairue, Iyamho, Edo State, Nigeria

Introduction

Essential oils (EOs) are now receiving increased attention in the food industry as natural inhibitors or bio preservatives due to customer concerns about artificial preservatives. The number of publications published about the use of EOs in foods can be used to gauge this growing interest. The term "essential oils and foods" produced 5559 results in a basic search of the Web of Science database from 1950 to 2017, was chosen as the search subject and "article" as the document type. Despite the fact that the first article ever found was written in 1953 (Ooi et al., 2006) and that a steady stream of publications followed, it wasn't until 1990 that the number of articles produced each year was recorded. The main reasons why EOs are used to extend the shelf life of foods are their antioxidant and antimicrobial properties. The number of publications found when the search keywords "antioxidant" (1920 papers), "antimicrobial" (2473 papers), or both (973 papers) were included serves as another example of this. Given the types of foods that were primarily used in these studies, it can be said that EOs have been used as bio preservatives in all kinds of foods, but fruits and vegetables have seen the largest recorded usage of them. These include fruits (657 papers), vegetables (403 papers), fish products (415 papers), meat products (410 papers), milk and dairy products (216 papers), and bread and baked foods (97 papers).

For humans, smells have important symbolic implications. For instance, unpleasant odors serve as a warning sign for spoiled food, but pleasant aromas are employed for relaxation, medication, and fragrance. In the midst of our stressful schedules, it is sometimes beneficial to pause and smell the roses, as the phrase goes. But what happens if there aren't enough roses for everyone who wants to take a sniff? EOs are one of the many remedies that have been created to address the problem. For thousands of years, they have been utilized in a variety of ways for medical purposes, in scents, and in commercial goods. The sector is expected to develop significantly over the next several years because of rising demand on a global scale. It is now robust. The definition of EOs in the dictionary is that they are hydrophobic liquids, or not water soluble, and include volatile fragrance molecules from the plants from which they were harvested. They are not necessary, despite their name, despite what the industry would undoubtedly claim. The liquids that carry the "essence," or the distinctive smell, of the plant they come from, are what are meant when we use the word "essential," instead. Volatile oils, ethereal oils, or the essence or oil of the plant, such as bergamot oil, are some other names that are frequently used for these substances. Archeologists may have discovered apparatus for producing EOs around the third millennium BC that was maybe constructed of ceramic. Zosimus of Panopolis, a Greek alchemist, described producing "holy water and remedy" in his writings from the fifth century AD. In the Roman Empire and Asia, there was a large commerce in odoriferous materials. Floral and fragrant liquids were employed in cooking, commerce, and perfumery during medieval times. They were also utilized as digestive aids. The works of Arab scholar Ibn al-Baitar from the 12th century provide the first documented details of how to manufacture EOs. Modern chemistry made it possible to isolate the chemical constituents of EOs, which allowed for a more streamlined production process for EOs. The vocabulary also changed as a result, with scientific publications starting to refer to substances like "methyl salicylate" instead of "oil of wintergreen." Modern EO mass manufacturing was made possible because of this improved understanding.

Applications of Essential Oils in the Food Industry. DOI: https://doi.org/10.1016/B978-0-323-98340-2.00013-4

Distillation and expressiveness have both been used for thousands of years as the two main techniques for producing EOs. Since then, more contemporary techniques like solvent and carbon dioxide (CO_2) extraction have joined these traditional technologies. Distillation is by far the most popular of the four. Most widely used EOs, such as eucalyptus, peppermint, lavender, and tea tree oils, are produced using the distillation process. Water, steam, a combination of water and steam, or percolation distillation are all methods for extracting EOs. All of these processes have the advantage of extracting the volatile compounds at lower temperatures than the boiling points of the individual components, according to the US National Association for Holistic Aromatherapy. Then, it would be simple to separate these components from the condensed water. Plant material that has not been processed, such as bark, flowers, leaves, roots, and/or seeds, is placed in an alembic, or still chamber, above water for the distillation process. The still is closed, and as the water warms, steam or water rises to the top, gently puncturing the plant matter and vaporizing the volatile compounds. As the compounds ascend with the steam, they contact the condenser coils. Due to the fact that EOs and water do not mix, after time, and depending on the oil's weight, the oils may ultimately be sucked off the top or bottom of the collecting dish. Hydrosol, hydrolat, or floral water is the term for the leftover water that still contains certain water-soluble components and traces of the essential oil. Applications for hydrosol include skincare items and perfumes that are comparable to those for EOs. A comparable extraction process to the manufacturing of olive oil is an expression or cold pressing. Orange, tangerine, lemon, and lime oils are typically extracted using it expression, sometimes known as cold pressing, which is an extraction technique used in the same way as olive oil is made. Citrus oils, including those from orange, tangerine, lemon, and lime, are what it is generally used for. The citrus fruit's rind is put into a container that has spikes or needles to puncture it as part of the current production process. The EO is continuously released from the rind with a probing motion, and it is then gathered beneath the container, in a smaller dish. Tradition dictates that when the extraction liquid has been allowed to stand to separate into its oil and water/juice components, the oil is then siphoned off. But in commercial facilities, centrifuges are now frequently used to perform the separation, which speeds up the process and minimizes the need for manual work. The expression of citrus oils is often less expensive than the distillation of oils since citrus peels are frequently easily accessible in large quantities from the food industry.

Commercially available essential oil with higher food application

Orange oil

By cold pressing the rind, a by-product of the orange processing industry, the orange oil from the citrus sinensis plant is obtained. About 0.3%−0.5% of orange oil is produced during the extraction process. The oil is yellow to orange in color, viscous almost like water, and has a sweet and acidic smell. The hydrocarbon that gives citrus fruits their distinctive aroma, d-limonene, makes up about 90% of the oil. Orange EO is frequently used as a flavoring and coloring ingredient in the food and beverage sector, for instance in fruit juices, jams, bakery goods, and liqueurs similar to Curacao. Because of its pleasing aroma, it is often used in cleaning products and perfumes. Due to its alleged depressive, anti-inflammatory, antibacterial, and aphrodisiac effects, orange oil is expected to see an increase in demand across a variety of applications, according to Grand View Research. Additionally, orange oil has little effect on termites but may be used as a natural insecticide to eradicate whole ant colonies. Skin irritation is one of the possible handling risks associated with orange oil since the high limonene concentration destroys the natural oils of the skin. It should not be applied directly to the skin since it might have a phototoxic impact, especially before spending a lot of time in the sun. According to studies, orange oil causes cancer in male rats, nonetheless, the International Agency for Research on Cancer rates it as safe for humans and there is no evidence that it causes cancer in people. According to market statistics from Ultra International, orange oil prices climbed to all-time highs in 2016, and neither it nor other firms anticipate that trend to change very soon. The market was impacted by the lowest Florida season output in ten years, which was 17% lower than the previous season and the lowest since 1963, as well as drought conditions in South Africa. According to Ultra International's spring 2017 market assessment, "there is no question we're still in for a rough period" because local juice supplies are expected to hit an all-time low by June 2017 and prices reached record highs in January 2017.

Corn mint and peppermint oil

Plants in the mint family include Mentha x piperita, sometimes known as peppermint, and Mentha arvensis, also known as corn mint, field mint, or wild mint.. Usually, the plant's leaves are steam distilled to produce an EO with up to 80% menthol. However, the bulk of commercially available maize mint oil has been fractionated and dementholized, removing a significant amount of the menthol component. Despite this, both oils are thin, clear liquids that smell strongly bittersweet. Similar uses exist for both mint oils. Natural insecticides like pulegon and menthone are present in high concentrations in them. The stimulation of the digestive system, relief from headaches, muscular pains, colds, and sinusitis are only a few of its medical applications. Because of their capacity to destroy microorganisms and chill the body, they are also widely utilized in dental care products. However, because maize mint oil is less expensive, it is more prevalent in these applications—fragrance and food production—where they both are used. They may also be utilized in plumbing, where their strong aroma can be used to find leaks. Future market insights predict that the global increase in interest in aromatherapy will significantly drive the market for peppermint EO (because corn mint is less popular in aromatherapy). Natural mosquito repellents have become more popular as a result of the current Zika virus outbreak, and mint oils are expected to continue to see record growth rates in the fragrance industry. Peppermint oil production is mostly based in North America, although India is the global leader in corn mint production. According to Ultra International, both industries, however, experience difficulty with the restricted amount of raw materials that are accessible. To keep prices in check, producers are still optimistic that favorable weather conditions in 2017 would result in greater harvests.

Eucalyptus oil

The phrase "eucalyptus oil" is used to describe a variety of eucalyptus plant EOs. Traded eucalyptus oils often fall into one of three groups: industrial, medical, and fragrance-related types. The "typical" cineole-containing oil, an opaque liquid with a strong, woody scent, is the most common of the three, even though they are all produced by steam distillation. Pharmaceutical uses call for oil to have a minimum cineole content of 70%, while several types, including Eucalyptus kochii, generate oil with a cineole content of 80% to 95%. The amount of cineole in low-grade oil can be increased. In items like cough syrup, lozenges, ointments, and inhalants, Eucalyptus oil is used medicinally to treat the flu and cough symptoms. Eucalyptus vapor is breathed as a bronchitis treatment because of its antibacterial and decongestant properties. Eucalyptus oil is sometimes found in personal hygiene items like soaps and toothpaste. As an insect repellant and biopesticide, the oil is also employed. It works well as a fungal inhibitor, according to a Chinese study from 2016. Eucalyptus oil is used in small amounts to flavor food and beverages because of its antibacterial properties, which help to keep food from spoiling. Eucalyptus oil has also demonstrated its promise as a feedstock for biofuels that reduce harmful emissions, notably in the aviation sector, via research and practical demonstrations. Chinese producers account for roughly 75% of the world's eucalyptus oil supply, displacing Australia, which has historically led the market. However, Ultra International observes that since 2014, output in China has decreased as firms have waited in anticipation of rising crude oil prices before beginning the distillation process. Thus, production is increasing in Australia, where new plantations are being planted, and increased demand, which is also aided by aromatherapy, is driving up production in Spain and Portugal.

Citronella oil

One of the EOs produced from lemongrass is citronella oil, which is split into two chemotypes based on the chemical make-up of the plant and the plant variety. The Ceylon type, from the Cymbopogon nardus variety, consists mostly of 18%−20% geraniol and 5%−15% citronellal, in contrast to the Java type, which is derived from the Cymbopogon winterianus variety and contains 11%−13% geraniol and 32%−45% citronellal. Distillation is used to extract the oil, which is a thin, transparent liquid with a lemony, sweet scent. According to Transparency Market Research, citronella has a long history of usage in perfumes and as a flavor enhancer, with the latter accounting for the majority of sales of the product. Citronella oil is a component of detergents, industrial polishes, soaps, and cleaning agents in addition to foods and fragrances. Since 1948, the oil has been approved for use as a natural insect repellent in the USA, which is another common use for it. Citronella has encountered opposition in other places when used as repellents. Citronella oil is not allowed to be sold as an insect repellent in the EU or the UK, and Canada prohibited it in this application in 2012-although the prohibition was quickly lifted. Although some studies have found a possibility of a medication interaction, Citronella oil is safe when used as indicated, according to the US Food and Drug Administration. The Asia-Pacific area, which includes China, India, and Indonesia, is the world's top producer of citronella oil. The market may, however, be affected, Ultra International and Global Essence believe, as a result of recent low harvests and price increases. In China, there is not an excess of, thus prices have remained consistent at about US$17.5/kg despite a 15% increase in prices in Indonesia. By the end of the year, however, prices may rise due to sluggish output and decreasing supply.

Tea tree oil

Tea tree oil is a liquid with a yellow hue that is created by distilling the leaves of the Melaleuca alternifolia tree. It is also known as melaleuca oil or tea tree oil. The ISO 4730–2004 standard is used to standardize commercially traded tea tree oil, which has a woody, herbal scent. It concludes that the required levels of terpinen-4-ol, gamma-terpinene, and six additional compounds in the oil are 30%–48%, 10%–28%, and 6%, respectively. Acne, nail fungus infections, and sports foot are among the conditions that tea tree oil is used to cure, however, there isn't yet solid proof of its efficacy in treating any of these conditions. In low concentrations, it is also widely used in skincare and cosmetics products, as well as in haircare products. The exploding interest in aromatherapy is raising as well, much like the interest in other EOs in the sales of tea tree oil. Recent investigations have also raised the possibility of some antibacterial and antiseptic properties. Although tea tree oil is generally harmless, there are occasional issues. When consumed, tea tree oil is poisonous and has a broad range of potential side effects, including nausea, hallucinations, and coma, according to a 2006 study. Tea tree oil is irritating because it should be diluted before applying to the skin. Tea tree oil has several components that oxidize quickly when exposed to air, and this oil may become allergenic. Northern Africa, Australia, China, and the Asia Pacific are among the regions that produce tea tree oil. In Australia, where the entire crop may be decreased by up to 30%, and in other regions of Asia, this year's harvests are in jeopardy due to heavy rainfall in early 2017. According to Ultra International, crop loss may prevent these regions from meeting demand this year. Due to this, Australian tea tree oil now costs about $50 USD. South Africa's output might potentially increase by up to fourfold over the next several years, whereas China's production is either steady or increasing. This increase keeps prices in these locations at or near $25 and $45 per unit, respectively, and keeps worldwide demand satisfied.

Market trends of essential oil

The market value of EOs is expected to reach USD 16.0 billion by 2026 with a CAGR of 9.3%. Due to its numerous uses and end-use applications, the market for EOs is expanding.

Oils that are extracted for their taste and smell are known as EOs because of their high concentration. They come from plant components such as leaves, stems, flowers, bark, roots, and volatile aromatic chemicals. They have qualities that are beneficial to sustaining health, including antioxidant, antibacterial, and anti-infective characteristics. They assist in the treatment of a multitude of illnesses, including immune system issues, sleep issues, disorders linked to stress, and depression. After EOs are removed, producers need a processor to whom they may sell the final product. Pharmacies, cosmetics, food and beverage, soaps, and household cleaning goods all employ EOs. While they are mostly used in the later sectors for their color, flavor, and aroma, they are sought after in the first two for their medical and pharmacological properties. By 2024, the market, which exceeded US$6 billion in value in 2015, may be worth US$13.94 billion, according to Grand View Research's projections because of the continually increasing demand. In order to defend themselves opposed dangers like insects and fungus, plants create several EOs that serve as natural biocides. Due in part to these qualities, they have long been used to make pesticides and for medical purposes. For instance, the antibacterial compound carvacrol, which is present in Oregano oil, which includes *E. coli*, stops the development of many bacteria. thymol, while *E. coli*, which is present in thyme oil, lowers bacterial resistance to conventional antibiotics like penicillin. The fungicidal qualities of eucalyptus oil, on the other hand, have been discovered, and this means that it may be a powerful, organic pesticide. Antidepressant, antibacterial, aphrodisiac, and diuretic therapies are a few other "medical" applications for different oils. Grand View Research reports that the food and beverage industry is the largest market for EOs, accounting for roughly 34.6% of the total worldwide demand of 178,800 tons in 2015. According to the market consultant, the rise of developing countries and the increased desire for natural products in Western markets will keep EOs in high demand. Spa and relaxation, which had 29% of the market in 2015 and is projected to grow at an 8.7% CAGR, is another rapidly expanding market. The growing acceptance of aromatherapy and other complementary therapies is driving the relaxing effects of essential oils, when used in massages, and is increasing customer demand. There is a chance that EO supplies could run out despite rising demand. Both US-based Global Essence and Dutch producer of aromatherapy and EO products Ultra International predict that rising demand and scarcity will push up the price of EOs. Additional worries for the sector include those related to the environment brought on by the production of EOs, climate change that may have an impact on how many aromatic plants thrive, and globalization that may forcibly compel farmers to switch to alternate crops. The demand, however, should continue to rise over the period of at least the next five to six years, if not longer, assuming output can be raised. Synthesis of EOs levels is difficult to determine since they vary greatly depending on the plant used, the location, and the manufacturing processes. However, according to Grand View Research, orange oil, which accounted for 29.4% of the worldwide market's total volume in 2015, is the most popular product on the market. The top five EOs also include tea tree, corn, peppermint, eucalyptus, and citronella.

Market dynamics for essential oils

Driver: growing interest in aromatherapy

Use of EOs for medicinal and cosmetic reasons is known as aromatherapy. As a result, EOs are frequently employed in aromatherapy. They help both physiological and psychological activity in the body at the same time. Aromatherapy was once thought of as an alternative therapy; nevertheless, it is now recognized as a mainstream therapy. A number of variables, the most significant of them being the last decade's overall growth in wealth, might be responsible for the rising demand for aromatherapy. The tastes and inclinations of customers have also changed in favor of more upscale goods as earnings have grown. According to the International Spa Association spas see an increase in visitors every year (ISPA). According to the 2019 US Spa Industry Study, the spa industry saw improvements across all key performance indicators. In 2018, the spa industry achieved records for overall revenue, spa visits, spa locations, revenue per visit, and total employees. The overall spa sales reached USD 18.3 billion in 2018, an increase of 4.7% from USD 17.5 billion in 2017. A total of 190 million people visited spas in 2018, an increase of 1.6%. From 21,770 in 2017, there are now 22,160 spa locations in the US, a rise of 1.8%. The revenue per visit increased to USD 96.50, a 3% increase over 2017.

Natural resource depletion is a constraint

A sound environment offers a variety of advantages that serve as cornerstone of human welfare. Because they balance the environment by supplying food and shelter, plants are a crucial component of this ecosystem. Through the use of therapeutic goods like EOs, they also promote physical and mental health. But with time, this source runs out. This could be related to a rise in population and industry, which have caused deforestation.

As the population grows, more trees are chopped down to make room for the extra people. Up to 28,000 species might become extinct in the next 25 years owing to deforestation, according to The World Counts. Natural catastrophes have caused the loss of 13 million hectares of forest, or they have been converted to other uses.

EOs are extraordinarily potent substances with a very limited yield. One pound of lavender EO, for instance, uses about 220 pounds of lavender flowers. A significant element that creates a barrier in the market for EOs is deforestation, which also contributes to the poor yield of EOs.

Possibility: growing interest in natural components

Recently, there has been a growth in the demand for "all-natural components," primarily as a result of consumers being more aware of the advantages of natural ingredients and having more money available to spend on pricey natural goods. As people's awareness of their own health increases, they are more willing to spend money on high-quality natural goods that would have a lasting impact, which further feeds the market for these items.

For their formulations to offer the most benefits to the customers, many producers from various sectors are coming up with novel ways to use organic and healthy components. "Natural claims" have a significant role in consumer decisions to buy food, drinks, and cosmetic products. The "all natural" taste and aroma that EOs give is used in a variety of sectors, including food and beverage, cosmetics, feed, and home care, among others. This opens up a number of prospects for this industry.

Issues: the prevalence of artificial and contaminated products

Contamination of EOs is a significant worry that has grown to be a significant obstacle for this industry. Not every EO is the same, and some could be adulterated with synthetic substances. These tampered-with EOs serve only as fragrance products and lack the qualities of true EOs.

EOs that have been tampered with are exceedingly dangerous. They might have a wide range of negative consequences and harm the organs permanently. The high cost of EOs is mostly to blame for the prevalence of such tainted and manufactured items. While the price of a liter of pure rose EO is over USD 78000, a lesser-quality version of the same volume would cost between $200 and $300. The predominance of synthetic/adulterated items is due to the fact that adulteration reduces the price of the product and makes it accessible to the general public.

According to predicted market share by applications throughout the forecast period, the EOs market's food and beverage sector will dominate

EOs are known for their health benefits since they are all-natural substances that carry the flavor of the plant or animal from which they were extracted. They are used as artificial ingredient substitutes in food and beverage products to give the finished goods more flavor. The demand for foods and beverages that use EOs as additives has increased as more people are becoming aware of the health benefits of these oils. Useful substances are recognized for their health advantages since they are all-natural compounds that transmit the flavor of the source from which they were produced. In place of artificial components, they are used to flavor food and beverages. Customers' preferences for food and beverage items that include EOs as additives have grown as they are becoming more aware of the health advantages of doing so.

The market for distillation segment is projected to account for the largest share during the forecast period

The distillation industry held the highest proportion of the market for EOs when it came to the method of extraction. By evaporating a mixture, turning the vapor into liquid, and then condensing it again, distillation is the process of separating its constituent parts. Temperature-sensitive molecules, such as aromatic compounds, are processed using the distillation technique. The water-insoluble chemicals are separated from nonvolatile elements using distillation in this process, which is just mildly volatile. For the extraction of oils, many types of distillation techniques are utilized, including steam distillation, water distillation, and water and steam distillation. Various EOs, including eucalyptus, lavender, and jasmine, are the principal uses for it. The fragrance and EO businesses frequently employ distillation as an extraction technique.

The essential oils market is driven by clover leaf throughout the anticipated era

The segment with the greatest anticipated CAGR by type throughout the projection period is the clover leaf sector. Cloves are dried flower buds that come from a tree that is native to Indonesia. Minerals, vitamins C and E, calcium, and magnesium are present in trace levels, and it also contains a lot of fiber. Clove leaf oil is a popular fragrance ingredient in cosmetics, toothpaste, soaps, toiletries, lotions, massage oils, and fragrances due to its potent germicidal properties, use in dental preparations, and use in these products.

The region with the quickest rate of growth in the worldwide market for essential oils is Asia Pacific

It is anticipated that the EO market would expand most quickly in the Asia Pacific area. EO demand is rising as a result of consumers choosing natural EO constituents like pepper and spearmint. In 2020, the biggest market for natural antioxidants was Asia Pacific, which had a market share of almost 16%. One of the major markets for food antioxidants is the Asia Pacific area, which includes both the largest and most populous continent in the world.

Conclusion

The commercially available EOs with a greater focus on the food business are covered in this chapter along with comprehensive information on market trends. To lessen the influence of microbial activity in food items, a variety of EO classes and their individual components are employed as natural antibacterial agents. Distillation techniques are frequently used to extract the natural chemicals known as EOs from aromatic and therapeutic plants. These substances play a significant part in the safety and extension of food product shelf life. As a result of the development of spoilage microorganisms being suppressed, there has been an increase in food safety as well as a decrease in biogenic amines, mostly in beef and meat and dairy products (Adejumo, Adetunji, et al., 2017; Adetunji & Varma, 2020; Adetunji, 2008, 2019; Adetunji et al., 2021a, 2021b, 2022; Adetunji, Adetunji, Olaniyan, et al., 2021; Adetunji, Ajani, et al., 2013; Adetunji, Ajayi, et al., 2021; Adetunji, Akram, et al., 2021; Adetunji, Anani, et al., 2021; Adetunji, Arowora, et al., 2013; Adetunji, Egbuna, et al., 2020; Adetunji, Kumar, et al., 2019; Adetunji, Michael, Kadiri, et al., 2021; Adetunji, Michael, Nwankwo, et al., 2021; Adetunji, Michael, Varma, et al., 2021; Adetunji, Nwankwo, et al., 2021; Adetunji, Ojediran, et al., 2019; Adetunji, Olaniyan, et al., 2021; Adetunji, Oloke, et al., 2020; Adetunji, Palai, et al., 2021; Adetunji, Panpatte, et al., 2019; Adetunji, Phazang, et al., 2017; Adetunji, Roli, et al., 2020; Bello et al., 2019; Egbuna et al., 2020; Olaniyan & Adetunji, 2021; Rauf et al., 2021; Thangadurai et al., 2021; Ukhurebor, & Adetunji, 2021). Since volatile oils may be used to improve food safety and shelf life and can be viewed as a natural substitute for conventional food preservatives, the usage of plant extracts and commercially available EOs in consumer goods is anticipated to rise in the coming years.

References

Adejumo, I. O., Adetunji, C. O., & Adeyemi, O. S. (2017). Influence of UV light exposure on mineral composition and biomass production of myco-meat produced from different agricultural substrates. *Journal of Agricultural Sciences, Belgrade*, *62*(1), 51−59.

Adetunji, C.O. (2008). The antibacterial activities and preliminary phytochemical screening of vernoniaamygdalina and Aloe vera against some selected bacteria, pp 40−43 [Msc Thesis University of Ilorin].

Adetunji, C. O., Arowora, K., Fawole, O., & Adetunji, J. B. (2013). Effects of coatings on storability of carrot under evaporative coolant system. *Albanian Journal of Agricultural Sciences*, *12*(3).

Adetunji, C. O., Michael, O. S., Rathee, S., Singh, K. R. B., Ajayi, O. O., Adetunji, J. B., Ojha, A., Singh, J., & Singh, R. P. (2022). Potentialities of nanomaterials for the management and treatment of metabolic syndrome: A new insight. *Materials Today Advances*, *13*, 100198.

Adetunji, C. O., Palai, S., Ekwuabu, C. P., Egbuna, C., Adetunji, J. B., Ehis-Eriakha, C. B., Kesh, S. S., & Mtewa, A. G. (2021). *General principle of primary and secondary plant metabolites: Biogenesis, metabolism, and extraction. Preparation of Phytopharmaceuticals for the Management of Disorders* (pp. 3−23). Publisher Academic Press.

Adetunji, J. B., Ajani, A. O., Adetunji, C. O., Fawole, O. B., Arowora, K. A., Nwaubani, S. I., Ajayi, E. S., Oloke, J. K., & Aina, J. A. (2013). Postharvest quality and safety maintenance of the physical properties of Daucus carota L. fruits by Neem oil and Moringa oil treatment: A new edible coatings. *Agrosearch*, *13*(1), 131−141.

Adetunji, C. O. (2019). Environmental impact and ecotoxicological influence of biofabricated and inorganic nanoparticle on soil activity. In D. Panpatte, & Y. Jhala (Eds.), *Nanotechnology for agriculture*. Singapore: Springer. Available from https://doi.org/10.1007/978-981-32-9370-0_12.

Adetunji, C. O., Ajayi, O. O., Akram, M., Olaniyan, O. T., Chishti, M. A., Inobeme, A., Olaniyan, S., Adetunji, J. B., Olaniyan, M., & Awotunde, S. O. (2021). Medicinal plants used in the treatment of influenza A virus infections. In K. Dua, S. Nammi, D. Chang, D. K. Chellappan, G. Gupta, & T. Collet (Eds.), *Medicinal plants for lung diseases*. Singapore: Springer. Available from https://doi.org/10.1007/978-981-33-6850-7_19.

Adetunji, C. O., Akram, M., Olaniyan, O. T., Ajayi, O. O., Inobeme, A., Olaniyan, S., Hameed, L., & Adetunji, J. B. (2021). Targeting SARS-CoV-2 novel Corona (COVID-19) virus infection using medicinal plants. In K. Dua, S. Nammi, D. Chang, D. K. Chellappan, G. Gupta, & T. Collet (Eds.), *Medicinal plants for lung diseases*. Singapore: Springer. Available from https://doi.org/10.1007/978-981-33-6850-7_21.

Adetunji, C. O., Anani, O. A., Olaniyan, O. T., Inobeme, A., Olisaka, F. N., Uwadiae, E. O., & Obayagbona, O. N. (2021). Recent trends in organic farming. In R. Soni, D. C. Suyal, P. Bhargava, & R. Goel (Eds.), *Microbiological activity for soil and plant health management*. Singapore: Springer. Available from https://doi.org/10.1007/978-981-16-2922-8_20.

Adetunji, C. O., Inobeme, A., Olaniyan, O. T., Ajayi, O. O., Olaniyan, S., & Adetunji, J. B. (2021a). Application of nanodrugs derived from active metabolites of medicinal plants for the treatment of inflammatory and lung diseases: Recent advances. In K. Dua, S. Nammi, D. Chang, D. K. Chellappan, G. Gupta, & T. Collet (Eds.), *Medicinal plants for lung diseases*. Singapore: Springer. Available from https://doi.org/10.1007/978-981-33-6850-7_26.

Adetunji, C. O., Inobeme, A., Olaniyan, O. T., Ajayi, O. O., Olaniyan, S., & Adetunji, J. B. (2021b). Application of nanodrugs derived from active metabolites of medicinal plants for the treatment of inflammatory and lung diseases: Recent advances. In K. Dua, S. Nammi, D. Chang, D. K. Chellappan, G. Gupta, & T. Collet (Eds.), *Medicinal plants for lung diseases*. Singapore: Springer. Available from https://doi.org/10.1007/978-981-33-6850-7_26.

Adetunji, C. O., Kumar, D., Raina, M., Arogundade, O., & Sarin, N. B. (2019). Endophytic microorganisms as biological control agents for plant pathogens: A panacea for sustainable agriculture. In A. Varma, S. Tripathi, & R. Prasad (Eds.), *Plant biotic interactions*. Cham: Springer. Available from https://doi.org/10.1007/978-3-030-26657-8_1.

Adetunji, C. O., Michael, O. S., Varma, A., Oloke, J. K., Kadiri, O., Akram, M., Bodunrinde, R. E., Imtiaz, A., Adetunji, J. B., Shahzad, K., Jain, A., Ubi, B. E., Majeed, N., Ozolua, P., & Olisaka, F. N. (2021). Recent advances in the application of biotechnology for improving the production of secondary metabolites from Quinoa. In A. Varma (Ed.), *Biology and biotechnology of Quinoa*. Singapore: Springer. Available from https://doi.org/10.1007/978-981-16-3832-9_17.

Adetunji, C. O., Michael, O. S., Kadiri, O., Varma, A., Akram, M., Oloke, J. K., Shafique, H., Adetunji, J. B., Jain, A., Bodunrinde, R. E., Ozolua, P., & Ubi, B. E. (2021). Quinoa: From farm to traditional healing, food application, and phytopharmacology. In A. Varma (Ed.), *Biology and biotechnology of Quinoa*. Singapore: Springer. Available from https://doi.org/10.1007/978-981-16-3832-9_20.

Adetunji, C. O., Michael, O. S., Nwankwo, W., Ukhurebor, K. E., Anani, O. A., Oloke, J. K., Varma, A., Kadiri, O., Jain, A., & Adetunji, J. B. (2021). Quinoa, the next biotech plant: Food security and environmental and health hot spots. In A. Varma (Ed.), *Biology and biotechnology of Quinoa*. Singapore: Springer. Available from https://doi.org/10.1007/978-981-16-3832-9_19.

Adetunji, C. O., Nwankwo, W., Ukhurebor, K., Olayinka, A. S., & Makinde, A. S. (2021). Application of biosensor for the identification of various pathogens and pests mitigating against the agricultural production: Recent advances. In R. N. Pudake, U. Jain, & C. Kole (Eds.), *Biosensors in agriculture: Recent trends and future perspectives. Concepts and strategies in plant sciences*. Cham: Springer. Available from https://doi.org/10.1007/978-3-030-66165-6_9.

Adetunji, C. O., Ojediran, J. O., Adetunji, J. B., & Owa, S. O. (2019). Influence of chitosan edible coating on postharvest qualities of Capsicum annuum L. during storage in evaporative cooling system. *Croatian journal of Food Science and Technology*, *11*(1), 59−66.

Adetunji, C. O., Olaniyan, O. T., Akram, M., Ajayi, O. O., Inobeme, A., Olaniyan, S., Khan, F. S., & Adetunji, J. B. (2021). Medicinal plants used in the treatment of pulmonary hypertension. In K. Dua, S. Nammi, D. Chang, D. K. Chellappan, G. Gupta, & T. Collet (Eds.), *Medicinal plants for lung diseases*. Singapore: Springer. Available from https://doi.org/10.1007/978-981-33-6850-7_14.

Adetunji, C. O., Oloke, J. K., & Prasad, G. (2020). Effect of carbon-to-nitrogen ratio on eco-friendly mycoherbicide activity from Lasiodiplodia pseudotheobromae C1136 for sustainable weeds management in organic agriculture. *Environment, Development and Sustainability*, *22*, 1977−1990. Available from https://doi.org/10.1007/s10668-018-0273-1.

Adetunji, C. O., Panpatte, D. G., Bello, O. M., & Adekoya, M. A. (2019). Application of nanoengineered metabolites from beneficial and eco-friendly microorganisms as a biological control agents for plant pests and pathogens. In D. Panpatte, & Y. Jhala (Eds.), *Nanotechnology for agriculture: Crop production & protection*. Singapore: Springer. Available from https://doi.org/10.1007/978-981-32-9374-8_13.

Adetunji, C. O., Phazang, P., & Sarin, N. B. (2017). Significance of rhamnolipids as a biological control agent in the management of crops/plant pathogens. *Current Trends in Biomedical Engineering & Biosciences*, *10*(3), 54−55.

Adetunji, C. O., Roli, O. I., & Adetunji, J. B. (2020). Exopolysaccharides derived from beneficial microorganisms: Antimicrobial, food, and health benefits. In P. Mishra, R. R. Mishra, & C. O. Adetunji (Eds.), *Innovations in food technology*. Singapore: Springer. Available from https://doi.org/10.1007/978-981-15-6121-4_10.

Adetunji, C. O., & Varma, A. (2020). Biotechnological application of *Trichoderma*: A powerful fungal isolate with diverse potentials for the attainment of food safety, management of pest and diseases, healthy planet, and sustainable agriculture. In C. Manoharachary, H. B. Singh, & A. Varma (Eds.), *Trichoderma: Agricultural applications and beyond. Soil biology* (Vol. 61). Cham: Springer. Available from https://doi.org/10.1007/978-3-030-54758-5_12.

Adetunji, C. O., Egbuna, C., Tijjani, H., Adom, D., Al-Ani, L. K. T., & Patrick-Iwuanyanwu, K. C. (2020). *Homemade preparations of natural biopesticides and applications. Natural Remedies for Pest, Disease and Weed Control* (pp. 179−185). Publisher Academic Press.

Adetunji, J. B., Adetunji, C. O., & Olaniyan, O. T. (2021). African walnuts: A natural depository of nutritional and bioactive compounds essential for food and nutritional security in Africa. In O. O. Babalola (Ed.), *Food security and safety*. Cham: Springer. Available from https://doi.org/10.1007/978-3-030-50672-8_19.

Bello, O. M., Ibitoye, T., & Adetunji, C. (2019). Assessing antimicrobial agents of Nigeria flora. *Journal of King Saud University-Science*, *31*(4), 1379−1383.

Egbuna, C., Gupta, E., Ezzat, S. M., Jeevanandam, J., Mishra, N., Akram, M., Sudharani, N., Adetunji, C. O., Singh, P., Ifemeje, J. C., Deepak, M., Bhavana, A., Walag, A. M. P., Ansari, R., Adetunji, J. B., Laila, U., Olisah, M. C., & Onyekere, P. F. (2020). Aloe species as valuable sources of functional bioactives. In C. Egbuna, & G. Dable Tupas (Eds.), *Functional foods and nutraceuticals*. Cham: Springer. Available from https://doi.org/10.1007/978-3-030-42319-3_18.

Olaniyan, O. T., & Adetunji, C. O. (2021). Biological, biochemical, and biodiversity of biomolecules from marine-based beneficial microorganisms: Industrial perspective. In C. O. Adetunji, D. G. Panpatte, & Y. K. Jhala (Eds.), *Microbial rejuvenation of polluted environment. Microorganisms for sustainability* (Vol. 27). Singapore: Springer. Available from https://doi.org/10.1007/978-981-15-7459-7_4.

Ooi, L. S., Li, Y., Kam, S. L., Wang, H., Wong, E. Y., & Ooi, V. E. (2006). Antimicrobial activities of cinnamon oil and cinnamaldehyde from the Chinese medicinal herb Cinnamomum cassia Blume. *The American Journal of Chinese Medicine*, *34*, 511−522.

Rauf, A., Akram, M., Semwal, P., Mujawah, A. A. H., Muhammad, N., Riaz, Z., Munir, N., Piotrovsky, D., Vdovina, I., Bouyahya, A., Adetunji, C. O., Shariati, M. A., Almarhoon, Z. M., Mabkhot, Y. N., & Khan, H. (2021). Antispasmodic potential of medicinal plants: A comprehensive review. *Oxidative Medicine and Cellular Longevity*, *2021*, Article ID 4889719, 12 pages. Available from https://doi.org/10.1155/2021/4889719.

Thangadurai, D., Naik, J., Sangeetha, J., Al-Tawaha, A. R. M. S., Adetunji, C. O., Islam, S., David, M., Shettar, A. K., & Adetunji, J. B. (2021). Nanomaterials from agrowastes: Past, present, and the future. In O. V. Kharissova, L. M. Torres-Martínez, & B. I. Kharisov (Eds.), *Handbook of nanomaterials and nanocomposites for energy and environmental applications*. Cham: Springer. Available from https://doi.org/10.1007/978-3-030-36268-3_43.

Ukhurebor, K. E., & Adetunji, C. O. (2021). Relevance of biosensor in climate smart organic agriculture and their role in environmental sustainability: What has been done and what we need to do? In R. N. Pudake, U. Jain, & C. Kole (Eds.), *Biosensors in agriculture: Recent trends and future perspectives. Concepts and strategies in plant sciences*. Cham: Springer. Available from https://doi.org/10.1007/978-3-030-66165-6_7.

Chapter 22

Socioeconomic factors and major setbacks involved in the application of essential oil in food systems: what we need to do

Benjamin Olusola Abere[1] and Charles Oluwaseun Adetunji[2]

[1]*Department of Economics, Edo State University Uzairue, Iyamho, Edo State, Nigeria,* [2]*Applied Microbiology, Biotechnology and Nanotechnology Laboratory, Department of Microbiology, Edo State University Uzairue, Iyamho, Edo State, Nigeria*

Introduction

Secondary metabolites from aromatic plants that are volatile in nature include essential oils (EOs). They frequently result from a single cell's metabolic processes or the coordinated action of a group of specialized cells (Pott et al., 2019). However, monoterpenes, which have the largest hydrocarbon content, make up the majority of these complex formulations of molecules that make up EOs (Ben Salha et al., 2019). Specialized structures next to channel cavities, glandular trichomes, epidermal cells, secretory cells, or some spreading sections are where EOs are created and collected. Important storage components such as mint leaves, vetiver roots, nutmeg seeds, wild bergamot flower, anise fruit, cinnamon bark, ginger rhizome, and sandalwood are frequently linked to EOs (Hanif et al., 2019). On the composition of EOs, environmental strictures such as photoperiod, temperature, growing methods, irradiance, and relative humidity have a major influence. In addition to all of these other aspects, the method used to extract the EO affects not just its makeup but also how likely it is for its elements to occur (Mickiene et al., 2011). EOs are complex combinations of terpenoids and aromatic chemicals that are very volatile and exceedingly complicated in composition. The principal flavor ingredients in drinks are these metabolic by-products (Huang et al., 2009). Since EOs can prevent food deterioration and the growth of harmful microbes, Food processors and researchers are concentrating on using processed food items. Numerous studies in the field have shown that intricate bioactive extraordinary properties of EOs are due to compounds including glycosides, phenols, alkaloids, steroids, tannins, and coumarins (Huang et al., 2009). Thanks to a variety of bioactive substances, such as carvacrol and thymol in oregano, vanillin in vanilla, eugenol in cloves, allyl isothiocyanate in mustard, menthol in peppermint, and allicin in garlic, among others, various herbs and spices and their functional properties are distinctive in their own right (Shaaban et al., 2012).

Researchers now have clear scientific proof, thanks to this data, that these bioactive chemicals can protect against a wide range of concerns with food deterioration and intoxication. What are secondary metabolites? They are viewed as the plants' strategic weapons in battle, allowing them to defend themselves from intruders like bacteria, fungus, insects, and herbivorous animals. Secondary metabolites, such as saponins, anthraquinones, glucose neighboring inolates, glycosides, tannins, polyacetylenes, anthraquinones, cyanogen, and terpenes, function as allelochemicals and negatively impact the physiological condition of competing species (Mambanzulua Ngoma et al., 2015). For the purpose of pollination, these EOs also draw insects, aiding the neighboring plants' ability to reproduce (Satyal et al., 2019). The goal of the present analysis is to examine if using EOs in industrial food processing may replace dangerous chemical additions in a safer way.

Applications of Essential Oils in the Food Industry. DOI: https://doi.org/10.1016/B978-0-323-98340-2.00014-6

Socioeconomic factors and major setbacks facing the application of essential oil in food system

Investigations of societal factors affecting health and socioeconomic variations in health frequently appear. A lower socioeconomic level (SES) is often associated with worse health and an earlier death rate than a higher SES. Socioeconomic health inequities persist even among older people (Fors & Thorslund, 2015; Fors et al., 2007; Geyer et al., 2006; Hoffmann, 2008; Huisman et al., 2003). However, health disparities were seen in all nations across all age categories, according to the findings of a comparative research that looked at eleven European countries' morbidity rates (Huisman et al., 2003). These disparities may have a significant impact on a sizable section of the older population due to the fact that the majority of sickness and age-related deaths do occur, putting more of a financial strain on public expenditure when people get older. SES is generally operationalized in studies of health disparities in later life as either education, social class, or income, and frequently without providing a justification for the indicator's choice (Geyer et al., 2006; Goldthorpe, 2009; Ploubidis et al., 2014).

This study's primary goal was to investigate the relationships between old age health and the three most prevalent SES indicators: education, social status, and income. Occupational complexity was also added as a substitute for SES since, according to recent research (Le Grand & Tåhlin, 2013; Tåhlin, 2011, 2007), complexity is a major factor in the stratification of the labor market. Although education, social standing, the intricacy of one's job, and money have certain commonalities, they may also have different connections to aging health. We thus examined the association between these characteristics, a composite measure of the variables, and changes in daily activity restrictions, psychological distress, and mobility constraints from working ages through old life.

Occupational complexity, income, health, social class, and education

The majority of people reach their educational peak in their early adult years, which aids in bridging socioeconomic generational gaps (Mirowsky & Ross, 2003). According to studies, variables that affect the relationship between education and health include advancements in terms of human capital, mental resources, housing conditions, access to better healthcare, and a healthier way of life, and choosing (both direct and indirect) (Hayward et al., 2015). Low-educated individuals, according to studies, age more quickly and have a faster deterioration in health (Dupre, 2007). However, as compared to other indices like wealth, income, tenure, and deprivation, education typically has a lesser relationship with aging health (Avlund et al., 2003; Duncan et al., 2002; Grundy & Holt, 2001). The most common method for class stratification is the use of profession. According to several class schemas, ownership is the main basis for discrimination in occupations between employers and employees, for example). Then, several employer groupings and workers stand divided according to things like organizational size and kind, necessary skills, power dynamics, and workplaces (Goldthorpe, 2007; Rose & Harrison, 2010). Social class influences current income as well as income security, stability, and increase over the long term (Goldthorpe, 2004; Watson et al., 2009). The correlation between socioeconomic position and health in later life has mixed results. While more studies have discovered (Enroth et al., 2013; Fors et al., 2007) a connection between mortality among older people and socioeconomic status, Duncan et al. suggested that productivity, efficiency, and skill needs may be able to explain social stratification in the job market more effectively than traditional class theories (Le Grand & Tåhlin, 2013; Tåhlin, 2011, 2007). Thlin discovered using data from Sweden that occupational complexity levels and earned income had a stronger relationship than conventional social class schemas (Tåhlin, 2011, 2007).

Further, Thlin makes the case that productivity and efficiency may be gauged by looking at the degree of occupational complexity related to particular employment. In light of this, we have added an additional SES indicator called an occupational complexity measure. Income is frequently seen as an unambiguous sign of financial wealth, and it has a strong and favorable correlation with lifespan (Chetty et al., 2016; Rehnberg & Fritzell, 2016). People's health may be impacted by their wealth since it makes it easier for those at the top of the economic scale to live healthy lives and deprives those at the bottom of these resources (Marmot, 2002). According to the Andersen health behavioral model (Aday & Andersen, 1974; Andersen, 1995), income is a crucial enabling resource for access to and use of healthcare. Conversely, declining health can also have a negative impact on revenue (Muennig, 2008). Therefore, bi-directional causal processes are probably what determine how wealth and health are related. Compared to factors like educational attainment and social status, for example, the financial condition is frequently found to be most significantly related to health and mortality in old age in research (Avlund et al., 2003; Duncan et al., 2002; Grundy & Holt, 2001).

Studies on the health of the elderly may include several SES markers for a variety of reasons. It might be to keep track of or to comprehend how the distribution of social resources affects health. In this instance, the myriad ways, and methods, it is important to take into account how factors like education, social standing, wealth, and occupational complexity may impact health. Without having a special interest in the causes behind the health disparities, another justification would be to simply alter a model to account for as much socioeconomic variance as possible. According to the findings of certain research, this application might be appropriate for a composite measure of individual-level markers (Galobardes et al., 2007). A prior research's findings, which obtained identical data to the current investigation, found that none of the individual variables were as suitable for this purpose as a composite measure of person-level SES characteristics (Darin-Mattsson et al., 2015). Using a composite measurement of SES variables at the individual level, on the other hand, may conceal the underlying mechanisms and halt progress in our knowledge of the various ways that SES affects health.

Due to the fact that learning is correlated with position and difficulty of the occupation, which are correlated with income (Le Grand & Tåhlin, 2013; Mirowsky & Ross, 2005), education is inextricably linked in relation to social class, intricacy of the job, and income. Additionally, socioeconomic disadvantage typically builds up over the course of a lifetime, spanning across and dividing socioeconomic levels (Dannefer, 2003). Evidence suggests that exposure to socioeconomic disadvantages may have a greater and more long-lasting impact on health across the course of a person's life during certain, vulnerable times (Ben-Shlomo & Kuh, 2002).

In various studies, many individuals of working age have demonstrated that various SES variables are differentially linked through various health outcomes, which suggests that various underlying processes may be responsible for these correlations (Geyer et al., 2006; Torssander & Erikson, 2009). These results imply that when analyzing socioeconomic disparities in health, the indicator chosen may have an impact on the conclusions drawn and the way they are interpreted. The relationship between various SES markers and aging health is less well understood. A thorough examination of how the link between SES and health in old age varies by an indicator of SES may lead to a considerable knowledge of the mechanisms generating socioeconomic differences in good health in old age. Because this is when people frequently reach their highest levels of education, social status, and wealth, we choose to study these aspects later in life when working. The most popular SES metrics are based on stratification based on employment and education. In the current study, in order to evaluate late-life health, mobility restrictions, ADL restrictions, and psychological discomfort are employed.

By way of social costs, these disorders place a heavy load (Ratigan et al., 2016). These are all common areas where older people experience health problems. It is uncommon for adults under the age of 40 to have mobility or ADL difficulties, but as people become older, the incidence of these conditions rises. Distress, limits on mobility and ADLs, and the requirement for institutionalization are all well-predicted by these factors (Ahacic et al., 2000; Larsson et al., 2006; Molarius et al., 2009). Furthermore, psychological distress also predicts various health outcomes, particularly relevant to the elderly population, such as dementia, mortality, cardiovascular illnesses, and cerebrovascular disorders.

Major factors affecting essential oil

The external elements of the environment is often connected to exogenous variables like light, precipitation, growth place (altitude, latitude), and composition of the soil (pH, components). Light, precipitation, a growth environment, and soil light is the source of rising monoterpene and phenyl propane concentrations in Ocimum basilicum L (Johnson et al., 1999), as well as the source of rising monoterpene concentrations in Satureja douglasii Benth (Peer & Langenheim, 1998). These findings generally concur with the notion that several secondary pathway enzymes depend on UV-B (Kun et al., 1984). Because biosynthetic activities occur in an aquatic environment, the availability of water might have an impact on secondary metabolism. Monoterpenes are produced more often when there is more water available, as is known (Palà-Paùl et al., 2001; Taveira et al., 2003).

Thymus piperella L.'s EO., on the other hand, has been shown to have more monoterpenes when there is less water available (Boira & Blanquer, 1998). The impact of altitude on the proximal latitudes affects the chemical makeup of *Citrus bergamia* Risso's EO (Satta et al., 1999). Low-latitude *Haptics suaveolens* L. EO had a greater sesquiterpene concentration, while *Satria montana* L. ssp. Montana EO had findings that were equivalent (Azavedo et al., 2001). When viewed in the context of environmental conditions, the nature of the soil, including pH and its structure, can sometimes be held responsible for changes in the chemical makeup of the EOs. The aromatic plant extracts *Helichrysum italicism* ssp. pH, however, had little impact. Microphyll (Slavkovska et al., 2001) has a significant impact on Rosmarinus officialism Officinalis (Manunta, 1985−1986) growing in the same environment and at the same elevation. *Eryngium champetre* L.'s EO's varying chemical composition. was examined by Plalà-Paul and colleagues (2008) when they looked at the compositional variation of the EO. Growing in both alkaline and acidic soil, they discovered that the pH of the soil had a substantial impact on the dispersion of terpenes.

Myrcene levels were markedly higher and -cur cumene levels were markedly lower in the population developing on acidic soil. According to the presence of calcareous and siliceous soils, the chemical composition of EOs may change, as discussed in a few papers. High monoterpene concentration was a distinguishing feature of the Ten Spineless EOs of Thymus (De Feo et al., 2003). Growing the soil is calcareous, whereas sesquiterpenes were a hallmark of plant oils from siliceous soil. Comparing the oils produced by plants cultivated in calcareous soil to those produced by *Cistus monspeliensis* L.'s EOs, it was discovered that the latter had higher quality (Robles & Garzino, 1999). Because of this, it is unclear how different external stimuli affect the chemical makeup of EOs. A number of variables that may simultaneously affect the metabolism are responsible for this impact, which is dependent on the plants.

Variation by season

Numerous factors, such as temperature, radiation, and precipitation, have an impact on seasonal chemical variations. Recent developments in this area primarily concentrate on improving yields and harvesting times in relation to biologically active, harmful, or valuable chemicals. According to a study on the effects period for harvesting *Thymus pulegioides* L., the phenol content of the EO was greater at the start of the blooming stage (Satta et al., 1999). Winter saw a decrease in oxygenated monoterpenes in *Salvia officinalis* L. EO, but the same season saw an increase in monoterpene hydrocarbons (Santos-Gomes & Fernandes-Ferreira, 2001). In this regard, it was noted that linalool and the primary combination of *Artemisia pallens* L. EO. Both sharply decreased from the beginning to the end of the blooming period, whereas certain chemicals increased (Mallavarapu et al., 1999). The EO of Salvia libanotica Boiss is also superior to the spring and has been demonstrated to have more camphene, thujone, and camphor in the winter (Farhat et al., 2001). The metabolic pathways that produce the main compounds in *Lavandula stoechas* L. exhibit a significant seasonal variation, according to recent research. ssp. EO of stoechas (Angioni et al., 2006).

Seasonal changes in temperate climates were studied, however seasonal variations in tropical climates should also be included. *Guarea macrophylla* Vahl, a tropical plant, is described in a publication (Lago et al., 2006). ssp. similar vegetative stage (flowering time), in both dry (October) and rainy (November) seasons, tuberculoma's EO composition was practically the same (June). Thus, rather than just one component, it appears that a number of variables are incorporated into the Eos' chemical composition. If the metabolic pathways were altered as a result of various circumstances, their effects may be explained. In the same plant species, external causes may produce ecotypes and chemotypes.

Conclusion

Based on the plant's economic performance as described in this paper, it can be inferred that the commercial production of EO, which is highly profitable and ought to be appealing to any potential investor, may not be feasible if its various challenges—discussed in this chapter—are not effectively addressed. Aside from the fact that the home market is woefully underserved, the product has excellent export potential, especially in Europe, Asia, and the West Africa Sub-Saharan area, with the goal of bringing in foreign revenue for our nation. By offering work chances to unemployed young people, this would aid in reducing poverty (Adejumo et al., 2017; Adetunji, 2008, 2019; Adetunji & Varma, 2020; Adetunji et al., 2013a, 2013b, 2017, 2019, 2020a, 2020b, 2020c, 2021a, 2021b, 2021c, 2021d, 2021e, 2021f, 2021g, 2021h, 2021i, 2021j, 2021k, 2021l, 2022; Bello et al., 2019; Egbuna et al., 2020; Olaniyan & Adetunji, 2021; Rauf et al., 2021; Thangadurai et al., 2021; Ukhurebor & Adetunji, 2021).

References

Aday, L. A., & Andersen, R. (1974). A framework for the study of access to medical care. *Health Services Research, 9*, 208.

Adejumo, I. O., Adetunji, C. O., & Adeyemi, O. S. (2017). Influence of UV light exposure on mineral composition and biomass production of myco-meat produced from different agricultural substrates. *Journal of Agricultural Sciences, 62*(1), 51−59, Belgrade.

Adetunji, C. O. (2008). *The antibacterial activities and preliminary phytochemical screening of vernoniaamygdalina and Aloe vera against some selected bacteria* [M. Sc. thesis]. University of Ilorin, pp. 40−43.

Adetunji, C. O., Ajayi, O. O., Akram, M., Olaniyan, O. T., Chishti, M. A., Inobeme, A., Olaniyan, S., Adetunji, J., Olaniyan, M., & Awotunde, S. O. (2021l). Medicinal plants used in the treatment of influenza A virus infections. In K. Dua, S. Nammi, D. Chang, D. K. Chellappan, G. Gupta, & T. Collet (Eds.), *Medicinal plants for lung diseases*. Singapore: Springer. Available from https://doi.org/10.1007/978-981-33-6850-7_19.

Adetunji, C. O., Akram, M., Olaniyan, O. T., Ajayi, O. O., Inobeme, A., Olaniyan, S., Hameed, L., & Adetunji, J. B. (2021a). Targeting SARS-CoV-2 novel Corona (COVID-19) virus infection using medicinal plants. In K. Dua, S. Nammi, D. Chang, D. K. Chellappan, G. Gupta, & T. Collet (Eds.), *Medicinal plants for lung diseases*. Singapore: Springer. Available from https://doi.org/10.1007/978-981-33-6850-7_21.

Adetunji, C. O., Anani, O. A., Olaniyan, O. T., Inobeme, A., Olisaka, F. N., Uwadiae, E. O., & Obayagbona, O. N. (2021f). Recent trends in organic farming. In R. Soni, D. C. Suyal, P. Bhargava, & R. Goel (Eds.), *Microbiological activity for soil and plant health management*. Singapore: Springer. Available from https://doi.org/10.1007/978-981-16-2922-8_20.

Adetunji, C. O., Egbuna, C., Tijjani, H., Adom, D., Al-Ani, L. K. T., & Patrick-Iwuanyanwu, K. C. (2020b). *Homemade preparations of natural biopesticides and applications. Natural Remedies for Pest, Disease and Weed Control* (pp. 179–185). Publisher Academic Press.

Adetunji, C. O., Inobeme, A., Olaniyan, O. T., Ajayi, O. O., Olaniyan, S., & Adetunji, J. B. (2021d). Application of nanodrugs derived from active metabolites of medicinal plants for the treatment of inflammatory and lung diseases: Recent advances. In K. Dua, S. Nammi, D. Chang, D. K. Chellappan, G. Gupta, & T. Collet (Eds.), *Medicinal plants for lung diseases*. Singapore: Springer. Available from https://doi.org/10.1007/978-981-33-6850-7_26.

Adetunji, C. O., Inobeme, A., Olaniyan, O. T., Ajayi, O. O., Olaniyan, S., & Adetunji, J. B. (2021e). Application of nanodrugs derived from active metabolites of medicinal plants for the treatment of inflammatory and lung diseases: Recent advances. In K. Dua, S. Nammi, D. Chang, D. K. Chellappan, G. Gupta, & T. Collet (Eds.), *Medicinal plants for lung diseases*. Singapore: Springer. Available from https://doi.org/10.1007/978-981-33-6850-7_26.

Adetunji, C. O., Kayode, A., Bolajoko, F. O., & Bunmi, A. J. (2013b). Effects of coatings on storability of carrot under evaporative coolant system. *Albanian Journal of Agricultural Sciences*, *12*(3).

Adetunji, C. O., Kumar, D., Raina, M., Arogundade, O., & Sarin, N. B. (2019). Endophytic microorganisms as biological control agents for plant pathogens: A panacea for sustainable agriculture. In A. Varma, S. Tripathi, & R. Prasad (Eds.), *Plant biotic interactions*. Cham: Springer. Available from https://doi.org/10.1007/978-3-030-26657-8_1.

Adetunji, C. O., Michael, O. S., Kadiri, O., Varma, A., Akram, M., Oloke, J. K., Shafique, H., Adetunji, J. B., Jain, A., Bodunrinde, R. E., Ozolua, P., & Ubi, B. E. (2021g). Quinoa: From farm to traditional healing, food application, and phytopharmacology. In A. Varma (Ed.), *Biology and biotechnology of Quinoa*. Singapore: Springer. Available from https://doi.org/10.1007/978-981-16-3832-9_20.

Adetunji, C. O., Michael, O. S., Nwankwo, W., Ukhurebor, K. E., Anani, O. A., Oloke, J. K., Varma, A., Kadiri, O., Jain, A., & Adetunji, J. B. (2021h). Quinoa, The next biotech plant: Food security and environmental and health hot spots. In A. Varma (Ed.), *Biology and biotechnology of Quinoa*. Singapore: Springer. Available from https://doi.org/10.1007/978-981-16-3832-9_19.

Adetunji, C. O., Michael, O. S., Rathee, S., Singh, K. R. B., Ajayi, O. O., Adetunji, J. B., Ojha, A., Singh, J., & Singh, R. P. (2022). Potentialities of nanomaterials for the management and treatment of metabolic syndrome: A new insight. *Materials Today Advances*, *13*, 100198.

Adetunji, C. O., Michael, O. S., Varma, A., Oloke, J. K., Kadiri, O., Akram, M., Bodunrinde, R. E., Imtiaz, A., Adetunji, J. B., Shahzad, K., Jain, A., Ubi, B. E., Majeed, N., Ozolua, P., & Olisaka, F. N. (2021i). Recent advances in the application of biotechnology for improving the production of secondary metabolites from Quinoa. In A. Varma (Ed.), *Biology and biotechnology of Quinoa*. Singapore: Springer. Available from https://doi.org/10.1007/978-981-16-3832-9_17.

Adetunji, C. O., Nwankwo, W., Ukhurebor, K., Olayinka, A. S., & Makinde, A. S. (2021b). Application of biosensor for the identification of various pathogens and pests mitigating against the agricultural production: Recent advances. In R. N. Pudake, U. Jain, & C. Kole (Eds.), *Biosensors in agriculture: Recent trends and future perspectives. Concepts and strategies in plant sciences*. Cham: Springer. Available from https://doi.org/10.1007/978-3-030. Available from 66165-6_9.

Adetunji, C. O., Olaniyan, O. T., Akram, M., Ajayi, O. O., Inobeme, A., Olaniyan, S., Khan, F. S., & Adetunji, J. B. (2021k). Medicinal plants used in the treatment of pulmonary hypertension. In K. Dua, S. Nammi, D. Chang, D. K. Chellappan, G. Gupta, & T. Collet (Eds.), *Medicinal plants for lung diseases*. Singapore: Springer. Available from https://doi.org/10.1007/978-981-33-6850-7_14.

Adetunji, C. O., Oloke, J. K., & Prasad, G. (2020a). Effect of carbon-to-nitrogen ratio on eco-friendly mycoherbicide activity from Lasiodiplodia pseudotheobromae C1136 for sustainable weeds management in organic agriculture. *Environment, Development and Sustainability*, *22*, 1977–1990. Available from https://doi.org/10.1007/s10668-018-0273-1.

Adetunji, C. O., Palai, S., Ekwuabu, C. P., Egbuna, C., Adetunji, J. B., Ehis-Eriakha, C. B., Kesh, S. S., & Mtewa, A. G. (2021c). *General principle of primary and secondary plant metabolites: Biogenesis, metabolism, and extraction. Preparation of phytopharmaceuticals for the management of disorders* (pp. 3–23). Publisher Academic Press.

Adetunji, C. O. (2019). Environmental impact and ecotoxicological influence of biofabricated and inorganic nanoparticle on soil activity. In D. Panpatte, & Y. Jhala (Eds.), *Nanotechnology for agriculture*. Singapore: Springer. Available from https://doi.org/10.1007/978-981-32-9370-0_12.

Adetunji, C. O., Phazang, P., & Sarin, N. B. (2017). Significance of rhamnolipids as a biological control agent in the management of crops/plant pathogens. *Current Trends in Biomedical Engineering & Biosciences*, *10*(3), 54–55.

Adetunji, C. O., Roli, O. I., & Adetunji, J. B. (2020c). Exopolysaccharides derived from beneficial microorganisms: Antimicrobial, food, and health benefits. In P. Mishra, R. R. Mishra, & C. O. Adetunji (Eds.), *Innovations in food technology*. Singapore: Springer. Available from https://doi.org/10.1007/978-981-15-6121-4_10.

Adetunji, C. O., & Varma, A. (2020). Biotechnological application of trichoderma: A powerful fungal isolate with diverse potentials for the attainment of food safety, management of pest and diseases, healthy planet, and sustainable agriculture. In C. Manoharachary, H. B. Singh, & A. Varma (Eds.), *Trichoderma: Agricultural applications and beyond. Soil biology* (vol 61). Cham: Springer. Available from https://doi.org/10.1007/978-3-030-54758-5_12.

Adetunji, J. B., Adetunji, C. O., & Olaniyan, O. T. (2021j). African walnuts: A natural depository of nutritional and bioactive compounds essential for food and nutritional security in Africa. In O. O. Babalola (Ed.), *Food security and safety*. Cham: Springer. Available from https://doi.org/10.1007/978-3-030-50672-8_19.

Adetunji, J. B., Ajani, A. O., Adetunji, C. O., Fawole, O. B., Arowora, K. A., Nwaubani, S. I., Ajayi, E. S., Oloke, J. K., & Aina, J. A. (2013a). Postharvest quality and safety maintenance of the physical properties of *Daucus carota* L. fruits by Neem oil and Moringa oil treatment: A new edible coatings. *Agrosearch, 13*(1), 131−141.

Ahacic, K., Parker, M., & Thorslund, M. (2000). Mobility limitations in the Swedish population from 1968 to 1992: Age, gender and social class differences. *Aging Milan Italy, 12*, 190−198.

Andersen, R. M. (1995). Revisiting the behavioral model and access to medical care: Does it matter? *Journal of Health and Social Behavior, 36*, 1−10.

Angioni, A., Barra, A., Coroneo, V., Dessi, S., & Cabras, P. (2006). Seasonal, plant part chemical variability and antifungal activity investigation of *Lavandula stoechas* L ssp stoechas essential oils. *Journal of Agricultural and Food Chemistry, 54*, 4364−4370.

Avlund, K., Holstein, B. E., Osler, M., Damsgaard, M. T., Holm-Pedersen, P., & Rasmussen, N. K. (2003). Social position and health in old age: The relevance of different indicators of social position. *Scandinavian Journal of Social Medicine, 31*, 126−136.

Azavedo, N. R., Campos, I. F. P., Ferreira, H. D., Portes, T. A., Santos, S. C., Seraphin, J. C., Paula, J. R., & Ferri, P. H. (2001). Chemical variability in the essential oil of *Hyptis suaveolens*. *Phytochemistry, 57*, 733−736.

Bello, O. M., Ibitoye, T., & Adetunji, C. (2019). Assessing antimicrobial agents of *Nigeria flora*. *Journal of King Saud University-Science, 31*(4), 1379−1383.

Ben Salha, G., Herrera Díaz, R., Lengliz, O., Abderrabba, M., & Labidi, J. (2019). Effect of the chemical composition of free-terpene hydrocarbons essential oils on antifungal activity. *Molecules (Basel, Switzerland), 24*(19), 3532. Available from https://doi.org/10.3390/molecules24193532.

Ben-Shlomo, Y., & Kuh, D. (2002). A life course approach to chronic disease epidemiology: Conceptual models, empirical challenges and interdisciplinary perspectives. *International Journal of Epidemiology, 31*(2).

Boira, H., & Blanquer, A. (1998). Environmental factors affecting chemical variability of essential oils in *Thymus piperella* L. *Biochemical Systematics and Ecology, 26*, 811−822.

Chetty, R., Stepner, M., Abraham, S., Lin, S., Scuderi, B., Turner, N., et al. (2016). The association between income and life expectancy in the United States, 2001- 2014. *JAMA: The Journal of the American Medical Association, 315*, 1750−1766.

Dannefer, D. (2003). Cumulative advantage/disadvantage and the life course: Crossfertilizing age and social science theory. *The Journals of Gerontology. Series B, Psychological Sciences and Social Sciences, 58*, S327−S337.

Darin-Mattsson, A., Andel, R., Fors, S., & Kåreholt, I. (2015). Are occupational complexity and socioeconomic position related to psychological distress 20 years later? *Journal of Aging and Health, 27*, 1266−1285.

Duncan, G. J., Daly, M. C., McDonough, P., & Williams, D. R. (2002). Optimal indicators of socioeconomic status for health research. *American Journal of Public Health, 92*, 1151−1157.

Dupre, M. E. (2007). Educational differences in age-related patterns of disease: Reconsidering the cumulative disadvantage and age-as-leveler hypotheses. *Journal of Health and Social Behavior, 48*, 1−15.

Egbuna, C., Gupta, E., Ezzat, S. M., Jeevanandam, J., Mishra, N., Akram, M., Sudharani, N., Adetunji, C. O., Singh, P., Ifemeje, J. C., Deepak, M., Bhavana, A., Mark, A., Walag, P., Ansari, R., Adetunji, J. B., Laila, U., Olisah, M. C., & Onyekere, P. F. (2020). Aloe species as valuable sources of functional bioactives. In C. Egbuna, & G. Dable Tupas (Eds.), *Functional foods and nutraceuticals*. Cham: Springer. Available from https://doi.org/10.1007/978-3-030-42319-3_18.

Enroth, L., Raitanen, J., Hervonen, A., & Jylhä, M. (2013). Do socioeconomic health differences persist in nonagenarians? *The Journals of Gerontology. Series B, Psychological Sciences and Social Sciences, 68*, 837−847.

Farhat, G. N., Affara, N. I., & Gali-Muhtasib, H. U. (2001). Seasonal changes in the composition of the essential oil extract of East Mediterranean sage (*Salvia libanotica*) and its toxicity in mice. *Toxicon, 39*, 1601−1605.

De Feo, V., Bruno, M., Tahiri, B., Napolitano, F., & Senatore, F. (2003). Chemical composition and antibacterial activity of essential oils from *Thymus spinulosus* Ten (Laminaceae). *Journal of Agricultural and Food Chemistry, 51*, 3849−3853.

Fors, S., Lennartsson, C., & Lundberg, O. (2007). Health inequalities among older adults in Sweden 1991−2002. *European Journal of Public Health, 18*, 138−143.

Fors, S., & Thorslund, M. (2015). Enduring inequality: Educational disparities in health among the oldest old in Sweden 1992−2011. *International Journal of Public Health, 60*, 91.

Galobardes, B., Lynch, J., & Smith, G. D. (2007). Measuring socioeconomic position in health research. *British Medical Bulletin, 1*, 21−37.

Geyer, S., Hemström, Ö., Peter, R., & Vågerö, D. (2006). Education, income, and occupational class cannot be used interchangeably in social epidemiology. Empirical evidence against a common practice. *Journal of Epidemiology Community Dental Health, 60*, 804−810.

Goldthorpe, J. H. (2004). *The economic basis of social class*.

Goldthorpe, J. H. (2007). *On sociology. 2. Illustration and retrospect*. Standford: Stanford University Press.

Goldthorpe, J. H. (2009). Analysing social inequality: A critique of two recent contributions from economics and epidemiology. *European Sociological Review, 26*, 731−744.

Le Grand, C., & Tåhlin, M. (2013). Class, occupation, wages, and skills: The iron law of labor market inequality. *Cl. Stratif. Anal. 30*, 3−46, Emerald Group Publishing Limited.

Grundy, E., & Holt, G. (2001). The socioeconomic status of older adults: How should we measure it in studies of health inequalities? *Journal of Epidemiology and Community Health, 55*, 895−904.

Hanif, M. A., Nisar, S., Khan, G. S., Mushtaq, Z., & Zubair, M. (2019). *Essential oils. Essential oil research malik. Sonia* (pp. 3−17). Cham: Springer.

Hayward, M. D., Hummer, R. A., & Sasson, I. (2015). Trends and group differences in the association between educational attainment and US adult mortality: Implications for understanding education's causal influence. *Social Science & Medicine*, *127*, 8–18.

Hoffmann, R. (2008). *Socioeconomic differences in old age mortality*. Springer Science & Business Media.

Huang, W. Y., Cai, Y. Z., & Zhang, Y. (2009). Natural phenolic compounds from medicinal herbs and dietary plants: Potential use for cancer prevention. *Nutrition and Cancer*, *62*(1), 1–20. Available from https://doi.org/10.1080/01635580903191585.

Huisman, M., Kunst, A. E., & Mackenbach, J. P. (2003). Socioeconomic inequalities in morbidity among the elderly; a European overview. *Social Science & Medicine*, *57*, 861–873.

Johnson, C. B., Kirby, J., Naxakis, G., & Pearson, S. (1999). Substantial UV-mediated induction of essential oils in sweet basil (*Ocimum basilicum* L.). *Phytochemistry*, *51*, 507–510.

Kun, D. N., Chappell, J., Boudet, A., & Hahlbrock, K. (1984). Induction of phenylalanine ammonia-lyase and 4-coumarate:CoA ligase mRNAs in cultured plant cells by UV light or fungal elicitor. *Proceedings of the National Academy of Sciences*, *81*, 1102–1106, USA.

Lago, J. H. G., Soares, M. G., Batista-Pereira, L. G., Silva, M. F. G. F., Correa, A. G., Fernandes, J. B., Vieira, P. C., & Roque, N. F. (2006). Volatile oil from *Guarea macrophylla* ssp. tubercolata: Seasonal variation and electroantennographic detection by *Hypsipyla grandella*. *Phytochemistry*, *67*, 589–594.

Larsson, K., Thorslund, M., & Kåreholt, I. (2006). Are public care and services for older people targeted according to need? Applying the behavioural model on longitudinal data of a Swedish urban older population. *European Journal of Ageing*, *3*, 22–33.

Mallavarapu, G. R., Kulkarni, R. N., Baskaran, K., Rao, L., & Ramesh, S. (1999).) Influence of plant growth stage on the essential oil content and composition in Davana (*Artemisia pallens* Wall.). *Journal of Agricultural and Food Chemistry*, *47*, 254–258.

Mambanzulua Ngoma, P., Hiligsmann, S., Sumbu Zola, E., Ongena, M., & Thonart, P. (2015). Impact of different plant secondary metabolites addition: Saponin, tannic acid, salicin and aloin on glucose anaerobic co-digestion. *Fermentation Technology*, *4*(1), 1–11.

Manunta, A. (1985–1986). Influenza del pH del substrato sulla composizione dell'olio essenziale di *Rosmarinus officinalis* L. *Studi Sassaresi*, *32*, 111–118.

Marmot, M. (2002). The influence of income on health: Views of an epidemiologist. *Health Affairs (Millwood)*, *21*, 31–46.

Mickiene, R., Bakutis, B., & Baliukoniene, V. (2011). Antimicrobial activity of two essential oils. *Annuals of Agricultural and Environmental Medicine*, *18*(1), 139–144.

Mirowsky, J., & Ross, C. E. (2003). *Education, social status, and health*. New York: Transaction Publishers.

Mirowsky, J., & Ross, C. E. (2005). Education, cumulative advantage, and health. *Ageing International*, *30*, 27.

Molarius, A., Berglund, K., Eriksson, C., Eriksson, H. G., Lindén-Boström, M., Nordström, E., et al. (2009). Mental health symptoms in relation to socio-economic conditions and lifestyle factors—a population-based study in Sweden. *BMC Public Health*, *9*, 302.

Muennig, P. (2008). Health selection vs. causation in the income gradient: What can we learn from graphical trends? *Journal of Health Care for the Poor and Underserved*, *19*, 574–579.

Olaniyan, O. T., & Adetunji, C. O. (2021). Biological, biochemical, and biodiversity of biomolecules from marine-based beneficial microorganisms: Industrial perspective. In C. O. Adetunji, D. G. Panpatte, & Y. K. Jhala (Eds.), *Microbial rejuvenation of polluted environment. Microorganisms for sustainability* (vol 27). Singapore: Springer. Available from https://doi.org/10.1007/978-981-15-7459-7_4.

Palà-Paul, J., Usano-Alemany, J., Soria, A. C., Pèrez-Alonso, M., & Brophy, J. J. (2008). Essential oil composition of *Erygium campestre* L. growing in different soil types. A preliminary study. *Natural Product Communications*, *3*, 1121–1126.

Palà-Paùl, J., Perèz-Alonso, M. J., Velasco-Neguerela, A., Palà-Paùl, R., Sanz, J., & Conejero, F. C. O. (2001). Seasonal variation in chemical constituents of *Santolina rosmarinifolia* L. ssp. rosmarinifolia. *Biochemical Systematics and Ecology*, *29*, 663–672.

Peer, W. A., & Langenheim, J. H. (1998). Influence of phytochrome on leaf monoterpene variation in Satureja douglasii. *Biochemical Systematics and Ecology*, *26*, 25–34.

Ploubidis, G. B., Benova, L., Grundy, E., Laydon, D., & DeStavola, B. (2014). Lifelong socio economic position and biomarkers of later life health: Testing the contribution of competing hypotheses. *Social Science & Medicine*, *119*, 258–265.

Pott, D. M., Osorio, S., & Vallarino, J. G. (2019). From central to specialized metabolism: An overview of some secondary compounds derived from the primary metabolism for their role in conferring nutritional and organoleptic characteristics to fruit. *Frontiers in Plant Science*, *10*. Available from https://doi.org/10.3389/fpls.2019.00835.

Ratigan, A., Kritz-Silverstein, D., & Barrett-Connor, E. (2016). Sex differences in the association of physical function and cognitive function with life satisfaction in older age: The Rancho Bernardo Study. *Maturitas*, *89*, 29–35.

Rauf, A., Akram, M., Semwal, P., Mujawah, A. A. H., Muhammad, N., Riaz, Z., Munir, N., Piotrovsky, D., Vdovina, I., Bouyahya, A., Adetunji, C. O., Shariati, M. A., Almarhoon, Z. M., Mabkhot, Y. N., & Khan, H. (2021). Antispasmodic potential of medicinal plants: A comprehensive review. *Oxidative Medicine and Cellular Longevity*, Article ID 4889719, 12 pages, https://doi.org/10.1155/2021/4889719.

Rehnberg, J., & Fritzell, J. (2016). The shape of the association between income and mortality in old age: A longitudinal Swedish national register study. *SSM Population Health*, *2*, 750–756.

Robles, C., & Garzino, S. (1999). Infraspecific variability in the essential oil composition of Cistus monspeliensis leaves. *Phytochemistry*, *53*, 71–75.

Rose, D., & Harrison, E. (2010). *Social class in Europe: An introduction to the European socio-economic classification*. NewYork: Routledge.

Santos-Gomes, P. C., & Fernandes-Ferreira, M. (2001). Organ and season-dependent variation in the essential oil composition of Salvia officinalis cultivated at two different sites. *Journal of Agricultural and Food Chemistry*, *49*, 2908–2916.

Satta, M., Tuberoso, C. I. G., Angioni, A., Pirisi, F. M., & Cabras, P. (1999). Analysis of the essential oil of Helichrysum italicum G.Don. ssp. microphyllum (willd.) Nym. *Journal of Essential Oil Research*, *11*, 711–715.

Satyal, P., Hieu, H. V., Chuong, N. T. H., Hung, N. H., Tai, T. A., Hien, V. T., & Setzer, W. N. (2019). Chemical composition, Aedes mosquito larvicidal activity, and repellent activity against *Triatoma rubrofasciata* of *Severinia monophylla* leaf essential oil. *Parasitology Research*, *118*(3), 733–742. Available from https://doi.org/10.1007/s00436-019-06212-1.

Shaaban, H. A., El-Ghorab, A. H., & Shibamoto, T. (2012). Bioactivity of essential oils and their volatile aroma components. *Journal of Essential Oil Research*, *24*(2), 203–212. Available from https://doi.org/10.1080/10412905.2012.659528.

Slavkovska, V., Jancic, R., Bojovic, S., Milosavljevic, S., & Djokovic, D. (2001). Variability of essential oils of *Satureja montana* L and *Satureja kitaibelii* Wierzb. ex Heuff. from the central part of the Balkan peninsula. *Phytochemistry*, *57*, 71–76.

Taveira, F. S. N., de Lima, W. N., Andrade, E. H. A., & Maia, J. G. S. (2003). Seasonal essential oil variation of *Aniba canelilla*. *Biochemical Systematics and Ecology*, *31*, 69–75.

Thangadurai, D., Naik, J., Sangeetha, J., Al-Tawaha, A. R. M. S., Adetunji, C. O., Islam, S., David, M., Shettar, A. K., & Adetunji, J. B. (2021). Nanomaterials from agrowastes: Past, present, and the future. In O. V. Kharissova, L. M. Torres-Martínez, & B. I. Kharisov (Eds.), *Handbook of nanomaterials and nanocomposites for energy and environmental applications*. Cham: Springer. Available from https://doi.org/10.1007/978-3-030-36268-3_43.

Torssander, J., & Erikson, R. (2009). Stratification and mortality—A comparison of education, class, status, and income. *European Sociological Review*, *26*, 465–474.

Tåhlin, M. (2007). Class clues. *European Sociological Review*, *23*, 557–572.

Tåhlin, M. (2011). Vertical differentiation of work tasks: Conceptual and measurement issues. *Empirical Research in Vocational Education and Training*, *3*, 55–70.

Ukhurebor, K. E., & Adetunji, C. O. (2021). Relevance of biosensor in climate smart organic agriculture and their role in environmental sustainability: What has been done and what we need to do? In R. N. Pudake, U. Jain, & C. Kole (Eds.), *Biosensors in agriculture: Recent trends and future perspectives. Concepts and strategies in plant sciences*. Cham: Springer. Available from https://doi.org/10.1007/978-3-030-66165-6_7.

Watson, D., Whelan, C. T., & Maître, B. (2009). *Class and poverty: Cross-sectional and dynamic analysis of income poverty and life-style deprivation. Social class in Europe: An introduction to the European socio-economic classification*. London: Routledge.

Chapter 23

Circular, bioeconomy, and gross domestic product perspective of essential oils in the food industry

Benjamin Olusola Abere[1] and Charles Oluwaseun Adetunji[2]

[1]Department of Economics, Edo State University Uzairue, Iyamho, Edo State, Nigeria, [2]Applied Microbiology, Biotechnology and Nanotechnology Laboratory, Department of Microbiology, Edo State University Uzairue, Iyamho, Edo State, Nigeria

Introduction

The linear value chains that underpin our current economic systems depend on continual, increased exploitation of raw resources and disdain for them after usage. To close our resource loops, just 8.65 billion tonnes of raw materials, or 8.6% of all materials harvested, are now recycled back into the system. The remaining basic minerals totaling 92 billion tonnes that are needed to power our economy are taken out of the ground, processed, put to use, and then thrown away with no chance of material recovery (Galadima, 2019). Pollution of our environment by non-biodegradable items and materials is one effect of poor recycling rates. Due to ongoing population expansion, growing wealth and purchasing power, as well as technological improvements, it is anticipated that the overexploitation of natural resources would get worse. Despite the rise in environmental awareness, the levels of circularity are declining globally, falling by an average of 0.25% points over the last two years.

Our linear systems have catastrophic effects on environmental services, climate change, and biodiversity loss. According to estimates, between 1000 and 10,000 times greater than the rate of extinction it would naturally be (Daw et al., 2018). The mining and processing of materials results in an overall 62% increase in greenhouse gas emissions, and this tendency is rising (Sridhar et al., 2021). By 2050, the 1.5°C goal of the Paris Agreement must be met by achieving net zero emissions. We must move quickly away from the conventional "take-make-waste" economic paradigm and toward one that is regenerative by nature in order to cut carbon emissions and enable sustainable production cycles. To establish a system that supports a circular economy, it is necessary to preserve as much value as possible from resources, goods, components, and materials. This will enable systems with extended lifespans, optimal reuse, refurbishment, remanufacturing, and recycling/recovery.

A natural, 100% pure EO has been extracted through distillation or another process from a variety of sources, such as aromatic substances produced in plant parts such as the organs, flowers, buds, leaves, peel, bark, roots, seeds, or roots of various animals that exude fragrant compounds. Essential oils (EOs) are used in aromatherapy, a natural remedy for mental stress and poor physical conditions, in addition to their use as cosmetic smells for perfume products or toiletries and as culinary flavorings to be added to beverages, confections, and other processed meals. Since aromatherapy has become more and more popular in recent years, there has been an increase in the usage of EOs for aromatherapy. EOs have long been used for scenting purposes, such as food flavorings and cosmetic scents.

Secondary metabolites used by plants to attract pollinators and provide cover from predators are proven to be a suitable sort of secondary metabolites. Plant extracts and oils have been utilized for a huge range of uses for many thousands of years. In the 16th century, Paracelsus von Hohenheim used the term "essential oil," Quinta, to characterize it for the first time (Garry & Chalchat, 2019). EOs are blends of volatile substances created by the secondary metabolism of aromatic and other plant varieties (Mustapha et al., 2020). An EO is a fragrant blend of chemicals that is created by isolating volatile metabolites from plant material using steam- or hydro-distillation techniques (Daw et al., 2018). Hydrocarbons and volatile terpenes make up the majority of the components found in EOs (Tongnuanchan & Benjakul 2019). EOs have a long history of usage as therapeutics (Peter & Timmerhaus, 2019). The antibacterial qualities of

Applications of Essential Oils in the Food Industry. DOI: https://doi.org/10.1016/B978-0-323-98340-2.00015-8

extracts and EOs produced from medicinal plants, according to EQ de Lima et al., have been empirically acknowledged for millennia, but only lately have they been scientifically verified (Rawat, 2020). According to Almeidia's 2010 addition, numerous studies on the biological activity of medicinal plants from various parts of the world, inspired by the widespread use of local species, have demonstrated that the extracts and EOs of these plants are effective at suppressing the growth of a wide range of microorganisms, including fungi, yeasts, and bacteria (Burt, 2021).

The same goes for EOs, which have analgesic, anti-inflammatory, antiprotozoal, anticarcinogenic, medicinal, inhibitor, gastro-protecting, and acetylcholinesterase characteristics. The latter trait is very important for managing Alzheimer's disease, a neurodegenerative illness that typically affects the elderly and is responsible for 50% to 60% of dementia cases in those over 65 (Tajkarimi et al., 2019). As a result of the food industry's recent success in using EOs to control foodborne bacteria, plants' extracts are now receiving a lot of interest as potential alternatives to traditional antimicrobials. Methods of EO extraction and isolation EOs have been separated in a variety of ways, all of which improve their bioactive and therapeutic properties. Among others, the most efficient techniques used in these extraction procedures are freeze drying, rotary evaporation, steam distillation, hydrolysis, and GC chromatography tests. According to Karen et al., using GC to extract the EOs from the leaves of perennial (*Arachis glabrata* Benth.) and edible (*Arachis hypogaea* L.) peanut plants is successful (Solórzano-Santos & Miranda-Novales, 2022). The process of distilling substances in such a way that steam is injected into the raw material is known as steam distillation, and it is used to isolate substances that breakdown at high temperatures. The amount of EOs extracted by steam distillation is 93%, with other techniques being able to obtain the remaining 7% (Willem & Tjakko, 2019). Plant materials are submerged entirely in water, then boiled in order to hydrolyze them. Because the water in the area works as a barrier to keep it from heating up, this approach somewhat protects the extracted oils from damage (Bello et al., 2019).

Due to the rising popularity of aromatherapy, the number of imports of EOs began to rise yearly in the first half of the 1990s, but it peaked in 2004, and since then, import volumes have begun to stabilize. With 8801 tons imported in 2010, which was more than 70% of all imports, oil of orange has a disproportionately large percentage of the overall volume of imports. In contrast, EOs other than those from citrus fruit, omitting oils from mints, make up a significant portion of the market in terms of money, accounting for 4061 million yen, or around 30% of the total. Oil of orange has a high quantity of cheap items imported from price-competitive nations like Brazil, which has led to a relatively low volume of imports on a monetary basis among other things, although having a dominant volume of imports on a numerical basis. Other than citrus fruit EOs and without mint EOs, the unit prices of these oils, while being imported in lesser quantities than oil of orange and other products, have increased sharply as a result of increased product branding, particularly among French and British brands. Many different EOs aren't from citrus fruit; instead, they can be extracted from citronella, eucalyptus, rosemary, ylang-ylang, sandalwood, etc. The significant category consolidation in the 1999 international trade figures prevented the availability of comprehensive breakdowns. Due to a large variety of plant materials being used as the raw materials for EOs, import trends for specific items are also unclear because of the significant price variations even among the same item depending on the extraction site, method, plant growing environments, country of origin, and product grade.

EOs are mostly imported from Brazil and the United States for orange, lemon, and other citrus fruit oils; France and the United States for oils other than those of citrus fruit, such as oils of mint; and India and China for other EOs. While increasing steadily in recent years, particularly from South Africa (oil of orange, etc.) and Morocco (oil of roses, etc.), imports from Africa remain at modest levels in comparison to other continents, with the 2010 results at 0.6% of the total on a numerical basis and 1.1% on a monetary basis (Mustapha et al., 2020).

Japan grows few useful plants for making EOs, hence the majority of its supply comes from imports. Domestic EOs contain a very small number of species, including cypress, fir, and lavender, and the amount of production is quite small, even though volumes of domestically produced EOs are unclear due to a lack of precise information. Many Japanese manufacturers buy EOs in large quantities from foreign vendors and then pack or bottle them there to market as local goods under their own trademarks. When it comes to room deodorizers, items may frequently be passed off as local by importing certain raw components from abroad and blending them with others made domestically.

Up until about 2004, imports of EOs saw a surge in value due to the rising popularity of aromatherapy and their expanding usage as flavorings in a range of goods, including cosmetics and processed foods. A full circle has subsequently been reached in this pattern, and no significant upward movement has been seen since. Additionally, a number of occurrences during 2008, particularly the Lehman Shock, led to pullbacks in consumer spending, which resulted in a trend toward lower import values. There have been small ups and downs in import amounts, with a significant upward shift occurring during the recession from 2008 to 2009. However, import values have been declining. There are a number of potential causes for this increase, including the impact of the recession; more consumers may have chosen to buy less expensive aroma oils rather than expensive oils from luxury British or French brands; or they may have decided to go out less frequently to save money on entertainment and dining out, instead attempting to spend more quality time at home, for which the use of EOs was a tool (Mahato et al., 2019).

The demand for EOs has been rising in Japan as a result of the rising popularity of aromatherapy. As of March 2010, the Aroma Environment Association of Japan (AEAJ) had 51,312 individual members and 241 corporate members who dealt with aromatherapy, herbs, EOs, and other associated products and services (including aromatherapy schools and salons). Since 2005, individual membership has increased by double (about an eightfold increase from 2000). The number of successful candidates for the Aromatherapy Proficiency Examination has increased by a factor of four since 2005, approaching 200,000. One reason why EO imports have increased even during the recession is that the number of people using EOs has continuously increased along with the popularity of aromatherapy.

Uses of essential oil

Aromatherapy

More people are eager to enjoy fragrance in their daily lives as a consequence of the rising understanding that aroma may assist people's lives be enhanced, which leads to an increase in the number of people who love aromatherapy. The popularity of aromatherapy has spread widely due to a number of factors, including its urban and stylish image, media influence, and the fact that Japan has a foundation for the spread of aromatherapy because there was already a culture of herb enjoyment there before aromatherapy was introduced to the country. However, a more basic reason is that aromatherapy is now seen as a useful therapeutic treatment in a culture that is becoming more stressed. Making use of the natural ingredients of EOs by using them in housekeeping as well as in making homemade cosmetics, EOs are now used more and more in everyday life. From enjoying aroma in the air at home by adding drops of one's choice of EOs of the day to aroma pots, aroma lamps, or aroma diffusers (devices used to diffuse the oil in the form of atomized particles that are growing in popularity), to relaxing in an aroma bath by adding oils to the water, EOs are now used more. Numerous scents that fit their own lives, tastes, and moods are becoming more and more popular among consumers (Asuquo et al., 2019). As a result of the rising demand for EOs, markets are expanding for additional essential oil-related enterprises, such as services that give customers access to therapeutic and comforting spaces by utilizing the aromatherapy properties of EOs.

Recently, EOs began to be disseminated for use in aromatherapy in Japan, and as their history is relatively young, this is in part the reason why aromatherapy has not yet been acknowledged as a legitimate medical treatment (i.e., medical aromatherapy). Only recently academic medical societies have been formed, and more intensive educational and research endeavors have begun to attempt to apply the practice to medical care. As a result, markets for other essential oil-related industries, such as services to create healing and comfortable spaces for customers, are growing. In order to promote and advance aromatherapy in the medical sectors, the Japanese Society of Aromatherapy (JSA) was founded in 1997. As a result of the JSA's efforts to integrate aromatherapy into clinical settings, more medical facilities are being accredited to offer outpatient aromatherapy treatment. The Japan Medical Aromatherapy Association (JMAA), which was established in 2001 as a specific nonprofit company, carries out initiatives to promote medical aromatherapy among the general public through media like the Internet, contributing to a constant growth in medical awareness. A few examples of medicinal aromatherapy indications include: (1) influenza and other common respiratory illnesses; (2) allergy disorders; (3) atopic dermatitis and other skin conditions; (4) both menopause and dysmenorrhea; (5) psychosomatic illnesses; (6) psychological disorders like sleeplessness and panic disorder; (7) lifestyle-related illnesses; (8) backache, joint discomfort, and tight shoulders; and (9) constipation. Medical aromatherapy is being utilized in nursing and geriatric care settings in addition to being employed for therapeutic purposes. It should be mentioned that the majority of EOs used in aromatherapy and marketed on the open market are offered as everyday items. These goods might include not just natural oils that are 100% pure but also inferior oils that contain artificial preservatives and scents like fragrance oils, aroma oils, or potpourri oils that are mistakenly marketed as EOs. The Pharmaceutical Affairs Act prohibits describing the therapeutic effects and effectiveness of EOs, as well as giving specific instructions on how to use them because they are general merchandise rather than medications (or quasi-drugs). The key prerequisite is self-care on a user-responsibility basis; it is impossible to say that common consumers have a solid understanding of EO fundamentals like the characteristics or effects of scents, the optimal application, usage precautions, etc (De Lima et al., 2020).

Scenting purposes

It is difficult to say that common consumers have a solid understanding of EO fundamentals such as characteristics or effects of scents, the optimal application, use precautions, etc., mainly because self-care on a user-responsibility basis is seen as the essential prerequisite. The Japan Taste and Scent Association is the source of the data on flavor and fragrance cited here. Materials Association and pertain to the amount of domestic fragrance manufacturing. Food flavorings are manufactured in Japan in the greatest quantity, according to flavor and fragrance data, and a significant amount of EOs are thought to be employed in them. Food flavorings may also use synthetic tastes in addition to EOs; however, due to the recent increase in health consciousness, EOs have become increasingly well-liked as a 100% natural component. Their usage in food flavorings has been increasing annually. The usage of EOs in cosmetic scents for cosmetics and toiletries has increased in comparison to synthetic perfumes, however, it is unclear exactly how big of a number of EOs are utilized in fragrances.

There are many distribution methods for EOs used in aromatherapy and for scenting. The distribution of EOs used for scenting purposes occurs directly between foreign flavorings and fragrances, cosmetics, processed food supplies, and domestic manufacturers. Then, as flavorings or scents, these products are employed in processed foods, cosmetics, and other products. In contrast to scenting, numerous EOs are given through various pathways for use in aromatherapy. Foreign suppliers or importers provide EOs used for scenting directly to domestic makers of flavorings and perfumes, cosmetics, and processed foods, which are subsequently utilized as flavorings or smells in those goods. There are many distribution methods for EOs used in aromatherapy and for scenting. Those that make flavorings and perfumes, cosmetics, or processed foods in the United States get their EOs directly from foreign suppliers or importers to utilize as flavorings or scents in these products. Along with current over-the-counter sales, an increasing number of importers, distributors, and specialty shops put up websites and engage in online business. The avenues for distributing EOs are getting more complicated since some retailers only conduct business online (Başer & Demirci, 2020).

Literature review

Circular bioeconomy and essential oil

Recently, customers have demanded safe, all-natural foods with a longer shelf life, devoid of synthetic chemical preservatives, and less processing. It's difficult to get these things while also ensuring their safety and well-being. As a result, it is advised to add organic substances from plants and herbs that have antibacterial properties, either separately or in combination. Recent studies have demonstrated that a wide variety of species and herbs exhibit antimicrobial activity as a result of preventing the growth of harmful microbes, its EO fractions can be employed in food systems as antifungal, antibacterial, and antioxidant agents, assuring the microbiological safety of food items. In accordance with the French Pharmacopeia's Eighth Edition (1965), the term "EOs" (EOs) refers to products having a complicated overall composition made up of volatile plant-based principles that have undergone some processing during production. After being collected from different plant parts, including flowers, leaves, seeds, bark, fruits, and roots, EOs are stored in secretory cells, cavities, canals, epidermic cells, or glandular trichomes. The composition of EOs from a certain plant species may vary depending on harvesting seasons and geographical sources. EOs are liquid, volatile, limpid, infrequently colored, lipid-soluble, and soluble in organic solvents with a density that is frequently lower than that of water (Jones, 2019). A complex blend of several chemicals makes up the constitution of EOs. The primary category is made up of terpenes and terpenoids, while the other is made up of elements that are aromatic and aliphatic, all of which have low molecular weights. Compared to the other components, which are present at minimal levels, volatile oils have two or three primary components at high concentrations (20%−70%). EOs are phenolic in nature and mostly include mono- and sesquiterpenoids. Terpenes occur more often than aromatic compounds; nonetheless, the presence of phenolic components is necessary for EOs to have antibacterial action. Additional chemical elements found in plant-derived antimicrobials include flavonoids, thiosulfinates, glucosinolates, and saponins. Eugenol, carvacrol, and thymol are the EOs that have received the most investigation, which is a key component of thyme and *Origanum compactum* EOs (found in *Eugenia caryophylata*).

EOs work well to combat some types of organisms, including bacteria, viruses, fungus, protozoa, parasites, acarids, larvae, and mollusks. Steam distillation and hydrodistillation are the most common methods for removing antimicrobials from plants, whereas other methods like supercritical fluid extraction (SFE) provide greater solubility and faster mass transfer rates. There are various methods known for extracting EOs, for as using liquid carbon dioxide or microwaves (Maggi et al., 2019).

Biomass fuel for biomethane synthesis may be found in abundance in the food supply chain. In a Malaysian study of the economic viability of offering feed-in tariff for injecting biomethane into the natural gas grid, advocate for a rationalization of the price of natural gas along with the imposition of a carbon tax to make investments in the production, upgrading, and injection of biomethane derived from food waste appealing. Investing in the production of biomethane from indigenous renewable feedstock sources decreases reliance on imports, ensures energy security, and provides a practical means of addressing geopolitical challenges. Pellets, long fiber, and biofertilizer increase the profitability of the biorefinery. *Fusarium heterosporum* produces a lipase which is an enzyme that hydrolyzes fats, and palm oil mill wastes such as palm kernel shells may successfully transesterify palm oil mill effluents into biodiesel (Mancini et al., 2020). Chinese researchers have developed a unique method for producing both biogas and biodiesel by anaerobically digesting chicken manure and rapeseed straw that has been pre-treated with the liquid component of the digestate (from black soldier fly larvae grown on the solid fraction of the digestate). Alternatively, fish farms might receive the larvae as a source of a protein-rich diet (Amiri et al., 2018). It is possible to extract a significant amount of hydrogen from organic solid wastes and wastewater, and this might very well be the future's fuel. In an Italian example study, concentrate on the anaerobic digestion and dark fermentation of bovine dung and increased value is added to the bioeconomy by using grass silage to generate the H2-CH4 mixture (known as bio-hythane) and volatile fatty acids that serve as precursors to bioplastics. In two Italian studies (Mancini et al., 2020), the authors employed whey (dairy wastes) and molasses (waste from sugar mills) to make biohydrogen and bioplastics, namely polyhydroxybutyrate or PHB, by dark fermentation and photo-fermentation in that order.

Another biopolymer used to produce packaging "bio-paper" with exceptional mechanical and barrier properties is poly(3-hydroxybutyrate-co-3-hydroxyvalerate), or PHBV. Studies, once they are published, will establish the strength and shelf-life of naturally biodegradable bioplastics, such as PHB, used for packaging, which will serve as a barometer for their entrenchment as alternatives to petroleum-based plastics. Elephant dung paper, for example, is produced from non-wood materials and is a specialist item with a small market. When the paper product's brightness requirements aren't very high, Agro-food wastes may be possible to replace wood, according to Spanish experts that looked into the matter. Nanotechnologists from Spain have recommended employing biowaste-based nanomaterials for the purpose, but found that co-ground wood fibers and tannins in aqueous media fit the bill extremely well. Hydrophobicity and antioxidant effects are necessary properties of packaging films. Synthetic adhesives' bonding qualities can be enhanced by tannins and cellulose nanocrystals, according to research. A circular bioeconomy must have sustainable bio-packaging choices in place, while also extending the shelf life of packaged foods to decrease food waste. The three main traits of sustainable packaging are compatibility with the circular production-consumption system, the capacity to meet varied consumer wants, and the potential to support sustainable lifestyles by extending material life cycles (Braca et al., 2020). Braca et al. (2020) emphasizes the potential for viticulture to maximize its environmental friendliness while increasing its economic value through the diversification of products through the valorization of wastes, Additionally, the similar marketing approach for sugar beet, lemon, and mango has been advocated in other publications. Tomato plants under biotic/abiotic stress are rich sources of rutin and solanesol, which are prized for their therapeutic benefits. In a different application, tomato pomace is used as mulch to trap solar radiation and heat the soil to control nematodes, fungi, bacteria, and crop-damaging pests. The common practice in India is the open burning of agricultural garbage, contributes significantly to the country's GHG emissions, but it is simply ignored because this waste has a great deal of potential to be converted into biogas. Fossil fuels might be substituted for open burning if that were to occur on a big scale, which would result in a reduction in the energy sector's overall GHG emissions. To overcome technological, financial, policy interventions, and regulatory barriers and hasten the transition to a circular bioeconomy, better communication between decision-makers and potential biogas producers in the agriculture sector is needed (Aizpurua-Olaizola et al., 2021).

Application of essential oils in food safety

The leaves, bark, flowers, buds, seeds, roots, stems, and fruits of various aromatic plants have been used to extract EOs, which are important secondary metabolic products. The phrase "essential oil" is derived from the word "essence," which denotes the taste and scent it possesses. EOs include a variety of structurally similar terpenoids, phenylpropanoids, and low molecular weight, lipophilic, short-chain aliphatic hydrocarbons. Aldehydes, ketones, esters, oxides, and alcohols are only a few examples of oxygenated molecules that actively contribute to the synthesis of EOs in some aromatic plants. Hydrodistillation is a common extraction method for separating oils from plant sources. The bioactive components of EOs are volatile by nature, have a strong odor, and can emit a fragrance or flavor. Different plant species, including those in the families Asteraceae, Lamiaceae, Cyperaceae, Zingerberaceae, Piperaceae, Apiaceae, Myrtaceae, Solanaceae, and Lauraceae, have EOs isolated from them. One way that EOs contribute to plant defense in the environment is by being able to withstand environmental stress, having better antifungal and antibacterial capabilities, and maybe being valuable in the food and pharmaceutical industries. Eugenol has been discovered to be a significant bioactive component of *Cinnamomum verum* EO, and it has been shown to have potent anticancer and antibacterial properties. Antifungal, anticancer, antibacterial, and antiviral agents are all actively used with it (Guan et al., 2019). A notable member of the Lamiaceae family, *Thymus persicus* is used medicinally. A few of the volatile ingredients that make up the EO of

T. persicus are carvacrol (27.01%), thymol (11.86%), p-cymene (10.16%), and -terpineol. The two primary components of *Origanum compactum* EO, thymol (27%) and carvacrol (30%) have been discovered. The proportion of camphor and thuoyne in *Artemisia herba-alba* EO was found to be substantial, but the primary elements of *Anethum graveolens* EO were carvone (58%), limonene (37%) and linalool (68%).

Carvacrol (27.01%), thymol (11.86%), p-cymene (10.16%), and -terpineol are only a few of the volatile compounds that make up *T. persicus'* EO. As the two main ingredients of *Origanum compactum* EO, thymol (27%) and carvacrol (30%) have been identified. *Artemisia herba-alba* EO was found to include camphor, and thuoyne, among other ingredients. Carvone (58%) and limonene (37%) from the EO of A. graveolens, and linalool (68%) from the EO of *Coriandrum sativum* are the most abundant compounds. EOs exhibit broad-spectrum antibacterial, fungitoxic, and antimycotoxigenic actions in food products that have been postharvest stored. Its variable bioefficacy, which includes its antiviral, antidepressant, antibacterial, and toxin (bacterial toxins and mycotoxins) detoxification activities, was the inspiration for its application in the management of stored food in order to ensure its use as a green substitute for synthetic preservatives). On the use of EOs as a natural antibacterial, antifungal, and antimycotoxigenic agent in the model and food system, there is a plethora of knowledge accessible. Studies by Dwivedy et al. suggest that *Mentha cardiaca* EO might be used as a green substitute to prevent fungal and AFB1 contamination in dried fruits being kept (2017). The in vitro antiaflatoxigenic and antifungal activities of *M. cardiaca* EO were each 1.0 l/mL. The EO demonstrated high antioxidant activity (DPPH IC50 = 15.89 l/mL), which served to shield dry fruits from lipid peroxidation by scavenging biodegrading free radicals. It also worked to stop a variety of fungus that may contaminate food and AFB1. Additionally, the EO showed a good LD50 value (7133.70 mg/kg), indicating that it is not hazardous to animals and suggesting that it might be used widely. The efficiency of *Artemisia nilagirica* EO in stopping fungus invasion and contamination with AFB1 was examined in vitro and in vivo to extend the shelf life of stored millets. AFB1 secretion and toxic *Aspergillus flavus* growth were both totally suppressed during in vitro tests at concentrations of 1 and 1, respectively, l/mL. By scavenging free radicals, the EO's potent antioxidant activity (DPPHIC50 = 2.51 and ABTSIC50 = 1.07 l/mL) successfully inhibited the production of the AFB1 protein and decreased the possibility that food components would oxidize, hence lowering the nutritional quality (Elshafie et al., 2021).

Numerous in vitro tests have shown the antibacterial and antimycotoxigenic efficiency of EOs and their constituents, however, EO must be widely used in food-based industrial sectors for its in vivo practical usefulness in food systems. It is claimed that *P. aeruginosa*, *E. coli*, *Shigella dysanteriae*, *S. typhimurium*, *Salmonella enteritidis*, and *Yersinia enterocolitica* are six gram-negative bacteria, while *Lactococcus lactis*, *S. aureus*, *L. monocytogenes*, and *Brochotrix thermosphacta* are seven distinct gram-positive bacteria. Clove EO performed better in the research than rosemary essential oil, and the application of combinations of clove and rosemary EO showed greater antibacterial activity for the preservation and extension of the shelf life of poultry meat. The EO was shown to be particularly effective against *L. monocytogenes*, with minimum inhibitory concentrations of 0.10 and 0.15 l/mL and minimum bactericidal concentrations of 0.15, respectively. The authors also observed an increase in pH in the control beef samples due to the breakdown of meat proteins into free amino acids and the production of various amines as a result of alkaline reactions, but the pH was maintained in the beef samples treated with *M. alternifolia* EO, which was supported by the elimination of spoilage bacteria infestations and inhibition of toxin production in ground beef. According to a recent research, cinnamon and clove EOs can prevent *Penicillium spp* and *Aspergillus spp* infection in baked goods without degrading their organoleptic properties. In a research, the EO of *Allium sativum* significantly decreased the infestation of A. *Niger* and A. *flavus* in plum fruit while also preserving the fruit's quality for ten days of storage. A. *Niger* and A. *flavus* growth was completely inhibited in an in vitro trial at concentrations of 7.5 and 6.5 g/mL, although an in vivo application of A. *sativum* EO could require a fivefold higher concentration than it did for in vitro antifungal efficacy. A. *sativum* EO in plum fruit decreased A. *Niger* infestation by 50%−80%, whereas A. *flavus* infection was suppressed to varied degrees (20%−55%). Dwivedy (2019) proved *I. verum* EO's ability to entirely shield Pistacio seeds from fungal and aflatoxin B1 contamination in situ. The EOs of lemongrass, oregano, and thyme fully suppressed the growth of A. *flavus*, A. *parasiticus*, and A. *clavatus* in oats (2017). The ability of thyme, cinnamon, and lemongrass EOs to prevent *Penicillium citrinum* and A. *flavus* infection in peanut kernels was investigated at the lowest minimum inhibitory concentration of 40 l/mL.

Gross domestic product perspective of essential oil in food industry

A linear economic model has been the basis for all of society since the industrial revolution, which can be summed up as taking the resources you need, producing the items you want to sell and profit from, and discarding the rest

(Ludwiczuk et al., 2019). For many years, this extractive industrial paradigm has helped mankind by generating tangible prosperity. However, it has resulted in a circumstance that cannot maintain current generations indefinitely. The results of scientific modeling point to a level of resource consumption that is already over what is considered to be sustainable. According to the Global Footprint Network, about 1.7 Earths would be needed to supply all the people on the planet with the natural resources they need. With a rising global population and a commensurate growth in our consumption needs, it can be concluded from these facts that the linear economic model will not hold up (Ozel & Kaymaz, 2020). The population of the globe, which was predicted to be 7.4 billion in 2017, is projected to rise to 9.7 billion by 2050 and to 11.2 billion by 2100 as a result of the more than 140 million births that take place annually. By 2050, the Food and Agriculture Organization of the United Nations (FAO) projects that the caloric demand will increase by 70% and that there would be a requirement for double the current crop production for both human consumption and animal feed. In low income countries of the world food prices would rise as a result of the imbalance between supply and demand, which would disproportionately affect the poor. By 2050, 2.5 billion people are expected to live in Africa, which would make up nearly 27% of the world's population. By 2060, 1.1 billion Africans, or 42% of the people in the continent, are expected to belong to the continent's middle class, which is defined as the group of individuals who make between $4 and $20 each day. It is projected that the middle class would boost demand for food, fiber, fuel, and pharmaceuticals and change patterns of material and food consumption (e.g., increase intake of livestock-based products). In Sub-Sahara Africa [SSA], increasing population pressure is already causing farmers to use more and more marginal land (Ozel & Kaymaz, 2020). The majority of African farming techniques are still unsustainable due to land degradation. Approximately 25% of the deterioration of productive lands is attributed to the loss of nutrients and soil organic carbon brought on by continuous cropping. The significant waste creation trend is causing greenhouse gas (GHG) emissions in a similar way to how the land is degrading. The breakdown of trash at open landfills contributes to over 7% of Africa's GHG emissions. Climate change has a relationship to GHG emissions. SSA is particularly susceptible because of its inadequate capacity for adaptation to climate change, even if the negative effects of climate change are widespread. It is anticipated that the SSA will lose $3.33 billion in GDP, or 0.2% of its GDP, by 2050 because of climate change (Jirovetz et al, 2020). These effects imply that additional goods that respect the constraints of the available resources are required and should be produced sustainably. Concerns that SSA is now dealing with low economic development, climate change, and environmental degradation. The term "circular bioeconomy," which is a combination of the terms "bioeconomy" and "circular economy" (Ayoade & Daniel, 2020), has significant promise for addressing these issues. A good example is the circular bioeconomy strategy, which reduces the amount of resources needed to produce one unit of economic output. The circular bioeconomy's minimal reliance on chemical inputs may help to slow global warming (Ayoade & Daniel, 2020). By eliminating garbage that is deteriorating, the reuse of waste, particularly biowaste, into industrial systems would slow environmental deterioration. Because of the abundance of fertile land and longer daylight hours in SSA, there are plenty of opportunities for biomass production that may be utilized as sources of renewable energy, helping to slow down global warming (Ayoade & Daniel, 2020). The central idea of the circular bioeconomy is "the sustainable, resource-efficient valorization of biomass in integrated, multi-output production chains (e.g., biorefineries), while also using leftovers and wastes and cascading the value of biomass through time". One can focus on economic, social, or environmental aspects of biomass in order to increase its value over time, but ideally, all three sustainability pillars should be taken into consideration (Chunhui et al., 2020). Sustainable biomass sourcing, circular and long-lasting product design, the use of waste and residues, integrated, multiple-output production chains, bioenergy and biofuels, bio-based products, food, and feed, prolonged and shared use, energy recovery and composting, recycling and cascading, and are the main elements of the circular bioeconomy. The bio-based economy in Africa is largely driven by public research for development (R4D), which tackles a variety of issues, such as (1) raising agricultural productivity, (2) reducing resource demands and environmental pressures, (3) adjusting to the effects of climate change, (4) expanding opportunities for bioresource value addition and turning waste into useful products like energy, and (5) revitalizing rural communities and enhancing rural livelihoods (Constanza et al., 2015).

Conclusion

This chapter covers EOs, their important components, applications, and distribution methods. It also covers the circular bioeconomy and EOs, their role in ensuring the safety of food, and their impact on GDP from the viewpoint of the food sector. In accordance with the United Nations' Sustainable Development Goals (SDGs), food sustainability and security has been a significant problem (Adetunji, 2008, 2019; Adetunji et al., 2013, 2019, 2022; Adetunji, Egbuna et al., 2020; Adetunji, Oloke et al., 2020; Adetunji, Roli et al., 2020; Adejumo, Adetunji et al., 2017; Adejumo, Phazang et al., 2017; Adetunji et al., 2013; Adetunji, Adetunji et al., 2021; Adetunji, Ajayi et al., 2021; Adetunji, Akram et al., 2021; Adetunji, Anani et al., 2021;

Adetunji, Inobeme et al., 2021; Adetunji, Michael, kadiri et al., 2021; Adetunji, Michael, Nwankwo et al., 2021; Adetunji, Michael, Varma et al., 2021; Adetunji, Nwankwo et al., 2021; Adetunji, Olaniyan et al., 2021; Adetunji, Palai et al., 2021; Adetunji & Varma, 2020; Bello et al., 2019; Egbuna et al., 2020; Olaniyan & Adetunji, 2021; Rauf et al., 2021; Thangadurai et al., 2021; Ukhurebor & Adetunji, 2021). Food preservation is therefore a major concern as it is necessary to ensure year-round food availability. It is currently necessary for the food industries to utilize active packaging so as to lower the initial microbial load and/or restrict the growth of the leftover bacteria during manufacturing and storage. If the bioactive chemicals found in EOs have the ability to fulfill the specified goal, then EO development and application will proceed accordingly. A larger degree of resistance to foodborne pathogens and other microbes has been seen as a result of EO research and use in a variety of industries, including the food, pharmaceutical, and cosmetic industries.

Aside from the fact that the home market is woefully underserved, the product has excellent export potential, especially to Europe, Asia, and the West Africa Sub-Saharan area, with the goal of bringing in foreign revenue for our nation. Therefore, it is highly advised that the state and local governments of Nigeria assist in the establishment of such a pilot plant in their respective areas, particularly where the raw materials are plentiful. In addition to such attempts, offering work chances to unemployed young people would also aid in reducing poverty.

References

Adejumo, I. O., Adetunji, C. O., & Adeyemi, O. S. (2017). Influence of UV light exposure on mineral composition and biomass production of myco-meat produced from different agricultural substrates. *Journal of Agricultural Sciences, Belgrade, 62*(1), 51−59.

Adetunji, C.O. (2008). The antibacterial activities and preliminary phytochemical screening of vernoniaamygdalina and Aloe vera against some selected bacteria, pp. 40−43 [M. sc Thesis University of Ilorin].

Adetunji, C. O. (2019). Environmental impact and ecotoxicological influence of biofabricated and inorganic nanoparticle on soil activity. In D. Panpatte, & Y. Jhala (Eds.), *Nanotechnology for agriculture*. Singapore: Springer. Available from https://doi.org/10.1007/978-981-32-9370-0_12.

Adetunji, J. B., Adetunji, C. O., & Olaniyan, O. T. (2021). African walnuts: A natural depository of nutritional and bioactive compounds essential for food and nutritional security in Africa. In O. O. Babalola (Ed.), *Food security and safety*. Cham: Springer. Available from https://doi.org/10.1007/978-3-030-50672-8_19.

Adetunji, J. B., Ajani, A. O., Adetunji, C. O., Fawole, O. B., Arowora, K. A., Nwaubani, S. I., Ajayi, E. S., Oloke, J. K., & Aina, J. A. (2013). Postharvest quality and safety maintenance of the physical properties of Daucus carota L. fruits by Neem oil and Moringa oil treatment: A new edible coatings. *Agrosearch, 13*(1), 131−141.

Adetunji, C. O., Ajayi, O. O., Akram, M., Olaniyan, O. T., Chishti, M. A., Inobeme, A., Olaniyan, S., Adetunji, J. B., Olaniyan, M., & Awotunde, S. O. (2021). Medicinal plants used in the treatment of influenza a virus infections. In K. Dua, S. Nammi, D. Chang, D. K. Chellappan, G. Gupta, & T. Collet (Eds.), *Medicinal plants for lung diseases*. Singapore: Springer. Available from https://doi.org/10.1007/978-981-33-6850-7_19.

Adetunji, C. O., Akram, M., Olaniyan, O. T., Ajayi, O. O., Inobeme, A., Olaniyan, S., Hameed, L., & Adetunji, J. B. (2021). Targeting SARS-CoV-2 novel corona (COVID-19) virus infection using medicinal plants. In K. Dua, S. Nammi, D. Chang, D. K. Chellappan, G. Gupta, & T. Collet (Eds.), *Medicinal plants for lung diseases*. Singapore: Springer. Available from https://doi.org/10.1007/978-981-33-6850-7_21.

Adetunji, C. O., Anani, O. A., Olaniyan, O. T., Inobeme, A., Olisaka, F. N., Uwadiae, E. O., & Obayagbona, O. N. (2021). Recent trends in organic farming. In R. Soni, D. C. Suyal, P. Bhargava, & R. Goel (Eds.), *Microbiological activity for soil and plant health management*. Singapore: Springer. Available from https://doi.org/10.1007/978-981-16-2922-8_20.

Adetunji, C. O., Egbuna, C., Tijjani, H., Adom, D., Al-Ani, L. K. T., & Patrick-Iwuanyanwu, K. C. (2020). *Homemade preparations of natural biopesticides and applications. Natural remedies for pest, disease and weed control* (pp. 179−185). Publisher Academic Press.

Adetunji, C. O., Inobeme, A., Olaniyan, O. T., Ajayi, O. O., Olaniyan, S., & Adetunji, J. B. (2021). Application of nanodrugs derived from active metabolites of medicinal plants for the treatment of inflammatory and lung diseases: Recent advances. In K. Dua, S. Nammi, D. Chang, D. K. Chellappan, G. Gupta, & T. Collet (Eds.), *Medicinal plants for lung diseases*. Singapore: Springer. Available from https://doi.org/10.1007/978-981-33-6850-7_26.

Adetunji, C. O., Kumar, D., Raina, M., Arogundade, O., & Sarin, N. B. (2019). Endophytic microorganisms as biological control agents for plant pathogens: A panacea for sustainable agriculture. In A. Varma, S. Tripathi, & R. Prasad (Eds.), *Plant biotic interactions*. Cham: Springer. Available from https://doi.org/10.1007/978-3-030-26657-8_1.

Adetunji, C. O., Michael, O. S., Kadiri, O., Varma, A., Akram, M., Oloke, J. K., Shafique, H., Adetunji, J. B., Jain, A., Bodunrinde, R. E., Ozolua, P., & Ubi, B. E. (2021). Quinoa: From farm to traditional healing, food application, and phytopharmacology. In A. Varma (Ed.), *Biology and biotechnology of Quinoa*. Singapore: Springer. Available from https://doi.org/10.1007/978-981-16-3832-9_20.

Adetunji, C. O., Michael, O. S., Nwankwo, W., Ukhurebor, K. E., Anani, O. A., Oloke, J. K., Varma, A., Kadiri, O., Jain, A., & Adetunji, J. B. (2021). Quinoa, The next biotech plant: Food security and environmental and health hot spots. In A. Varma (Ed.), *Biology and biotechnology of Quinoa*. Singapore: Springer. Available from https://doi.org/10.1007/978-981-16-3832-9_19.

Adetunji, C. O., Michael, O. S., Rathee, S., Singh, K. R. B., Ajayi, O. O., Adetunji, J. B., Ojha, A., Singh, J., & Singh, R. P. (2022). Potentialities of nanomaterials for the management and treatment of metabolic syndrome: A new insight. *Materials Today Advances, 13*100198.

Adetunji, C. O., Michael, O. S., Varma, A., Oloke, J. K., Kadiri, O., Akram, M., Bodunrinde, R. E., Imtiaz, A., Adetunji, J. B., Shahzad, K., Jain, A., Ubi, B. E., Majeed, N., Ozolua, P., & Olisaka, F. N. (2021). Recent advances in the application of biotechnology for improving the production of secondary metabolites from Quinoa. In A. Varma (Ed.), *Biology and biotechnology of Quinoa*. Singapore: Springer. Available from https://doi.org/10.1007/978-981-16-3832-9_17.

Adetunji, C. O., Nwankwo, W., Ukhurebor, K., Olayinka, A. S., & Makinde, A. S. (2021). Application of biosensor for the identification of various pathogens and pests mitigating against the agricultural production: Recent advances. In R. N. Pudake, U. Jain, & C. Kole (Eds.), *Biosensors in agriculture: Recent trends and future perspectives. Concepts and strategies in plant sciences*. Cham: Springer. Available from https://doi.org/10.1007/978-3-030-66165-6_9.

Adetunji, C. O., Olaniyan, O. T., Akram, M., Ajayi, O. O., Inobeme, A., Olaniyan, S., Khan, F. S., & Adetunji, J. B. (2021). Medicinal plants used in the treatment of pulmonary hypertension. In K. Dua, S. Nammi, D. Chang, D. K. Chellappan, G. Gupta, & T. Collet (Eds.), *Medicinal plants for lung diseases*. Singapore: Springer. Available from https://doi.org/10.1007/978-981-33-6850-7_14.

Adetunji, C. O., Oloke, J. K., & Prasad, G. (2020). Effect of carbon-to-nitrogen ratio on eco-friendly mycoherbicide activity from Lasiodiplodia pseudotheobromae C1136 for sustainable weeds management in organic agriculture. *Environ Dev Sustain, 22*, 1977−1990. Available from https://doi.org/10.1007/s10668-018-0273-1.

Adetunji, C. O., Palai, S., Ekwuabu, C. P., Egbuna, C., Adetunji, J. B., Ehis-Eriakha, C. B., Kesh, S. S., & Mtewa, A. G. (2021). *General principle of primary and secondary plant metabolites: Biogenesis, metabolism, and extraction. Preparation of Phytopharmaceuticals for the Management of Disorders* (pp. 3−23). Publisher Academic Press.

Adetunji, C. O., Phazang, P., & Sarin, N. B. (2017). Significance of rhamnolipids as a biological control agent in the management of crops/plant pathogens. *Current Trends in Biomedical Engineering & Biosciences, 10*(3), 54−55.

Adetunji, C. O., Roli, O. I., & Adetunji, J. B. (2020). Exopolysaccharides derived from beneficial microorganisms: Antimicrobial, food, and health benefits. In P. Mishra, R. R. Mishra, & C. O. Adetunji (Eds.), *Innovations in food technology*. Singapore: Springer. Available from https://doi.org/10.1007/978-981-15-6121-4_10.

Adetunji, C. O., & Varma, A. (2020). Biotechnological application of trichoderma: A powerful fungal isolate with diverse potentials for the attainment of food safety, management of pest and diseases, healthy planet, and sustainable agriculture. In C. Manoharachary, H. B. Singh, & A. Varma (Eds.), *Trichoderma: Agricultural applications and beyond. Soil biology* (vol 61). Cham: Springer. Available from https://doi.org/10.1007/978-3-030-54758-5_12.

Aizpurua-Olaizola, O., Ormazabal, M., Vallejo, A., Olivares, M., Navarro, P., et al. (2021). Optimization of supercritical fluid consecutive extractions of fatty acids and polyphenols from Vitis vinifera grape wastes. *Journal of food science, 80*, E101−E107.

Amiri, A., Dugas, R., Pichot, A. L., & Bompeix, G. (2018). In vitro and in vitro activity of eugenol oil (Eugenia caryophylata) against four important postharvest apple pathogens. *International Journal of Food Microbiology, 126*, 13−19.

Asuquo, O. R., Fischer, C. E., Mesembe, O. E., Igiri, O. E., & Ekom, J. E. (2019). Comparative study of aqueous and ethanolic leaf extracts of Spondias mombin on neurobehaviour in male rats. *IOSR Journal of Pharmacy and Biological Sciences New York, USA, 5*, 29−35.

Ayoade, K., & Daniel, B. A. (2020). Basic concept in process equipment and plant design. *Publication of Raw Materials Research and Development Council, Abuja*, 233−240.

Başer, K. H. C., & Demirci, F. (2020). Chemistry of essential oils. In R. G. Berger (Ed.), *Flavours and fragrances: chemistry, bioprocessing and sustainability* (pp. 43−86). New York: Springer, 2007.

Bello, O. M., Ibitoye, T., & Adetunji, C. (2019). Assessing antimicrobial agents of Nigeria flora. *Journal of King Saud University-Science, 31*(4), 1379−1383.

Braca, A., Siciliano, T., D'Arrigo, M., & Germanò, M. P. (2020). Chemical composition and antimicrobial activity of Momordica charantia seed essential oil. *Fitoterapia, 79*, 123−125.

Burt, S. (2021). Essential oils: their antibacterial properties and potential applications in foods-a review. *International journal of food microbiology, 94*, 223−253.

Chunhui, D., Ning, Y., Aiqin, W., & Xiangmin, Z. (2020). Determination of essential oil in a traditional Chinese medicine, Fructus amomi by pressurized hot water extraction followed by liquid-phase microextraction and gas chromatography−mass spectrometry. *Analytica Chimica Acta, 536*, 237−244.

Constanza, K., Tallury, S., Whaley, J., Sanders, T., & Dean, L. (2015). Chemical composition of the essential oils from leaves of edible (Arachis hypogaea L.) and Perennial (Arachis glabrata Benth.) peanut plants. *Journal of Essential Oil Bearing Plants, 18*, 605−612.

Daw, Y., El- Barty, G. E., & Muhammad, E. A. (2018). Inhibition of *Aspergellius* parasitices growth and Aflatoxin production by some essential oils. *J. Afr. Crop Sci. 3, 4*, 511−517.

De Lima, E. Q., de Oliveira, E., & de Brito, H. R. (2020). Extraction and characterization of the essential oils from Spondias mombin L.(Caj). *Spondias purpurea L.(Ciriguela) and Spondia ssp (Cajarana do serto). African Journal of Agricultural Research, 11*, 105−116.

Dwivedy, M. (2019). Isolation, chemical characterization and evaluation of insecticide property of essential oil Piper amplumKunt, in Masters IDissertação (chemistry) R.U.o. Blumenau, Editor. Blumenau-SC, p. 88.

Egbuna, C., Gupta, E., Ezzat, S. M., Jeevanandam, J., Mishra, N., Akram, M., Sudharani, N., Adetunji, C. O., Singh, P., Ifemeje, J. C., Deepak, M., Bhavana, A., Walag, A. M. P., Ansari, R., Adetunji, J. B., Laila, U., Olisah, M. C., & Onyekere, P. F. (2020). Aloe species as valuable sources of functional bioactives. In C. Egbuna, & G. Dable Tupas (Eds.), *Functional foods and nutraceuticals*. Cham: Springer. Available from https://doi.org/10.1007/978-3-030-42319-3_18.

Elshafie, H. S., Mancini, E., Camele, I., De Martino, L., & De Feo, V. (2021). In vivo antifungal activity of two essential oils from Mediterranean plants against postharvest brown rot disease of peach fruit. *Industrial Crops and Products, 66*, 11−15.

Galadima, M.S. (2019). Design and fabrication of pilot plant for steam distillation of essential oils. [M.sc Thesis. Ahmadu Bello University Zaria − Nigeria]. (Unpublished).

Garry, R. P., & Chalchat, J. C. (2019). Essential oils and their development. *Journal of Fruits, 50*(6), 453—458.

Guan, W., Li, S., Yan, R., Tang, S., & Quan, C. (2020). Comparison of essential oils of clove buds extracted with supercritical carbon dioxide and other three traditional extraction methods. *Food Chemistry, 101*, 1558—1564.

Jirovetz, L., Buchbauer, G., Ngassoum, M. B., & Geissler, M. (2020). Aroma compound analysis of Piper nigrum and Piper guineense essential oils from Cameroon using solid-phase microextraction—gas chromatography, solid-phase microextraction—gas chromatography—mass spectrometry and olfactometry. *Journal of Chromatography A, 976*, 265—275.

Jones, M. (2019). The complete guide to creating oils, soaps, creams, and herbal gels for your mind and body: 101 natural body care recipes. Atlantic Publishing Company.

Ludwiczuk, A., Skalicka-Woźniak, K., & Georgiev, M. (2019). Terpenoids. In S. Badal, & R. Delgoda (Eds.), *Pharmacognosy: Fundamentals, applications and Strategied* (pp. 233—266). London: Elsevier.

Maggi, F., Bramucci, M., Cecchini, C., Coman, M. M., & Cresci, A. (2019). Composition and biological activity of essential oil of Achillea ligustica All. (Asteraceae) naturalized in central Italy: Ideal candidate for anti-cariogenic formulations. *Fitoterapia, 80*, 313—319.

Mahato, N., Sharma, K., Koteswararao, R., Sinha, M., & Baral, E. (2019). Citrus essential oils: Extraction, authentication and application in food preservation. *Critical Reviews in Food Science and Nutrition New York USA, 59*, 611—625.

Mancini, E., Camele, I., Elshafie, H. S., De Martino, L., Pellegrino, C., et al. (2020). Chemical composition and biological activity of the essential oil of Origanum vulgare ssp. hirtum from different areas in the Southern Apennines (Italy). *Chemistry & biodiversity, 11*, 639—651.

Mustapha, Z. O., Fahreltin, G., & Alistiar, C. L. (2020). Subcritical water of essential oils from Thymbra spicata. *Journal of Food Chemistry America, 82*, 381—386.

Olaniyan, O. T., & Adetunji, C. O. (2021). Biological, biochemical, and biodiversity of biomolecules from marine-based beneficial microorganisms: Industrial perspective. In C. O. Adetunji, D. G. Panpatte, & Y. K. Jhala (Eds.), *Microbial rejuvenation of polluted environment. Microorganisms for sustainability* (vol. 27). Singapore: Springer. Available from https://doi.org/10.1007/978-981-15-7459-7_4.

Ozel, M. Z., & Kaymaz, H. (2020). Superheated water extraction, steam distillation and Soxhlet extraction of essential oils of Origanum onites. *Analytical and bioanalytical chemistry, 379*, 1127—1133.

Peter, M. S., & Timmerhaus, K. D. (2019). *s. Plant design and economics for chemical engineer* (4th ed.). NewYork, U.S.A: McGraw-Hill.

Rauf, A., Akram, M., Semwal, P., Mujawah, A. A. H., Muhammad, N., Riaz, Z., Munir, N., Piotrovsky, D., Vdovina, I., Bouyahya, A., Adetunji, C. O., Ali Shariati, M., Almarhoon, Z. M., Mabkhot, Y. N., & Khan, H. (2021). Antispasmodic potential of medicinal plants: A comprehensive review. *Oxidative Medicine and Cellular Longevity*12. Available from https://doi.org/10.1155/2021/4889719, vol. 2021, Article ID 4889719.

Rawat, S. (2020). Food spoilage: Microorganisms and their prevention. *Asian Journal of Plant Science and Research, 5*, 47—56.

Solórzano-Santos, F., & Miranda-Novales, M. G. (2022). Essential oils from aromatic herbs as antimicrobial agents. *Current opinion in biotechnology, 23*, 136—141.

Sridhar, S. R., Velusamy, R. R., Rejavel, R., Selladumai, M., & Srinivasan, N. (2021). Antifungal activity of some essential oils. *Journal of Agricultural and Food Chemistry, 51*, 7596—7599.

Tajkarimi, M., Ibrahim, S. A., & Cliver, D. (2019). Antimicrobial herb and spice compounds in food. *Food control, 21*, 1199—1218.

Thangadurai, D., Naik, J., Sangeetha, J., Al-Tawaha, A. R. M. S., Adetunji, C. O., Islam, S., David, M., Shettar, A. K., & Adetunji, J. B. (2021). Nanomaterials from agrowastes: Past, present, and the future. In O. V. Kharissova, L. M. Torres-Martínez, & B. I. Kharisov (Eds.), *Handbook of nanomaterials and nanocomposites for energy and environmental applications*. Cham: Springer. Available from https://doi.org/10.1007/978-3-030-36268-3_43.

Tongnuanchan, P., & Benjakul, S. (2019). Essential oils: Extraction, bioactivities, and their uses for food preservation. *Journal of food science, 79*, R1231—R1249.

Ukhurebor, K. E., & Adetunji, C. O. (2021). Relevance of biosensor in climate smart organic agriculture and their role in environmental sustainability: What has been done and what we need to do? In R. N. Pudake, U. Jain, & C. Kole (Eds.), *Biosensors in agriculture: Recent trends and future perspectives. Concepts and strategies in plant sciences*. Cham: Springer. Available from https://doi.org/10.1007/978-3-030-66165-6_7.

Willem, V., & Tjakko, M. (2019). The role of sB in the stress response of Gram-positive bacteria-targets for food preservation and safety. *Current Opinion in Biotechnology, 16*, 218—224.

Chapter 24

Application of essential oil in livestock production

Nyejirime Young Wike[1], Olugbemi T. Olaniyan[2], Charles Oluwaseun Adetunji[3], Juliana Bunmi Adetunji[4], Olalekan Akinbo[5], Babatunde Oluwafemi Adetuyi[6], Abel Inobeme[7], Oloruntoyin Ajenifujah-Solebo[8], Yovwin D. Godwin[9], Majolagbe Olusola Nathaniel[10], Ismail Ayoade Odetokun[11], Oluwabukola Atinuke Popoola[8], Olatunji Matthew Kolawole[12] and Mohammed Bello Yerima[13]

[1]Department of Human Physiology, Faculty of Basic Medical Sciences, Rhema University, Aba, Abia State, Nigeria, [2]Laboratory for Reproductive Biology and Developmental Programming, Department of Physiology, Faculty of Basic Medical Sciences, Rhema University, Aba, Abia State, Nigeria, [3]Applied Microbiology, Biotechnology and Nanotechnology Laboratory, Department of Microbiology, Edo State University Uzairue, Iyamho, Edo State, Nigeria, [4]Department of Biochemistry, Osun State University, Osogbo, Osun State, Nigeria, [5]Centre of Excellence in Science, Technology, and Innovation, AUDA-NEPAD, Johannesburg, Gauteng, South Africa, [6]Department of Natural Sciences, Faculty of Pure and Applied Sciences, Precious Cornerstone University, Ibadan, Oyo State, Nigeria, [7]Department of Chemistry, Edo State University Uzairue Iyamho, Auchi, Edo State, Nigeria, [8]Genetics, Genomics and Bioinformatics Department, National Biotechnology Development Agency, Abuja, FCT, Nigeria, [9]Department of Family Medicine, Faculty of Clinical Sciences, Delta State University, Abraka, Delta State, Nigeria, [10]Microbiology Unit, Department of Pure and Applied Biology, Ladoke Akintola University of Technology, Ogbomoso, Oyo State, Nigeria, [11]Department of Veterinary Public Health and Preventive Medicine, University of Ilorin, Ilorin, Kwara State, Nigeria, [12]Department of Microbiology Faculty of Life Sciences, University of Ilorin, Ilorin, Kwara State, Nigeria, [13]Department of Microbiology, Sokoto State University, Sokoto, Sokoto State, Nigeria

Introduction

The increasing expansion of livestock farming and the internationalization of commerce contributed to the widespread and ineffective utilization of antimicrobials for viral infections. Ever since the conclusion of the 1990s, consciousness of such a concern has grown, leading to a progressive involvement of the global audience (Lesage, 2021). This resulted in an essential concern as it has been identified as a strain promoting the faster development of resistant pathogens globally (Goossens et al., 2005; Wall et al., 2016). Medications must be used with caution in the long term. It ought to be handled from a single medical viewpoint, considering that drug resistance in people, nutrition, the ecosystem, and wildlife are linked, and interchange can happen continually (Forsberg et al., 2012; Hu et al., 2017).

The European Union prohibited all medications as livestock growth enhancers in 2006 and recommended replacements (Simitzis, 2017). Organic goods like edible herbs and essential oils, including herbal preparations, were viewed as viable replacement treatments (Mushtaq et al., 2018). Essential oils are some of the most commercially significant organic compounds, and they are usually accountable for the well-being qualities of various animals. Such substances were taken from a variety of plants that are mostly found in moderate to warm climates, such as the Middle East and tropical areas, and are a significant element of conventional medicinal products. Essential oils are a combination of reduced compounds such as hydrocarbons, acetaldehyde, alkynes, and esters, which, in combination with their active ingredients, are responsible for the fragrant scent that such products commonly show (Jean, 2008).

Currently, about 3000 essential oils are recognized, with 10% of them being technically and financially significant. Such goods have the ability to be repositories of several pharmacological activities having a variety of therapeutic qualities, and they are consistent with recent consumption patterns for organic ingredients (Brenes & Roura, 2010). These also have the benefit of becoming more tolerable in the nervous system and diminishing negative impacts (Liu et al., 2017). Essential oils have also been studied for their taste qualities (Gutierrez et al., 2009; Shi et al., 2017) as well as

Applications of Essential Oils in the Food Industry. DOI: https://doi.org/10.1016/B978-0-323-98340-2.00030-4

their ability to be employed as functional ingredients since they increase the storage period (Fernandes et al., 2017) or lower the amount of Escherichia bacteria (Moro et al., 2015). The biggest barrier to employing them as feed additives is that they are frequently insufficiently powerful and produce sensory property modifications when used in normal concentrations to elicit the bactericidal property. Nonetheless, more research into chemical interactions is required to discover the much more therapeutic benefits that remain elusive.

Over the last few years, essential oils have gained importance in the livestock and healthcare fields. Unfortunately, its oxidative and antimutagenic properties in the livestock industry have received little attention (Castillo et al., 2005; Lykkesfeldt & Svendsen, 2007). Essential oils have been shown in research to enhance wildlife productivity and effectiveness by raising digestion (Jang et al., 2007), raising the quantity of probiotics like *Lactobacillus* spp. (Adaszýnska-Skwirzýnska & Szczerbínska, 2019; Cetin et al., 2016), arousing resistance to infection and the gastric microbes and reducing cell damage (Li et al., 2012; Zeng, Zhang, Wang, et al., 2015; Zeng, Zhang, Zhao, et al., 2015). Nevertheless, certain variations in essential oil efficacy remain, owing to the structure of the phytochemicals and various internal and external variables like illness, nutrient intake, surroundings, and, especially, diet composition (Omonijo et al., 2018).

According to the Developmental Origins of Health and Disease paradigm, using essential oils as novel nutritional additions during early development (pregnancy or breastfeeding) could be a very viable strategy for reducing antiviral drug use in cattle. Barker and Osmond (1986), Boyle (1955), pioneered this notion in individuals, and it has now been applied to zoology (Funston & Summers, 2013; Luebbe et al., 2019). This approach could produce young livestock in an antibiotic-free manufacturing environment, resulting in higher productivity and effectiveness at a later age, thereby helping to decrease resistant bacteria. Essential oils, generally known as "unstable" or "intangible" oils (Guenther, 1948), are fragrant lipid-soluble solutions derived from natural substances like different parts of a plant. They can be generated through translation, polymerization, enfleurage, or separation; however, simple distillation is the most extensively utilized method for commercialization of essential oils (Van de Braak & Leijten, 1999). The term "essential oil" is thought to have originated in the Middle Ages with the Swiss scientist and inventor Paracelsus von Hohenheim, who named the main contributor to a medication Quinta crucial (Guenther, 1948). As previously stated, approximately three thousand derivatives are recognized, of which approximately three hundred are well-known around the world and are primarily aimed at the tastes and perfumes industry (Van de Braak & Leijten, 1999).

Certain essential oils have been known to possess antibacterial characteristics (Guenther, 1948), which were reviewed previously (Nychas, 1995; Shelef et al., 1984), like the antibacterial activities of ingredients (Shelef et al., 1984). Nonetheless, the sudden rise in "eco-friendly" consumer culture has rekindled great interest in such substances (Nychas, 1995; Tuley de Silva, 1996). Aside from antimicrobial activity capabilities (Mari et al., 2003; Nychas, 1995; Ultee & Smid, 2001), extracts and their constituents were demonstrated to possess antiviral, antitumor (Mari et al., 2003), anti-toxigenic (Juglal et al., 2002; Ultee & Smid, 2001), antiprotozoal (Pandey et al., 2000; Pessoa et al., 2002), and insecticidal (Karpouhtsis et al., 1998; Konstantopoulou et al., 1992) qualities. Such properties may be linked to the role of these chemicals in crops (Guenther, 1948; Mahmoud & Croteau, 2002).

Essential oil in the diet of livestock

The impacts of essential oils in the diet on the flesh and stage of lactation of livestock are conflicting, most likely because of the varying concentrations of the essential oils delivered within each research study. Nevertheless, there is an agreement that gives low dosages of essential oils (from 1.33 and 4 g per liverstock each day) could enhance beef quality and longevity in certain circumstances (decreased discoloration, high antioxidant effect, and reduced oxidative damage in the meat). In crossbred cows, a regular intake of 3.5 g per head of livestock could be advised (De Oliveira Monteschio et al., 2017; Rivaroli et al., 2016). Such small concentrations have demonstrated antimicrobial capacity that might decrease oxidative damage (and hence enhance hue) and oxidative peptidases such as caspases, thereby improving textural characteristics (Harris et al., 2001).

Nevertheless, large concentrations may have a pro-oxidant impact on the livestock, with negative repercussions for medical issues and delivery performance. This is due to the fact that high quantities of essential oil can permeabilize the organelles, altering the transfer of electrons and releasing additional oxidative stress like superoxide anion which damages fatty acids and proteins. Certain compounds may be deposited at the muscle stage, as feeding sheep with lavender essential oils enhanced the textural properties of mutton (e.g., taste and palatability) (Smeti et al., 2018). Nevertheless, it is worth noting that the majority of reported research investigating the impacts of essential oils on carcass characteristics when given to livestock does not assess the propagation of breakdown products to the item; that could vary depending on the uptake, dispersion, digestion, and efflux of the multiple elements contained in the structure of the essential oil administered.

El-Essawy et al. (2021) recently looked into the impact of an essential oil mixture (licorice flavor, garlic, and rosemary) on 8 breastfeeding Sahara goats. The addition of these essential oils enhanced digestion without any impact on consumption or dairy production. They discovered that adding essential oils enhanced the brewing process. When opposed to the normal control, introducing essential oils increased milk protein output, amount of fat, and percentages of mixed and monounsaturated saturated fats. Benchaar (2021) discovered that adding Thymus essential oils to milk production diets did not enhance fermentation. Soltan et al. (2018) looked into the impact on Santa Inês sheep of a nanoparticulate combination of essential oil containing methyl esters, tocopherol, curcumin, and pepper aqueous extracts. They discovered that feeding this combination reduces methane generation.

Essential oil application in poultry production

The substitute of antimicrobials stimulants with other natural and healthy compounds is a essential element for the animal production. Because of their flexibility, active compounds can be utilized as synthetic chemicals in poultry farming. The utilization of natural ingredients and other organic ingredients as production supplements has yielded a few impressive outcomes. Classic effectiveness variables for poultry production are muscle mass, development, nutrition, and improved feed proportion. Different writers have documented the beneficial effect of natural ingredients on poultry farming efficiency (Al-Kassie, 2010; Calislar et al., 2009; Aguilar et al., 2014; Azadegam Mehr et al., 2014; Erhan et al., 2012; Hong et al., 2012; Karadas et al., 2014; Roofchaee et al., 2011; Saleh et al., 2014; Vukić-Vranješ et al., 2013; Zeng, Zhang, Wang, et al., 2015; Zeng, Zhang, Zhao, et al., 2015).

It seems that using natural ingredients as developmental enhancer replacements in poultry feed does not always enhance, and sometimes intensifies, quality parameters (Brenes & Roura, 2010; Demir et al., 2008; Kirkpinar et al., 2011; Ocak et al., 2008; Saleh et al., 2014; Zeng, Zhang, Wang, et al., 2015; Zeng, Zhang, Zhao, et al., 2015). This is most likely caused by an incorrect apparent viscosity or a brief application period. The discrepancies in the observed outcomes could be due to engaging frail chicks, infringements of biocontainment regulations, or the influence of external variables, such as bedsheets, illumination, devices, the existence of rats, and so on. This imbalance may also be a consequence of nutritional "flaws" during the research projects, e.g., an uneven diet or polluted nourishment or liquid.

Khosravinia (2015) demonstrated that adding savory oil to drinkable water not only improves productivity and quality but also lessens outcomes and experiences for the dead animal, which is connected with a decrease in garbage humidity and the oil's antimicrobial activities. Essential oils could indeed assist in increasing the sanitation standard at manufacturing sites. According to Witkowska and Sowinska (2013), peppermint and thyme oil may be employed to cloud development at all levels. Both oils appear to be effective, though thyme oil was more effective against Coli microbes and peppermint oil was more effective against the emergence of bacteria. Other uses for essential oils include combining them with alpha-tocopherol at the end of the chicken's developmental period. As a consequence of this diagnosis, the antioxidant capacity of mutton and its consumer items improves. Color fastness and moisture resistance are both improved. Chicken, furthermore, actually tastes nicer (Tongnuanchan & Benjakul, 2014). Much more recent research shows that essential oils may be utilized as organic antimicrobial drugs in raw meat conservation and protection. For instance, using 2% rosemary oil strengthened the storability of chicken breast flesh (Petrova et al., 2013; Ramos et al., 2011; Tongnuanchan & Benjakul, 2014). Spicy seasoning and onion essential oils increase the flow of blood, enabling the quicker elimination of dangerous substances from the broiler's skin. Nevertheless, it should be considered that onion or garlic oils have a detrimental impact on the aroma and flavor of animal flesh, so allow for a prolonged waiting time frame.

Effect of application of oregano essential oil on poultry and egg production

The consequences of oregano essential oil on chickens and laying hens cannot be viewed as definitive or constant because they may vary slightly between many studies conducted by similar researchers (Arpášová, 2011; Arpášová et al., 2015). Oregano essential oil has been shown in research to enhance effectiveness (Khattak et al., 2014), regular and ultimate muscle mass in chicken meat (Peng et al., 2016; Suchý et al., 2010), dietary utilization (Amrik & Bilkei, 2004; Basmacioğlu et al., 2010), egg mass and development (Abd et al., 2008; Arpášová, 2011; Suchý et al., 2010), and food consumption (Angelovičová et al., 2010) Improvements in feed utilization, yield, and quality may be linked to modifications in digestive shape, such as an increase in goblet cell length to increase crept intensity proportion (Peng et al., 2016) or biochemical processes, such as intestinal absorption as a result of chymotrypsin involvement (Basmacioğlu et al., 2010) and parasitic infection mitigation (Mohiti-Asli & Ghanaatparast-Rashti, 2015). Several studies have suggested that antioxidants from the oregano essential oil may move into the muscle of the laying hen, inhibiting the cascade of events linked with peroxidation and thereby reducing oxidation in albumen (Florou-Paneri et al., 2006). Conversely, many researchers have asserted that oregano essential oil has no impact on its effectiveness (Basmacioğlu et al., 2010; Jang et al., 2007; Lee et al., 2003). For instance, Arpášová et al. (2015) discovered that the inclusion of thyme and oregano essential oils had no impact on the muscle mass, feed intake, and transformation of laying hens' mass and muscle mass in egg production.

Essential oil on sow production

With respect to optimum productivity, the impacts of including oregano essential oil in foods are conflicting, as they are in broiler farming (Simitzis et al., 2010; Zeng, Zhang, Wang, et al., 2015; Zeng, Zhang, Zhao, et al., 2015). Nevertheless, beneficial benefits linked to including oregano essential oil in their food may be more obvious in other places, such as antitumor (Zeng, Zhang, Wang, et al., 2015; Zeng, Zhang, Zhao, et al., 2015) and alterations in blood count (Stelter et al., 2013) in pigs, it improves reproductive capacity (Allan & Bilkei, 2005), reduces cell damage, and improves litter performance (Tan et al., 2015). Because it improves cellular function, oregano essential oil can also be utilized to reduce anxiety caused by mass transit (Zhang et al., 2015; Zou et al., 2016). Lastly, pigs fed 1000 ppm oregano essential oil produced high-quality animal flesh with low peroxidation (Alarcón-Rojo et al., 2013).

Application of essential oil on piglets

The majority of studies on essential oils in sows were already aimed at nursery sows, due to the modifications in their food intake and other challenges they display at this important moment, which often have a detrimental effect on their well-being and efficiency. According to various researches, consuming essential oils throughout this time causes alterations in the intestinal epithelium which support a safer microbial population (Franz et al., 2010; Huang et al., 2010; Li et al., 2012). In certain instances, the emergence of safer microbes seems to outnumber the detrimental pathogenic microorganisms that carry disease and inadequate nutrient consumption and efficiency in the initial days or months of breastfeeding. Li et al. (2012) discovered that coated essential oils (thymol and cinnamaldehyde evaluated in these research findings) increased quality, invulnerability, and intestinal microbiomes in 240 piglets that were a month and few days old (at the beginning of investigation) more than a prolonged time frame; findings revealed a decrease in Escherichia coli numbers in excrement, enhanced lymphocyte transition as well as decreased gastroenteritis. Huang et al. (2010) observed that feeding mashed-up essential oils to 90 exclusively breastfed piglets for a month and 14 days caused an increased quality in post-weaning total mass with no obvious adverse impacts on well-being or other success factors. In a piglet survey performed beyond a period of three weeks after breastfeeding at day 21, Neill et al. (2006) found that in-feed bacteriocins enhanced productive efficiency more efficiently than a meal with natural ingredients.

Conclusion

Essential oils have been studied both internally and externally in livestock production. There seems to be significant proof that essential oils improve livestock production quality, as demonstrated by decreased food consumption, greater muscle fat mass, and improved resistance as well as wellness. As a result, innovative livestock components that consist of essential oils with potent, effective antibacterial characteristics are continuously being provided. Without a doubt, one of the most significant advantages of essential oils is that no microorganisms with drug resistance have previously been published as a result of their components. Essential oils are employed at a variety of concentration levels, and an added benefit is that they may be applied in relation to treatment. Unlike chemotherapy drugs, phytogenic food additives do not harm the livestock and do not necessitate a time frame prior to actually killing, ensuring consumer protection.

References

Abd, E. L., Motaal, A. M., Ahmed, A. M. H., Bahakaim, A. S. A., & Fathi, M. M. (2008). Productive performance and immunocompetence of commercial laying hens given diets supplemeted with eucalyptus. *International Journal of Poultry Science*, 7, 445–449.

Adaszýnska-Skwirzýnska, M., & Szczerbínska, D. (2019). The effect of lavender (*Lavandula angustifolia*) essential oil as a drinking water supplement on the production performance, blood biochemical parameters, and ileal microflora in broiler chickens. *Poultry Science*, 98, 358–365. Available from https://doi.org/10.3382/ps/pey385.

Aguilar, C. A. L., Lima, K. R. S., Manno, M. C., Maia, J. G. S., Fernandes Neto, D. L., Tava-res, F. B., Roque, T. J. L., Mendonca, R. A., & Carmo, E. S. N. (2014). Rosewood (*Aniba rosaeodora* Ducke) oil in broiler chickens diet. *Revista Brasileira de Saúde e Produção Animal, 15.* Available from http://doi.org/10.1590/S1519-99402014000100014.

Alarcón-Rojo, A. D., Peña-Gonzalez, E., Janacua-Vidales, H., Santana, V., & Ortega, J. A. (2013). Meat quality and lipid oxidation of pork after die-tary supplementation with oregano essential oil. *World Applied Sciences Journal, 21*(5), 665–673.

Al-Kassie, G. A. M. (2010). The role of peppermint (*Mentha piperita*) on performance in broiler diets. *Agriculture and Biology Journal of North America, 1,* 1009–1013.

Allan, P., & Bilkei, G. (2005). Oregano improves reproductive performance of sows. *Theriogenology, 63*(3), 716–721. Available from https://doi.org/10.1016/j.theriogenology.2003.06.010.

Amrik, B., & Bilkei, G. (2004). Influence of farm application of oregano on performances of sows. *Canadian Veterinary Journal, 45,* 674–677.

Angelovičová, M., Kačaniová, M., Angelovič, M., & Lopašovský, Ľ. (2010). Per os use of Thymi aetheroleum for growth performance of the broiler chickens. *Scientific Journal for Food Industry, 4,* 127–132, Special Issue 2010.

Arpášová, H. (2011). *Phytobiotics to replace antibiotic growth promoters and their impact on the usefulness and quality of the eggs of laying hens type.* Nitra: SPU, 2011. 101 s. ISBN 978-80-552–0555-7.

Arpášová, H., Branislav, G., Hrnčár, C., Fik, M., Herkel, R., & Pistová, V. (2015). The effect of essential oils on performance of laying hens. *Animal Science and Biotechnologies, 48,* 8–14.

Azadegam Mehr, M., Hassanabadi, A., Nassiri Moghaddam, H., & Kermanshahi, H. (2014). Supplementation of clove essential oils and probiotic to the broiler's diet on performance, carcass traits and blood components. *Iranian Journal of Applied Animal Science, 1,* 117–122.

Barker, D. J. P., & Osmond, C. (1986). Infant mortality, childhood nutrition, and ischaemic heart disease in england and wales. *Lancet, 327,* 1077–1081. Available from https://doi.org/10.1016/S0140-6736(86)91340-1.

Basmacioğlu, M. H., Baysal, S., Misirlioğlu, Z., Polat, M., Yilmaz, H., & Turan, N. (2010). Effects of oregano essential oil with or without feed enzymes on growth performance, digestive enzyme, nutrient digestibility, lipid metabolism and immune response of broilers fed on wheat-soybean meal diets. *British Poultry Science, 51*(1), 67–80.

Benchaar, C. (2021). Diet supplementation with thyme oil and its main component thymol failed to favorably alter rumen fermentation, improve nutri-ent utilization, or enhance milk production in dairy cows. *Journal of Dairy Science, 104,* 324–336. Available from https://doi.org/10.3168/jds.2020-18401.

Boyle, W. (1955). *The American perfumer and essential oil review 66. Spices and essential oils as presservatives* (pp. 25–28). https://www.scirp.org/(S(351jmbntvnsjt1aadkposzje))/reference/ReferencesPapers.aspx?ReferenceID = 902045 (Accessed 30.01.21).

Brenes, A., & Roura, E. (2010). Essential oils in poultry nutrition: Main effects and modes of action. *Animal Feed Science and Technology, 158,* 1–14. Available from https://doi.org/10.1016/j.anifeedsci.2010.03.007.

Calislar, S., Gemci, I., & Kamalak, A. (2009). Effects of Orego-Stim® on broiler chick performance and some blood parameters. *Journal of Animal and Veterinary Advances, 8,* 2617–2620.

Castillo, C., Hernandez, J., Bravo, A., Lopez-Alonso, M., Pereira, V., & Benedito, J. L. (2005). Oxidative status during late pregnancy and early lacta-tion in dairy cows. *Veterinary Journal (London, England: 1997), 169,* 286–292. Available from https://doi.org/10.1016/j.tvjl.2004.02.001.

Cetin, E., Yibar, A., Yesilbag, D., Cetin, I., & Cengiz, S. S. (2016). The effect of volatile oil mixtures on the performance and ilio-caecal microflora of broiler chickens. *British Poultry Science, 57,* 780–787. Available from https://doi.org/10.1080/00071668.2016.1214682.

De Oliveira Monteschio, J., de Souza, K. A., Vital, A. C. P., Guerrero, A., Valero, M. V., Kempinski, E. M. B. C., Barcelos, V. C., Nascimento, K. F., & do Prado, I. N. (2017). Clove and rosemary essential oils and encapsuled active principles (eugenol, thymol and vanillin blend) on meat quality of feedlot-finished heifers. *Meat Science, 130,* 50–57. Available from https://doi.org/10.1016/j.meatsci.2017.04.002.

Demir, E., Kilinc, K., Yildirim, Y., Dincer, F., & Esecel, H. (2008). *Comparative effects of mint, sage, thyme and flavomycin in wheat-based broiler diets,* . Archivos de Zootecnia (11, pp. 54–63). .

El-Essawy, A. M., Anele, U. Y., Abdel-Wahed, A. M., Abdou, A. R., & Khattab, I. M. (2021). Effects of anise, clove and thyme essential oils supple-mentation on rumen fermentation, blood metabolites, milk yield and milk composition in lactating goats. *Animal Feed Science and Technology, 271,* 114760. Available from https://doi.org/10.1016/j.anifeedsci.2020.114760.

Erhan, M. K., Bolukbas, S. C., & Urusan, H. (2012). Biological activities of pennyroyal (*Mentha pulegium* L.) in broilers. *Livestock Science, 146,* 189–192.

Fernandes, R. V., de, B., Guimarães, I. C., Ferreira, C. L. R., Botrel, D. A., Borges, S. V., & de Souza, A. U. (2017). Microencapsulated rosemary (*Rosmarinus officinalis*) essential oil as a biopreservative in minas frescal cheese. *Journal of Food Processing and Preservation, 41,* e12759. Available from https://doi.org/10.1111/jfpp.12759.

Florou-Paneri, P., Dostas, D., Mitsopoulos, I., Dostas, V., Botsoglou, E., Nikolakakis, I., & Botsoglou, N. (2006). Effect of feeding rosemary and α-tocopheryl acetate on hen performance and egg quality. *Poultry Science, 465,* 143–149.

Forsberg, K. J., Reyes, A., Wang, B., Selleck, E. M., Sommer, M. O., & Dantas, G. (2012). The shared antibiotic resistome of soil bacteria and human pathogens. *Science (New York, N.Y.), 337,* 1107–1111. Available from https://doi.org/10.1126/science.1220761.

Franz, C., et al. (2010). Essential oils and aromatic plants in animal feeding–a European perspective. A review. *Flavour and Fragrance Journal, 25* (5), 327–340.

Funston, R. N., & Summers, A. F. (2013). Effect of prenatal programming on heifer development. *Veterinary Clinics Food Animal Practice, 29,* 517–536. Available from https://doi.org/10.1016/j.cvfa.2013.07.001.

Goossens, H., Ferech, M., Vander Stichele, R., & Elseviers, M. (2005). Outpatient antibiotic use in Europe and association with resistance: A cross-national database study. *Lancet, 365,* 579–587. Available from https://doi.org/10.1016/S0140-6736(05)70799-6.

Guenther, E. (1948). The essential oils. *D.* New York, NY, USA: Van Nostrand.

Gutierrez, J., Barry-Ryan, C., & Bourke, P. (2009). Antimicrobial activity of plant essential oils using food model media: Efficacy, synergistic potential and interactions with food components. *Food Microbiology, 26,* 142−150. Available from https://doi.org/10.1016/j.fm.2008.10.008.

Harris, S. E., Huff-Lonergan, E., Lonergan, S. M., Jones, W. R., & Rankins, D. (2001). Antioxidant status affects color stability and tenderness of calcium chloride-injected beef. *Journal of Animal Science, 79,* 666−677. Available from https://doi.org/10.2527/2001.793666x.

Hong, J. C., Steiner, T., Aufy, A., & Lien, T. F. (2012). Effects of supplemental essential oil on growth performance, lipid metabolites and immunity, intestinal characteristics, microbiota and carcass traits in broilers. *Livestock Science, 144,* 253−262.

Hu, Y., Gao, G. F., & Zhu, B. (2017). The antibiotic resistome: Gene flow in environments, animals and human beings. *Frontiers in Medicine, 11,* 161−168. Available from https://doi.org/10.1007/s11684-017-0531-x.

Huang, Y., et al. (2010). Effects of dietary supplementation with blended essential oils on growth performance, nutrient digestibility, blood profiles and fecal characteristics in weanling pigs. *Asian-Australasian Journal of Animal Sciences, 23*(5), 607.

Jang, I. S., Ko, Y. H., Kang, S. Y., & Lee, C. Y. (2007). Effect of a commercial essential oil on growth performance, digestive enzyme activity and intestinal microflora population in broiler chickens. *Animal Feed Science and Technology, 134,* 304−315. Available from https://doi.org/10.1016/j.anifeedsci.2006.06.009.

Jean, B. (2008). *Pharmacognosy phytochemistry, medicinal plants, 2nd ed* (Volume 934). Paris, France: Lavoisier.

Juglal, S., Govinden, R., & Odhav, B. (2002). Spice oils for the control of co-occurring mycotoxin-producing fungi. *Journal of Food Protection, 65,* 683−687. Available from https://doi.org/10.4315/0362-028X-65.4.683.

Karadas, F., Pirgozliev, V., Rose, S. P., Dimitrov, D., Oduguwa, O., & Bravo, D. (2014). Dietary essential oils improve the hepatic antioxidative status of broiler chickens. *British Poultry Science, 3,* 329−334.

Karpouhtsis, I., Pardali, E., Feggou, E., Kokkini, S., Scouras, Z. G., & Mavragani-Tsipidou, P. (1998). Insecticidal and genotoxic activities of oregano essential oils. *Journal of Agricultural and Food Chemistry, 46,* 1111−1115. Available from https://doi.org/10.1021/jf970822o.

Khattak, F., Ronchi, A., Castelli, P., & Sparks, N. (2014). Effects of natural blend of essential oil on growth performance blood biochemistry, cecal morphology, and carcass quality of broiler chickens. *Poultry Science, 93,* 132−137.

Khosravinia, H. (2015).). Litter quality and external carcass defects in broiler chicken influenced by supplementation of drinking water with savory (Satureja khuzistanica) essential oils. *Global Journal of Animal Scientific Research, 1,* 247−252.

Kirkpinar, F., Unulu, H. B., & Ozdemir, G. (2011). Effects of oregano and garlic essential oils on performance, carcase, organ and blood characteristics and intestinal microfora of broilers. *Livestock Science, 137,* 219−225.

Konstantopoulou, I., Vassilopoulou, L., Mavragani-Tsipidou, P., & Scouras, Z. G. (1992). Insecticidal effects of essential oils. A study of the effects of essential oils extracted from eleven Greek aromatic plants on *Drosophila auraria. Experientia, 48,* 616−619. Available from https://doi.org/10.1007/BF01920251.

Lee, K. W., Everts, H., Lankhorst, A. E., Kappert, H. J., & Beynen, A. C. (2003). Addition of β-ionone to the diet fails to affect growth performance in female broiler chickens. *Animal Feed Science and Technology, 106,* 219−223.

Lesage, M. (2021). Les Antibiorésistances en Élevage: Vers des Solutions Intégrées. MONTREUIL SOUS BOIS. Available from https://agriculture.gouv.fr/sites/minagri/files/cep_analyse82_antibioresistances_en_elevage.pdf (Accessed 13.01.21).

Li, S. Y., Ru, Y. J., Liu, M., Xu, B., Péron, A., & Shi, X. G. (2012). The effect of essential oils on performance, immunity and gut microbial population in weaner pigs. *Livestock Science, 145*(1−3), 119−123. Available from https://doi.org/10.1016/j.livsci.2012.01.005.

Liu, Q., Meng, X., Li, Y., Zhao, C.-N., Tang, G.-Y., & Li, H.-B. (2017). Antibacterial and antifungal activities of spices. *International Journal of Molecular Sciences, 18,* 1283. Available from https://doi.org/10.3390/ijms18061283.

Luebbe, K. M., Stalker, L. A., Klopfenstein, T. J., & Funston, R. N. (2019). Influence of weaning date and late gestation supplementation on beef system productivity I: Animal performance. *Translational Animal Science, 3,* 1492−1501. Available from https://doi.org/10.1093/tas/txz116.

Lykkesfeldt, J., & Svendsen, O. (2007). Oxidants and antioxidants in disease: Oxidative stress in farm animals. *Veterinary Journal 2007, 173,* 502−511. Available from https://doi.org/10.1016/j.tvjl.2006.06.005.

Mahmoud, S. S., & Croteau, R. B. (2002). Strategies for transgenic manipulation of monoterpene biosynthesis in plants. *Trends in Plant Science, 7,* 366−373. Available from https://doi.org/10.1016/S1360-1385(02)02303-8.

Mari, M., Bertolini, P., & Pratella, G. C. (2003). Non-conventional methods for the control of post-harvest pear diseases. *Journal of Applied Microbiology, 94,* 761−766. Available from https://doi.org/10.1046/j.1365-2672.2003.01920.x.

Mohiti-Asli, M., & Ghanaatparast-Rashti, M. (2015). Dietary oregano essential oil alleviates experimentally induced coccidiosis in broilers. *Preventive Veterinary Medicine, 120*(2), 195−202.

Moro, A., Librán, C. M., Berruga, M. I., Carmona, M., & Zalacain, A. (2015). Dairy matrix effect on the transference of rosemary (*Rosmarinus officinalis*) essential oil compounds during cheese making. *Journal of the Science of Food and Agriculture, 95,* 1507−1513. Available from https://doi.org/10.1002/jsfa.6853.

Mushtaq, S., Shah, A. M., Shah, A., Lone, S. A., Hussain, A., Hassan, Q. P., & Ali, M. N. (2018). Bovine mastitis: An appraisal of its alternative herbal cure. *Microbial Pathogenesis, 114,* 357−361. Available from https://doi.org/10.1016/j.micpath.2017.12.024.

Neill, C. R., et al. (2006). Effects of oregano oil on growth performance of nursery pigs. *Journal of Swine Health and Production, 14*(6), 312−316.

Nychas, G. J. E. (1995). Natural antimicrobials from plants. In G. W. Gould (Ed.), *New methods of food preservation* (pp. 58−89). London, UK: Blackie Academic and Professional.

Ocak, N., Erener, G., Ak, F. B., Sungu, M., Altop, A., & Ozmen, A. (2008). Performance of broilers fed diets supplemented with dry peppermint (*Mentha piperita* L.) or thyme (*Thymus vulgaris* L.) leaves as growth promoter source. *Czech Journal of Animal Science, 53,* 169−175.

Omonijo, F. A., Ni, L., Gong, J., Wang, Q., Lahaye, L., & Yang, C. (2018). Essential oils as alternatives to antibiotics in swine production. *Animal Nutrition, 4*, 126–136. Available from https://doi.org/10.1016/j.aninu.2017.09.001.

Pandey, R., Kalra, A., Tandon, S., Mehrotra, N., Singh, H. N., & Kumar, S. (2000). Essential oils as potent source of nematicidal compounds. *Journal of Phytopathology, 148*, 501–502. Available from https://doi.org/10.1046/j.1439-0434.2000.00493.x.

Peng, Q. Y., Li, J. D., Li, Z., Duanb, Z. Y., & Wua, Y. P. (2016). Effects of dietary supplementation with oregano essential oil on growth performance, carcass traits and jejunal morphology in broiler chickens. *Animal Feed Science and Technology, 214*, 148–153.

Pessoa, L. M., Morais, S. M., Bevilaqua, C. M. L., & Luciano, J. H. S. (2002). Anthelmintic activity of essential oil of *Ocimum gratissimum* Linn. and eugenol against *Haemonchus contortus*. *Veterinary Parasitology, 109*, 59–63. Available from https://doi.org/10.1016/S0304-4017(02)00253-4.

Petrova, J., Pavelkova, A., Hleba, L., Pochop, J., Rovna, K., & Kaniova, M. (2013). Microbiological quality of fresh chicken breast meat after rosemary essential oil treatment and vacuum packaging. *Journal of Animal Science and Biotechnology, 1*, 140–144.

Ramos, F. A., Martinez, A. P., Montes, E. S., Garcia, J. M., Perez, C. M., Valesco, J. L., & Gaytran, C. N. (2011). Effects of dietary oregano essential oil and vitamin E on the lipid oxidation stability of cooked chicken breast meat. *Poultry Science, 2*, 505–511.

Rivaroli, D. C., Guerrero, A., Valero, M. V., Zawadzki, F., Eiras, C. E., del Mar Campo, M., Sañudo, C., Jorge, A. M., & do Prado, I. N. (2016). Effect of essential oils on meat and fat qualities of crossbred young bulls finished in feedlots. *Meat Science, 121*, 278–284. Available from https://doi.org/10.1016/j.meatsci.2016.06.017.

Roofchaee, A., Mehradad, I., Ebrahimzadeeh, M. A., & Akbari, M. R. (2011). Effect of dietary oregano (*Origanum vulgare* L.) essential oil on growth performance, cecal microflora and serum antioxidant activity of broiler chickens. *African Journal of Biotechnology, 32*, 6177–6183.

Saleh, N., Allam, T., El-latif, A. A., & Ghazy, E. (2014). The effects of dietary supplementation of different levels of thyme (*Thymus vulgaris*) and ginger (*Zingiber officinale*) essential oils on performance, hematological, biochemical and immunological parameters of broiler chickens. *Global Veterinary, 6*, 736–744.

Shelef, L. A., Jyothi, E. K., & Bulgarellii, M. A. (1984). Growth of enteropathogenic and spoilage bacteria in sage-containing broth and foods. *Journal of Food Science, 49*, 737–740. Available from https://doi.org/10.1111/j.1365-2621.1984.tb13198.x.

Shi, C., Zhang, X., Zhao, X., Meng, R., Liu, Z., Chen, X., & Guo, N. (2017). Synergistic interactions of nisin in combination with cinnamaldehyde against *Staphylococcus aureus* in pasteurized milk. *Food Control, 71*, 10–16. Available from https://doi.org/10.1016/j.foodcont.2016.06.020.

Simitzis, P. E. (2017). Enrichment of animal diets with essential oils—A great perspective on improving animal performance and quality characteristics of the derived products. *Medicines, 4*, 35. Available from https://doi.org/10.3390/medicines4020035.

Simitzis, P. E., Symeon, G. K., Charismiadou, M. A., Bizelis, J. A., & Deligeorgis, S. G. (2010). The effects of dietary oregano oil supplementation on pig meat characteristics. *Meat Science, 84*(4), 670–676. Available from https://doi.org/10.1016/j.meatsci.2009.11.001.

Smeti, S., Hajji, H., Mekki, I., Mahouachi, M., & Atti, N. (2018). Effects of dose and administration form of rosemary essential oils on meat quality and fatty acid profile of lamb. *Small Ruminant Research: The Journal of the International Goat Association, 158*, 62–68. Available from https://doi.org/10.1016/j.smallrumres.2017.10.007.

Soltan, Y. A., Natel, A. S., Araujo, R. C., Morsy, A. S., & Abdalla, A. L. (2018). Progressive adaptation of sheep to a microencapsulated blend of essential oils: Ruminal fermentation, methane emission, nutrient digestibility, and microbial protein synthesis. *Animal Feed Science and Technology, 237*, 8–18. Available from https://doi.org/10.1016/j.anifeedsci.2018.01.004.

Stelter, K., Frahm, J., Paulsen, J., Berk, A., Kleinwächter, M., Selmar, D., & Dänicke, S. (2013). Effects of oregano on performance and immunmodulating factors in weaned piglets. *Archives of Animal Nutrition, 67*(6), 461–476.

Suchý, P., Strakova, E., Mas, N., Serman, V., Vecerek, V., Bedrica, L., Lukac, Z., & Horvat, Z. (2010). The effect of a herbal additive on performance parameters in layers. *Tierarztliche Umschau, 65*, 74–78.

Tan, Ch, Wei, H., Sun, H., Ao, J., Long, G., Jiang, S., & Peng, J. (2015). Effects of dietary supplementation of oregano essential oil to sows on oxidative stress status, lactation feed intake of sows, and piglet performance. *Journal of Nutrition, 1*, 1–9.

Tongnuanchan, P., & Benjakul, S. (2014). Essential oils: Extraction, bioactivities, and their uses for food preservation. *Journal of Food Science, 7*, 1231–1249.

Tuley de Silva, K. (1996). *A manual on the essential oil industry*. Vienna, Austria: United Nations Industrial Development Organization.

Ultee, A., & Smid, E. J. (2001). Influence of carvacrol on growth and toxin production by Bacillus cereus. *International Journal of Food Microbiology, 64*, 373–378. Available from https://doi.org/10.1016/S0168-1605(00)00480-3.

Van de Braak, S.A. A.J., & Leijten, G.C. J.J. (1999). *Essential oils and oleoresins: A survey in the Netherlands and other major markets in the European Union* (p. 116). CBI, Centre for the Promotion of Imports from Developing Countries; Rotterdam, The Netherlands.

Vukić-Vranješ, N., Tolimir, D., Vukmirović, R., Čolović, V., Stanaćev, P., & Ikonić, S. (2013). Effect of phytogenic additives on performance, morphology and caecal microflora of broiler chicken. *Biotechnology in Animal Husbandry, 2*, 311–319.

Wall, B. A., Mateus, A. L. P., Marshall, L., Pfeiffer, D. U., Lubroth, J., Ormel, H. J., Otto, P., & Patriarchi, A. (2016). *Drivers, dynamics and epidemiology of antimicrobial resistance in animal production*. Rome, Italy: Food and Agriculture Organization of the United Nations accessed on 14 February 2021. Available from http://www.fao.org/publications/card/fr/c/d5f6d40d-ef08-4fcc-866b-5e5a92a12dbf/.

Witkowska, D., & Sowinska, J. (2013). The effectiveness of peppermint and thyme essential oil mist in reducing bacterial contamination in broiler houses. *Poultry Science, 11*, 2834–2843.

Zeng, Z., Xu, X., Zhang, Q., Li, P., Zhao, P., Li, Q., Liu, J., & Piao, X. (2015). Effects of essential oil supplementation of a low-energy diet on performance, intestinal morphology and microflora, immune properties and antioxidant activities in weaned pigs. *Animal Science Journal, 86*, 279–285. Available from https://doi.org/10.1111/asj.12277.

Zeng, Z., Zhang, S., Wang, H., & Paio, X. (2015). Essential oil and aromatic plants as feed additives in non-ruminant nutrition: A review. *Journal of Animal Science and Biotechnology, 6*, 7−15.

Zhang, T., Zhou, Y. F., Zou, Y., Hu, X. M., Zheng, L. F., Wei, H. K., Giannenas, I., Jinc, L. Z., Peng, J., & Jiang, S. W. (2015). Effects of dietary oregano essential oil supplementation on the stress response, antioxidative capacity, and HSPs mRNA expression of transported pigs. *Livestock Science, 180*, 143−149.

Zou, Y., Xiang, Q., Wang, J., Wei, H., & Peng, J. (2016). Effects of oregano essential oil or quercetin supplementation on body weight loss, carcass characteristics, meat quality and antioxidant status in finishing pigs under transport stress. *Livestock Science, 192*, 33−38. Available from https://doi.org/10.1016/j.livsci.2016.08.005.

Chapter 25

Stability of essential oil during different types of food processing and storage and their role in postharvest management of fruits and vegetables

Nyejirime Young Wike[1], Olugbemi T. Olaniyan[2], Charles Oluwaseun Adetunji[3], Juliana Bunmi Adetunji[4], Olalekan Akinbo[5], Babatunde Oluwafemi Adetuyi[6], Abel Inobeme[7], Yovwin D. Godwin[8], Oloruntoyin Ajenifujah-Solebo[9], Majolagbe Olusola Nathaniel[10], Ismail Ayoade Odetokun[11], Oluwabukola Atinuke Popoola[9], Olatunji Matthew Kolawole[12] and Mohammed Bello Yerima[13]

[1]Department of Human Physiology, Faculty of Basic Medical Sciences, Rhema University, Aba, Abia State, Nigeria, [2]Laboratory for Reproductive Biology and Developmental Programming, Department of Physiology, Faculty of Basic Medical Sciences, Rhema University, Aba, Abia State, Nigeria, [3]Applied Microbiology, Biotechnology and Nanotechnology Laboratory, Department of Microbiology, Edo State University Uzairue, Iyamho, Edo State, Nigeria, [4]Department of Biochemistry, Osun State University, Osogbo, Osun State, Nigeria, [5]Centre of Excellence in Science, Technology, and Innovation, AUDA-NEPAD, Johannesburg, Gauteng, South Africa, [6]Department of Natural Sciences, Faculty of Pure and Applied Sciences, Precious Cornerstone University, Ibadan, Oyo State, Nigeria, [7]Department of Chemistry, Edo State University Uzairue Iyamho, Auchi, Edo State, Nigeria, [8]Department of Family Medicine, Faculty of Clinical Sciences, Delta State University, Abraka, Delta State, Nigeria, [9]Genetics, Genomics and Bioinformatics Department, National Biotechnology Development Agency, Abuja, FCT, Nigeria, [10]Microbiology Unit, Department of Pure and Applied Biology, Ladoke Akintola University of Technology, Ogbomoso, Oyo State, Nigeria, [11]Department of Veterinary Public Health and Preventive Medicine, University of Ilorin, Ilorin, Kwara State, Nigeria, [12]Department of Microbiology, Faculty of Life Sciences, University of Ilorin, Ilorin, Kwara State, Nigeria, [13]Department of Microbiology, Sokoto State University, Sokoto, Sokoto State, Nigeria

Introduction

Fresh produce is high in phytoconstituents, vitamins, and minerals, all of which are vital to the body. A daily intake of 400 g of fresh produce is recommended for the prevention of long-term illnesses (like myocardial infarction, cancer, insulin resistance, and overweight) and the treatment of many nutritional problems (World Health Organization (WHO), 2003). With the world's population nearing nine billion people, 70%−100% more fresh produce is required to sustain humanity (Godfray et al., 2010). As a result, reducing damage along the fresh produce distribution network is critical to ensuring their availability. Fresh produce, on the other hand, is inherently perishable, with a brief postharvest existence. These agricultural products are susceptible to mechanical damage, fluctuations in managing thermal environments, and pollutants throughout collection and processing, all of which may result in postharvest deterioration. Microbes are estimated to deteriorate 20%−25% of cultivated produce throughout postharvest operations (Droby, 2006; Singh & Sharma, 2007; Zhu, 2006).

Reasonable precautions are being investigated in order to avoid this. Treatment processes, such as the use of authorized antimicrobials, are the most frequently used because of the convenience and low cost of implementation. Notwithstanding, their detrimental consequences on buyers' well-being, like byproducts and a lengthy deterioration timeframe, have piqued the interest of buyers, producers, and operators. This is due to the brief timeframe between fungicidal usage and intake, and the impact may be more disastrous than those used in the pre-harvest phase (Amin et al.,

Applications of Essential Oils in the Food Industry. DOI: https://doi.org/10.1016/B978-0-323-98340-2.00027-4

2012). As a result, it is essential to devise secure, efficient, and environmentally sustainable methods to circumvent the constraints of antimicrobials in lowering yield loss.

Plant-derived essential oils have antimicrobial activities (Bosquez-Molina et al., 2010), lesser side effects, less ecological consequences (Burt, 2004), and are eco-friendly and compostable (Tzortzakis & Economakis, 2007). As a result, utilizing essential oils postharvest is an efficient method.

Cleaning fruits after cultivation and during preservation is discouraged by typical small fruit manufacturing systems. Because small fruits are eaten fresh, it might encourage the development of food-borne microbes and pose a threat of food safety to buyers. Furthermore, because they are susceptible to deteriorative microbes, myopathies, and cell damage, tiny fruit and vegetables possess an extremely short lifespan (Duduk et al., 2015). This is owed in part to their high nutritional value and moisture, as well as their neutral affinity, which makes them vulnerable to microbial pathogens. As a result, damages throughout cultivation, management, and advertising can reach up to 50% (Agrios, 2000). Microbes may also generate mycotoxins, which end up making the fruit unsafe for consumption (Tripathi et al., 2008).

Essential oil in postharvest management of fruit and vegetables

The "Rongrien" rambutan (*Nephelium lappaceum* Linn. cv. Rongrien) is well recognized in Thailand as an economically significant fruit. Nevertheless, rambutan fruits are extremely vulnerable to synthetic sealant harm, particularly to their spinners. Physical stress might hasten spinneret discoloration and create an injury for microbes to enter after cultivation. Dehydration from the spinterns, as well as the epidermis, can cause seed coat discoloration, with the impacts on the spinterns being more prominent compared to those on the epidermis (Yingsanga et al., 2006). More so, a greater concentration can cause water loss and increased vulnerability to charring. Furthermore, the greater discoloration rates of the spinturns may be due to increased polyphenol oxidase and peroxide actions in the spinturns compared to the husk (Yingsanga et al., 2008). Low-temperature storage (10C−13C) and reformed air wrapping with low-density polyethylene have been suggested to extend the shelf life of the product of rambutan fruit since these decrease weight reduction as well as epidermis charring (Julianti et al., 2012).

In recent history, buyers have become more security- and health-conscious regarding nutrition. As a result, studies have concentrated on ecologically responsible substances for handling fresh produce, as well as postharvest therapies for preventive measures and storage period advancement. Because of their beneficial properties, essential oils have been extensively researched among many substitute naturally derived isolates. Combustible components in essential oils are generally a combination of terpenes, terpenoids, as well as other aliphatic and aromatic molecules (Bakkali et al., 2008) that are frequently utilized as artificial flavors in the food market. These substances also show antimicrobial, fungicidal, and antioxidant activities (Sanchez-Gonzalez et al., 2011; Schaneberg & Khan, 2002). Nevertheless, a few research findings have shown that the effectiveness of essential oils is primarily determined by their components, which are strongly impacted by the separation process used (Desai & Parikh, 2015; Schaneberg & Khan, 2002).

Numerous researchers have examined how natural oils, like basil oil (Juven et al., 1994) cassia oil (Tzortzakis, 2007), cymbopogon oil (Raybaudi-Massilia et al., 2008), and thyme oil (Juliano et al., 2008), can help with postharvest preventive measures. Several more studies have found that essential oils possess a favorable impact on fruit value and disease management, like on rambutans using cinnamaldehyde and curcumin oil (Phuoc, 2020; Sivakumar et al., 2002), chilies using Cymbopogon oil (Ali, et al., 2015), papayas using thyme and Mexican lime oils (Bosquez-Molina et al., 2010), and avocados using thyme and cinnamon oils (Bill et al., 2014; Combrinck et al., 2011). The high concentration of active chemicals in natural ingredients, on the other hand, may restrict their utilization in postharvest therapies, primarily since they influence the textural properties of the encased fruits or veggies (Majewska et al., 2019; Serrano et al., 2006). As a result, optimizing the optimum dose of an active ingredient in its implementation as a preservation method is crucial.

Essential oils' impact on the protective coating of ground beef

Ground-beef protection and storage period extension are critical research goals. Built-in safety containers, contemporary forms of handling (pasteurization, high intensity), and dietary supplements with antioxidant and antimicrobial properties are all methods for increasing the storage period. The combustion of cofactor pigments has a significant effect on the total consumer acceptance of the final piece, causing the appealing coloration of meat and its derivatives to fade. Ground beef appearance is a crucial concern for people, so hue evaluation is a vital component of a ground beef qualitative inquiry.

Several variables influence fatty acids and protein denaturation in beef products, including the extent of milling, storage conditions, availability of illumination, plastic wrapping category, storage period, availability of antioxidants, and others. Quite enough exploration has recently been devoted to the investigation of antioxidant properties, particularly those of naturally derived and prepared compounds utilizing various separation techniques (Oswell et al., 2018; Ribeiro et al., 2019). Plant food products frequently have a diverse set of biological properties: for example, they can inhibit peroxidation modifications and microbe development while providing the desired taste and appearance. The application of essential oils as antioxidants can prevent fatty acid and hemoglobin oxidative damage. Phytochemicals can be found in a variety of foods, including fruits, vegetables, herbs, and spices (Ahn et al., 2007; Naveena et al., 2008; Nunez et al., 2008; Shah et al., 2014).

The impact of natural ingredients on ground beef appearance could be attributable to their ability to reduce myoglobin and fatty acid oxidative damage. According to Hashemi Gahruie et al. (2017), cinnamon and rosemary concentrates reduced reactive thiobarbituric acid compounds and metmyoglobin in beef burgers. Yu et al. (2002) reported similar findings, indicating that rosemary extracts significantly reduced peroxidation and hue transformation in cooked chicken. Belantine et al. (2006) found that rosemary-treated minced beef stayed red for longer periods of time and had lower Tbars substance results than the untreated control. The introduction of rosemary oil to mined meat impeded discoloration throughout storage, according to Formanek et al. (2003). Finely ground rosemary enhanced the tenderness of meat patties, according to Sanchez-Escalante et al. (2001).

Conclusion

Numerous essential oils and their constituents are utilized as organic bacteriocins to decrease the effects of microbes in food products. Generally, the phenolic composition of essential oils functions as cell wall permeabilizers. Gram-positive microbes are more responsive to essential oils as well as their biologically active components than gram-negative bacteria. The primary constituents occurring in essential oils are thymol, carvacrol, and cinnamaldehyde, which are accountable for optimizing antimicrobial effects through numerous means such as altering permeation, changing cell wall fatty acids, and inhibiting electron orbital driving power. Essential oils derived from plant species that are either well-known or unknown to humanity are effective postharvest fungicide replacements. Once introduced to fresh produce, the oil may not only inhibit the growth of microorganisms but also sustain consumer characteristics. The sector needs such characteristics. Unfortunately, there have been few efforts to utilize essential oils to reduce yield loss.

References

Agrios, G. N. (2000). *Significance of plant diseases* (pp. 25–37). London, UK: Academic Press, Plant pathology.

Ahn, J., Grun, I. U., & Mustapha, A. (2007). *Food microbiology, 24*, 7–14.

Ali, A., Noh, N. M., & Mustafa, M. A. (2015). Antimicrobial activity of chitosan enriched with lemongrass oil against anthracnose of bell pepper. *Food Package Shelf Life, 3*, 56–61.

Amin, M., Ding, P., Jugah, K., & Hasanah, M. G. (2012). Effect of hot water dip treatment on control of postharvest anthracnose disease of banana var. Berangan. *African Journal of Agricultural Research, 7*(1), 6–10.

Bakkali, F., Averbeck, S., Averbeck, D., & Idaomar, I. (2008). Biological effects of essential oils-a review. *Food and Chemical Toxicology: An International Journal Published for the British Industrial Biological Research Association, 46*, 446–475.

Belantine, C. W., Crandall, P. G., O'Bryan, C. A., Duong, D. Q., & Pohlman, F. W. (2006). *Meat Science, 73*, 413–421.

Bill, M., Sivakumar, D., Korsten, L., & Thompson, A. K. (2014). The efficacy of combined application of edible coatings and thyme oil in inducing resistance components in avocado (Persea americana Mill.) against anthracnose during post-harvest storage. *Crop Protection (Guildford, Surrey), 64*, 159–167.

Bosquez-Molina, E., Ronquillo-de Jesús, E., BautistaBanos, S., Verde-Calvoa, J. R., & Morales-López, J. (2010). Inhibitory effect of essential oils against Colletotrichum gloeosporioides and Rhizopus stolonifer in stored papaya fruit and their possible application in coatings. *Postharvest Biology and Technology, 57*, 132–137.

Burt, S. (2004). Essential oils: Their antibacterial properties and potential applications in foods-a review. *International Journal of Food Microbiology, 94*, 223–253.

Combrinck, S., Regnier, T., & Kamatou, G. P. P. (2011). In vitro activity of eighteen essential oils and some major components against common post-harvest fungal pathogens of fruit. *Industrial Crops and Products, 33*, 344–349.

Desai, M. A., & Parikh, J. (2015). Extraction of essential oil from leaves of lemongrass using microwave radiation: Optimization, comparative, kinetic, and biological studies. *ACS Sustainable Chemistry & Engineering, 3*, 421–431.

Droby, S. (2006). Improving quality and safety of fresh fruit and vegetables after harvest by the use of biocontrol agents and natural materials. *Acta Horticulturae, 709*, 45–51.

Duduk, N., Markovic, T., Vasic, M., Duduk, B., Vico, I., & Obradovic, A. (2015). Antifungal activity of three essential oils against *Colletotrichum acutatum*, the causal agent of strawberry anthracnose. *Journal of Essential Oil Bearing Plants, 18*, 529–537. Available from https://doi.org/10.1080/0972060X.2015.1004120.

Formanek, Z., Lynch, A., Galvin, K., Farkas, J., & Kerry, J. P. (2003). *Meat Science, 63*(4), 433–440.

Godfray, H. C. J., Beddington, J. R., Crute, J. I., Haddad, L., Lawrence, D., Muir, J. F., ... Toulmin, C. (2010). Food security: The challenge of feeding 9 billion people. *Science (New York, N.Y.), 327*, 812–818.

Hashemi Gahruie, H., Hosseini, S. M. H., Taghavifard, M. H., Eskandari, M. H., Golmakani, M. T., & Shad, E. (2017). *Journal of Food Quality*, 6350156.

Juliano, C., Demurtas, C., & Piu, L. (2008). In vitro study on the anticandidal activity of Melaleuca alternifolia (tea tree) essential oil combined with chitosan. *Flavour and Fragrance Journal, 23*, 227–231.

Julianti, E., Ridwansyah., Yusraini, E., & Suhaidi, I. (2012). Effect of modified atmosphere packaging on postharvest quality of Rambutan cv. Binjai. *Journal of Food Science and Engineering, 2*, 111–117.

Juven, B. J., Kanner, J., Schved, F., & Weisslowicz, H. (1994). Factors that interact with the antibacterial action of thyme and its active constituents. *The Journal of Applied Bacteriology, 76*, 626–631.

Majewska, E., Kozowska, M., Gruczyska-Skowska, E., Kowalska, D., & Tarnowska, K. (2019). Lemongrass (Cymbopogon citratus) essential oil: Extraction, composition, bioactivity and uses for food preservation-a review. *Polish Journal of Food and Nutrition Sciences, 69*, 327–341.

Naveena, B. M., Sen, A. R., Kingsly, R. P., Singh, D. B., & Kondaiah, N. (2008). *International Journal of Food Science & Technology, 43*, 1807–1812.

Nunez, M. T., de Gonzalez., Boleman, R. M., Miller, R. K., Keeton, J. T., & Rhee, K. S. (2008). *Journal of Food Science, 73*, 63–71.

Oswell, N. J., Thippareddi, H., & Pegg, R. B. (2018). *Meat Science, 145*, 469–479.

Phuoc, M. N. (2020). Incorporation of turmeric oil into chitosan edible coating in preservation of rambutan fruit, Nephelium lappaceum. *Journal of the Entomological Research, 44*, 179–182.

Raybaudi-Massilia, R. M., Mosqueda-Melgar, J., & Martin-Belloso, O. (2008). Edible alginate-based coating as carrier of antimicrobials to improve shelf-life and safety of fresh-cut melon. *International Journal of Food Microbiology, 121*, 313–327.

Ribeiro, J. S., Santos, M. J. M. C., Silva, L. K. R., Pereira, L. C. L., Santos, I. A., da Silva Lannes, S. C., & da Silva, M. V. (2019). *Meat Science, 148*, 181–188.

Sanchez-Escalante, A., Djenane, D., Torrescano, G., Beltran, J. A., & Roncales, P. (2001). *Meat Science., 58*(4), 421–429.

Sanchez-Gonzalez, L., Vargas, M., Gonzalez-Martinez, C., Chiralt, A., & Chafer, M. (2011). Use of essential oils in bioactive edible coatings. *Food Engineering Reviews, 3*, 1–16.

Schaneberg, B. T., & Khan, I. A. (2002). Comparison of extraction methods for marker compounds in the essential oil of lemon grass by GC. *Journal of Agricultural and Food Chemistry, 50*, 1345–1349.

Serrano, M., Valverde, J. M., Guillen, F., Castillo, S., Martinez-Romero, D., & Valero, D. (2006). Use of Aloe vera gel coating preserves the functional properties of table grapes. *Journal of Agricultural and Food Chemistry, 54*, 3882–3886.

Shah, M. A., Bosco, S. J. D., & Mir, S. A. (2014). *Meat Science., 98*(1), 21–33.

Singh, D., & Sharma, R. R. (2007). Postharvest diseases of fruit and vegetables and their management. In D. Prasad (Ed.), *Sustainable pest management* (pp. 273–313). New Delhi: Daya Publishing House.

Sivakumar, D., Wijeratnama, R. S. W., Wijesunderab, R. L. C., & Abeyesekere, M. (2002). Control of postharvest diseases of rambutan using cinnamaldehyde. *Crop Protection (Guildford, Surrey), 21*, 847–852.

Tripathi, P., Dubey, N. K., & Shukla, A. K. (2008). Use of some essential oils as post-harvest botanical fungicides in the management of grey mold of grapes caused by *Botrytis cinereal*. *World Journal of Microbiology and Biotechnology., 24*, 39–46. Available from https://doi.org/10.1007/s11274-007-9435-2.

Tzortzakis, N. G. (2007). Maintaining postharvest quality of fresh produce with volatile compounds. *Innovative Food Science and Emerging Technologies, 8*, 111–116.

Tzortzakis, N. G., & Economakis, C. D. (2007). Antifungal activity of lemongrass (Cympopogon citratus L.) essential oil against key postharvest pathogens. *Innovative Food Science and Emerging Technologies, 8*, 253–258.

World Health Organization (WHO). (2003). *Diet, nutrition and the prevention of chronic diseases. WHO Technical Report Series, No. 916*. Geneva: World Health Organization.

Yingsanga, P., Srilaong, V., Kanlayanarat, S., Noichinda, S., & McGlasson, W. B. (2008). Relationship between browning and related enzymes (PAL, PPO and POD) in rambutan fruit (Nephelium lappaceum Linn.) cvs. Rongrien and See-Chompoo. *Postharvest Biology and Technology, 50*, 164–168.

Yingsanga, P., Srilaong, V., McGlasson, W. B., Kabanoff, E., Kanlayanarat, S., & Noichinda, S. (2006). Morphological differences associated with water loss in Rambutan fruit cv. Rongrien and See-chompoo. *Acta Horticulturae, 712*, 453–459.

Yu, L., Scanlin, L., Wilson, J., & Schmidt, G. (2002). *Journal of Food Science, 67*(2), 582–585.

Zhu, S. J. (2006). Non-chemical approaches to decay control in postharvest fruit. In B. Noureddine, & S. Norio (Eds.), *Advances in postharvest technologies for horticultural crops* (pp. 297–313). India: Research Signpost.

Chapter 26

Role of essential oil in feed production over the food chain in animal nutrition

Nyejirime Young Wike[1], Charles Oluwaseun Adetunji[2], Olugbemi T. Olaniyan[3], Babatunde Oluwafemi Adetuyi[4], Abel Inobeme[5], Juliana Bunmi Adetunji[6], Oluwabukola Atinuke Popoola[7], Oloruntoyin Ajenifujah-Solebo[7], Yovwin D. Godwin[8], Olatunji Matthew Kolawole[9], Olalekan Akinbo[10], Joan Imah-Harry[4] and Mohammed Bello Yerima[11]

[1]Department of Human Physiology, Faculty of Basic Medical Sciences, Rhema University, Aba, Abia State, Nigeria, [2]Applied Microbiology, Biotechnology and Nanotechnology Laboratory, Department of Microbiology, Edo State University Uzairue, Iyamho, Edo State, Nigeria, [3]Laboratory for Reproductive Biology and Developmental Programming, Department of Physiology, Faculty of Basic Medical Sciences, Rhema University, Aba, Abia State, Nigeria, [4]Department of Natural Sciences, Faculty of Pure and Applied Sciences, Precious Cornerstone University, Ibadan, Oyo State, Nigeria, [5]Department of Chemistry, Edo State University Uzairue Iyamho, Auchi, Edo State, Nigeria, [6]Department of Biochemistry, Osun State University, Osogbo, Osun State, Nigeria, [7]Genetics, Genomics and Bioinformatics Department, National Biotechnology Development Agency, Abuja, FCT, Nigeria, [8]Department of Family Medicine, Faculty of Clinical Sciences, Delta State University, Abraka, Delta State, Nigeria, [9]Department of Microbiology, Faculty of Life Sciences, University of Ilorin, Ilorin, Kwara State, Nigeria, [10]Centre of Excellence in Science, Technology, and Innovation, AUDA-NEPAD, Johannesburg, Gauteng, South Africa, [11]Department of Microbiology, Sokoto State University, Sokoto, Sokoto State, Nigeria

Introduction

For many decades, antimicrobial agents were employed as feed supplements in poultry diets, particularly in promotional agriculture. Antimicrobials to prevent infection, as well as a developmental stimulator, are becoming increasingly important in order to maintain animal production development (Roura et al., 1992; Waldroup et al., 2003). According to Thomke and Elwinger (1998), resistance bacteria commonly found in the gastrointestinal system may slow growth, which is linked to the harmful bacteria of such an animal's surroundings. Vaccines work on infective gut microbes, which contain harmful chemicals that damage the animal's livelihood or production performance. Antimicrobial agents used in this manner accumulate in animal organs, possibly contributing to bacterial resistance in humans via the food chain and, causing medical issues, according to Levy and Marshall (2004).

Several nations prohibited antimicrobial utilization as supplemental feed in animal farming. As a result, the urgent requirement is to develop a substitute for antimicrobials. As a result, the WHO has urged the quest for antimicrobial options to regulate intestinal mucosa epidemics as reported by Fritts and Waldroup (2003); Ayed et al. (2004) and Humphrey et al. (2002). Numerous researchers, including Langhout (2000), Mellor (2000), Wenk (2000), and Humphrey et al. (2002), have attempted to find alternatives to antimicrobials. Humphrey et al. (2002) discovered that lactoferrin as well as lysozyme have antimicrobial properties, which makes them suitable for use as nutritional supplements in animals instead of antimicrobials. Plant oils have been discovered to possess antibacterial properties, as well as free radical scavenging, anti-inflammatory, neuroprotective, absorption of nutrients, and hypolipidemic properties (Viuda–Martos et al., 2009; 2011). Because of their flexibility, EOs may serve as feed additives in animal farming.

Natural oils are a significant class of plant-derived food supplements (Bakkali et al., 2008; Banerjee et al., 2015; Banik et al., 2016; Bartoš et al., 2016; Bourgaud et al., 2001; Broom & Reefmann, 2005; Champagne & Boutry, 2016; Hristov et al., 2013; Jacela et al., 2010; Jamwa et al., 2018; Khan et al., 2013; Kis & Bilkei, 2003; Liu et al., 2014; Nasir & Grashorn 2010; Saito & Matsuda, 2010; Verpoorte, 1998; Yang et al., 2015; Zeng et al., 2015). According to Prakash et al. (2012), Prakash et al. (2015), Pavela and Benelli (2016), and Ambrosio et al. (2017), EOs have been widely used in complementary therapies, as flavoring and perfume in beauty products and ingredients, and more recently as prescription medicines, organic artificial flavors, emulsifiers, and biological pest control because of their potent fragrant properties and biocompatibility. According to Bakkali et al. (2008) and Prakash et al. (2012), the quality

Applications of Essential Oils in the Food Industry. DOI: https://doi.org/10.1016/B978-0-323-98340-2.00022-5

attributes of EOs are defined by the unique combination of aromatic and medicinal organic substances produced by plant biosynthetic pathways. Flowering plants, farming practices, cultivation conditions, vegetation chemotypes, and other variables affect the functional properties of EOs, irrespective of the intended use (Figueiredo et al., 2008). Because EOs are dynamic and responsive, their efficiency in animals may be influenced by a number of factors during production (Maenner et al., 2011), volatile oils storage (Nguyen et al., 2009), and gastrointestinal system conditions in the animals (Piva et al., 2007).

EOs have an impact on livestock nutrition and manufacturing. They can have rapid onset or prolonged, repairable or irreparable, poisonous, compensatory mechanisms, precautionary, or restorative impacts on animal life, as with any other active substance (Durmic & Blache, 2012). Durmic and Blache (2012) conducted studies to investigate the consequences of substances gotten from plants on various gastrointestinal systems and processes (nutrient intake, abdominal digestion, small bowel, hepatic, heart and lungs, renal tubular tract, skin, wool, hematological parameters, inflammatory processes, fertilization, and pressure and feelings in the nervous system), all of which have implications for animal welfare.

Enhanced nutritional intake is a consequence of enhanced nutrient intestinal absorption by EOs, as shown in poultry and porcine animals (Emami et al., 2012; Ahmed et al., 2013). Herbal remedies may greatly affect intestinal absorption and the pace of nutrient movement through the gastrointestinal tract, with effects on bile biosynthetic pathways, secretions, bile, fluid overload, and enzymatic activities (Platel & Srinivasan, 2000, 2001), but information is incongruent (Muhl & Liebert, 2007) and mostly derived from human healthcare (Jacela et al., 2010). According to Hong et al. (2012) and Namkung et al. (2004), phytogenic food additives increase nutrient uptake by increasing the absorbent interfacial interaction. Plant-based ingredients are known to boost the small intestinal maximum level in chickens and farm animals.

Supplementing feed and water in animal nutrition with essential oil

Plant oils are introduced to water and diets to boost flavor, and they are furthermore given to livestock to strengthen immune function and achievement (Kirkpinar et al., 2014; Witkowska & Sowi'nska, 2013). Since in vitro experiments also affirmed EOs' antimicrobial effectiveness, they may be employed as strength development promoters in farm animal feeding, particularly in poultry and pig farming (Zhai et al., 2018). Natural oils improve the taste as well as acceptability of diets, particularly those devoid of such characteristics, thereby increasing nutrient utilization. Nevertheless, because of their pungent smell, EOs must be administered with caution (Zhai et al., 2018). EOs encourage the production of proteolytic juices, influence metabolic activity and intestinal microbiota, and enhance food consumption, nutrient absorption, and accessibility (Zhai et al., 2018). The dosages and composition of the primary bioactive components in natural medicines determine their efficacy (Giannenas et al., 2013).

Numerous authors (Awaad et al., 2010, 2016; Jamroz et al., 2005; Bölükbaşı et al., 2008; Castillo et al., 2006; Cetin et al., 2016; Cross et al., 2007; Denli et al., 2004; De Souza et al., 2019; Firmino et al., 2021; Frankič-Korošec et al., 2009; Florou-Paneri et al., 2005; Hashemipour et al., 2013; Jerzsele et al., 2021; Lee et al., 2003; Tekippe et al., 2013; Tiihonen et al., 2010) provide the findings of various research studies looking into the impact of diets that include vital oils or their bioactive constituents on animal well-being and efficiency. Plant oils are progressively utilized to avoid and cure disease outbreaks, particularly in non-ruminants such as poultry and pigs (Zhai et al., 2018). EO neuroprotective compounds have been recognized as a possible treatment approach in ruminant and non-ruminant husbandry by Nehme et al. (2021). In poultry production, EOs function as natural coccidiostats, alleviating gastrointestinal clinical manifestations and reducing the movement of coccidia oocysts in animal waste. Natural oils are also used in pig farms to avoid piglet digestive problems due to their bactericidal properties. Plant oils are believed to enhance farm animals' immune functions. Pig nutrients augmented with EOs improve piglet immune reactions at weaning (Zhai et al., 2018; Franz et al., 2010).

EO in fish meal can sometimes actively promote bowel protection via the host-microbial co-metabolism (Firmino et al., 2021). Ruminal metabolism can be affected by EO therapeutic potential (Tekippe et al., 2013). Ruminants reported decreased methane production (Belanche et al., 2020), ammonia secretions, and absorption of nutrients (Franz et al., 2010).

As recorded by Frankič-Korošec et al. (2009), natural oils (such as sage, peppermint, and rosemary EOs) boost lipids and bile as well as pancreatic enzyme efflux and have a beneficial impact on the alimentary canal. A combination of carvacrol, cinnamaldehyde, and paprika oleoresin, according to the researchers, has protective effects and proficiently protects immune cells in the blood of swine from combustion (Frankič-Korošec et al., 2009).

Thyme oil strengthened the intestinal epithelium, which defends against the movement of harmful chemicals from poultry diets, for example. The inclusion of carvacrol in diets may improve bowel surface characteristics by enhancing the dimension of poultry GI tract villi (IRTA, 2015). The addition of nanoparticulate garlic, carvacrol, as well as thymol flavorings to a fish meal improved the transcription factors, immunologic profile, and microbiome composition of the intestinal wall (Firmino et al., 2021). In contrast with the untreated population, Bölükbaşı et al. (2008) discovered an increase in the feed-to-food ratio, a decrease in plasma levels of triglycerides and cholesterol, an increase in egg mass, and an increase in eggshell percentage in laying chickens whose nutrition was supplemented with thyme, sage, and rosemary EOs.

Denli et al. (2004) discovered that thymol enhanced quail growth in body weight, nutrient digestibility, and feed utilization. Those fed with thyme oil forage seemed to have a higher dead animal mass, compact size, and a lower proportion of belly obesity than the control group. Tiihonen et al. (2010) found more *Lactobacillus* and *E. coli* in the cecum of broiler birds given a mixture of EOs containing thymol and cinnamic aldehyde. Cetin et al. (2016) obtained comparable results in poultry nutrition using oregano, rosemary, and fennel EO combinations. Cross et al. (2007) discovered that thyme and yarrow EOs improved lean mass and thyme EO increased feed consumption in broilers; however, none of the EOs investigated had probiotic effects.

Lippens et al. (2005) added cinnamon, thyme, and oregano, as well as phytoconstituents and phenolic compounds, to broiler chicken nutrition. The efficiency of these broilers was compared to that of the treatment group, and the birds received enzyme inhibitor feed. When weighed against the control group, the exchange rate of feed was 2.9% lower in the EO group and 0.4% lower in the enzyme inhibitor group. Hashemipour et al. (2013), on the other hand, observed a boost in broiler chicken effectiveness in reaction to diets supplemented with thymol and carvacrol. Furthermore, the findings indicate that chemopreventive and gastrointestinal proteases had a beneficial impact on the birds' immunogenicity. Awaad et al. (2010, 2016), presented comparable findings (immunostimulatory impacts and enhanced efficiency) in a study that included broilers given peppermint as well as eucalyptus EOs in water. In a study carried out by De Souza et al., 2019, they discovered in heifer nutrition a mixture of natural clove oil as well as defensive preservatives like eugenol, thymol, and vanillin that enhanced efficiency (body fat mass and the effectiveness of feed conversion). Rosemary EO used alone had a negative effect on herb effectiveness, but it had a favorable effect when combined with other EOs.

Giannenas et al., 2013 observed that in a few surveys, a particular EO or a mixture of EOs failed to enhance lean mass improvements, nutrient digestibility, or the amount of feed converted. Tekippe et al., 2013 examined what impact an ingredient having EO (incorporating polyphenols and geraniol) would have on the digestion process, fiber content, and lactating milk production effectiveness. The referenced authors noted that the evaluated EO prototype had only a small effect on digestion and dairy cattle efficiency, but a consistent increase in overall NDF nutritional value. It ought to be mentioned that the investigated EO product enhanced manure's total combined nitrogen energy output.

Maintenance of food and infection control

The preceding studies would have confirmed that a diet supplemented with natural ingredients that affect the lipid makeup of animal flesh as well as the browning of free fatty acids could, in fact, define meat tenderness during farm animal raising (Zhai et al., 2018). Giannenas et al. (2013), discovered that plant substances with antioxidant capacity could be introduced to feed to optimize meat tenderness throughout stockpiling. Plant oils may also be incorporated into meat and meat-related items explicitly (Zhai et al., 2018). Because of their medicinal uses and antioxidant capacity, EO compounds inhibit the quality deterioration of ground beef, and they can be utilized as efficient and, most notably, unprocessed meat artificial flavors (Nehme et al., 2021; Silva et al., 2019). Adding plant oils to animal-based commodities, such as uncooked and high-temperature-analyzed ground beef, could enhance their performance and bacteriological security (Kirkpinar et al., 2014; Giannenas et al., 2013; Witkowska & Sowińska, 2013, Witkowska et al., 2016, 2019).

The inclusion of natural ingrdient serves as healthy flavor enhancers, influencing the flavor and the smell of processed meat as well as improving their textural characteristics as well as palatability. Nevertheless, as powerful antioxidants, EOs must be cautiously administered as they may convey a nasty taste or exhibit poisonous consequences at elevated concentrations (Zhai et al., 2018). The outcomes of research examining the impact of EOs as flavor enhancers are displayed by several researchers (Bonilla et al., 2014; Gómez-Estaca et al., 2010; Ibrahim et al., 2018; Karoui & Hassoun, 2017; Kostova, et al., 2016; Sarıcaoglu, & Turhan, 2020; Shaltout et al., 2017; Wrona et al., 2021). Amino acids were tested from birds whose bones were automatically removed from the meat, new ground chicken fillet, nonfermented pork fat (butter), ground pork meat, raw beef meat, minced beef, uncooked, pasteurized, and fermented cow's milk, milk with microbes, fish-cod, and East Coast sardines steaks (Bonilla et al., 2014; Gómez-Estaca et al., 2010; Kostova et al., 2016; Ibrahim et al., 2018; Shaltout et al., 2017). Among the key difficulties, according to da Silva (Silva et al., 2019), there will be the advancement of innovative techniques for integrating EOs into complicated food supply chains in order to reduce sensory changes, boost antibacterial properties, and assist with food quality assurance. In the coming years, complementary mixtures of various EOs and quality parameters, in addition to the interplay among EO components and farming practices, must be fully examined. Implementation also represents an innovative method of delivering animal nutrition components. Various nanoparticulate methods have been suggested to safeguard the active substances and excellent biocompatibility of EOs from deterioration and corrosion mechanisms throughout nutrient preparation and preservation, in addition to various situations in the gastrointestinal tract, and also to regulate their discharge and blending with the basal feed additives (Stevanovíc et al., 2018).

Conclusion

Plant oils, a type of plant-derived food supplement, have been assumed to be a less expensive and better substitute for anabolic steroids like antibiotics. Natural oils are a substitute for antimicrobials in animal nutrition. Utilizing natural ingredients has increased the dependability of raw feedstock while also having a variety of positive effects on the well-being and effectiveness of domesticated animals. EOs are anticipated to be a standard part of livestock feed, which will be important for the growth of the livestock production and farm animal businesses. Numerous variables, nevertheless, impact their consequences for animals. As a result, paying attention to details and maximizing the EO composition and reliability to be used as a feed nutrient are required, which are influenced by variables influencing secondary plant metabolic pathways during manufacturing, accompanied by the production preprocessing stage of food supplements, as well as their interrelationships with other substances from the food material to their intestinal absorption.

References

Ahmed, S. T., Hossain, M. E., Kim, G. M., Hwang, J. A., Ji, H., & Yang, C. J. (2013). Effects of resveratrol and essential oils on growth performance, immunity, digestibility and fecal microbial shedding in challenged piglets. *Asian-Australasian Journal of Animal Sciences*, 26, 683–690. Available from https://doi.org/10.5713/ajas.2012.12683.

Ambrosio, C. M. S., de Alencar, M., de Sousa, R. L. M., Moreno, A. M., & Da Gloria, E. M. (2017). Antimicrobial activity of several essential oils on pathogenic and beneficial bacteria. *Industrial Crops and Products*, 97, 128–136. Available from https://doi.org/10.1016/j.indcrop.2016.11.045.

Awaad, M. H. H., Abdel-Alim, G. A., Sayed Kawkab, K. S. S., Ahmed, A., Nada, A. A., Metwalli, A. S. Z., & Alkhalaf, A. N. (2010). Immunostimulant effects of essential oils of peppermint and eucalyptus in chickens. *Pakistan Veterinary Journal*, 30, 61–66.

Awaad, M. H. H., Afify, M. A. A., Zoulfekar, S. A., Mohammed, F. F., Elmenawy, M. A., & Hafez, H. M. (2016). Modulating effect of peppermint and eucalyptus essential oils on vVND infected chickens. *Pakistan Veterinary Journal*, 36, 350–355.

Ayed, M. H., Laamari, Z., & Rekik, B. (2004). Effects of incorporating an antibiotic Avilamycin and a probiotic Activis in broiler diets. *Western section ASAS, Am. Society Ani. Sci. Champaign, IL*, 55, 237–240.

Bakkali, F., Averbeck, S., Averbeck, D., & Idaomar, M. (2008). Biological effects of essential oils—A review. *Food and Chemical Toxicology*, 46, 446–475. Available from https://doi.org/10.1016/j.fct.2007.09.106.

Banerjee, J., Erehman, B. O., Gohlke, T., Wilhelm, R., Preissner, M., & Dunkel, D. (2015). Super natural II-a database of natural products. *Nucleic Acids Research*, 43, 935–939. Available from https://doi.org/10.1093/nar/gku886.

Banik, B. K., Durmic, Z., Erskine, W., Revell, C. K., Vadhanabhuti, J., McSweeney, C. S., Padmanabha, J., Flematti, G. R., Algreiby, A. A., & Vercoe, P. E. (2016). Bioactive fractions from the pasture legume *Biserrulapelecinus* L. have an anti-methanogenic effect against key rumen methanogens. *Anaerobe.*, 39, 173–182. Available from https://doi.org/10.1016/j.anaerobe.2016.04.004.

Bartoš, P., Dolan, A., Smutný, L., Šístková, M., Celjak, I., Šoch, M., & Havelka, Z. (2016). Effects of phytogenic feed additives on growth performance and on ammonia and greenhouse gases emissions in growing-finishing pigs. *Animal Feed Science and Technology*, 212, 143–148. Available from https://doi.org/10.1016/j.anifeedsci.2015.11.003.

Belanche, A., Newbold, C. J., Morgavi, D. P., Bach, A., Zweifel, B., & Yáñez-Ruiz, D. R. (2020). A Meta-analysis describing the effects of the EOs blend Agolin Ruminant on performance, rumen fermentation and methane emissions in dairy cows. *Animals*, 10, 620.

Bölükbąsı, S. C., Erhan, M. K., & Kaynar, Ö. (2008). The effect of feeding thyme, sage and rosemary oil on laying hen performance, cholesterol and some proteins ratio of egg yolk and Escherichia coli count in feces. *European Poultry Science*, 72, 231–237.

Bonilla, J., Vargas, M., Atarés, L., & Chiralt, A. (2014). Effect of chitosan essential oil films on the storage-keeping quality of pork meat products. *Food Bioprocess Tech*, 7, 2443–2450.

Bourgaud, F., Gravot, A., Milesi, S., & Gontier, E. (2001). Production of plant secondary metabolites: A historical perspective. *Plant Science*, 161, 839–851. Available from https://doi.org/10.1016/S0168-9452(01)00490-3.

Broom, D. M., & Reefmann, N. (2005). Chicken welfare as indicated by lesions on carcases in supermarkets. *British Poultry Science*, 46, 407–414. Available from https://doi.org/10.1080/00071660500181149.

Castillo, M., Martin-Orue, S. M., Roca, M., Manzanilla, E. G., Badiola, I., Perez, J. F., & Gasa, J. (2006). The response of gastrointestinal microbiota to avilamycin, butyrate, and plant extracts in early-weaned pigs. *Journal of Animal Science*, 84, 2725–2734.

Cetin, E., Yibar, A., Yesilbag, D., Cetin, I., & Cengiz, S. S. (2016). The effect of volatile oil mixtures on the performance and ilio-caecal microflora of broiler chickens. *British Poultry Science*, 57, 780–787.

Champagne, A., & Boutry, M. (2016). Proteomics of terpenoid biosynthesis and secretion in trichomes of higher plant species. *Biochimica et Biophysica Acta, 1864*, 1039–1049. Available from https://doi.org/10.1016/j.bbapap.2016.02.010.

Cross, D. E., McDevitt, R. M., Hillman, K., & Acamovic, T. (2007). The effect of herbs and their associated essential oils on performance, dietary digestibility and gut microflora in chickens from 7 to 28 days of age. *British Poultry Science, 48*, 496–506.

De Souza, K. A., de Oliveira Monteschio, J., Mottin, C., Ramos, T. R., de Moraes Pinto, L. A., Eiras, C. E., Guerrero, A., & do Prado, I. N. (2019). Effects of diet supplementation with clove and rosemary essential oils and protected oils (eugenol, thymol and vanillin) on animal performance, carcass characteristics, digestibility, and ingestive behavior activities for Nellore heifers finished in feedlot. *Livestock Science, 220*, 190–195.

Denli, M., Okan, F., & Uluocak, A. N. (2004). Effect of dietary supplementation of herb essential oils on the growth performance, carcass and intestinal characteristic of quail. *South African Journal of Animal Science, 34*, 174–179.

Durmic, Z., & Blache, D. (2012). Bioactive plants and plant products: Effects on animal function, health and welfare. *Animal Feed Science and Technology, 176*, 150–162. Available from https://doi.org/10.1016/j.anifeedsci.2012.07.018.

Emami, N. K., Samie, A., Rahmani, H. R., & Ruiz-Feria, C. A. (2012). The effect of peppermint essential oil and fructooligosaccharides, as alternatives to virginiamycin, on growth performance, digestibility, gut morphology and immune response of male broilers. *Animal Feed Science and Technology, 175*, 57–64. Available from https://doi.org/10.1016/j.anifeedsci.2012.04.001.

Figueiredo, A. C., Barroso, J. G., Pedro, L. G., & Scheffer, J. J. C. (2008). Factors affecting secondary metabolite production in plants: Volatile components and essential oils. *Flavour and Fragrance Journal, 23*, 213–226. Available from https://doi.org/10.1002/ffj.1875.

Firmino, J. P., Vallejos-Vidal, E., Balebona, M. C., Ramayo-Caldas, Y., Cerezo, I. M., Salomón, R., Tort, L., Estevez, A., Moriñigo, M. Á., Reyes-López, F. E., et al. (2021). Diet, immunity, and microbiota interactions: An integrative analysis of the intestine transcriptional response and microbiota modulation in gilthead seabream (Sparus aurata) fed an essential oils-based functional diet. *Frontiers in Immunology, 12*, 356.

Florou-Paneri, P., Nikolakakis, I., Giannenas, I., Koidis, A., Botsoglou, E., Dotas, V., & Mitsopoulos, I. (2005). Hen performance and egg quality as affected by dietary oregano essential oil and alpha-tocopheryl acetate supplementation. *International Journal of Poultry Science, 4*, 449–454.

Frankič-Korošec, T., Voljc, M., Salobir, J., & Rezar, V. (2009). Use of herbs and spices and their extracts in animal nutrition. *Acta Agriculturae Slovenica, 94*, 95–102.

Franz, C., Baser, K. H. C., & Windisch, W. (2010). EOs and aromatic plants in animal feeding—A European perspective. A review. *Flavour and Fragrance Journal, 25*, 327–340.

Fritts, C. A., & Waldroup, P. W. (2003). Evaluation of bio—mos mannanoligosaccharide as a replacement for growth promoting antibiotics in diets for turkeys. *International Journal of Poultry Science, 2*, 19–22.

Giannenas, I., Bonos, E., Christaki, E., & Florou-Paneri, P. (2013). Essential oils and their applications in animal nutrition. *Medicinal & Aromatic Plants, 2*, 1–12.

Gómez-Estaca, J., López de Lacey, A., López-Caballero, M. E., Gómez-Guillén, M. C., & Montero, P. (2010). Biodegradable gelatin-chitosan films incorporated with essential oils as antimicrobial agents for fish preservation. *Food Microbiology, 27*, 889–896.

Hashemipour, H., Kermanshahi, H., Golian, A., & Veldkamp, T. (2013). Effect of thymol and carvacrol feed supplementation on performance, antioxidant enzyme activities, fatty acid composition, digestive enzyme activities, and immune response in broiler chickens. *Poultry Science, 92*, 2059–2069.

Hong, J. C., Steiner, T., Aufy, A., & Lien, T. F. (2012). Effects of supplemental essential oil on growth performance, lipid metabolites and immunity, intestinal characteristics, microbiota and carcass traits in broilers. *Livestock Science, 144*, 253–262. Available from https://doi.org/10.1016/j.livsci.2011.12.008.

Hristov, A. N., Lee, C., Cassidy, T., Heyler, K., Tekippe, J. A., Varga, G. A., Corl, B., & Brandt, R. C. (2013). Effect of *Origanum vulgare* L. leaves on rumen fermentation, production, and milk fatty acid composition in lactating dairy cows. *Journal of Dairy Science, 96*, 1189–1202. Available from https://doi.org/10.3168/jds.2012-5975.

Humphrey, B. D., Huang, N., & Klasing, K. C. (2002). Rice expressing lactoferrin and lysozyme has antibiotic—like properties when fed to chicks. *Journal of Nutrition, 132*, 1214–1218.

Ibrahim, H. M., Hassan, M. A., Amin, R. A., Shawqy, N. A., & Elkoly, R. L. (2018). Effect of some essential oils on the bacteriological quality of some chicken meat products. *Journal of Benha Veterinary Medical, 35*, 42–49.

IRTA. (2015). Review of immune stimulator substances/agents that are susceptible of being used as feed additives: Mode of action and identification of end-points for efficacy assessment; EN-905 (vol. 12, p. 266). Catalonia, Spain: EFSA Supporting Publication.

Jacela, J. Y., DeRouchey, J. M., Tokach, M. D., Goodband, R. D., Nelssen, J. L., Renter, D. G., & Dritz, S. S. (2010). Feed additives for swine: Fact sheets-prebiotics and probiotics, and phytogenics. *Journal of Swine Health and Production, 18*, 132–136. Available from https://doi.org/10.4148/2378-5977.7067.

Jamroz, D., Wiliczkiewicz, A., Wertelecki, T., Orda, J., & Skorupińska, J. (2005). Use of active substances of plant origin in chicken diets based on maize and locally grown cereals. *British Poultry Science, 46*, 485–493.

Jamwa, K., Bhattacharya, S., & Puri, S. (2018). Plant growth regulator mediated consequences of secondary metabolites in medicinal plants. *Journal of Applied Research on Medicinal and Aromatic Plants, 9*, 26–38.

Jerzsele, A., Szeker, K., Csizinszky, R., Gere, E., Jakab, C., Mallo, J. J., & Galfi, P. (2021). Efficacy of protected sodium butyrate, a protected blend of essential oils, their combination, and Bacillus amyloliquefaciens spore suspension against artificially induced necrotic enteritis in broilers. *Poultry Science, 91*, 837–843, [CrossRef] Molecules 26, 3798 19 of 20.

Karoui, R., & Hassoun, A. (2017). Efficiency of rosemary and basil essential oils on the shelf-life extension of Atlantic mackerel (Scomber scombrus) fillets stored at 2°C. *Journal of AOAC International, 100*, 335–344.

Khan, S. H., Anjum, M. A., Parveen, A., Khawaja, T., & Ashraf, N. M. (2013). Effects of black cumin seed (*Nigella sativa* L.) on performance and immune system in newly evolved crossbred laying hens. *Veterinary Quarterly, 33*, 13–19. Available from https://doi.org/10.1080/01652176.2013.782119.

Kirkpinar, F., Ünlü, H. B., Serdaroˇglu, M., & Turp, G. Y. (2014). Effects of dietary oregano and garlic essential oils on carcass characteristics, meat composition, colour, pH and sensory quality of broiler meat. *British Poultry Science, 55*, 157–166.

Kis, R. K., & Bilkei, G. (2003). Effect of a phytogenic feed additive on weaning-to-estrus interval and farrowing rate in sows. *Journal of Swine Health and Production, 11*, 296–299.

Kostova, I., Damyanova, S. T., Ivanova, N., Stoyanova, A., Ivanova, M., & Vlaseva, R. (2016). Use of essential oils in dairy products. Essential oil of basil (Ocimum basilicum L.). *Indian Journal of Applied Research, 6*, 211–213.

Langhout, P. (2000). New additives for broiler chickens. *World Poultry, 16*, 22–27.

Lee, K. W., Everts, H., Kappert, H. J., Yeom, K. H., & Beynen, A. C. (2003). Dietary carvacrol lowers body weight gain but improves feed conversion in female broiler chickens. *Journal of Applied Poultry Research, 12*, 394–399.

Levy, S. B., & Marshall, B. (2004). Antibacterial resistance worldwide: Causes, challenges and responses. *Nature Medicine, 10*, 122–129.

Lippens, M., Huyghebaert, G., & Cerchiari, E. (2005). Effect of the use of coated plant extracts and organic acids as alternatives for antimicrobial growth promoters on the performance of broiler chickens. *European Poultry Science, 69*, 261–266.

Liu, Y., Song, M., Che, T. M., Bravo, D., Maddox, C. W., & Pettigrew, J. E. (2014). Effects of capsicum oleoresin, garlic botanical, and turmeric oleoresin on gene expression profile of ileal mucosa in weaned pigs. *Journal of Animal Science, 92*, 3426–3440. Available from https://doi.org/10.2527/jas.2013-6496.

Maenner, K., Vahjen, W., & Simon, O. (2011). Studies on the effects of essential-oil-based feed additives on performance, ileal nutrient digestibility, and selected bacterial groups in the gastrointestinal tract of piglets. *Journal of Animal Science, 89*, 2106–2112. Available from https://doi.org/10.2527/jas.2010-2950.

Mellor, S. (2000). Nutraceuticals – alternatives to antibiotics. *World Poultry, 16*(2), 30–33.

Muhl, A., & Liebert, F. (2007). No impact of a phytogenic feed additive on digestion and unspecific immune reaction in piglets. *Journal of Animal Physiology and Animal Nutrition, 91*, 426–431. Available from https://doi.org/10.1111/j.1439-0396.2006.00671.x.

Namkung, H., Li, M., Gong, J., Yu, H., Cottrill, M., & Lange, C. F. M. (2004). Impact of feeding blends of organic acids and herbal extracts on growth performance, gut microbiota and digestive function in newly weaned pigs. *Canadian Journal of Animal Science, 84*, 697–704. Available from https://doi.org/10.4141/A04-005.

Nasir, Z., & Grashorn, M. A. (2010). Effects of Echinacea purpurea and Nigella sativa supplementation on broiler performance, carcass and meat quality. *Journal of Animal and Feed Sciences, 19*, 94–104. Available from https://doi.org/10.22358/jafs/66273/2010.

Nehme, R., Andrés, S., Pereira, R. B., Ben Jemaa, M., Bouhallab, S., Ceciliani, F., López, S., Rahali, F. Z., Ksouri, R., Pereira, D. M., et al. (2021). Essential oils in livestock: From health to food quality. *Antioxidants, 10*, 330.

Nguyen, H., Campi, E. M., Jackson, W. R., & Patti, A. F. (2009). Effect of oxidative deterioration on flavour and aroma components of lemon oil. *Food Chemistry, 112*, 388–393. Available from https://doi.org/10.1016/j.foodchem.2008.05.090.

Pavela, R., & Benelli, G. (2016). Essential oils as ecofriendly biopesticides? Challenges and constraints. *Trends Plant Science, 21*, 1000–1007. Available from https://doi.org/10.1016/j.tplants.2016.10.005.

Piva, A., Pizzamiglio, V., Morlacchini, M., Tedeschi, M., & Piva, G. (2007). Lipid microencapsulation allows slow release of organic acids and natural identical flavors along the swine intestine. *Journal of Animal Science, 85*, 486–493. Available from https://doi.org/10.2527/jas.2006-323.

Platel, K., & Srinivasan, K. (2000). Stimulatory influence of select spices on bile secretion in rats. *Nutrition Research, 20*, 1493–1503. Available from https://doi.org/10.1016/S0271-5317(00)80030-5.

Platel, K., & Srinivasan, K. (2001). Studies on the influence of dietary spices on food transit time in experimental rats. *Nutrition Research, 21*, 1309–1314. Available from https://doi.org/10.1016/S0271-5317(01)00331-1.

Prakash, B., Kedia, A., Mishra, P. K., & Dubey, N. K. (2015). Plant essential oils as food preservatives to control moulds, mycotoxin contamination and oxidative deterioration of agri-food commodities—Potentials and challenges. *Food Control, 47*, 381–391. Available from https://doi.org/10.1016/j.foodcont.2014.07.023.

Prakash, B., Singh, P., Kedia, A., & Dubey, N. K. (2012). Assessment of some essential oils as food preservatives based on antifungal, antiaflatoxin, antioxidant activities and in vivo efficacy in food system. *Food Research International, 49*, 201–208. Available from https://doi.org/10.1016/j.foodres.2012.08.020.

Roura, E., Homedes, J., & Klasing, K. C. (1992). Prevention of immunologic stress contributes to the growth–permitting ability of dietary antibiotics in chicks. *Journal of Nutrition, 122*, 2383–2390.

Saito, K., & Matsuda, F. (2010). Metabolomics for functional genomics, systems biology, and biotechnology. *Annual Review of Plant Biology, 61*, 463–489. Available from https://doi.org/10.1146/annurev.arplant.043008.092035.

Sarıcaoglu, F. T., & Turhan, S. (2020). Physicochemical, antioxidant and antimicrobial properties of mechanically deboned chicken meat protein films enriched with various essential oils. *Food Package Shelf Life, 25*, 1–11.

Shaltout, F. A., Thabet, M. G., & Koura, H. A. (2017). Impact of some essential oils on the quality aspect and shelf life of meat. *Journal of Nutrition & Food Sciences, 7*, 1–7.

Silva, R. S., Lima, A. S., da Silva, L. P., Silva, R. N., Pereira, E. M., de Oliveira, F. L. N., & Azerêdo, G. A. (2019). Addition of essential oils and inulin for production of reduced salt and fat ham. *Australian Journal of Crop Science, 13*, 1031–1036.

Stevanović, Z. D., Bošnjak-Neumüller, J., Pajić-Lijaković, I., & Raj, J. (2018). Essential oils as feed additives—Future perspectives. *Molecules, 23*, 1717.

Tekippe, J. A., Tacoma, R., Hristov, A. N., Lee, C., Oh, J., Heyler, K. S., Cassidy, T. W., Varga, G. A., & Bravo, D. (2013). Effect of essential oils on ruminal fermentation and lactation performance of dairy cows. *Journal of Dairy Science, 96*, 7892−7903.

Thomke, S., & Elwinger, K. (1998). Growth promotants in feeding pigs and poultry. III. Alternatives to antibiotic growth promotants. *Annales de Zootechnie, 47*, 245−271.

Tiihonen, K., Kettunen, H., Bento, M. H. L., Saarinen, M., Lahtinen, S., Ouwehand, A. C., Schulze, H., & Rautonen, N. (2010). The effect of feeding essential oils on broiler performance and gut microbiota. *British Poultry Science, 51*, 381−392.

Verpoorte, R. (1998). Exploration of nature's chemodiversity: The role of secondary metabolites as leads in drug development. *Drug Discovery Today, 3*, 232−238. Available from https://doi.org/10.1016/S1359-6446(97)01167-7.

Viuda-Martos, M., Ruiz-Navajas, Y., Fernandez-Lopez, J., & Perez-Alvarez, J. A. (2011). Spices as functional foods. *Critical Reviews in Food Science and Nutrition, 51*(1), 13−28.

Viuda-Martos, M., Ruiz-Navajas, Y., Fernandez-Lopez, J., & Perez-Alvarez, J. A. (2009). Effect of adding citrus waste water, thyme and oregano essential oil on the chemical, physical and sensory characteristics of a bologna sausage. *Innovative Food Science and Emerging Technologies, 10*, 655−660.

Waldroup, P. W., Oviedo-Rondo, E. O., & Fritts, C. A. (2003). Comparison of Bio−MOS and antibiotic feeding programme in broiler diets containing copper sulphate. *Journal of Poultry Science, 2*, 28−31.

Wenk, C. (2000). Recent advances in animal feed additives such as metabolic modifiers, antimicrobial agents, probiotics, enzymes and highly available minerals. *Asian-Australasian Journal of Animal Sciences, 13*, 86−95.

Witkowska, D., & Sowi'nska, J. (2013). The effectiveness of peppermint and thyme essential oil mist in reducing bacterial contamination in broiler houses. *Poultry Science, 92*, 2834−2843.

Witkowska, D., Sowi'nska, J., Murawska, D., Matusevicius, P., Kwiatkowska-Stenzel, A., Mituniewicz, T., & Wójcik, A. (2019). Effect of peppermint and thyme essential oil mist on performance and physiological parameters in broiler chickens. *South African Journal of Animal Science, 49*, 29−39.

Witkowska, D., Sowi'nska, J., Zebrowska, J., & Mituniewicz, E. (2016). The antifungal properties of peppermint and thyme essential oils misted in broiler houses. *Brazilian Journal of Poultry Science, 18*, 629−638.

Wrona, M., Silva, F., Salafranca, J., Nerina, C., Alfonso, M. J., & Caballero, M. A. (2021). Design of new natural antioxidant active packaging: Screening flowsheet from pure essential oils and vegetable oils to ex vivo testing in meat samples. *Food Container, 120*, 1−18.

Yang, C., Chowdhury, M. A. K., Hou, Y., & Gong, J. (2015). Phytogenic compounds as alternatives to in-feed antibiotics: Potentials and challenges in application. *Pathogens., 4*, 137−156. Available from https://doi.org/10.3390/pathogens4010137.

Zeng, Z., Zhang, S., Wang, H., & Piao, X. (2015). Essential oil and aromatic plants as feed additives in non-ruminant nutrition: A review. *Journal of Animal Science and Biotechnology, 6*. Available from https://doi.org/10.1186/s40104-015-0004-5.

Zhai, H., Liu, H., Wang, S., Wu, J., & Kluenter, A. M. (2018). Potential of essential oils for poultry and pigs. *Animal Nutrition, 4*, 179−186.

Chapter 27

Biopreservative effects of essential oils in the food industry: oils and nuts, seeds and, seed products

Osarenkhoe Omorefosa Osemwegie[1], Adeyemi Ayotunde Adeyanju[1], Damilare Emmanuel Rotimi[2], Fisayo Yemisi Daramola[3], Charles Oluwaseun Adetunji[4], Francis Bayo Lewu[3] and A.T. Odeyemi[1]

[1]Department of Food Science and Microbiology, Landmark University, Omu-Aran, Kwara State, Nigeria, [2]Department of Biochemistry, Landmark University, Omu-Aran, Kwara State, Nigeria, [3]Department of Agriculture, Cape Peninsula University of Agriculture, Cape Town, Western Cape, South Africa, [4]Applied Microbiology, Biotechnology and Nanotechnology Laboratory, Department of Microbiology, Edo State University Uzairue, Iyamho, Edo State, Nigeria

Introduction

Current consumers' expectations regarding foods include a reduction in overprocessing foods while encouraging the availability of minimally processed foods, and limiting the application of conventional chemical antimicrobials like sulfur oxides, nitrates, benzoate, citric acid, and sorbates as preservatives. These conventional practices are intended to prevent the outbreaks of foodborne diseases, guarantee food safety, reduce food wastage, and extend the shelf of foods as well as pharmaceuticals. They were recently observed to cause a range of public health concerns that include the proliferation of microbial multidrug resistance or tolerance to food processing and artificial preservation methods, hypersensitivity, toxicity due to antimicrobial residues, allergies (asthma, skin irritation, and flu), and cancer (Mani-López et al., 2018; Onyeaka & Nwabor, 2022). These are aside from their long-term effects on the environment and ecosystem functions. Their application is popular and transcends diverse food production and processing aspects (meat, poultry, fish, dairy, canned foods, preharvest, postharvest, and beverages). Furthermore, these conventional or artificial (chemical and physical) methods suffered as one of the early inventions used to preserve agricultural produce and hunts. This contextually suggests that they are methods that predate modern science and form the foundation for the development of newer technologies including the invention of biopreservative options, nanotechnology (or green technology), cryopreservation technology, optimization of preservative factors, and antimicrobial-delivery systems for food products (Dey & Nagababu, 2022; Yee et al., 2022).

Sequel to the emerging health, environmental and resistant concerns associated with the chronic application of chemical preservatives, a radical shift to exploring naturally derived enzymes, spices as well as microbial, plants (phytometabolites), and animal products as alternatives to conventional chemical preservatives with quality organoleptic efficiency are now experienced globally. This, along with other biotechnological means that involved varying or manipulated critical requirements to inhibit or inactivate the activities of potential and innate food spoilage, pathogenic microbiota or metabolites, or products associated with foods in all its ramifications, is considered as biopreservation. Methods intended to enhance the biochemical products of biological origin including the fermentaters, and the natural capacity of biota to produce antimicrobials, or their amenability to genetic modification for optimizing preservative status are also categorized as biopreservation (Elsser-Gravesen & Elsser-Gravesen, 2013; Rendueles et al., 2022). The dramatic attraction to biopreservative gents is underscored by the preponderance, sustainability, tractability or amenability, environmentally safe, and non-toxic nature of natural agents or their derived products. However, their sensitivity to local environmental eccentricities and agent-food component interactions may limit their potency as biopreservative agents (Premanath et al., 2022). Also, the effect of biopreservatives on various food processing techniques, preservation methods, food biofunctions, palatability, and sensory have only been moderately studied while little information abounds in the literature on their underlying antimicrobial and ecophysiological actions in food matrices (Gálvez et al., 2011).

Applications of Essential Oils in the Food Industry. DOI: https://doi.org/10.1016/B978-0-323-98340-2.00016-X

The increasing human craving for the consumption of minimally processed foods and vegetables as well as raw fruits or extracted juices with fresh-tasting attributes as recommended by international agencies (World Health Organization—WHO, Food and Agriculture Organization—FAO, United State Department of Agriculture—USDA, European Food Safety Authority—EFSA) has fueled the application of biopreservation. Therefore biopreservation techniques, when applied to food singly or integrative sphere (hurdle technology), are required to limit or prevent the colonization of potential pathogens and foodborne illnesses, decontaminate raw or fresh foods or beverages, extend the shelf life of highly perishable foods/beverages, and enhanced the foods' bioactivity, nutrient, nutritional, sensory, and palate qualities. In addition, the choice of a natural preservative in the food preservation process or system depends on the versatility of the agent or its ability to act in an unbiased manner. The biochemical and physical nature of food, prevailing environmental factors, native biological composition, and microbial density as well as the diversity of the food are the other factors that could influence the activity of biopreservative agents. The growing popularity of industrial biopreservation in the maintenance of microbiological food standards in food processing and production is stimulating the need for a better understanding of the science of biopreservation of foods (Fig. 27.1) involving diverse biological-derived agencies with phenomenal boon propensity.

Selected live nonpathogenic microorganisms including phages that possess natural antibiosis tendencies, and diverse biogenic (microorganisms, plants, and animals) derivatives such as organic acids, peptides (reuterin, pediocin, protamine, ovotransferrin, lactoferrins, natamycin, pleurocidin, nisin, enterocins, and bacteriocins), enzymes (endolysins and lysozymes), polysaccharides, and essential oils (EOs) have been listed as biopreservative agents (Pateiro et al., 2021; Tiwari & Dubey, 2022; Wu et al., 2021; Yusuf, 2018).

Although more and more biogenic derivatives are emerging from studies across the world and evaluated for their innate ability to eliminate pathogenic microbes or for their antimicrobial application in food processing, there is a paucity of information on the comparative studies of their efficiency or the limitations of each application in the literature. However, the study by Drioiche et al. (2022) suggested that EOs, particularly those from botanical and microbial origins, may possess eco-sustainable, antimicrobial, antioxidant, and thermolabile attributes. In addition, Mukurumbira et al. (2022) alluded to their potential against multidrug-resistant strains and their broad safety, aroma/flavoring, and absorbances characteristics. Notwithstanding the slight restriction posed by their volatility and climate-factor sensitivity, EOs have a more versatile application advantage over other preservative agents, particularly when used in large-scale. Tiwari and Dubey (2022) also recognized their amenability to nanotechnology which may be used to improve their bioefficacy in maintaining the nutritional and sensory parameters of food items and overcoming their application limitations, including that which is related to their chemical dynamics. While biogenic EOs rank closer to polysaccharides in applicability, little is known about their target and nontarget actions or their natural potential in quorum sensing (signal transduction) and how this impacts their application in the biopreservation of different food items. Hence, this review aims at exploring the diverse knowledge sphere and gaps in biogenic EOs' applications in food processing. This is with particular emphasis on the science of the EOs derived from the oils, nuts, seeds, and seed products of botanicals of varying ecogeographical locations.

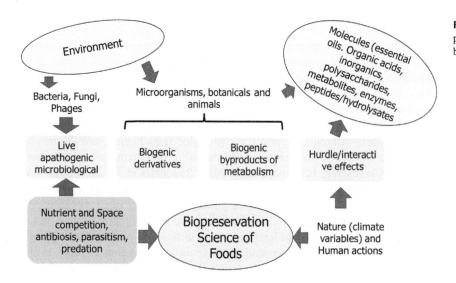

FIGURE 27.1 Schematic illustration of a phenomenal boon propensity of agencies of biopreservation in a food system.

Sources of essential oils used for biopreservation

From an evolutionary perspective, EOs were logically assumed to have coevolved along with other primary and secondary metabolites to support plants' efficient colonization of dry lands or to drive plants' overall fitness (Duplais et al., 2020). This is attested to by the emerging reports on the contributions of various EOs to the mediation of diverse naturally occurring biological phenomena like pollination, phytophagous infestations or herbivory, resiliency to stresses that are due to climate change, protection against pathogen attacks, yield quality, and growth. necessary for plants' wellness well as communication (Polturak et al., 2022). Suffices it to say that EOs and their derivatives are more associated with plants' survival. Plants form the most studied organismic reservoir of the complex mixture of EOs and hydrocarbons compared to all other organisms including microorganisms. Notwithstanding their bioactive, ecophysiological, chemoecological, and phytotherapeutic (herbivory deterrent) roles, the understanding of their spatiotemporal evolution, distribution, synthesizing genes and transposal dynamics, molecular mechanisms, and conditions that stimulate their production is still sketchy (Li et al., 2022). Also, the characterization of the diversity of beneficial phytochemicals including peptides, enzymes, oils, polysaccharides, and hormones inherent in plants is still inexhaustive, and the capacity of these plants to synthesize, acquire, or spin-off products of these phytometabolomes for survival in a constantly changing environment demands further study. Therefore, many other organisms equally possess the inherent potential to express survival genes and biochemicals required for weathering the diverse stresses exerted by the natural environment. Consequently, the capacity to produce natural EOs directly or indirectly has been found among plants, microorganisms (bacteria, fungi, algae), and an insignificant number of animals (Chávez-González et al., 2016; Ferranti & Velotto, 2023).

Plants' sources of essential oils used as biopreservatives

Scientific literature is dominated by different accounts of EOs' extraction for biopreservation, aromatherapy, medicine, and mysticism. Ríos (2016) relates aromatic, and volatile lipophilic metabolites with varying concentrations of monoterpenes, sesquiterpenes, fatty acids, allyl, isoallyl, phenol, aldehyde, esters, nitrogen, sulfur, and ketones compositions to EOs from plants (Chávez-González et al., 2016). Plant parts or organs (flower, seed, fruit, leaf, stem, barks, and root) are implicated in the synthesis, accumulation, and spritzing of EOs, and contain them in varying concentrations (Ferranti & Velotto, 2023; Karanicola et al., 2021). The popularity of EOs of plant origins and their identified presence in over 3000 species out of the vast 17,000 genera belonging to 642 families showed that plant screening for EOs dated many decades back. Since then, over 400 kinds of EOs have been profiled from less than 18% of the plant population with an even lesser number found suitable for commercial extraction (Hinks, 2014). The plant families such as Compositae, Labiatae, Lauraceae, Myrtaceae, Poaceae (Graminae), Pinaceae, Rutaceae, and Zingiberaceae are some of the richest source of EOs (Table 27.1). Plants-derived EOs have unique compositions of chemical variables that make them tractably important in apothecary, massage, food (flavoring, spices, cooking oils, preservatives), aromatherapy, personal care, traditions, and mythical rites (Estores & Frye, 2015). Like other secondary metabolites of plant origins, EOs are uniquely produced in the plastids or cytoplasm through different biochemical (malonic acid, mevalonic acid, methyl-D-erythritol-4-phosphate, and isoprenoid) pathways and squirted from specialized glandular trichomes to the environment. These biochemical processes are triggered and accelerated by ecologically induced stresses and are well-studied as biopreservatives of foods (Mani-López et al., 2018). Depending on their chemical architecture or the part of the plant material from which they are sourced, they are extracted by distillation (steam and water), extrusion, fat absorption or enfleurage, carbon dioxide extraction, or solvent extraction methods. These methods which may impact the quality of their bioactivity could be facilitated or assisted by other advanced technologies like ultrasound, pressurized or supercritical, microwave, nanocomposite, enzymes, and mathematical simulations techniques (Adeyemi et al., 2022; Herrera et al., 2022; Liu et al., 2021; Saleh et al., 2022). Although most EOs may not be suitable for the biopreservation of all foods because of dose requirements, application (emulsion, capsule, spray, encapsulation) compatibility, the effect of the application on organoleptic and market qualities, and antimicrobial spectrum (Ponce et al., 2011). This may suggest a variation in the chemical compositions of EOs due to the source's geographical location, season, cultivars, tissue material, soil type, growth stage, nature of input (fertilizer, soil amendment, hormones), cultivation conditions, and practice (Li & Chang, 2016).

Plants' EOs have witnessed more interesting research attention as a natural preservative than many other phytochemicals or EOs from non-plant bioresources because they offered certain desirable properties and benefits that make them suitable for food biopreservation (D'Almeida et al., 2022). Its abundant presence in a diverse phytoresources, versatile applications beyond food preservation, ability to remain active or labile over a wide range of temperatures, pH,

TABLE 27.1 Some plant sources of essential oils (EOs) with prospects in food preservation.

S/n	Plant source		Plant parts	Chemical composition of EO	EO impact on food	References
	Common name	Botanical name				
1	Agarwood (Balsaminaceae)	*Aquilaria* spp.	Bark, leaf	Agarospirol, jinkohol, agarofuran, vanillin, benzylacetone	Flavor, aroma, antimicrobial (*E. coli, Staphylococcus aureus, Bacillus subtilis*), antioxidant	Ali et al. (2016), Kadir et al. (2020)
2	Balsam (Fabaceae)	*Impatiens balsamina* L., *Myroxylon pereira* (Royle) Klotsch	Pericarp, leaf, flower	Linalool, benzaldehyde, cinnamate, cinnamyl, peruviol, vanillin, cinnamic acid	Antimicrobial (coliforms), antioxidant	Hintz et al. (2015), Pires et al. (2021)
3	Basil (Lamiaceae)	*Ocimum* spp.	Leaf	Eugenol, linalool, anethole, estragole	Foodborne fungi and mycotoxins, *Salmonella typhi, E. coli, B. subtilis, S. aureus*	Li and Chang (2016), Erol et al. (2022)
4	Bay or Laurel leaf (Lauraceae)	*Laurus nobilis* L.,	Leaf	Chavicol, limonene, geranyl acetate, neral, myrcene, terpineol	Foodborne and spoilage microorganisms	Li et al. (2022)
5	Bergamot (Rutaceae)	*Monarda didyma* L. (Syn. *Citrus bergamia*)	Fruit, leaf	Linalyl acetate, pinene (α, β), geraniol, myrcene, limonene, furan	*Listeria monocytogenes, Pseudomonas putida, Enterococcus faecium, Arcobacter butzleri*; pathogenic yeasts	
6	Cannabis (Cannabaceae)	*Cannabis sativa* L.	All parts of the plant but more in the flower and leaf	Caryophyllene, α-humulene, decane, α-pinene, myrcene, D-linalool.	Additive in starch-based food packaging, insect repellant, *S. aureus, Clostridium* spp. *Pseudomonas* spp., Pectobacterium, pathogenic yeasts	Li et al. (2022)
7	Carrot (Apiaceae)	*Daucus carota* L.	Seed, flower	Carotol, Dauca-4,8-diene, Caryophyllene	*Fusarium oxysporum, Monilinia fructigena, Trichoderma* spp.	
8	Cinnamon (Lauraceae)	*Cinnamon* spp.	Leaf, bark, fruits	*Cinnamaldehyde*, cinnamic acid, eugenol, cinnamyl acetate, thujene, cubebene	*Salmonella enterica, E. coli*, coliforms, molds	Valdivieso-Ugarte et al. (2021)
9	Citrus (Ruraceae-oranges) (sweet and bitter), tangarine, grapefruit, and lemon	*Citrus* spp.	Peels, fruit, seed, leaf, bark	Limonene, sitosterol, Sabinen, Ocimene, germacrene B, farnesene	*S. aureus, E. coli, Aspergillus niger, Pseudomonas aeruginosa*	Brahmi et al. (2021), Singh et al. (2021)

No.	Plant (family)	Scientific name	Parts used	Compounds	Mesophilic, psychrotrophic, and coliform bacteria	References
10	Clove/Clovos (Myrtaceae)	*Syzygium aromaticum* (Spring.) Spraque (Syn: *Caryophyllus aromaticum, Eugenia caryophyllata, Dianthus* sp.)	Leaf and flower buds	Caryophyllene, humulene, eugenol. terpineol, limonene	Mesophilic, psychrotrophic, and coliform bacteria	Valdivieso-Ugarte et al. (2021), Ponce et al. (2011)
11	Eucalyptus/gum tree (Myrtaceae)	*Eucalyptus* spp.	Leaf, flower, fruits, bark	Cymene, eucalyptol, pinene, myrcene, citronellyl, cineole	*S. aureus, Streptococcus pyogenes, Pseudomonas aeruginosa, Candida albicans*	Limam et al. (2020)
12	Garlic (Amaryllidaceae)	*Allium* spp.	Leaf, flower, bulb, stalk	Cymene, carvacrol, terpinene	*Colletotrichum* sp., *Trichophyton* spp., *Sclerotium cepivorum, Enterobacter cloacae, Klebsiella pneumoniae, Shewanella putrefaciens*, spoilage yeasts	Ezeorba et al. (2022); Putnik et al. (2019)
13	Ginger (Zingiberaceae)	*Zingiber officinale* Rosc.	Rhizome, root, leaf	Linalool, camphene, pinene, nerol, borneol. Terpineol, geranial, eucalyptol	*Fusarium oxysporum, Pyricularia oryzae, Xanthomonas* sp., *Ralstonia solanacearum, Bacillus* spp., *Klebsiella* sp.	Halimin et al. (2022), Abdullahi et al. (2020)
14	Jasmine (Olive family or Oleaceae)	*Jasminum* spp.	Flower, leaf	Indole, linalool, Jasmone, farnesene, nerol	*Trichophyton* spp. *E. coli, Aspergillus flavus, Shigella* spp. *Streptococcus* spp., *Fusobacterium nucleatum, Porphyromonas gingivalis*	Issa et al. (2020), Ahmed et al. (2016)
17	Lemongrass (Poaceae)	*Cymbopogon citratus* (DC. ex Nees) Stapf.	Leaf	Citral, neral, geranial, myrcene, verbenol, copaene, caryophyllene	*Bacillus amyloliquefaciens, Candida* spp., *S. aureus.*	Gao et al. (2020)
18	Moringa (Moringaceae)	*Moringa oleifera* Lam	Leaf, flower, seed, fruit, bark	Neral, linalool, camphene, carvacrol	*L. monocytogenes, P. putida, E. faecium, Arcobacter butzleri*, pathogenic yeasts	Abd El-Hack et al. (2022)
19	Neem (Meliaceae)	*Azadirachta indica* A. Juss	Leaf, seed, bark	Azadirachtin, margolonone, gedunin, nimbin	*Aeromonas hydrophila, P. aeruginosa, Salmonella enteritidis*	Ghosh et al. (2016)
20	Nutmeg (Myristicaceae)	*Myristica fragrans* Houtt.	Seed (nut), kernel, leaf, mace	Camphene, dipentene, limonene sabinene, eugenol, terpene myrcene, pinene	*Aspergillus* spp., *Proteus* spp., *Bacillus* spp., *Shigella dysenteriae*	Ashokkumar et al. (2022); (Periasamy et al., 2016)
21	Pepper (pink, green, white, black)	*Pipper* spp. (Syn. *Capsicum* spp.)	Fruit, leaf, seed	Pinene, Limonene, caryophyllene, bisabolene, phellandrene	*E. coli, S. aureus, L. monocytogenes*	Wang et al. (2022); Li et al. (2020)

(Continued)

TABLE 27.1 (Continued)

S/n	Plant source		Plant parts	Chemical composition of EO	EO impact on food	References
	Common name	Botanical name				
22	Tea tree (Myrtaceae)	*Camellia sinensis* (L) *O. Kuntze, Melaleuca alternifolia* (Maiden & Betche) Cheel, *Leptospermum scoparium* Forster & G. Forster L.	Leaf, bark	Pinene, sabinene, myrcene, cineole, cymene	Mesophilic, psychrotrophic, and coliform bacteria	Tian et al. (2023); Jiang et al. (2021)
23	Thyme (Lamiaceae)	*Thymus vulgaris*	Flower, leaf	Thymol, Carvacrol, borneol, cymene, thujanol, camphene, linalool	*Salmonella* sp. *Listeria monocytogenes, E. coli, Pseudomonas aeruginosa, Enterococcus* spp.	Drioiche et al. (2022)
24	Tumeric	*Curcuma longa*	Rhizome or underground root,	Turmerone, curlone, caryophyllene, eucalyptol, phellandrene, zingiberene		
25	Hemp	*Cannabis sativa*	Leaf, seed, flower, bark, root	Terpene, linalool, humulene, pinene, caryophyllene, myrcene	*Pseudomonas* spp., *Clostridium* spp., *Enterococcus* spp., yeasts,	Nissen et al. (2010); Li et al. (2022)

and pressure, non-toxic or inactive to nontarget in vivo, and relatively broad antimicrobial (bacteria, yeast, and fungi) and pesticidal spectrum may be the major attractants to this class of EOs (Drioiche et al., 2022; Han & Bhat, 2014; Khalili et al., 2015; Lin et al., 2022). These notwithstanding, the benefits-cost ratio, the effect of chronic long-term exploitation of EOs on natural plants' biodiversity, forestation, and conservation, genetic determinants of the preservative capability of EOs which is the Deoxyribonucleic acid (DNA), and their eventual transposition may suffice as limitations to their use as preservatives in the food industry. Understanding how EOs used as a biopreservative impacts the dynamics of food-beneficial microbes, particularly the gut normal flora, the development of multiresistant capability, bioactivity (biofunction and nutrient bioavailability) of the ingested food, and foodborne disease agents' dispersal across trophic levels may be critical to the success of this category of biopreservatives. Therefore this may suggest the need for further research on other bioresources for EOs with bioequivalent food preservative quality to complement those derived from plants.

Sources of microbial and animal essential oils used as biopreservatives

Even though only a few microorganisms (bacteria, fungi, algae, and yeasts) have been implicated in EO extraction studies compared to plants, a selection of microorganisms, their enzymes, bacteriocins, polysaccharides, and biosurfactants are found to be valuable to biopreservation systems. Many microorganisms, by reason of possessing natural abilities such as chemical antibiosis, superior competitive capacity, and predatory tendency, may be beneficial in systematically constraining the activities of food spoilage and pathogenic microbes (Barcenilla et al., 2022). The innate or acquired antibiosis traits of several beneficial microorganisms vary across species. These may include the production of antimicrobials or other toxigenic compounds (peptides, organic acids, endolysins, bacteriocins), alteration of the local pH of the system, secretion of proteolytic, nucleolytic, and lipolytic enzymes, and the release of other secondary metabolites that can disorient ligand-receptor systems or stimulate the competitiveness of native microbes to resource utilization (Margalho et al., 2021; Mouafo et al., 2020; Rendueles et al., 2022). Also, one of the important abilities of microorganisms and bacteriophages that are explored in food processing and preservation is their aggressive but greedy sequestration of shared resources (food, oxygen, space, nutrients, and moisture) and intricate tendency to outcompete resident potential foodborne pathogenic as well as food-spoilage agents (Barcenilla et al., 2022). The numerous biotechnological applications of microorganisms and their products in different aspects of the food value chain are proof of their versatility in nanosystems and as a synergistic composite of other antimicrobial systems of similar or divergent origins.

Conversely, there is a relative paucity of literature on the extraction of microbial as well as animal EOs and their direct applications as natural preservatives in food processing systems. However, the pieces of literature available were found to focus more on the usage of EOs for the decapacitation of food spoilage and foodborne pathogenic microbes in food manufacturing. This may be attributed to the non-adaptability of some of the widely employed conventional EO extraction technologies like distillation (stem, solvent, cold, hot), treatments (microwave, ultrasound, ohmic-assisted, laser), and pyrolysis to microorganisms and animals. Notwithstanding, de Souza et al. (2022) reported the indirect use of rhizobacteria, mycorrhizal and filamentous fungi in the improvement of EO extraction efficiency, selectivity, time, content, and yield from tea tree (*Melaleuca alternifolia*) as well as lemongrass (*Cymbopogon citratus*). A micropropagated biomass oregano (*Origanum vulgare*) induced with mycorrhizal *Glomus viscosum* showed increased glandular density and a boost in the EO parameters of the plant (Morone-Fortunato & Avato, 2008). Also, Formighieri & Melis (2018) reported that a cyanobacterium, *Synechocystis* shares β-phellandrene synthesis genes of different plants which qualifies it as a potential alternative for variable monoterpene EO extraction in addition to its ecological contribution to the aquatic food chain (Fawcett et al., 2022). Even though it could be hypothetically assumed that plants' beneficial and endophytic microbes have links to the generation of isopentenyl diphosphate (IPP), dimethylallyl diphosphate (DMAPP) and larger isoprenoid precursors critical to EOs biosynthesis, their mechanistic role in biosynthetic pathways (mevalonic acid—MVA, 1-deoxy-d-xylulose-5-phosphate—DXP, 2-C-methylerythritol-4-phosphate—MEP) yielding EOs in plants remains unclear (Cheng et al., 2022). This inextricably gave support to the logic that suggests that some members of the rhizospheric microbial communities could acquire EO biosynthetic traits through horizontal gene transfer, genetic transformation, and long years of association with plants (Kaur et al., 2019). Suffices to say that the genes for the biosynthesis of EOs may yet be latent or cryptic in endophytic and rhizospheric microorganisms. Consequently, more microorganisms should be screened for EO biosynthetic potential toward expanding the spectra of biopreservative sources beyond the current options. Therefore a better understanding of how to optimally harness plant growth-promoting microorganisms to produce EOs and innovating appropriate systematic extraction technologies are required for propelling healthy food systems.

Similarly, the literature has a relative bias toward the multifaceted use (aromatherapy, feed supplements, wound healing, anesthetics, immunomodulator, antibiotics, biopreservation of meat, fish, pork, and poultry) of essentials from plants for animals' well-being (Hou et al., 2022). This suggests a huge knowledge gap in the spectra of biological resources that can service the demand for natural biopreservatives in the food value chain and that animals or their products are currently unpopular for EO production. However, Vázquez et al. (2022) demonstrated that tuna head wastes obtained from consuming tuna saltwater finfish can be valorized for high-quality oil production. The facilitating effect of the antioxidant activity of thyme and rosemary EOs on tilapia fillet preservation has been documented (Albarracín et al., 2012), and the use of EOs to improve the safety and shelf life of meat and dairy products in Mediterranean diets was noted in the work of Tokur et al. (2021). In another study, Karanicola et al. (2021) used ultrasound-assisted dilute acid hydrolysis of orange peel waste (*Citrus*) and hydrolysate from bacterial fermentation designed experiment to optimize EO production. Therefore research innovations are required in the application of EOs for the preservation of ruminant animals' meat, poultry, pork, and seafood.

Trends in essential oil applications in the food industry and benefits in other industries

Considering recent concerns associated with using chemical preservatives in processed foods to prevent spoilage, the food manufacturing sector is awakened to developing healthy foods that are free of artificial preservatives. The food business is under more pressure now than ever to provide products that would satisfy consumers' growing interest in natural foods and palliate health concerns arising from chemically processed foods. As a result, many food-producing companies have shifted focus to using diverse plant-based food additives of organic origins to boost food storage stability. Many bioactive and functional compounds found in these plant-based food additives have been linked to a variety of food-related health-promoting qualities (Benkhoud et al., 2022). Therefore, due to the extensive spectrum of biological activities (antiinflammatory, antinociceptive, antibacterial, anticarcinogenic, and antioxidant) linked to natural food additives including biopreservatives, EOs have in recent years drawn significant attention in the food industry. EOs are a complex composition of volatile components such as terpenes, terpenoids, and aromatic components (Fig. 27.2). Their phenolic components, which have strong antioxidant properties, informed their use as preservatives in the food sector to halt product deterioration and enhance product shelf life. Therefore this section explores the different applications of EOs in the food industry.

Fruits and vegetables

Bacterial enzymes are the primary cause of deterioration in pasteurized fruit juices. *Salmonella* spp., *Listeria* spp., and *E. coli* can all be effectively eliminated by incorporating EOs into fruit juice (Raybaudi-Massilia et al., 2006). Further investigation by Raybaudi-Massilia et al. (2009) revealed that the quality of fresh "Piel de Sapo" melon can be retained longer by combining malic acid with gingerol, citral, and eugenol. In a 21-day trial, edible food coatings with EOs integrated into an alginate base were found to be effective against *Salmonella enteritidis* (Ayala-Zavala et al., 2009). Also, an edible coating comprised of apple puree and alginate that contained EOs of oregano, lemongrass, and vanillin increased the shelf life of Fuji apples. Alfalfa seeds were protected from *Salmonella* contamination by thymol and cinnamaldehyde (Friedman, 2017). According to a different study, minimally processed eggplant salad treated with oregano EO (7–21 L/g) lowered the risk of *E. coli* O157:H7 contamination (Burt, 2007).

Dairy products

The use of D-limonene, terpeneless lime, geraniol, and terpinol was found to reduce the growth of bacteria in dairy products while milk was preserved longer in an in vitro investigation when treated with a combination of terpineol and orange essential oil. Although terpineol antibacterial effectiveness was observed to be counteracted by the milk's fat content, it is still most effective in skim milk (7 log CFU/mL), low-fat milk (4 log CFU/mL), and full cream milk (3 log CFU/mL). Gram-positive bacteria (*Enterococcus hirae*, *S. aureus*, and *Bacillus licheniformis*) were successfully controlled in contaminated milk samples with the use of nanoemulsions made from thyme essential oil. The EO also improved the physical and chemical properties of milk (Perricone et al., 2015). Furthermore, it was demonstrated that the growth of *S. enteritidis* was suppressed in low-fat cheese treated with thyme clove, and cinnamon EOs but altered the sensory profile of the cheese. Yangilar and Yildiz (2018) investigated the blending of yogurt with ginger and chamomile EOs at 0.2% and 0.4% concentrations, respectively, and did chemical, microbiological, and sensory

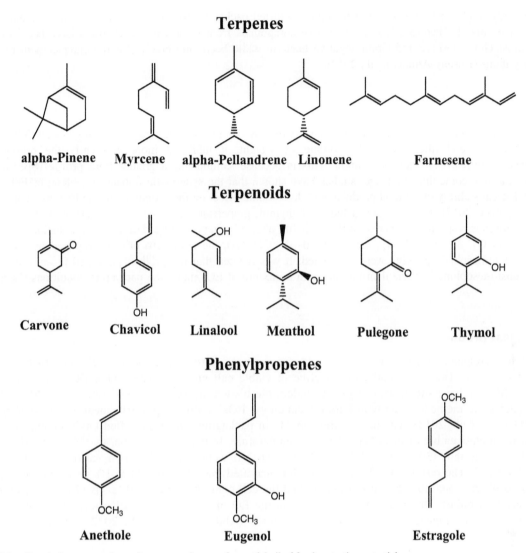

FIGURE 27.2 Chemical representations of some constituents of essential oil of food protective potential.

studies. Their results showed that 8.01 CFU/mL were generated by yogurt with 0.4% ginger essential oil, compared to 8.32 CFU/mL when essential oil-free control samples were used. It was however noticed that chamomile and ginger EOs significantly increased the shelf life of the yogurt due to their stronger antibacterial and antioxidant properties. Incorporating *Thymus capitatus* EO nanoemulsions in semi-skimmed ultra-high-temperature (UHT)-treated milk and the use of used nanotechnology to preserve milk resulted in products with improved physiological and microbiological stabilities, and reduced growth of spoilage organisms (Jemaa et al., 2018). Furthermore, the oxidative and fermentative stability of UHT-treated semi-skimmed milk was enhanced. This study infers EOs could be more effective as preservatives than their widely used synthetic counterparts.

Cereal-based products

The primary problem limiting bread's shelf life is fungal deterioration due to molds. Bakery products are therefore supplemented with various natural ingredients to improve their quality. Bread's sensory characteristics and microbial development are impacted by citrus peel EOs according to Salim-ur-Rehman et al. (2007). Spraying bread with orange peel EO improve its ability to stay fresh. Research on pasta made from amaranth (*Amaranthus hypochondriacus*) demonstrated that natural bioactive compounds such as limonene, chitosan, and thymol improve pasta's microbiological stability at concentrations between 2000 and 4000 ppm in a refrigerator (Lucera et al., 2012). Similarly, some plant-derived compounds were also found to retard the activity of *Bacillus cereus* in rice-based food products (Budka & Khan, 2010).

Another study implicated cinnamon EO in active modified packaging for increased sensory and nutrient quality of gluten-free sliced bread. Therefore EO and bioactive compounds from some herbs and spices have been found to be efficient against Gram-positive and Gram-negative bacteria while becoming crucial for the enhancement of the shelf life of bakery items (Basuny-Amany et al., 2012).

Seafoods

The biopreservative capacity of EOs in shellfish is decreased by fats likewise in meat and meat products. When oregano EO was applied to cod fillets at 0.05% volume by weight, it inhibited more efficiently the proliferation of *Photobacterium phosphoreum* than on salmon. This is because salmon has a greater lipid profile (Speranza et al., 2010). Hence, to overcome this challenge, studies have shown that the synergistic impact of a blend of EOs improves the microbiological stability of seafood products with less countereffect on the efficiency of the EOs by their fat content (Hassoun & Çoban, 2017). For example, a blend of thymol, grapefruit seed, and lemon EOs at 0.11%, 0.10%, and 0.12% concentrations respectively, increased the shelf life of fish burgers by 40%. Utilizing three distinct gas combinations ($5:95$—$O_2:CO_2$, $50:50$—$CO_2:O_2$, and $30:40:30$—$N_2:CO_2:O_2$), the combined effects produced by these on EOs under modified atmospheric packing were conducted. It was noticed that the product packaged in an airbag was less effective against mesophilic bacteria than the one with a mixture of EOs under the packaging conditions (Lucera et al., 2012).

Meat and meat products

EOs like thyme, oregano, eugenol, clove, and coriander can be used to reduce the microflora (autochthonous) that causes meat to decay. There is a noticeable decline in viable cell numbers when these EOs are used (Speranza et al., 2010). Studies have shown that meat fat hinders the bioactivity of EOs (Perricone et al., 2015). By using innovative packaging methods along with hurdle technology, EOs' ability to preserve meat against deterioration is enhanced (Degala et al., 2018). The use of oregano EO in packaging with a modified environment increases the shelf life of fresh chicken breast meat kept at a cool temperature. In this investigation, two distinct modified atmosphere packaging designs ($30:70$, $CO_2:N_2$ and $70:30$, $CO_2:N_2$) were assessed at two various EO concentrations (0.1% and 1% w/w). The results of a 25-day study demonstrated that oregano EO at 1% performed better in both modified atmospheric conditions because it maintained the cell count below crucial levels (7 log cfu/g) throughout the study period. Similarly, thymol and p-cymene from the EO of *Satureja horvattii* successfully protected meat against *Listeria monocytogenes* in a 96-hour storage study (Abd El-Hack et al., 2022; Bukvički et al., 2014) (Table 27.2).

Pharmaceutical and other applications

Numerous research works abound on the medicinal use of EOs of plant origin. The nano-drug delivery system has made it possible to continuously regulate drug release, facilitate deep tissue penetration of pharmaceuticals with nano-particle size, cellular absorption, membrane permeability, and protection of medication therapy at the extracellular and intracellular levels (Mehdizadeh & Moghaddam, 2018). Pharmaceutical products such as mouthwashes or gargles and inhalants have been cited as some of the most efficient health-related applications of EOs. Oral usage of EOs is rare even though they have been frequently proven to be safe for intake. In this instance, milk and soy milk are typically used to dilute them before oral administration. Even while EOs are frequently safe to use locally, there might be unexpected dangers involved with their unregulated usage. For example, there might be a chance that the uncontrolled use of citrus oils on the skin couldcause some negative responses due to the sensitivity of the oils to UV (Chouhan et al., 2017).

The indirect delivery of EOs through the consumption of edible carrier nuts (hazelnuts, walnuts, peanuts, pine nuts, cashew nuts, breadnuts, Brazil nuts, coconuts) and seeds (moringa, mustard, sesame, hemp, watermelon, pumpkin, melon) can be used to supplement food matrixes in the milled form to stabilize spoilage (Nazir & Wani, 2022). This supports the perception that nuts and seeds rich in EO content have great ethnopharmacological and food applications and might even be classified as functional foods (Selli et al., 2022). Most EOs that are derived from plant nuts and seeds have often been implicated in the formulation of different modifications of bioactive edible coating and films or biodegradable packaging (Behbahani et al., 2020; Tran et al., 2021). Suffices it to say that the nuts and seeds form a

TABLE 27.2 Roles of essential oils (EOs) and applications in across different aspects of the food industry

Product	EOs	Modern technology	Examples of food products	References
Cereal-based products	Citrus peel, thyme, cinnamon, sage, rosemary, and oregano essential oils	Active packaging	Bakery products, rice-based food products, gluten-free sliced bread, fresh amaranth-based pasta	Salim-ur-Rehman et al. (2007), Lucera et al. (2012), Basuny-Amany et al. (2012); Budka and Khan (2010)
Dairy products	Thyme, cinnamon, dove, and orange essential oils	Nanoencapsulation	Low-fat milk, full cream milk, skimmed milk; low-fat cheese	Perricone et al. (2015), Kavas et al. (2015)
Seafoods	Oregano, grapefruit seed, thymol, and lemon essential oil	Modified atmosphere packaging	Salmon, cod fillets	Speranza et al. (2010), Hassoun and Çoban (2017), Lucera et al. (2012)
Meat and meat products	Eugenol, oregano, thyme, clove, and coriander essential oils, winter savory essential oil	Modified atmosphere packaging	Fresh chicken, breast meat, pork preservation	Speranza et al. (2010), Perricone et al. (2015), Degala et al. (2018), Chouliara et al. (2007) Bukvički et al. (2014), Al-Hijazeen and Al-Rawashdeh (2017), Hassoun and Çoban (2017), Lucera et al. (2012)
Fruits and vegetables	Ginger, oregano, lemongrass, and vanillin essential oils	Edible food coatings	Salad, Fuji apples, eggplant, Pear apple juice, and Piel de Sapo melon	Raybaudi-Massilia et al. (2006), Raybaudi-Massilia et al. (2009), Ayala-Zavala et al. (2009), Friedman. (2017), Burt. (2007)

cost-effective means of harnessing the benefits of EOs in plant sources without concerns associated with the extraction process.

Several investigations on EOs encapsulation of food products for the slow release of bioactivity are being undertaken, a few of which include studying their microcapsule products. The main goals of EO encapsulation are to regulate the rate of bioactivity release, protect active ingredients from the environment, reduce their volatility, and boost their biological activity. It has been proven to be an effective and practical strategy. In this regard, numerous EO-loaded delivery systems with good biodegradability and biocompatibility have been reported; these systems often contain nanoparticles made of polymers and lipids as well as molecular complexes containing cyclodextrin inclusions. As an illustration, lipid nanoparticles were developed to load various EOs and were used in medicine to hasten the healing of skin lesions (Saporito et al., 2017). Additionally, these nanostructured lipid carriers appeared to speed up the healing process in vivo (Tanha et al., 2017). In a different study, oregano EO was enclosed in chitosan nanoparticles using a two-step procedure. The results of the experiment showed that nanoparticles were successfully used to encapsulate oregano EO suggesting that nanoparticles have an effective encapsulation rate and great loading capacity. According to in vitro release experiments, the active components in oregano EO were immediately released at the start of the nanoparticle formation, commencing to release gradually. A nanocomposite film consisting of chitosan and silver has also been found to have a huge potential for usage in water purification, antibacterial applications, and wound dressing (Hosseini et al., 2013). It is therefore reported by a few studies that nano-formulations containing EOs hold promise for the prospect of biomedical and nutraceutical applications even though additional studies may be required to sufficiently validate the extent of their benefits in those contexts.

Strategies of biological preservation by essential oils

The use of EOs has increased in food processing systems due to customers' search for less expensive and more "natural" alternative inhibitors of food spoilage as well as foodborne pathogens. In food, crop protection, and cosmetic sectors, EOs are widely used because they are considered biodegradable, abundant, sustainable, and "less harmful" than synthetic preservatives. This relative advantage has stimulated the need to examine the safety and efficacy of EOs in various biological systems to better comprehend their mechanisms, biopreservative functions, and health enhancement. Some of the documented mechanisms underscoring the use of EOs as effective biopreservatives in the food matrix include bioactivities related to antioxidant, antimicrobial, antiviral, antiinflammatory, antiangiogenic, antitumor, immunomodulatory, and antiapoptotic functions.

Antioxidant activity

Essetial oils have been studied extensively for their ability to protect organisms against free radicals. The cellular damage and disorders related to metabolism are two of the numerous consequences of the oxidation of many biological molecules that are palliated by natural EOs of high antioxidant properties (Lui & Mori, 2005; Moreira et al., 2005; Naito et al., 2006). EOs include volatile components that not only alter the oxidative stress pattern of the food medium but also serve as prooxidants by targeting the cellular redox potential of potential spoilage agents. It may also cause relatively significant damage to the DNA and cellular proteins which are critical drivers of the food biodeterioration process by microorganisms (Elsayed et al., 2020). Therefore it is important to carry out more investigations on how both the antioxidant and prooxidant properties of EOs optimize their biopreservative function. It has long been known that EOs have strong phenolic content which possibly enhances a direct interaction between their excellent antioxidant and antimicrobial actions. The phenolic compounds are linked to the behavior of the EOs as hydrogen donors, and singlet oxygen quencher/scavengers/reducing agents (Rice-Evans et al., 1997). Also, their volatile elements may cause disorientation of microbial activities and facilitate microbicidal effects (Naito et al., 2006). Thyme and clove leaf oils are two EOs that have good antioxidant properties in various studies. Aside from thymol, several other major constituents of thyme oil, including carvacrol, aldehyde, terpinene, myrcene, and linalool, have been found to inhibit ferric ion-induced lipid peroxidation in the food matrix and rat brain homogenates even though their individual effectiveness was not as high as that of thyme oil as a whole (Youdim et al., 2002). This, therefore, implies that more studies are needed to establish comprehensively the synergic relationship between the role of antioxidants in establishing EOs' preservative functions in different food processing systems.

Antiviral activity

Drug-resistant viral strains have sparked research into the use of EOs as an alternative to synthetic preservatives. Although less information exists on the role of EOs in the minimization of foodborne viruses, they are still a concern to public health and food safety for the past 200 years (Bosch et al., 2016). Schnitzler et al. (2001) observed that eucalyptus EO has the potential to cause a decrease in plaque formation due to a virus suggesting that the oil may have antiviral effects before or during virus adsorption to the host cell. Similar inhibition of viral reproduction was observed when eucalyptus oil was employed in vivo. This was attributed to the damage caused by the oil to the viral envelope structure which is necessary for its adsorption to host cells. The oil may have also limited viral multiplication by interfering with the host DNA polymerase (Limam et al., 2020). *Melaleuca* species, lemongrass, sandalwood, and *Juniperus* species have also been evaluated for their virucidal property (Burt, 2004; Wei & Shibamoto, 2007). The antiviral activity of ginger, thyme, hyssop, and sandalwood oils has also been documented in the healthcare sector (Schnitzler et al., 2001). Viruses have the tendency to spread through the food value chain due to their ubiquity and often escape due to their peculiar acellular nature. Their detection or conventional control measures facilitate their potential impact on illnesses across the food industry (Olaimat et al., 2022). Furthermore, food safety experts, mindful of the eminent danger of foodborne viral outbreaks, are working tirelessly across the globe foreffective measures of control of these viruses in different aspects of the food value chain even though the understanding of essential oil's effectiveness in achieving this feat remains vague (Bosch et al., 2018).

Antimicrobial activity

The time-kill test is the best approach for measuring bactericidal impact and a powerful tool for learning about the antimicrobial agent's dynamic interaction with a microbial strain. The time-kill test also reveals an antibacterial action that is time-dependent or concentration-dependent (Li et al., 2014). The EO from the leaf of *Forsythia koreana* has been shown to harm the cytoplasmic membrane of foodborne and harmful bacteria. The composition of bioactive constituents, functional groups contained in the active components, and their synergistic interactions or counteractive effect are crucial to determining the potency of EOs (Dorman & Deans, 2000). Consequently, some strains of microorganisms may become more susceptible to the antimicrobial constituents of EO than others. Gram-positive bacterial contaminants of food are more vulnerable to EO than Gram-negative contaminants (Azhdarzadeh & Hojjati, 2016; Huang et al., 2014). Due to the rigidity of the outer membrane of gram-negative bacteria, and the thickness of the lipopolysaccharide (LPS), the diffusion of hydrophobic antimicrobial contents of EOs is limited. Contrary to this, the thin nature of the peptidoglycan walls of Gram-positive bacteria increases the surface area of small antimicrobial molecules' permeability making it easier to access the cell membrane (Mangalagiri et al., 2021). Furthermore, the lipophilic endings of lipoteichoic acid found in the cell membrane of Gram-positive bacteria may facilitate the entry of hydrophobic components of EO (Diao et al., 2014). Studies reveal that bioactive components in EOs may adhere to cells' surfaces and then pass through to their phospholipid bilayer. It may lead to cell death if the structural integrity of the cell membrane is compromised by their build-up (Bajpai et al., 2013; Lv et al., 2011). However, the understanding of antimicrobial impacts of EOs on some Gram-positive biodeteriorants possibly acquired resistant strategies such as the evolution of β-lactamase production, aminoglycoside-modifying enzymes, target site modification, active efflux, and penicillin-binding proteins is vital to their wide applications in preventing foodborne infections. *E. coli* and *S. aureus* cell membranes were damaged by cinnamon EOs at the minimum inhibitory concentration (MIC) level while their cells were inhibited at the minimum bactericidal concentration (MBC) level (Zhang et al., 2016). According to El Kolli et al. (2016), cinnamon EOs also increased the electrical conductivity in the first few hours due to a rapid leakage of small electrolytes, the concentration of proteins, and nucleic acids in the cell suspension, and a 3−fivefold decrease in bacterial metabolic activity. Turmeric EO is derived from *Curcuma longa* and was found to inhibit *Aspergillus flavus* (Hu et al., 2017). Similarly, the EO acted by disrupting the fungal cell endomembrane system, including the plasma membrane and mitochondria.

Antitoxigenic activity

Many foodborne toxigenic pathogens and spoilage agents are reckoned with through the food because they produce various low molecular weight toxins. This may have life-long implications for food safety or public health when foods intoxicated by these toxins are consumed. This is in addition to their inherent ability to cause the deterioration of food nutritional quality and disrupt the guts and intestinal normal floral balance to the extent of provoking opportunistic infections. Although EOs are generally recognized as safe (GRAS) food additives for many nontarget organisms including food microbes by the FDA and cosmetics acts of many nations, Bibiano et al. (2022) and Zerkani et al. (2022) noted that the innate secondary metabolite constituents of EOs can selectively slow down food intoxications or neutralize their toxigenic effects if consumed. EOs have anticancer properties in several investigations and have become valuable in preventing the long-term consequences (cancers, tumors, neurodegenerative disorders, congenital cases) of consuming foods contaminated with microbial toxins. While the process by which EOs decapacitates toxigenic pathogens or their toxins is not fully understood, it may be philosophically linked to the stalling of toxin synthesis, adsorption of the toxin to food molecules due to encapsulation effect, and the efflux by either blocking the transport pathway or transfigurate the toxins into non-toxic molecules (Krstić et al., 2021; Lasram et al., 2019). Being lipohilic in nature, EOs easily transverse food pathogenic organisms, chelate, bind critical protein receptors, and inhibit spoilage microbes. For example, Lasram et al. (2019) noted that carvone and linalool constituents of *Carum carvi* and *Coriandrum sativum* EOs showed a significant decrease of 77.9% and 0.1% respectively in aflatoxin B1 synthesis. Also, the efficiency of essential oil-based nanoencapsulation improved the stability of herbal raw materials (HRMs) and the controlled release of bioactive principles that inhibits the biodeterioration activity of postharvest filamentous fungi (*Aspergillus flavus*, *Penicillium*, and *Fusarium*) and mycotoxins.

Conclusion

The growing application of EOs in the food industry is widely linked to their multidimensional benefits. This could be why EO market across the globe has become a billion-dollar investment that is valued at around USD 18.6 billion in 2020 and still appreciating in value add. The market is also highly diversified and priced based on their utility and extraction profile. Their utility-based market categorization into mutually exclusive financial investment options where each option parades its own prospective EOs have shaped their demand structure for decades. Food and beverages, aromatherapy, healthcare, homecare, daycare, cosmetics, toiletries, and crop protection constitute the various market categories. EOs are sold as aromatherapeutic, healthcare and homecare products, dominating the global market. However, there is little empirical evidence to support the theoretical notion that medicinal herbs and trees have high contents of versatile EOs than the nonherbal plants. Also, this could be the reason why most edible seeds and nuts that are rich in EO are classified as functional foods. Similarly, more investigations evaluating the toxicity of plants' EOs are still ongoing even though popular accounts in the literature affirmed them as safe for use particularly against nontarget organisms. This only suggests that there is need for the establishment of standard dosage/concentration and exercise of control in the use of EOs in the food industry to forestall the emergence of resistance strains. Therefore the understanding of how the bioactive principles of EOs disarm foodborne pathogens and stabilize the process of food spoilage/biodeterioration is crucial for establishing standardization in their application. Furthermore, more studies are recommended to improve understanding on how EO molecules interact with food molecules to deliver optimal health, taste, and metabolic properties or detect components of the food molecules that can counteract the preservative effects of the EOs. Some species of algae, fungi, and bacteria have equally shown promise as biogenic sources of EOs even though they only produce limited quantities and are not amenable to current conventional technologies employed for EO extraction from plant. Therefore the science and application of EOs in the food industry could be advanced with technological innovations that optimize EO extraction from bacteria, algae, and fungi at a commercial scale. This, aside from expanding the options of molecules with biopreservative/food additive properties, may offer partial relief to the overexploitation vegetation, especially the food crops. Suffices it to say that edible plants are left to primarily satisfy the hunger demands of the increasing global population.

References

Abd El-Hack, M. E., Alqhtani, A. H., Swelum, A. A., El-Saadony, M. T., Salem, H. M., Babalghith, A. O., Taha, A. E., Ahmed, O., Abdo, M., & El-Tarabily, K. A. (2022). Pharmacological, nutritional and antimicrobial uses of *Moringa oleifera* Lam. leaves in poultry nutrition: An updated knowledge. *Poultry Science* 102031. Available from https://doi.org/10.1016/j.psj.2022.102031.

Abdullahi, A., Khairulmazmi, A., Yasmeen, S., Ismail, I. S., Norhayu, A., Sulaiman, M. R., Ahmed, O. H., & Ismail, M. R. (2020). Phytochemical profiling and antimicrobial activity of ginger (*Zingiber officinale*) essential oils against important phytopathogens. *Arabian Journal of Chemistry*, *13*(11), 8012–8025. Available from https://doi.org/10.1016/j.arabjc.2020.09.031.

Adeyemi, I., Meribout, M., & Khezzar, L. (2022). Recent developments, challenges, and prospects of ultrasound-assisted oil technologies. *Ultrasonics Sonochemistry*, *82* 105902. Available from https://doi.org/10.1016/j.ultsonch.2021.105902.

Ahmed, N., Hanani, Y. A., Ansari, S. Y., & Anwar, S. (2016). Jasmine (*Jasminum sambac* L., Oleaceae) oils. In V. R. Preedy (Ed.), *Essential oils in food preservation, flavor and safety* (pp. 487–494). Academic Press. Available from https://doi.org/10.1016/B978-0-12-416641-7.00055-9.

Albarracín, H., William., Alfonso, A., Christian., Sánchez, B., & Iván, C. (2012). Application of essential oils as a preservative to improve the shelf life of Nile tilapia (*Oreochromis niloticus*). *Vitae*, *19*(1), 34–40.

Ali, N. A. M., Jin, C. B., & Jamil, M. (2016). Agarwood (*Aquilaria malaccensis*) oils. In V. R. Preedy (Ed.), *Essential oils in food preservation, flavor and safety* (pp. 173–180). , Academic Press. Available from https://doi.org/10.1016/B978-0-12-416641-7.00018-3.

Al-Hijazeen, M., & Al-Rawashdeh, M. (2017). Preservative effects of rosemary extract (*Rosmarinus officinalis* L.) on quality and storage stability of chicken meat patties. *Food Science and Technology*, *39*(1), 27–34.

Ashokkumar, K., Vellaikumar, S., Muthusamy, M., Dhanya, M. K, & Aiswarya, S. (2022). Compositional variation in the leaf, mace, kernel, and seed essential oil of nutmeg (*Myristica fragrans* Houtt.) from the Western Ghats, India. *Natural Product Research*, *36*(1), 432–435. Available from https://doi.org/10.1080/14786419.2020.1771713.

Ayala-Zavala, J. F., González-Aguilar, G. A., & Del-Toro-Sánche, L. (2009). Enhancing safety and aroma appealing of fresh-cut fruits and vegetables using the antimicrobial and aromatic power of essential oils. *Journal of Food Science*, *74*(7), R84–R91.

Azhdarzadeh, F., & Hojjati, M. (2016). Chemical composition and antimicrobial activity of leaf, ripe and unripe peel of bitter orange (*Citrus aurantium*) essential oils. *Nutrition and Food Sciences Research*, *3*(1), 43–50.

Bajpai, V. K., Sharma, A., & Baek, K.-H. (2013). Antibacterial mode of action of *Cudrania tricuspidata* fruit essential oil, affecting membrane permeability and surface characteristics of food-borne pathogens. *Food Control*, *32*(2), 582–590.

Barcenilla, C., Ducic, M., López, M., Prieto, M., & Álvarez-Ordóñez, A. (2022). Application of lactic acid bacteria for the biopreservation of meat products: A systematic review. *Meat Science*, *183* 108661. Available from https://doi.org/10.1016/j.meatsci.2021.108661.

Basuny-Amany, M., Nasef, S. L., Mahmoud, E. A. M., & Shaker, M. A. (2012). Use of medicinal and aromatic plants for increasing quality of some bakery products. *International Science and Investigation: The Journal*, *54*, 820–828. (2012).

Behbahani, B. A., Noshad, M., & Jooyandeh, H. (2020). Improving oxidative and microbial stability of beef using Shahri Balangu seed mucilage loaded with Cumin essential oil as a bioactive edible coating. *Biocatalysis and Agricultural Biotechnology*, *24* 101563. Available from ttps://doi.org/10.1016/j.bcab.2020.101563.

Benkhoud, H., M'Rabet, Y., GaraAli, M., Mezni, M., & Hosni, K. (2022). Essential oils as flavoring and preservative agents: Impact on volatile profile, sensory attributes, and the oxidative stability of flavored extra virgin olive oil. *Journal of Food Processing and Preservation*, *46*(5) e15379.

Bibiano, C. S., Alves, D. S., Freire, B. C., Bertolucci, S. K. V., & Carvalho, G. A. (2022). Toxicity of essential oils and pure compounds of Lamiaceae species against *Spodoptera frugiperda* (Lepidoptera: Noctuidae) and their safety for the nontarget organism *Trichogramma pretiosum* (Hymenoptera: Trichogrammatidae). *Crop Protection*, *158*, 0261–2194. Available from https://doi.org/10.1016/j.cropro.2022.106011.

Bosch, A., Pintó, R. M., & Guix, S. (2016). Foodborne viruses. *Current Opinion in Food Science*, *8*, 110–119. Available from https://doi.org/10.1016/j.cofs.2016.04.002.

Bosch, A., Gkogka, E., Guyader, F. S. L., Loisy-Hamon, F., Lee, A., van Lieshout, L., Marthi, B., Myrmel, M., Sansom, A., Schultz, A. C., Winkler, A., Zuber, S., & Phister, T. (2018). Foodborne viruses: Detection, risk assessment, and control options in food processing. *International Journal of Food Microbiology*, *285*, 110−128. Available from https://doi.org/10.1016/j.ijfoodmicro.2018.06.001.

Brahmi, F., Mokhtari, O., Legssyer, B., Hamdani, I., Asehraou, A., Hasnaoui, I., Rokni, Y., Diass, K., Oualdi, I., & Tahani, A. (2021). Chemical and biological characterization of essential oils extracted from citrus fruits peels. *Materials Today: Proceedings*, *45*(8), 7794−7799. Available from https://doi.org/10.1016/j.matpr.2021.03.587.

Budka, D., & Khan, N. A. (2010). The effect of *Ocimum basilicum*, *Thymus vulgaris*, *Origanum vulgare* essential oils on *Bacillus cereus* in rice-based foods. *European Journal of Behavioral Sciences*, *2*(1), 17−20.

Bukvički, D., Stojković, D., Soković, M., Vannini, L., Montanari, C., Pejin, B., Savić, A., Velijić, M., Grujić, S., & Marin, P. D. (2014). *Satureja horvatii* essential oil: In vitro antimicrobial and antiradical properties and in situ control of *Listeria monocytogenes* in pork meat. *Meat Science*, *96*(3), 1355−1360.

Burt, S. (2004). Essential oils: Their antibacterial properties and potential applications in foods—A review. *International Journal of Food Microbiology*, *94*(3), 223−253.

Burt, S.A. (2007). Antibacterial activity of essential oils: potential applications in food. https://dspace.library.uu.nl/bitstream/handle/1874/24273/full.pdf? sequence = 6.

Chávez-González, M. L., Rodríguez-Herrera, R., & Aguilar, C. N. (2016). Essential oils: a natural alternative to combat antibiotics resistance. In K. Kon, & M. Rai (Eds.), *Antibiotic resistance*. Academic Press, 227-237. Available from https://doi.org/10.1016/B978-0-12-803642-6.00011-3.

Cheng, X., Bi, L.-W., Li, S.-N., Lu, Y.-J., Wang, J., Xu, S.-C., Gu, Y., Zhao, Z.-D., & Chen, Y.-X. (2022). Succession of endophytic bacterial community and its contribution to cinnamon oil production during cinnamon shade-drying process. *Food Chemistry: Molecular Sciences*, *4*100094. Available from https://doi.org/10.1016/j.fochms.2022.100094.

Chouhan, S., Sharma, K., & Guleria, S. (2017). Antimicrobial activity of some essential oils - Present status and future perspectives. *Medicines*, *4*(3), 58.

Chouliara, E., Karatapanis, A., Savvaidis, I. N., & Kontominas, M. G. (2007). Combined effect of oregano essential oil and modified atmosphere packaging on shelf-life extension of fresh chicken breast meat, stored at 40°C. *Food Microbiology*, *24*(6), 607−617.

D'Almeida, R. E., Sued, N., & Arena, M. E. (2022). *Citrus paradisi* and *Citrus reticulata* essential oils interfere with *Pseudomonas aeruginosa* quorum sensing in vivo on *Caenorhabditis elegans*. *Phytomedicine Plus*, *2*(1)100160. Available from https://doi.org/10.1016/j.phyplu.2021.100160.

de Souza, B. C., da Cruz, R. M. S., Lourenço, E. L. B., Pinc, M. M., Dalmagro, M., da Silva, C., Faria Nunes, M. G. L., de Souza, L. G. H., & Alberton, O. (2022). Inoculation of lemongrass with arbuscular mycorrhizal fungi and rhizobacteria alters plant growth and essential oil production. *Rhizosphere*, *22*100514. Available from https://doi.org/10.1016/j.rhisph.2022.100514.

Degala, H. L., Mahapatra, A. K., Demirci, A., & Kannan, G. (2018). Evaluation of non-thermal hurdle technology for ultraviolet-light to inactivate *Escherichia coli* K12 on goat meat surfaces. *Food Control*, *90*, 113−120. (2018).

Dey, S., & Nagababu, B. H. (2022). Applications of food colour and bio-preservatives in the food and its effect on the human health. *Food Chemistry Advances*, *1*100019. Available from https://doi.org/10.1016/j.focha.2022.100019.

Diao, W.-R., Hu, Q.-P., Zhang, H., & Xu, J.-G. (2014).). Chemical composition, antibacterial activity and mechanism of action of essential oil from seeds of fennel (*Foeniculum vulgare* Mill.). *Food Control*, *35*(1), 109−116.

Dorman, H. D., & Deans, S. G. (2000). Antimicrobial agents from plants: Antibacterial activity of plant volatile oils. *Journal of Applied Microbiology*, *88*(2), 308−316.

Drioiche, M., Radi, F. Z., Ailli, A., Bouzoubaa, A., Boutakiout, A., Mekdad, S., Kamaly, O. A. L., Saleh, A., Maouloua, M., Bousta, D., Sahpaz, S., Makhoukhi, F. E. L., & Zair, T. (2022). Correlation between the chemical composition and the antimicrobial properties of seven samples of essential oils of endemic Thymes in Morocco against multi-resistant bacteria and pathogenic fungi. *Saudi Pharmaceutical Journal*. Available from https://doi.org/10.1016/j.jsps.2022.06.022.

Duplais, C., Papon, N., & Courdavault, V. (2020). Tracking the origin and evolution of plant metabolites. *Trends in Plant Science*, *25*(12), 1182−1184. Available from https://doi.org/10.1016/j.tplants.2020.08.010.

El Kolli, M., Laouer, H., El Kolli, H., Akkal, S., & Sahli, F. (2016). Chemical analysis, antimicrobial and anti-oxidative properties of *Daucus gracilis* essential oil and its mechanism of action. *Asian Pacific Journal of Tropical Biomedicine*, *6*(1), 8−15.

Elsayed, E. A., Farooq, M., Sharaf-Eldin, M. A., El-Enshasy, H. A., & Wadaan, M. (2020). Evaluation of developmental toxicity and anti-angiogenic potential of essential oils from *Moringa oleifera* and *Moringa peregrina* seeds in zebrafish (*Danio rerio*) model. *South African Journal of Botany*, *129*, 229−237. Available from https://doi.org/10.1016/j.sajb.2019.07.022.

Elsser-Gravesen, D., & Elsser-Gravesen, A. (2013). Biopreservatives. In H. Zorn, & P. Czermak (Eds.), *Biotechnology of food and feed additives. Advances in biochemical engineering/biotechnology* (p. 143). Berlin, Heidelberg: Springer. Available from https://doi.org/10.1007/10_2013_234.

Estores, I. M., & Frye, J. (2015). Healing environments: Integrative medicine and palliative care in acute care settings. *Critical Care Nursing Clinics of North America*, *27*(3), 369−382. Available from https://doi.org/10.1016/j.cnc.2015.05.002.

Erol, N. D., Erdem, O. A., Yilmaz, S. T., Kayalar, H., & Cakli, S. (2022). Effects of the BHA and basil essential oil on nutritional, chemical, and sensory characteristics of sunflower oil and sardine (*Sardina pilchardus*) fillets during repeated deep-frying. *LWT*, *163*113557. Available from https://doi.org/10.1016/j.lwt.2022.113557.

Ezeorba, T. P. C., Chukwudozie, K. I., Ezema, C. A., Anaduaka, E. G., Nweze, E. J., & Okeke, E. S. (2022). Potentials for health and therapeutic benefits of garlic essential oils: Recent findings and future prospects. *Pharmacological Research - Modern Chinese Medicine*, *3*100075. Available from https://doi.org/10.1016/j.prmcm.2022.100075.

Fawcett, C. A., Senhorinho, G. N. A., Laamanen, C. A., & Scott, J. A. (2022). Microalgae as an alternative to oil crops for edible oils and animal feed. *Algal Research*, *64*102663. Available from https://doi.org/10.1016/j.algal.2022.102663.

Ferranti, P., & Velotto, S. (2023). Edible oil from algal sources: Characteristics and properties as novel food ingredient. *Reference Module in Food Science*. Available from https://doi.org/10.1016/B978-0-12-823960-5.00007-X.

Formighieri, C., & Melis, A. (2018). Cyanobacterial production of plant essential oils. *Planta, 248*(4), 933–946. doi:10.1007/s00425-018-2948-0. Epub 2018 Jul 4. PMID: 29974209.

Friedman, M. (2017). Chemistry, antimicrobial mechanisms, and antibiotic activities of cinnamaldehyde against pathogenic bacteria in animal feeds and human foods. *Journal of Agricultural and Food Chemistry, 65*(48), 10406–10423.

Gálvez, A., Abriouel, H., López, R. L., & Omar, N. B. (2011). Biological control of pathogens and post-processing spoilage microorganisms in fresh and processed fruit and vegetables. In C. Lacroix (Ed.), *Woodhead publishing series in food science, technology and nutrition, protective cultures, antimicrobial metabolites and bacteriophages for food and beverage biopreservation* (pp. 403–432). Woodhead Publishing. Available from https://doi.org/10.1533/9780857090522.3.403.

Gao, S., Liu, G., Li, J., Chen, J., Li, L., Li, Z., Zhang, X., Zhang, S., Thorne, R. F., & Zhang, S. (2020). Antimicrobial activity of lemongrass essential oil (*Cymbopogon flexuosus*) and its active component citral against dual species biofilms of Staphylococcus aureus and *Candida* species. *Frontier in Cellular and. Infection. Microbiology, 10*603858. Available from https://doi.org/10.3389/fcimb.2020.603858.

Ghosh, V., Sugumar, S., Mukherjee, A., & Chandrasekaran, N. (2016). Neem (*Azadirachta indica*) oils. In V. R. Preedy (Ed.), *Essential oils in food preservation, flavor and safety* (pp. 593–599). Academic Press. Available from https://doi.org/10.1016/B978-0-12-416641-7.00067-5.

Halimin, N. M. S., Abdullah, M. O., Wahab, N. A., Junin, R., Husaini, A. A. S. A., & Agi, A. (2022). Oil extracts from fresh and dried Iban ginger. *Chinese Journal of Analytical Chemistry, 50*(8)100119. Available from https://doi.org/10.1016/j.cjac.2022.100119.

Han, C. V., & Bhat, R. (2014).). In vitro control of food-borne pathogenic bacteria by essential oils and solvent extracts of underutilized flower buds of *Paeonia suffruticosa* (Andr.). *Industrial Crops and Products, 54*, 203–208. Available from https://doi.org/10.1016/j.indcrop.2014.01.014.

Hassoun, A., & Çoban, Ö. E. (2017). Essential oils for antimicrobial and antioxidant applications in fish and other seafood products. *Trends in Food Science and Technology, 68*, 26–36.

Herrera, J. G., Ramos, M. P., Albuquerque, B. N. L., de Aguiar, J. C. R. O. F., Neto, A. C. A., Paiva, P. M. G., Navarro, D. M. A. F., & Pinto, L. (2022). Multivariate evaluation of process parameters to obtain essential oil of *Piper corcovadensis* using supercritical fluid extraction. *Microchemical Journal, 181*107747. Available from https://doi.org/10.1016/j.microc.2022.107747.

Hinks, V. (2014). Essential oil safety, a guide for health care professionals. *Journal of Herbal Medicine, 4*(3), 172–173. Available from https://doi.org/10.1016/j.hermed.2014.05.002.

Hintz, T., Matthews, K. K., & Di, R. (2015). The use of plant antimicrobial compounds for food preservation. *Biomed Research International, 2015*246264. Available from https://doi.org/10.1155/2015/246264.

Hosseini, S. F., Zandi, M., Rezaei, M., & Farahmandghavi, F. (2013). Two-step method for encapsulation of oregano essential oil in chitosan nanoparticles: Preparation, characterization and in vitro release study. *Carbohydrate Polymers, 95*(1), 50–56.

Hou, T., Sana, S. S., Li, H., Xing, Y., Nanda, A., Netala, V. R., & Zhang, Z. (2022). Essential oils and its antibacterial, antifungal and antioxidant activity applications: A review. *Food Bioscience, 47*(ISSN 2212-4292)101716. Available from https://doi.org/10.1016/j.fbio.2022.101716.

Hu, Y., Zhang, J., Kong, W., Zhao, G., & Yang, M. (2017). Mechanisms of antifungal and anti-aflatoxigenic properties of essential oil derived from turmeric (*Curcuma longa* L.) on *Aspergillus flavus*. *Food Chemistry, 220*, 1–8.

Huang, D., Xu, J.-G., Liu, J.-X., Zhang, H., & Hu, Q. (2014). Chemical constituents, antibacterial activity and mechanism of action of the essential oil from *Cinnamomum cassia* bark against four food-related bacteria. *Microbiology (Reading, England), 83*(4), 357–365.

Issa, M. Y., Mohsen, E., Younis, I. Y., Nofal, E. S., & Farag, M. A. (2020). Volatiles distribution in jasmine flowers taxa grown in Egypt and its commercial products as analyzed via solid-phase microextraction (SPME) coupled to chemometrics. *Industrial Crops and Products, 144*112002. Available from https://doi.org/10.1016/j.indcrop.2019.112002.

Jemaa, M. B., Falleh, H., Saada, M., Oueslati, M., Snoussi, M., & Ksouri, R. (2018). *Thymus capitatus* essential oil ameliorates pasteurization efficiency. *Journal of Food Science and Technology, 55*(9), 3446–3452.

Jiang, S., Zhao, T., Wei, J., Cao, Z., Xu, Y., Wei, J., Xu, F., Wang, H., & Shao, X. (2021). Preparation and characterization of tea tree oil/ hydroxypropyl-β-cyclodextrininclusion complex and its application to control brown rot in peach fruit. *Food Hydrocolloids, 121*107037. Available from https://doi.org/10.1016/j.foodhyd.2021.107037.

Kadir, F. A. A., Azizan, K. A., & Othman, R. (2020). Datasets of essential oils from naturally formed and synthetically induced *Aquilaria malaccensis* agarwoods. *Data in Brief, 28*104987. Available from https://doi.org/10.1016/j.dib.2019.104987.

Khalili, S. T., Mohsenifar, A., Beyki, M., Zhaveh, S., Rahmani-Cherati, T., Abdollahi, A., Bayat, M., & Tabatabaei, M. (2015). Encapsulation of Thyme essential oils in chitosan-benzoic acid nanogel with enhanced antimicrobial activity against *Aspergillus flavus*. *LWT - Food Science and Technology, 60*(1), 502–508. Available from https://doi.org/10.1016/j.lwt.2014.07.054.

Karanicola, P., Patsalou, M., Stergiou, P. Y., Kavallieratou, A., Evripidou, N., Christou, P., Panagiotou, G., Damianou, C., Papamichael, E. M., & Koutinas, M. (2021). Ultrasound-assisted dilute acid hydrolysis for production of essential oils, pectin and bacterial cellulose via a citrus processing waste biorefinery. *Bioresource Technology, 342*126010. Available from https://doi.org/10.1016/j.biortech.2021.126010.

Kaur, G., Arya, S. K., Singh, B., Singh, S., Dhar, Y. V., Verma, P. C., & Ganjewala, D. (2019). Transcriptome analysis of the palmarosa *Cymbopogon martinii* inflorescence with emphasis on genes involved in essential oil biosynthesis. *Industrial Crops and Products, 140*, 111602. Available from https://doi.org/10.1016/j.indcrop.2019.111602.

Kavas, G., Kavas, N., & Saygili, D. (2015). The effects of thyme and clove essential oil fortified edible films on the physical, chemical and microbiological characteristics of Kashar cheese. *Journal of Food Quality, 38*(6), 405–412.

Krstić, M., Stupar, M., Đukić-Ćosić, D., Baralić, K., & Mračević, S. D. (2021). Health risk assessment of toxic metals and toxigenic fungi in commercial herbal tea samples from Belgrade, Serbia. *Journal of Food Composition and Analysis, 104*104159. Available from https://doi.org/10.1016/j.jfca.2021.104159.

Lasram, S., Zemni, H., Hamdi, Z., Chenenaoui, S., Houissa, H., Tounsi, M. S., & Ghorbel, A. (2019). Antifungal and antiaflatoxinogenic activities of Carum carvi L., Coriandrum sativum L. seed essential oils and their major terpene component against Aspergillus flavus. *Industrial Crops and Products, 134*, 11−18. Available from https://doi.org/10.1016/j.indcrop.2019.03.037.

Li, Q. X., & Chang, C. L. (2016). Chapter 25 - Basil (*Ocimum basilicum* L.) oils. In V. R. Preedy (Ed.), *Essential oils in food preservation, flavor and safety* (pp. 231−238). Academic Press. Available from https://doi.org/10.1016/B978-0-12-416641-7.00025-0.

Li, W., Liu, J., Zhang, H., Liu, Z., Wang, Y., Xing, L., He, Q., & Du, H. (2022). Plant pan-genomics: Recent advances, new challenges, and roads ahead. *Journal of Genetics and Genomics, 1673−8527.* Available from https://doi.org/10.1016/j.jgg.2022.06.004, ISSN.

Li, Y.-X., Zhang, C., Pan, S., Chen, L., Liu, M., Yang, K., Zeng, X., & Tian, J. (2020). Analysis of chemical components and biological activities of essential oils from black and white pepper (*Piper nigrum* L.) in five provinces of southern China. *LWT, 117*108644. Available from https://doi.org/10.1016/j.lwt.2019.108644.

Limam, H., Jemaa, M. B., Tammar, S., Ksibi, N., Khammassi, S., Jallouli, S., DelRe, G., & Msaada, K. (2020). Variation in chemical profile of leaves essential oils from thirteen Tunisian Eucalyptus species and evaluation of their antioxidant and antibacterial properties. *Industrial Crops and Products, 158*112964. Available from https://doi.org/10.1016/j.indcrop.2020.112964.

Lin, H.-J., Lin, Y.-L., Huang, B.-B., Lin, Y.-T., Li, H.-K., Lu, W.-J., Lin, T.-C., Tsui, Y.-C., & Lin, H.-T. V. (2022). Solid- and vapour-phase antifungal activities of six essential oils and their applications in postharvest fungal control of peach (*Prunus persica* L. Batsch). *LWT, 156*113031. Available from https://doi.org/10.1016/j.lwt.2021.113031.

Liu, Z., Li, H., Cui, G., Wei, M., Zou, Z., & Ni, H. (2021). Efficient extraction of essential oil from *Cinnamomum burmannii* leaves using enzymolysis pretreatment and followed by microwave-assisted method. *LWT, 147*111497. Available from https://doi.org/10.1016/j.lwt.2021.111497.

Li, W.-R., Shi, Q.-S., Liang, Q., Xie, X.-B., Huang, X.-M., & Chen, Y.-B. (2014). Antibacterial activity and kinetics of *Litsea cubeba* oil on *Escherichia coli. PLoS One, 9*(11)e110983.

Lucera, A., Del Nobile, M. A., Costa, C., & Conte, A. (2012). Food applications of natural antimicrobial compounds. *Frontiers in Microbiology, 3*, 287.

Lui, J., & Mori, A. (2005). Oxidative damage hypothesis of stress associated aging acceleration: Neuroprotective effects of natural and nutritional antioxidants. *Research Communication in Biology Psychology, Psychiatry and Neuroscience,* 30−31.

Lv, F., Liang, H., Yuan, Q., & Li, C. (2011). In vitro antimicrobial effects and mechanism of action of selected plant essential oil combinations against four food-related microorganisms. *Food Research International, 44*(9), 3057−3064.

Mangalagiri, N. P., Panditi, S. K., & Jeevigunta, N. L. L. (2021). Antimicrobial activity of essential plant oils and their major components. *Heliyon, 7* (4)e06835. Available from https://doi.org/10.1016/j.heliyon.2021.e06835.

Mani-López, E., Palou, E., & López-Malo, A. (2018). Biopreservatives as agents to prevent food spoilage. In A. Maria Holban, & A. Mihai Grumezescu (Eds.), *Handbook of food bioengineering, microbial contamination and food degradation* (pp. 235−270). Academic Press, 2018. Available from https://doi.org/10.1016/B978-0-12-811515-2.00008-1.

Margalho, L. P., Kamimura, B. A., Brexó, R. P., Alvarenga, V. O., Cebeci, A. S., Janssen, P. W. M., Dijkstra, A., Starrenburg, M. J. C., Sheombarsing, R. S., Cruz, A. G., Alkema, W., Bachmann, H., & Sant'Ana, A. S. (2021). High throughput screening of technological and biopreservation traits of a large set of wild lactic acid bacteria from Brazilian artisanal cheeses. *Food Microbiology, 100*103872. Available from https://doi.org/10.1016/j.fm.2021.103872.

Mehdizadeh, L., & Moghaddam, M. (2018). Essential oils: Biological activity and therapeutic potential. *Therapeutic, Probiotic, and Unconventional Foods,* 167−179.

Moreira, P., Smith, M., Zhu, X., Honda, K., Lee, H., Aliev, G., & Perry, G. (2005). Since oxidative damage is a key phenomenon in Alzheimer's disease, treatment with antioxidants seems to be a promising approach for slowing disease progression. Oxidative damage and Alzheimer's disease: Are antioxidant therapies useful. *Drug News Perspective, 18,* 13−19.

Morone-Fortunato, I., & Avato, P. (2008). Plant development and synthesis of essential oils in micropropagated and mycorrhiza inoculated plants of *Origanum vulgare* L. ssp. *hirtum* (Link) Ietswaart. *Plant Cell Tiss Organ Cult, 93,* 139−149. Available from https://doi.org/10.1007/s11240-008-9353-5.

Mouafo, H. T., Mbawala, A., Tanaji, K., Somashekar, D., & Ndjouenkeu, R. (2020). Improvement of the shelf life of raw ground goat meat by using biosurfactants produced by lactobacilli strains as biopreservatives. *LWT, 133*110071. Available from https://doi.org/10.1016/j.lwt.2020.110071.

Mukurumbira, A. R., Shellie, R. A., Keast, R., Palombo, E. A., & Jadhav, S. R. (2022). Encapsulation of essential oils and their application in antimicrobial active packaging. *Food Control, 136*108883. Available from https://doi.org/10.1016/j.foodcont.2022.108883.

Naito, Y., Uchiyama, K., & Yoshikawa, T. (2006). Oxidative stress involvement in diabetic nephropathy and its prevention by astaxanthin. *Oxidative Stress and Disease, 21,* 235.

Nazir, S., & Wani, I. A. (2022). Development and characterization of an antimicrobial edible film from basil seed (*Ocimum basilicum* L.) mucilage and sodium alginate. *Biocatalysis and Agricultural Biotechnology*102450. Available from https://doi.org/10.1016/j.bcab.2022.102450.

Nissen, Lorenzo, Zatta, Alessandro, Stefanini, Ilaria, Grandi, Silvia, Sgorbati, Barbara, Bruno, Biavati, & Monti, Andrea (2010). Characterization and antimicrobial activity of essential oils of industrial hemp varieties (*Cannabis sativa* L.). *Fitoterapia, 81*(5), 413−419. (ISSN 0367-326X). Available from https://doi.org/10.1016/j.fitote.2009.11.010.

Olaimat, A. N., Al-Nabulsi, A. A., & Osaili, T. M. (2022). Control of foodborne pathogens in common foods in the Middle East. In N. Ioannis, S. Tareq, & M. Osaili (Eds.), *Food safety in the Middle East* (pp. 187−226). Academic Press. Available from https://doi.org/10.1016/B978-0-12-822417-5.00003-9.

Onyeaka, H. N., & Nwabor, O. F. (2022). Conventional preservation and preservative. In N. N. Helen, F. Ozioma, & Nwabor (Eds.), *Food preservation and safety of natural products* (pp. 51–56). Academic Press. Available from https://doi.org/10.1016/B978-0-323-85700-0.00008-3.

Pateiro, M., Munekata, P. E. S., Sant'Ana, A. S., Domínguez, R., Rodríguez-Lázaro, D., & Lorenzo, J. M. (2021). Application of essential oils as antimicrobial agents against spoilage and pathogenic microorganisms in meat products. *International Journal of Food Microbiology, 337*108966. Available from https://doi.org/10.1016/j.ijfoodmicro.2020.108966.

Periasamy, G., Karim, A., Gibrelibanos, M., Gebremedhin, G., & Gilani, A. (2016). Chapter 69 - Nutmeg (*Myristica fragrans* Houtt.) oils. In Victor R. Preedy (Ed.), *Essential oils in food preservation, flavor and safety* (ISBN 978012416641, pp. 607–616). Academic Press. Available from https://doi.org/10.1016/B978-0-12-416641-7.00069-9.

Perricone, M., Arace, E., Corbo, M. R., Sinigaglia, M., & Bevilacqua, A. (2015). Bioactivity of essential oils: A review on their interaction with food components. *Frontiers in Microbiology, 6,* 76.

Pires, E. O., Caleja, C., Garcia, C. C., Ferreira, I. C. F. R., & Barros, L. (2021). Current status of genus Impatiens: Bioactive compounds and natural pigments with health benefits. *Trends in Food Science & Technology, 117,* 106–124. Available from https://doi.org/10.1016/j.tifs.2021.01.074.

Premanath, R., James, J. P., Karunasagar, I., Vaňková, E., & Scholtz, V. (2022). Tropical plant products as biopreservatives and their application in food safety. *Food Control, 141*109185. Available from https://doi.org/10.1016/j.foodcont.2022.109185.

Polturak, G., Liu, Z., & Osbourn, A. (2022). New and emerging concepts in the evolution and function of plant biosynthetic gene clusters. *Current Opinion in Green and Sustainable Chemistry (Weinheim an der Bergstrasse, Germany), 33*100568. Available from https://doi.org/10.1016/j.cogsc.2021.100568.

Ponce, A., Roura, S. I., & Moreira, M. R. (2011). Essential oils as biopreservatives: Different methods for the technological application in lettuce leaves. *Journal of Food Science, 76*(1), M34–M40. Available from https://doi.org/10.1111/j.1750-3841.2010.01880.x.

Putnik, P., Gabrić, D., Roohinejad, S., Barba, F. J., Granato, D., Mallikarjunan, K., Lorenzo, J. M., & Kovačević, D. B. (2019). An overview of organosulfur compounds from *Allium* spp.: From processing and preservation to evaluation of their bioavailability, antimicrobial, and anti-inflammatory properties. *Food Chemistry, 276,* 680–691. Available from https://doi.org/10.1016/j.foodchem.2018.10.068.

Raybaudi-Massilia, R. M., Mosqueda-Melgar, J., & Martin-Belloso, O. (2006). Antimicrobial activity of essential oils on *Salmonella enteritidis*, *Escherichia coli*, and *Listeria innocua* in fruit juices. *Journal of Food Protection, 69*(7), 1579–1586.

Raybaudi-Massilia, R. M., Mosqueda-Melgar, J., Soliva-Fortuny, R., & Martín-Belloso, O. (2009). Control of pathogenic and spoilage microorganisms in fresh-cut fruits and fruit juices by traditional and alternative natural antimicrobials. *Comprehensive Reviews in Food Science and Food Safety, 8*(3), 157–180.

Rendueles, C., Duarte, A. C., Escobedo, S., Fernández, L., Rodríguez, A., García, P., & Martínez, B. (2022). Combined use of bacteriocins and bacteriophages as food biopreservatives. A review. *International Journal of Food Microbiology, 368*109611. Available from https://doi.org/10.1016/j.ijfoodmicro.2022.109611.

Rice-Evans, C., Miller, N., & Paganga, G. (1997). Antioxidant properties of phenolic compounds. *Trends in Plant Science, 2*(4), 152–159.

Ríos, J.-L. (2016). Essential oils: What they are and how the terms are used and defined. In V. R. Preedy (Ed.), *Essential oils in food preservation, flavor and safety* (pp. 3–10). , Academic Press. Available from https://doi.org/10.1016/B978-0-12-416641-7.00001-8.

Saleh, A., Pirouzifard, M., Khaledabad, M. A., & Almasi, H. (2022). Optimization and characterization of *Lippia citriodora* essential oil loaded niosomes: A novel plant-based food nano preservative. *Colloids and Surfaces A: Physicochemical and Engineering Aspects, 650*129480. Available from https://doi.org/10.1016/j.colsurfa.2022.129480, ISSN 0927-7757.

Salim-ur-Rehman, S. H., Nawaz, H., Ahmad, M. M., Murtaza, M. A., & Rizvi, A. J. (2007). Inhibitory effect of citrus peel essential oils on the microbial growth of bread. *Pakistan Journal of Nutrition, 6*(6), 558–561.

Saporito, F., Sandri, G., Bonferoni, M. C., Rossi, S., Boselli, C., Cornaglia, A. I., Mannucci, B., Grisoli, B. V., & Ferrari, F. (2017). Essential oil-loaded lipid nanoparticles for wound healing. *International Journal of Nanomedicine, 13,* 175–186.

Schnitzler, P., Schön, K., & Reichling, J. (2001). Antiviral activity of Australian tea tree oil and eucalyptus oil against herpes simplex virus in cell culture. *Die Pharmazie, 56*(4), 343–347.

Selli, S., Guclu, G., Sevindik, O., & Kelebek, H. (2022). Biochemistry, antioxidant, and antimicrobial properties of hazelnut (*Corylus avellana* L.) oil. In A. Adam Mariod (Ed.), *Multiple biological activities of unconventional seed oils* (pp. 397–412). Academic Press. Available from https://doi.org/10.1016/B978-0-12-824135-6.00012-X.

Singh, B., Singh, J. P., Kaur, A., & Yadav, M. P. (2021). Insights into the chemical composition and bioactivities of citrus peel essential oils. *Food Research International, 143*110231. Available from https://doi.org/10.1016/j.foodres.2021.110231.

Speranza, B., Corbo, M. R., & Sinigaglia, M. (2010). Essential oils for preserving perishable foods: Possibilities and limitations. In A. Bevilacqua, M. R. Corbo, & M. Sinigaglia (Eds.), *Application of alternative food-preservation technologies to enhance food safety and stability* (23, pp. 35–37). Sharjah: United Arab Emirates:Bentham Publisher. (eds).

Tanha, S., Rafiee-Tehrani, M., Abdollahi, M., Vakilian, S., Esmaili, Z., Naraghi, Z. S., Seyedjafari, E., & Javar, H. A. (2017). G-CSF-loaded nanofiber/nanoparticle composite coated with collagen promotes wound healing in vivo. *Journal of Biomedical Materials Research. Part A, 105*(10), 2830–2842.

Tian, Y., Zhou, L., Liu, J., Yu, K., Yu, W., Jiang, H., Zhong, J., Zou, L., & Liu, W. (2023). Effect of sustained-release tea tree essential oil solid preservative on fresh-cut pineapple storage quality in modified atmospheres packaging. *Food Chemistry, 417*135898. Available from https://doi.org/10.1016/j.foodchem.2023.135898.

Tiwari, S., & Dubey, N. K. (2022). Nanoencapsulated essential oils as novel green preservatives against fungal and mycotoxin contamination of food commodities. *Current Opinion in Food Science, 45*100831. Available from https://doi.org/10.1016/j.cofs.2022.100831.

Tokur, B., Korkmaz, K., & Uçar, Y. (2021). Enhancing sunflower oil by the addition of commercial thyme and rosemary essential oils: The effect on lipid quality of Mediterranean horse mackerel and anchovy during traditional pan-frying. *International Journal of Gastronomy and Food Science*, *26*100428. Available from https://doi.org/10.1016/j.ijgfs.2021.100428.

Tran, V. T., Kingwascharapong, P., Tanaka, F., & Tanaka, F. (2021). Effect of edible coatings developed from chitosan incorporated with tea seed oil on Japanese pear. *Scientia Horticulturae*, *288*110314. Available from https://doi.org/10.1016/j.scienta.2021.110314.

Valdivieso-Ugarte, M., Plaza-Diaz, J., Gomez-Llorente, C., Gómez, E. L., Sabés-Alsina, M., & Gil, Á. (2021). In vitro examination of antibacterial and immunomodulatory activities of cinnamon, white thyme, and clove essential oils. *Journal of Functional Foods*, *81*104436. Available from https://doi.org/10.1016/j.jff.2021.104436.

Vázquez, J. A., Pedreira, A., Durán, S., Cabanelas, D., Souto-Montero, P., Martínez, P., Mulet, M., Pérez-Martín, R. I., & Valcarcel, J. (2022). Biorefinery for tuna head wastes: Production of protein hydrolysates, high-quality oils, minerals and bacterial peptones. *Journal of Cleaner Production*, *357*131909. Available from https://doi.org/10.1016/j.jclepro.2022.131909.

Wang, Y., Luo, J., Hou, X., Wu, H., Li, Q., Li, S., Luo, Q., Li, M., Liu, X., Shen, G., Cheng, A., & Zhang, Z. (2022). Physicochemical, antibacterial, and biodegradability properties of green Sichuan pepper (*Zanthoxylum armatum* DC.) essential oil incorporated starch films. *LWT*, *161*113392. Available from https://doi.org/10.1016/j.lwt.2022.113392.

Wei, A., & Shibamoto, T. (2007). Antioxidant activities and volatile constituents of various essential oils. *Journal of Agricultural and Food Chemistry*, *55*(5), 1737–1742.

Wu, K., Jin, R., Bao, X., Yu, G., & Yi, F. (2021). Potential roles of essential oils from the flower, fruit and leaf of Citrus medica L. var. sarcodactylis in preventing spoilage of Chinese steamed bread. *Food Bioscience*, *43*101271. Available from https://doi.org/10.1016/j.fbio.2021.101271.

Yee, Q. Y., Hassim, M. H., Chemmangattuvalappil, N. G., Ten, J. Y., & Raslan, R. (2022). Optimization of quality, safety and health aspects in personal care product preservative design. *Process Safety and Environmental Protection*, *157*, 246–253. Available from https://doi.org/10.1016/j.psep.2021.11.025.

Yangilar, F., & Yildiz, P. O. (2018). Effects of using combined essential oils on quality parameters of bio-yogurt. *Journal of Food Processing and Preservation*, *42*(1)e13332.

Youdim, K., Deans, S., & Finlayson, H. (2002). The antioxidant properties of thyme (*Thymus zygis* L.) essential oil: An inhibitor of lipid peroxidation and a free radical scavenger. *Journal of Essential Oil Research*, *14*(3), 210–215.

Yusuf, M. (2018). Natural antimicrobial agents for food biopreservation. In A. Mihai Grumezescu, & A. Maria Holban (Eds.), *Handbook of food bioengineering, food packaging and preservation* (pp. 409–438). Academic Press. Available from https://doi.org/10.1016/B978-0-12-811516-9.00012-9.

Zhang, Y., Liu, X., Wang, Y., Jiang, P., & Quek, S. (2016). Antibacterial activity and mechanism of cinnamon essential oil against Escherichia coli and Staphylococcus aureus. *Food Control*, *59*, 282–289.

Zerkani, K., Amalich, S., Tagnaout, I., Bouharroud, R., & Zair, T. (2022). Chemical composition, pharmaceutical potential and toxicity of the essential oils extracted from the leaves, fruits and barks of *Pistacia atlantica*. *Biocatalysis and Agricultural Biotechnology*, *43*102431. Available from https://doi.org/10.1016/j.bcab.2022.102431.

Chapter 28

Antibiofilm effect of essential oils in food industry

Blessing Itohan Omo-Omorodion[1] and Charles Oluwaseun Adetunji[2]

[1]Applied Microbiology, Biotechnology and Nanotechnology Laboratory, Department of Microbiology, Edo University Iyamho, Auchi, Edo State, Nigeria, [2]Applied Microbiology, Biotechnology and Nanotechnology Laboratory, Department of Microbiology, Edo State University Uzairue, Iyamho, Edo State, Nigeria

Introduction

Essential oils are naturally occurring compounds found in plant parts such as seeds, stem, bark, flower, or in the whole plant. They have been reported to be used for several purposes including food preservation (Martinengo, Arunachalam, & Shi, 2021). Because essential oils (EOs) are secondary metabolites, they have shown a wide range of biological properties and usefulness in different sectors including the food industry as preservatives. For example, EOs extracted from citrus plants, such as sesquiterpenes, monoterpenes, and oxygenated derivatives, possess strong antimicrobial properties against food-causing bacteria such as *Escherichia coli, Listeria monocytogenes, Pseudomonas fluorescens, Bacillus thermosphacta, Campylobacter jejuni, Salmonella* spp., and *Bacillus cereus* (Lin, Briandet, & Kovács, 2022). However, microbial adherence on food-grade surfaces causes the formation of biofilms which results in significant in great economic loss in the food industry. Surfaces in food industries such as stainless steel, aluminum, and glasses conveyor belts, and wastewater pipes, as well as food processing units like heating exchangers, are good sources of substrates for biofilms by pathogenic and food-spoiling bacteria. These biofilms can persist for long and cause cross-contamination of food industries which can lead to food poisoning and food deterioration (Maia et al. 2020).

Microbial biofilms

Biofilms are a complex matrix of microbes that contain sticky gels of polysaccharides (glycocalyx), proteins, and organic components that are usually attached to biotic and abiotic surfaces such as food processing environments, drinking water stations, clinics, and industrial surfaces. The glycocalyx is strongly anionic and acts as a physical barrier that protects microbes from external agents. Accordingly, the formation of biofilms usually takes different stages of cyclical and dynamic process. At first, there is a reversible binding of planktonic cells to the surface which is then followed by an irreversible attachment to surfaces with the production of more proteins and polysaccharides in addition to DNAs, lipids, and exopolysaccharides. This primary biofilm structure grows and matures with time to provide mechanical and chemical stability to cells (Hall & Mah, 2017). It has earlier been reported that biofilms are heterogenous in nature comprising multispecies and different microbes communities. They either cooperate or compete. This feature influences the formation of biofilms, attachment to surfaces, and resistance to harsh conditions (Rossi et al., 2022). Kerekes et al. (2019) noted that biofilms formed from monocultures are usually less fitted and easy to target when compared to multispecies (Al-Shabib et al., 2017).

Foodborne pathogens

Characteristically, most microbes related to the food industry are predominantly in a mixed biofilm state. Some of which include *Enterohemorrhagic E. coli, L. monocytogenes, S. enterica,* and *B. cereus. B. cereus* is an endospore-forming bacteria that can stand very high temperatures of 80°C and pasteurization and make rich biofilms. They are present in fat-rich foods like cheese, butter, and milk and have been indicated in enterotoxin illnesses like diarrhea and

Applications of Essential Oils in the Food Industry. DOI: https://doi.org/10.1016/B978-0-323-98340-2.00029-8

abdominal pain when ingested (Yang, Lin, Aljuffali, & Fang, 2017). *B. cereus* has been reported to form endospores and its toxins can pass through high-temperature short-time pasteurization (Wu, Liu, Nakamoto, Wall, & Li, 2022). Similarly, enterohemorrhagic *E. coli* strain 0157:H7 are thick-forming biofilms (Rossi et al., 2022) found on food stainless steels and borosilicate glasses in food processing industries. It has been found that little to no amount of *E. coli* strain O157:H7 are found in polypropylene surfaces. *E. coli* is found in contaminated meat, milk, water sources, fruits, and veggies. It produces shiga toxins (STEC) and enterohaemorrhagic gastroenteritis, and are among the major causes of blood-watery diarrhea. Furthermore, *S. aureus* and *S. enterica* have been reported in cases of biofilm formation (Rodrigues et al., 2018) and causing agents of most mediated diseases. From farm animals, methicillin-resistant *S. aureus* (MRSA) has been found to form biofilms on animal surfaces (Wen, Wang, Zhao, Chen, & Zeng, 2021).

Nevertheless, *L. monocytogenes* are Gram-positive rod-shaped bacterium usually present in soil, water, and air. They are commonly associated with food products such as meats, processed food, and vegetables. *L. monocytogenes* have been indicated in listeriosis, a serious public health concern that primarily affects pregnant women, newborns, and people with weakened immunity. This route of infection is the oral ingestion of contaminated food. *L. monocytogenes*, are colonizers of biotic and abiotic surfaces where they form biofilms. Biofilms formed by *L. monocytogenes* when compared to other bacteria have been revealed to form resistant biofilms that are resistant to antibiotics, biocides, and detergents (Rodrigues et al., 2018). It is not clear why this is so but some schools of thought believe that *L. monocytogenes* require unusual growth conditions and survival mechanisms that make control and eradication from surfaces difficult. This is still a big food and hygiene safety concern in most countries.

However, as concerns are raised over the use of synthetic chemicals on food products, alternative and preferential use of natural products like EOs have become important. The knowledge into biofilms formations and resilience has made researchers delve into developing and understanding compounds that have antibiofilm properties to reduce illness caused by foodborne pathogens (Alvarez-Ordóñez, Coughlan, Briandet, & Cotter, 2019).

Antibiofilm property of essential oils in food industry

EOs have been reported to have antibiofilm activities on microorganisms. They inhibit the formation of biofilms by exerting their bioactive constituents which interfere with the mechanism of biofilm formation at different stages. This interference occurs in the form of inhibition, removal, and prevention of biofilm spread on food contact surfaces, thus serving as sanitizers in food industries.

It has also been documented that high biofilm-controlling ability occurs through the regulation of genes and proteins that code for motility, exopolysaccharides matrix formation, and quorum sensing in microbes. To increase the antibiofilm activities in EOs, recent researches are directed toward incorporating nanosized and blended formulations and combinations of EOs in the food sector. This has drawn much attention from food scientists as synthetic preservatives used in food industries have documented cases of intoxications, allergies, cancer, and degenerative diseases (Carter, Feng, & Li, 2019). Arousing interest in an alternate source has stimulated research on EOs. EOs are biodegradable and required in low quantities without toxic effects (Uddin Mahamud, Nahar, Ashrafudoulla, Park, & Ha, 2022). Additionally, because of the lipophilic oily nature of EOs, solutions containing EOs are easily washed off without residual odor problems. Food industries use extracts of EOs because of their ability to grow on pathogenic microbes (Nové et al., 2020).

Interestingly, EOs such as thyme oil, cinnamon oils, and oregano have shown antibiofilm activities against food-spoiling bacteria such as *L. monocytogenes, E. coli, Bacillus thermosphacta,* and *Pseudomonas fluorescens*. More recently, some bacteria have shown tolerance to minute quantities of EOs examples of which include *E. coli,* and *S. aureus*. These bacteria are ubiquitous and are leading causes of infection in clinical, environmental, and food industries (Alvarez-Ordóñez et al., 2019).

Essential oils that target food spoilage–causing microbes

Essential oils are complex with over 50 compounds in a mixture. EOs like lemon oil, tea tree oil, cinnamon oil, thyme oil, mustard oil, lavender oil, clove oil, peppermint, oregano, and eucalyptus oils have been implicated in food preservation; for example, cinnamon oil is a volatile compound that contains three important compounds namely *trans*-cinnamaldehyde, eugenol, and linalool. Cinnamaldehyde, a potent compound in cinnamon oil has been reported to actively inhibit the growth of Gram-positive and Gram-negative bacteria and fungi such as molds (Prakash, Kedia, Mishra, & Dubey, 2015). The free radical scavenging properties in Cinnamon attach to the cell wall of pathogenic microbes (Kim & Kang, 2020). EOs cause cell wall degradation, membrane damage, and disruption of proton motive force (Carter et al., 2019).

Similarly, tea tree oil has been implicated to be widely used against molds (Prakash et al., 2015) and has been used as antimold agents in bakeries (Alvarez-Ordóñez et al., 2019). Clove oils; however, it contains major compounds of phenylpropanoids which include thymol, carvacrol, cinnamaldehyde, 2-heptanone eugenyl acetate, eugenol, and β-caryophyllene. These biological compounds could inhibit germ tube formation of *C. albicans, Candida* spp. such as *C. albicans, C. tropicalis*, and *C. glabrata*, and inhibits the growth of multi-resistant Staphylococcus species.

More so, active ingredients in lavender oil are linalool, camphor, linalool (3,7-dimethylocta-1,6-dien-3-ol), l, 1,8-cineole, lavandulyl acetate, *B*-ocimene, terpinen-4-ol, l-fenchone, viridiflorol, avandulol and linalyl acetate (3,7-dimethyl-1,6-octa-dien-3-yl acetate). These compounds have shown strong antibiofilm forming properties again biofilm-producing bacteria and fungi like yeasts, *Aspergillus niger, Aspergillus flavus, Cryptococcus neoformans*, dermatophytes, and *C.* species. Likewise, Eucalyptus oil that contains spathulenol, terpinen-4-ol, α-terpineol, aromadendrene *trans*-pinocarveol, phellandral, globulol, limonene, cuminal, and cryptone, α-pinene, and *p*-cymene. They are used mostly in food companies for their strong flavors and antibiofilm properties in food processing companies (Bai, Nakatsu, & Bhunia, 2021).

However, the concentration of these EOs necessary to cause inhibition of biofilm formation varies from one EOs to the other and species to species (Kerekes et al., 2019).

Conclusion

EOs are important in the food industry as they play major roles in inhibiting the growth of biofilms formation of food pathogens. Knowing of the basis and mechanism with which EOs inhibit the formation of biofilms is necessary. Synergistic studies have shown desirable aroma and strong preservatory properties of EOs. The use of essential oil is expected to grow in the food industries in the future.

Implication for further studies

There are several types of EOs with varying compositions that serve as antibiofilms in the food industry. It is suggested that more studies that understand and compare the effect of EOs on Gram-negative and Gram-positive bacteria at varying conditions be understudied.

Nanotechnology is a promising technology. The use of nanotechnology in fabricating EOs should be considered. For example, dairy products and equipment are rich sources of biofilms, and the inherent antimicrobial properties of nanoparticles have been largely studied. Very little is known about antibiofilms.

Microencapsulated edible films or coated EOs have been reported to enhance safe and quality food and cereals. The characteristics and biosafety of nanoparticles used should be explored. In some cases, the microencapsulated edible film or coating with EOs along with antimicrobial agents effectively enhances the safety and quality of cereals and food products, which need to be investigated in future studies.

References

Al-Shabib, N. A., Husain, F. M., Ahmad, I., Khan, M. S., Khan, R. A., & Khan, J. M. (2017). Rutin inhibits mono and multi-species biofilm formation by foodborne drug resistant *Escherichia coli* and Staphylococcus aureus. *Food Control, 79*, 325–332.

Alvarez-Ordóñez, A., Coughlan, L. M., Briandet, R., & Cotter, P. D. (2019). Biofilms in food processing environments: Challenges and opportunities. *Annual Review of Food Science and Technology, 10*, 173–195, [CrossRef] [PubMed].

Bai, X., Nakatsu, C. H., & Bhunia, A. K. (2021). Bacterial biofilms and their implications in pathogenesis and food safety. *Foods, 10*, 2117, [CrossRef] [PubMed].

Carter, M. Q., Feng, D., & Li, H. H. (2019). Curli fimbriae confer shiga toxin-producing Escherichia coli a competitive trait in mixed biofilms. *Food Microbiology, 82*, 482–488.

Hall, C. W., & Mah, T. F. (2017). Molecular mechanisms of biofilm-based antibiotic resistance and tolerance in pathogenic bacteria. *FEMS Microbiology Reviews, 41*, 276–301.

Kerekes, E. B., Vidács, A., Takó, M., Petkovits, T., Vágvölgyi, C., Horváth, G., Balázs, V. L., & Krisch, J. (2019). Anti-biofilm effect of selected essential oils and main components on mono- and polymicrobic bacterial cultures. *Microorganisms, 7*, 345.

Kim, S.-Y., & Kang, S.-S. (2020). Anti-biofilm activities of Manuka honey against Escherichia coli O157:H7. *Food Science of Animal Resources, 40*, 668–674.

Lin, Y., Briandet, R., & Kovács, Á. T. (2022). *Bacillus cereus* sensu lato biofilm formation and its ecological importance. *Biofilm, 4*100070.

Maia, D. S. V., Haubert, L., Kroning, I. S., Soares, K. D. S., Oliveira, T. L., & da Silva, W. P. (2020). Biofilm formation by Staphylococcus aureus isolated from food poisoning outbreaks and effect of *Butia odorata* Barb. Rodr. Extract on planktonic and biofilm cells. *LWT, 117*108685.

Martinengo, P., Arunachalam, K., & Shi, C. J. F. (2021). Polyphenolic antibacterials for food preservation: Review, challenges, and current applications. *Foods, 10*, 2469, 14. Chitlapilly Dass, S.; Wang, R. Biofilm through the Looking Glass: A Microbial Food Safety Perspective. Pathogens 2022, 11, 346. [CrossRef].

Nové, M., Kincses, A., Szalontai, B., Rácz, B., Blair, J. M. A., González-Prádena, A., Benito-Lama, M., Domínguez-Álvarez, E., & Spengler, G. (2020). Biofilm eradication by symmetrical selenoesters for food-borne pathogens. *Microorganisms, 8*, 566.

Prakash, B., Kedia, A., Mishra, P. K., & Dubey, N. K. (2015). Plant essential oils as food preservatives to control moulds, mycotoxin contamination and oxidative deterioration of agri-food commodities—Potentials and challenges. *Food Control, 47*, 381—391.

Rodrigues, J. B. D. S., Souza, N. T. D., Scarano, J. O. A., Sousa, J. M. D., Lira, M. C., Figueiredo, R. C. B. Q. D., de Souza, E. L., & Magnani, M. (2018). Efficacy of using oregano essential oil and carvacrol to remove young and mature Staphylococcus aureus biofilms on food-contact surfaces of stainless steel. *LWT, 93*, 293—299.

Rossi, C., Chaves-López, C., Serio, A., Casaccia, M., Maggio, F., & Paparella, A. (2022). Effectiveness and mechanisms of essential oils for biofilm control on food-contact surfaces: An updated review. *Critical Reviews in Food Science and Nutrition, 62*, 2172—2191.

Uddin Mahamud, A.G.M.S.; Nahar, S.; Ashrafudoulla, M.; Park, S.H.; Ha, S.-D. Insights into antibiofilm mechanisms of phytochemicals: Prospects in the food industry. Critical Reviews in Food Science and Nutrition 2022.

Wen, Q. H., Wang, R., Zhao, S. Q., Chen, B. R., & Zeng, X. A. (2021). Inhibition of biofilm formation of foodborne Staphylococcus aureus by the citrus flavonoid naringenin. *Foods, 10*, 2614.

Wu, B., Liu, X., Nakamoto, S. T., Wall, M., & Li, Y. (2022). Antimicrobial activity of Ohelo Berry (*Vaccinium calycinum*) Juice against *Listeria monocytogenes* and its potential for milk preservation. *Microorganisms, 10*, 548, [CrossRef] [PubMed].

Yang, S.-C., Lin, C.-H., Aljuffali, I. A., & Fang, J.-Y. (2017). Current pathogenic *Escherichia coli* foodborne outbreak cases and therapy development. *Archives of Microbiology, 199*, 811—825.

Chapter 29

Significance of essential oils for the treatment of infectious diseases

Blessing Itohan Omo-Omorodion[1] and Charles Oluwaseun Adetunji[2]

[1]Applied Microbiology, Biotechnology and Nanotechnology Laboratory, Department of Microbiology, Edo University Iyamho, Auchi, Edo State, Nigeria,
[2]Applied Microbiology, Biotechnology and Nanotechnology Laboratory, Department of Microbiology, Edo State University Uzairue, Iyamho, Edo State, Nigeria

Introduction

Aromatic plants contain significant properties and derivatives implicated the human health. One such derivative that has been reported is essential oils (Bakkali et al., 2008). Essential oils from plant sources contain antibacterial, antiviral, and antifungal properties and have been considered useful as they serve as natural alternatives to treat and manage infectious diseases (Soltani & Madadlou, 2015) such as flu, common cold, influenza, Hepatitis A and B, tuberculosis (TB), pneumonia, HIV, malaria including severe acute respiratory syndrome (SARS) (Lv et al., 2011; Prabhu & Poulose, 2012; Soltani & Madadlou, 2015). Besides, other applications have been reported in cancer, nervous system disorders, and cardiovascular diseases (Lv et al., 2011). More so, essential oils have featured in the reduction of cholesterol and regulation of glucose levels, in cosmetic and food production sectors (Soltani & Madadlou, 2015). This chapter focuses on the significance of essential oils to treat infectious diseases.

Infectious diseases

Typically, many microorganisms live inside or on our bodies and are generally harmless but helpful. These microorganisms such as bacteria, fungi, viruses, and parasites are referred to as micro floras, but under certain conditions, they can become infectious and cause disease especially if a host immune system is compromised or overwhelmed.

Infectious diseases are prominent causes of global morbidity and mortality. These diseases are spread from person to person through various factors such as skin contact, saliva droplets in the air, contaminated foods and drinks, insect bites, blood, bodily fluids, contaminated needles and syringes, and unsafe sexual intercourse. They are usually the major reasons why people visit hospitals frequently, especially in developing countries (Prabhu & Poulose, 2012). The severity of infectious diseases can range from mild to severe cases with symptoms such as fever, muscle and body aches, headache, fatigue, chills, vomiting dehydration, and diarrhea.

Brief historic facet of infectious diseases

The medical history of infectious diseases in humans is enthralling. The history of infectious disease is believed not to be a result of geographical, economic, or political factors (Rai et al., 2015). It is the history of people who have defined infections, differentiated infections, and isolated, characterizing the etiological agents associated with an infection (Rai et al., 2015). Additionally, infectious disease history has also looked at the history of individuals who have developed diagnostic tests, engineered treatment plans, proposed preventive measures to reduce acquisition of the infectious agents, and developed chemo-preventive measures and vaccines to address infectious agents (Danilcauk et al., 2006).

Infectious disease history is a story of how early physicians and scientists worked on the effects of toxins produced by microbes such as bacterial, fungi, chlamydia, mycoplasma, rickettsia and understood the life cycle of parasites (Ghosh et al., 2013). As a result of this understanding, improvements in public health prevention and control measures for infectious diseases were made (Feng et al., 2008). In time past, infectious diseases have impacted the human population negatively, making civilization altered (Matsumura et al., 2003). From the plague that caused a change in power

Applications of Essential Oils in the Food Industry. DOI: https://doi.org/10.1016/B978-0-323-98340-2.00017-1

between Athens and Spartans to the downfall of Greece and ancient Rome, down to the plague that befell the Justinian, infectious diseases have caused a significant change in the history of humanity (Ravi Kumar, 2000). During the invasion of explorers, conquerors, and invading armies of Europeans in Latin America, Asia, India, and Africa, insects and rodent vectors that harbored infectious agents were introduced and sustained into nonendemic areas (Moghimi et al., 2016). Human populations became exposed to virulent microbes and this changed the history of native populations.

Also, infectious disease history can also be analyzed by evolutionary biology. The interaction between microbes and humans (parasitic or symbiotics) is also an interesting historic part of infectious diseases. Microorganisms strive for survival through nutrient uptake, reproduction, and stability (Bajera et al., 2017). To have a successful host-parasitic interaction, transmission is either direct or indirect requiring an intermediary host dependence (Bajera et al., 2017). For example, the association of parasite with the human host in the case of malaria requires an intermediate host that is a key for the pathogen to complete its life cycle (Bajera et al., 2017). Therefore, infectious diseases are characterized by either human-to-human transmission, for example, typhoid fever, measles, chickenpox, tuberculosis, or vector-definite host transmission such as malaria, trypanosomiasis, and plague. Before the invention of modern medicines, herbs were used to manage most infectious diseases, with little or no knowledge of the bioactive compounds present in plant parts.

Despite many advances in chemotherapy, the use of plants for the treatment and management of infectious diseases is still in use to date (Pinto et al., 2006). Medicinal plants have been reported to have low toxicity, good pharmacological activities, and good sustainability. Because of these properties, many researchers have focused on the beneficiary properties of medicinal plants used in treating several diseases (Zhang et al., 2011). These plant medicines are used in their crude form or their active compounds extracted as liquids or oils which are used in the treatment of diseases.

Conventional treatment of infectious diseases

There has been an upsurge in the demand for aromatic and medicinal plants' essential oils because of their rich antioxidative properties and antimicrobial effects on pathogens (Jess Vergis et al., 2015). This is because essential oils have been reported to contain radical scavenging properties on pathogens (Bilia et al., 2014). More so, the antimicrobial properties of essential oils have also played significant roles in reducing antibiotic resistance since there is a massive abuse of synthesized antibiotics which is a huge global health problem.

Essential oils

Essential oils, also referred to as aromatic oily liquids, are volatile derivatives of plant parts such as flowers, leaves, barks, buds, fruits, stems, roots, and seeds (Wagnera & Ulrich-Merzenich, 2009). It was first discovered by Arabs in the mid-ages and it is obtained by steam or hydro-distillation methods (Bassolé et al., 2010). The density of essential oils is generally lower than water. These characteristics allow essential oils and their constituents to split lipids available in bacteria mitochondria and cell membrane, making them permeable, as a result, this causes the cells to have molecule and ions leakage thereby resulting in bacterial cell death (García-García et al., 2011). Essential oils are soluble in organic solvents, scarcely colored, volatile liquid, and limpid. They contain combinations of diverse lipophilic and volatile constituents, including sesquiterpenes, monoterpenes, and/or phenylpropanoids, usually with a pleasing smell (Nguefack et al., 2012). They are natural mixtures of compounds and consist of nearly 20−60 bioactive ingredients in different amounts. They are usually characterized by 2−3 major components present to as high as 20%−70% in comparison to other trace constituents that are present in minute amounts (Bag & Chattopadhyay, 2015). The amount of these key components varies from different plant parts and species. They have been reported to protect plants from microbial contamination because of their antibacterial, antifungal, antiviral, and insecticidal properties (Hossain et al., 2016). Evidence has shown that plant-sourced medicines contain a rich array of bioactive metabolites (Carson et al., 2006). Researchers have screened and understudied a variety of medicinal plants with essential oils for antimicrobial activities (Safaei-Ghomi & Ahd, 2010).

Mechanism of actions of essential oils

Reports have revealed essential oils are rich in in vitro pharmacological activities (Zhang et al., 2017). These plant-natural derived substances have potential and promising properties in the treatment of several infections. They possess antiinflammatory and immunostimulatory properties (Lorian & Gemmel, 1980). Essential oils with antimicrobial properties act on target sites of pathogens (Stevic et al., 2014); for instance, *Litsea cubeba* oil destroys the inner and outer membrane of *E. coli*, *Foeniculum vulgare* (fennel oil) destroys the membrane integrity of *Shigella dysenteriae*, and

Forsythia koreana essential oil acts on the cytoplasmic membrane of most foodborne pathogens. These oils are either bactericidal or static in their mode of action targeting a specific site on the pathogen (Table 29.1).

Abundant sources of novel bioactive secondary metabolites have been reported and separated for use from therapeutic plants like peppermint (*Mentha piperita*) (Abreu et al., 2012; Donsì et al., 2012). This is basically because of their antimicrobial properties (Donsì et al., 2012). Other essential oils are used against Gram-positive and Gram-negative bacteria, fungi, and yeast, along with some viruses such as fennel (*Foeniculum vulgare*), thyme (*Thymus vulgaris*), etc.

Significance of essential oils

The abuse of antimicrobials is a major reason for the emergence of drug-resistant microbes. To address this increasing resistance, an effective and sustainable antimicrobial source is commendable. These antimicrobials in the form of essential oils form the basis of food preservatives and additives, aroma therapeutics, and complementary medicine. Some examples include celery, lemongrass, rosemary, mint, angelica, bergamot, mandarin, orange, peppermint, and caraway. Other reported plants with medicinal essential oils also include citronella, coriander, eucalyptus, geranium, petitgrain, pine, juniper, lavandin, lavender, lemon, sage, and thyme (Knezevic et al., 2016; Zhang et al., 2014). This chapter will explain the phytochemical properties and uses of commonly used essential oils.

Eucalyptus oil

Eucalyptus oil: Its generic name is *Eucalypti aetheroleum*, and is characterized by two main bioactive components namely: oxygenated monoterpene (β-fenchol) and oxygenated sesquiterpene (α-eudesmol), 1,8-cineole. These bioactive compounds are present in the fresh leaves of eucalyptus plants (Beykia et al., 2014). Other minor compounds are α-pinene (2%−8%) and camphor (less than 0.1%) (Beykia et al., 2014). The terminal branchlets or fresh leaves are usually processed by steam distillation to get their essential oil. The most common species used are the *Eucalyptus globulus Labill* followed by *E. polybractea* R.T. Baker and then *E. smithii* R.T. Baker (Beykia et al., 2014). Eucalyptus oil

TABLE 29.1 Plant-based essential oils and the pathogen target site.

Essential oils	Target sites	Pathogenic microorganisms	References
Essential oils from *Coriaria nepalensis*	Inhibition of ergosterol biosynthesis and destruction of membrane integrity	*Candida* spp.	Palaniappan and Holley (2010)
Turmeric (*Curcuma longa* L.) against	Stop the production of ergosterol, malate dehydrogenase. Also, inhibits mitochondrial ATPase activation, and succinate dehydrogenase activities	*Aspergillus flavus*	Moon et al. (2011)
Dipterocarpus gracilis	Inhibition of growth that act on the cytoplasmic membrane of the plant	*Bacillus cereus* and *Proteus mirabilis*	Aumeeruddy-Elalfi et al. (2015)
Black pepper essential oil (BPEO)	Causes deformation, leakage of plant by breaking cell membrane	*E. coli*	Dorman and Deans (2000)
Litsea cubeba oil	Penetration of inner membrane to create holes and gaps and eventual damage and cell death	*E. coli*	Sousa et al. (2010)
Fennel oil from *Foeniculum vulgare*	destruction of the integrity of pathogen membrane	*Shigella dysenteriae*	Feyzioglu and Tornuk (2016)
Peppermint (*Mentha piperita*)	Bacterial cell lysis	*Streptococcus pneumoniae*	Herculano et al. (2015)
Thyme (*Thymus vulgaris*)	Causes leakage of ions in bacteria thus reducing glucose availability and disrupts fungi vesicles and cell membrane. Thyme also impair the buildup of ergosterol in bacteria cell	*Candida* spp.	Ahmad et al. (2011)

appears pale yellow or as a colorless liquid with an aromatic smell described as camphoraceous, including a spicy taste (Salvia-Trujillo et al., 2015). In the production process, to reduce undesirable aldehydes, the oil after steam distillation is fixed using alkaline treatment followed by fractional distillation (Herculano et al., 2015). Eucalyptus oil is used to treat common cold, flu, catarrh, bronchitis, and upper respiratory tract infections (Fabio et al., 2007). The oils can also be used as ointments in both children and adults, but studies have not been reported on pregnant women and lactation without a doctor's prescription (Herculano et al., 2015).

Peppermint oil

Peppermint oil (*Menthae piperitae aetheroleum*) appears as a transparent, pale yellowish, or greenish-yellow pale liquid. The oil is gotten from fresh aerial flowering parts of the peppermint plant by steam distillation. It is characterized by methanolic smell and taste and a cold sensation. Peppermint essential oils contain about 1.2%−3% of peppermint and about 30%−55% of menthol, 14%−32% of menthon 1.5%−10% of isomenthone, 2.8%−10%, of menthyl acetate, 1%−9% of menthofuran, 3.5%−14% of 1,8-cineole, 1%−5% of limonene, and less or equal to 3% of pulegone. Peppermint oils also contain less than 1% carvone and a higher proportion of cineole than limonene compound (Fabio et al., 2007).

Peppermint oil is useful for managing digestive disorders such as irritable bowel syndrome, flatulence, and symptoms of upper respiratory tract infections. Evidence abounds that peppermint oil is contraindicated in children under the age of 2 years because the menthol compound present can make laryngospasm/constriction and reflex apnea in susceptible people. It is advisable that the use of peppermint oil directly should be discouraged in children because of the high risk of bronchial spasms and laryngeal it causes when used as nasal drops or inhalations (Imai et al., 2001). Menthol has been reported to cause jaundice in neonates and is discouraged in pregnant women or lactating mothers except under a doctor's prescription (Fabio et al., 2007).

Thymol oil

Thymol oil is extracted from the commonly used thyme (*T. vulgaris*). It contains 2-isopropyl-5-methyl phenol also referred to as Thymol. Thymol is a phenol derivative of cymene which is a natural form of monoterpene. More so, thymol is an isomer of carvacrol (Imai et al., 2001) and is characterized by a clear, yellow, and sometimes very dark reddish-brown coloration with a spicy smell and aromatic taste (Imai et al., 2001). Thymol oil has antimicrobial activities by impairing the citrate metabolic pathway and misfolding bacterial membrane proteins in pathogens (Horne et al., 2001). In fungi, thymol oil interacts with cell envelop and intracellular components. It disrupts the fungi vesicles and cell membrane and impairs the biosynthesis of ergosterol in Candida species (Guerra-Rosas et al., 2017).

In other pathogenic microbes, thymol oil increases the absorptivity of both *P. aeruginosa* and *S. aureus* microbes by dissipating gradients pH regardless of glucose disposal and causes outflow of inorganic ions in the pathogens (Ahmad et al., 2011). Thymol oil is used in the treatment of diarrhea (Hosseini et al., 2013).

Cloves

The germicidal property of cloves (*Syzygium aromaticum*) oil makes it a very useful oil in the treatment of yeast infection, sexually transmitted diseases, asthma, bronchitis, toothache, and mouth ulcer (Iannitelli et al., 2011). It contains eugenol, as an active component and appears as a transparent to pale yellowish oily liquid. Eugenol acts as an antifungal, antiinflammatory bioactive molecule. Eugenol acts by destroying pathogen cell walls and preventing enzymatic activity in microbes (Aijaz Ahmad et al., 2011).

Tea tree

Essential oils from the tea tree (*Melaleucae aetheroleum*) are recovered from the foliage and terminal leaves of the plant by steam distillation. It is usually a clear, colorless liquid and sometimes pale yellowish oil with a sweet characteristic smell. The essential oils of the tea tree are made up mainly of monoterpenes including terpinen-4-ol, γ-terpinene, and 1,8-cineole (Lambert et al., 2001). Tea tree oil is used in the management of respirational infections such as common cold, bronchitis, and influenza. It is also implicated in the treatment of fungi infections and dandruff, and in removing lice from the hair and furs of animals. It scarcely results in skin irritation and infection. However, its use in pregnancy has not been established (Martuccia et al., 2015)

Oreganol

Oreganol a phenolic monoterpenoid essential oil is a well-studied essential oil with major constituent carvacrol. It is remarkable for its antimicrobial activity. The action site of a pathogen is the cytoplasmic membrane. The outer part of the cytoplasmic membrane is affected by carvacrol which results in a passive transport of ions across the Gram-negative bacteria membrane. Carvacrol is a fast-acting compound that deactivates *E. coli*, and *Salmonella* spp. in a very short time, within five minutes (Stefanakis et al., 2013). With fungal isolates, Oreganol acts to down-regulate gene transcription causing Ca^{2+} stress and reducing nutrient uptake in *Candida* spp. Other researchers have also stated that oregano has an antibacterial effect on microbes like *L. monocytogenes* involved in abortion, *A. hydrophila*, *E. coli*, *Brochothrix thermosphacta*, and *Pseudomonas fragi* (Rosato et al., 2017).

Anise oil

Anise oil (*Anisi aetheroleum*) appears as a transparent colorless or pale yellowish liquid acquired by steam distillation from ripened fruits of *Pimpinella anisum* L. It usually appears in a jelly form between 14°C and 16°C because it contains trans-anethole as a major component and lesser amounts of anisaldehyde, and methyl-cavicol (Diao et al., 2014). In cases of respiratory disorders and colds, Anise oil is used as an expectorant in cough (Guleria et al., 2008). Its dosage is between 50 and 200 μL taken three times per day for 14 days. Its use in infants and teenagers below 18 years is not advisable because there is no sufficient data for its safety in that age range; moreover, it contains estragole, which is contraindicative for the age range.

Anise and its oil should be avoided by anethole and hypersensitive people (Aligiannis et al., 2001; Bagamboula et al., 2004). Alcoholic extracts of anise oil are also not advisable for use in pregnancy and for lactating women because mild estrogenic activity and nonfertility effects of anethole have been proven in rats (Rattanachaikunsopon & Phumkhachorn, 2010). Hypersensitive reactions may occur in the skin and the respirational system.

Bitter fennel fruit oil

Foeniculi amari fructus aetheroleum also known as bitter fennel fruit oil is a clear, transparent, or pale-yellow liquid with a characteristic smell. The oil is obtained by steam distillation from the ripe fruits of *Foeniculum vulgare* Miller, ssp. vulgare var. vulgare. The main components are fenchone (12.0%–25.0%) and trans-anethole (55.0%–75.0%) (Mann et al., 2000). Therapeutic products containing bitter fennel fruit oil are used for colds and coughs as an expectorant. However, its use in infants and teenagers under 18 years of age is still contraindicated because not enough data to support its use in this population. There is the possibility that allergic reactions to the active substance (e.g., trans-anethole) arise because of the hormonal (estrogenic) activity of trans-anethole in taking too many doses of fennel oil. This effect may affect hormone therapy, oral contraceptive pill, and hormone replacement therapy. The possibility of hypersensitivity to fennel oil affecting the respiratory system is also very likely in children, but their frequency is not known (Mann et al., 2000).

Essential oils as antibacterial

Essential oils have been used as antibacterial to counter an array of Gram-positive and Gram-negative bacteria isolates. These essential oils have shown beyond doubt that they are promising against organisms such as Salmonella and Staphylococci associated with gastrointestinal diseases and oral pathogens (Burt et al., 2007). They can be used as alternative medicines and antibiotics because of their antibacterial properties and extensive study (MCetin et al., 2011). A typical example of essential oil that has shown good bactericidal properties is Basil essential oil. It has been reported to kill bacterial isolates such as *Aeromonas*, *P. fluorescens* and *Hydrophila* (Reichling, 1999). Eucalyptus oil, rosemary oil, lavender oils, and tea tree oils have been positive to kill oral bacterial isolates recovered from dental caries issues. Examples of isolates are *Actinobacillus actinomycetemcomitans Streptococcus mutans*, *Porphyromonas gingivalis and Fusobacterium nucleatum*, and *Streptococcus sobrinus*.

Essential oils as antifungal

Essential oils from the tea tree (*Melaleuca alternifolia*) have displayed antifungal activity using in vitro testing. Carson et al. 2006 reported that tea trees had a wide range of fungicidal properties on filamentous and dermatophyte fungi (Carson et al., 2006). The authors reported that germinated *Aspergillus niger* conidia were also more sensitive to tea trees compared to nongerminated fungi. However, the fresh leaves of the tea tree plant contained more fungicidal bioactive compounds. Other essential oils that contain antifungal properties include black mustard seed oils, Cuminum oils, etc. (Auddy et al., 2003).

Essential oils as antiviral

As an antiviral, essential oils have been reported to have antiviral properties. Evaluating the effect of *M. armillaris*, *M. ericifolia*, *M. leucadendron*, and *Melaleuca styphelioides* oils on Herpes simplex virus on kidney cells of African green monkey showed that up to 99% reduction of plaque was recorded using *M. armillaris* essential oils and up to 91.5% for *M. ericifolia* essential oil treatment (Costa et al., 2015).

Essential oils as antiinflammatory

Tea tree essential oils have been shown to reduce weal and flare reactions by histamine in humans. It was seen that topical application resulted in reduced inflammation caused by histamine diphosphate after about 10 minutes. Essential oils have shown noncytotoxic concentrations exertion by increasing interleukin-10 production and inducing antiinflammation (Caballero et al., 2003).

Essential oils as antilice

Tea tree oils have also been implicated in insecticidal activity because of their anticholinesterase potential. Evidence has shown that tea tree oils when applied to the head infested with lice, there was a drastic reduction and clearance of lice with continuous use of tea tree essential oils. It is interesting to know that most hair shampoos include tea trees as key ingredients (Astani et al., 2010).

Essential oils as antidandruff

It has been observed that tee tree oil and jojoba seed oils are well tolerated in reducing mild to moderate dandruff (Auddy et al., 2003; De Sousa Barros et al., 2015; Hale et al., 2008). However, this area of research still needs more data.

Essential oils as antitumor

Terpinen-4-ol and tea tree oil have been recorded to cause very slow development of human melanoma M14 WT cells and adriamycin-resistant M14 cells too. They do so causing cell death through the caspase-dependent mechanism in melanoma cells. Geraniol is a vital component of plant essential oil and has been recorded to be useful in the management of human colon cancer cells (Fisher & Phillips, 2008). Although not much is known about antitumor activities in plant oils, few studies have shown promising results (Fisher & Phillips, 2008).

Essential oils as antioxidants

Essential oils such as citrus plant oils have also shown antioxidative properties. The seeds of *Nigella sativa* L have effective hydroxyl radical scavenging activity. Others include Manuka (*Leptospermum scoparium*), *M. armillaris*. Kanuka (*Kunzea ericoides*), and *Leptospermum petersonii*. They act by altering the parameters of superoxide dismutase and improve vitamins C and E concentrations which help to remove or mop up injurious radicals in the body system. *T. vulgaris*, *C. limon*, *E. globulus*, and *Cupressus sempervirens* essential oils have also been proven to have antiinflammatory activities (Kaloustian et al., 2008).

Essential oils as insect/mosquito repellant

Essential oils of *Nepeta parnassica*, on the *Culex pipiens molestus* strains showed repellency on the insects (Burt, 2004).

Essential oils having spasmodic action

L. scoparium and *K. ericoides* essential oils have strong spasmogenic and spasmolytic activity respectively when tested in the ileum of rats. More so, *Ferula gummosa* essential oil is a good relaxing oil. It relaxes the contractile overactivity of the ileum which is mostly the primary cause of gastrointestinal disorders (Devi et al., 2010).

Essential oils having hormonal action

Some essential oils are good stimulators of estrogens such as geraniol, nerol, geranial, and trans-anethole oils, unlike eugenol oils which have antiestrogenic activity (Djenane et al., 2012).

Conclusion

This chapter has provided detailed information on the following application of essential oil such as eucalyptus oil, peppermint oil, thymol oil, cloves, tea tree, oregano, anise oil, and bitter fennel fruit oil derived from many medicinal plants in the management of numerous diseases. This chapter also highlights the food and pharmacological benefits that could be derived from these essential oils such as antibacterial, antifungal, antiviral, antiinflammatory, antilice, antidandruff, antitumor, antioxidant, insect/mosquito repellant action, spasmodic action, and hormonal action.

References

Abreu, F. O. M. S., Oliveira, E. F., Paula, H. C. B., & De Paula, R. C. M. (2012). Chitosan/cashew gum nanogels for essential oil encapsulation. *Carbohydrate Polymers*, 89, 1277−1282, [CrossRef] [PubMed].

Ahmad, A., Khan, A., Akhtar, F., Yousuf, S., Xess, I., Khan, L., & Manzoor, N. (2011). Fungicidal activity of thymol and carvacrol by disrupting ergosterol biosynthesis and membrane integrity against Candida. *European Journal of Clinical Microbiology & Infectious Diseases*, 30, 41−50, [CrossRef] [PubMed].

Aijaz Ahmad, A., Khan, A., Kumar, P., Bhatt, R. P., & Manzoor, N. (2011). Antifungal activity of *Coriaria nepalensis* essential oil by disrupting ergosterol biosynthesis and membrane integrity against Candida. *Yeast (Chichester, England)*, 28, 611−617, [CrossRef] [PubMed].

Aligiannis, N., Kalpoutzakis, E., Mitaku, S., & Chinou, I. B. (2001). Composition and antimicrobial activity of the essential oils of two *Origanum* species. *Journal of Agricultural and Food Chemistry*, 49, 4168−4170, [CrossRef] [PubMed].

Astani, A., Reichling, J., & Schnitzler, P. (2010). Comparative study on the antiviral activity of selected monoterpenes derived from essential oils. *Phytotherapy Research: PTR*, 24, 673−679, [CrossRef] [PubMed].

Auddy, B., Ferreira, M., Blasina, F., Lafon, L., Arredondo, F., Dajas, F., Tripathi, P. C., Seal, T., & Mukherjee, B. (2003). Screening of antioxidant activity of three Indian medicinal plants, traditionally used for the management of neuro-degenerative diseases. *Journal of Ethnopharmacology*, 84, 131−138, [CrossRef].

Aumeeruddy-Elalfi, Z., Gurib-Fakim, A., & Mahomoodally, F. (2015). Antimicrobial, antibiotic potentiating activity and phytochemical profile of essential oils from exotic and endemic medicinal plants of Mauritius. *Industrial Crops and Products*, 71, 197−204, [CrossRef].

Bag, A., & Chattopadhyay, R. R. (2015). Evaluation of synergistic antibacterial and antioxidant efficacy of essential oils of spices and herbs in combination. *PLoS One*, 10, e0131321, [CrossRef] [PubMed] Medicines **2017**, 4, 58 21 of 21.

Bagamboula, C. F., Uyttendaele, M., & Debevere, J. (2004). Inhibitory effect of thyme and basil essential oils, carvacrol, thymol, estragol, linalool and p-cymene towards *Shigella sonnei* and *S. flexneri*. *Food Microbiology*, 21, 33−42.

Bajera, T., Silha, D., Ventura, K., & Bajerov, P. (2017). Composition and antimicrobial activity of the essential oil, distilled aromatic water and herbal infusion from *Epilobium parviflorum* Schreb. *Industrial Crops and Products*, 100, 95−105.

Bakkali, F., Averbeck, S., Averbeck, D., & Idaomar, M. (2008). Biological effects of essential oils—A review. *Food and Chemical Toxicology: An International Journal Published for the British Industrial Biological Research Association*, 46, 446−475.

Bassolé, I. H. N., Lamien-Meda, A., Bayala, B., Tirogo, S., Franz, C., Novak, J., Nebié, R. C., & Dicko, M. H. (2010). Composition and antimicrobial activities of Lippia multi-flora Moldenke, *Mentha piperita* L. and *Ocimum basilicum* L. essential oils and their major monoterpene alcohols alone and in combination. *Molecules (Basel, Switzerland)*, 15, 7825−7839, [CrossRef] [PubMed].

Beykia, M., Zhaveha, S., Khalilib, S. T., Rahmani-Cheratic, T., Abollahic, A., Bayatd, M., Tabatabaeie, M., & Mohsenifar, A. (2014). Encapsulation of *Mentha piperita* essential oils in chitosan−cinnamic acid nanogel with enhanced antimicrobial activity against *Aspergillus flavus*. *Industrial Crops and Products*, 54, 310−319, [CrossRef].

Bilia, A. R., Guccione, C., Isacchi, B., Righeschi, C., Firenzuoli, F., & Bergonzi, M. C. (2014). Essential oils loaded in nanosystems: A developing strategy for a successful therapeutic approach. *Evidence-Based Complementary and Alternative Medicine*, 651593, 1−14, [CrossRef] [PubMed].

Burt, S. (2004). Essential oils: Their antibacterial properties and potential applications in foods—A review. *International Journal of Food Microbiology*, 94, 223−253, [CrossRef] [PubMed].

Burt, S. A., Van Der Zee, R., Koets, A. P., De Graaff, A. M., Van Knapen, F., Gaastra, W., Haagsman, H. P., & Veldhuizen, E. J. A. (2007). Carvacrol induces heat shock protein 60 and inhibits synthesis of flagellin in *Escherichia coli* O157:H7. *Applied and Environmental Microbiology*, 73, 4484−4490, [CrossRef] [PubMed].

Caballero, B., Trugo, L. C., & Finglas, P. M. (2003). *Encyclopedia of food sciences and nutrition*. Amsterdam, The Netherlands: Academic Press.

Carson, C. F., Hammer, K. A., & Riley, T. V. (2006). *Melaleuca alternifolia* (tea tree) oil: A review of antimicrobial and other medicinal properties. *Clinical Microbiology Reviews*, 19, 50−62.

Costa, D. C., Costa, H. S., Albuquerque, T. G., Ramos, F., Castilho, M. C., & Sanches-Silva, A. (2015). Advances inphenolic compounds analysis of aromatic plants and their potential applications. *Trends in Food Science and Technology*, 45, 336−354, [CrossRef].

Danilcauk, M., Lund, A., Saido, J., Yamada, H., & Michalik, J. (2006). Conduction electron spin resonance of small silver particles. *Spectrochimica Acta Part A, 63*, 189–191, [CrossRef] [PubMed].

De Sousa Barros, A., de Morais, S. M., Ferreira, P. A. T., Vieira, Í. G. P., Craveiro, A. A., de Santos Fontenelle, R. O., de Menezes, J. E. S. A., da Silva, F. W. F., & de Sousa, H. A. (2015). Chemical composition and functional properties of essential oils from Mentha species. *Industrial Crops and Products, 76*, 557–564, [CrossRef].

Devi, K. P., Nisha, S. A., Sakthivel, R., & Pandian, S. K. (2010). Eugenol (an essential oil of clove) acts as an antibacterial agent against *Salmonella typhi* by disrupting the cellular membrane. *Journal of Ethnopharmacology, 130*, 107–115.

Diao, W. R., Hua, Q. P., Zhang, H., & Xu, J. G. (2014). Chemical composition, antibacterial activity and mechanism of action of essential oil from seeds of fennel (*Foeniculum vulgare* Mill). *Food Control, 35*, 109–116, [CrossRef].

Djenane, D., Yangueela, J., Gomez, D., & Roncales, P. (2012). Perspectives on the use of essential oils as antimicrobials against *Campylobacter jejuni* CECT 7572 in retail chicken meats packaged in micro aerobic atmosphere. *Journal of Food Safety, 32*, 37–47.

Donsì, F., Annunziata, M., Vincensi, M., & Ferrari, G. (2012). Design of nanoemulsion-based delivery systems of natural antimicrobials: Effect of the emulsifier. *Journal of Biotechnology, 159*, 342–350, [CrossRef] [PubMed].

Dorman, H. J. D., & Deans, S. G. (2000). Antimicrobial agents from plants: Antibacterial activity of plant volatile oils. *Journal of Applied Microbiology, 88*, 308–316, [CrossRef] [PubMed].

Fabio, A., Cermelli, C., Fabio, G., Nicoletti, P., & Quaglio, P. (2007). Screening of the antibacterial effects of a variety of essential oils on microorganisms responsible for respiratory infections. *Phytotherapy Research: PTR, 21*, 374–377.

Feng, Q. L., Wu, J., Chen, G. Q., Cui, F. Z., & Kim, J. O. (2008). A mechanistic study of the antibacterial effect of silver ions on *Escherichia coli* and *Staphylococcus aureus*. *Journal of Biomedical Materials Research, 52*, 662–668, [CrossRef].

Feyzioglu, G. C., & Tornuk, F. (2016). Development of chitosan nanoparticles loaded with summer savory (*Satureja hortensis* L.) essential oil for antimicrobial and antioxidant delivery applications. *LWT - Food Science and Technology, 70*, 104–110, [CrossRef].

Fisher, K., & Phillips, C. (2008). Potential antimicrobial uses of essential oils in food: Is citrus the answer? *Trends in Food Science and Technology, 19*, 156–164, [CrossRef].

García-García, R., López-Malo, A., & Palou, E. (2011). Bactericidal action of binary and ternary mixtures of carvacrol, thymol, and eugenol against *Listeria innocua*. *Journal of Food Science, 76*, M95–M100, [CrossRef] [PubMed].

Ghosh, I. N., Patil, S. D., Sharma, T. K., Srivastava, S. K., Pathania, R., & Navani, N. K. (2013). Synergistic action of cinnamaldehyde with silver nanoparticles against spore-forming bacteria: A case for judicious use of silver nanoparticles for antibacterial applications. *International Journal of Nanomedicine, 8*, 4721–4731.

Guerra-Rosas, M. I., Morales-Castro, J., Cubero-M_arquez, M. A., & Salvia-Trujillo, L. (2017). Antimicrobial activity of nanoemulsions containing essential oils and high methoxyl pectin during long-term storage. *Food Control, 77*, 131–138, [CrossRef].

Guleria, S., Kumar, A., & Tiku, A. K. (2008). Chemical composition and fungitoxic activity of essential oil of *Thuja orientalis* L. grown in the North-Western Himalaya. *Zeitschrift für Naturforschung C, 63*, 211–214, [CrossRef].

Hale, A. L., Reddivari, L., Nzaramba, M. N., Bamberg, J. B., & Miller, J. C., Jr. (2008). Interspecific variability for antioxidant activity and phenolic content among *Solanum* species. *American Journal of Potato Research, 85*, 332–341.

Herculano, E. D., de Paula, H. C. B., de Figueiredo, E. A. T., Dias, F. G. B., de, A., & Pereira, V. (2015). Physicochemical and antimicrobial properties of nanoencapsulated *Eucalyptus staigeriana* essential oil. *LWT - Food Science and Technology, 6*, 484–491, [CrossRef].

Horne, D., Holm, M., Oberg, D. G., Chao, S., & Young, D. G. (2001). Antimicrobial effects of essential oils on *Staphylococcus pneumoniae*. *Journal of Essential Oil Research, 13*, 387–392.

Hossain, F., Follett, P., Dang Vu, K., Harich, M., Salmieri, S., & Lacroix, M. (2016). Evidence for synergistic activity of plant-derived essential oils against fungal pathogens of food. *Food Microbiology, 53*, 24–30, [CrossRef].

Hosseini, S. F., Zandi, M., Rezaei, M., & Farahmandghavi, F. (2013). Two-Step method for encapsulation of oregano essential oil in chitosan nanoparticles: Preparation, characterization and in vitro release study. *Carbohydrate Polymers, 95*, 50–56, [CrossRef] [PubMed].

Iannitelli, A., Grande, R., Stefano, A. D., Giulio, M. D., Sozio, P., Bessa, L. J., Laserra, S., Paolini, C., Protasi, F., & Cellini, L. (2011). Potential antibacterial activity of carvacrol-loaded poly(DL-lactide-co-glycolide) (PLGA) nanoparticles against microbial biofilm. *International Journal of Molecular Sciences, 12*, 5039–5051, [CrossRef] [PubMed].

Imai, H., Osawa, K., Yasuda, H., Hamashima, H., Arai, T., & Sasatsu, M. (2001). Inhibition by essential oils of peppermint and spearmint of the growth of pathogenic bacteria. *Microbios, 106*, 31–39.

Jess Vergis, J., Gokulakrishnan, P., Agarwal, R. K., & Kumar, A. (2015). Essential oils as natural food antimicrobial agents: A review. *Critical Reviews in Food Science and Nutrition, 55*, 1320–1323.

Kaloustian, J., Chevalier, J., Mikail, C., Martino, M., Abou, L., & Vergnes, M. F. (2008). Étude de six huiles essentielles: Composition chimique et activité antibactérienne. *Phytothérapie, 6*, 160–164, [CrossRef].

Knezevic, P., Verica Aleksic, V., Simin, N., Svircev, E., Petrovic, A., & Mimica-Dukic, N. (2016). Antimicrobial activity of *Eucalyptus camaldulensis* essential oils and their interactions with conventional antimicrobial agents against multi-drug resistant *Acinetobacter baumannii*. *Journal of Ethnopharmacology, 178*, 125–136, [CrossRef] [PubMed].

Lambert, R. J. W., Skandamis, P. N., Coote, P. J., & Nychas, G. J. E. (2001). A study of the minimum inhibitory concentration and mode of action of oregano essential oil, thymol and carvacrol. *Journal of Applied Microbiology, 91*, 453–462, [CrossRef] [PubMed].

Lorian, V., & Gemmel, C. G. (1980). *Effects of low antibiotic concentrations on bacteria: Effects on ultrastructure, virulence, and susceptibility to immunodefenses* (p. 493) Baltimore: Williams and Wilkins.

Lv, F., Liang, H., Yuan, Q., & Li, C. (2011). *In Vitro* antimicrobial effects and mechanism of action of selected plant essential oil combinations against four food related microorganisms. *Food Research International, 44,* 3057–3064.

Mann, C. M., Cox, S. D., & Markham, J. L. (2000). The outer membrane of *Pseudomonas aeruginosa* NCTC 6749 contributes to its tolerance to the essential oil of *Melaleuca alternifolia* (tea tree oil). *Letters in Applied Microbiology, 30,* 294–297.

Martuccia, J. F., Gendeb, L. B., Neiraa, L. M., & Ruseckaite, R. A. (2015). Oregano and lavender essential oils as antioxidant and antimicrobial additives of biogenic gelatin films. *Industrial Crops and Products, 71,* 205–213.

Matsumura, Y., Yoshikat, K., Kunisaki, S., & Tsuchido, T. (2003). Mode of action of silver zeolite and its comparison with that of silver nitrate. *Applied and Environmental Microbiology, 16,* 4278–4281, [CrossRef].

MCetin, B., Cakmakci, S., & Cakmakci, R. (2011). The investigation of antimicrobial activity of thyme and oregano essential oils. *Turkish Journal of Agriculture and Forestry, 35,* 145–154.

Moghimi, R., Ghaderi, L., Rafati, H., Aliahmadi, A., & Mcclements, D. J. (2016). Superior antibacterial activity of nanoemulsion of *Thymus daenensis* essential oil against E. coli. *Food Chemistry, 194,* 410–415.

Moon, S. E., Kim, H. Y., & Cha, J. D. (2011). Synergistic effect between clove oil and its major compounds and antibiotics against oral bacteria. *Archives of Oral Biology, 56,* 907–916, [CrossRef] [PubMed].

Nguefack, J., Tamgue, O., Dongmo, J. B. L., Dakole, C. D., Leth, V., Vis-mer, H. F., Zollo, P. H., & Nkengfack, A. E. (2012). Synergistic action between fractions of essential oils from *Cymbopogon citratus, Ocimum gratissimum* and *Thymus vulgaris* against *Penicillium expansum. Food Control, 23,* 377–383, [CrossRef].

Palaniappan, K., & Holley, R. A. (2010). Use of natural antimicrobials to increase antibiotic susceptibility of drug resistant bacteria. *International Journal of Food Microbiology, 140,* 164–168, [CrossRef] [PubMed].

Pinto, E., Pina-Vaz, C., Salgueiro, L., Goncalves, M. J., Costa-de-Oliveira, S., Cavaleiro, C., Palmeira, A., Rodrigues, A., & Martinez-de-Oliveira, J. (2006). Antifungal activity of the essential oil of *Thymus pulegioides* on *Candida, Aspergillus* and dermatophyte species. *Journal of Medical Microbiology, 55,* 1367–1373.

Prabhu, S., & Poulose, E. K. (2012). Silver nanoparticles: Mechanism of antimicrobial action, synthesis, medical applications, and toxicity effects. *International Nano Letters, 2,* 32, [CrossRef].

Rai, M., Ingle, A. P., Gade, A. K., Duarte, M. C. T., & Duran, N. (2015). Three *Phoma* spp: Synthesized novel silver nanoparticles that possess excellent antimicrobial efficacy. *IET Nanobiotechnology / IET, 9,* 280–287.

Rattanachaikunsopon, P., & Phumkhachorn, P. (2010). Assessment of factors influencing antimicrobial activity of carvacrol and cymene against *Vibrio cholera* in food. *Journal of Bioscience and Bioengineering, 110,* 614–619, [CrossRef] [PubMed].

Ravi Kumar, M. N. (2000). Nano and microparticles as controlled drug delivery devices. *Journal of Pharmaceutical Sciences, 3,* 234–258.

Reichling, J. (1999). Plant-microbe interaction and secondary metabolites with antiviral, antibacterial and antifungal properties. In M. Wink (Ed.), *Functions of plant secondary metabolites and their exploitation in biotechnology* (pp. 187–273). Sheffield: Sheffield Academic Press.

Rosato, A., Piarulli, M., Corbo, F., Muraglia, M., Carone, A., Vitali, M. E., & Vitali, C. (2017). In vitro synergistic action of certain combinations of gentamicin and essential oils. *Current Medicinal Chemistry,* 3289–3295, [CrossRef].

Safaei-Ghomi, J., & Ahd, A. A. (2010). Antimicrobial and antifungal properties of the essential oil and methanol extracts of *Eucalyptus largiflorens* and *Eucalyptus intertexta. Pharmacognosy Magazine., 6,* 172–175, [CrossRef] [PubMed].

Salvia-Trujillo, L., Rojas-Graü, A., Soliva-Fortuny, R., & Martín-Belloso, O. (2015). Physicochemical characterization and antimicrobial activity of foodgrade emulsions and nanoemulsions incorporating essential oils. *Food Hydrocolloids, 43,* 547–556, [CrossRef].

Soltani, S., & Madadlou, A. (2015). Gelation characteristics of the sugar beet pectin solution charged with fish oil-loaded zein nanoparticles. *Food Hydrocolloids, 43,* 664–669.

Sousa, E. O., Silva, N. F., Rodrigues, F. F., Campos, A. R., Lima, S. G., & Costa, J. G. (2010). Chemical composition andresistance-modifying effect of the essential oil of Lantana camara Linn. *Pharmacognosy Magazine., 6,* 79–82.

Stefanakis, M. K., Touloupakis, E., Anastasopoulos, E., Ghanotakis, D., Katerinopoulos, H. E., & Makridis, P. (2013). Antibacterial activity of essential oils from plants of the genus *Origanum. Food Control, 34,* 539–546.

Stevic, T., Beric, T., Savikin, K., Sokovic, M., GoCevac, D., Dimkic, I., & Stankovic, S. (2014). Antifungal activity of selected essential oils against fungi isolated from medicinal plant. *Industrial Crops and Products, 55,* 116–122.

Wagnera, H., & Ulrich-Merzenich, G. (2009). Synergy research: Approaching a new generation of phytopharmaceuticals. *Phytomedicine: International Journal of Phytotherapy and Phytopharmacology, 16,* 97–110, [CrossRef] [PubMed].

Zhang, D., Hu, H., Rao, Q., & Zhao, Z. (2011). Synergistic effects and physiological responses of selected bacterial isolates from animal feed to four natural antibacterials and two antibiotics. *Foodborne Pathogens and Disease, 8,* 1055–1062, [CrossRef] [PubMed].

Zhang, J., Ye, K. P., Zhang, X., Pan, D. D., Sun, Y. Y., & Cao, J. X. (2017). Antibacterial activity and mechanism of action of black pepper essential oil on meat-borne *Escherichia coli. Frontiers in Microbiology, 7,* 2094, [CrossRef].

Zhang, Z., Vriesekoop, F., Yuan, Q., & Liang, H. (2014). Effects of nisin on the antimicrobial activity of d-limonene and its nano-emulsion. *Food Chemistry, 150,* 307–312, [CrossRef] [PubMed].

Chapter 30

Antihyperlipidemic and antioxidant properties of medicinal attributes of essential oil

Muhammad Akram[1], Rabia Zahid[1], Babatunde Oluwafemi Adetuyi[2], Olalekan Akinbo[3], Juliana Bunmi Adetunji[4], Charles Oluwaseun Adetunji[5], Mojisola Christiana Owoseni[6], Majolagbe Olusola Nathaniel[7], Ismail Ayoade Odetokun[8], Oluwabukola Atinuke Popoola[9], Joan Imah Harry[10], Olatunji Matthew Kolawole[11] and Mohammed Bello Yerima[12]

[1]Department of Eastern Medicine, Government College University, Faisalabad, Pakistan, [2]Department of Natural Science, Faculty of Pure and Sciences, Precious Cornerstone University, Ibadan, Oyo State, Nigeria, [3]Centre of Excellence in Science, Technology, and Innovation, AUDA-NEPAD, Johannesburg, Gauteng, South Africa, [4]Department of Biochemistry, Osun State University, Osogbo, Osun State, Nigeria, [5]Applied Microbiology, Biotechnology and Nanotechnology Laboratory, Department of Microbiology, Edo State University Uzairue, Iyamho, Edo State, Nigeria, [6]Department of Microbiology, Federal University of Lafia, Lafia, Nasarawa State, Nigeria, [7]Microbiology Unit, Department of Pure and Applied Biology, Ladoke Akintola University of Technology, Ogbomoso, Oyo State, Nigeria, [8]Department of Veterinary Public Health and Preventive Medicine, University of Ilorin, Ilorin, Kwara State, Nigeria, [9]Genetics, Genomics and Bioinformatics Department, National Biotechnology Development Agency, Abuja, FCT, Nigeria, [10]Department of Natural Sciences, Faculty of Pure and Applied Sciences, Precious Cornerstone University, Ibadan, Oyo State, Nigeria, [11]Department of Microbiology, Faculty of Life Sciences, University of Ilorin, Ilorin, Kwara State, Nigeria, [12]Department of Microbiology, Sokoto State University, Sokoto, Sokoto State, Nigeria

Introduction

Paracelsus von Hohenheim was the pioneer to utilize the word "basic oil" without precedent in the sixteenth century, named the viable segment of medication, "Quinta fundamental". Fundamental oils are sweet-smelling substances that are present in particular cells or organs of specific plants, utilized to shield themselves from killers and irritations, in addition to pulling in pollinators. At the end of the day, fundamental oils are a piece of the resistant arrangement of the plant. Fundamental oils are profoundly focused on unpredictable substances extricated from different pieces of plant species, all with explicit remedial and lively effects. Essential oils' major uses in pharmaceutical agents had declined by the middle of the twentieth century, leaving only those used for fragrances, grooming, and dietary supplements (Batiha et al., 2021). In any case, fundamental oils are deprived of contraindications and unfriendly impacts that regularly confuse some medical issues.

Basic oils are moderate across the board in the kingdom Plantae, a few families being exceptionally wealthy in such oils, in number and amount. Regularly, fundamental oils are detected in unrivaled plants (around 50 families) having a place with two sets of plants; one is angiosperms that have been explored (James-Okoro et al., 2021). Due to medicinal and antimicrobial properties such as blockage, loose bowels, measles, intestinal sickness, onchocerciasis, yellow fever, stomach torment, and so forth, the stems and thinner roots of the plant are used to make toothpicks and other oral healthcare tools (Adetunji, 2008, 2019; Adetunji & Anani, 2020; Adetunji & Ugbenyen, 2019; Adetunji, Adetunj, Olaniyan, et al., 2021; Adetunji, Adetunji, Michael, et al., 2021; Adetunji, Ajayi, et al., 2021; Adetunji, Akram, et al., 2021; Adetunji, Inobeme, et al., 2021; Adetunji, Michael, et al., 2021; Adetunji, Olaniyan, et al., 2021; Bello et al., 2019; Hameed et al., 2022; Ogunlana et al., 2020). Second, ginsenosides have been shown to have a wide variety of beneficial effects, including analgesic, bactericidal, anticataleptic, antidiabetic, antihyperlipidemic, fungicidal, antihypercholesterolemic, antimicrobial, cancer preventative, relaxant, sedative, cardioprotective, neuroprotective, and other uses (Adetuyi, Adebayo, et al., 2022; Adetuyi, Adebisi, et al., 2022; Adetuyi, Odine, et al., 2022; Adetuyi, Ogundipe, et al., 2022; Adetuyi, Olajide, Oluwatosin, et al., 2022; Adetuyi, Olajide, Omowumi, et al., 2022).

Combinations with a lower subatomic mass, such as those present in volatile oil, can readily cross the cell membrane and set off a cascade of physiological responses. Antioxidant, alleviating, lipid-lowering, antidiabetic, and hepatic-protective effects have

Applications of Essential Oils in the Food Industry. DOI: https://doi.org/10.1016/B978-0-323-98340-2.00018-3

been the primary focal points of research into the health benefits of basic oils. While there are around 3,000 different types of essential oils, only around 10% are really used in things like medicine, food additives, cosmetics, and perfumes (Bakkali, Averbeck et al., 2008). The union and gathering of fundamental oils happen in glandular brushes and papillae, or in the secretory cells. Fundamental oils can collect in all plant organs, yet in differing sums. In this way we can found them in roots, blooms, leaves, natural products, stems, or bark wood. The substance in basic plant soils is regularly beneath 1%, once in a while arriving at 15% or much more, in the dry result of certain plants (Didunyemi et al., 2019). Likewise, different species, albeit naturally smell, contain restorative substances that included basic oils. Extraction of fundamental oils is costly a direct result of the huge measure of crude material needed to create oil in a couple of milliliters. This explains why authorized essential oils are so darn expensive. For example, it takes roughly 60 roses to extract one droplet of rose essential oil (Ouis & Hariri, 2018). Be that as it may, there are likewise more affordable oils because of the wealth of economical crude ingredients and great efficiency. Such oils included citrus—lemon, bergamot, orange, lime, tea tree oil, and lemongrass oil. Along these lines, the basic oil is valuable, however just one drop is adequate for helpful outcomes, besides, surpassing a 2% measurements is lethal and produces antagonistic effects (Barazesh, Oloumi et al., 2017). A few methods can be utilized to remove fundamental oils from various pieces of the fragrant plant, containing water/steam refining, dissolvable extraction, articulation under strain, supercritical liquid, and subcritical water extractions. The concoction piece of basic oils is exceptionally fluctuated and the principle segments can be a piece of the aliphatic, sweet-smelling, and terpenic arrangement. For the most part, fundamental oils contain ternary, once-in-a-while quaternary substances. Unpredictable items are comprised of terpenes, aldehydes, aromatics, phenols, ketones, unstable acids, esters, and so on. The plant substantial liable to hydrodynamics isn't constantly prepared in the wake of reaping (Zerkaoui, Benslimane et al., 2018). Fresh plants produce more desirable perfume mixtures and more medicinal action than dried ones, with the exception of cinnamon, lime blossoms, and lavender. The lower inconsistent pee is sometimes obtained from dry plants due to morphology and composition modifications brought on by atmospheric action, heat, granule gathering, and maybe modification (Hariri et al., 2017). Certain oils, for example, citrus offer a lower future than others, for example, sandalwood and patchouli, the last appearing to show signs of improvement over the long haul. The normal life expectancy of basic oils fluctuates, contingent upon both the assembling procedure and the preservation strategies (Olajide, Adetuyi, et al., 2022; Olajide, Omowumi, Odine, 2022; Olajide, Omowumi, Okunlola, et al., 2022). Hermetic and secure glass containers should be used to store oils between 15°C and 20°C. Some oils have a minimum shelf life of 4 years if properly stored (Bozhanska, 2018).

Basic oils as antioxidant agents

Oxidation of biological molecules including amino acid residues, unsaturated fat, and Genetic material is caused by oxidative stress and reactive forms of oxygen, leading to atomic changes like atherosclerosis, aging, breathing problems, and metabolic syndrome (Zarkovic, 2003). In most cases, the reactive species found in electronic devices can be neutralized by the body's built-in defensive measure (Halliwell & Gutteridge, 1990). A discrepancy between formation of free radicals and elimination by the body's cell membrane fortification system results in oxidative stress. To restore equilibrium between oxidative stress and chemopreventive drugs, additional cellular reinforcements from the outside world are required. Reactive oxygen species (ROS) and free radicals (FR) have been linked to a variety of disorders, according to medical data (Adetuyi, Adebisi, et al., 2020; Adetuyi, Okeowo, et al., 2020; Adetuyi, Olajide, et al., 2020). Plants can be a source of new cancer-preventing mixtures due to the high volume of potent antioxidants they create to combat the oxidative stress of exposure to sunlight and oxygen. Basic oils, as common wellsprings of phenolic parts, pull in agents to assess their action as cancer prevention agents or free extreme foragers.

Antihyperlipidemic effects of essential oils

A few parts of current ways of life, for example, high-fat weight control plans and inadequate exercise, are ordinarily connected with hyperlipidemia, which has a quickly expanding predominance (Adetuyi, Adebisi, et al., 2020; Adetuyi, Okeowo, et al., 2020; Adetuyi, Olajide, et al., 2020). The real intricacy of hyperlipidemia is atherosclerosis, basically ascribed to elevated amounts of blood cholesterol and cholesterol-rich plasma lipoproteins, for example, low-thickness lipoprotein (LDL), which instigate ectopic fat aggregation in major blood vessel dividers. People with a raised LDL cholesterol level are at expanded hazard for CHD. Cell oxidative pressure frequently quickens the procedure of atherosclerosis, as oxidized LDL is framed (Awoyelu et al., 2020). These atherogenic LDL particles are great substrates for macrophage forager receptors, for example, CD36 and SR-An, and accordingly advance the arrangement of froth cells, which can prompt atherosclerosis. In like manner, treatment to bring down cholesterol levels just as fitting cancer prevention agent admission may enhance lipid profiles in people with hyperlipidemia (Adetuyi et al., 2020). Thus, broad investigations have been led on the improvement of restorative medications to anticipate hyperlipidemia; nonetheless, tranquilize reactions are a major concern (Salvamani, Gunasekaran et al., 2016).

Utilitarian sustenance or elective meds, for example, basic oils, have attracted consideration as choices to decrease the issues from medication use in expansion to staple products of the soil, different phytochemicals from restorative plants and herbs contain bioactive intensifies that may tweak the outflow of qualities and proteins basic for are effectively examining these hypocholesterolemic phytochemicals (Adewale et al., 2022).

Aromatic ingredients and essential oils

Essential oils are partially responsible for the medicinal benefits associated with aromatic plants. Since there is no fatty ingredient in essential oil, it cannot be considered oil (Ogunlana et al., 2020). The vesicles of these specialized plant parts release an essence rich in distinctive tastes and active ingredients. The secretory glands are pressed or refined to obtain useful substances. Cold-pressing citrus peels, and dehydrating other plant parts like leaves, flowers, roots, and wood are just a few examples (Bozhanska, 2018). The end result of these treatments is a concentrated product with a pleasant aroma and a verified source of active ingredients. Depending on context, you might also refer to essential oil as volatile oil or vaporous oil. The volatile aromatic components of essential oils have been shown to have strong antioxidant and antihyperlipidemic effect, which is at odds with LDL oxidation. Amonoterpene compound terpinolene effectively inhibits the breakdown of the lipid and protein components of low-density lipoprotein. This obstruction is due to a blockage in the degradation of endogenous LDL carotenoids rather than to the safety of endogenous β-tocopherol as is the case with some flavonoids. Oils with high concentrations of total phenolics, including thymol and eugenol, have the highest antioxidant activities and can also change the propensity of LDL particles for their active site (Adeleke et al., 2022). It was shown that the number and type of phenol moieties in oil were related to their resistance to LDL oxidation. As an illustration, eugenol was reported to reduce copper-catalyzed peroxidation of human LDL in vitro by a whopping 50%. Essential oils with low to medium concentrations of thymol, phenolics, cuminol, or carvacrol saw just a 10% reduction in effectiveness (Olajide, Adetuyi, et al., 2022; Olajide, Omowumi, Odine, 2022; Olajide, Omowumi, Okunlola, et al., 2022). The monoterpene molecule γ-terpinene was found to suppress Lipid oxidation even in the exponential growth phase, which is in addition to the overall phenolic content. Terpinene blocked Cu^{2+}-activated and AAP H-induced oxidation of human LDL in vitro (Shuaib, Rohit et al., 2016).

Plants with antioxidant and hypolipidemic essential oil

Oils can be derived from almost every component of a plant, such as the foliage, blossoms, stalks, nuts, and even the roots. Fragrant plants typically carry volatile oil across their full extents, in a number of different fixes. In contrast to ginger, which produces more pleasant-smelling oil in the rhizome, the rose typically secretes its etheric oil at the level of blossoms (Adetuyi, Dairo, Didunyemi, 2015; Adetuyi, Dairo, Oluwole, 2015; Adetuyi, Oluwole, Dairo, 2015). Each kind of volatile oil is unique, with its own characteristics even among members of the same plant, and it has great potential uses and is as variable as Nature itself. An unstable oil's distinctive aroma is the result of a complex interplay between the plant's variety, harvest time, atmospheric conditions, and the section of the plant that was used to extract the oil (Adetuyi, Omolabi, et al., 2021; Adetuyi, Toloyai, et al., 2021). Different exacerbates that are components of basic oils were identified. About 200 distinct components make up the mint oil alone. It turns out that many of the combinations are isomers of very simple molecules (Hariri et al., 2017; Hariri et al., 2018). In these ever-evolving plant varieties, we find compounds like cineol, fenchone, limonene, menthol, mentone, pinene, and sabinen, some of which are present only in trace levels, making it extremely challenging to simulate in a research facility the structure of naturally occurring fragrant oils (Didunyemi et al., 2020). Nature is without a doubt the best scientific expert, for the fragrant forces of plants in the whole realized verdure couldn't be incorporated in the a large number of long periods of joint endeavors of all physicists on the planet (Jahan, Chowdhury et al., 2015).

Hypercholesterolemia assumes a significant job in the improvement of infections, which are driving reason for death (Nazir et al., 2022). In spite of the fact that the lipid-bringing down impacts of bioactive mixes in basic oils are accounted for despite everything it need research and assessment. As a result of the abundance of the corresponding fruit skins, citrus-derived oils are widely used in manufacturing (Adetuyi et al., 2020). Actual natural goods include lemon, orange, harsh orange, bergamot, mandarin, and grapefruit. In particular, Limonene, a major component of lemon essential oil, has been shown to have lipid-lowering effects by increasing expression of the transcription factors peroxisome proliferator-activated receptor alpha (PPARα) and liver X receptor beta (LXRβ) (Jing, Zhang et al., 2013). Terpinene showed a lipid-lowering effect and was present in high concentrations in the essential oils of citrus fruits like bergamot (14%), mandarin (17%), and lemon (10%). The γ-terpinene content of tea tree essential oil is high, coming in at 23.0%.

Cancer-fighting activity has been observed in Indian gooseberry, tumeric, mango, bitter melon, Indian sandalwood, chirata, and Winter Cherry (Scartezzini & Speroni, 2000). The DPPH radical test performed at ambient temperature reveals that the essential oils of basil, cinnamon, clove, nutmeg, oregano, and thyme have characteristics that act as scavenging free radicals and cellular reinforcers. In β-carotene/linoleic acid structure, the basic oil of Thymus serpyllum showed a reactive oxygen species scavenging activity similar to that of the produced butylatedhydroxytoluene (BHT) (Ogunlana, Adetuyi, Esalomi, et al., 2021; 2021; Ogunlana, Adetuyi, Rotimi, et al., 2021). The elevated concentrations of the polyphenols thymol and carvacrol (20.5% and 58.1%, respectively) are responsible for the cancer-preventing compound activity. The high concentrations of thymol (36.5%) and carvacrol (29.8%) in Thymus spathulifolius essential oil gave it a cell-reinforcing effect. Oregano basic oil had a cell-reinforcing effect that was on par with those of α-tocopherol and BHT but weaker compared to ascorbic acid. Thymol (at 35%) and carvacrol (32%, individually) have been implicated as the active ingredients in this case. Oregano oil administration delayed lipid oxidation in rabbits, but not as effectively as treatment with a similar concentration of α-tocopheryl acetic acid. However, when tested on turkeys, it showed equivalent efficacy to a similar concentration of β-tocopheryl acetic acid in delaying iron-initiated, peroxidation. The primary oils of both Salvia cryptantha and Salvia multicaulis have antioxidant properties that allow them to seek out and destroy free radicals. In terms of activity, these oils outperformed curcumin, ascorbic acid, and BHT (Adetuyi, Adebayo, et al., 2022; Adetuyi, Adebisi, et al., 2022; Adetuyi, Odine, et al., 2022; Adetuyi, Ogundipe, et al., 2022; Adetuyi, Olajide, Oluwatosin, et al., 2022; Adetuyi, Olajide, Omowumi, et al., 2022). The nonenzymatic lipid peroxidation of rat liver homogenate was slowed by the essential oil of Achillea millefolium, which also exhibited a hydroxyl radical scavenging effect in the Fe^{3+}—EDTA-H_2O_2 deoxyribose structure. It was also shown that the essential oil of the plant Curcuma zedoaria was a fantastic scavenger of the DPPH radical. Neral/geranial, citronellal, isomenthone, and menthone are the most abundant scavengers in the essential oil of lemon salve, demonstrating cell reinforcement and reactive species scavenging action (Adetuyi, Adebayo, et al., 2022; Adetuyi, Adebisi, et al., 2022; Adetuyi, Odine, et al., 2022; Adetuyi, Ogundipe, et al., 2022; Adetuyi, Olajide, Oluwatosin, et al., 2022; Adetuyi, Olajide, Omowumi, et al., 2022). It is not just the phenolic contents in essential oils that contribute to their cancer-preventative properties; other compounds such monoterpene alcohols, ketones, aldehydes, hydrocarbons, and ethers also play a role in neutralizing free radicals. For instance, the cancer-preventive activity of the essential oil of thymus caespititius, thymus camphoratus, and thymus mastichina was sometimes similar to that of β-tocopherol. Surprisingly, thymol and carvacrol are hardly present, although linalool and 1, 8-cineole are abundant in all 3 varieties (Shuaib, Rohit et al., 2016).

Conclusion

It's completely possible that their motion can't be attributed to a given criterion, but rather to the existence of numerous factual zones in the phone, due to the large number of different groups of the chemical mixtures included in the essential oils synthesis. Low solubility is a crucial property of essential oils and their mixtures, as it allows them to alter the bilayer composition of the bacterial membrane and increase its permeability, leading to a greater loss of particulates as well as other cellular components. Naturally dynamic exacerbates may be found in abundance in essential oils, and the need for cancer-fighting extracts derived from medicinal plants, specifically essential oils, is on the rise. There is a possibility that the hypocholesterolemic impact of the plant mixtures included in these oils will be felt. It has been suggested as an expansion of such underutilized and disregarded therapeutic herbs to sustenance and refreshments to ensure against LDL oxidation and to decrease oxidative misfortune.

References

Adeleke, D. A., Olajide, P. A., Omowumi, O. S., Okunlola, D. D., Taiwo, A. M., & Adetuyi, B. O. (2022). Effect of monosodium glutamate on the body system. *World News of Natural Sciences*, 44, 1−23.

Adetunji, C.O. (2008). *The antibacterial activities and preliminary phytochemical screening of vernoniaamygdalina and Aloe vera against some selected bacteria* (pp. 40−43) (Msc thesis). University of Ilorin.

Adetunji, C. O. (2019). Environmental impact and ecotoxicological influence of biofabricated and inorganic nanoparticle on soil activity. In D. Panpatte, & Y. Jhala (Eds.), *Nanotechnology for agriculture*. Singapore: Springer. Available from https://doi.org/10.1007/978-981-32-9370-0_12.

Adetunji, C. O., Adetunji, C. O., Michael, O. S., Kadiri, O., Varma, A., Akram, M., Oloke, J. K., Shafique, H., Adetunji, J. B., Jain, A., Bodunrinde, R. E., Ozolua, P., & Ubi, B. E. (2021). Quinoa: From farm to traditional healing, food application, and phytopharmacology. In A. Varma (Ed.), *Biology and biotechnology of quinoa*. Singapore: Springer. Available from https://doi.org/10.1007/978-981-16-3832-9_20.

Adetunji, C. O., Adetunji, C. O., Olaniyan, O. T., Akram, M., Ajayi, O. O., Inobeme, A., Olaniyan, S., Khan, F. S., & Adetunji, J. B. (2021). Medicinal plants used in the treatment of pulmonary hypertension. In K. Dua, S. Nammi, D. Chang, D. K. Chellappan, G. Gupta, & T. Collet (Eds.), *Medicinal plants for lung diseases*. Singapore: Springer. Available from https://doi.org/10.1007/978-981-33-6850-7_14.

Adetunji, C. O., Ajayi, O. O., Akram, M., Olaniyan, O. T., Chishti, M. A., Inobeme, A., Olaniyan, S., Adetunji, J. B., Olaniyan, M., & Awotunde, S. O. (2021). Medicinal plants used in the treatment of influenza a virus infections. In K. Dua, S. Nammi, D. Chang, D. K. Chellappan, G. Gupta, & T. Collet (Eds.), *Medicinal plants for lung diseases*. Singapore: Springer. Available from https://doi.org/10.1007/978-981-33-6850-7_19.

Adetunji, C. O., Akram, M., Olaniyan, O. T., Ajayi, O. O., Inobeme, A., Olaniyan, S., Hameed, L., & Adetunji, J. B. (2021). Targeting SARS-CoV-2 novel Corona (COVID-19) virus infection using medicinal plants. In K. Dua, S. Nammi, D. Chang, D. K. Chellappan, G. Gupta, & T. Collet (Eds.), *Medicinal plants for lung diseases*. Singapore: Springer. Available from https://doi.org/10.1007/978-981-33-6850-7_21.

Adetunji, C. O., & Anani, O. A. (2020). Bio-fertilizer from *Trichoderma*: Boom for agriculture production and management of soil- and root-borne plant pathogens. In P. Mishra, R. R. Mishra, & C. O. Adetunji (Eds.), *Innovations in food technology*. Singapore: Springer. Available from https://doi.org/10.1007/978-981-15-6121-4_17.

Adetunji, C. O., Inobeme, A., Olaniyan, O. T., Ajayi, O. O., Olaniyan, S., & Adetunji, J. B. (2021). Application of nanodrugs derived from active metabolites of medicinal plants for the treatment of inflammatory and lung diseases: Recent advances. In K. Dua, S. Nammi, D. Chang, D. K. Chellappan, G. Gupta, & T. Collet (Eds.), *Medicinal plants for lung diseases*. Singapore: Springer. Available from https://doi.org/10.1007/978-981-33-6850-7_26.

Adetunji, C. O., Michael, O. S., Varma, A., Oloke, J. K., Kadiri, O., Akram, M., Bodunrinde, R. E., Imtiaz, A., Adetunji, J. B., Shahzad, K., Jain, A., Ubi, B. E., Majeed, N., Ozolua, P., & Olisaka, F. N. (2021). Recent advances in the application of biotechnology for improving the production of secondary metabolites from quinoa. In A. Varma (Ed.), *Biology and biotechnology of quinoa*. Singapore: Springer. Available from https://doi.org/10.1007/978-981-16-3832-9_17.

Adetunji, C. O., Olaniyan, O. T., Anani, O. A., Olisaka, F. N., Inobeme, A., Bodunrinde, R. E., Adetunji, J. B., Singh, K. R. B., Palnam, W. D., & Singh, R. P. (2021). *Current scenario of nanomaterials in the environmental, agricultural, and biomedical fields. Nanomaterials in bionanotechnology* (1st Edition, p. 30)Imprint CRC Press, Book.

Adetunji, C. O., & Ugbenyen, M. A. (2019). Mechanism of action of nanopesticide derived from microorganism for the alleviation of abiotic and biotic stress affecting crop productivity. In D. Panpatte, & Y. Jhala (Eds.), *Nanotechnology for agriculture: Crop production & protection*. Singapore: Springer. Available from https://doi.org/10.1007/978-981-32-9374-8_7.

Adetuyi, B., Dairo, J., & Oluwole, E. (2015). Biochemical effects of shea butter and groundnut oils on white albino rats. *International Journal of Chemistry and Chemical Processes, 1*(8), 1−17.

Adetuyi, B. O., Adebayo, P. F., Olajide, P. A., Atanda, O. O., & Oloke, J. K. (2022). Involvement of free radicals in the ageing of cutaneous membrane. *World News of Natural Sciences, 43*, 11−37.

Adetuyi, B. O., Adebisi, O. A., Adetuyi, O. A., Ogunlana, O. O., Toloyai, P. E., Egbuna, C., & Patrick-Iwuanyanwu, K. C. (2022). Ficus exasperata attenuates acetaminophen-induced hepatic damage via NF-κB signaling mechanism in experimental rat model. *BioMed Research International, 2022*.

Adetuyi, B. O., Adebisi, O. A., Awoyelu, E. H., Adetuyi, O. A., & Ogunlana, O. O. (2020). Phytochemical and toxicological effect of ethanol extract of heliotropium indicum on liver of male albino rats. *Letters in Applied NanoBioscience, 10*(2), 2085−2095.

Adetuyi, B. O., Dairo, J. O., & Didunyemi, O. M. (2015). Anti-hyperglycemic potency of jatropha gossypiifolia in alloxan induced diabetes. *Biochem Pharmacol (Los Angel), 4*(193), 2167, 0501.

Adetuyi, B. O., Odine, G. O., Olajide, P. A., Adetuyi, O. A., Atanda, O. O., & Oloke, J. K. (2022). Nutraceuticals: Role in metabolic disease, prevention and treatment. *World News of Natural Sciences, 42*, 1−27.

Adetuyi, B. O., Ogundipe, A. E., Ogunlana, O. O., Egbuna, C., Estella, O. U., Mishra, A. P., & Achar, R. R. (2022). *Banana peel as a source of nutraceuticals. Food and agricultural byproducts as important source of valuable nutraceuticals* (pp. 243−250). Cham: Springer.

Adetuyi, B. O., Okeowo, T. O., Adetuyi, O. A., Adebisi, O. A., Ogunlana, O. O., Oretade, J. O., & Batiha, G. E. S. (2020). Ganoderma lucidum from red mushroom attenuates formaldehyde-induced liver damage in experimental male rat model. *Biology, 9*(10), 313.

Adetuyi, B. O., Olajide, P. A., Awoyelu, E. H., Adetuyi, O. A., Adebisi, O. A., & Oloke, J. K. (2020). Epidemiology and treatment options for COVID-19: A review. *African Journal of Reproductive Health, 24*(2), 142−153.

Adetuyi, B. O., Olajide, P. A., Oluwatosin, A., & Oloke, J. K. (2022). Preventive phytochemicals of cancer as speed breakers in inflammatory signaling. *Research Journal of Life Sciences, Bioinformatics, Pharmaceutical and Chemical Sciences, 8*(1), 30−61.

Adetuyi, B. O., Olajide, P. A., Omowumi, O. S., Odine, G. O., Okunlola, D. D., Taiwo, A. M., & Opayinka, O. D. (2022). Blockage of Alzheimer's gene: Breakthrough effect of Apolipoprotein E4. *African Journal of Advanced Pure and Applied Sciences (AJAPAS)*, 26−33.

Adetuyi, B. O., Oluwole, E. O., & Dairo, J. O. (2015). Chemoprotective potential of ethanol extract of ganoderma lucidum on liver and kidney parameters in plasmodium beghei-induced mice. *International Journal of Chemistry and Chemical Processes (IJCC), 1*(8), 29−36.

Adetuyi, B. O., Omolabi, F. K., Olajide, P. A., & Oloke, J. K. (2021). Pharmacological, biochemical and therapeutic potential of milk thistle (silymarin): A review. *World News of Natural Sciences, 37*, 75−91.

Adetuyi, B. O., Toloyai, P. E. Y., Ojugbeli, E. T., Oyebanjo, O. T., Adetuyi, O. A., Uche, C. Z., & Egbuna, C. (2021). Neurorestorative roles of microgliosis and astrogliosis in neuroinflammation and neurodegeneration. *Scicom Journal of Medical and Applied Medical Sciences, 1*(1), 1−5.

Adewale, G. G., Olajide, P. A., Omowumi, O. S., Okunlola, D. D., Taiwo, A. M., & Adetuyi, B. O. (2022). Toxicological significance of the occurrence of selenium in foods. *World News of Natural Sciences, 44*, 63−88.

Awoyelu, E. H., Oladipo, E. K., Adetuyi, B. O., Senbadejo, T. Y., Oyawoye, O. M., & Oloke, J. K. (2020). Phyloevolutionary analysis of SARS-CoV-2 in Nigeria. *New Microbes and New Infections, 36*, 100717.

Bakkali, F., Averbeck, S., Averbeck, D., & Idaomar, M. (2008). Biological effects of essential oils—A review. *Food and chemical toxicology, 46*(2), 446–475.

Barazesh, F., Oloumi, H., Nasibi, F., & Kalantari, K. M. (2017). Effect of spermine, epibrassinolid and their interaction on inflorescence buds and fruits abscission of pistachio tree (Pistacia vera L.),"Ahmad—Aghai" cultivar. *Banat's Journal of Biotechnology, 8*(16), 105–115.

Batiha, G. E., Awad, D. A., Algamma, A. M., Nyamota, R., Wahed, M. I., Shah, M. A., Amin, M. N., Adetuyi, B. O., Hetta, H. F., Cruz-Marins, N., Koirala, N., Ghosh, A., & Sabatier, J.-M. (2021). Diary-derived and egg white proteins in enhancing immune system against COVID-19. *Frontiers in Nutritionr. (Nutritional Immunology), 8*, 629440. Available from https://doi.org/10.3389/fnut.2021629440.

Bello, O. M., Ibitoye, T., & Adetunji, C. (2019). Assessing antimicrobial agents of Nigeria flora. *Journal of King Saud University-Science, 31*(4), 1379–1383.

Bozhanska, T. (2018). Botanical and morphological composition of artificial grassland of bird's-foot-trefoil (*Lotus corniculatus* L.) treated with lumbrical and lumbrex. *Banat's Journal of Biotechnology, 9*(18). Available from https://doi.org/10.7904/2068-4738-IX(18)-12.

Didunyemi, M. O., Adetuyi, B. O., & Oyebanjo, O. O. (2019). Morinda lucida attenuates acetaminophen-induced oxidative damage and hepatotoxicity in rats. *Journal of Biomedical. Sciences, 8*(2), 0.

Didunyemi, M., Adetuyi, B., & Oyewale, I. (2020). Inhibition of lipid peroxidation and in-vitro antioxidant capacity of aqueous, acetone and methanol leaf extracts of green and red acalypha wilkesiana muell arg. *International Journal of Biological & Medical Research, 11*(3), 7089–7094.

Halliwell, B., & Gutteridge, J. M. (1990). The antioxidants of human extracellular fluids. *Archives of Biochemistry and Biophysics, 280*(1), 1–8.

Hameed, A., Condò, C., Tauseef, I., Idrees, M., Ghazanfar, S., Farid, A., Muzammal, M., Al Mohaini, M., Alsalman, A. J., Al Hawaj, M. A., Adetunji, C. O., Dauda, W. P., Hameed, Y., Alhashem, Y. N., & Alanazi, A. A. (2022). Isolation and characterization of a cholesterol-lowering bacteria from *Bubalus bubalis* raw milk. *Fermentation, 8*(4), 163. Available from https://doi.org/10.3390/fermentation8040163.

Hariri, A., Ouis, N., Bouhadi, D., & Ould Yerou, K. (2017). Evaluation of the quality of the date syrups enriched by cheese whey during the period of storage. *Banat's Journal of Biotechnology, 8*(16).

Hariri, A., Ouis, N., Bouhadi, D., & Benatouche, Z. (2018). Characterization of the quality of the steamed yoghurts enriched by dates flesh and date powder variety H'loua. *Banat's Journal of Biotechnology, 9*(17).

Jahan, S., Chowdhury, S. F., Mitu, S. A., Shahriar, M., & Bhuiyan, M. A. (2015). Genomic DNA extraction methods: A comparative case study with gram-negative organisms. *Banat's Journal of Biotechnology, 6*(11), 61–68.

James-Okoro, P. P. O., Iheagwam, F. N., Sholeye, M. I., Umoren, I. A., Adetuyi, B. O., Ogundipe, A. E., & Ogunlana, O. O. (2021). Phytochemical and in vitro antioxidant assessment of Yoyo bitters. *World News of Natural Sciences, 37*, 1–17.

Jing, L., Zhang, Y., Fan, S., Gu, M., Guan, Y., Lu, X., Huang, C., & Zhou, Z. (2013). Preventive and ameliorating effects of citrus D-limonene on dyslipidemia and hyperglycemia in mice with high-fat diet-induced obesity. *European Journal of Pharmacology, 715*(1–3), 46–55.

Nazir, A., Itrat, N., Shahid, A., Mushtaq, Z., Abdulrahman, S. A., Egbuna, C., & Toloyai, P. E. Y. (2022). *Orange peel as source of nutraceuticals. Food and agricultural byproducts as important source of valuable nutraceuticals* (pp. 97–106). Cham: Springer.

Ogunlana, O. O., Adetuyi, B. O., Esalomi, E. F., Rotimi, M. I., Popoola, J. O., Ogunlana, O. E., & Adetuyi, O. A. (2021). Antidiabetic and antioxidant activities of the twigs of andrograhis paniculata on streptozotocin-induced diabetic male rats. *BioChem, 1*(3), 238–249.

Ogunlana, O. O., Adetuyi, B. O., Rotimi, M., Adeyemi, A., Akinyele, J., Ogunlana, O. E., & Batiha, G. E. S. (2021). Hypoglycemic and antioxidative activities of ethanol seed extract of Hunteria umbellate (Hallier F.) on streptozotocin-induced diabetic rats. *Clinical Phytoscience, 7*(1), 1–9.

Ogunlana, O. O., Babatunde, O. A., Tobi, S. A., Adegboye, B. E., Iheagwam, F. N., & Oluseyi, E. (2021c). Ogunlana. Ruzu bitters ameliorates high-fat diet induced non-alcoholic fatty liver disease in male Wistar rats. *Journal of Pharmacy and Pharmacognosy Research, 9*(3), 251–260.

Ogunlana, O. O., Ogunlana, O. E., Adekunbi, T. S., Adetuyi, B. O., Adegboye, B. E., & Iheagwam, F. N. (2020). Anti-inflammatory mechanism of ruzu bitters on diet-induced nonalcoholic fatty liver disease in male wistar rats. *Evidence-Based Complementary and Alternative Medicine, 2020*.

Olajide, P. A., Adetuyi, O. A., Omowumi, O. S., & Adetuyi, B. O. (2022). Anticancer and antioxidant phytochemicals as speed breakers in inflammatory signaling. *World News of Natural Sciences, 44*, 231–259.

Olajide, P. A., Omowumi, O. S., & Odine, G. O. (2022). Pathogenesis of reactive oxygen species: A review. *World News of Natural Sciences, 44*, 150–164.

Olajide, P. A., Omowumi, O. S., Okunlola, D. D., & Adetuyi, B. O. (2022). Deadly pandemia: Monkeypox disease, a case study. *African Journal of Advanced Pure and Applied Sciences (AJAPAS)*, 34–37.

Ouis, N., & Hariri, A. (2018). Antioxidant and antibacterial activities of the essential oils of Ceratonia siliqua. *Banat's Journal of Biotechnology, 9*(17).

Salvamani, S., Gunasekaran, B., Shukor, M. Y., Bakar, M. Z. A., & Ahmad, S. A. (2016). Phytochemical investigation, hypocholesterolemic and anti-atherosclerotic effects of Amaranthus viridis leaf extract in hypercholesterolemia-induced rabbits. *RSC Advances, 6*(39), 32685–32696.

Scartezzini, P., & Speroni, E. (2000). Review on some plants of Indian traditional medicine with antioxidant activity. *Journal of Ethnopharmacology, 71*(1–2), 23–43.

Shuaib, A., Rohit, A., & Piyush, M. (2016). A review article on essential oils. *Journal of Medicinal Plants Studies, 4*, 237–240.

Zarkovic, N. (2003). 4-Hydroxynonenal as a bioactive marker of pathophysiological processes. *Molecular Aspects of Medicine, 24*(4–5), 281–291.

Zerkaoui, L., Benslimane, M., & Hamimed, A. (2018). The purification performances of the lagooning process, case of the Beni Chougrane region in Mascara (Algerian NW). *Banat's Journal of Biotechnology, 9*(18), 20–28.

Index

Note: Page numbers followed by "*f*" and "*t*" refer to figures and tables, respectively.